NASA'S CONTRIBUTIONS TO AERONAUTICS

VOLUME 1

AERODYNAMICS

STRUCTURES

PROPULSION

CONTROLS

DR. RICHARD P. HALLION, EDITOR

National Aeronautics and Space Administration
Headquarters
300 E St SW
Washington, DC 20546

2010

NASA/SP-2010-570-Vol 1

www.nasa.gov

Contents

Foreword .... *vii*

CASE 1
Sweep and Swing: Reshaping the Wing for the Jet and Rocket Age
Richard P. Hallion 2

CASE 2
Richard Whitcomb and the Quest for Aerodynamic Efficiency
Jeremy Kinney 88

CASE 3
NACA–NASA and the Rotary Wing Revolution
John F. Ward 134

CASE 4
Softening the Sonic Boom: 50 Years of NASA Research
Lawrence R. Benson 180

CASE 5
Toward Transatmospheric Flight: From V-2 to the X-51
T.A. Heppenheimer 276

CASE 6
Physical Problems, Challenges, and Pragmatic Solutions
Robert G. Hoey 360

CASE 7
NASA and the Evolution of Computational Fluid Dynamics
John D. Anderson, Jr. 426

CASE 8
NASA and Computational Structural Analysis
David C. Aronstein 460

CASE 9
High-Temperature Structures and Materials
T.A. Heppenheimer 568

CASE 10
Fly-by-Wire: Making the Electric Jet
Albert C. Piccirillo 630

CASE 11
Advancing Propulsive Technology
James Banke 734

CASE 12
Leaner and Greener: Fuel Efficiency Takes Flight
Caitlin Harrington 784

CASE 13
Good Stewards: NASA's Role in Alternative Energy
Bruce I. Larrimer 822

Index 879

Foreword

AS THIS BOOK GOES TO PRESS, the National Aeronautics and Space Administration (NASA) has passed beyond the half century mark, its longevity a tribute to how essential successive Presidential administrations—and the American people whom they serve—have come to regard its scientific and technological expertise. In that half century, flight has advanced from supersonic to orbital velocities, the jetliner has become the dominant means of intercontinental mobility, astronauts have landed on the Moon, and robotic spacecraft developed by the Agency have explored the remote corners of the solar system and even passed into interstellar space.

Born of a crisis—the chaotic aftermath of the Soviet Union's space triumph with Sputnik—NASA rose magnificently to the challenge of the emergent space age. Within a decade of NASA's establishment, teams of astronauts would be planning for the lunar landings, first accomplished with Neil Armstrong's "one small step" on July 20, 1969. Few events have been so emotionally charged, and none so publicly visible or fraught with import, as his cautious descent from the spindly little Lunar Module Eagle to leave his historic boot-print upon the dusty plain of Tranquillity Base.

In the wake of Apollo, NASA embarked on a series of space initiatives that, if they might have lacked the emotional and attention-getting impact of Apollo, were nevertheless remarkable for their accomplishment and daring. The Space Shuttle, the International Space Station, the Hubble Space Telescope, and various planetary probes, landers, rovers, and flybys speak to the creativity of the Agency, the excellence of its technical personnel, and its dedication to space science and exploration.

But there is another aspect to NASA, one that is too often hidden in an age when the Agency is popularly known as America's space agency and when its most visible employees are the astronauts who courageously rocket into space, continuing humanity's quest into the unknown. That hidden aspect is aeronautics: lift-borne flight within the atmosphere, as distinct from the ballistic flight of astronautics, out into space. It is the first "A" in the Agency's name and the oldest-rooted of the Agency's technical competencies, dating to the formation, in 1915, of NASA's lineal predecessor, the National Advisory Committee for Aeronautics (NACA). It was the NACA that largely restored America's aeronautical primacy in the interwar years after 1918, deriving the airfoil profiles and configuration concepts that defined successive generations of ever-more-capable aircraft as America progressed from the subsonic piston era into the transonic and supersonic jet age. NASA, succeeding the NACA after the shock of Sputnik, took American aeronautics across the hypersonic frontier and onward into the era of composite structures, electronic flight controls, and energy-efficient flight.

This volume, the first of a two-volume set, traces contributions by NASA and the post–Second World War NACA to the field of aeronautics. It was that work that enabled the exploitation of the turbojet and high-speed aerodynamic revolution that led to the gas-turbine-powered jet age that followed, within which we still live. The subjects covered in this first volume are an eclectic mix of surveys, case studies, and biographical examinations ranging across multiple disciplines and technical competencies residing within the National Aeronautics and Space Administration. The topics are indicative of the range of Agency work and the capabilities of its staff. They include:

- The advent of the sharply swept-back wing, which enabled taking fullest advantage of the turbojet revolution and thereby launched the era of high-speed global mass mobility, becoming itself the iconic symbol of the jet age.
- The contributions and influence of Richard T. Whitcomb, a legendary NACA–NASA researcher who gave to aeronautics some of the key methods of reducing drag and improving flight efficiencies in the challenging transonic region, between subsonic and supersonic flight.
- The work of the NACA and NASA in furthering the rotary wing revolution via research programs on a range of rotorcraft from autogiros through helicopters, convertiplanes, ducted fan, tilt wing, and tilt rotor craft.

- How NASA worked from the earliest days of the supersonic revolution to mitigate the shock and disturbing effects of the sonic boom, developing creative test approaches to evaluate boom noise and overpressures, and then methods to alleviate boom formation and impingement, leading to novel aircraft shaping and methods that are today promising to revolutionize the design of transonic and supersonic civil and military aircraft.
- How the NACA and NASA, having mastered the transonic and supersonic regions, took on the challenge of extending lift-borne flight into the hypersonic region and thence into space, using exotic "transatmospheric" vehicles such as the legendary X-15, various lifting bodies, and the Space Shuttle, and extending the frontiers of air-breathing propulsion with the Mach 9+ scramjet-powered X-43.
- The physical problems and challenges that forced NASA and other researchers to study and find pragmatic solutions for such thorny issues as aeroelasticity, oscillatory instabilities forcing development of increasingly sophisticated artificial stability systems, flight simulation for high-performance aerospace vehicles, and aerothermodynamic structural deformation and heating.
- NASA's role in advancing and maturing computational fluid dynamics (CFD) and applying this new tool to aeronautical research and aerospace vehicle design.
- The exploitation of materials science and development of high-temperature structures to enable design of practical high-speed military and civil aircraft and spacecraft.
- The advent of computerized structural loads prediction, modeling, and simulation, which, like CFD, revolutionized aerospace design practices, enhancing both safety and efficiency.
- NASA's pioneering of electronic flight control ("fly-by-wire"), from rudimentary testbeds evolved from Apollo-era computer architectures and software, to increasingly sophisticated systems integrating aerodynamic and propulsion controls.

- How the NACA and NASA advanced the gas turbine revolution, producing more efficient engine concepts and technology for application to new generations of military and civilian aircraft.
- How NASA has contributed to the quest for fuel-efficient and environmentally friendly aircraft technology, studying combustion processes, alternative fuels, and pollutant transfer into the upper atmosphere, searching for appropriate technological solutions, and resulting in less polluting, less wasteful, and more efficient aircraft designs.
- The Agency's work in promoting global environmental good stewardship by applying its scientific and technical competencies to wind and solar energy, resulting in more efficient energy-producing wind turbines and high-altitude solar-powered long-endurance unpiloted aerial vehicles.

The record of NACA–NASA accomplishments in aeronautics demonstrates the value of consistent investment in aeronautical research as a means of maintaining the health and stability of America's aerospace industrial base. That base has generated an American predominance in both civil and military aeronautics, but one that is far from assured as the Nation enters the second century of winged flight. It is hoped that these studies, offering a glimpse at the inner workings of the Agency and its personnel, will prove of value to the men and women of NASA, to those who benefit across the United States and overseas from their dedicated work, and to students of aeronautics and members of the larger aerospace community. It is to the personnel of NASA, and the NACA before them, that this volume is dedicated, with affection and respect.

Dr. Richard P. Hallion
August 4, 2010

THIS PAGE INTENTIONALLY BLANK

The X-48B subscale demonstrator for the Blended Wing-Body (BWB). The BWB may represent the next extension of the swept and delta wing, to transform flight away from the rule of the "tube and wing" jetliner. NASA.

CASE 1

Sweep and Swing: Reshaping the Wing for the Jet and Rocket Age

Richard P. Hallion

The development of the swept and delta wing planform enabled practical attainment of the high speeds promised by the invention of the turbojet engine and the solid-and-liquid-fueled rocket. Refining the swept and delta planforms from theoretical constructs to practical realities involved many challenges and problems requiring creative analysis and study by NACA and NASA researchers. Their insight and perseverance led to the swept wing becoming the iconic symbol of the jet age.

THE PROGRESSIVE EVOLUTION OF AIRCRAFT DESIGN HAS WITNESSED continuous configuration changes, adaptations, and reinterpretations. The canard wood-and-fabric biplane launched the powered flight revolution and gave way to the tractor biplane and monoplane, and both gave way to the all-metal monoplane of the interwar era. The turbojet engine set aside the piston engine as the primary motive power for long-range commercial and military aircraft, and it has been continually refined to generate the sophisticated bypass turbofans of the present era, some with afterburning as well. The increasing airspeed of aircraft drove its own transformation of configuration, measurable in the changed relationship between aspect and fineness ratios. Across the primacy of the propeller-driven era, from the beginning of the 20th century to the end of the interwar era, wingspan generally far exceeded fuselage length. That changed early in the jet and rocket era. By the time military and test pilots from the National Advisory Committee for Aeronautics (NACA) first probed the speed of sound with the Bell XS-1 and Douglas D-558-1 Skystreak, wingspan and fuselage length were roughly equal. Within a decade, as aircraft speed extended into the supersonic regime, the ratio of wingspan to fuselage length dramatically reversed, evidenced by aircraft such as the Douglas X-3, the Lockheed F-104 Starfighter, and the Anglo-French Concorde Supersonic Transport (SST). Nicknames handily captured the transformation: the rakish X-3 was known informally as the "Stiletto" and the only slightly less sleek F-104 as the "Missile with a Man in It."

There was as well another manifestation of profound design transformation, one that gave to the airplane a new identity that swiftly became a global icon: the advent of the swept wing. If the biplane constituted the normative airplane of the first quarter century of flight and the straight wing cantilever monoplane that of the next quarter century, by the time of the golden anniversary of Kitty Hawk, the swept wing airplane had supplanted both, its futuristic predominance embodied by the elegant North American F-86 Sabre that did battle in "MiG Alley," high over North Korea's blue-gray hills bordering the Yalu River. In the post-Korean era, as swept wing Boeing 707 and Douglas DC-8 jet airliners replaced what historian Peter Brooks termed the "DC-4 generation" of straight wing propeller-driven transports, the swept wing became the iconic embodiment of the entire jet age.[1] Today, 75 years since its enunciation at an international conference, the high-speed swept wing is the commonly accepted global highway symbol for airports, whether an intercontinental center such as Los Angeles, Frankfurt, or Heathrow; regional hubs such as Dallas, Copenhagen, or Charlotte; or any of the myriad general aviation and business aviation airfields around the world, even those still primarily populated, ironically, by small, straight wing propeller-and-piston-driven airplanes.

The Tailless Imperative: The Early History of Swept and Delta Wings

The high-speed swept wing first appeared in the mid-1930s and, like most elements in aircraft design, was European by birth. But this did not mark the swept wing's first appearance in the world's skies. The swept wing dated to before the First World War, when John Dunne had developed a series of tailless flying wing biplanes using the swept planform as a means of ensuring inherent longitudinal stability, imparting "self-correcting" restoration of any gust-induced pitching motions. Dunne's aircraft, while freakish, did enjoy some commercial success. He sold manufacturing

---

1. Peter W. Brooks, *The Modern Airliner: Its Origins and Development* (London: Putnam & Co., Ltd., 1961), pp. 91–111. Brooks uses the term to describe a category of large airliner and transport aircraft defined by common shared design characteristics, including circular cross-section constant-diameter fuselages, four-engines, tricycle landing gear, and propeller-driven (piston and turbo-propeller), from the DC-4 through the Bristol Britannia, and predominant in the time period 1942 through 1958. Though some historians have quibbled with this, I find Brooks's reasoning convincing and his concept of such a "generation" both historically valid and of enduring value.

rights to the Burgess Company in the United States, which subsequently produced two "Burgess-Dunne" seaplanes for the U.S. Navy. Lt. Holden C. Richardson, subsequently one of the first members of the NACA, had urged their purchase "so that the[ir] advantages and limitations can be thoroughly determined . . . as it appears to be only the beginning of an important development in aeronautical design."[2]

That it was, though not in the fashion Richardson expected. The swept wing remained an international staple of tailless self-stabilizing design, typified in the interwar years by the various Westland Pterodactyl aircraft designed by Britain's G.T.R. Hill, the tailless aircraft of Boris Ivanovich Cheranovskiy, Waldo Waterman's Arrowplane, and a series of increasingly sophisticated sailplanes and powered aircraft designed by Germany's Alexander Lippisch. However, it would not become the "mainstream" element of aircraft design its proponents hoped until applied to a very different purpose: reducing transonic aerodynamic effects.[3] The transonic swept wing effectively increased a wing's critical Mach number (the "drag divergence Mach number"), delaying the onset of transonic drag rise and enabling an airplane to fly at higher transonic and supersonic speeds for the same energy expenditure and drag penalty that a straight wing airplane would expend and experience at much lower subsonic speeds.

In 1935, leading aerodynamicists gathered in Rome for the Volta Congress on High Speeds in Aviation, held to coincide with the opening of Italy's impressive new Guidonia laboratory complex. There, a young German fluid dynamicist, Adolf Busemann, unveiled the concept of using the swept wing as a means of attaining supersonic flight.[4] In his presentation, he

---

2. Quoted in Roy A. Grossnick, et al., *United States Naval Aviation 1910–1995* (Washington: U.S. Navy, 1997), p. 15; Gordon Swanborough and Peter M. Bowers, *United States Navy Aircraft Since 1911* (New York: Funk & Wagnalls, 1968), p. 394.
3. Alexander Lippisch, "Recent Tests of Tailless Airplanes," NACA TM-564 (1930), a NACA translation of his article *"Les nouveaux essays d'avions sans queue," l'Aérophile* (Feb. 1–15, 1930), pp. 35–39.
4. For Volta, see Theodore von Kármán and Lee Edson, *The Wind and Beyond: Theodore von Kármán, Pioneer in Aviation and Pathfinder in Space* (Boston: Little, Brown and Co., 1967), pp. 216–217, 221–222; Adolf Busemann, "Compressible Flow in the Thirties," *Annual Review of Fluid Mechanics*, vol. 3 (1971), pp. 6–11; Carlo Ferrari, "Recalling the Vth Volta Congress: High Speeds in Aviation," *Annual Review of Fluid Mechanics*, vol. 28 (1996), pp. 1–9; Hans-Ulrich Meier, "Historischer Rückblick zur Entwicklung der Hochgeschwindigkeitsaerodynamik," in H.-U. Meier, ed., *Die Pfeilflügelentwicklung in Deutschland bis 1945* (Bonn: Bernard & Graefe Verlag, 2006), pp. 16–36; and Michael Eckert, *The Dawn of Fluid Dynamics: A Discipline Between Science and Technology* (Weinheim: Wiley-VCH Verlag, 2006), pp. 228–231.

demonstrated the circulation pattern around a swept wing that, essentially, "fooled" it into "believing" it was flying at lower velocities. As well, he presented a sketch of an aircraft with such a "Pfielförmiges Tragwerk" ("Arrow-Shaped Lifting Surface"), though one that had, by the standards of subsequent design, very modest sweep and very high aspect ratio.[5]

Theodore von Kármán recalled not quite two decades later that afterward, at the conference banquet, "General [Arturo] Crocco, the organizer of the congress and a man of far-reaching vision, went further while doodling on the back of the menu card, drawing a plane with swept-back wings and tail, and even swept propeller blades, laughingly calling it 'Busemann's airplane.'"[6] Evidence exists that Crocco took the concept beyond mere dinner conversation, for afterward, Guidonia researchers evaluated a design blending modestly swept wings with a "push-pull" twin-engine fuselage configuration. However, Guidonia soon returned to the more conventional, reflecting the Italian air ministry's increasing emphasis upon building a large and powerful air arm incorporating already proven and dependable technology.[7]

Delegates from other nations present at Busemann's briefing missed its significance altogether, perhaps because his gently swept configuration—in the era of the DC-2 and DC-3, which had pronounced leading edge taper—looked far less radical than the theory and purpose behind it implied. NACA Langley Memorial Aeronautical Laboratory researchers had already evaluated far more sharply swept planforms at Langley for a seminal wing taper study the laboratory issued the next year.[8] Thus, at first glance, Busemann's design certainly did not look like a shape that would transform aviation from the firmly subsonic to the transonic, making possible the potential of the jet engine, and the jet age (with its jet set) that followed.

---

5. Adolf Busemann, "Aerodynamische Auftrieb bei Überschallgeschwindigkeit," Luftfahrtforschung, vol. 12, No. 6 (Oct. 3, 1935), pp. 210–220, esp. Abb. 4–5 (Figures 4–5).
6. Theodore von Kármán, Aerodynamics (New York: McGraw-Hill Book Company, Inc., 1963 ed.), p. 133.
7. Ministero dell'Aeronautica, 1° Divisione, Sezione Aerodinamica Resultati di Esperienze (Rome: Guidonia, 1936); the swept "double-ender" wind tunnel study (anticipating the layout of Dornier's Do 335 Pfeil ["Arrow"] of the late wartime years) was designated the J-10; its drawing is dated March 7, 1936. I thank Professor Claudio Bruno of the Università degli Studi di Roma "La Sapienza"; and Brigadier General Marcello di Lauro and Lieutenant Colonel Massimiliano Barlattani of the Stato Maggiore dell'Aeronautica Militare (SMdAM), Rome, for their very great assistance in enabling me to examine this study at the Ufficio Storico of the SMdAM in June 2009.
8. Raymond F. Anderson, "Determination of the Characteristics of Tapered Wings," NACA Report No. 572 (1936); see in particular Figs. 15 and 16, p. 11.

Therefore, for the United States and most other nations, over the next decade, the normative airplane remained one having straight (if tapered) wings and piston propulsion. For Germany, however, the future belonged to increasingly sharply swept and delta wings, and jet and rocket propulsion as well. Within 5 years of the Volta conference, with Europe engulfed in a new war, its engineers had already flown their first jet and rocket-powered aircraft, had expanded beyond Busemann's initial conception to derive shapes more closely anticipating subsequent high-speed aircraft and missile designs, and were busily testing models of swept wing transonic airplanes and supersonic missiles. Lippisch's swept wing sailplanes had presaged a new Messerschmitt rocket-propelled interceptor, the Me 163 Komet ("Comet"), and his broad, high aspect ratio deltas had given way to a rounded triangular planform that he envisioned as meeting the needs for transonic and supersonic flight. While many of these concepts by Lippisch and other German designers were impracticable, or unrelated to Germany's more immediate military needs, others possessed significant military or research potential. Only flawed decisions by the Third Reich's own leadership and the Allies' overrunning of Germany would prevent them from being developed and employed before the collapse of the Hitler regime in May 1945.[9]

Birthing the American Delta and Swept Wing
The extent to which the swept wing permeated German aeronautical thought understandably engendered tremendous postwar interest

---

9. For an example of such work, see Dr. Richard Lehnert, "Bericht über Dreikomponentenmessungen mit den Gleitermodellen A4 V12/a und A4 V12/c," Archiv Nr 66/34 (Peenemünde: Heeres-Versuchsstelle, Nov. 27, 1940), pp. 6-10, Box 674, "C10/V-2/History" file, archives of the National Museum of the United States Air Force, Dayton, OH. Re: German research deficiencies, see Adolf Baeumker, *Ein Beitrag zur Geschichte der Führung der deutschen Luftfahrttechnik im ersten halben Jahrhundert, 1900–1950* (Bad Godesberg: Deutschen Forschungs – und Versuchsanstalt für Luft – und Raumfahrt e. V., 1971), pp. 61–74; Col. Leslie E. Simon, *German Scientific Establishments* (Washington: Office of Technical Services, Department of Commerce, 1947), pp. 7–9; Helmuth Trischler, "Self-Mobilization or Resistance? Aeronautical Research and National Socialism," and Ulrich Albrecht, "Military Technology and National Socialist Ideology," in Monika Renneberg and Mark Walker, eds., *Science, Technology, and National Socialism* (Cambridge: Cambridge University Press, 1994), pp. 72–125. For science and the Third Reich more generally, see Alan D. Beyerchen, *Scientists Under Hitler: Politics and the Physics Community in the Third Reich* (New Haven: Yale University Press, 1977); and Kristie Macrakis, *Surviving the Swastika: Scientific Research in Nazi Germany* (New York: Oxford University Press, 1993).

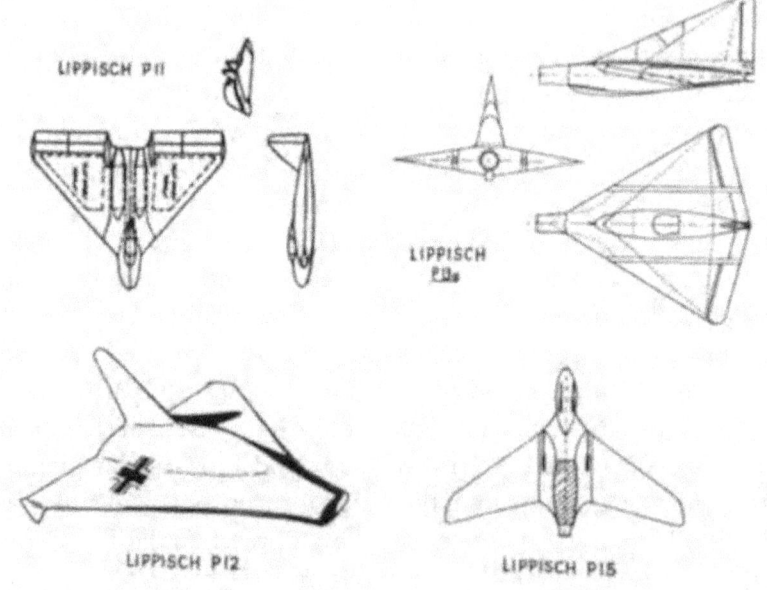

A sampling of various design concepts for Lippisch swept wing and delta aircraft. These original Lippisch sketches were incorporated in "German Aircraft: New and Projected Types," a 1946 Allied technical intelligence summary. USAF.

in the benefits of swept planforms for transonic and supersonic flight within the American, European, and Soviet aeronautical communities.[10] However, for America, uncovering German swept wing research and development furnished the *confirmation* of its value, not its *discovery*, for Robert T. Jones, an aerodynamicist at the Langley Memorial Aeronautical Laboratory, had independently discovered its benefits in 1944, a year before the Allies first entered Germany's shattered and shuttered research laboratories and design shops.[11]

The Gluhareff-Griswold Nexus

In 1936, Michael E. Gluhareff, an emigree Russian engineer who was chief of design for the Vought-Sikorsky Aircraft Division of United

---

10. USAAF, "German Aircraft, New and Projected Types" (1946), Box 568, "A-1A/Germ/1945" file, NMUSAF Archives; and J. McMasters and D. Muncy, "The Early Development of Jet Propelled Aircraft," AIAA Paper 2007-0151, Pts. 1–2 (2007).
11. See Richard P. Hallion, "Lippisch Gluhareff, and Jones: The Emergence of the Delta Planform and the Origins of the Sweptwing in the United States," *Aerospace Historian*, vol. 26, no. 1 (Mar. 1979), pp. 1–10.

Case 1 | Sweep and Swing: Reshaping the Wing for the Jet and Rocket Age

Aircraft Corporation, began examining various tailless aircraft configurations. By July 1941, his study had spawned a proposed interceptor fighter powered by a piston engine driving a contra-rotating pusher propeller. It had a rounded delta planform resembling an arrowhead, with leading edges swept aft at 56 degrees. It featured a tricycle retractable landing gear, twin ventral vertical fins, an extremely streamlined and rounded configuration, provisions for six heavy machine guns, and elevons (combined ailerons and elevators) for roll and pitch control. Gluhareff informed company founder Igor I. Sikorsky that its sharp sweep would delay the onset of transonic compressibility, noting, "The general shape and form of the aircraft is, therefore, outstandingly adaptable for extremely high speeds."[12]

In retrospect, Gluhareff's design was a remarkable achievement, conceived at just the right time to have been completed with turbojet propulsion (for which its configuration and internal layout was eminently suited) though circumstances conspired against its development. Sikorsky was then perfecting the first practical helicopter—the VS-300, another revolutionary development, of course—and chose understandably to concentrate on rotary wing flight. He did authorize Gluhareff to solicit support from inventor-entrepreneur Roger W. Griswold, president of the Ludington-Griswold Company, about building a wind tunnel model of the configuration.[13] Tests by United Aircraft proved so encouraging that Griswold approached the engineering staff of the Army Air Forces (AAF) at its Wright Field Aircraft Laboratory about sponsoring what was now called the "Dart."[14] But having their fill of visitors

---

12. Memo, Michael Gluhareff to I.I. Sikorsky, July 1941, copy in the Gluhareff Dart accession file, National Air and Space Museum, Smithsonian Institution, Washington, DC. Gluhareff's Dart appeared contemporaneously with a remarkably similar (though with a tractor propeller) Soviet design by Alexandr Sergeevich Moskalev. Though unclear, it seems Gluhareff first conceived the planform. It is possible that an informal interchange of information between the two occurred, as Soviet aeronautics and espionage authorities kept close track of American developments and the activities of the emigree Russian community in America.
13. Griswold is best known as coinventor (with Hugh De Haven) of the three-point seat restraint, which formed the basis for the modern automotive seat belt; Saab then advanced further, building upon their work. See "Three-Point Safety Belt is American, not Swedish, Invention," *Status Report*, vol. 35, no. 9 (Oct. 21, 2000), p. 7.
14. Vought-Sikorsky, "Aerodynamic Characteristics of the Preliminary Design of a 1/20 Scale Model of the Dart Fighter," Vought-Sikorsky Wind Tunnel Report No. 192 (Nov. 18, 1942), copy in the Gluhareff Dart accession file, National Air and Space Museum, Smithsonian Institution, Washington, DC.

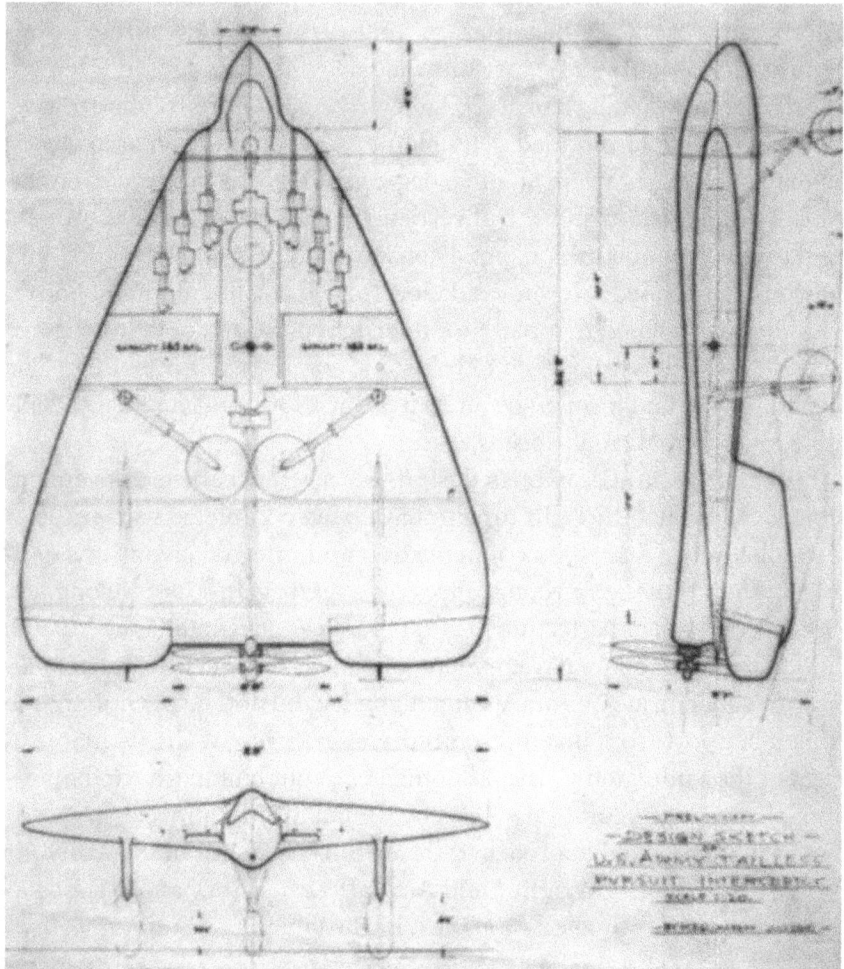

The proposed Gluhareff Dart fighter of 1941, showing both its novel layout and, for the time, nearly as novel tricycle landing gear layout. National Air and Space Museum, Smithsonian Institution.

bringing a series of the weird and unconventional, and charged with ensuring that the AAF acquired large numbers of aircraft, and quickly, the AAF's engineers did not pursue the project.[15]

So the Gluhareff-Griswold Dart never reached the hardware stage, the failure to build it counting as a loss to American midcentury aeronautics. As for Gluhareff, though he had made notable contributions to Sikorsky's large flying boats (and would, as well, to his helicopters), he continued

---

15. Letter, Roger W. Griswold to Maj. Donald R. Eastman, Oct. 22, 1946, Gluhareff Dart accession file, NASM.

to explore the basic design of his intriguing if abortive configuration, proposing a variety of derivatives, including in 1959 a Mach 2+ supersonic transport with a small canard wing and double-deck fuselage.[16] If the Dart never saw development, its configuration nevertheless proved significant. In 1944, Griswold resurrected the Dart shape for a proposed 2,000-pound guided glide bomb, or "glomb." The Army Air Forces recommended he obtain the NACA's opinion of its aerodynamics, and for this, Griswold turned to Langley Memorial Aeronautical Laboratory. There, on August 19, he met with the NACA's resident aerodynamic expert on "pilotless missiles," Robert T. Jones. Out of that contact would emerge both the American delta and swept wing.

Enter Robert T. Jones . . .

"R.T." Jones was a brilliant, flight-obsessed, and largely self-taught fluid dynamicist, having dropped out of the University of Missouri to join a flying circus, then working as a designer for Nicholas-Beazley, a small Missouri aircraft company. When the Great Depression collapsed the firm, his father used political connections as Chairman of the local Democratic Party to secure Jones a job running elevators in the U.S. Capitol. In his spare time and evenings, he studied mathematics and aerodynamics with Albert Zahm, the aeronautics Chair at the Library of Congress, and with Max Munk at Catholic University. Despite his lack of a formal engineering degree, through the efforts of Representative David Lewis (a homespun Maryland progressive with a strong interest in self-improvement who had taken math instruction from the young elevator operator), Jones received a temporary appointment as a "scientific aide" to the NACA. There, he quickly proved such a gifted and insightful researcher that he soon secured a coveted permanent position at Langley, consorting with the likes of John Stack, Eastman Jacobs, and Theodore Theodorsen.[17]

As he considered Griswold's "glomb," Jones recognized that its extremely low aspect ratio shape (that is, a shape having a very long

---

16. M.E. Gluhareff, "Tailless Airplane," U.S. patent No. 2,511,502, issued June 13, 1950; "Sikorsky Envisions Supersonic Airliner," *Aviation Week* (May 4, 1959), pp. 67–68; M.E. Gluhareff, "Aircraft with Retractable Auxiliary Airfoil," U.S. patent No. 2,941,752, issued June 21, 1960.
17. See William Sears's biographical introduction to the "Collected Works of Robert T. Jones," NASA TM-X-3334 (1976), pp. vii–ix; and Walter G. Vincenti, "Robert Thomas Jones," in *Biographical Memoirs*, vol. 86 (Washington: National Academy of Sciences, 2005), pp. 3–21.

wing root in relation to its total wingspan) could not be adequately analyzed using conventional Prandtl-rooted "lifting line" theory. Instead, Jones drew on the work of his mentor Munk, using papers that Munk had written on the flow of air around inclined airship hulls and swept wings, and one by the Guggenheim Aeronautical Laboratory's Hsue-shen Tsien, a von Kármán associate at the California Institute of Technology (Caltech), on airflow around inclined bodies of revolution. He analyzed it using linear equations governing two-dimensional incompressible flow, considering his results of little practical value, recalling three decades later, "I thought, well, this is so crude, nobody would be interested. So I just hid it in my desk."[18]

But it sparked his curiosity, and in January 1945, by which time he was busy thinking about nonlinear compressible flows, he had a revelation: the equations he had developed months earlier for the glomb analysis could be applied to a low aspect triangular wing operating in supersonic flow, one whose wing-leading edges were so sharply swept as to place them within the shock cone formed around the vehicle and hence operating in subsonic flow. In these conditions, the wing was essentially "fooled" into behaving as if it were operating at a much lower Mach number. As Jones recalled, "It finally dawned on me that the slender wing theory would hold for compressible flow and even at supersonic speed if it were near the center of the Mach cone. So, I immediately got the paper out and I added the compressible flow parts to it, which was really the important part, and then I wondered well, why is it that this slender wing doesn't have an effect on compressibility? Then I realized that it was because the obliquity of the edge and that this is the simple sweep theory and would work in spite of the compressibility effect. So, I wrote a paper which incorporated the slender wing theory and also sweep theory."[19] Jones then moved from considering a slender triangular delta [Δ] to the sharply sweptback wing [^], the reverse of

---

18. Transcript of interview of R.T. Jones by Walter Bonney, Sept. 24, 1974, p. 5, in Jones biographical file, No. 001147, Archives of the NASA Historical Division, National Aeronautics and Space Administration, Washington, DC.
19. Transcript of Jones-Bonney interview, p. 5; Hallion conversation with Dr. Robert T. Jones at NASA Ames Research Center, Sunnyvale, CA, July 14, 1977; Max M. Munk, "The Aerodynamic Forces on Airship Hills, NACA Report No. 184 (1923); Max M. Munk, "Note on the Relative Effect of the Dihedral and the Sweep Back of Airplane Wings," NACA TN-177 (1924); H.S. Tsien, "Supersonic Flow Over an Inclined Body of Revolution," *Journal of the Aeronautical Sciences*, vol. 5, no. 2 (Oct. 1938), pp. 480–483.

Germany, where the high-speed swept wing had preceded, not followed, the delta.[20]

Jones's delta and swept wing utilized, for their time, very thin airfoil sections, ones typical of supersonic aircraft to come. In contrast, German swept and delta wing developer Alexander Lippisch had employed much thicker sections that proved unsuitable for transonic flight. His tailless rocket-propelled swept wing Me 163 Komet ("Comet") interceptor, for example, essentially became uncontrollable at speeds slightly above Mach 0.82 thanks to stability changes induced by shock wave formation on its relatively thick wing. His design for a rocket-boosted, ramjet-powered delta fighter, the P 13, had such thick wing and tail sections—the pilot actually sat within the leading edge of the vertical fin—that it could never have achieved its desired transonic performance. As discussed subsequently, postwar NACA tests of a captured glider configuration of this design, the DFS DM-1, confirmed that transonic delta wings should be far thinner, with sharper leading edges. As a consequence, NACA researchers rejected the Lippisch approach, and, though some of them tried extrapolations of his designs (but with lower thickness-chord ratios and sharper leading edges), the NACA (and industry as well) adapted instead the thin slender delta, à la Jones.[21]

Dissemination, Deliberation, and Confirmation

In February 1945, Jones showed his notes on sweep to Jean Roché, the Army Air Forces technical liaison at Langley, and informed others as well, including Maj. Ezra Kotcher of the AAF's Air Technical Service

---

20. Note that although Lippisch called his tailless aircraft "deltas" as early as 1930, in fact they were generally broad high aspect ratio wings with pronounced leading edge taper, akin to the wing planform of America's classic DC-1/2/3 airliners. During the Second World War, Lippisch did develop some concepts for sharply swept deltas (though of very thick and impracticable wing section). Taken all together, Lippisch's deltas, whether of high or low aspect ratio planform, were not comparable to the thin slender and sharply swept (over 60 degrees) deltas of Jones, and Gluhareff before him, or Dietrich Küchemann at the Royal Aircraft Establishment afterwards, which were more akin to high-supersonic and hypersonic shapes of the 1950s–1960s.

21. For DM-1 and extrapolative tests, see Herbert A. Wilson, Jr., and J. Calvin Lowell, "Full-Scale Investigation of the Maximum Lift and Flow Characteristics of an Airplane Having Approximately Triangular Plan Form," NACA RM-L6K20 (1947); J. Calvin Lovell and Herbert A. Wilson, Jr., "Langley Full-Scale-Tunnel Investigation of Maximum Lift and Stability Characteristics of an Airplane Having Approximately Triangular Plan Form (DM-1 Glider)," NACA RM-L7F16 (1947); and Edward F. Whittle, Jr., and J. Calvin Lovell, "Full-Scale Investigation of an Equilateral Triangular Wing Having 10-Percent-Thick Biconvex Airfoil Sections," NACA RM-L8G05 (1948).

Command, and NACA colleagues Arthur Kantrowitz and Hartley A. Soulé.[22] Kotcher passed it along to von Kármán and Tsien—then working as scientific advisers to Gen. Henry H. "Hap" Arnold, the Army Air Forces' Chief of Staff—and Soulé and Kantrowitz urged Jones to inform the Agency's Director of Research, George W. Lewis, of his discovery.[23] Accordingly, on March 5, 1945, Jones informed Lewis, "I have recently made a theoretical analysis which indicates that a V-shaped wing travelling point foremost would be less affected by compressibility than other planforms. In fact, if the angle of the V is kept small relative to the Mach angle [the angle of the shockwave], the lift and center of pressure remain the same at speeds both above and below the speed of sound."[24] Jones subsequently undertook tests in the Langley 9-inch supersonic tunnel of a small, 4-inch-long daggerlike sheet-steel triangular wing with rounded leading edges and a span of only 1.5 inches, tests that complemented other trials at Aberdeen, MD, arranged by von Kármán and Tsien.

The Langley tests, through the transonic region and up to Mach 1.75, confirmed his expectations, and Jones published his first test results May 11, 1945, noting, "The lift distribution of a pointed airfoil travelling point-foremost is relatively unaffected by the compressibility of the air below or above the speed of sound."[25] This was almost 2 weeks before Lippisch informed von Kármán, then leading an AAF European study team, of his high-speed delta concepts (during a technical intelligence interrogation at St. Germain, France, on May 23), not quite a month before von Kármán assistant Clark Millikan visited the Messerschmitt advanced projects group at Oberammergau on June 9–10 and interrogated Waldemar Voigt about his swept wing fighter concepts, and well over a month before Millikan journeyed to Völkenrode to inter-

---

22. In 1944, Kotcher had conceived a rocket-powered "Mach 0.999" transonic research airplane (a humorous reference to the widely accepted notion of an "impenetrable" sonic "barrier") that subsequently inspired the Bell Aircraft Corporation to undertake design of the XS-1, the world's first supersonic manned airplane.
23. Kantrowitz would pioneer high-Mach research facilities design, and Soulé would serve the NACA as research airplane projects leader, supervising the Agency's Research Airplane Projects Panel (RAPP), a high-level steering group coordinating the NACA's X-series experimental aircraft programs.
24. Memo, Jones to Lewis, Mar. 5, 1945; see also ltr., Jones to Ernest O. Pearson, Jr., Feb. 2, 1960, and Navy/NACA Record of Invention Sheet, Apr. 10, 1946, Jones biographical file, NASA.
25. Robert T. Jones, "Properties of Low-Aspect-Ratio Pointed Wings at Speeds Below and Above the Speed of Sound," NACA TN-1032 (1946), p. 11 [first issued at NACA LMAL on May 11, 1945].

rogate German swept wing inventor Adolf Busemann, on June 20–21.[26]

Langley's peer reviewers and senior Agency official Theodore Theodorsen did not immediately accept Jones's assumption that a unified slender wing theory could apply to both compressible and incompressible flows and even questioned the evidence of sweep's benefits. Fortunately, Jones was greatly assisted in confounding skeptics by the timely results of NACA tunnel tests and falling body experiments, which left little doubt that sweep worked. As well, an associate of Jones made a most helpful discovery: locating a 1942 British translation of Busemann's 1935 paper. Evidence of an enemy's interest coincident with one's own work always heightens its perceived value, and undoubtedly, the Busemann paper, however dated, now strongly bolstered Jones's case. When it became time to assemble a bibliography for his swept wing report, Jones added Busemann's paper and other German sources by Albert Betz, H.G. Küssner, Ludwig Prandtl, and Hermann Schlichting, though it is unclear whether this reflected a collegial respect across the chasm of war or simply a shrewd appreciation of their persuasive value.[27]

Langley released his report in late June 1945.[28] In it, Jones noted: "the attachment of plane waves to the airfoil at near-sonic or supersonic speeds (Ackeret theory) may be avoided and the pressure drag may be reduced by the use of planforms in which the angle of sweepback is greater than the Mach angle. The analysis indicates that for aerodynamic efficiency, wings designed for flight at supersonic speeds should be swept back at an

---

26. For Millikan visit to Germany, see Millikan Diary 6, Box 35, Papers of Clark B. Millikan, Archives, California Institute of Technology, Pasadena, CA; Alexander Lippisch, ltr. to editor, *Aviation Week and Space Technology* (Jan. 6, 1975); in 1977, while curator of science and technology at the National Air and Space Museum, the author persuaded Jones to donate his historic delta test model to the museum; he had been using it for years as a letter opener!

27. Jones noted afterward that at Volta, Busemann "didn't have the idea of getting the wing inside the Mach cone so you got subsonic flow. The real key to [the swept wing] was to get subsonic flow at supersonic speed by getting the wing inside the Mach cone . . . the development of what I would say [was] the really correct sweep theory for supersonic speeds occurred in Germany in '43 or '44, and with me in 1945." (See transcript of Jones-Bonney interview, p. 6). But German researchers had mastered it earlier, as evident in a series of papers and presentations in a then-"*Geheim*" ("Secret") conference report by the Lilienthal-Gesellschaft für Luftfahrtforschung, *Allgemeine Strömungsforschung: Bericht über die Sitzung Hochgeschwindigkeitsfragen am 29 und 30 Oktober 1942 in Berlin* (Berlin: LGF, 1942).

28. For his report, see Robert T. Jones, "Wing Planforms for High-Speed Flight," NACA TN-1033 (1946) [first issued at LMAL on June 23, 1945, as Confidential Memorandum Report L5F21]. Jones's tortuous path to publication is related in James R. Hansen's *Engineer in Charge: A History of the Langley Aeronautical Laboratory, 1917–1958*, SP-4305 (Washington: NASA, 1987), pp. 284–285.

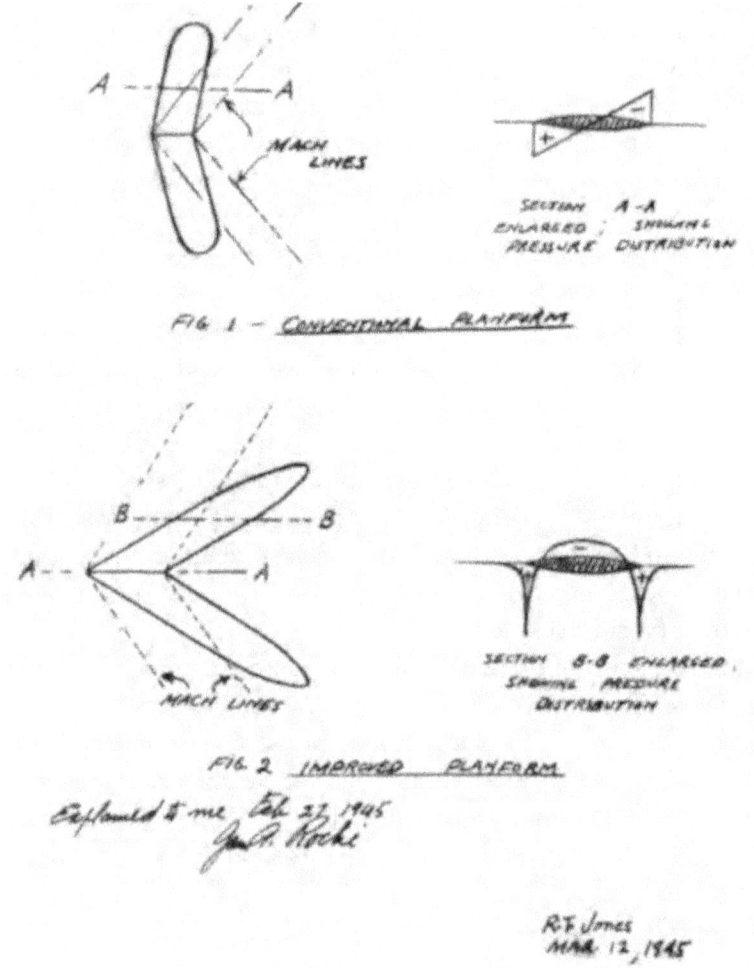

Jones showed these notes on the concept of high-speed wing sweep to Langley's AAF technical liaison representative Jean Roché on February 27, 1945. NASA.

angle greater than the Mach angle and the angle of sweepback should be such that the component of velocity normal to the leading edge is less than the critical speed of the airfoil sections. This principle may also be applied to wings designed for subsonic speeds near the speed of sound, for which the induced velocities resulting from the thickness might otherwise be sufficiently great to cause shock waves."[29] Such marked the effective birth of

---

29. Jones, "Wing Planforms for High-Speed Flight," NACA TN-1033, p. 1.

Case 1 | Sweep and Swing: Reshaping the Wing for the Jet and Rocket Age

the high-speed swept wing airplane in the United States, as his report weeks earlier had marked the birth of the American high-speed delta.

By the time Jones's report appeared, Germany's aeronautical establishment was already under the microscope of Allied technical intelligence, whose teams swiftly focused on its intensive investment in swept wing aerodynamics for its missiles and aircraft. Replicating reaction to the earlier "discovery" of "Göttingen aerodynamics" after the First World War, the post–Second World War influence of German example and practice was even more profound. Indeed, it affected the entire postwar course of European, Soviet, and American high-speed aerodynamic research, development, test, evaluation, and acquisition. In the increasingly tense national security environment of the burgeoning Cold War, the national intelligence services of the various advanced aeronautical nations understandably maintained very active technical collection efforts to learn what they did not already know.[30]

---

30. For the United States, this meant that Soviet intelligence collectors increasingly focused on American high-speed research. Bell Aircraft Corporation, manufacturer of the first American jet airplane, the first supersonic airplane, and advanced swept wing testbeds (the X-2 and X-5), figured prominently as a Soviet collection target as did the NACA. NACA engineer William Perl (born Mutterperl), a member of the Rosenberg spy ring who passed information on aviation and jet engines to Soviet intelligence, worked as a postwar research assistant for Caltech's Theodore von Kármán, director of the Guggenheim Aeronautical Laboratory of the California Institute of Technology (GALCIT), the Nation's premier academic aero research facility. He cultivated a close bond with TvK's sister Josephine ("Pipa") and TvK himself. Perl had almost unique access to the highest-level NACA and GALCIT reports on high-speed flight, and the state of advanced research and facilities planning for them and the U.S. Air Force. He associated as well with NACA notables, including Arthur Kantrowitz, Eastman Jacobs, and Robert T. Jones. So closely was he associated with von Kármán that he once helpfully reminded him where to find the combination to an office safe! He helped screen sensitive NACA data for a presentation TvK was making on high-speed stability and control, and TvK recommended Perl for consultation on tunnel development at the proposed new Arnold Engineering Development Center (AEDC) in Tennessee. Perl was unmasked by the Venona signals intelligence decryption program, interrogated on his associations with known Communists, and subsequently arrested and convicted of perjury. (He had falsely denied knowing the Rosenbergs.) More serious espionage charges were not brought, lest court proceedings compromise the ongoing *Venona* collection effort. The Papers of Theodore von Kármán, Box 31, Folder 31.38, Archives of the California Institute of Technology, and the Federal Bureau of Investigations' extensive Perl documentation contain much revealing correspondence on Perl and his associates. I thank Ernest Porter and the FBI historical office for arranging access to FBI material. See also Katherine A.S. Sibley, *Red Spies in America: Stolen Secrets and the Dawn of the Cold War* (Lawrence: University Press of Kansas, 2004); and John Earl Haynes and Harvey Klehr's *Early Cold War Spies: The Espionage Trials that Shaped American Politics* (Cambridge: Cambridge University Press, 2006) for further details on the Perl case.

The North American XP-86, prototype of the F-86 Sabre family, represented an amalgam of German and American swept wing and streamlined aerodynamics. USAF.

Swept Wing Challenges

The NACA so rapidly focused its attention on swept planforms that, within 2 years of the end of the Second World War, George Gray, author of a popular yet surprisingly detailed study of the Agency, could already write: "Just how far the sweepback principle can be applied with resulting advantage is a question. . . . At about 90 percent of the speed of sound both sweepback and low aspect ratio begin to be of value, and wings that combine the two features seem to offer a promising choice. At about Mach number 1.50, a sweepback of 60 degrees seems necessary to escape the backward flare of the Mach angle. . . . At Mach number 2.00, the angle is so acute that it is impossible to avoid it and still preserve the wings. It may be that designers preparing for flight at this speed will return to wings of low angles of sweep, and place their main dependence for drag reduction on thinning the profiles, lowering the aspect ratio, and sharpening the edges of wings."[31] By 1950, this grow-

---

31. George W. Gray, *Frontiers of Flight: The Story of NACA Research* (New York: Knopf, 1948), p. 348.

ing confidence in the old-new swept planform had resulted in transonic and supersonic research airplanes, a variety of military prototypes, and two operational jet fighters that would shortly clash over North Korea: the American F-86 Sabre (first flight in October 1947) and, in the Soviet Union, the MiG-15 (first flight in December 1947).[32]

Swept wing aircraft, for all their high-speed advantages, posed daunting stability, control, and handling qualities challenges. Foremost of these was pitch-up at low and high speeds, resulting from deteriorating longitudinal stability.[33] A swept wing airplane's lateral-directional stability was compromised as well by so-called "dihedral effects." Swept wing aircraft with excessive dihedral experienced pronounced combined rolling and yawing "Dutch roll" motions, which would be unacceptable on both production civil and military designs.[34] Such motions would induce airsickness in passengers on large aircraft and, on bomber, fighter, and attack aircraft, prevent accurate tracking of a maneuvering target or accurate bomb release. (Indeed, it was largely because of this kind of behavior that the U.S. Air Force did not proceed with production of Northrop's YB-49 flying wing jet bomber.) Adverse yaw posed another problem. At higher speeds, as a swept wing plane rolled from aileron

---

32. Re: German high-speed influence in the U.S., Britain, and Russia, see H.S. Tsien, "Reports on the Recent Aeronautical Developments of Several Selected Fields in Germany and Switzerland," in Theodore von Kármán, ed., *Where We Stand: First Report to General of the Army H.H. Arnold on Long Range Research Problems of the Air Forces with a Review of German Plans and Developments* (Washington: HQ AAF, Aug. 22, 1945), Microfilm Reel 194, Papers of Gen. Henry H. Arnold, Manuscript Division, U.S. Library of Congress, Washington, DC; Ronald Smelt, "A Critical Review of German Research on High-Speed Airflow," *Journal of the Royal Aeronautical Society,* vol. 50, No. 432 (Dec. 1946), pp. 899–934; Andrew Nahum, "I Believe the Americans Have Not Yet Taken Them All!" in Helmuth Trischler, Stefan Zeilinger, Robert Bud, and Bernard Finn, eds., *Tackling Transport* (London: Science Museum, 2003), pp. 99–138; Matthew Uttley, "Operation 'Sturgeon' and Britain's Post-War Exploitation of Nazi German Aeronautics," *Intelligence and National Security,* vol. 17, no. 2 (Sum. 2002), pp. 1–26; M.I. Gurevich, "O Pod'emnoi Sile Strelovidnogo Kryla v Sverkhzvukovom Potoke," *Prikladnaya Matematika i Mekhanika,* vol. 10 (1946), translated by the NACA as "Lift Force of an Arrow-Shaped Wing," NACA TM-1245 (1949). Gurevich, cofounder of the MiG bureau (he is the "G" in "MiG") was subsequently principal aerodynamicist of the MiG-15, the Soviet Union's swept wing equivalent to the American F-86. For a detailed examination of F-86 wing development and the influence of German work (particularly Göthert's) upon it, see Morgan M. Blair, "Evolution of the F-86," AIAA Paper 80-3039 (1980).
33. Pitch-up was of such significance that it is discussed subsequently in greater detail within this essay.
34. First comprehensively analyzed by Max M. Munk in his "Note on the Relative Effect of the Dihedral and the Sweep Back of Airplane Wings," NACA TN-177 (1924).

deflection, it experienced higher drag and loss of lift involving the lowered wing, generating a tendency of the airplane to turn (reverse) into the direction of the raised wing, effectively doing the opposite of what the pilot intended. Adverse yaw could be caused by aeroelastic effects as well. That swept wing aircraft would possess behavior characteristics significantly different than conventional straight wing designs did not come as a surprise to the NACA or other aerodynamic researchers in America and overseas. But all recognized the need to complement theory and ground-test methodologies with flight research.

The peculiarities of swept wing aircraft, at a time when early jet aircraft lacked the power-to-weight advantages of later designs, could—and often did—prove fatal. For example, Boeing designed the B-47, America's first large swept wing aircraft, with pod-mounted engines and a broad, highly tapered, thin swept wing. During flight-testing at higher speeds, test pilots found aileron input to roll the aircraft would twist the wing, the aileron effectively acting as a trim-tab does on a control surface. The twisted wing would overcome the rolling moment produced by the aileron, rolling the aircraft in the opposite direction. Aeroelastic structural divergence caused several accidents of the B-47 during its flight-testing and service introduction, forcing the Air Force to limit its permissible airspeed to 425 knots, as high as it could be safely flown if roll reversal were to be avoided. As a result, Boeing built its successors, the XB-52 and the Model 367-80 (prototype for the KC-135 family and inspiration for the civil 707), with much thicker wing roots and structures that were torsion resistant but that could still flex vertically to absorb structural loads and gust-induced loads during flight.[35]

Confronting Pitch-Up

But the most serious swept wing problem in the early jet era was pitch-up, a condition affecting both low- and high-speed flight, reflecting stall onset either from decreasing speed (low-speed pitch-up) or from trim changes during high-speed flight, particularly during accelerated

---

35. See John E. Steiner, "Transcontinental Rapid Transit: The 367-80 and a Transport Revolution—The 1953–1978 Quarter Century," AIAA Paper 78-3009 (1978), p. 93; John E. Steiner, "Jet Aviation Development: A Company Perspective," in Walter J. Boyne and Donald H. Lopez, eds., *The Jet Age: Forty Years of Jet Aviation* (Washington: Smithsonian Institution Press, 1979), pp. 145–148; and William H. Cook, *The Road to the 707: The Inside Story of Designing the 707* (Bellevue, WA: TYC Publishing Co., 1991), pp. 145–205.

maneuvers, such as "wind-up" turns that rapidly increased g-loading and angle of attack. Pitch-up occurred at the breakpoint in a lift curve, immediately beyond the peak point where the airplane's wing was operating at its highest lift-producing angle of attack, with its lift coefficient at maximum value. At the breakpoint, the wing would begin stalling, with flow separation from the airfoil, breaking the circulatory flow pattern around the wing. In ideal circumstances with a straight wing aircraft, the change in lift would occur simultaneously spanwise across the wing and would typically trigger a nose drop. But in a swept wing aircraft, the stall would first begin at the tips and progress inward, the center of lift shifting forward. As the plane's longitudinal (nose-up nose-down) stability decreased, the shifting center of lift would abruptly rotate the nose upward (hence the use of the expression "pitch-up"), even at a rate of onset beyond the capabilities of its elevator control surfaces to correct. As well, of course, since the ailerons that governed lateral control (roll control) were typically located outboard on a wing, a swept wing airplane could lose its lateral control authority precisely at a point when the pilot needed as much control capability and reserve as possible. Because stall onset is not always triggered uniformly, a swept wing airplane nearing the pitch-up point could experience sudden loss of lift on one wing, inducing abrupt rolling motions (called "wing dropping"), complicating its already dangerous low-speed behavior.

There was, of course, the possibility of overcoming such problems by sweeping a wing forward, not aft. A forward-swept wing (FSW) had both desirable high – and low-speed aerodynamic characteristics. Since the spanwise flow would run from the tips to the fuselage, the outer portions of the wing would stall last, thus preserving lateral control. As well, it would have more desirable pitching characteristics. Already, in the midst of the Second World War, the Germans had flown an experimental bomber, the Junkers Ju 287, featuring a forward swept wing, and a number of aircraft and missile projects were forecast for such planforms as well. The forward swept wing, and combined-sweep "M," "W," and even "X" planforms, received a great deal of postwar attention, both in America and abroad. Researchers at Langley modified wind tunnel and configuration models of both the XS-1 and D-558 to employ forward-swept wing planforms, and tested conceptual planforms with both aft and forward sweep to develop comparison data. But while the FSW undoubtedly had better low-speed behavior, it had higher profile drag and posed difficult structural problems for designers. In the precomposite structure era, an FSW had to be necessarily heavier than an aft-swept wing to avoid aeroelastic flexing that could inhibit both good flight performance and even flight safety. Further, the structural and weight limitations also limited the sweep angles that an FSW could then have; even as late as the 1960s, when Germany produced a business aircraft (the Hamburger Flugzeugbau HFB-320 Hansa Jet), it possessed only modest forward sweep and, though flown successfully and built in small numbers, was not a commercial success. It would take over three decades before the advent of computerized flight control, composite structures, and a more radical vision of forward sweep application would result in experimental planforms like the Rockwell Sabrebat FSW design concept, the piloted Air Force–DARPA–NASA Grumman X-29 (and, in Russia, an X-29-like experimental aircraft, the Sukhoi Su-37). Even so, and even though forward sweep would be applied to some weapon systems (for example, the AGM-129 stealthy cruise missile, where it contributed to its low radar reflectivity), forward wing sweeping would remain the exception to "normative" aft-swept wing design practice.[36]

Pitch-up was profoundly dangerous. At low speeds in proximity to the ground, it could—and often did—trigger a disastrous departure and crash. The recognition of such problems had caused the U.S. Navy to procure two modified Bell P-63 Kingcobra fighters (designated L-39), which had their wing panels replaced with 35-degree swept wing sections, and a fuselage extension to accommodate their now-changed center of lift. Not intended for high speeds, these two low-speed swept wing research aircraft were extensively flown by various contractor, Navy, and NACA research pilots to assess the basic behavior of the swept wing, with and without lift-and-control-augmenting devices such as wing slats and flaps. They quickly encountered its limitations. On one flight with the plane in "clean" (i.e., slat-free) configuration, Bell Company test pilot A.M. "Tex" Johnston gradually raised the nose of the plane while retarding power. After just "a slight tremor" indicating the onset of asymmetric tip stall, it "instantaneously rolled to an almost inverted

---

36. See, for example, Richard T. Whitcomb, "An Investigation of the Effects of Sweep on the Characteristics of a High-Aspect-Ratio Wing in the Langley 8-Ft. High Speed Tunnel," NACA RM-L6J01a (1947), conclusion 4, p. 19; Stephen Silverman, "The Next 25 Years of Fighter Aircraft," AIAA Paper No. 78-3013 (1978); Glen Spacht, "X-29 Integrated Technology Demonstrator and ATF," AIAA Paper No. 83-1058 (1983).

position."[37] Grumman test pilot Corwin "Corky" Meyer recalled that while the L-39 was "docile" with leading edge slats, without them it "cavorted like a cat on catnip."[38] The two L-39 aircraft furnished vital insight into the low-speed performance and limitations of swept wing aircraft, but they also clearly demonstrated that such aircraft could, in fact, be safely flown if their wings incorporated careful design and safety devices such as fixed leading edge slots or movable slats.[39]

In military aircraft, pitch-up could prevent a pilot from maneuvering effectively against a foe, could lead to loss of control of the airplane, and could result in such excessive airframe loadings that an airplane would break up. It was no respecter of designs, even outstanding ones such as North American's evocative F-86 Sabre, generally considered the finest jet fighter of its time by both American and foreign test pilots. First flown in October 1947, the Sabre quickly became an internationally recognized symbol of aeronautical excellence and advancement. When British test pilot Roland Beamont, a distinguished Royal Air Force fighter ace of the Second World War, evaluated the Sabre at Muroc Dry Lake in May 1948

---

37. A.M. "Tex" Johnston with Charles Barton, *Tex Johnston: Jet-Age Test Pilot* (Washington: Smithsonian Institution Press, 1991), p. 105. The designation "L-39" could be taken to imply that the swept wing testbeds were modifications of Bell's earlier and smaller P-39 Airacobra. In fact, it was coincidence; the L-39s were P-63 conversions, as is evident from examining photographs of the two L-39 aircraft.

38. Corwin H. Meyer, *Corky Meyer's Flight Journal: A Test Pilot's Tales of Dodging Disasters—Just in Time* (North Branch, MN: Specialty Press, 2006), p. 193.

39. NACA's L-39 trials are covered in three reports by S.A. Sjoberg and J.P. Reeder: "Flight Measurements of the Lateral and Directional Stability and Control Characteristics of an Airplane Having a 35° Sweptback Wing with 40-Percent-Span slots and a Comparison with Wind-Tunnel Data," NACA TN-1511 (1948); "Flight Measurements of the Longitudinal Stability, Stalling, and Lift Characteristics of an Airplane Having a 35° Sweptback Wing Without Slots and With 40-Percent-Span Slots and a Comparison with Wind-Tunnel Data," NACA TN-1679 (1948); and "Flight Measurements of the Stability, Control, and Stalling Characteristics of an Airplane Having a 35° Sweptback Wing Without Slots and With 80-Percent-Span Slots and a Comparison with Wind-Tunnel Data," NACA TN-1743 (1948). The American L-39s were matched by foreign equivalents, most notably in Sweden, where the Saab company flew a subscale swept wing variant of its conventional *Safir* light aircraft, designated the Saab 201, to support development of its J29 fighter, Western Europe's first production swept wing jet, which first flew in Sept. 1948. Like both the F-86 and MiG-15, it owed its design largely to German inspiration. Saab researchers were so impressed with what they had learned from the 201 that they subsequently flew another modified Safir, the Saab 202, with a more sharply swept wing planform intended for the company's next jet fighter, the J32 *Lansen* (Lance). See Hans G. Andersson, *Saab Aircraft Since 1937* (Washington: Smithsonian Institution Press, 1989), pp. 106, 117.

(a month after it had dived past the speed of sound, becoming the world's first supersonic turbojet airplane), he likewise dived it through Mach 1, thus becoming the first supersonic British pilot. Afterward, he noted approvingly in his test report, "The P-86 is an outstanding aircraft."[40] The Sabre's reputation was such that British authorities (frustrated by the slow development pace of Albion's own swept wing aircraft) tellingly referred to it simply as "That Aircraft." Vickers-Supermarine test pilot David Morgan recalled, "No British fighter of the day could match the handling of the North American F-86."[41] Indeed, designers from his company, frustrated by their slow progress turning the experimental Swift into a decent airplane, even resorted to crude subterfuge in an effort to unlock the Sabre's secrets. When a pair of Canadian pilots landed at the Supermarine plant in their Canadair-built Sabres, company officials, with apparent generosity, laid on a fancy lunch, driving them off to a local hotel. While the visiting airmen dined and chatted with solicitous Supermarine representatives, another team of engineers "swarmed over the Sabres to study their construction," marveling at "this splendid aircraft."[42]

Yet however "splendid" "That Aircraft" might otherwise have been, the Sabre killed unwary pilots by the dozens in accidents triggered by its low-speed pitch-up tendencies. Apollo 11 astronaut Michael Collins recalled his introduction to the F-86 at Nellis Air Force Base as "a brutal process. . . . In the eleven weeks I was there, twenty-two people were killed. In retrospect it seems preposterous to endure such casualty rates without help from the enemy, but at the time the risk appeared perfectly acceptable. . . . I'm surprised to have survived. I have never felt quite so threatened since."[43] In over a decade of tests with various Sabre variants

---

40. XP-86 test report, May 21, 1948, reprinted in Roland Beamont, *Testing Early Jets: Compressibility and the Supersonic Era* (Shrewsbury: Airlife, 1990), p. 36. Beamont's achievement remained largely secret; the first British pilot to fly through the speed of sound in a British airplane was John Derry, who did so in Sept. 1948.
41. Quote from Nigel Walpole, *Swift Justice: The Full Story of the Supermarine Swift* (Barnsley, UK: Pen & Sword Books, 2004), p. 38.
42. Charles Burnet, *Three Centuries to Concorde* (London: Mechanical Engineering Publications Ltd., 1979), pp. 121, 123.
43. Michael Collins, *Carrying the Fire: An Astronaut's Journeys* (New York: Farrar, Straus, and Giroux, 1974), p. 9. Another Sabre veteran who went through Nellis at the same time recalled to the author how he once took off on a training sortie with ominous columns of lingering smoke from three earlier Sabre accidents.

to improve their low-speed handling qualities, NACA Ames researchers assessed a variety of technical "fixes." The most beneficial was the combination of artificial feel (to give the pilot more reassuring higher maneuvering control forces during the approach-to-landing, combined with greater inherent stability than possible with a non–artificial-feel system), coupled with leading-edge suction to draw off the boundary layer airflow.[44] First evaluated on a test rig installed in the Ames 40-foot by 80-foot full-scale wind tunnel, the boundary layer control (BLC) experiment on the F-86 proved most valuable. Ames researchers concluded: "Leading edge boundary-layer control was most effective in providing a large reduction in both stalling speed and approach speed together with an increased margin of lift for flare and maneuvering during the [landing] approach," an important point, particularly for swept wing naval aircraft, which had to be controllable down to a landing on the confined deck of an aircraft carrier.[45] The trials benefitted not only future swept wing studies but, more generally, studies of BLC applications for Vertical/Short Take-Off and Landing (V/STOL) aircraft systems as well.[46]

Nor were the Sabre's high-speed pitch-up characteristics innocuous. The NACA flew extensive Sabre evaluations at its High-Speed Flight Research Station and at Ames to refine understanding of

---

44. For development of control boost, artificial feel, and control limiting, see Robert G. Mungall, "Flight Investigation of a Combined Geared Unbalancing-Tab and Servotab Control System as Used with an All-Movable Horizontal Tail," NACA TN-1763 (1948); William H. Phillips, "Theoretical Analysis of Some Simple Types of Acceleration Restrictors," NACA TN-2574 (1951); R. Porter Brown, Robert G. Chilton, and James B. Whitten, "Flight Investigation of a Mechanical Feel Device in an Irreversible Elevator Control System of a Large Airplane," NACA Report No. 1101 (1952); James J. Adams and James B. Whitten, "Tests of a Centering Spring Used as an Artificial Feel Device on the Elevator of a Fighter Airplane," NACA RM-L52G16; and Marvin Abramovitz, Stanley F. Schmidt, and Rudolph D. Van Dyke, Jr., "Investigation of the Use of a Stick Force Proportional to Pitching Acceleration for Normal-Acceleration Warning," NACA RM-A53E21 (1953).
45. George E. Cooper and Robert C. Innis, "Effect of Area-Suction-Type Boundary-Layer Control on the Landing-Approach Characteristics of a 35° Swept-Wing Fighter," NACA RM-A55K14 (1957), p. 11. Other relevant Ames F-86 studies are: George A. Rathert, Jr., L. Stewart Rolls, Lee Winograd, and George E. Cooper, "Preliminary Flight Investigation of the Wing-Dropping Tendency and Lateral-Control Characteristics of a 35° Swept-Wing Airplane at Transonic Mach Numbers," NACA RM-A50H03 (1950); and George A. Rathert, Jr., Howard L. Ziff, and George E. Cooper, "Preliminary Flight Investigation of the Maneuvering Accelerations and Buffet Boundary of a 35° Swept-Wing Airplane at High Altitude and Transonic Speeds," NACA RM-A50L04 (1951).
46. Edwin P. Hartman, *Adventures in Research: A History of the Ames Research Center, 1940–1965*, SP-4302 (Washington: NASA 1970), p. 252.

its transonic pitch-up behavior, which test pilot A. Scott Crossfield recalled as "violent and dangerous."[47] It could easily exceed its design load factors, sometimes pitching as high as 10 g. At 25,000 feet, in the very midst of its combat operating envelope (and at lift coefficients less than its maximum attainable lift) the Sabre's pitch-up onset was so severe that g forces once momentarily "blacked out" the test pilot. Overall, after extensive Ames tests, the early slatted F-86A with a conventional fixed horizontal stabilizer and movable elevator was judged "unsatisfactory" by a group of highly experienced fighter test pilots, thanks to its "severe pitch-up tendencies." The same group found the later slat-less "6-3" F-86F (so-called because its wing extended forward 6 inches at the root and 3 inches at the tip, a modification made by North American based on Korean war experience) had "moderate" pitch-up tendencies. Because of this, and because it had an adjustable (not fixed) horizontal stabilizer in addition to its elevator, the pilots judged the F-86F's pitch-up behavior "unsatisfactory but acceptable."[48]

Worse swept wing problems plagued the Sabre's great adversary, the Soviet MiG-15. Unlike the Sabre, the MiG-15 had a less aerodynamically pleasing configuration, and its fixed horizontal stabilizer and elevator combination, located midway up the vertical fin, made it more susceptible to aerodynamic "blanketing" of the tail by the wing and, hence severe pitch-up problems, as well as limiting its transonic maneuverability (to the Sabre's advantage). During the Korean war, Sabre pilots frequently saw MiG pilots eject from otherwise perfectly sound aircraft that had pitched up during turns, stalled, and entered flat, unrecoverable spins. Nearly five decades later, Soviet pilot Stepan Mikoyan (nephew of Anushavan "Artem" Mikoyan, cofounder of the MiG design bureau) conceded that high-speed accelerated stalls often triggered unrecoverable spins, leading to "a number of ejections and fatal accidents."[49] Postwar American testing of a MiG-15 delivered by a defecting North Korean

---

47. A. Scott Crossfield with Clay Blair, *Always Another Dawn: The Story of a Rocket Test Pilot* (Cleveland: World Publishing Co., 1960), pp. 193–194. See also W.C. Williams and A.S. Crossfield, "Handling Qualities of High-Speed Airplanes," NACA RM-L52A08 (1952), p. 3; Melvin Sadoff, John D. Stewart, and George E. Cooper, "Analytical Study of the Comparative Pitch-Up Behavior of Several Airplanes and Correlation with Pilot Opinion," NACA RM-A57D04 (1957).
48. Sadoff, Stewart, and Cooper, "Analytical Study of Comparative Pitch-Up Behavior," p. 12.
49. S.A. Mikoyan, *Stepan Anastasovich Mikoyan: An Autobiography* (Shrewsbury: Airlife, 1999), p. 289.

pilot confirmed the MiG's marked vulnerability to pitch-up-induced stalls and spins; indeed, the defector's own instructor had been lost in one such accident. Not surprisingly, when Mikoyan produced the MiG-17—the lineal successor to the MiG-15—it had a very different outer wing configuration giving it more benign behavior.[50]

Western European swept wing aircraft exhibited similar problems as their American and Soviet counterparts. For a brief while, influenced by the Messerschmitt Me 163 Komet and a variety of other German projects, designers were enthralled with the swept wing tailless configuration, believing it could resolve both the challenges of high-speed flight and also furnish inherent stability.[51] Then, in September 1946, British test pilot Geoffrey de Havilland perished in an experimental tailless transonic research aircraft, the de Havilland D.H. 108 Swallow, when it began an undamped violently divergent longitudinal pitching oscillation at Mach 0.875, breaking up over the Thames estuary and proving that the "sound barrier" could bite.[52] Subsequently the NACA evaluated the Northrop X-4, a generally similar American configuration. Tested at high altitude (and hence, at low dynamic pressure), the X-4 fortunately never "diverged" as violently as the ill-fated D.H. 108. Instead, as NACA pilot A. Scott Crossfield remembered, at Mach 0.88 "it broke

---

50. Maj. Gen. H.E. "Tom" Collins, USAF (ret.), "Testing the Russian MiG," in Ken Chilstrom, ed., *Testing at Old Wright Field* (Omaha: Westchester House Publishers, 1991), p. 46.

51. Two not so taken with the swept tailless configuration were Douglas aerodynamicist L. Eugene Root and former Focke-Wulf aerodynamicist Hans Multhopp. After an inspection trip to Messerschmitt in August 1945, Root wrote "a tailless design suffers a disadvantage of small allowable center of gravity travel. . . . Although equally good flying qualities can be obtained in either [tailless or conventional] case, the tailless design is considered more dangerous at very high speeds. For example, the Me 262 has been taken to a Mach number of 0.86 without serious difficulty, whereas the Me 163 could not exceed M = 0.82. For the Me 163 . . . it was not considered possible fundamentally to control the airplane longitudinally past M = 0.82 in view of a sudden diving moment and complete loss of elevator effectiveness." See L.E. Root, "Information of Messerschmitt Aircraft Design," Item Nos. 5, 25, File No. XXXII-37, Copy 079 (Aug. 1945), p. 3, Catalog D52.1Messerschmitt/144, in the Wright Field Microfilm Collection, National Air and Space Museum Archives, Paul E. Garber Restoration Facility, Silver Hill, MD. Focke-Wulf's Hans Multhopp, designer of the influential T-tail sweptwing Ta 183, was even more dismissive. After the war, while working at the Royal Aircraft Establishment, he remarked that it constituted an "awful fashion;" see Nahum, "I Believe . . .," in Trischler, et al., ed., *Tackling Transport*, p. 118. Multhopp later came to America, joining Martin and designing the SV-5 reentry shape that spawned the SV-5D PRIME, the X-24A, and the X-38.

52. Burnet, *Three Centuries to Concorde*, p. 102.

into a steady porpoising motion, like an automobile cushioning over a washboard road."[53] Conventional tailed European swept wing designs followed the same steep learning curve as American ones. Britain's Supermarine Swift, a much-touted design from the builder of the legendary Spitfire, had a "vicious" transonic pitch-up. By the time it entered service, it was years late, obsolescent, and useless for any other role save low-level tactical reconnaissance.[54]

The Skyrocket: The NACA's Pitch-Up Platform

Pitch-up afflicted a wide range of early transonic and supersonic jet fighters, and the NACA was fortunate in having an available research airplane that could study swept wing behavior across the transonic regime. This aircraft was the Douglas D-558-2 Skyrocket, "Phase II" of the larger D-558 research aircraft program, a Douglas Company venture begun in 1945 and sponsored by the U.S. Navy and the NACA. The D-558 program had begun as a companion to the XS-1 effort and represented a different design approach. Where the XS-1 was rocket powered, the D-558 Skystreak used a turbojet; where the XS-1 employed an ogival projectile shape with a midwing of 8-percent thickness-chord ratio, the D-558 used a constant-diameter tube wrapped around an axial-flow turbojet engine and a low wing of 10-percent thickness-chord ratio; and where the XS-1 was air launched, the D-558 took off from the ground as a conventional airplane. Both were straight wing designs, with their adjustable stabilizers and movable elevators placed midway up their vertical fins. All together, the Navy ordered six D-558 aircraft from the firm.[55]

Originally, swept wings had not featured in the D-558 program. Then the discovery by Douglas engineers of a plethora of German technical reports (coupled with the work of Jones and others in the United States) caused the Navy, the NACA, and Douglas to modify the D-558

---

53. Crossfield with Blair, *Always Another Dawn*, p. 39; Melvin Sadoff and Thomas R. Sisk, "Longitudinal-Stability Characteristics of the Northrop X-4 Airplane (USAF No. 46-677)," NACA RM-A50D27 (1950); and Williams and Crossfield, "Handling Qualities of High-Speed Airplanes."
54. Quote from Walpole, *Swift Justice*, pp. 58, 66. Walpole, a former Swift pilot, writes affectionately but frankly of its strengths and shortcomings. See also Burnet, *Three Centuries to Concorde*, pp. 127–128. Burnet was involved in analyzing Swift performance, and his book is an excellent review of Supermarine and other British efforts at this time.
55. For the origins of the D-558 program, see Richard P. Hallion, *Supersonic Flight: Breaking the Sound Barrier and Beyond—The Story of the Bell X-1 and Douglas D-558* (New York: The Macmillan Co. in association with the Smithsonian Institution, 1972).

program.[56] The last three aircraft were completed as a new swept wing design. Initially, the planned modification seemed straightforward: replace the straight wing and tail surfaces with swept ones. In anticipation, Langley tested models of the D-558 with a variety of swept wings. But the possibility of giving the swept wing D-558 supersonic performance—something the D-558 straight wing lacked—resulted in a more radical redesign. Gone was the simple Pitot intake inlet. Instead, designer Edward "Ed" Heinemann and his team chose an ogival body shape resembling the XS-1. The new 35-degree slat-equipped swept wing was relocated to midfuselage position and given anhedral (droop), with the landing gear relocated into the fuselage. In contrast to the original single-engine D-558s, the new swept wing design featured both a 6,000-pound thrust rocket engine and a small turbojet. Thus recast, it received the designation D-558-2 and the name Skyrocket, to distinguish it from the straight wing Skystreak, itself redesignated D-558-1. The result was one of the most elegant and significant aircraft of all time.

The first D-558-2 flew in February 1948, though initial flight tests gave little hint of how remarkably versatile and successful it would prove. At max takeoff weight, it was so underpowered (and thus so sluggish) that it needed four solid-fuel jettisonable assistance takeoff (JATO) rockets to help kick it aloft. Eventually, the Navy and the NACA would arrange to take the second and third Skyrockets and modify them for air launch from a modified PB2-1S (Navy B-29) Superfortress, dramatically improving both their safety and high-speed performance; fuel previously

---

56. Particularly Bernard Göthert's "Hochgeschwindigkeitsmessungen an einem Pfeilflügel (Pfeilwinkel ϕ = 35°)," in the previously cited Lilienthal-Gesellschaft, *Allgemeine Strömungsforschung*, pp. 30–40, subsequently translated and issued by the NACA as "High-Speed Measurements on a Swept-Back [sic] Wing (Sweepback Angle ϕ = 35°)," NACA TM-1102 (1947), which directly influenced design of the 35-degree swept wings employed on the F-86, the B-47, and the D-558-2. Göthert, incidentally, used NACA airfoil sections for his studies, another example of the Agency's pervasive international influence. At war's end he was in Berlin; when ordered to report to Russian authorities, he instead fled the city, making his way back to Göttingen, where he met Douglas engineer Apollo M.O. Smith, with the Naval Technical Mission to Europe. Smith arranged for him to immigrate to the United States, where he had a long and influential career, rising to Chief Scientist of Air Force Systems Command, a position he held from 1964 to 1966. See Tuncer Cebeci, ed., *Legacy of a Gentle Genius: The Life of A.M.O. Smith* (Long Beach: Horizons Publishing, Inc., 1999), p. 32. I acknowledge with grateful appreciation notes and correspondence received from members of the D-558 design team in 1971–1972, including the late Edward Heinemann, L. Eugene Root, A.M.O. Smith, Kermit Van Every, and Leo Devlin, illuminating the origins of the Skystreak and Skyrocket programs.

spent climbing aloft could now be more profitably expended exploring the transonic and supersonic regimes. While the third aircraft retained its jet and rocket engine, the second had its jet engine removed and additional tanks for rocket propellant and oxidizer installed. Thus modified, the second aircraft reached Mach 2.01 in November 1953, flown by Scott Crossfield, the first piloted Mach 2 flight, having earlier attained an altitude of 83,235 feet, piloted by Lt. Col. Marion Carl, a noted Marine aviator. Eventually, the NACA received the first D-558-2 as well (which Douglas had employed for contractor testing). The Agency modified it as an all-rocket aircraft, though it only completed a single check flight before being retired.

Before all-rocket modification, the second Skyrocket introduced Agency pilots to the hazards of pitch-up. On August 8, 1949, during its seventh flight, pilot Robert Champine banked into a 4 g turn at Mach 0.6, and the Skyrocket violently pitched up, reaching 6 g. It responded rapidly to full-down elevator, and Champine made an uneventful (if prudently precautionary) landing. Thereafter, until returning the airplane to Douglas for all-rocket modification in 1950, the NACA flew extensive pitch-up investigations with it. In November, pilot John Griffith replicated the 4 g and Mach 0.6 pitch-up that Champine had experienced earlier. This time, however, he attempted to continue flying to more fully assess the Skyrocket's behavior. Thus challenged, it snap-rolled on its back. After recovering, Griffith probed its low-speed behavior, gradually slowing, with flaps and gear extended and wing slats closed. At 14,000 feet and 130 mph, the Skyrocket pitched up, rolling into a spin, and losing 7,000 feet of altitude before its pilot could recover.[57] Clearly its ugly behavior did not match its alluring form.

Focused on extending the Skyrocket's performance into the supersonic regime by modifying the second aircraft as a pure rocket plane, the NACA turned to the third aircraft, which retained its jet engine as well as its rocket, for future pitch-up research. Air-launched, the jet-and-rocket Skyrocket had tremendous research productivity; it could accelerate

---

57. Hallion, *Supersonic Flight*, pp. 151–152, based upon D-558 biweekly progress reports. As well, I thank the late Robert Champine for his assistance to my research. See also S.A. Sjoberg and R.A. Champine, "Preliminary Flight Measurements of the Static Longitudinal Stability and Stalling Characteristics of the Douglas D-558-II Research Airplane (BuAero No. 37974)," NACA RM-L9H31a (1949); W.H. Stillwell, J.V. Wilmerding, and R.A. Champine, "Flight Measurements with the Douglas D-558-II (BuAero No. 37974) Research Airplane Low-Speed Stalling and Lift Characteristics," NACA RM-L50G10 (1950).

Leading-edge wing chord extensions tested on the third D-558-2 Skyrocket, one of many combinations of flaps, slats, fences, and extensions evaluated in the NACA's 6-year-long study of the Skyrocket's pitch-up behavior. NASA.

into the supersonic regime, above Mach 1.1, and its jet engine enabled it to "loiter" in the transonic region, making repeated data-gathering runs. Its comprehensive instrumentation package enabled assessment of loads, pressure distributions, and accelerations, evaluated against background data on flight conditions, aircraft attitude, and control surface position and forces. Between the end of 1950 and the fall of 1956, it completed 66 research flights on pitch-up and associated transonic phenomena, including the evaluation of the effects external wing stores—tanks and bomb shapes—had on aircraft performance. It evaluated a variety of proposed aerodynamic solutions and fixes to resolve the pitch-up problem, including various wing fence designs to "channel" airflow and inhibit the characteristic spanwise-flow (flow toward the wingtips) found with swept wing planforms, various combinations of slat and flap position, changes to leading edge shape, and "sawtooth" leading edge extensions on its outer wing panels. All of this testing reinforced what engineers suspected, namely that no one overall technical fix existed that could resolve the pitch-up challenge. Rather, swept wing aircraft design was clearly situational, and, depending on the mission of the aircraft and its resulting design, combinations of approaches worked best, chief among them being low placement of the horizontal tail, below the chord-line of the wing, coupled with provision of stability augmentation and pitch-damping flight control technology.[58]

Ensuring Longitudinal Control: Transforming the Horizontal Tail

Though not seemingly connected to the swept wing, the researching and documenting of the advantages of low-placed horizontal tail surfaces constituted one of the major NACA postwar contributions to flight, one dramatically improving both the safety and flight performance of swept wing designs. As a consequence, the jet fighter and attack aircraft of 1958 looked very different than did the initial jet (and rocket) aircraft of the immediate postwar era. Then, high-speed aircraft designers had emphasized tailless planforms, or ones in which the horizontal tail was well up the vertical fin (for example, both the XS-1 and the D-558 families). A decade later, aircraft introduced into test or service—such as the Vought F8U-1 Crusader, the Republic F-105B Thunderchief, the Grumman F11F-1 Tiger, the McDonnell F4H-1 Phantom II, the North American A3J-1 Vigilante, and the Northrop N-156 (progenitor of both the T-38 supersonic trainer and the F-5 lightweight fighter)—shared common characteristics: irreversible power-operated flight controls, stability augmentation, and damping, large vertical fins for enhanced directional stability, area-ruling, and low-placed, all-moving tails. Foreign aircraft exhibited similar features: for example, the MiG-21, Folland Gnat, and English Electric Lightning.

Aircraft lacking such features manifested often-perilous behavior. The Douglas XF4D-1 Skyray, a graceful rounded delta, had a sudden transonic pitch change reflecting its legacy of Messerschmitt-inspired tailless aerodynamic design. During one test run to Mach 0.98, it pitched

---

58. Jack Fischel and Jack Nugent, "Flight Determination of the Longitudinal Stability in Accelerated Maneuvers at Transonic Speeds for the Douglas D-558-II Research Airplane Including the Effects of an Outboard Wing Fence," NACA RM-L53A16 (1953); Jack Fischel, "Effect of Wing Slats and Inboard Wing Fences on the Longitudinal Stability Characteristics of the Douglas D-558-II Research Airplane in Accelerated Maneuvers at Subsonic and Transonic Speeds," NACA RM-L53L16 (1954); Jack Fischel and Cyril D. Brunn, "Longitudinal Stability Characteristics in Accelerated Maneuvers at Subsonic and Transonic Speeds of the Douglas D-558-II Research Airplane Equipped with a Leading-Edge Wing Chord-Extension," NACA RM-H54H16 (1954); M.J. Queijo, Byron M. Jaquet, and Walter D. Wolmart, "Wind-Tunnel Investigation at Low Speed of the Effects of Chordwise Wing Fences and Horizontal-Tail Position on the Static Longitudinal Stability Characteristics of an Airplane Model with a 35° Sweptback Wing," NACA Report 1203 (1954); Jack Fischel and Donald Reisert, "Effect of Several Wing Modifications on the Subsonic and Transonic Longitudinal Handling Qualities of the Douglas D-558-II Research Airplane," NACA RM-H56C30 (1956).

up so violently that test pilot Robert Rahn blacked out, becoming one of the first pilots to experience sudden g-induced loss-of-consciousness (g-loc). Fortunately, he recovered and returned safely, the battered plane now marred by prominent stress-induced wrinkles, giving it a prunelike appearance.[59] When Grumman entered the transonic swept wing era, it did so by converting its conventional straight wing F9F-5 Panther into a swept wing design, spawning the F9F-6 Cougar. (The use of an identical prefix—"F9F"—indicates just how closely the two aircraft were related.) But the Cougar's swept wing, midplaced horizontal tail, and thick wing section (inherited from the firmly subsonic Panther) were ill matched. The new Cougar had serious pitch-up and departure characteristics at low and high speeds, forcing redesign of its wing before it could be introduced into fleetwide service. Even afterward, however, it retained some unpleasant characteristics, particularly a restricted angle-of-attack range during carrier landing approaches that gave the pilot only a small maneuver margin before the Cougar would become unstable. Well aware of the likely outcome of stalling and pitching up in the last seconds of flight prior to "trapping" on a carrier, pilots opted to fly faster, though the safety they gained came at the price of less-precise approaches with greater risk of "wave-offs" (aborted landings) and "bolters" (touching down beyond the cables and having to accelerate back into the air).[60]

McDonnell's XF-88, a beefy twin-engine jet fighter prototype from the late 1940s, was placed on hold while more powerful engines were developed. When finally ordered into development in the early 1950s as the F-101 Voodoo, it featured a T-tail, a most unwise choice. Acceptable on airliners and transports, the T-tail was anathema for high-performance jet fighters. The Voodoo experienced serious pitch-up problems, and the cure was less a "fix" than simply a "patch": McDonnell installed

---

59. Robert O. Rahn, "XF4D Skyray Development: Now It Can Be Told," 22nd Symposium, Society of Experimental Test Pilots, Beverly Hills, CA, Sept. 30, 1978; and Edward H. Heinemann and Rosario Rausa, *Ed Heinemann: Combat Aircraft Designer* (Annapolis: Naval Institute Press, 1980), p. 192. Years later, another Skyray pilot at the Naval Air Test Center experienced a similar mishap, likewise making a near-miraculous recovery; the plane was so badly stressed that it never flew again.
60. Meyer, *Flight Journal*, pp. 196–198; he was nearly killed on one low-altitude low-speed pitch-up that ended in a near-fatal spin. The Cougar's approach behavior resulted in a Langley research program flown using a F9F-7 variant, which highlighted the need for more powerful, responsive, and controllable aircraft, such as the later McDonnell F4H-1 Phantom II. See Lindsay J. Lina, Garland J. Morris, and Robert A. Champine, "Flight Investigation of Factors Affecting the Choice of Minimum Approach Speed for Carrier-Type Landings of a Swept-Wing Jet Fighter Airplane," NACA RM-L57F13 (1957).

a stick-kicker that would automatically push the stick forward as angle of attack increased and the Voodoo approached the pitch-up point. Wisely, for their next fighter project (the superlative F4H-1 Phantom II), McDonnell designers lowered the horizontal tail location to the base of the fin, giving it a characteristically distinctive anhedral (droop).[61]

Better yet, however, was placing the horizontal tail below the line of the wing chord, which, in practical terms, typically meant at the base of the rear fuselage, and making it all-moving as well. In 1947, even before the first supersonic flights of the XS-1, NACA Langley researchers had evaluated a wind tunnel model of the proposed Bell XS-2 (later X-2) with a low-placed horizontal tail and a ventral fin, though (unfortunately, given its history as related subsequently) Bell completed it with a more conventional layout mirroring the XS-1's midfin location.[62] The now-classic jet age low, all-moving "stabilator" tail was first incorporated on the North American YF-100A Super Sabre, the first of the "Century series" of American fighters.

The low all-moving tail reflected extensive NACA research dating to the midst of the Second World War. While the all-moving tail surface had been a standard feature of early airplanes such as the German Fokker Eindecker ("Monoplane") and French Morane Bullet fighters of the "Great War," the near constant high workload it made for a pilot caused it to fall

---

61. Robert C. Little, "Voodoo! Testing McAir's Formidable F-101," *Air Power History*, vol. 41, no. 1 (spring 1994), pp. 6–7. In Britain, designer George Edwards likewise added anhedral (though more modest than the Phantom's) to the Supermarine Scimitar, another pitch-up plagued swept wing fighter. See Robert Gardner, *From Bouncing Bombs to Concorde: the Authorised Biography of Aviation Pioneer Sir George Edwards OM* (Stroud, UK: Sutton Publishing, 2006), p. 125. Though not per se a swept wing aircraft, the Lockheed F-104 Starfighter, another T-tail design, likewise experienced tail-blanketing and consequent pitch-up, necessitating installation of a stick-kicker and imposing of limitations on high angle-of-attack maneuvering. At the time of its design, the benefits of a low-placed tail were already recognized, and it is surprising that Clarence "Kelly" Johnson, Lockheed's legendary designer, did not incorporate one. Certainly afterward, he recognized its value, for when, in 1971, he proposed a lineal derivative of the F-104, the CL-1200 Lancer (subsequently designated the X-27 but never built and flown), as a lightweight NATO export fighter, it featured a low, not high, all-moving horizontal tail. For X-27 see Jay Miller, *The X-Planes: X-1 to X-45* (Hinckley, UK: Midland Publishing, 2001), pp. 284–289.

62. See Joseph Weil, Paul Comisarow, and Kenneth W. Goodson, "Longitudinal Stability and Control Characteristics of an Airplane Model Having a 42.8° Sweptback Circular-Arc Wing with Aspect Ratio 4.00, Taper Ratio 0.60, and Sweptback Tail Surfaces," NACA RM-L7G28 (1947). Considerable debate likewise existed on whether the XS-2 should have a shoulder-mounted wing with anhedral, a midwing (like the XS-1) without any dihedral or anhedral, or a low wing with or without dihedral. Bell opted for a low wing with slight dihedral.

from grace, in favor of a fixed stabilizer and movable elevator surface. But by the early 1940s, NACA researchers recognized "its possible advantages as a longitudinal control for flight at high Mach numbers."[63] Accordingly, researchers at the Langley Memorial Aeronautical Laboratory modified an experimental Curtiss XP-42 fighter on loan from the Army Air Forces by removing its conventional horizontal tail surfaces and replacing them with an all-moving tail plane hinged at its aerodynamic center and controlled by a trailing edge servotab. Initial tests during turns at 200 mph proved disappointing, with pilots finding the all-moving surface too sensitive and its control forces too light (and thus dangerous, for they could easily subject the airplane to excessive maneuvering loads) and demanding continuous attention particularly in choppy air. So the XP-42 was modified yet again, this time with a geared, not servotab, control mechanism. If not perfect, the results were much better and more encouraging, with pilots now having the kind of variation in stick force to give them feedback on how effectively they were controlling the airplane.[64] Recognizing that the all-moving tail could substantially increase longitudinal control authority in the transonic region, NACA researchers continued their study efforts into the postwar years, encouraged by initial flight-test results of the Bell XS-1, which began approaching higher transonic Mach numbers in the fall of 1946. Though its adjustable horizontal stabilizer with a movable elevator constituted an admittedly interim step on the path to an all-moving surface, the XS-1's excursions through the speed of sound generated convincing proof that designers could dramatically increase transonic longitudinal control authority via an all-moving tail.[65]

---

63. Harold F. Kleckner, "Preliminary Flight Research on an All-Movable Horizontal Tail as a Longitudinal Control for Flight at High Mach Numbers," NACA ARR-L5C08 (Mar. 1945), p. 1.
64. Harold F. Kleckner, "Flight Tests of an All-Movable Horizontal Tail with Geared Unbalancing Tabs on the Curtiss XP-42 Airplane," NACA TN-1139 (1946).
65. Hubert M. Drake and John R. Carden, "Elevator-Stabilizer Effectiveness and Trim of the X-1 Airplane to a Mach Number of 1.06," NACA RM-L50G20 (1950). Despite the 1950 publication date, this report covers the results of XS-1 testing from Oct. 1946 through the first supersonic flight to M = 1.06 on Oct. 14, 1947. European designers recognized the value of such a tail layout as well. The Miles M.52, a jet-powered supersonic research airplane intemperately canceled by the British Labour government, would have incorporated similar surfaces; "This unfortunate decision," Sir Roy Fedden wrote a decade later, "cost us at least ten years in aeronautical progress." See his *Britain's Air Survival: An Appraisement and Strategy for Success* (London: Cassell & Co., Ltd., 1957), p. 20.

Tail location—midfin (as in the XS-1 and D-558), at the base of the fin (as with the F-86 and most other jet aircraft), or high (as with the T-tail F-101)—was another significant issue. German wartime research had favored no tail surfaces or, at the other extreme, high T-tails—for example, the DFS 346 supersonic research aircraft under development at war's end or the proposed Focke-Wulf Ta 183 swept wing jet fighter (which influenced the design of the MiG-15 and early Lavochkin swept wing jet fighters and a proposed British supersonic research aircraft). But the pitch-up problems encountered by the Skyrocket and even the F-86, as angle of attack increased, argued powerfully against such locations. In 1949, coincident with the Air Force and North American beginning development of the Sabre 45, a 45-degree swept wing successor to the F-86, Jack D. Brewer and Jacob H. Lichtenstein, two researchers at Langley, undertook a series of studies of tail size, length, and vertical location using the Langley stability tunnel and a model having 45-degree swept wing and tail surfaces. Their research demonstrated that placing a tail well aft of the wing and along the fuselage centerline (as viewed from the side) improved longitudinal stability and control.[66] Building upon their work, Langley researchers William Alford, Jr., and Thomas Pasteur, Jr., ran an investigation in the Langley 7-foot by 10-foot high-speed tunnel to determine aspect ratio and location effects on the longitudinal stability of a swept wing model across the transonic regime from Mach 0.80 to Mach 0.93. "The results," they subsequently reported in 1953, "indicted that, within the range of variables considered, the most favorable pitching-moment characteristics at a Mach number of 0.90 were obtained by locating the tail below the wing-chord plane."[67] Compared to this, other changes were inconsequential.

Flight tests at Ames in 1952 of a North American YF-86D (an interceptor variant of the F-86) specially modified with a low-placed horizontal tail, confirmed the Langley test results. As researchers noted, "The

---

66. Jack D. Brewer and Jacob H. Lichtenstein, "Effect of Horizontal Tail on Low-Speed Static Lateral Stability Characteristics of a Model Having 45° Sweptback Wing and Tail Surfaces," NACA TN-2010 (1950); and Jacob H. Lichtenstein, "Experimental Determination of the Effect of Horizontal-Tail Size, Tail Length, and Vertical Location on Low-Speed Static Longitudinal Stability and Damping in Pitch of a Model Having 45° Sweptback Wing and Tail Surfaces," NACA Report 1096 (1952).
67. William J. Alford, Jr., and Thomas B. Pasteur, Jr., "The Effects of Changes in Aspect Ratio and Tail Height on the Longitudinal Stability Characteristics at High Subsonic Speeds of a Model with a Wing Having 32.6° Sweepback," NACA RM-L53L09 (1953), p. 1.

test airplane, while having essentially the same unstable airplane static pitching moments as another version of this airplane [the F-86A] with an uncontrollable pitch-up, had only a mild pitch-up which was easily controllable," and had a nearly 40-percent increase in stabilizer and elevator effectiveness at transonic speeds.[68] The prototype YF-100 Super Sabre, first flown in May 1953, incorporated the fruits of this research. Next came the Vought XF8U-1 Crusader and the Republic YF-105 Thunderchief, and thereafter a plethora of other types. Aviation had returned full circle to the technology with which powered, controlled flight had begun: back to pivoted all-moving pitch-control surfaces of a kind the Wrights and other pioneers would have immediately recognized and appreciated.

Inertial Coupling: Dangerous Byproduct of High-Speed Design

The progression of aircraft flight speeds from subsonic to transonic and on into the supersonic changed the proportional relationship of wing to fuselage. As speed rose, the ratio of span to fuselage length decreased. At the onset of the subsonic era, the Wright Flyer had a wingspan-to-fuselage length ratio of 1.91. The SPAD XIII fighter of World War I was 1.30. The Second World War's P-51D decreased to 1.14. Then came the supersonic era: the XS-1 was 0.90. In 1953, the F-100A, lowered the ratio to 0.80, and the F-104A of 1954 cut this in half, to 0.40. The radical X-3 had a remarkably slender wingspan-to-fuselage length ratio of just 0.34: not without reason was it nicknamed the "Stiletto." But while the dramatic increase in fuselage length at the expense of span spoke to the need to reduce wing-aspect ratio and increase fuselage fineness ratio to achieve idealized supersonic shaping, any resulting aerodynamic benefit came only at the price of significant performance limitations and risk.

Increasing fuselage length while reducing span dramatically changed the mass distribution of these new designs: whereas earlier airplanes had most of their mass concentrated along the span of their wings, as the wing-fuselage ratio changed from well above 1.0 to well below this figure, the distribution of mass shifted to along the fuselage. Since a long forward fuselage inherently reduces directional stability, and since the small low aspect ratio wings of these airplanes reduced their roll stability, a potentially deadly mix of technical circumstances existed to

---

68. Norman M. McFadden and Donovan R. Heinle, "Flight Investigation of the Effects of Horizontal-Tail Height, Moment of Inertia, and Control Effectiveness on the Pitch-up Characteristics of a 35° Swept-Wing Fighter Airplane at High Subsonic Speeds," NACA RM-A54F21 (1955).

produce a major crisis: the onset of transonic and supersonic inertial coupling, also termed roll-coupling.

William Hewitt Phillips of the NACA's Langley laboratory had first forecast inertial coupling. His pronouncement sprang from a fortuitous experience while supervising tests of a large XS-1 "falling body" model in the summer of 1947. The model (dropped from a high-flying B-29 over a test range near Langley to assess XS-1 elevator control effectiveness as it approached Mach 1) incorporated a simple autopilot and was intended to rotate slowly as it fell, so as to maintain a "predictable trajectory."[69] But after the drop, things went rapidly awry. The model experienced violent pitching and rapid rolling "well below" the speed of sound and fell so far from its planned impact point that it literally disappeared from history. But optical observations, coupled with telemetric data, led Phillips to conclude that "some kind of gyroscopic effect" had taken place. Intrigued, he drew upon coursework from Professors Manfred Rauscher and Charles Stark Draper of the Massachusetts Institute of Technology, using the analogy of the coupling dynamics of a rotating rod. He substituted the values obtained from the falling XS-1 model, discovering that "the results clearly showed the possibility of a divergent motion. . . . The instability was likely to occur when the values of longitudinal stability and directional stability were markedly different and when a large amount of the weight was distributed along the fuselage."[70] Hewitt subsequently published a seminal NACA Technical Note in 1948, which presciently concluded: "Design trends of very high-speed aircraft, which include short wing spans, fuselages of high density, and flight at high altitude, all tend to increase the inertia forces due to rolling in comparison with the aerodynamic restoring forces provided by the longitudinal and directional stabilities. It is therefore desirable to investigate the effects of rolling on the longitudinal and directional stabilities of these aircraft. . . . The rolling motion introduces coupling between the longitudinal and lateral motion of the aircraft."[71] Out of

---

69. W. Hewitt Phillips, *Journey in Aeronautical Research: A Career at NASA Langley Research Center*, No. 12 in the *Monographs in Aerospace History Series* (Washington: NASA, 1998), p. 70; for an excellent survey, see Richard E. Day, *Coupling Dynamics in Aircraft: A Historical Perspective*, SP-532 (Washington: NASA, 1997).

70. Phillips, *Journey in Aeronautical Research*, p. 72.

71. William H. Phillips, "Effect of Steady Rolling on Longitudinal and Directional Stability," NACA TN-1627 (1948), pp. 1–2.

Case 1 | Sweep and Swing: Reshaping the Wing for the Jet and Rocket Age

this came the expression "inertial coupling" and its more descriptive equivalent, "roll-coupling." Phillips continued his research on roll-coupling and rolling maneuvers in accelerated flight, noting in 1949 that high-speed rolls could generate "exceptionally large" sideslip loads on a vertical fin that might risk structural failure. He concluded: "The provision of adequate directional stability, especially at small angles of sideslip, in order to prevent excessive sideslipping in rolls at high speed is therefore important from structural considerations as well as from the standpoint of providing desirable flying qualities."[72]

In the summer of 1952, as part of an investigation effort studying coupled lateral and longitudinal oscillations, researchers at the NACA's Pilotless Aircraft Research Division at Wallops Island, VA, fired a series of large rocket-boosted swept wing model airplanes. Spanning over 3 feet, but with a length of nearly 6 feet, they had the general aerodynamic shape of the D-558-2 as originally conceived: with a slightly shorter vertical fin. These models accelerated to supersonic speed and then, after rocket burnout and separation, glided onward while onboard telemetry instrumentation relayed a continuous stream of key performance and behavior parameters as they decelerated through the speed of sound before diving into the sea. On August 6, 1952, technicians launched one equipped with a small pulse rocket to deliberately destabilize it with a timely burst of rocket thrust. After booster burnout, as the model decelerated below Mach 1, the small nose thruster fired, inducing combined yawing, sideslip, and rolling motions. But instead of damping out, the model swiftly went out of control, as if a replay of the XS-1 falling body test 5 years previously. It rolled, pitched, and yawed until it plunged into the Atlantic, its death throes caught by onboard instrumentation and radioed to a NACA ground station. If dry, the summary words of the resulting test report held ominous import for future flight-testing of full-size piloted aircraft: "From the flight time history of a rocket-propelled model of a representative 35° sweptback wing airplane, it is indicated that coupled longitudinal motions were excited and sustained by pure lateral oscillations. The resulting longitudinal motions had twice the frequency of the lateral oscillations and rapidly developed lift loads of appreciable magnitude. The longitudinal moments are attributed to two sources, aerodynamic

---

72. William H. Phillips, "Appreciation and Prediction of Flying Qualities," NACA Report No. 927 (1949), p. 32.

moments due to sideslip and inertial cross-coupling. The roll characteristics are indicated to be the predominating influence in the inertial cross-coupling terms."[73]

Two model tests, 5 years apart, had shown that roll coupling was clearly more than a theoretical possibility. Shortly thereafter it turned into an alarming reality when the Bell X-1A, North American YF-100 Super Sabre, and Douglas X-3 entered flight-testing. Each of these encountered it with varying degrees of severity. The first to do so was the Bell X-1A, a longer, more streamlined, and more powerful derivative of the original XS-1.[74] The X-1A arrived at Edwards in early 1953, flew a brief contractor program, and then entered Air Force evaluation in November. On December 12, 1953, test pilot Charles E. "Chuck" Yeager nearly died when it went out of control at Mach 2.44 at nearly 80,000 feet. In the low dynamic pressure ("low q" in engineering parlance) of the upper atmosphere, a slight engine thrust misalignment likely caused it to begin a slow left roll. As Yeager attempted to control it, the X-1A rolled rapidly to the right, then violently back to the left, tumbling completely out of control and falling over 50,000 feet before the badly battered Yeager managed to regain control. Gliding back to Edwards, he succinctly radioed: "You know, if I'd had a seat, you wouldn't still see me in this thing."[75] Afterward, NACA engineers concluded that "lateral stability difficulties were encountered which resulted in uncontrolled rolling motions of the airplane at Mach numbers near 2.0. Analysis indicates that this behavior apparently results from a combination of low directional stability

---

73. James H. Parks, "Experimental Evidence of Sustained Coupled Longitudinal and Lateral Oscillations from a Rocket-Propelled Model of a 35° Swept Wing Airplane Configuration," NACA RM-L54D15 (1954). For more on Wallops testing, see Joseph A. Shortal, *A New Dimension: Wallops Island Flight Test Range: The First Fifteen Years*, RP-1028 (Washington: NASA, 1978), pp. 256–257. For the record, the wingspan-to-fuselage ratio of the model was 0.59, significantly lower than the XS-1.

74. It is worth noting that the advanced X-1A (and X-1B and X-1D) had a wingspan-to-fuselage length ratio of 0.79, compared to the 0.90 XS-1, the drop model of which first encountered inertial coupling. Their longer fuselage forebody likewise contributed even further to their tendency toward lateral-directional instability.

75. Yeager pilot report and attached transcript, Dec. 23, 1953; J.L. Powell, Jr., "X-1A Airplane Contract W33-038-ac-20062, Flight Test Progress Report No. 15, Period From 9 December through 20 December 1953," Bell Aircraft Corporation Report No. 58-980-019 (Feb. 3, 1954), both from AFFTC History Office archives. I thank the staff of the AFFTC History Office and the NASA DFRC Library and Archives for locating these and other documents.

Case 1 | Sweep and Swing: Reshaping the Wing for the Jet and Rocket Age

and damping in roll."[76] The predictions made in Phillips's 1948 NACA Technical Note had come to life, and even worse would soon follow.

By the time of Yeager's harrowing X-1A flight, the prototype YF-100, having first flown in May 1953, was well into its flight-test program. North American and the Air Force were moving quickly to fulfill ambitious production plans for this new fighter. Yet all was not well. The prototype Super Sabre had sharply swept wings, a long fuselage, and a small vertical fin. While fighter pilots, entranced by its speed, were enthusiastic about the new plane, Air Force test pilots were far less sanguine, noting its lateral-directional stability was "unsatisfactory throughout the entire combat speed range," with lateral-directional oscillations showing "no tendency to damp at all."[77] Even so, in the interest of reducing weight and drag, North American actually *shrank* the size of the vertical fin for the production F-100A, lowering its height, reducing its area and aspect ratio, and increasing its taper ratio. The changes further cut the directional stability of the F-100A, by some estimates as much as half, over the YF-100.[78] The first production F-100As entered service in the late summer of 1954. Inertial coupling now struck with a vengeance. In October and November, two accidents claimed North American's chief test pilot, George "Wheaties" Welch, and Royal Air Force Air Commodore Geoffrey Stephenson, commander of Britain's Central Fighter Establishment. Others followed. The accidents resulted in an immediate grounding while the Air Force, North American, and the NACA crafted complementary research programs to analyze and fix the troubled program.[79]

Then, in the midst of the F-100's travail, inertial coupling struck the Douglas X-3. First flown in October 1952, the X-3 had vestigial straight wings and tail surfaces joined to a missile-like fuselage. Though it was

---

76. Hubert M. Drake and Wendell H. Stillwell, "Behavior of the Bell X-1A Research Airplane During Exploratory Flights at Mach Numbers Near 2.0 and at Extreme Altitude," NACA RM-H55G25 (1955), p. 10.
77. Alfred D. Phillips and Lt. Col. Frank K. Everest, USAF, "Phase II Flight Test of the North American YF-100 Airplane USAF No. 52-5754," AFFTC TR-53-33 (1953), Appendix I, p. 8. For the difference between fighter pilots and test pilots in regarding the F-100, see Brig. Gen. Frank. K. Everest, Jr., with John Guenther, *The Fastest Man Alive* (New York: Bantam, 1990 ed.), pp. 6, 11–13.
78. James R. Peele, "Memorandum for Research Airplane Projects Leader [Hartley A. Soulé, hereafter RAPL]: Results of Flights 2, 3, and 4 of the F-100A (52-5778) airplane" (Nov. 19, 1954), DFRC Archives.
79. Ronald-Bel Stiffler, *The History of the Air Force Flight Test Center: 1 January 1954–30 June 1954* (Edwards AFB: AFFTC, July 13, 1955), vol. 1, pp. 66–67, copy in AFFTC History Office archives.

the most highly streamlined airplane of its time, mediocre engines confounded hopes it might achieve Mach 2 speeds, and it never flew faster than Mach 1.21, and that only in a dive. The NACA acquired it for research in December 1953, following contractor flights and a brief Air Force evaluation. On October 27, 1954, during its 10th NACA flight, test pilot Joseph A. Walker initiated an abrupt left aileron roll at Mach 0.92 at 30,000 feet. The X-3 pitched up as it rolled, sideslipping as well. After it returned to stable flight, Walker initiated another left roll at Mach 1.05. This time, it responded even more violently. Sideslip angle exceeded 21 degrees, and it reached –6.7 g during a pitch-down, immediately pitching up to over 7 g. Fortunately, the wild motions subsided, and Walker, like Yeager before him, returned safely to Earth.[80] With the example of the X-1A, the F-100A, and the X-3, researchers had conclusive proof of a newly emergent crisis imperiling the practical exploitation of the high-speed frontier.

The F-100A raised the most concern, for it was the first of an entire new class of supersonic fighter aircraft, the "Century series," with which the United States Air Force and at least some of its allies hoped to reequip. Welch's F-100A had sideslipped and promptly disintegrated during a diving left roll initiated at Mach 1.5 at 25,000 feet. As Phillips had predicted in 1949, the loads had proven too great for the fin to withstand (afterward, North American engineers "admitted they had been naive in estimating the effects of reducing the aspect ratio and area of the YF-100 prototype tail").[81] Curing the F-100's inertial coupling problems took months of extensive NACA and Air Force flight-testing, much of it very high-risk, coupled with analytical studies by Langley personnel using a Reeves Electronic Analogue Computer (REAC), an early form of a digital analyzer. During one roll at Mach 0.7 (and only using two-thirds of available aileron travel), NACA test pilot A. Scott Crossfield experienced "a large yaw divergence accompanied by a violent pitch-down . . . which subjected the airplane to approximately –4.4g vertical acceleration."[82] Clearly the F-100A needed significant redesign: the Super

---

80. NACA High-Speed Flight Station, "Flight Experience with Two High-Speed Airplanes Having Violent Lateral-Longitudinal Coupling in Aileron Rolls," NACA RM-H55A13 (1955), p. 4.
81. Joseph Weil, "Memo to RAPL: Visit of HSFS personnel to North American Aviation, Inc. on Nov. 8, 1954" (Nov. 19, 1954), DFRC Archives.
82. Peele memo to RAPL, Nov. 19, 1954; for Langley REAC studies, see Charles J. Donlan [NACA LRC], "Memo for Associate Director [Floyd Thompson]: Industry-Service-NACA Conference on F-100, Dec. 16, 1954" (Dec. 28. 1954), DFRC Archives.

Case 1 | Sweep and Swing: Reshaping the Wing for the Jet and Rocket Age

Sabre's accidents and behavior (and that of the X-3 as well) highlighted that streamlined supersonic aircraft needed greatly increased tail area, coupled with artificial stability and motion damping, to keep sideslip from developing to dangerous values. North American subsequently dramatically increased the size of the F-100's vertical fin, increased its wingspan by 2 feet (to shift the plane's center of gravity forward), and incorporated a yaw damper to control sideslip. Though the F-100 subsequently became a reliable fighter-bomber (it flew in American service for almost a quarter century and longer in foreign air arms), it remained one that demanded the constant attention and respect of its pilots.[83]

Inertial coupling was not, of course, a byproduct of conceptualizing the swept and delta wings, nor was it limited (as the experience of the XS-1 falling model, X-1A, and X-3 indicated) just to aircraft possessing swept or delta planforms. Rather, it was a byproduct of the revolution in high-speed flight, reflecting the overall change in the parametric relationship between span and length that characterized aircraft design in the jet age. Low aspect ratio straight wing aircraft like the X-3 and the later Lockheed F-104 were severely constrained by the threat of inertial coupling, even more than many swept wing aircraft were.[84] But for swept wing and delta designers, inertial coupling became a particular challenge they had to resolve, along with pitch-up. As the low-placed horizontal tail reflected the problem of pitch-up, the increasing size of vertical fins (and the addition of ventral fins and strakes as well) incorporated on new aircraft such as the Navy's F8U-1 and the Air Force's

---

83. Joseph Weil and Walter C. Williams, "Memo for RAPL: Meeting of NACA and Air Force personnel at North American Aviation, Inc on Monday, Nov. 22, 1954, to discuss means of expediting solution of stability and control problems on the F-100A airplane" (Nov. 26, 1954), DFRC Archives; Thomas W. Finch, "Memo for RAPL: Progress report for the F-100A (52-5778) airplane for the period Nov. 1 to Nov. 30, 1954" (Dec. 20, 1954), DFRC Archives; Hubert M. Drake, Thomas W. Finch, and James R. Peele, "Flight Measurements of Directional Stability to a Mach Number of 1.48 for an Airplane Tested with Three Different Vertical Tail Configurations," NACA RM-H55G26 (1955); Marion H. Yancey, Jr., and Maj. Stuart R. Childs, USAF, "Phase IV Stability Tests of the F-100A Aircraft, USAF S/N 52-5767," AFFTC TR-55-9 (1955); 1st Lt. David C. Leisy, USAF, and Capt. Hugh P. Hunerwadel, USAF, "ARDC F-100D Category II Performance Stability and Control Tests," AFFTC TR-58-27 (1958).

84. See, for example, Robert G. Hoey and Capt. Iven C. Kincheloe, USAF, "ARDC F-104A Stability and Control," AFFTC TR-58-14 (1958); and Capt. Slayton L. Johns, USAF, and Capt. James W. Wood, USAF, "ARDC F-104A Stability and Control with External Stores," AFFTC TR-58-14, Addendum 1 (July 1959).

F-105B (and the twin-fins that followed in the 1970s on aircraft such as the F-14A, F-15A, and F/A-18A) spoke to the serious challenge the inertial coupling phenomenon posed to aircraft design. Not visible were such "under the skin" systems as yaw dampers and the strict limitations on abrupt transonic and supersonic rolling taught to pilots transitioning into these and many other first-generation supersonic designs.[85]

The story of the first encounters with inertial coupling is a salutary, cautionary tale. A key model test had resulted in analysis leading to the issuance of a seminal report but one recognized as such only in retrospect. A half decade after the report's release, pilots died because the significance of the report for future aircraft design and behavior had been missed. Even within the NACA, recognition of seriousness of reduced transonic and supersonic lateral-directional stability had been slow. When, in August 1953, NACA engineers submitted thoughts for a tentative research plan for an F-100A that the Agency would receive, attention focused on longitudinal pitch-up, assessing its handling qualities (particularly its suitability as a gun platform, something seemingly more appropriately done by the Air Force Flight Test Center or the Air Proving Ground at Eglin), and the correlation of flight and wind tunnel measurements.[86] Even after the experience of the X-1A, F-100A, and X-3, even after all the fixes and training, it is disturbing how inertial coupling stilled claimed the unwary.[87] Over time, the combination of refined design, advances in stability augmentation (and eventually the advent of computer-controlled fly-by-wire flight) would largely render

---

85. For example, Thomas R. Sisk and William H. Andrews, "Flight Experience with a Delta-Wing Airplane Having Violent Lateral-Longitudinal Coupling in Aileron Rolls," NACA RM-H55H03 (1955).
86. William H. Phillips [NACA LRC], "Memo for Associate Director: Flight program for F-100A airplane" (Aug. 10, 1953), DFRC Archives. Even odder, it was Phillips who had identified inertial coupling in TN-1627 in 1948!
87. The best known was Capt. Milburn "Mel" Apt, who died in late 1956. His Bell X-2 went out of control as he turned back to Edwards after having attained Mach 3.2, possibly because of lagging instrumentation readings leading him to conclude he was flying at a slower speed. Undoubtedly the nearly decade-old design of the X-2 contributed to its violent coupling tendencies. It is sobering that in 1947 NACA had evaluated some design options (tail location, vertical fin design) that, had Bell incorporated them on the X-2, might have turned Apt's accident into an incident. See Ronald Bel Stiffler, *The Bell X-2 Rocket Research Aircraft: The Flight Test Program* (Edwards AFB: Air Force Flight Test Center, 1957); and Richard E. Day and Donald Reisert, "Flight Behavior of the X-2 Research Airplane to a Mach Number of 3.20 and a Geometric Altitude of 126,200 Feet," NACA TM-X-137 (1959).

inertial coupling a curiosity. But for pilots of a certain age—those who remember aircraft such as the X-3, F-100, F-101, F-102, and F-104—the expression "inertial coupling," like "pitch-up," will always serve to remind that what is an analytical curiosity in the engineer's laboratory is a harsh reality in the pilot's cockpit.

Implementing the Delta Planform

While swept wing adaptation in Europe, Russia, and America followed a generally similar pattern, the delta wing underwent markedly different international development. Generally, European designers initially emulated the Lippisch approach, resulting in designs with relatively thick wing sections (exemplified by the Avro Vulcan bomber and the "tailed" Gloster Javelin interceptor) that inhibited their ability to operate beyond the transonic. Only after the practical demonstration of Convair's emerging family of thin-wing delta designs—the XF-92A research aircraft, the F-102 interceptor, the XF2Y-1 experimental naval fighter, the B-58 supersonic bomber, and the F-106 interceptor—did they conceptualize more "supersonic friendly" designs, typified by the Swedish Saab J35 Draken ("Dragon"), the British Fairey F.D.2 research airplane, the French Dassault Mirage I (progenitor of the Mirage fighter and bomber family). By the late 1950s, British and French aerodynamicists had so completely "closed" any "delta gap" that might have existed between Europe and America that they were already conceptualizing development of a Mach 2 supersonic transatlantic transport using a shapely "ogee" reflexive delta planform, a study effort that would, a decade later, spawn the Anglo-French Concorde.[88] Not so taken with the pure delta, Soviet designers joined American-like thin delta wings to the low-placed horizontal tail, generating advanced MiG and Sukhoi fighters and interceptors. These "tailed deltas" (particularly the MiG-21) possessed far better transonic and supersonic turning performance than could be attained by a conventional delta with its high induced drag onset at the increasing angles of attack characteristic of hard-maneuvering. (An American equivalent was the Douglas Company's superlative A4D-1 Skyhawk,

---

88. Keneth Owen, *Concorde: Story of a Supersonic Pioneer* (London: Science Museum, 2001), pp. 21–60; Andrew Nahum, "The Royal Aircraft Establishment from 1945 to Concorde," in Robert Bud and Philip Gummett, eds., *Cold War, Hot Science: Applied Research in Britain's Defence Laboratories, 1945–1990* (London: Science Museum, 1999), pp. 29–58; and Andersson, *Saab Aircraft*, pp. 124–129.

a light attack bomber with maneuvering performance better than most fighters.)

Although it is commonly accepted that American delta aircraft owe their inspiration to the work of Lippisch—Convair's delta aircraft repeatedly being cited as the products of his influence—in fact, they do not.[89] Unlike, say, the swept wing F-86 and B-47, which directly reflected German aerodynamic thought and example, America's delta wing aircraft reflected indigenous, not foreign, research and inspiration. By the time that Lippisch first met with Allied technical intelligence experts, American aerodynamicists were already advancing along a very different path than the one he had followed. Jones had already enunciated his thin, sharply swept delta theory and undertaken his first tunnel tests of it. In June 1946—a full year after the German collapse—Convair engineers developing the experimental delta XP-92 interceptor had their chance to meet with Lippisch at Wright Field. By then, however, they had already independently decided upon a thin delta planform. "We had heard about Dr. Lippisch's work and this gave us some moral support," Convair designer Adolph Burstein recalled, adding: "but not much else. . . . We did not go along with many of his ideas, such as a very thick airfoil."[90] Burstein and his colleagues arrived at their delta shape by beginning with a 45-degree swept wing, gradually increasing its sweepback angle, and then "filling in" the ever-closing trailing edges, until they arrived at the classic 60-degree triangular delta planform the company incorporated on all its subsequent delta aircraft. With a 6.5 thickness-chord ratio—less than half that of Lippisch's DM-1—it was an altogether different-looking airplane.[91] Nor was Convair alone in going its own way; Douglas naval aircraft designer Edward Heinemann acknowledged that "At the close of World War II the work with delta planforms accomplished by

---

89. Even the official Air Force history of the service's postwar fighter development repeats the canard, though it does acknowledge that "low-aspect-ratio wing forms were also studied by the U.S. National Advisory Committee for Aeronautics." See Marcelle Size Knaack, *Post-World War II Fighters 1945–1973*, vol. 1 of *Encyclopedia of U.S. Air Force Aircraft and Missile Systems* (Washington: Office of Air Force History, 1978), p. 159, no. 1.

90. Letter, Adolph Burstein to Richard P. Hallion, Jan. 25, 1972. Despite his "Germanic" name, Burstein, one of the XF-92A's designers, was not a German scientist or engineer who came to America after 1945. Rather, he was a Russian emigree from St. Petersburg who had come to the United States in 1925.

91. See Hallion, "Lippisch Gluhareff, and Jones," and R.P. Hallion, "Convair's Delta Alpha," *Air Enthusiast Quarterly*, No. 2 (1976).

Case 1 | Sweep and Swing: Reshaping the Wing for the Jet and Rocket Age

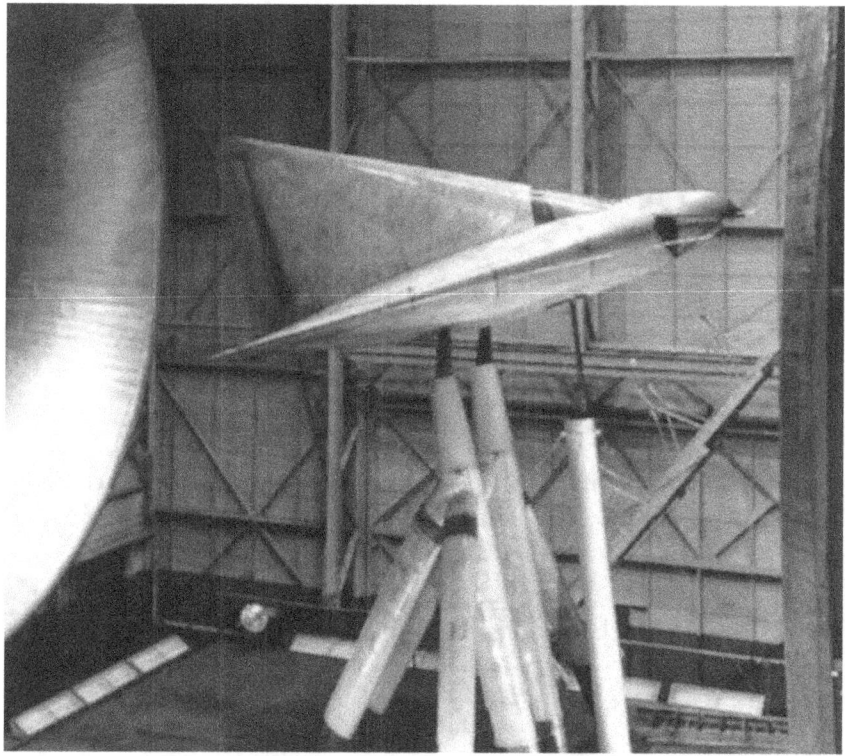

Lippisch DM-1 glider in the Langley Full Scale Tunnel, 1946. The thick-wing section is readily apparent, as is the oversize vertical fin, both of which rendered the concept unsuitable for transonic flight. NASA.

Dr. Lippisch in Germany became generally known and appreciated," but that "Extensive wind tunnel tests showed there was no special merit to an equilateral triangle planform—especially those designed with thicker airfoils."[92]

The chronology of American delta development, and the technical choices and paths followed by American engineers, supports both statements. At war's end, advancing ground forces at Prien, Austria, had discovered a thick-wing wooden delta glider, the DM-1, which Lippisch had intended as a low-speed testbed for a proposed supersonic fighter, the P 13. At Army Air Forces' request, it was shipped back to America in January 1946 for comprehensive testing in the Full-Scale Tunnel at the NACA's Langley Aeronautical Laboratory. Had the tests gone well,

---

92. Edward H. Heinemann, "Design of High-Speed Aircraft," a paper presented at the Fifth International Aeronautical Conference, Royal Aeronautical Society-Institute of the Aeronautical Sciences, Los Angeles, CA, June 20–24, 1955, p. 3. Copy from the Boeing-McDonnell Douglas Archives.

the possibility existed as that, as the Germans had intended, it might be flown as a glider. But the tunnel tests quickly disabused delta enthusiasts of these hopes. As the AAF's Langley liaison officer subsequently reported, the "Initial test results were very disappointing; the lift coefficient was low, the drag was high, the directional stability was unsatisfactory, and the craft was considered unsafe for flight tests."[93]

Afterward, Langley engineers undertook a comprehensive study of the DM-1 configuration, not in the spirit of emulation but rather attempting to find a way to fix it. After giving its wings sharp leading edges, sealing all slots and gaps around control surfaces, and removing the thick vertical fin and replacing it with a thin one (relocating the pilot under a streamlined bubble canopy), they had markedly improved its performance, doubling its lift coefficient, from 0.6 to over 1.2. But it remained an unsatisfactory design, proof enough that the Lippisch concept of deltas was hardly one that could serve—or did serve—as a veritable template (as has been so often alleged) for the supersonic American, Swedish, and French delta fighters and bombers that flew over the next decade.[94] Subsequently, NACA engineers looked to far thinner and more streamlined configurations that, if not yet as extreme as Robert T. Jones's original daggerlike concept, were even more amenable to the rigors of transonic and supersonic flight than the generously rounded contours of Lippisch's thick wings and awkward pilot-enclosing vertical fins. By the beginning of 1947, they were already examining the technical requirements of slender, low aspect ratio delta configurations

---

93. Ltr., Maj. Howard C. Goodell, USAF, to Paul E. Garber, "DM-1 Glider Disposal," Nov. 28, 1949, in Gluhareff Dart accession file, National Air and Space Museum.

94. For Langley's progressive evaluation and modification of the DM-1, see two reports by Herbert A. Wilson, Jr., and J. Calvin Lovell, "Full Scale Investigation of the Maximum Lift and Flow Characteristics of an Airplane Having Approximately Triangular Plan Form," NACA RM-L6K20 (1947); and "Langley Full-Scale Tunnel Investigation of Maximum Lift and Stability Characteristics of an Airplane Having Approximately Triangular Plan Form (DM-1 Glider), NACA RM-L7F16 (1947). Changes are detailed in RM L7F16, Fig. 4. The closest expression of Germanic delta philosophy in America was not a Convair delta, but a Douglas one: the Navy-Marine F4D-1 Skyray fighter. Its design was greatly influenced by German tailless and swept wing reports Douglas engineers L. Eugene Root and Apollo M.O. Smith had discovered while assigned to an Allied technical intelligence team examining the Messerschmitt advanced projects office at Oberammergau and interviewing its senior personnel, particularly chief designer Woldemar Voigt; I wish to acknowledge with gratitude notes on their experiences received in 1972 from both the late L. Eugene Root and A.M.O. Smith. See also Cebeci, ed., *Legacy of a Gentle Genius*, pp. 30–36.

Case 1 | Sweep and Swing: Reshaping the Wing for the Jet and Rocket Age

The Convair XF-92A, the world's first delta jet airplane, at the NACA High-Speed Flight Research Station, now the Dryden Flight Research Center, in 1953. NASA.

to meet emerging military specifications for a Mach 1.5, 60,000-foot bomber interceptor.[95]

First Flight Experiences

Out of this mutually reinforcing climate of thought emerged the world's first delta jet airplane, the Convair XF-92A, first flown in September 1948. This technology explorer (for despite its "fighter" designation, it was always intended for research purposes) demonstrated the potential of the delta wing and encouraged Convair and Air Force authorities to pursue a delta planform for a future interceptor design. Originally, that design had been the "XP-92," an impractical barrel-shaped rocket-boosted ramjet with the pilot sitting in a conical nose within the ramjet's

---

95. R.M. Cross, "Characteristics of a Triangular-Winged Aircraft: 2: Stability and Control," in NACA, *Conference on Aerodynamic Problems of Transonic Airplane Design* (1947), pp. 163–186, and Figs. 6 and 12. See also Edward F. Whittle, Jr., and J. Calvin Lovell, "Full-Scale Investigation of an Equilateral Triangular Wing Having 10-percent-Thick Biconvex Airfoil Sections," NACA RM-L8G05 (1948), Fig. 2.

circular inlet, similar to René Leduc's straight wing air-launched French ramjet designs of the same period. Following its cancellation, work on the XF-92A continued, supporting the Air Force's "1954 Interceptor" initiative, which Convair hoped to win with, essentially, a bigger and more powerful version of the XF-92A. Aside from greater power, the interceptor would have to have a nose radar and thus "cheek" inlets rather than the simple Pitot nose inlet of the smaller testbed. The "1954 Interceptor" eventually became two: the "interim" Mach 1+ F-102 Delta Dagger and the "ultimate" Mach 2+ F-106 Delta Dart.

The XF-92A contributed markedly to delta understanding but was far from a trouble-free design. Deltas evinced a variety of quirks and performance deficiencies, some of which they shared with their swept wing brethren. Deltas manifested the same tendency to persistent combined lateral-directional Dutch roll motions, as well as pitch-up, from Mach number effects as they entered further into the transonic regime. The extreme sweep of their wings accentuated spanwise flow tendencies, making wing fences almost mandatory in all cases. Their high angle-of-attack ("hi AoA") landing approaches highlighted potentially serious control deficiencies, for, unlike a conventional fighter, the delta lacked separate elevators and ailerons. It relied instead on elevons—combined elevator-ailerons—for pitch and roll control. Thus, with the stick pulled back on final approach, the nose would rise, and if the plane encountered a sudden gust that induced a rolling motion, the pilot might lack sufficient remaining reserve "travel" from the deflected elevon to correct for the rolling motion. Further complicating landing approaches and turn performance was the delta's inherently high-induced drag as it turned or was at higher angles of attack. Deltas needed lots of power. The high-induced drag of the delta led to a rapid bleeding off of airspeed during turns and thus inhibited its holding altitude during turning maneuvers. Tests with the little XF-92A in 1953 by NACA research pilot Scott Crossfield indicated that as much as 3,000 feet of altitude could be lost trying to maintain constant speed in a turning maneuver—and this was after it had been modified to incorporate an afterburner for greater power. "Every time I took off in that plane I held my brief until I reached sufficient altitude to use the ejection seat," Crossfield recollected later. "The pilot never really flew that airplane, he corralled it."[96]

---

96. Crossfield with Blair, *Always Another Dawn*, p. 167; Thomas R. Sisk and Duane O. Muhleman, "Longitudinal Stability Characteristics in Maneuvering Flight of the Convair XF-92A Delta-Wing Airplane Including the Effects of Wing Fences," NACA RM-H4J27 (1955).

All together, the NACA completed 25 flights in the XF-92A before a landing gear collapse brought its research career to an end.

Tests of the XF-92A foreshadowed similar challenges with the next Convair delta, the prototype YF-102 interceptor. The YF-102 is infamous for having suffered from such high transonic drag rise that it could not accelerate through the speed of sound, a discovery that led, as Air Force test pilot Lt. Col. Frank K. "Pete" Everest recalled, to "surprise and concern.... The National Advisory Committee for Aeronautics had claimed all along that the airplane would not go supersonic, and now their predictions came true."[97] (How the YF-102 was transformed from embarrassing failure to operational success, thanks to Richard Whitcomb's "area rule" theory and its practical application to the F-102 design, is covered elsewhere in this volume in a case study on Whitcomb's contributions to aeronautics, by historian Jeremy Kinney.) But more than reshaping of its fuselage was required before the F-102 became a success. Instead, its wing underwent fundamental aerodynamic redesign reflecting the second stage in American delta development and its third stage overall.

Reshaping the Delta: Deriving Conical Camber

Having preceded the explication of the swept wing in Jones's original research, the roots of the delta's redesign now lay, somewhat ironically, in his expanding upon the slender swept wing research he had first begun at Langley. After the war, Jones had left Virginia's Tidewater region for the equally pleasant Bay area environment of Sunnyvale, CA, and there had continued his swept wing studies. By 1947, he had evolved a sharply swept symmetrical airfoil planform he considered suitable for a supersonic jet transport. Such a planform, with the leading edges of the wings within the shock cone formed around the vehicle and thus in a region of subsonic flow, could perhaps have a lift-to-drag (L/D) ratio as high as 10, though at the price of much higher landing speeds.[98] Tests of a small model in the Ames 1-foot by 3-foot supersonic tunnel and a larger one in the Ames 40-foot by 80-foot tunnel encouraged Jones and inspired fellow Ames researchers Charles F. Hall and John C. Heitmeyer to build upon his work. Hall and Heitmeyer considered the behavior of the combined wing-body, with the wing twisted and

---

97. Everest with Guenther, *Fastest Man Alive*, p. 109.
98. R.T. Jones, "Characteristics of a Configuration with a Large Angle of Sweepback," in NACA, *Conference on Aerodynamic Problems of Transonic Airplane Design* (1947), pp. 165–168, Figs. 1–6.

given camber (curvature) to evenly distribute the flight loads, deriving a sharply swept and tapered wing configuration that demonstrated an L/D of 8.9 during tunnel tests to Mach 1.53.[99] In the refinement of its planform, it called to mind the shape (though, of course, not the airfoil section) of Whitcomb's later supercritical transonic transport wing conceptualizations.[100]

Hall and Heitmeyer next broadened their research to examine slender deltas likewise featuring aerodynamic twist and camber. In 2 years, 1951–1952, they coauthored a dozen reports, culminating in the issuance of a seminal study by Hall in the spring of 1953 that summarized the lift, drag, pitching moment, and load distribution data on a variety of thin delta wings of varying aspect ratios operating from Mach 0.25 (touchdown velocity) to Mach 1.9. Out of this came the concept of leading edge "conical camber": twisting and rounding the leading edge of a delta wing to minimize performance-robbing drag generated by the wing's own lifting force. The modified delta had minimal camber at the wing root and maximum camber at the tip, the lineal development of the camber along the leading edge effectively representing the surface of a steadily expanding cone nestled under the leading edge of the wing.[101]

Hall's conical camber, like Whitcomb's area rule, came just in time to save the F-102 program. Both were necessary to make it a success: Whitcomb's area rule to get it through the sound barrier, and Hall's to give it acceptable transonic and supersonic flying qualities. If overshadowed by Whitcomb's achievement—which resulted in the young Langley aerodynamicist receiving the Robert J. Collier Trophy, American aviation's most prestigious award, in 1954—Hall's conical camber concept was nevertheless a critical one. Comparative flight-testing of the YF-102 at the NACA High-Speed Flight Station at Edwards from late

---

99. Charles F. Hall and John C. Heitmeyer, "Aerodynamic Study of a Wing-Fuselage Combination Employing a Wing Swept Back 63°—Characteristics at Supersonic Speeds of a Model with the Wing Twisted and Cambered for Uniform Load," NACA RM-A9J24 (1950).
100. Though no transport or military aircraft ever flew with such a slender swept wing, just such a configuration was subsequently employed on the largest swept wing tailless vehicle ever flown, the Northrop Snark intercontinental cruise missile. Though the Snark did not enter operational service for a variety of other reasons, it did demonstrate that, aerodynamically, such a wing configuration was eminently suitable for long-range transonic cruising flight.
101. Charles F. Hall, "Lift, Drag, and Pitching Moment of Low-Aspect Ratio Wings at Subsonic and Supersonic Speeds," NACA RM-A53A30 (1953). For the views of an Ames onlooker, see Hartman, Adventures in Research, pp. 202–207.

1954 to mid-1955 with and without conical camber indicated that conical camber gave it lower drag and increased its maximum lift-to-drag ratio by approximately 20 percent over a test Mach number range of 0.6 to 1.17, at altitudes of 25,000, 40,000, and 50,000 feet. Transonic stability of the cambered versus symmetrical YF-102 more than doubled, and "no severe pitch-up tendencies were exhibited, except when accelerating or decelerating through the trim-change region."[102]

With the advent of conical camber, the age of the practical transonic-supersonic delta wing had arrived. By mid-decade, the F-102's aerodynamic deficiencies had been cured, and it was well on its way to service use.[103] Convair designers were refining the delta planform to generate the F-102's successor, the superlative F-106, and a four-engine Mach 2+ bomber, the delta wing B-58 Hustler. Overseas, Britain's Fairey Company had under test a delta of its own, the F.D.2, which would shortly establish an international speed record, while, in France, Dassault engineers were conceptualizing a design that would spawn the Mirage family and be responsible, in 1967, for one of the most remarkable aerial victories of all time. Jones's supersonic delta vision from over a decade previously had become reality, thanks in part to Whitcomb's interference studies (which Jones himself would expand at Ames) and Hall's conceptualization of conical camber.

Extending the Delta into the Hypersonic and Orbital Frontier

The next stage in delta development took it from the realm of the transonic and supersonic into the hypersonic, again thanks to a healthy rivalry and differing technical perspective between those two great research centers, Ames and Langley. The area was hypersonics: flight at speeds higher than Mach 5, an area of intense inquiry in the mid-1950s following upon the success of the supersonic Round One research

---

102. Quoted in William E. Andrews, Thomas R. Sisk, and Robert W. Darville, "Longitudinal Stability Characteristics of the Convair YF-102 Airplane Determined from Flight Tests," NACA RM-H56I17 (1956), p. 1; see also Edwin J. Saltzman, Donald R. Bellman, and Norman T. Musialowski, "Flight-Determined Transonic Lift and Drag Characteristics of the YF-102 Airplane With Two Wing Configurations," NACA RM-H56E08 (1956).

103. Although it still experienced some troubled sailing: like most of the Century series fighters, the F-102 had other, more tortuous acquisition and program management problems unrelated to its aerodynamics that contributed to its delayed service entry. See Thomas A. Marschak, *The Role of Project Histories in the Study of R&D*, Rand report P-2850 (Santa Monica: The Rand Corporation, 1965), pp. 66–81; and Knaack, *Fighters*, pp. 163–167.

aircraft. Already a Round Two hypersonic test vehicle, the soon-to-emerge North American X-15 was underway. But what of high-hypersonics, the hypersonics of flight at Mach 10 to orbital velocity?

Hypersonics constituted a natural application for the low aspect ratio delta planform. Before the Second World War, Austrian engineer Eugen Sänger and his mathematician wife, Irene Sänger-Bredt, had conceptualized the Silbervogel ("Silver Bird"), a flat-bottom, half ogive body shape as a potential Earth-girdling hypersonic boost-glider. It had, for its time, a remarkable advanced aerodynamic profile, introducing the flat bottom and ogival configuration that did, in fact, come to characterize hypersonic aerothermodynamic design. But in one respect it did not: Sänger-Bredt's "antipodal aircraft" had a conventional wing (though of low aspect planform and with supersonic wedge airfoils). Although it proved very influential on the course of postwar hypersonics, by the mid-1950s, as high-speed aerodynamic thinking advanced beyond the supersonic and into the hypersonic realm, attention increasingly turned toward the sharply swept delta planform.

In 1951, Ames researchers H. Julian Allen and Alfred Eggers, Jr., had postulated the blunt-body reentry theory that led to the advent of the practical reentry shape used subsequently both for missile warheads and the first human presence in space.[104] (Their work, and the emergence of the hypersonics field generally, are discussed in greater detail in T.A. Heppenheimer's accompanying essay on transatmospherics.) While blunt-body theory enabled safely transiting the atmosphere, it did not furnish the flexibility of a large landing "footprint"; indeed, in practice, blunt-body reentry was limited to "throwaway" reentry shapes and programs such as Mercury, Gemini, and Apollo that necessitated a large and cumbersome investment in oceanic recovery of returning spacecraft. Some sort of lifting vehicle that could fly at hypersonic velocities would have far greater flexibility.

Related to the problem of hypersonic flight was the challenge of increasing lift-to-drag ratios at high supersonic speeds. Eggers, working with Ames researcher Clarence A. Syvertson, now turned away from blunt-body theory to examine thin, slender deltas. The two recognized that "the components of the aircraft should be individually

---

104. H. Julian Allen and A.J. Eggers, Jr., "A Study of the Motion and Aerodynamic Heating of Ballistic Missiles Entering the Earth's Atmosphere at High Supersonic Speeds," NACA TR-1381 (1953); Hartman, *Adventures in Research*, pp. 215–218.

and collectively arranged to impart the maximum downward and the minimum forward momentum to the surrounding air."[105] Out of this emerged the hypersonic "flattop" delta, a high-wing concept having the wing perched above the body (in this case, surmounting the classic half-ogive hypersonic shape), incongruously much like a general aviation light airplane such as a Cessna 152. At mid-span, its tips would angle sharply downward, capturing the momentum of flow imparted laterally outward from the body and deflecting it into downward momentum, thus greatly increasing lift. The tips as well furnished directional stability. This flattop concept, which Eggers and Syvertson enunciated in 1956, spawned an Ames concept for a hypersonic "beyond X-15" Round Three research vehicle that could be air-launched from a modified Convair B-36 bomber for initial trials to Mach 6 and, once proven, could then be launched vertically as the second stage of a two-stage system capable of reaching Mach 10 and transiting the United States. The Ames vehicle, with an overall length of 70 feet and a span of just 25 feet, represented a bold concept that seemed likely to spawn the anticipated Round Three hypersonic boost-glider.[106]

But the flattop delta was swiftly undone by a rival Round Three Langley concept that echoed more the earlier work of Sänger-Bredt. A 1957 study by Peter Korycinski and John Becker demonstrated that a flat-bottom (that is, low-wing) delta boost-glider would have better cooling characteristics (a vital concern at hypersonic velocities) and thus require less weight for thermal protection systems. Any lift-to-drag advantages of the Ames flattop high-wing concept were thus nullified. Round Three went forward, evolving into the abortive Air Force–NASA X-20 Dyna-Soar program, which employed the Langley

---

105. A.J. Eggers, Jr., and Clarence A. Syvertson, "Aircraft Configurations Developing High Lift-Drag Ratios at High Supersonic Speeds," NACA RM-A55L05 (1956), p. 1.
106. Ames staff, "Preliminary Investigation of a New Research Airplane for Exploring the Problems of Efficient Hypersonic Flight," (Jan. 18, 1957), copy in the archives of the Historical Office, NASA Johnson Space Center, Houston, TX. Drawings and more data on this concept can be found in Richard P. Hallion, ed., *From Max Valier to Project PRIME (1924–1967)*, vol. 1 of *The Hypersonic Revolution: Case Studies in the History of Hypersonic Technology* (Washington: USAF, 1998), pp.II-vi–II-x. Round One, in NACA parlance, was the original X-1 and D-558 programs. Round Two was the X-15. Round Three was what eventually emerged as the X-20 Dyna-Soar development effort.

flat-bottom approach, not the high-wing flattop delta of Ames.[107] Ames and Langley contested a decade later, this time in rival lifting bodies, with the Ames half-cone flattop M-2 (the product of Allen, Eggers, Syvertson, George Edwards, and George Kenyon) competing against Langley's HL-10 fattened flat-bottom delta (by Eugene S. Love). Again, it was the flat-bottom delta that proved superior, confirmed by tests in the mid-1970s with an even more refined flat-bottom Air Force-derived slender delta body shape, the Martin X-24B.[108]

When orbital cross range proved even of greater significance, Shuttle proponents from the National Aeronautics and Space Administration (NASA) and the Air Force in the 1970s looked away from flattop and lifting body approaches and more toward blended bodies, modified delta planforms, and exotic delta "wave riders." Though NASA's Spacecraft Design Division briefly considered a conventionally tailed, straight and swept wing Shuttle concepts, reflecting an influential study by Johnson's Maxime Faget, it moved rapidly toward deltas after analysis indicated such designs had a tendency of hypersonic spins, suspect aerothermal survivability, and too small a cross range during return from orbit. Between mid-1971 and the late summer of 1972, the Spacecraft Design Division evaluated no less than 37 separate delta configurations, ranging from simple triangular shapes echoing the early days of Jones to much more complex ogee shape reflecting the refinement of the delta as exemplified by the Anglo-French Concorde. Aside from continuous review by the Manned Spacecraft Center (MSC; subsequently the NASA Lyndon B. Johnson Space Center), these evaluations benefitted greatly from aerodynamic analysis by NASA's Ames and Langley hypersonic

---

107. See John V. Becker, "The Development of Winged Reentry Vehicles, 1952–1963," in Hallion, ed., *Hypersonic Revolution*, vol. 1, pp. 379–448. It is worth noting that one significant aircraft project did use the Eggers-Syvertson wing but in a modified form: the massive North American XB-70A Valkyrie Mach 3+ experimental bomber. The XB-70 had its six engines, landing gear, and weapons bays located under the wing in a large wedge-shaped centerbody. The long, cobralike nose ran forward from the wing and featured canard control surfaces. Its sharply swept delta wing had outer wing panels that could entrap the lateral momentum off the ventral centerbody and transfer it downward to furnish compression lift.

108. For further detail, see R. Dale Reed with Darlene Lister, *Wingless Flight: The Lifting Body Story*, SP-4220 (Washington: NASA 1997); Milton O. Thompson and Curtis Peebles, *Flying Without Wings: NASA Lifting Bodies and The Birth of the Space Shuttle* (Washington: Smithsonian Institution Press, 1999); and Johnny G. Armstrong, "Flight Planning and Conduct of the X-24B Research Aircraft Flight Test Program," Air Force Flight Test Center TR-76-11 (1977).

communities, the practical low lift-to-drag-ratio flight-test experience of researchers at the NASA Flight Research Center, and the rocketry and space flight expertise of the Marshall Space Flight Center, whose experts assessed each proposal from the standpoint of technical feasibility and launch vehicle practicality. This multi-Center review strongly endorsed development of a modified delta planform, in part because the delta had inherently better stability characteristics during the high angle-of-attack reentry profile that any returning Shuttle would have to experience. Two families emerged as finalists: The 036 series, with small payload bays and three engines, and the 040 family, of similar planform but with larger payload bays and four engines. Then, in late January 1972, MSC engineers evolved the 040C configuration: a three-engine design using new high-pressure engines. The 040C design became the baseline for subsequent Orbiter studies. While many questions remained over the final form that Shuttle's launch system would take, with the 040C study, the shape of the orbiter, and its all-important wing, was essentially fixed. Again, the flat-bottom delta had carried the day.[109]

Swing Wing: The Path to Variable Geometry

The notion of variable wing-sweeping dates to the earliest days of aviation and, in many respects, represents an expression of the "bird imitative" philosophy of flight that gave the ornithopter and other flexible wing concepts to aviation. Varying the sweep of a wing was first conceptualized as a means of adjusting longitudinal trim. Subsequently,

---

109. Spacecraft Design Division, *Summary of MSC Shuttle Configurations (External HO Tanks)* (Houston: Manned Spacecraft Center, June 30, 1972, rev. ed.), passim. I thank the late Dr. Edward C. Ezell for making a copy of this document available for my research. The range of configurations and wind tunnel testing done in support of Shuttle development is in A. Miles Whitnah and Ernest R. Hillje, "Space Shuttle Wind Tunnel Testing Summary," NASA Reference Publication 1125 (1984), esp. pp. 5–7. See also Alfred C. Draper, Melvin L. Buck, and William H. Goesch, "A Delta Shuttle Orbiter." *Astronautics & Aeronautics*, vol. 9, No. 1 (Jan. 1971), pp. 26–35 (I acknowledge with gratitude the assistance and advice of the late Al Draper, while we both worked at Aeronautical Systems Division, Wright-Patterson AFB, in 1986–1987); Joseph Weil and Bruce G. Powers, "Correlation of Predicted and Flight Derived Stability and Control Derivatives with Particular Application to Tailless Delta Wing Configurations," NASA TM-81361 (July 1981); and J.P. Loftus, Jr., et al. "The Evolution of the Space Shuttle Design," a reference paper prepared for the Rogers Commission, 1986 (copy in NASA JSC History Office archives). The evolution of Shuttle configuration evolution is examined more broadly in Richard P. Hallion and James O. Young, "Space Shuttle: Fulfillment of a Dream," in Hallion, ed., *From Scramjet to the National Aero-Space Plane (1964–1986)*, vol. 2 of The Hypersonic *Revolution: Case Studies in the History of Hypersonic Technology* (Washington: USAF, 1998) pp. 947–1173.

A time-lapse photograph of the Bell X-5, showing the range of its wing sweep. Note how the wing roots translated fore and aft to accommodate changes in center of lift with varying sweep angles. NASA.

variable-geometry advocates postulated possible use of asymmetric sweeping as a means of roll control. Lippisch, pioneer of tailless and delta design, likewise filed a patent in 1942 for a scheme of wing sweeping, but it was another German, Waldemar Voigt (the chief of advanced design for the Messerschmitt firm) who triggered the path to modern variable wing-sweeping. Ironically, at the time he did so, he had no plan to make use of such a scheme himself. Rather, he designed a graceful midwing turbojet swept wing fighter, the P 1101. The German air ministry rejected its development based upon assessments of its likely utility. Voigt decided to continue its development, planning to use the airplane as an in-house swept wing research aircraft, fitted with wings of varying sweep and ballasted to accommodate changes in center of lift.[110]

---

110. The best survey of v-g origins remains Robert L. Perry's *Innovation and Military Requirements: A Comparative Study*, Rand Report RM-5182PR (Santa Monica: The Rand Corporation, 1967), upon which this account is based.

Case 1 | Sweep and Swing: Reshaping the Wing for the Jet and Rocket Age

By war's end, when the Oberammergau plant was overrun by American forces, the P 1101 was over 80-percent complete. A technical team led by Robert J. Woods, a member of the NACA Aerodynamics Committee, moved in to assess the plant and its projects. Woods immediately recognized the value of the P 1101 program, but with a twist: he proposed to Voigt that the plane be finished with a wing that could be variably swept in flight, rather than with multiple wings that could be installed and removed on the ground. Woods's advocacy, and the results of NACA variable-sweep tests by Charles Donlan of a modified XS-1 model in the Langley 7-foot by 10-foot wind tunnel, convinced the NACA to support development of such an aircraft. In May 1949, the Air Force Air Materiel Command issued a contract covering development of two Bell variable sweep airplanes, to be designated X-5. They were effectively American-built versions of the P 1101, but with American, not German, propulsion, larger cockpit canopies for greater pilot visibility, and, of course, variable sweep wings that could range from 20 to 60 degrees.[111]

The first X-5 flew in June 1951 and within 5 weeks had demonstrated variable in-flight wing sweep to its maximum 60-degree aft position. Slightly over a year later, Grumman flew a prototype variable wing-sweep naval fighter, the XF10F-1 Jaguar. Neither aircraft represented a mature application of variable sweep design. The mechanism in each was heavy and complex and shifted the wing roots back and forth down the centerline of the aircraft to accommodate center of lift changes as the wing was swept and unswept. Each of the two had poor flying qualities unrelated to the variable-sweep concept, reflecting badly on their design. The XF10F-1 was merely unpleasant (its test pilot, the colorful Corwin "Corky" Meyer, tellingly recollected later "I had never attended a test pilots' school, but, for me, the F10F provided the complete curriculum"), but the X-5 was lethal.[112] It had a vicious pitch-up at higher-sweep angles, and its aerodynamic design ensured that it would have very great difficulty when it departed into a spin. The combination of the two led to the death of Air Force test pilot Raymond Popson in the crash of the second X-5

---

111. The history of the X-5 is examined minutely in Warren E. Green's *The Bell X-5 Research Airplane* (Wright-Patterson AFB: Wright Air Development Center, March 1954). For NACA work, see LRC staff, "Summary of NACA/NASA Variable-Sweep Research and Development Leading to the F-111 (TFX)," Langley Working Paper LWP-285 (Dec. 22, 1966).

112. Corwin H. Meyer, "Wild, Wild Cat: The XF10F," 20th Symposium, The Society of Experimental Test Pilots, Beverly Hills, CA, Sept. 15, 1976.

in 1953. More fortunate, NACA pilots completed 133 research flights in the first X-5 before retiring it in 1955.

The X-5 experience demonstrated that variable geometry worked, and the potential of combining good low-speed performance with high-speed supersonic dash intrigued military authorities looking at future interceptor and long-range strike aircraft concepts. Coincidentally, in the late 1950s, Langley developed increasingly close ties with the British aeronautical community, largely a result of the personal influence of John Stack of Langley Research Center, who, in characteristic fashion, used his forceful personality to secure a strong transatlantic partnership. This partnership, best known for its influence upon Anglo-American V/STOL research leading to the Harrier strike fighter, influenced as well the course of variable-geometry research. Barnes Wallis of Vickers had conceptualized a sharply swept variable-geometry tailless design, the Swallow, but was not satisfied with the degree of support he was receiving for the idea within British aeronautical and governmental circles. Accordingly, he turned to the United States. Over November 13–18, 1958, Stack sponsored an Anglo-American meeting at Langley to craft a joint research program, in which Wallis and his senior staff briefed the Swallow design.[113] As revealed by subsequent Langley tunnel tests over the next 6 months, Wallis's Swallow had many stability and control deficiencies but one significant attribute: its outboard wing-pivot design. Unlike the X-5 and Jaguar and other early symmetrical-sweep v-g concepts, the wing did not adjust for changing center of lift position by translating fore and aft along the fuselage centerline using a track-type approach and a single pivot point. Rather, slightly outboard of the fuselage centerline, each wing panel had its own independent pivot point. This permitted elimination of the complex track and allowed use of a sharply swept forebody to address at least some of the changes in center-of-lift location as the wings moved aft and forward. The remainder could be accommodated by control surface deflection and shifting fuel. Studies in Langley's 7-foot by 10-foot tunnel led to refinement of the outboard pivot concept and, eventually, a patent to William J. Alford and E.C. Polhamus for its concept, awarded in September 1962. Wallis's inspiration, joined with insightful research by Alford and Polhamus and

---

113. For meeting, see LRC staff, "Summary of NACA/NASA Variable-Sweep Research and Development," p. 8; and J.E. Morpurgo, *Barnes Wallis: a Biography* (Harmondsworth, UK: Penguin Books, 1973), p. 423. NASA Langley Photograph L58-771a, dated Nov. 13, 1958, documents the Stack-Wallis meeting; it is also catalogued as NASA LaRC image EL-2008-00001.

followed by adaptation of a conventional "tailed" configuration (a critical necessity in the pre-fly-by-wire computer-controlled era), made variable wing sweep a practical reality.[114] (Understandably, after returning to Britain, Wallis had mixed feelings about the NASA involvement. On one hand, he had sought it after what he perceived as a "go slow" approach to his idea in Britain. On the other, following enunciation of outboard wing sweep, he believed—as his biographer subsequently wrote—"The Americans stole his ideas.")[115]

Thus, by the early 1960s, multiple developments—swept wings, high-performance afterburning turbofans, area ruling, the outboard wing pivot, low horizontal tail, advanced stability augmentation systems, to select just a few—made possible the design of variable-geometry combat aircraft. The first of these was the General Dynamics Tactical Fighter Experimental (TFX), which became the F-111. It was a troubled program, though, like most of the Century series that had preceded it (the F-102 in particular), this had essentially nothing to do with the adaptation of a variably swept wing. Instead, a poorly written specification emphasizing joint service over practical, attainable military utility resulted in development of a compromised design. The result was a decade of lost fighter time for the U.S. Navy, which never did receive the aircraft it sought, and a constrained Air Force program that resulted in the eventual development of a satisfactory strike aircraft—the F-111F—but years late and at tremendous cost. Throughout the evolution of the F-111, NASA research proved of crucial importance to saving the program. NASA Langley, Ames, and Lewis researchers invested over 30,000

---

114. LRC, "Summary of NACA/NASA Variable-Sweep Research;" see also William J. Alford, Jr., and William P. Henderson, "An Exploratory Investigation of Variable-Wing-Sweep Airplane Configurations," NASA TM-X-142 (1959); William J. Alford, Jr., Arvo A. Luoma, and William P. Henderson, "Wind-Tunnel Studies at Subsonic and Transonic Speeds of a Multiple-Mission Variable-Wing-Sweep Airplane Configuration," NASA TM-X-206 (1959); and Gerald V. Foster and Odell A. Morris, "Aerodynamic Characteristics in Pitch at a Mach Number of 1.97 of Two Variable-Wing-Sweep V/STOL Configurations with Outboard Wing Panels Swept Back 75°," NASA TM-X-322 (1960).

115. Morpurgo, *Wallis*, p. 422, and Derek Wood, *Project Cancelled: British Aircraft that Never Flew* (Indianapolis: The Bobbs-Merrill Company, Inc., 1975), pp. 182–195. After the Nov. 1958 meeting, NASA tunnel tests revealed very great deficiencies attending his tailless concept that Stack and others reported back to Vickers in June 1959. In short, the outboard pivot was but one element necessary for making a successful v-g aircraft. Others were provision for a conventional tail and design of a practicable airframe. In short, Wallis had an idea, but it took Alford and Polhamus and other NASA researchers to refine it and render it achievable.

hours of wind tunnel test time in the F-111 (over 22,000 at Langley alone), addressing various shortcomings in its design, including excessive drag, lack of transonic and supersonic maneuverability, deficient directional stability, and inlet distortion that plagued its engine performance. As a result, the Air Force F-111 became a reliable weapon system, evidenced by its performance in Desert Storm, where it flew long-range strike missions, performed electronic jamming, and proved the war's single most successful "tank plinker," on occasion destroying upward of 150 tanks per night and 1,500 over the length of the 43-day conflict.[116]

From the experience gained with the F-111 program sprang the Grumman F-14 Tomcat naval fighter and the Rockwell B-1 bomber, both of which experienced fewer development problems, benefitting greatly from NASA tunnel and other analytical research.[117] Emulating American variable-geometry development, Britain, France, and the Soviet Union undertook their own development efforts, spawning the experimental Dassault Mirage G (test-flown, though never placed in service), the multipartner NATO Tornado interceptor and strike fighter program, and a range of Soviet fighter and bomber aircraft, including the MiG-23/27 Flogger, the Sukhoi Su-17/22 Fitter, the Su-24 Fencer, the Tupolev Tu-22M Backfire, and the Tu-160 Blackjack.[118]

Variable geometry has had a mixed history since; in the heyday of the space program, many proposals existed for tailored lifting body shapes deploying "switchblade" wings, and the variable-sweep wing was a prominent feature of the Boeing SST concept before its subsequent rejection. The tailored aerodynamics and power available with modern aircraft have rendered variable-geometry approaches less attractive than they once were, particularly because, no matter how well thought out, they invari-

---

116. NASA F-111 tunnel research, analysis, and support is detailed in *Testimony of Edward C. Polhamus, in U.S. Senate, TFX Contract Investigation (Second Series): Hearings Before the Permanent Subcommittee on Investigations of the Committee on Government Operations, United States Senate, 91st Congress, 2nd Session, Part 2* (Washington: GPO, 1970), pp. 339–363; for the F-111 in Desert Storm, see Tom Clancy with Gen. Chuck Horner (New York: G.P. Putnam's Sons, 1999), pp. 318, 417, 424, and 450.

117. See Joseph R. Chambers, *Partners in Freedom: Contributions of the Langley Research Center to U.S. Military Aircraft of the 1990s*, SP-2000-4519 (Washington; NASA, 2000), which treats these and other programs in great and authoritative detail.

118. Robert W. Kress, "Variable Sweep Wing Design," AIAA Paper No. 83-1051 (1983) is an excellent survey. The Su-24 was clearly F-111 inspired, and the Tu-160 was embarrassingly similar in configuration to the American B-1.

Case 1 | Sweep and Swing: Reshaping the Wing for the Jet and Rocket Age

The Grumman F-14A Tomcat naval fighter marked the maturation of the variable wing-sweep concept. This is one was assigned to Dryden for high angle of attack and departure flight-testing. NASA.

ably involve greater cost, weight, and structural complexity. In 1945–1946, John Campbell and Hubert Drake undertook tests in the Langley Free Flight Tunnel of a simple model with a single pivot, so that its wing could be skewed over a range of sweep angles. This concept, which German aerodynamicists had earlier proposed in the Second World War, demonstrated "that an airplane wing can be skewed as a unit to angles as great as 40° without encountering serious stability and control difficulties."[119] This concept, the simplest of all variable-geometry schemes, returned to the fore in the late 1970s, thanks to the work of Robert T. Jones, who adopted and expanded upon it to generate the so-called "oblique wing" design concept. Jones conceptualized the oblique wing as a means of producing a transonic transport that would have minimal drag and a minimal sonic boom; he even foresaw possible twin fuselage transports with a skewed wing shifting their relative position back and forth. Tests with a subscale turbojet demonstrator, the **AD-1** (for Ames-Dryden), at the Dryden Flight Research Center confirmed what Campbell and Drake had discovered

---

119. John P. Campbell and Hubert M. Drake, "Investigation of Stability and Control Characteristics of an Airplane Model with Skewed Wing in the Langley Free-Flight Tunnel," NACA TN-1208 (May 1947), p. 10.

nearly four decades previously, namely that at moderate sweep angles the oblique wing possessed few vices. But at higher sweep angles near 60 degrees, its deficits became more pronounced, calling into question whether its promise could ever actually be achieved.[120] On the whole, the variable-geometry wing has not enjoyed the kind of widespread success that its adherents hoped. While it may be expected that, from time to time, variable sweep aircraft will be designed and flown for particular purposes, overall the fixed conventional planform, outfitted with all manner of flaps and slats and blowing, sucking, and perhaps even warping technology, continues to prevail.

The Quest for Refinement

By the end of the 1960s, the "classic" era of aircraft design was arguably at an end. As exemplars of the highest state of aviation technology, the piston engine had given way to the gas turbine, the wood-and-fabric aircraft to the all-metal, the straight wing had given way to the swept and delta. Aircraft flight speeds had risen from a mere 40 mph at the time of the Wright brothers to over 100 times as fast, as the X-15A-2 demonstrated when it streaked to Mach 6.70 (4,520 mph) in October 1967, piloted by Maj. William J. Knight. Fighters, by that time, had been flying on a Mach 2 plateau for a decade and transports on a Mach 0.82 plateau for roughly the same amount of time. In space, Americans were basking in the glow of the recent Apollo triumph, where a team of astronauts, led by former NACA–NASA research pilot Neil Armstrong—a Round One and Round Two veteran whose experience included both the X-1 and the X-15—journeyed to the Moon, landed two of their number upon it, and then returned to Earth.

Such accomplishments hardly meant that the frontiers of the sky were closing, or that NASA had little to do. Indeed, in some respects, it was facing even greater challenges: conducting comprehensive aeronautical research at a time when, increasingly, more people identified it with space than aeronautics and when, in the aftermath of the Apollo success, monies were increasingly tight. Added to this was a dramatically transforming world situation: increasing tension in the Middle East, a growing Soviet threat, rising oil prices, open concern over environmental stewardship,

---

120. Richard P. Hallion and Michael H. Gorn, *On the Frontier: Experimental Flight at NASA Dryden* (Washington: Smithsonian Books, 2002), pp. 256–260, and personal recollections of the program from the time.

Case 1 | Sweep and Swing: Reshaping the Wing for the Jet and Rocket Age

and a national turning away from the reflexive perception that limitless technological progress was both a given and a good thing.

Within this framework, NASA work increasingly turned to achieving efficiencies: more fuel-efficient and energy-efficient civilian flight, and more efficient military systems. It was not NASA's business to, per se, design new aircraft, but, as NACA–NASA history amply demonstrated, the Agency's mark could be found on many aircraft and their innovations. Little things counted for much. When, for example, NACA High-Speed Flight Research Station pilots flew a Douglas D-558-1 Skystreak modified with a row of small vortex generators (little rectangular fins of 0.5-inch chord standing vertically like a row of razor blades) on its upper wing surface, they hardly expected that such a small energy-imparting modification would so dramatically improve its transonic handling qualities that rows of vortex generators would become a commonly recognized feature on many aircraft, including such "classics" as the B-52, the 707, and the A-4.[121] In the post-1970 period, NASA assiduously pursued three concepts related to swept wing and delta flight, in hopes that each would pay great dividends: the supercritical wing, the winglet, and the arrow wing.[122] All had roots embedded and nourished in the earliest days of the supersonic and swept/delta revolution. Each reflected Whitcomb's passion—indeed obsession, in its most positive sense—with minimizing interference effects and achieving the greatest possible aerodynamic efficiency without incurring performance-robbing complexity. Many had researched configurations approaching the purity of the arrow wing, but it was Whitcomb who first actually achieved such a configuration, as part of Langley's Supersonic Transport study effort.

Long a subject of individual research and thought, Langley's institutional SST studies had begun in 1958, when the ever-enthusiastic John Stack formed a Supersonic Transport Research Committee (STRC). It evaluated the maturity of various disciplines—particularly the "classics" of aerodynamics, structures, propulsion, and controls—and then forecast the overall feasibility of a Supersonic Transport. The Stack team presented the results of their studies to the head of the Federal Aviation

---

121. De E. Beeler, Donald R. Bellman, and John H. Griffith, "Flight Determination of the Effects of Wing Vortex Generators on the Aerodynamic Characteristics of the Douglas D-558-I Airplane," NACA RM-L51A23 (1951).
122. All three bore the imprint of Richard Whitcomb and thus, in this survey, are not examined in detail, since his work is more thoroughly treated in a companion essay by Jeremy Kinney.

Administration (FAA), Elwood Quesada, a retired Air Force general, in December 1959. Their report, issued the following year, concluded: "the state of the art appears sufficiently advanced to permit the design of an airplane at least marginally capable of performing the supersonic transport mission."[123] NASA swiftly ramped up to match growing interest in the FAA in such aircraft; within a decade, SST-focused research would constitute over a quarter of all NASA aeronautics research undertaken at the Langley, Ames, and Lewis Centers.[124]

Given that the British and French subsequently designed the Mach 2+ Concorde, and the Soviets the Tupolev Tu-144, NASA Langley's technological optimism in 1959–1960 was, within limits, technically well justified, and such optimism infused Washington's political community as well. In March 1966, President Lyndon Johnson announced that the first American SST, designed to cruise at Mach 2.7, would fly at decade's end and enter commercial service in 1974.[125] But such expectations would prove overly optimistic. As Mach number rose, so too did a number of daunting technical challenges encountered by the more ambitious aircraft American SST proponents favored. Assessing the technology alone did not address the serious questions—research and development investment, production costs, operating economics, and environmental concerns, for example—such aircraft would pose and would limit the airline acceptance (and, hence, market success) of even the "modest" Concorde and Tu-144. Air transport constitutes a system of systems, and excellence in some does not guarantee or imply excellence overall. Political support, strongly bipartisan over the Kennedy-Johnson era, withered in the Nixon

---

123. LRC staff, "The Supersonic Transport—A Technical Summary," NASA TN-D-423 (1960), p. 93; this was the summary report of the briefings presented the previous fall to Quesada. NASA research on supersonic cruise is the subject of a companion essay in this study, by William Flanagan, and Whitcomb's work is detailed in the previously cited Kinney study in this volume.

124. In FY 1968, NASA expended $10.8 million in then-year dollars on SST research at Langley, Ames, and Lewis, against a total aeronautics research expenditure of $42.9 million at those Centers. See *Testimony of James E. Webb in U.S. Senate, Aeronautical Research and Development Policy: Hearings Before the Committee on Aeronautical and Space Sciences, United States Senate, 90th Congress, 1st Session* (Washington: GPO, 1967), p. 39.

125. Lyndon B. Johnson, "President's Message on Transportation," Mar. 2, 1966, reprinted in *Legislative Reference Service of the Library of Congress, Policy Planning for Aeronautical Research and Development: Staff Report Prepared for Use of the Committee on Aeronautical and Space Sciences United States Senate by the Legislative Reference Service Library of Congress, Document No. 90, U.S. Senate, 89th Congress, 2nd Session* (Washington: GPO, 1966), pp. 50–51.

Case 1 | Sweep and Swing: Reshaping the Wing for the Jet and Rocket Age

years as technical and other challenges arose, and a re-action against the SST set in, fueled by questions over the value of high technology and reaction to the long and costly war in Southeast Asia.[126]

From the standpoint of aircraft design, from Langley's interest emerged a series of Supersonic Commercial Air Transport (SCAT) design studies, most of which incorporated variable-geometry planforms reflecting a growing popular wisdom that future military or civilian supersonic cruise designs would necessarily incorporate such wings. Whitcomb, focused on simplicity and efficiency, demurred, preferring instead a sharply swept arrow configuration, the SCAT-4, which he had derived. It drew upon a two-decade tradition of Langley swept and delta studies running through those of Clinton E. Brown and F. Edward McLean in the 1950s, back to the thin swept and delta research manifested in Robert T. Jones's original concepts in 1944–1945. Though he was not successful at the time at selling his vision of what such an aircraft should be (and, in fact, left the Stack SST study effort as a result), in time the fixed wing predominated. In 1964, a Langley team comprised of Harry Carlson, Roy Harris, Ed McLean, Wilbur Middleton, and A. Warner Robins derived a fixed wing variant of the variable-sweep SCAT-15, generating an elegant slender arrow wing called the SCAT-15F. SCAT-15F had an incredible lift-to-drag ratio of 9.3 at Mach 2.6, well beyond what previous analysis and thought had deemed possible, though it also had serious low-speed pitch-up and deep-stall tendencies that triggered intensive investigations by researchers using the Langley Full-Scale Tunnel.[127] Out of this came a revised SCAT-15F configuration, with leading edge flaps, wing notches, area-and-camber-increasing Fowler flaps, and a small, horizontal tail, all of which worked to make it a much more acceptable planform. The development

---

126. For various perspectives on Anglo-French-Soviet-American SST development, see Kenneth Owen, *Concorde: Story of a Supersonic Pioneer* (London: Science Museum, 2001); Howard Moon, *Soviet SST: The Technopolitics of the Tupolev Tu-144* (New York: Orion Books, 1989); R.E.G. Davies, *Supersonic (Airliner) Non-Sense: A Case Study in Applied Market Research* (McLean, VA: Paladwr Press, 1998); Mel Horwitch, *Clipped Wings: The American SST Conflict* (Cambridge: The MIT Press, 1982); and Eric M. Conway, *High-Speed Dreams: NASA and the Technopolitics of Supersonic Transportation, 1945–1999* (Baltimore: The Johns Hopkins Press, 2005).

127. Deep stall is a dangerous condition wherein an airplane pitches to a high angle of attack, stalls, and then descends in a stabilized stalled attitude, impervious to corrective control inputs. It is more typically encountered by swept wing T-tail aircraft, and one infamous British accident, to a BAC 1-11 airliner, claimed the life of a crack flight-test crew captained by the legendary Mike Lithgow, an early supersonic and sweptwing pioneer.

of the high supersonic L/D fixed wing eventually led Boeing (winner of the Government's SST design competition) to abandon variable-sweep in favor of a highly refined small-tailed delta, for its final SST proposal, though congressional refusal to furnish needed developmental monies brought the American SST development effort to a sorry end.[128] It did not, however, end interest in similar configurations for a range of other missions. Today, in an era of vastly different technology, with much higher-performing engines, better structures, and better means of modeling and simulating the aerodynamic and propulsive performance of such designs, tailored fixed arrow wing configurations are commonplace for future advanced high-speed civil and military aircraft applications.

As the American SST program, plagued by controversy and numerous wounds (many self-inflicted), died amid performance and environmental concerns, Whitcomb increasingly turned his attention to the transonic, thereby giving to aviation one of its most compelling images, that of the graceful supercritical wing and, of less aesthetic appeal but no less significance, the wingtip winglet. Both, in various forms, became standard design elements of future civil and military transport design and are examined elsewhere (by historian Jeremy Kinney) in this work.

---

128. Langley's SCAT studies are summarized in David A. Anderton, *Sixty Years of Aeronautical Research, 1917–1977, EP-145* (Washington: NASA, 1978), pp. 54–58. Relevant reports on specific configurations and predecessors include: Donald D. Baals, Thomas A. Toll, and Owen G. Morris, "Airplane Configurations for Cruise at a Mach Number of 3," NACA RM-L58E14a (1958); Odell A. Morris and A. Warner Robins, "Aerodynamic Characteristics at Mach Number 2.01 of an Airplane Configuration Having a Cambered and Twisted Arrow Wing Designed for a Mach Number of 3.0," NASA TM-X-115 (1959); Cornelius Driver, M. Leroy Spearman, and William A. Corlett, "Aerodynamic Characteristics at Mach Numbers From 1.61 to 2.86 of a Supersonic Transport Model With a Blended Wing-Body, Variable-Sweep Auxiliary Wing Panels, Outboard Tail Surfaces, and a Design Mach Number of 2.2," NASA TM-X-817 (1963); Odell A. Morris and James C. Patterson, Jr., "Transonic Aerodynamic Characteristics of Supersonic Transport Model With a Fixed, Warped Wing Having 74° Sweep," NASA TM-X-1167 (1965); Odell A. Morris, and Roger H. Fournier, "Aerodynamic Characteristics at Mach Numbers 2.30, 2.60, and 2.96 of a Supersonic Transport Model Having Fixed, Warped Wing," NASA TM-X-1115 (1965); A. Warner Robins, Odell A. Morris, and Roy V. Harris, Jr., "Recent Research Results in the Aerodynamics of Supersonic Vehicles," AIAA Paper 65-717 (1965); Donald D. Baals, A. Warner Robins, and Roy V. Harris, Jr., "Aerodynamic Design Integration of Supersonic Aircraft," AIAA Paper 68-1018 (1968); Odell A. Morris, Dennis E. Fuller, and Carolyn B. Watson, "Aerodynamic Characteristics of a Fixed Arrow-Wing Supersonic Cruise Aircraft at Mach Numbers of 2.30, 2.70, and 2.95," NASA TM-78706 (1978); and John P. Decker and Peter F. Jacobs, "Stability and Performance Characteristics of a Fixed Arrow Wing Supersonic Transport Configuration (SCAT 15F-9898) at Mach Numbers from 0.60 to 1.20," NASA TM-78726 (1978).

As for the arrow wing, military exigency and the Cold War combined to ensure that studies of this most promising configuration spawned the "cranked arrow wing" of the late 1970s. Following cancellation of the national SST effort, NASA researchers continued studying supersonic cruise for both military and civil applications, under the guise of a new study effort, the Advanced Supersonic Technology (AST) effort. AST was succeeded by another Langley-run cruise-focused effort, the Supersonic Cruise Aircraft Research (SCAR, later shortened to SCR) program. SCR lasted until 1982, when NASA terminated it to focus more attention and resources on the already troubled Shuttle program. But meantime, it had spawned the Supersonic Cruise and Maneuver Prototype (SCAMP), a derivative of the F-16 designed to cruise at supersonic speeds. Its "cranked arrow" wing, blending a 70-degree swept inboard leading edge and a 50-degree swept outboard leading edge, looked deceptively simple but embodied sophisticated shaping and camber (reflecting the long legacy of SCAT studies, particularly the refinement of the SCAT-15F), with leading edge vortex flaps to improve both transonic and low-speed performance. General Dynamics' F-16 designer Harry J. Hillaker adopted the planform for a proposed strike fighter version of the F-16 because it reduced supersonic wave drag, increasing the F-16's potential combat mission radius by as much as 65 percent and more than doubling its permissible angle-of-attack range as well. In the early 1980s, SCAMP, now designated the F-16XL, competed with the prototype F-15E Strike Eagle at Edwards Air Force Base for an Air Force deep-strike fighter contract. But the F-16XL was too small an airplane to win the completion; with greater internal fuel and volume, the larger Strike Eagle offered more growth potential and versatility. The two F-16XL aircraft, among the most beautiful ever flown, remained at Edwards, where they flew a variety of research missions at NASA Dryden, refining understanding of the complex flows around cranked arrow profiles and addressing such technical issues as the possibility of supersonic laminar flow control by using active suction. Interest in the cranked arrow has persisted, as it remains a most attractive design option for future supersonic cruise aircraft, whether piloted or not, both civil and military.[129]

---

129. Harry J. Hillaker, "The F-16: A Technology Demonstrator, a Prototype, and a Flight Demonstrator," AIAA Paper No. 83-1063 (1983). The "XL" designation for the cranked-arrow F-16 reflected Harry Hillaker's passionate interest in golf, for it echoed the name of a particularly popular long-distance golf ball, the Top Flite XL. See also Chambers, Innovation in Flight, pp. 42, 48, 58–59.

By the end of the 1980s, for military aircraft, concern over aerodynamic shaping of aircraft was beginning to take second place behind concern over their electromagnetic signature. Where something such as the blended wing-body delta SR-71 possessed an innate purity and beauty of form, inherent when aerodynamics is given the position of primacy in aircraft design, something such as the swept wing, V-tail F-117 stealth fighter did not: all angles and panels, it hardly looked aerodynamic, and, indeed, it had numerous deficits cured only by its being birthed in the electronic fly-by-wire and composites era. But in other aspects it performed with equal brilliance: not the brilliance of Mach 3+, but the quiet brilliance of penetrating a high-threat integrated air defense network, attacking a key target, and escaping without detection.

For the future of the swept surface, one had to look elsewhere, back to the transonic, where it could be glimpsed in the boldly imaginative lines of the Blended Wing-Body (BWB) transport. Conceived by Robert H. Liebeck, a gifted Boeing designer who had begun his career at Douglas, where he worked with the legendary A.M.O. Smith, the BWB represented a conception of pure aerodynamic efficiency predating NASA, the NACA that had preceded it, and even, indeed, Jack Northrop and the Horten brothers. It hearkened back to the earliest concepts for Nurflügeln (flying wings) by Hugo Junkers before the First World War, the first designer to appreciate how one could insightfully incorporate the cantilever all-metal structure to achieve a pure lifting surface.[130] Conceived while Liebeck worked for McDonnell-Douglas in the latter years before its own merger with Boeing, the graceful BWB was not strictly a flying wing but, rather, a hybrid wing-body combination whose elegant high aspect ratio wing blended smoothly into a wide, flat-bottom fuselage, the wings sprouting tall winglets at their tips for lateral control, thus differing significantly from earlier concepts such as the Boeing "Spanloader" and the Horten, Armstrong-Whitworth, and Northrop flying wings. Early design conceptions envisioned upward of 800 passen-

---

130. For Junkers, see Hugo Junkers, *Gleitflieger mit zur Aufnahme von nicht Auftrieg erzeugen Teilen dienenden Hohlkörpern, Patentschrift Nr. 253788, Klasse 77h, Gruppe 5* (Berlin: Reichspatentamt, Nov. 14, 1912). For Liebeck, see Robert H. Liebeck, Mark A. Page, Blaine K. Rawdon, Paul W. Scott, and Robert A. Wright, "Concepts for Advanced Subsonic Transports," NASA CR-4624 (1994); Robert H. Liebeck, "Design of the Blended Wing Body Subsonic Transport," *Journal of Aircraft*, vol. 41, no. 1 (Jan.–Feb. 2004). pp. 10–25; and Chambers, Innovation in Flight, pp. 86–92.

Case 1 | Sweep and Swing: Reshaping the Wing for the Jet and Rocket Age

gers flying in a three-engine, double-deck, 823,000-pound, manta-shaped BWB (spanning 289 feet with a length of 161 feet), cruising across the globe at Mach 0.85. Subsequent analysis resulted in a smaller design sized for 450 passengers, the BWB-450, which served as the baseline for later research and evaluation, which concluded that the most suitable role for the BWB might be for a range of global heavy-lift multi-purpose military missions rather than passenger-carrying.[131] Extensive studies by NASA Langley and Lewis researchers; McDonnell-Douglas (now Boeing) BWB team members; and academic researchers from Stanford University, the University of Southern California, Clark Atlanta University, and the University of Florida confirmed the aerodynamic and propulsive promise inherent in the BWB, particularly its potential to carry great loads at transonic speeds over global distances with unprecedented aerodynamic and energy efficiency, resulting in potentially 30-percent better fuel economy than that achievable by traditional "tube and wing" airlifters.[132]

These and many other studies, including tests by Boeing and the United States Air Force, encouraged the next logical step: developing a subscale unmanned aerial vehicle (UAV) to assess the low-speed flight-control characteristics of the BWB in actual flight. This became the X-48B, a 21-foot span, 8.5-percent scale UAV testbed of the BWB-450 configuration, powered by three 240-pound thrust Williams turbojets. Boeing had Cranfield Aerospace, Ltd., in Great Britain build two X-48Bs for the company's Phantom Works. After completion, the first X-48B completed 250 hours of tunnel tests in the Langley Full-Scale Tunnel (run by Old Dominion University) in May 2006. Readying the BWB for flight

---

131. These included heavy-lift cargo, air-refueling, and other military missions rather than use as a civil airliner. See NASA LRC, "The Blended-Wing-Body: Super Jumbo Jet Concept Would Carry 800 Passengers," NASA Facts, FS-1997-07-24-LaRC (July 1997); and NASA LRC, "The Blended Wing Body: A Revolutionary Concept in Aircraft Design," NASA Facts, FS-2001-04-24-LaRC (Apr. 2001). For an early appreciation of the military value of BWB designs, see Gene H. McCall, et al., *Aircraft & Propulsion*, a volume in the *New World Vistas: Air and Space Power for the 21st Century* series (Washington: HQ USAF Scientific Advisory Board, 1995), p. 6.

132. Robert H. Liebeck, Mark A. Page and Blaine K. Rawdon, "Blended-Wing-Body Subsonic Commercial Transport," AIAA Paper 98-0438 (1998); Sean Wakayama, "Multidisciplinary Design Optimization of the Blended-Wing-Body," AIAA Paper 98-4938 (1998); Dino Roman, J.B. Allen, and Robert H. Liebeck, "Aerodynamic Design Challenges of the Blended-Wing-Body Subsonic Transport," AIAA Paper 2000-4335 (2000). Fuel economy figure from Dryden Flight Research Center, "X-24B Blended Wing-Body" (Apr. 2, 2009).

The NASA F-16XL cranked-arrow research aircraft aloft over the Dryden Flight Research Center on December 16, 1997. NASA.

consumed another year until, on July 20, 2007, the second example took to the air at Dryden, becoming the first of the X-48B testbeds to fly. By the end of the year, it had completed five research flights. Subsequent testing explored its stability and control at increasing angles of attack (to as great as 16-degree AoA), pointing to possible ways of furnishing improved controllability at even higher angles of attack.[133] Time will tell if the world's skies will fill with blended wing-body shapes. But to those who follow the technology of the sky, if seemingly fantastic, it is well within the realm of the possible, given the history of the swept and delta wings—and NACA–NASA's role in furthering them.

In conclusion, the invention of the swept and delta wing blended creative and imaginative analysis and insight, great risk, and steadfast research. If in introspect their story has a clarity and a cohesiveness that was not necessarily visible to those at the time, it is because time has stripped the story to its essence. It is unfortunate that the perception that America was "given" (or "took") the swept and delta wing in full-blown maturity from the laboratories of the Third Reich possesses

---

133. DFRC, "X-24B Blended Wing-Body" (Apr. 2, 2009).

such persistency, for it obscures the complex roots of the swept and delta wing in both Europe and America, the role of the NACA and NASA in maturing them, and, at heart, the accomplishments of successive generations of Americans within the NACA–NASA and elsewhere who worked to take what were, in most cases, very immature concepts and turn them into practical reality. Doing so required achieving many other things, among which were securing a practical means of effective longitudinal control at transonic speeds (the low, all-moving, and powered tail), reducing transonic drag rise, developing stability augmentation systems, and refining aircraft handling qualities. Defeating the transonic drag "hump"; reducing pitch-up to nuance, not nuisance; and overcoming the danger of inertial coupling were all crucial to ensuring that the swept and delta wing could fulfill their transforming promise. Once achieved, that gave to the world the means to fulfill the promise of the jet engine. As a result, international security and global transportation patterns were dramatically altered and a new transnational global consciousness born. It is something that workers of the NACA past, and NASA past, present, and future, can look back upon with a sense of both pride and accomplishment.

Recommended Additional Readings
Reports, Papers, Articles, and Presentations:

Marvin Abramovitz, Stanley F. Schmidt, and Rudolph D. Van Dyke, Jr., "Investigation of the Use of a Stick Force Proportional to Pitching Acceleration for Normal-Acceleration Warning," NACA RM-A53E21 (1953).

James J. Adams and James B. Whitten, "Tests of a Centering Spring Used as an Artificial Feel Device on the Elevator of a Fighter Airplane," NACA RM-L52G16 (1952).

William J. Alford, Jr., and Thomas B. Pasteur, Jr., "The Effects of Changes in Aspect Ratio and Tail Height on the Longitudinal Stability Characteristics at High Subsonic Speeds of a Model with a Wing Having 32.6° Sweepback," NACA RM-L53L09 (1953).

William J. Alford, Jr., and William P. Henderson, "An Exploratory Investigation of Variable-Wing-Sweep Airplane Configurations," NASA TM-X-142 (1959).

William J. Alford, Jr., Arvo A. Luoma, and William P. Henderson, "Wind-Tunnel Studies at Subsonic and Transonic Speeds of a Multiple-Mission Variable-Wing-Sweep Airplane Configuration," NASA TM-X-206 (1959).

H. Julian Allen and A.J. Eggers, Jr., "A Study of the Motion and Aerodynamic Heating of Ballistic Missiles Entering the Earth's Atmosphere at High Supersonic Speeds," NACA TR-1381 (1953).

Ames Research Center staff, "Collected Works of Robert T. Jones," NASA TM-X-3334 (1976).

Raymond F. Anderson, "Determination of the Characteristics of Tapered Wings," NACA Report No. 572 (1936).

William E. Andrews, Thomas R. Sisk, and Robert W. Darville, "Longitudinal Stability Characteristics of the Convair YF-102 Airplane Determined from Flight Tests," NACA RM-H56I17 (1956).

Johnny G. Armstrong, "Flight Planning and Conduct of the X-24B Research Aircraft Flight Test Program," Air Force Flight Test Center TR-76-11 (1977).

Theodore G. Ayers and James B. Hallissy, "Historical Background and Design Evolution of the Transonic Aircraft Technology Supercritical Wing," NASA TM-81356 (1981).

Donald D. Baals, Thomas A. Toll, and Owen G. Morris, "Airplane Configurations for Cruise at a Mach Number of 3," NACA RM-L58E14a (1958).

Donald D. Baals, A. Warner Robins, and Roy V. Harris, Jr., "Aerodynamic Design Integration of Supersonic Aircraft," AIAA Paper 68-1018 (1968).

De E. Beeler, Donald R. Bellman, and John H. Griffith, "Flight Determination of the Effects of Wing Vortex Generators on the Aerodynamic Characteristics of the Douglas D-558-I Airplane," NACA RM-L51A23 (1951).

Morgan M. Blair, "Evolution of the F-86," AIAA Paper 80-3039 (1980).

Jack D. Brewer and Jacob H. Lichtenstein, "Effect of Horizontal Tail on Low-Speed Static Lateral Stability Characteristics of a Model Having 45° Sweptback Wing and Tail Surfaces," NACA TN-2010 (1950).

R. Porter Brown, Robert G. Chilton, and James B. Whitten, "Flight Investigation of a Mechanical Feel Device in an Irreversible Elevator Control System of a Large Airplane," NACA Report No. 1101 (1952).

Adolf Busemann, "Aerodynamische Auftrieb bei Überschallgeschwindigkeit," *Luftfahrtforschung*, vol. 12, no. 6 (Oct. 3, 1935).

Adolf Busemann, "Compressible Flow in the Thirties," *Annual Review of Fluid Mechanics*, vol. 3 (1971).

John P. Campbell and Hubert M. Drake, "Investigation of Stability and Control Characteristics of an Airplane Model with Skewed Wing in the Langley Free-Flight Tunnel," NACA TN-1208 (May 1947).

George E. Cooper and Robert C. Innis, "Effect of Area-Suction-Type Boundary-Layer Control on the Landing-Approach Characteristics of a 35° Swept-Wing Fighter," NACA RM-A55K14 (1957).

R.M. Cross, "Characteristics of a Triangular-Winged Aircraft: 2: Stability and Control," in NACA, *Conference on Aerodynamic Problems of Transonic Airplane Design* (1947).

Richard E. Day and Donald Reisert, "Flight Behavior of the X-2 Research Airplane to a Mach Number of 3.20 and a Geometric Altitude of 126,200 Feet," NACA TM-X-137 (1959).

John P. Decker and Peter F. Jacobs, "Stability and Performance Characteristics of a Fixed Arrow Wing Supersonic Transport Configuration (SCAT 15F-9898) at Mach Numbers from 0.60 to 1.20," NASA TM-78726 (1978).

Hubert M. Drake and John R. Carden, "Elevator-Stabilizer Effectiveness and Trim of the X-1 Airplane to a Mach Number of 1.06," NACA RM-L50G20 (1950).

Hubert M. Drake and Wendell H. Stillwell, "Behavior of the Bell X-1A Research Airplane During Exploratory Flights at Mach Numbers Near 2.0 and at Extreme Altitude," NACA RM-H55G25 (1955).

Hubert M. Drake, Thomas W. Finch, and James R. Peele, "Flight Measurements of Directional Stability to a Mach Number of 1.48 for an Airplane Tested with Three Different Vertical Tail Configurations," NACA RM-H55G26 (1955).

Alfred C. Draper, Melvin L. Buck, and William H. Goesch, "A Delta Shuttle Orbiter." *Astronautics & Aeronautics*, vol. 9, no. 1 (Jan. 1971).

Cornelius Driver, M. Leroy Spearman, and William A. Corlett, "Aerodynamic Characteristics at Mach Numbers From 1.61 to 2.86 of a Supersonic Transport Model With a Blended Wing-Body, Variable-Sweep Auxiliary Wing Panels, Outboard Tail Surfaces, and a Design Mach Number of 2.2," NASA TM-X-817 (1963).

Alfred J. Eggers, Jr., and Clarence A. Syvertson, "Aircraft Configurations Developing High Lift-Drag Ratios at High Supersonic Speeds," NACA RM-A55L05 (1956).

Carlo Ferrari, "Recalling the Vth Volta Congress: High Speeds in Aviation," *Annual Review of Fluid Mechanics*, vol. 28 (1996).

Jack Fischel and Jack Nugent, "Flight Determination of the Longitudinal Stability in Accelerated Maneuvers at Transonic Speeds for the Douglas D-558-II Research Airplane Including the Effects of an Outboard Wing Fence," NACA RM-L53A16 (1953).

Jack Fischel, "Effect of Wing Slats and Inboard Wing Fences on the Longitudinal Stability Characteristics of the Douglas D-558-II Research Airplane in Accelerated Maneuvers at Subsonic and Transonic Speeds," NACA RM-L53L16 (1954).

Jack Fischel and Cyril D. Brunn, "Longitudinal Stability Characteristics in Accelerated Maneuvers at Subsonic and Transonic Speeds of the Douglas D-558-II Research Airplane Equipped with a Leading-Edge Wing Chord-Extension," NACA RM-H54H16 (1954).

Jack Fischel and Donald Reisert, "Effect of Several Wing Modifications on the Subsonic and Transonic Longitudinal Handling Qualities of the Douglas D-558-II Research Airplane," NACA RM-H56C30 (1956).

Gerald V. Foster and Odell A. Morris, "Aerodynamic Characteristics in Pitch at a Mach Number of 1.97 of Two Variable-Wing-Sweep V/STOL Configurations with Outboard Wing Panels Swept Back 75°," NASA TM-X-322 (1960).

Bernard Göthert, "High-Speed Measurements on a Swept-Back Wing (Sweepback Angle $\phi = 35°$)," NACA TM-1102 (1947).

Charles F. Hall and John C. Heitmeyer, "Aerodynamic Study of a Wing-Fuselage Combination Employing a Wing Swept Back 63°—Characteristics at Supersonic Speeds of a Model with the Wing Twisted and Cambered for Uniform Load," NACA RM-A9J24 (1950).

Charles F. Hall, "Lift, Drag, and Pitching Moment of Low-Aspect Ratio Wings at Subsonic and Supersonic Speeds," NACA RM-A53A30 (1953).

Richard P. Hallion, "Convair's Delta Alpha," *Air Enthusiast Quarterly*, no. 2 (1976).

Richard P. Hallion, "Lippisch Gluhareff, and Jones: The Emergence of the Delta Planform and the Origins of the Sweptwing in the United States," *Aerospace Historian*, vol. 26, no. 1 (Mar. 1979).

Robert G. Hoey and Capt. Iven C. Kincheloe, USAF, "ARDC F-104A Stability and Control," AFFTC TR-58-14 (1958).

Capt. Slayton L. Johns, USAF, and Capt. James W. Wood, USAF, "ARDC F-104A Stability and Control with External Stores," AFFTC TR-58-14 Addendum 1 (July 1959).

Robert T. Jones, "Properties of Low-Aspect-Ratio Pointed Wings at Speeds Below and Above the Speed of Sound," NACA TN-1032 (1946).

Robert T. Jones, "Wing Planforms for High-Speed Flight," NACA TN-1033 (1946).

Robert T. Jones, "Characteristics of a Configuration with a Large Angle of Sweepback," in NACA, *Conference on Aerodynamic Problems of Transonic Airplane Design* (1947).

Harold F. Kleckner, "Preliminary Flight Research on an All-Movable Horizontal Tail as a Longitudinal Control for Flight at High Mach Numbers," NACA ARR-L5C08 (March 1945).

Harold F. Kleckner, "Flight Tests of an All-Movable Horizontal Tail with Geared Unbalancing Tabs on the Curtiss XP-42 Airplane," NACA TN-1139 (1946).

Robert W. Kress, "Variable Sweep Wing Design," AIAA Paper No. 83-1051 (1983).

1st Lt. David C. Leisy and Capt. Hugh P. Hunerwadel, "ARDC F-100D Category II Performance Stability and Control Tests," AFFTC TR-58-27 (1958).

Jacob H. Lichtenstein, "Experimental Determination of the Effect of Horizontal-Tail Size, Tail Length, and Vertical Location on Low-Speed Static Longitudinal Stability and Damping in Pitch of a Model Having 45° Sweptback Wing and Tail Surfaces," NACA Report 1096 (1952).

Robert H. Liebeck, "Design of the Blended Wing Body Subsonic Transport," *Journal of Aircraft*, vol. 41, no. 1 (Jan.–Feb. 2004). pp. 10–25.

Robert H. Liebeck, Mark A. Page, and Blaine K. Rawdon, "Blended-Wing-Body Subsonic Commercial Transport," AIAA Paper 98-0438 (1998).

Robert H. Liebeck, Mark A. Page, Blaine K. Rawdon, Paul W. Scott, and Robert A. Wright, "Concepts for Advanced Subsonic Transports," NASA CR-4624 (1994).

Lindsay J. Lina, Garland J. Morris, and Robert A. Champine, "Flight Investigation of Factors Affecting the Choice of Minimum Approach Speed for Carrier-Type Landings of a Swept-Wing Jet Fighter Airplane," NACA RM-L57F13 (1957).

Alexander Lippisch, "Recent Tests of Tailless Airplanes," NACA TM-564 (1930), a NACA translation of his article *"Les nouveaux essays d'avions sans queue," l'Aérophile* (Feb. 1–15, 1930).

J. Calvin Lovell and Herbert A. Wilson, Jr., "Langley Full-Scale-Tunnel Investigation of Maximum Lift and Stability Characteristics of an Airplane Having Approximately Triangular Plan Form (DM-1 Glider)," NACA RM-L7F16 (1947).

Norman M. McFadden and Donovan R. Heinle, "Flight Investigation of the Effects of Horizontal-Tail Height, Moment of Inertia, and Control Effectiveness on the Pitch-up Characteristics of a 35° Swept-Wing Fighter Airplane at High Subsonic Speeds," NACA RM-A54F21 (1955).

John McMasters and D. Muncy, "The Early Development of Jet Propelled Aircraft," AIAA Paper 2007-0151, Pts. 1–2 (2007).

Odell A. Morris and A. Warner Robins, "Aerodynamic Characteristics at Mach Number 2.01 of an Airplane Configuration Having a Cambered and Twisted Arrow Wing Designed for a Mach Number of 3.0," NASA TM-X-115 (1959).

Odell A. Morris and James C. Patterson, Jr., "Transonic Aerodynamic Characteristics of Supersonic Transport Model With a Fixed, Warped Wing Having 74° Sweep," NASA TM-X-1167 (1965).

Odell A. Morris and Roger H. Fournier, "Aerodynamic Characteristics at Mach Numbers 2.30, 2.60, and 2.96 of a Supersonic Transport Model Having Fixed, Warped Wing," NASA TM-X-1115 (1965).

Odell A. Morris, Dennis E. Fuller, and Carolyn B. Watson, "Aerodynamic Characteristics of a Fixed Arrow-Wing Supersonic Cruise Aircraft at Mach Numbers of 2.30, 2.70, and 2.95," NASA TM-78706 (1978).

Robert G. Mungall, "Flight Investigation of a Combined Geared Unbalancing-Tab and Servotab Control System as Used with an All-Movable Horizontal Tail," NACA TN-1763 (1948).

Max M. Munk, "The Aerodynamic Forces on Airship Hulls," NACA Report No. 184 (1923).

Max M. Munk, "Note on the Relative Effect of the Dihedral and the Sweep Back of Airplane Wings," NACA TN-177 (1924).

Andrew Nahum, "The Royal Aircraft Establishment from 1945 to Concorde," in Robert Bud and Philip Gummett, eds., *Cold War, Hot Science: Applied Research in Britain's Defence Laboratories, 1945–1990* (London: Science Museum, 1999), pp. 29–58.

Andrew Nahum, "I Believe the Americans Have Not Yet Taken Them All!" in Helmuth Trischler, Stefan Zeilinger, Robert Bud, and Bernard Finn, eds., *Tackling Transport* (London: Science Museum, 2003), pp. 99–138.

NASA Langley Research Center staff, "The Supersonic Transport—A Technical Summary," NASA TN-D-423 (1960).

NASA Langley Research Center staff, "Summary of NACA/NASA Variable-Sweep Research and Development Leading to the F-111 (TFX)," Langley Working Paper LWP-285 (Dec. 22, 1966).

Roy J. Niewald and Jack D. Brewer, "Bibliography of NACA Reports Related to Aircraft Control and Guidance Systems, January 1949–April 1954," NACA RM-54F01 (1954).

James H. Parks, "Experimental Evidence of Sustained Coupled Longitudinal and Lateral Oscillations from a Rocket-Propelled Model of a 35° Swept Wing Airplane Configuration," NACA RM-L54D15 (1954).

Paul Pellicano, Joseph Krumenacker, and David Van Hoy, "X-29 High Angle-of-Attack Flight Test Procedures, Results, and Lessons Learned," Society of Flight Test Engineers 21st Annual Symposium (1990).

Alfred D. Phillips and Lt. Col. Frank K. Everest, USAF, "Phase II Flight Test of the North American YF-100 Airplane USAF No. 52-5754," AFFTC TR-53-33 (1953).

William H. Phillips, "Effect of Steady Rolling on Longitudinal and Directional Stability," NACA TN-1627 (1948).

William H. Phillips, "Appreciation and Prediction of Flying Qualities," NACA Report No. 927 (1949).

William H. Phillips, "Theoretical Analysis of Some Simple Types of Acceleration Restrictors," NACA TN-2574 (1951).

M.J. Queijo, Byron M. Jaquet, and Walter D. Wolmart, "Wind-Tunnel Investigation at Low Speed of the Effects of Chordwise Wing Fences and Horizontal-Tail Position on the Static Longitudinal Stability Characteristics of an Airplane Model with a 35° Sweptback Wing," NACA Report 1203 (1954).

George A. Rathert, Jr., L. Stewart Rolls, Lee Winograd, and George E. Cooper, "Preliminary Flight Investigation of the Wing-Dropping Tendency and Lateral-Control Characteristics of a 35° Swept-Wing Airplane at Transonic Mach Numbers," NACA RM-A50H03 (1950).

George A. Rathert, Jr., Howard L. Ziff, and George E. Cooper, "Preliminary Flight Investigation of the Maneuvering Accelerations and Buffet Boundary of a 35° Swept-Wing Airplane at High Altitude and Transonic Speeds," NACA RM-A50L04 (1951).

A. Warner Robins, Odell A. Morris, and Roy V. Harris, Jr., "Recent Research Results in the Aerodynamics of Supersonic Vehicles," AIAA Paper 65-717 (1965).

Dino Roman, J.B. Allen, and Robert H. Liebeck, "Aerodynamic Design Challenges of the Blended-Wing-Body Subsonic Transport," AIAA Paper 2000-4335 (2000).

Melvin Sadoff and Thomas R. Sisk, "Longitudinal-Stability Characteristics of the Northrop X-4 Airplane (USAF No. 46-677)," NACA RM-A50D27 (1950).

Melvin Sadoff, John D. Stewart, and George E. Cooper, "Analytical Study of the Comparative Pitch-Up Behavior of Several Airplanes and Correlation with Pilot Opinion," NACA RM-A57D04 (1957).

Edwin J. Saltzman, Donald R. Bellman, and Norman T. Musialowski, "Flight-Determined Transonic Lift and Drag Characteristics of the YF-102 Airplane With Two Wing Configurations," NACA RM-H56E08 (1956).

Thomas R. Sisk and Duane O. Muhleman, "Longitudinal Stability Characteristics in Maneuvering Flight of the Convair XF-92A Delta-Wing Airplane Including the Effects of Wing Fences," NACA RM-H54J27 (1955).

Thomas R. Sisk and William H. Andrews, "Flight Experience with a Delta-Wing Airplane Having Violent Lateral-Longitudinal Coupling in Aileron Rolls," NACA RM-H55H03 (1955).

S.A. Sjoberg and J.P. Reeder: "Flight Measurements of the Lateral and Directional Stability and Control Characteristics of an Airplane Having a 35° Sweptback Wing with 40-Percent-Span slots and a Comparison with Wind-Tunnel Data," NACA TN-1511 (1948).

S.A. Sjoberg and J.P. Reeder, "Flight Measurements of the Longitudinal Stability, Stalling, and Lift Characteristics of an Airplane Having a 35° Sweptback Wing Without Slots and With 40-Percent-Span Slots and a Comparison with Wind-Tunnel Data," NACA TN-1679 (1948).

S.A. Sjoberg and J.P. Reeder, "Flight Measurements of the Stability, Control, and Stalling Characteristics of an Airplane Having a 35° Sweptback Wing Without Slots and With 80-Percent-Span Slots and a Comparison with Wind-Tunnel Data," NACA TN-1743 (1948).

S.A. Sjoberg and R.A. Champine, "Preliminary Flight Measurements of the Static Longitudinal Stability and Stalling Characteristics of the Douglas D-558-II Research Airplane (BuAero No. 37974)," NACA RM-L9H31a (1949).

Ronald Smelt, "A Critical Review of German Research on High-Speed Airflow," *Journal of the Royal Aeronautical Society,* vol. 50, no. 432 (Dec. 1946).

John E. Steiner, "Transcontinental Rapid Transit: The 367-80 and a Transport Revolution—The 1953–1978 Quarter Century," AIAA Paper 78-3009 (1978).

Ronald Bel Stiffler, *The Bell X-2 Rocket Research Aircraft: The Flight Test Program* (Edwards AFB: Air Force Flight Test Center, 1957).

W.H. Stillwell, J.V. Wilmerding, and R.A. Champine, "Flight Measurements with the Douglas D-558-II (BuAero No. 37974) Research Airplane Low-Speed Stalling and Lift Characteristics," NACA RM-L50G10 (1950).

H.S. Tsien, "Supersonic Flow Over an Inclined Body of Revolution," *Journal of the Aeronautical Sciences,* vol. 5, no. 2 (Oct. 1938).

United States Senate, TFX Contract Investigation (Second Series): Hearings Before the Permanent Subcommittee on Investigations of the Committee on Government Operations, United States Senate, 91st Congress, 2nd Session, Pts. 1–3 (Washington: GPO, 1970).

Walter G. Vincenti, "Robert Thomas Jones," in *Biographical Memoirs*, vol. 86 (Washington: National Academy of Sciences, 2005).

Marion H. Yancey, Jr., and Maj. Stuart R. Childs, USAF, "Phase IV Stability Tests of the F-100A Aircraft, USAF S/N 52-5767," AFFTC TR-55-9 (1955).

Sean Wakayama, "Multidisciplinary Design Optimization of the Blended-Wing-Body," AIAA Paper 98-4938 (1998).

Joseph Weil, Paul Comisarow, and Kenneth W. Goodson, "Longitudinal Stability and Control Characteristics of an Airplane Model Having a 42.8° Sweptback Circular-Arc Wing with Aspect Ratio 4.00, Taper Ratio 0.60, and Sweptback Tail Surfaces," NACA RM-L7G28 (1947).

A. Miles Whitnah and Ernest R. Hillje, "Space Shuttle Wind Tunnel Testing Summary," NASA Reference Publication 1125 (1984).

Richard T. Whitcomb, "An Investigation of the Effects of Sweep on the Characteristics of a High-Aspect-Ratio Wing in the Langley 8-Ft. High Speed Tunnel," NACA RM-L6J01a (1947).

Edward F. Whittle, Jr., and J. Calvin Lovell, "Full-Scale Investigation of an Equilateral Triangular Wing Having 10-Percent-Thick Biconvex Airfoil Sections," NACA RM-L8G05 (1948).

W.C. Williams and A.S. Crossfield, "Handling Qualities of High-Speed Airplanes," NACA RM-L52A08 (1952).

Herbert A. Wilson, Jr., and J. Calvin Lovell, "Full Scale Investigation of the Maximum Lift and Flow Characteristics of an Airplane Having Approximately Triangular Plan Form," NACA RM-L6K20 (1947).

Herbert A. Wilson, Jr., and J. Calvin Lovell, "Langley Full-Scale Tunnel Investigation of Maximum Lift and Stability Characteristics of an Airplane Having Approximately Triangular Plan Form (DM-1 Glider), NACA RM-L7F16 (1947).

Books and Monographs:

David A. Anderton, *Sixty Years of Aeronautical Research, 1917–1977*, EP-145 (Washington: NASA, 1978).

Roland Beamont, *Testing Early Jets: Compressibility and the Supersonic Era* (Shrewsbury: Airlife, 1990).

Walter J. Boyne and Donald H. Lopez, eds., *The Jet Age: Forty Years of Jet Aviation* (Washington: Smithsonian Institution Press, 1979).

Peter W. Brooks, *The Modern Airliner: Its Origins and Development* (London: Putnam & Co., Ltd., 1961).

Charles Burnet, *Three Centuries to Concorde* (London: Mechanical Engineering Publications Ltd., 1979).

Joseph R. Chambers, *Partners in Freedom: Contributions of the Langley Research Center to U.S. Military Aircraft of the 1990s,* SP-2000-4519 (Washington: NASA, 2000).

Joseph R. Chambers, *Innovation in Flight: Research of the NASA Langley Research Center on Revolutionary Advanced Concepts for Aeronautics*, SP-2005-4539 (Washington: NASA, 2005).

Eric M. Conway, *High-Speed Dreams: NASA and the Technopolitics of Supersonic Transportation, 1945–1999* (Baltimore: The Johns Hopkins Press, 2005).

William H. Cook, *The Road to the 707: The Inside Story of Designing the 707* (Bellevue, WA: TYC Publishing Co., 1991).

A. Scott Crossfield with Clay Blair, *Always Another Dawn: The Story of a Rocket Test Pilot* (Cleveland: World Publishing Co., 1960).

Richard E. Day, *Coupling Dynamics in Aircraft: A Historical Perspective*, SP-532 (Washington: NASA, 1997).

Michael Eckert, *The Dawn of Fluid Dynamics: A Discipline Between Science and Technology* (Weinheim: Wiley-VCH Verlag, 2006).

Brig. Gen. Frank K. Everest, Jr., with John Guenther, *The Fastest Man Alive* (New York: Bantam, 1990 ed.).

George W. Gray, *Frontiers of Flight: The Story of NACA Research* (New York: Knopf, 1948).

Roy A. Grossnick, et al., *United States Naval Aviation 1910–1995* (Washington: U.S. Navy, 1997).

Richard P. Hallion, *Supersonic Flight: Breaking the Sound Barrier and Beyond—The Story of the Bell X-1 and Douglas D-558* (New York: The Macmillan Co. in association with the Smithsonian Institution, 1972).

Richard P. Hallion, ed., *The Hypersonic Revolution: Case Studies in the History of Hypersonic Technology*, vols. 1–2 (Washington: USAF, 1998).

Richard P. Hallion and Michael H. Gorn, *On the Frontier: Experimental Flight at NASA Dryden* (Washington: Smithsonian Books, 2002).

James R. Hansen, *Engineer in Charge: A History of the Langley Aeronautical Laboratory, 1917–1958*, SP-4305 (Washington: NASA, 1987).

Edwin P. Hartman, *Adventures in Research: A History of the Ames Research Center, 1940–1965*, SP-4302 (Washington: NASA 1970).

Theodore von Kármán, *Aerodynamics* (New York: McGraw-Hill Book Company, Inc., 1963 ed.).

Theodore von Kármán and Lee Edson, *The Wind and Beyond: Theodore von Kármán, Pioneer in Aviation and Pathfinder in Space* (Boston: Little, Brown and Co., 1967).

Thomas A. Marschak, *The Role of Project Histories in the Study of R&D*, Rand Report P-2850 (Santa Monica: The Rand Corporation, 1965).

Hans-Ulrich Meier, ed., *Die Pfeilflügelentwicklung in Deutschland bis 1945* (Bonn: Bernard & Graefe Verlag, 2006).

Jay Miller, *The X-Planes: X-1 to X-45* (Hinckley, UK: Midland Publishing, 2001).

Kenneth Owen, *Concorde: Story of a Supersonic Pioneer* (London: Science Museum, 2001).

Robert L. Perry, *Innovation and Military Requirements: A Comparative Study*, Rand Report RM-5182PR (Santa Monica: The Rand Corporation, 1967).

W. Hewitt Phillips, *Journey in Aeronautical Research: A Career at NASA Langley Research Center*, No. 12 in *Monographs in Aerospace History* (Washington: NASA, 1998).

R. Dale Reed with Darlene Lister, *Wingless Flight: The Lifting Body Story*, SP-4220 (Washington: NASA 1997).

Joseph A. Shortal, *A New Dimension: Wallops Island Flight Test Range: The First Fifteen Years*, RP-1028 (Washington: NASA, 1978).

Milton O. Thompson and Curtis Peebles, *Flying Without Wings: NASA Lifting Bodies and The Birth of the Space Shuttle* (Washington: Smithsonian Institution Press, 1999).

Milton O. Thompson with J.D. Hunley, *Flight Research: Problems Encountered and What They Should Teach Us*, SP-2000-4522 (Washington: NASA, 2000).

Helmuth Trischler, Stefan Zeilinger, Robert Bud, and Bernard Finn, eds., *Tackling Transport* (London: Science Museum, 2003).

Whitcomb evaluates the shape of one of his area rule models in the 8-foot High Speed Tunnel. NASA.

CASE 2

Richard Whitcomb and the Quest for Aerodynamic Efficiency

Jeremy Kinney

Much of the history of aircraft design in the postwar era is encapsulated by the remarkable work of NACA–NASA engineer Richard T. Whitcomb. Whitcomb, a transonic and supersonic pioneer, gave to aeronautics the wasp-waisted area ruled transonic airplane, the graceful and highly efficient supercritical wing, and the distinctive wingtip winglet. But he also contributed greatly to the development of advanced wind tunnel design and testing. His life offers insights into the process of aeronautical creativity and the role of the genius figure in advancing flight.

ON DECEMBER 21, 1954, Convair test pilot Richard L. "Dick" Johnson flew the YF-102A Delta Dagger prototype to Mach 1, an achievement that marked the meeting of a challenge that had been facing the American aeronautical community. The Delta Dagger's contoured fuselage, shaped by a new design concept, the area rule, enabled an efficient transition from subsonic to supersonic via the transonic regime. Seventeen years later, test pilot Thomas C. "Tom" McMurtry made the first flight in the F-8 Supercritical Wing flight research vehicle on March 9, 1971. The flying testbed featured a new wing designed to cruise at near-supersonic speeds for improved fuel economy. Another 17 years later, the Boeing Company announced the successful maiden flight of what would be the manufacturer's best-selling airliner, the 747-400, on April 29, 1988. Incorporated into the design of the jumbo jet were winglets: small vertical surfaces that reduced drag by smoothing turbulent airflow at the wingtips to increase fuel efficiency.[1] All three of these revolutionary innovations originated with one person, Richard T.

---

1. David A. Anderton, "NACA Formula Eases Supersonic Flight," *Aviation Week* vol. 63 (Sept. 12, 1955): p. 15; Marvin Miles, "New Fighter Jet Gets Test," *Los Angeles Times,* Mar. 10, 1971, p. 26; "Boeing's 747-400 Jet Makes Maiden Flight," *Wall Street Journal,* May 2, 1988, p. 8.

"Dick" Whitcomb, an aeronautical engineer working for the National Advisory Committee for Aeronautics (NACA) and its successor, the National Aeronautics and Space Administration (NASA).

A major aeronautical revolution was shaping the direction and use of the airplane during the latter half of the 20th century. The invention of the turbojet engine in Europe and its incorporation into the airplane transformed aviation. The aeronautical community followed a basic premise—to make the airplane fly higher, faster, farther, and cheaper than ever before—as national, military, industrial, and economic factors shaped requirements. As a researcher at the Langley Memorial Aeronautical Laboratory in Hampton, VA, Dick Whitcomb was part of this movement, which was central to the missions of both the NACA and NASA.[2] His three fundamental contributions, the area rule fuselage, the supercritical wing, and the winglet, each in their own aerodynamic ways offered an increase in speed and performance without an increase in power. Whitcomb was highly individualistic, visionary, creative, and practical, and his personality, engineering style, and the working environment nurtured at Langley facilitated his quest for aerodynamic efficiency.

The Making of an Engineer

Richard Travis Whitcomb was born on February 21, 1921, in Evanston, IL, and grew up in Worcester, MA. He was the eldest of four children in a family led by mathematician-engineer Kenneth F. Whitcomb.[3] Whitcomb was one of the many air-minded American children building and testing aircraft models throughout the 1920s and 1930s.[4] At the age of 12, he created an aeronautical laboratory in his family's basement. Whitcomb spent the majority of his time there building, flying, and innovating rubberband-powered model airplanes, with the exception of reluctantly eating, sleeping, and going to school. He never had a desire to fly himself, but, in his words, he pursued aeronautics for the "fascination of

---

2. Whitcomb's story has been interpreted from the viewpoint of the NACA and NASA's overall contributions to aeronautics by several historians and engineers. This chapter depends heavily on the work of James Hansen, Richard Hallion, Michael Gorn, Lane Wallace, John Becker, Donald Baals, and William Corliss.

3. Clay Blair, Jr., "The Man Who Put the Squeeze on Aircraft Design," *Air Force Magazine*, vol. 39 (Jan. 1956): p. 50.

4. "Richard Travis Whitcomb: Distinguished Research Associate," NASA Langley Research Center, Apr. 1983, File CW-463000-01, National Air and Space Museum Archives.

making a model that would fly." One innovation Whitcomb developed was a propeller that folded back when it stopped spinning to reduce aerodynamic drag. He won several model airplane contests and was a prizewinner in the Fisher Body Company automobile model competition; both were formative events for young American men who would become the aeronautical engineers of the 1940s. Even as a young man, Whitcomb exhibited an enthusiastic drive that could not be diverted until the challenge was overcome.[5]

A major influence on Whitcomb during his early years was his paternal grandfather, who had left farming in Illinois to become a manufacturer of mechanical vending machines. Independent and driven, the grandfather was also an acquaintance of Thomas A. Edison. Whitcomb listened attentively to his grandfather's stories about Edison and soon came to idolize the inventor for his ideas as well as for his freethinking individuality.[6] The admiration for his grandfather and for Edison shaped Whitcomb's approach to aeronautical engineering.

Whitcomb received a scholarship to nearby Worcester Polytechnic Institute and entered the prestigious school's engineering program in 1939. He lived at home to save money and spent the majority of his time in the institute's wind tunnel. Interested in helping with the war effort, Whitcomb's senior project was the design of a guided bomb. He graduated with distinction with a bachelor's of science degree in mechanical engineering. A 1943 *Fortune* magazine article on the NACA convinced Whitcomb to join the Government-civilian research facility at Hampton, VA.[7]

Airplanes ventured into a new aerodynamic regime, the so-called "transonic barrier," as Whitcomb entered into his second year at Worcester. At speeds approaching Mach 1, aircraft experienced sudden changes in stability and control, extreme buffeting, and, most importantly, a dramatic increase in drag, which exposed three challenges to

---

5. Richard Witkin, "Air Scientist Got His Start When 12," *New York Times*, Oct. 3, 1955, p. 20 (quote); Ray Bert, "Winged Victory: Meet Richard Whitcomb," *Transformations* (fall 2002), http://www.wpi.edu/News/Transformations/2002Fall/whitcomb.html (Accessed Feb. 14, 2009); "Jet Pioneers—Richard T. Whitcomb," n.d., File CW-463000-01, National Air and Space Museum Archives.
6. Barbara Rowes, "When You Ride Tomorrow's Airplanes, You'll Thank Dick Whitcomb," *Washington Post-Times Herald*, Aug. 31, 1969, p. 165.
7. Bert, "Winged Victory"; Witkin, "Air Scientist Got His Start When 12"; Brian Welch, "Whitcomb: Aeronautical Research and the Better Shape," *Langley Researcher* (Mar. 21, 1980): p. 4.

the aeronautical community, involving propulsion, research facilities, and aerodynamics. The first challenge involved the propeller and piston-engine propulsion system. The highly developed and reliable system was at a plateau and incapable of powering the airplane in the transonic regime. The turbojet revolution brought forth by the introduction of jet engines in Great Britain and Germany in the early 1940s provided the power needed for transonic flight. The latter two challenges directly involved the NACA and, to an extent, Dick Whitcomb, during the course of the 1940s. Bridging the gap between subsonic and supersonic speeds was a major aerodynamic challenge.[8]

Little was known about the transonic regime, which falls between Mach 0.8 and 1.2. Aeronautical engineers faced a daunting challenge rooted in developing new tools and concepts. The aerodynamicist's primary tool, the wind tunnel, was unable to operate and generate data at transonic speeds. Four approaches were used in lieu of an available wind tunnel in the 1940s for transonic research. One way to generate data for speeds beyond 350 mph was through aircraft diving at terminal velocity, which was dangerous for test pilots and of limited value for aeronautical engineers. Moreover, a representative drag-weight ratio for a 1940-era airplane ensured that it was unable to exceed Mach 0.8. Another way was the use of a falling body, an instrumented missile dropped from the bomb bay of a Boeing B-29 Superfortress. A third method was the wing-flow model. NACA personnel mounted a small, instrumented airfoil on top of the wing of a North American P-51 Mustang fighter. The Mustang traveled at high subsonic speeds and provided a recoverable method in real-time conditions. Finally, the NACA launched small models mounted atop rockets from the Wallops Island facility on Virginia's Eastern Shore.[9] The disadvantages for these three methods were that they only generated data for short periods of time and that there were many variables regarding conditions that could affect the tests.

Even if a wind tunnel existed that was capable of evaluating aircraft at transonic speeds, there was no concept that guaranteed a successful

---

8. John Becker, *The High Speed Frontier: Case Histories of Four NACA Programs 1920–1950*, NASA SP-445 (Washington, DC: U.S. Government Printing Office, 1980), p. 61.

9. Becker, *High Speed Frontier*, p. 61; Lane E. Wallace, "The Whitcomb Area Rule: NACA Aerodynamics Research and Innovation," in Pam E. Mack, ed., *From Engineering Science to Big Science: The NACA and NASA Collier Trophy Research Project Winners*, (Washington, DC: National Aeronautics and Space Administration, 1998), p. 137.

transonic aircraft design. A growing base of knowledge in supersonic aircraft design emerged in Europe beginning in the 1930s. Jakob Ackeret operated the first wind tunnel capable of generating Mach 2 in Zurich, Switzerland, and designed tunnels for other countries. The international high-speed aerodynamics community met at the Volta Conference held in Rome in 1935. A paper presented by German aerodynamicist Adolf Busemann argued that if aircraft designers swept the wing back from the fuselage, it would offset the increase in drag beyond speeds of Mach 1. Busemann offered a revolutionary answer to the problem of high-speed aerodynamics and the sound barrier. In retrospect, the Volta Conference proved to be a turning point in high-speed aerodynamics research, 'especially for Nazi Germany. In 1944, Dietrich Küchemann discovered that a contoured fuselage resembling the now-iconic Coca-Cola soft drink bottle was ideal when combined with Busemann's swept wings. American researcher Robert T. Jones independently discovered the swept wing at NACA Langley almost a decade after the Volta Conference. Jones was a respected Langley aerodynamicist, and his five-page 1945 report provided a standard definition of the aerodynamics of a swept wing. The report appeared at the same time that high-speed aerodynamic information from Nazi Germany was reaching the United States.[10]

As the German and American high-speed traditions merged after World War II, the American aeronautical community realized that there were still many questions to be answered regarding high-speed flight. Three NACA programs in the late 1940s and early 1950s overcame the remaining aerodynamic and facility "barriers" in what John Becker characterized as "one of the most effective team efforts in the annals of aeronautics." The National Aeronautics Association recognized these NACA achievements three times through aviation's highest award, the Collier Trophy, for 1947, 1951, and 1954. The first award, for the achievement of supersonic flight by the X-1, was presented jointly to John Stack of the NACA, manufacturer Lawrence D. Bell, and Air Force test pilot Capt. Charles E. "Chuck" Yeager. The second award in 1952 recognized the slotted transonic tunnel development pioneered by John Stack and his associates at NACA Langley.[11] The third award recognized the direct byproduct of the development of a wind tunnel in which the visionary

---

10. John D. Anderson, Jr., *A History of Aerodynamics and its Impact on Flying Machines* (New York: Cambridge University Press, 1997), pp. 419, 424–425.

11. Becker, *High Speed Frontier*, p. 61.

mind of Dick Whitcomb developed the design concept that would enable aircraft to efficiently transition from subsonic to supersonic speeds through the transonic regime.

Dick Whitcomb and the Transonic-Supersonic Breakthrough

Whitcomb joined the research community at Langley in 1943 as a member of Stack's Transonic Aerodynamics Branch working in the 8-foot High-Speed Tunnel (HST). Initially, NACA managers placed him in the Flight Instrument Research Division, but Whitcomb's force of personality ensured that he would be working directly on problems related to aircraft design. As many of his colleagues and historians would attest, Whitcomb quickly became known for an analytical ability rooted in mathematics, instinct, and aesthetics.[12]

In 1945, Langley increased the power of the 8-foot HST to generate Mach 0.95 speeds, and Whitcomb was becoming increasingly familiar with transonic aerodynamics, which helped him in his developing investigation into the design of supersonic aircraft. The onset of drag created by shock waves at transonic speeds was the primary challenge. John Stack, Ezra Kotcher, and Lawrence D. Bell proved that breaking the sound barrier was possible when Chuck Yeager flew the Bell X-1 to Mach 1.06 (700 mph) on October 14, 1947. Designed in the style of a .50-caliber bullet with straight wings, the Bell X-1 was a successful supersonic airplane, but it was a rocket-powered research airplane designed specifically for and limited to that purpose. The X-1 would not offer designers the shape of future supersonic airplanes. Operational turbojet-powered aircraft designed for military missions were much heavier and would use up much of their fuel gradually accelerating toward Mach 1 to lessen transonic drag.[13] The key was to get operation aircraft through the transonic regime, which ranged from Mach 0.9 to Mach 1.1.

A very small body of transonic research existed when Whitcomb undertook his investigation. British researchers W.T. Lord of the Royal Aeronautical Establishment and G.N. Ward of the University of Manchester and American Wallace D. Hayes attempted to solve the problem of transonic drag through mathematical analyses shortly after World War II in 1946. These studies generated mathematical symbols

---

12. James R. Hansen, *Engineer in Charge: A History of the Langley Aeronautical Laboratory, 1917–1958*, NASA SP-4305 (Washington, DC: NASA, 1987), pp. 331–332.
13. Ibid., p. 332.

that did not lend themselves to the design and shape of transonic and supersonic aircraft.[14]

Whitcomb's analysis of available data generated by the NACA in ground and free-flight tests led him to submit a proposal for testing swept wing and fuselage combinations in the 8-foot HST in July 1948. There had been some success in delaying transonic drag by addressing the relationship between wing sweep and fuselage shape. Whitcomb believed that careful attention to arrangement and shape of the wing and fuselage would result in their counteracting each other. His goal was to reach a milestone in supersonic aircraft design. The tests, conducted from late 1949 to early 1950, revealed no significant decrease in drag at high subsonic (Mach 0.95) and low supersonic (Mach 1.2) speeds. The wing-fuselage combinations actually generated higher drag than their individual values combined. Whitcomb was at an impasse and realized he needed to refocus on learning more about the fundamental nature of transonic airflow.[15]

Just before Whitcomb had submitted his proposal for his wind tunnel tests, John Stack ordered the conversion of the 8-foot HST in the spring of 1948 to a slotted throat to enable research in the transonic regime. In theory, slots in the tunnel's test section, or throat, would enable smooth operation at very high subsonic speeds and at low supersonic speeds. The initial conversion was not satisfactory because of uneven flow. Whitcomb and his colleagues, physicist Ray Wright and engineer Virgil S. Ritchie, hand-shaped the slots based on their visualization of smooth transonic flow. They also worked directly with Langley woodworkers to design and fabricate a channel at the downstream end of the test section that reintroduced air that traveled through the slots. Their painstaking work led to the inauguration of transonic operations within the 8-foot HST 7 months later, on October 6, 1950.[16] Whitcomb,

---

14. Hansen, *Engineer in Charge*, p. 341. As James R. Hansen has suggested, these were certainly antecedents to Whitcomb's area rule, but it was his highly intuitive visual mind that resulted in something original.

15. Ibid., p. 332.

16. The NACA referred to the facility as the 8-foot Transonic Tunnel after Oct. 1950, but for the purposes of clarity and to avoid confusion with the follow-on 8-foot Transonic Pressure Tunnel, the original designation 8-foot High Speed Tunnel is used in this text. Hansen, *Engineer in Charge*, pp. 327–328, 454; Steven T. Corneliussen, "The Transonic Wind Tunnel and the NACA Technical Culture," in Pam E. Mack, ed., *From Engineering Science to Big Science: The NACA and NASA Collier Trophy Research Project Winners* (Washington, DC: NASA, 1998), p. 133.

The slotted-throat test section of the 8-foot High-Speed Tunnel. NASA.

as a young engineer, was helping to refine a tunnel configuration that was going to allow him to realize his potential as a visionary experimental aeronautical engineer.

The NACA distributed a confidential report on the new tunnel during the fall of 1948, which was distributed to the military services and select manufacturers. By the following spring, rumors had been circulating about the new tunnel throughout the industry. Initially, the call for secrecy evolved into outright public acknowledgement of the NACA's new transonic tunnels (including the 16-foot HST) with the awarding of the 1951 Collier Trophy to John Stack and 19 of his associates at Langley for the slotted wall. The Collier Trophy specifically recognized the importance of a research tool, which was a first in the 40-year history of the award. The NACA claimed that its slotted-throat transonic tunnels gave the United States a 2-year lead in the design of supersonic military aircraft.[17]

17. Hansen, *Engineer in Charge*, pp. 329, 330–331.

With the availability of the 8-foot HST and its slotted throat, the combined use of previously available wind tunnel components—the tunnel balance, pressure orifice, tuft surveys, and schlieren photographs—resulted in a new theoretical understanding of transonic drag. The schlieren photographs revealed three shock waves at transonic speeds. One was the familiar shock wave that formed at the nose of an aircraft as it pushed forward through the air. The other two were, according to Whitcomb, "fascinating new types" of shock waves never before observed, in which the fuselage and wings met and at the trailing edge of the wing. These shocks contributed to a new understanding that transonic drag was much larger in proportion to the size of the fuselage and wing than previously believed. Whitcomb speculated that these new shock waves were the cause of transonic drag.[18]

The Path to Area Rule

Conventional high-speed aircraft design emulated Ernst Mach's finding that bullet shapes produced less drag. Aircraft designers started with a pointed nose and gradually thickened the fuselage to increase its cross-sectional area, added wings and a tail, and then decreased the diameter of the fuselage. The rule of thumb for an ideal streamlined body for supersonic flight was a function of the diameter of the fuselage. Understanding the incorporation of the wing and tail, which were added for practical purposes because airplanes need them to fly, into Mach's ideal high-speed soon became the focus of Whitcomb's investigation.[19]

The 8-foot HST team at Langley began a series of tests on various wing and body combinations in November 1951. The wind tunnel models featured swept, straight, and delta wings, and fuselages with varying amounts of curvature. The objective was to evaluate the amount of drag generated by the interference of the two shapes at transonic speeds. The tests resulted in two important realizations for Whitcomb. First, variations in fuselage shape led to marked changes in wing drag. Second, and most importantly, he learned that the combination of fuselage and wing drag had to be considered together as a synergistic aerodynamic system rather than separately, as they had been before.[20]

---

18. Richard T. Whitcomb and Thomas C. Kelly, "A Study of the Flow Over a 45-degree Sweptback Wing-Fuselage Combination at Transonic Mach Numbers," NACA RM-L52D01 (June 25, 1952), p. 1; Hansen, *Engineer in Charge*, p. 333.
19. Ibid., p. 333.
20. Ibid., p. 334.

While Whitcomb was performing his tests, he took a break to attend a Langley technical symposium, where swept wing pioneer Adolf Busemann presented a helpful concept for imagining transonic flow. Busemann asserted that wind tunnel researchers should emulate aerodynamicists and theoretical scientists in visualizing airflow as analogous to plumbing. In Busemann's mind, an object surrounded by streamlines constituted a single stream tube. Visualizing "uniform pipes going over the surface of the configuration" assisted wind tunnel researchers in determining the nature of transonic flow.[21]

Whitcomb contemplated his findings in the 8-foot HST and Busemann's analogy during one of his daily thinking sessions in December 1951. Since his days at Worcester, he dedicated a specific part of his day to thinking. At the core of Whitcomb's success in solving efficiency problems aerodynamically was the fact that, in the words of one NASA historian, he was the kind of "rare genius who can see things no one else can."[22] His relied upon his mind's eye—the nonverbal thinking necessary for engineering—to visualize the aerodynamic process, specifically transonic airflow.[23] Whitcomb's ability to apply his findings to the design of aircraft was a clear indication that using his mind through intuitive reasoning was as much an analytical aerodynamic tool as a research airplane, wind tunnel, or slide rule.

With his feet propped up on his desk in his office a flash of inspiration—a "Eureka" moment, in the mythic tradition of his hero, Edison—led him to the solution of reducing transonic drag. Whitcomb realized that the total cross-sectional area of a fuselage, wing, and tail caused transonic drag or, in his words: "transonic drag is a function of the longitudinal development of the cross-sectional areas of the entire airplane."[24] It was simply not just the result of shock waves forming at the nose of the airplane, but drag-inducing shock waves formed just behind the wings. Whitcomb visualized in his mind's eye that if a designer narrowed the fuselage or reduced its cross section, where the wings attached, and enlarged the fuselage again at the trailing edge,

21. Ibid., p. 334.
22. Roger D. Launius, quoted in James Schultz, *Crafting Flight: Aircraft Pioneers and the Contributions of the Men and Women of NASA Langley Research Center* (Washington, DC: NASA, 2003), p. 183.
23. Eugene S. Ferguson, *Engineering and the Mind's Eye* (Boston: MIT Press, 1994), p. 41; Hansen, *Engineer in Charge*, p. 328.
24. Whitcomb quoted in Welch, "Whitcomb," p. 5.

then the fuselage would facilitate a smoother transition from subsonic to supersonic speeds. Pinching the fuselage to resemble a wasp's waist allowed for smoother flow of the streamlines as they traveled from the nose and over the fuselage, wings, and tail. Even though the fuselage was shaped differently, the overall cross section was the same along the length of the fuselage. Without the pinch, the streamlines would bunch and form shock waves, which created the high energy losses that prevented supersonic flight.[25] The removal at the wing of those "aerodynamic anchors," as historians Donald Baals and William Corliss called them, and the recognition of the sensitive balance between fuselage and wing volume were the key.[26]

Verification of the new idea involved the comparison of the data compiled in the 8-foot HST, all other available NACA-gathered transonic data, and Busemann's plumbing concept. Whitcomb was convinced that his area rule made sense of the questions he had been investigating. Interestingly enough, Whitcomb's colleagues in the 8-foot HST, including John Stack, were skeptical of his findings. He presented his findings to the Langley community at its in-house technical seminar.[27] After Whitcomb's 20-minute talk, Busemann remarked: "Some people come up with half-baked ideas and call them theories. Whitcomb comes up with a brilliant idea and calls it a rule of thumb."[28] The name "area rule" came from the combination of "cross-sectional area" with "rule of thumb."[29]

With Busemann's endorsement, Whitcomb set out to validate the rule through the wind tunnel testing in the 8-foot HST. His models featured fuselages narrowed at the waist. He had enough data by April 1952 indicating that pinching the fuselage resulted in a significant reduction in transonic drag. The resultant research memorandum, "A Study of the Zero Lift Drag Characteristics of Wing-Body Combinations near

---

25. Richard T. Whitcomb, "A Study of the Zero-Lift Drag-Rise Characteristics of Wing-Body Combinations Near the Speed of Sound," NACA TR-1273 (1956), pp. 519, 538–539; *Engineer in Charge*, pp. 334–335.
26. Donald D. Baals and William R. Corliss, *Wind Tunnels of NASA* (Washington, DC: Scientific and Technical Information Branch, National Aeronautics and Space Administration, 1981), p. 63.
27. Hansen, *Engineer in Charge*, p. 336.
28. Quoted in Richard P. Hallion, *Designers and Test Pilots* (Alexandria, VA: Time-Life Books, 1983), p. 143.
29. Michael Gorn, *Expanding the Envelope: Flight Research at NACA and NASA* (Lexington: University Press of Kentucky, 2001), p. 329.

the Speed of Sound," appeared the following September. The NACA immediately distributed it secretly to industry.[30]

The area rule provided a transonic solution to aircraft designers in four steps. First, the designer plotted the cross sections of the aircraft fuselage along its length. Second, a comparison was made between the design's actual area distribution, which reflected outside considerations, such as engine diameter and the overall size dictated by an aircraft carrier's elevator deck, and the ideal area distribution that originated in previous NACA mathematical studies. The third step involved the reconciliation of the actual area distribution with the ideal area distribution. Once again, practical design considerations shaped this step. Finally, the designer converted the new area distribution back into cross sections, which resulted in the narrowed fuselage that took into account the overall area of the fuselage and wing combination.[31] A designer that followed those four steps would produce a successful design with minimum transonic drag.

Validation in Flight

As Whitcomb was discovering the area rule, Convair in San Diego, CA, was finalizing its design of a new supersonic all-weather fighter-interceptor, began in 1951, for a substantial Air Force contract. The YF-102 Delta Dagger combined Mach's ideal high-speed bullet-shaped fuselage and delta wings pioneered on the Air Force's Convair XF-92A research airplane with the new Pratt & Whitney J57 turbojet, the world's most powerful at 10,000 pounds thrust. Armed entirely with air-to-air and forward-firing missiles, the YF-102 was to be the prototype for America's first piloted air defense weapon's system.[32] Convair heard of the NACA's transonic research at Langley and feared that its investment in the YF-102 and the payoff with the Air Force would come to naught if the new airplane could not fly supersonic.[33] Convair's reputation and a considerable Department of Defense contract were at stake.

---

30. Richard T. Whitcomb, "A Study of the Zero-Lift Drag-Rise Characteristics of Wing-Body Combinations Near the Speed of Sound," NACA RM-L52H08 (Sept. 3, 1952). RM-L52H08 was superseded by TR-1273 (see note 23) when the document became unclassified in 1956.
31. Anderton, "NACA Formula Eases Supersonic Flight," pp. 13–14.
32. Gordon Swanborough, *United States Military Aircraft Since 1909* (London: Putnam, 1963), pp. 151, 153.
33. Hansen, *Engineer in Charge*, p. 337.

Case 2 | Richard Whitcomb and the Quest for Aerodynamic Efficiency

A delegation of Convair engineers visited Langley in mid-August 1952, where the engineers witnessed a disappointing test of an YF-102 model in the 8-foot HST. The data indicated, according to the NACA at least, that the YF-102 was unable to reach Mach 1 in level flight. The transonic drag exhibited near Mach 1 simply counteracted the ability of the J57 to push the YF-102 through the sound barrier. They asked Whitcomb what could be done, and he unveiled his new rule of thumb for the design of supersonic aircraft. The data, Whitcomb's solution, and what was perceived as the continued skepticism on the part of his boss, John Stack, left the Convair engineers unconvinced as they went back to San Diego with their model.[34] They did not yet see the area rule as the solution to their perceived problem.

Nevertheless, Whitcomb worked with Convair's aerodynamicists to incorporate the area rule into the YF-102. New wind tunnel evaluations in May 1953 revealed a nominal decrease in transonic drag. He traveled to San Diego in August to assist Convair in reshaping the YF-102 fuselage. The NACA notified Convair that the modified design, soon be designated the YF-102A, was capable of supersonic flight in October.[35]

Despite the fruitful collaboration with Whitcomb, Convair was hedging its bets when it continued the production of the prototype YF-102 in the hope that it was a supersonic airplane. The new delta wing fighter with a straight fuselage was unable to reach its designed supersonic speeds during its full-scale flight evaluation and tests by the Air Force in January 1954. The disappointing performance of the YF-102 to reach only Mach 0.98 in level flight confirmed the NACA's wind tunnel findings and validated Whitcomb's research that led to his area rule. The Air Force realistically shifted the focus toward production of the YF-102A after NACA Director Hugh Dryden guaranteed that Chief of Staff of the Air Force Gen. Nathan F. Twining developed a solution to the problem and that the information had been made available to Convair and the rest of the aviation industry. The Air Force ordered Convair to stop production of the YF-102 and retool to manufacture the improved area rule design.[36]

It took Convair only 7 months to prepare the prototype YF-102A, thanks to the collaboration with Whitcomb. Overall, the new fighter-interceptor

---

34. Baals and Corliss, *Wind Tunnels of NASA*, p. 62; Hansen, *Engineer in Charge*, p. 337.
35. Hansen, *Engineer in Charge*, p. 337.
36. Hansen, *Engineer in Charge*, pp. 337–338; Richard P. Hallion, *On the Frontier: Flight Research at Dryden, 1946–1981* (Washington, DC: NASA, 1984), p. 90; Baals and Corliss, *Wind Tunnels of NASA*, p. 63.

was much more refined than its predecessor was, with sharper features at the redesigned nose and canopy. An even more powerful version of the J57 turbojet engine produced 17,000 pounds thrust with afterburner. The primary difference was the contoured fuselage that resembled a wasp's waist and obvious fairings that expanded the circumference of the tail. With an area rule fuselage, the newly re-designed YF-102A easily went supersonic. Convair test pilot Pete Everest undertook the second flight test on December 21, 1954, during which the YF-102A climbed away from Lindbergh Field, San Diego, and "slipped easily past the sound barrier and kept right on going." More importantly, the YF-102A's top speed was 25 percent faster, at Mach 1.2.[37]

The Air Force resumed the contract with Convair, and the manufacturer delivered 975 production F-102A air defense interceptors, with the first entering active service in mid-1956. The fighter-interceptors equipped Air Defense Command and United States Air Force in Europe squadrons during the critical period of the late 1950s and 1960s. The increase in performance was dramatic. The F-102A could cruise at 1,000 mph and at a ceiling of over 50,000 feet. It replaced three subsonic interceptor aircraft in the Air Force inventory—the North American F-86D Sabre, F-89 Scorpion, and F-94 Starfire—which were 600–650 mph aircraft with a 45,000-foot ceiling range. Besides speed and altitude, the F-102A was better equipped to face the Soviet Myasishchev Bison, Tupolev Bear, and Ilyushin Badger nuclear-armed bombers with a full complement of Hughes Falcon guided missiles and Mighty Mouse rockets. Convair incorporated the F-102A's armament in a drag-reducing internal weapons bay.

When the F-102A entered operational service, the media made much of the fact that the F-102 "almost ended up in the discard heap" because of its "difficulties wriggling its way through the sound barrier." With an area rule fuselage, the F-102A "swept past the sonic problem." The downside to the F-102A's supersonic capability was the noise from its J57 turbojet. The Air Force regularly courted civic leaders from areas near Air Force bases through familiarization flights so that they would understand the mission and role of the F-102A.[38]

---

37. Hansen, *Engineer in Charge*, 338; Swanborough, *United States Military Aircraft Since 1909*, p. 152; Hallion, *On the Frontier*, pp. 90, 144 [quote]; Baals and Corliss, *Wind Tunnels of NASA*, p. 63.
38. Swanborough, *United States Military Aircraft Since 1909*, pp. 152, 154–155; Richard Witkin, "Supersonic Jets Will Defend City," *New York Times*, Jan 3, 1957, p. 12.

The Air Force's F-102 got a whole new look after implementing Richard Whitcomb's area rule. At left is the YF-102 without the area rule, and at right is the new YF-102A version. NASA.

Convair produced the follow-on version, the F-106 Delta Dart, from 1956 to 1960. The Dart was capable of twice the speed of the Dagger with its Pratt & Whitney J75 engine.[39] The F-106 was the primary air defense interceptor defending the continental United States up to the early 1980s. Convair built upon its success with the F-102A and the F-106, two cornerstone aircraft in the Air Force's Century series of aircraft, and introduced more area rule aircraft: the XF2Y-1 Sea Dart and the B-58 Hustler.[40]

The YF-102/YF-102A exercise was valuable in demonstrating the importance of the area rule and of the NACA to the aviation industry and the military, especially when a major contract was at stake.[41] Whitcomb's revolutionary and intuitive idea enabled a new generation of supersonic military aircraft, and it spread throughout the industry. Like Convair, Chance Vought redesigned its F8U Crusader carrier-based interceptor with an area rule fuselage. The first production aircraft appeared in September 1956, and deliveries began in March 1957. Four months later, in July 1957, Marine Maj. John H. Glenn, Jr., as part of Project Bullet,

---

39. Swanborough, *United States Military Aircraft Since 1909*, pp. 152, 154–155.
40. Hallion, *On the Frontier*, p. 57.
41. Ibid., p. 96.

made a recordbreaking supersonic transcontinental flight from Los Angeles to New York in 3 hours 23 minutes. Crusaders served in Navy and Marine fighter and reconnaissance squadrons throughout the 1960s and 1970s, with the last airframes leaving operational service in 1987.[42]

Grumman was the first to design and manufacture from the ground up an area rule airplane. Under contract to produce a carrier-based supersonic fighter, the F9F-9 Tiger, for the Navy, Grumman sent a team of engineers to Langley, just 2 weeks after receiving Whitcomb's pivotal September 1952 report, to learn more about transonic drag. Whitcomb traveled to Bethpage, NY, in February 1953 to evaluate the design before wind tunnel and rocket-model tests were to be conducted by the NACA. The tests revealed that the new fighter was capable of supersonic speeds in level flight with no appreciable transonic drag. Grumman constructed the prototype, and in August 1954, with company test pilot C.H. "Corky" Meyer at the controls, the F9F-9 achieved Mach 1 in level flight without the assistance of an afterburner, which was a good 4 months before the supersonic flight of the F-102A.[43] The Tiger, later designated the F11F-1, served with the fleet as a frontline carrier fighter from 1957 to 1961 and with the Navy's demonstration team, the Blue Angels.[44]

Another aircraft designed from the ground up with an area rule fuselage represented the next step in military aircraft performance in the late 1950s. The legendary Lockheed "Skunk Works" introduced the F-104 Starfighter, "the missile with a man in it," in 1954. Characterized by its short, stubby wings and needle nose, the production prototype F-104, powered by a General Electric J79 turbojet, was the first jet to exceed Mach 2 (1,320 mph) in flight, on April 24, 1956. Starfighters joined operational Air Force units in 1958. An international manufacturing scheme and sales to 14 countries in Europe, Asia, and the Middle East ensured that the Starfighter was in frontline use through the rest of the 20th century.[45]

---

42. Gordon Swanborough and Peter M. Bowers, *United States Navy Aircraft Since 1911* (Annapolis: Naval Institute Press, 1990), pp. 456, 459; Barrett Tillman, *MiG Master: The Story of the F-8 Crusader* (Annapolis: Naval Institute Press, 2007), pp. 55–60.

43. Anderton, "NACA Formula Eases Supersonic Flight," 15; Hansen, *Engineer in Charge*, p. 339

44. René J. Francillon, *Grumman Aircraft Since 1929* (Naval Institute Press, 1989), p. 377; Swanborough and Bowers, *United States Navy Aircraft Since 1911*, pp. 256–257.

45. René J. Francillon, *Lockheed Aircraft Since 1913* (Annapolis: Naval Institute Press, 1987), pp. 329, 331, 342.

The area rule profile of the Grumman Tiger. National Air and Space Museum.

The area rule opened the way for the further refinement of supersonic aircraft, which allowed for concentration on other areas within the synergistic system of the airplane. Whitcomb and his colleagues continued to issue reports refining the concept and giving designers more options to design aircraft with higher performance. Working by himself and with researcher Thomas L. Fischetti, Whitcomb worked to refine high-speed aircraft, especially the Chance Vought F8U-1 Crusader, which evolved into one of the finest fighters of the postwar era.[46]

Spurred on by the success of the F-104, NACA researchers at the Lewis Flight Propulsion Laboratory in Cleveland, OH, estimated that innovations in jet engine design would increase aircraft speeds

---

46. Richard T. Whitcomb and Thomas L. Fischetti, "Development of a Supersonic Area Rule and an Application to the Design of a Wing-Body Combination Having High Lift-to-Drag Ratios," NACA RM-L53H31A (Aug. 18, 1953); and Richard T. Whitcomb, "Some Considerations Regarding the Application of the Supersonic Area Rule to the Design of Airplane Fuselages," NACA RM-L56E23a (July 3, 1956).

upward of 2,600 mph, or Mach 4, based on advanced metallurgy and the sophisticated aerodynamic design of engine inlets, including variable-geometry inlets and exhaust nozzles.[47] One thing was for certain: supersonic aircraft of the 1950s and 1960s would have an area rule fuselage.

The area rule gave the American defense establishment breathing room in the tense 1950s, when the Cold War and the constant need to possess the technological edge, real or perceived, was crucial to the survival of the free world. The design concept was a state secret at a time when no jets were known to be capable of reaching supersonic speeds, due to transonic drag. The aviation press had known about it since January 1954 and kept the secret for national security purposes. The NACA intended to make a public announcement when the first aircraft incorporating the design element entered production. *Aero Digest* unofficially broke the story a week early in its September 1955 issue, when it proclaimed, "The **SOUND BARRIER** has been broken for good," and declared the area rule the "first major aerodynamic breakthrough in the past decade." In describing the area rule and the Grumman XF9F-9 Tiger, *Aero Digest* stressed the bottom line for the innovation: the area rule provided the same performance with less power.[48]

The official announcement followed. Secretary of the Air Force Donald A. Quarles remarked on the CBS Sunday morning television news program *Face the Nation* on September 11, 1955, that the area rule was "the kind of breakthrough that makes fundamental research so very important."[49] *Aviation Week* declared it "one of the most significant military scientific breakthroughs since the atomic bomb."[50] These statements highlighted the crucial importance of the NACA to American aeronautics.

The news of the area rule spread out to the American public. The media likened the shape of an area rule fuselage to a "Coke bottle," a

---

47. Richard Witkin, "Aviation: 2,600 M.P.H.," *New York Times*, Oct. 20, 1957, p. X33.
48. "Aero News Digest," *Aero Digest* (Sept. 1955): p. 5. *Aero Digest* released the story without permission because publisher Fred Hamlin learned that the NACA had arranged, without his knowledge, to make the announcement in the rival journal, *Aviation Week*, on Sept. 19. "New Design Increasing Airplane Speeds Hailed," *Los Angeles Times*, Sept. 12, 1955, p. 10; Alvin Shuster, "'Pinch-Waist' Plane Lifts Supersonic Speed 25%," *New York Times*, Sept. 12, 1955, p. 15.
49. Quoted in Alvin Shuster, "'Pinch-Waist' Plane Lifts Supersonic Speed 25%," *New York Times*, Sept. 12, 1955, p. 15.
50. Commentary by Robert Hotz, "The Area-Rule Breakthrough," *Aviation Week* (Sept. 12, 1955), p. 152.

"wasp waist," an "hourglass," or the figure of actress Marilyn Monroe.[51] While the Coke bottle description of the area rule is commonplace today, the NACA contended that Dietrich Küchemann's Coke bottle and Whitcomb's area rule were not the same and lamented the use of the term. Küchemann's 1944 design concept pertained only to swept wings and tailored the specific flow of streamlines. Whitcomb's rule applied to any shape and contoured a fuselage to maintain an area equivalent to the entire stream tube.[52] Whitcomb actually preferred "indented."[53] One learned writer explained to readers of the *Christian Science Monitor* that an aircraft with an area rule slipped through the transonic barrier due to the "Huckleberry Finn technique," which the character used to suck in his stomach to squeeze through a hole in Aunt Polly's fence.[54]

Whitcomb quickly received just recognition from the aeronautical community for his 3-year development of the area rule. The National Aeronautics Association awarded him the Collier Trophy for 1954 for his creation of "a powerful, simple, and useful method" of reducing transonic drag and the power needed to overcome it.[55] Moreover, the award citation designated the area rule as "a contribution to basic knowledge" that increased aircraft speed and range while reducing drag and using the same power.[56] As Vice President Richard M. Nixon presented him the award at the ceremony, Whitcomb joined the other key figures in aviation history, including Orville Wright, Glenn Curtiss, and his boss, John Stack, in the pantheon of individuals crucial to the growth of American aeronautics.[57]

Besides the Collier, Whitcomb received the Exceptional Service Medal of the U.S. Air Force in 1955 and the inaugural NACA Distinguished Service Medal in 1956.[58] At the age of 35, he accepted an honorary doctor of engineering degree from his alma mater, Worcester Polytechnic

---

51. "New Plane Shape Increases Speed," *The Washington Post-Times Herald*, Sept. 12, 1955, p. 18; "New Design Increasing Airplane Speeds Hailed," *Los Angeles Times*, Sept. 12, 1955, p. 10; "Radial Shift in Air Design Bared by U.S.," *Christian Science Monitor*, Sept. 12, 1955, p. 1.
52. "Area Rule and Coke Bottle," *Aviation Week* (Sept. 12, 1955): p. 13. This source appeared as a sidebar in Anderton, "NACA Formula Eases Supersonic Flight."
53. Richard Witkin, "The 'Wasp-Waist' Plane," *New York Times*, Oct. 2, 1955, p. 20.
54. Maurice A. Garbell, "Transonic Planes Cut Drag with 'Wasp Waist,'" *Christian Science Monitor*, Oct. 14, 1955, p. 5.
55. James J. Hagerty, Jr., "The Collier Trophy Winner," *Collier's* (Dec. 9, 1955): n.p.
56. "Designer to Be Honored For Pinched-Waist Plane," *New York Times*, Nov. 23, 1955, p. 48.
57. Neal Stanford, "Wing Design Seeks Speed," *Christian Science Monitor*, Feb. 17, 1970, p. 5.
58. "Whitcomb Receives NACA's First DSM," *U.S. Air Services* (Oct. 1956): p. 20.

Institute, in 1956.⁵⁹ Whitcomb also rose within the ranks at Langley, where he became head of Transonic Aerodynamics Branch in 1958.

Whitcomb's achievement was part of a highly innovative period for Langley and the rest of the NACA, all of which contributed to the success of the second aeronautical revolution. Besides John Stack's involvement in the X-1 program, the NACA worked with the Air Force, Navy, and the aerospace industry on the resultant high-speed X-aircraft programs. Robert T. Jones developed his swept wing theory. Other NACA researchers generated design data on different aircraft configurations, such as variable-sweep wings, for high-speed aircraft. Whitcomb was directly involved in two of these major innovations: the slotted tunnel and the area rule.⁶⁰

Inventing the Supercritical Wing

Whitcomb was hardly an individual content to rest on his laurels or bask in the glow of previous successes, and after his success with area ruling, he wasted no time in moving further into the transonic and supersonic research regime. In the late 1950s, the introduction of practical subsonic commercial jetliners led many in the aeronautical community to place a new emphasis on what would be considered the next logical step: a Supersonic Transport (SST). John Stack recognized the importance of the SST to the aeronautics program in NASA in 1958. As NASA placed its primary emphasis on space, he and his researchers would work on the next plateau in commercial aviation. Through the Supersonic Transport Research Committee, Stack and his successor, Laurence K. Loftin, Jr., oversaw work on the design of a Supersonic Commercial Air Transport (SCAT). The goal was to create an airliner capable of outperforming the cruise performance of the Mach 3 North American XB-70 Valkyrie bomber. Whitcomb developed a six-engine arrowlike highly swept wing SST configuration that stood out as possessing the best lift-to-drag (L/D) ratio among the Langley designs called SCAT 4.⁶¹

---

59. Rowes, "When You Ride Tomorrow's Airplanes, You'll Thank Dick Whitcomb."
60. Joseph R. Chambers, *Innovation in Flight: Research of the NASA Langley Research Center on Revolutionary Advanced Concepts for Aeronautics*, NASA SP-2005-4539 (Washington, DC: NASA, 2005), p. 18.
61. Whitcomb also rejected the committee's emphasis on variable-geometry wings as too heavy, which led to his ejection from the design committee by Stack. Eric M. Conway, *High Speed Dreams: NASA and the Technopolitics of Supersonic Transportation, 1945–1999* (Baltimore: Johns Hopkins University Press, 2005), pp. 54–55; Becker, *High Speed Frontier*, pp. 55–56.

Manufacturers' analyses indicated that Whitcomb's SCAT 4 exhibited the lowest range and highest weight among a group of designs that would generate high operating and fuel costs and was too heavy when compared with subsonic transports. Despite President John F. Kennedy's June 1963 commitment to the development of "a commercially successful supersonic transport superior to that being built in any other country in the world," Whitcomb saw the writing on the wall and quickly disassociated himself from the American supersonic transport program in 1963.[62] Always keeping in mind his priorities based on practicality and what he could do to improve the airplane, Whitcomb said: "I'm going back where I know I can make things pay off."[63] For Whitcomb, practicality outweighed the lure of speed equated with technological progress.

Whitcomb decided to turn his attention back toward improving subsonic aircraft, specifically a totally new airfoil shape. Airfoils and wings had been evolving over the course of the 20th century. They reflected the ever-changing knowledge and requirements for increased aircraft performance and efficiency. They also represented the bright minds that developed them. The thin cambered airfoil of the Wright brothers, the thick airfoils of the Germans in World War I, the industry-standard Clark Y of the 1920s, and the NACA four- and five-digit series airfoils innovated by Eastman Jacobs exemplified advances in and general approaches toward airfoil design and theory.[64]

Despite these advances and others, subsonic aircraft flew at 85-percent efficiency.[65] The problem was that, as subsonic airplanes moved toward their maximum speed of 660 mph, increased drag and instability developed. Air moving over the upper surface of wings reached supersonic speeds, while the rest of the airplane traveled at a slower rate. The plane had to fly at slower speeds at decreased performance and efficiency.[66]

When Whitcomb returned to transonic research in 1964, he specifically wanted to develop an airfoil for commercial aircraft that delayed the onset of high transonic drag near Mach 1 by reducing air friction and turbu-

---

62. Conway, *High Speed Dreams*, p. 55; Gorn, *Expanding the Envelope*, p. 56; Quote from Chambers, *Innovation in Flight*, p. 28.
63. Gorn, *Expanding the Envelope*, p. 331.
64. For more information on the history of airfoils and their theorists and designers, see Anderson, *A History of Aerodynamics*.
65. Rowes, "When You Ride Tomorrow's Airplanes, You'll Thank Dick Whitcomb," p. 165.
66. Thomas Grubisich, "Fuel-Saver in Wings," *The Washington Post*, July 11, 1974, p. C1.

Whitcomb inspecting a supercritical wing model in the 8-Foot TPT. NASA.

lence across an aircraft's major aerodynamic surface, the wing. Whitcomb went intuitively against conventional airfoil design, in which the upper surface curved downward on the leading and trailing edges to create lift. He envisioned a smoother flow of air by turning a conventional airfoil upside down. Whitcomb's airfoil was flat on top with a downward curved rear section.[67] The shape delayed the formation of shock waves and moved them further toward the rear of the wing to increase total wing efficiency. The rear lower surface formed into deeper, more concave curve to compensate for the lift lost along the flattened wing top. The blunt leading edge facilitated better takeoff, landing, and maneuvering performance. Overall, Whitcomb's airfoil slowed airflow, which lessened drag and buffeting, and improved stability.[68]

With the wing captured in his mind's eye, Whitcomb turned it into mathematical calculations and transformed his findings into a wind tunnel model created by his own hands. He spent days at a time in the 8-foot Transonic Pressure Tunnel (TPT), sleeping on a nearby cot when needed, as he took advantage of the 24-hour schedule to confirm his findings.[69]

67. Gorn, *Expanding the Envelope*, p. 331.
68. Grubisich, "Fuel-Saver in Wings."
69. Gorn, *Expanding the Envelope*, p. 331.

Just as if he were still in his boyhood laboratory, Whitcomb stated that: "When I've got an idea, I'm up in the tunnel. The 8-foot runs on two shifts, so you have to stay with the job 16 hours a day. I didn't want to drive back and forth just to sleep, so I ended up bringing a cot out here."[70]

Whitcomb and researcher Larry L. Clark published their wind tunnel findings in "An Airfoil Shape for Efficient Flight at Supercritical Mach Numbers," which summarized much of the early work at Langley. Their investigation compared a supercritical airfoil with a NASA airfoil. They concluded that the former developed more abrupt drag rise than the latter.[71] Whitcomb presented those initial findings at an aircraft aerodynamics conference held at Langley in May 1966.[72] He called his new innovation a "supercritical wing" by combining "super" (meaning "beyond") with "critical" Mach number, which is the speed supersonic flow revealed itself above the wing. Unlike a conventional wing, where a strong shock wave and boundary layer separation occurred in the transonic regime, a supercritical wing had both a weaker shock wave and less developed boundary layer separation. Whitcomb's tests revealed that a supercritical wing with 35-degree sweep produced 5 percent less drag, improved stability, and encountered less buffeting than a conventional wing at speeds up to Mach 0.90.[73]

Langley Director of Aeronautics Laurence K. Loftin believed that Whitcomb's new supercritical airfoil would reduce transonic drag and result in improved fuel economy. He also knew that wind tunnel data alone would not convince aircraft manufacturers to adopt the new airfoil. Loftin first endorsed the independent analyses of Whitcomb's idea at the Courant Institute at New York University, which proved the viability of the concept. More importantly, NASA had to prove the value of the new technology to industry by actually building, installing, and flying the wing on an aircraft.[74]

---

70. Welch, "Whitcomb," p. 5.
71. Richard T. Whitcomb and Larry L. Clark, "An Airfoil Shape for Efficient Flight at Supercritical Mach Numbers," NASA TM-X-1109 (Apr. 20, 1965).
72. For a first-person account of the development of the supercritical wing, see Richard T. Whitcomb, "Research Associated with the Langley 8-Foot Tunnels Branch: Lecture at Ames Research Center, October 21, 1970," NASA TM-108686 (1970).
73. Richard T. Whitcomb, "The State of Technology Before the F-8 Supercritical Wing," in *Proceedings of the F-8 Digital Fly-By-Wire and Supercritical Wing First Flight's 20th Anniversary, May 27, 1992*, NASA CP-3256, vol. 1 (Washington, DC: NASA, 1996), p. 81; Gorn, *Expanding the Envelope*, pp. 331–332.
74. Gorn, *Expanding the Envelope*, pp. 332, 394, 401.

The major players met in March 1967 to discuss turning Whitcomb's concept into a reality. The practicalities of manufacturing, flight characteristics, structural integrity, and safety required a flight research program. The group selected the Navy Chance Vought F-8A fighter as the flight platform. The F-8A possessed specific attributes that made it ideal for the program. While not an airliner, the F-8A had an easily removable modular wing readymade for replacement, fuselage-mounted landing gear that did not interfere with the wing, engine thrust capable of operation in the transonic regime, and lower operating costs than a multi-engine airliner. Langley contracted Vought to design a supercritical wing for the F-8 and collaborated with Whitcomb during wind tunnel testing beginning during the summer of 1967. Unfortunately for the program, NASA Headquarters suspended all ongoing contracts in January 1968 and Vought withdrew from the program.[75]

SCW Takes to the Air

Langley and the Flight Research Center entered into a joint program outlined in a November 1968 memorandum. Loftin and Whitcomb lead a Langley team responsible for defining the overall objectives, determining the wing contours and construction tolerances, and conducting wind tunnel tests during the flight program. Flight Research Center personnel determined the size, weight, and balance of the wing; acquired the F-8A airframe and managed the modification program; and conducted the flight research program. North American Rockwell won the contract for the supercritical wing and delivered it to the Flight Research Center in November 1970 at a cost of $1.8 million. Flight Research Center technicians installed the new wing on a Navy surplus TF-8A trainer.[76] At the onset of the flight program, Whitcomb predicted the new wing design would allow airliners to cruise 100 mph faster and close to the speed of sound (nearly 660 mph) at an altitude of 45,000 feet with the same amount of power.[77]

NASA test pilot Thomas C. McMurtry took to the air in the F-8 Supercritical Wing flight research vehicle on March 9, 1971. Eighty-six

---

75. Thomas C. Kelly and Richard T. Whitcomb, "Evolution of the F-8 Supercritical Wing Configuration," in *Supercritical Wing Technology—A Progress Report on Flight Evaluations*, NASA SP-301 (1972), p. 35; Gorn, *Expanding the Envelope*, pp. 332–333.
76. Kelly and Whitcomb, "Evolution of the F-8 Supercritical Wing Configuration," in *Supercritical Wing Technology*, p. 35; Gorn, *Expanding the Envelope*, pp. 333–334.
77. Neal Stanford, "Wing Design Seeks Speed," *Christian Science Monitor*, Feb. 17, 1970, p. 5.

flights later, the program ended on May 23, 1973. A pivotal document generated during the program was *Supercritical Wing Technology—A Progress Report on Flight Evaluations*, which captured the ongoing results of the program. From the standpoint of actually flying the F-8, McMurtry noted that: "the introduction of the supercritical wing is not expected to create any serious problems in day-to-day transport operations." The combined flight and wind tunnel tests revealed increased efficiency of commercial aircraft by 15 percent and, more importantly, a 2.5-percent increase in profits. In the high-stakes business of international commercial aviation, the supercritical wing and its ability to increase the range, speed, and fuel efficiency of subsonic jet aircraft without an increase in required power or additional weight was a revolutionary new innovation.[78]

NASA went beyond flight tests with the F-8, which was a flight-test vehicle built specifically for proving the concept. The Transonic Aircraft Technology (TACT) program was a joint NASA–U.S. Air Force partnership begun in 1972 that investigated the application of supercritical wing technology to future combat aircraft. The program evaluated a modified General Dynamics F-111A variable-sweep tactical aircraft to ascertain its overall performance, handling qualities, and transonic maneuverability and to define the local aerodynamics of the airfoil and determine wake drag. Whitcomb worked directly with General Dynamics and the Air Force Flight Dynamics Laboratory on the concept.[79] NASA worked to refine the supercritical wing, and its resultant theory through continued comparison of wind tunnel and flight tests that continued the Langley and Flight Research Center collaboration.[80]

Whitcomb developed the supercritical airfoil using his logical cut-and-try procedures. Ironically, what was considered to be an unsophisticated research technique in the second half of the 20th century, a process John Becker called "Edisonian," yielded the complex super-

---

78. Thomas C. McMurtry, Neil W. Matheny, and Donald H. Gatlin, "Piloting and Operational Aspects of the F-8 Supercritical Wing Airplane," in *Supercritical Wing Technology—A Progress Report on Flight Evaluations*. NASA SP-301, (Washington, DC, NASA, 1972), p. 102; Gorn, *Expanding the Envelope*, pp. 335, 337.
79. Joseph Well, "Summary and Future Plans," in *Supercritical Wing Technology*, pp. 127–128.
80. See Jon S. Pyle and Louis L. Steers, "Flight-Determined Lift and Drag Characteristics of an F-8 Airplane Modified with a Supercritical Wing with Comparisons to Wind Tunnel Results," NASA TM-X-3250 (Jan. 16, 1975); and Lawrence C. Montoya and Richard D. Banner, "F-8 Supercritical Wing Flight Pressure, Boundary-Layer, and Wake Measurements and Comparisons with Wind Tunnel Data," NASA TM-X-3544 (June 1977).

critical airfoil. The key, once again, was the fact that the researcher, Whitcomb, possessed "truly unusual insights and intuitions."[81] Whitcomb used his intuitive imagination to search for a solution over the course of 8 years. Mathematicians verified his work after the fact and created a formula for use by the aviation industry.[82] Whitcomb received patent No. 3,952,971 for his supercritical wing in May 1976. NASA possessed the rights to granting licenses, and several foreign nations already had filed patent applications.[83]

The spread of the supercritical wing to the aviation industry was slow in the late 1970s. There was no doubt that the supercritical wing possessed the potential of saving the airline industry $300 million annually. Both Government experts and the airlines agreed on its new importance. Unfortunately, the reality of the situation in the mid-1970s was that the purchase of new aircraft or conversion of existing aircraft would cost the airlines millions of dollars, and it was estimated that $1.5 billion in fuel costs would be lost before the transition would be completed. The impetus would be a fuel crisis like the Arab oil embargo, during which the price per gallon increased from 12 to 30 cents within the space of a year.[84]

The introduction of the supercritical wing on production aircraft centered on the Air Force's Advanced Medium Short Take-Off and Landing (STOL) Transport competition between McDonnell-Douglas and Boeing to replace the Lockheed C-130 Hercules in the early 1970s. The McDonnell-Douglas design, the YC-15, was the first large transport with supercritical wings in 1975. Neither the YC-15 nor the Boeing YC-14 replaced the Hercules because of the cancellation of the competition, but their wings represented to the press an "exotic advance" that provided new levels of aircraft fuel economy in an era of growing fuel costs.[85]

During the design process of the YC-14, Boeing aerodynamicists also selected a supercritical airfoil for the wing. They based their decision on previous research with the 747 airliner wing, data from Whitcomb's research at Langley, and the promising performance of a Navy T-2C Buckeye that North American Aviation modified with a supercritical air-

---

81. Becker, *High Speed Frontier*, p. 59.
82. Grubisich, "Fuel-Saver in Wings."
83. Stacy V. Jones, "New Aircraft Wing Invented," *New York Times*, May 1, 1976, p. 46.
84. Grubisich, "Fuel-Saver in Wings."
85. Richard Witkin, "McDonnell Douglas Unveils New Cargo Jet," *New York Times*, Aug. 6, 1975, p. 65.

foil to gain experience for the F-8 wing project and undergoing flight tests in November 1969. Boeing's correlation of wind tunnel and flight test data convinced the company to introduce supercritical airfoils on the YC-14 and for all of its subsequent commercial transports, including the triumphant "paperless" airplane, the 777 of the 1990s.[86]

The business jet community embraced the supercritical wing in the increasingly fuel- and energy-conscious 1970s. Business jet pioneer Bill Lear incorporated the new technology in the Canadair Challenger 600, which took to the air in 1978. Rockwell International incorporated the technology into the upgraded Sabreliner 65 of 1979. The extensively redesigned Dassault Falcon 50, introduced the same year, relied upon a supercritical wing that enabled an over-3,000-mile range.[87]

The supercritical wing program gave NASA the ability to stay in the public eye, as it was an obvious contribution to aeronautical technology. The program also improved public relations and the stature of both Langley and Dryden at a time in the 1960s and 1970s when the first "A" in NASA—aeronautics—was secondary to the single "S"—space. For this reason, historian Richard P. Hallion has called the supercritical wing program "Dryden's life blood" in the early 1970s.[88]

Subsonic transports, business jets, STOL aircraft, and uncrewed aerial vehicles incorporate supercritical wing technology today.[89] All airliners today have supercritical airfoils custom-designed and fine-tuned by manufacturers with computational fluid dynamics software programs. There is no NASA supercritical airfoil family like the significant NACA four- and five-airfoil families. The Boeing 777 wing embodies a Whitcomb heritage. This revolutionary information appeared in NASA technical notes (TN) and other publications with little or no fanfare and through direct consultation with Whitcomb. A Lockheed engineer and former employee of Whitcomb in the late 1960s remarked on his days at NASA Langley:

When I was working for Dick Whitcomb at NASA, there was hardly a week that went by that some industry person did not come in to see him. It was a time when NASA was being constantly asked for technical

---

86. Chambers, *Innovation in Flight*, p. 183; Hallion, *On the Frontier*, p. 204.
87. Hallion, *On the Frontier*, pp. 206–207.
88. Ibid., p. 172.
89. For an overview of NASA development of supercritical airfoils up to 1990, see Charles D. Harris, "NASA Supercritical Airfoils—A Matrix of Family-Related Airfoils," NASA TP-2969 (1990).

advice, and Dick always gave that advice freely. He was always there when industry wanted him to help out. This is the kind of cooperation that makes industry want to work with NASA. As a result of that sharing, we have seen the influence of supercritical technology to go just about every corner of our industry.[90]

Whitcomb set the stage and the direction of contemporary aircraft design.

More accolades were given to Whitcomb by the Government and industry during the years he worked on the supercritical wing. From NASA, he received the Medal for Exceptional Scientific Achievement in 1969, and 5 years later, NASA Administrator James Fletcher awarded Whitcomb $25,000 in cash for the invention of the supercritical wing from NASA in June 1974. The NASA Inventions and Contributions Board recommended the cash prize to recognize individual contributions to the Agency's programs. It was the largest cash award given to an individual at NASA.[91] In 1969, Whitcomb accepted the Sylvanus Albert Reed Award from the American Institute of Aeronautics and Astronautics, the organization's highest honor for achievement in aerospace engineering. In 1973, President Richard M. Nixon presented him the highest honor for science and technology awarded by the U.S. Government, the National Medal of Science.[92] The National Aeronautics Association bestowed upon Whitcomb the Wright Brothers Memorial Trophy in 1974 for his dual achievements in developing the area rule and supercritical wing.[93]

Winglets — Yet Another Whitcomb Innovation

Whitcomb continued to search for ways to improve the subsonic airplane beyond his work on supercritical airfoils. The Organization of the Petroleum Exporting Countries (OPEC) oil embargo of 1973–1974 dramatically affected the cost of airline operations with high fuel prices.[94] NASA implemented the Aircraft Energy Efficiency (ACEE) program as part of

---

90. Blackwell, "Influence on Today's Aircraft," p. 114.
91. "Dr. Whitcomb to Receive $25,000 Award from NASA," NASA Release No. 74-148 (June 4, 1974): pp. 1, 3, File CW-463000-01, National Air and Space Museum Archives; Gorn, *Expanding the Envelope*, p. 337.
92. Grubisich, "Fuel-Saver in Wings."
93. "Richard Travis Whitcomb: Distinguished Research Associate," NASA Langley Research Center, Apr. 1983.
94. Welch, "Whitcomb," p. 5.

the national energy conservation effort in the 1970s. At this time, *Science* magazine featured an article discussing how soaring birds used their tip feathers to control flight characteristics. Whitcomb immediately shifted focus toward the wingtips of an aircraft—specifically flow phenomena related to induced drag—for his next challenge.[95]

Two types of drag affect the aerodynamic efficiency of a wing: profile drag and induced drag. Profile drag is a two-dimensional phenomenon and is clearly represented by the iconic airflow in the slipstream image that represents aerodynamics. Induced drag results from three-dimensional airflow near the wingtips. That airflow rolls up over the tip and produces vortexes trailing behind the wing. The energy exhausted in the wingtip vortex creates induced drag. Wings operating in high-lift, low-speed performance regimes can generate large amounts of induced drag. For subsonic transports, induced drag amounts to as much as 50 percent of the total drag of the airplane.[96]

As part of the program, Whitcomb chose to address the wingtip vortex, the turbulent air found at the end of an airplane wing. These vortexes resulted from differences in air pressure generated on the upper and lower surfaces of the wing. As the higher-pressure air forms along the lower surface of the wing, it creates its own airflow along the length of the wing. At the wingtip, the airflow curls upward and forms an energy-robbing vortex that trails behind. Moreover, wingtip vortexes create enough turbulent air to endanger other aircraft that venture into their wake.

Whitcomb sought a way to control the wingtip vortex with a new aeronautical structure called the winglet. Winglets are vertical wing-like surfaces that extend above and sometimes below the tip of each wing. A winglet designer can balance the relationship between cant, the angle the winglet bends from the vertical, and toe, the angle the winglet deviates from airflow, to produce a lift force that, when placed forward of the airfoils, generates thrust from the turbulent wingtip vortexes. This phenomenon is akin to a sailboat tacking upwind while, in the words of aviation observer George Larson: "the keel squeezes the boat forward like a pinched watermelon seed."[97]

---

95. Joseph R. Chambers, *Concept to Reality: Contributions of the Langley Research Center to U.S. Civil Aircraft of the 1990s* (Washington, DC: NASA, 2003), p. 35.
96. Ibid., p. 35.
97. George Larson, "Winglets," *Air & Space Magazine* (Sept. 01, 2001), http://www.airspacemag.com/flight-today/wing.html (Accessed Feb. 20, 2009).

There were precedents for the use of what Whitcomb would call a "nonplanar," or nonhorizontal, lifting system. It was known in the burgeoning aeronautical community of the late 1800s that the induced drag of wingtip vortexes degraded aerodynamic efficiency. Aeronautical pioneer Frederick W. Lanchester patented vertical surfaces, or "endplates," to be mounted at an airplane's wingtips, in 1897. His research revealed that vertical structures reduced drag at high lift. Theoretical studies conducted by the Army Air Service Engineering Division in 1924 and the NACA in 1938 in the United States and by the British Aeronautical Research Committee in 1956 investigated various nonplanar lifting systems, including vertical wingtip surfaces.[98] They argued that theoretically, these structures would provide significant aerodynamic improvements for aircraft. Experimentation revealed that while there was the potential of reducing induced drag, the use of simple endplates produced too much profile drag to justify their use.[99]

Whitcomb and his research team investigated the drag-reducing properties of winglets for a first-generation, narrow-body subsonic jet transport in the 8-foot TPT from 1974 to 1976. They used a semispan model, meaning it was cut in half and mounted on the tunnel wall to enable the use of a larger test object that would facilitate a higher Reynolds number and the use of specific test equipment. He compared a wing with a winglet and the same wing with a straight extension to increase its span. The constant was that both the winglet and extension exerted the same structural load on the wing. Whitcomb found that winglets reduced drag by approximately 20 percent and doubled the improvement in the lift-to-drag ratio to 9 percent compared with the straight wing extension. Whitcomb published his findings in "A Design Approach and Selected Wind-Tunnel Results at High Subsonic Speeds for Wing-Tip Mounted Winglets."[100] It was obvious that the reduction in drag generated by a pair of winglets boosted performance by enabling higher cruise speeds.

98. See F. Nagel, Wings With End Plates. Memo. Rep. 130, Eng. Div., McCook Field, Nov. 4, 1924; W. Mangler, "The Lift Distribution of Wings With End Plates," NACA TM-856 (1938); J. Weber, Theoretical Load Distribution on a Wing with Vertical Plates. R. & M. No. 2960, British A.R.C., 1956.
99. Richard T. Whitcomb, "A Design Approach and Selected Wind-Tunnel Results at High Subsonic Speeds for Wing-Tip Mounted Winglets," NASA TN-D-8260 (July 1976), p. 1; Chambers, Concept to Reality, p. 35.
100. Whitcomb, "A Design Approach and Selected Wind-Tunnel Results at High Subsonic Speeds for Wing-Tip Mounted Winglets," NASA TN-D-8260 (July 1976), pp. 13–14.

With the results, Whitcomb provided a general design approach for the basic design of winglets based on theoretical calculations, physical flow considerations, and emulation of his overall approach to aerodynamics, primarily "extensive exploratory experiments." What made a winglet rather than a simple vertical surface attached to the end of a wing was the designer's ability to use well-known wing design principles to incorporate side forces to reduce lift-induced inflow above the wingtip and outflow below the tip to create a vortex diffuser. The placement and optimum height of the winglet reflected both aerodynamic and structural considerations in which the designer had to take into account the efficiency of the winglet as well as its weight. For practical operational purposes, the lower portion of the winglet could not hang down far below the wingtip for fear of damage on the ground. The fact that the ideal airfoil shape for a winglet was NASA's general aviation airfoil made it even easier to incorporate winglets into an aircraft design.[101] Whitcomb's basic rules provided that foundation.

Experimental wind tunnel studies of winglets in the 8-foot TPT continued through the 1970s. Whitcomb and his colleagues Stuart G. Flechner and Peter F. Jacobs concentrated next on the effects of winglets on a representative second-generation jet transport—the semispan model vaguely resembled a Douglas DC-10—at high subsonic speeds, specifically Mach 0.7 to 0.83. They concluded that winglets significantly reduced the induced drag coefficient while lowering overall drag. The smoothing out of the vortex behind the wingtip by the winglet accounted for the reduction in induced drag. As in the previous study, they saw that winglets generated a small increase in lift. The researchers calculated that winglets reduced drag better than simple wingtip extensions did, despite a minor increase in structural bending moments.[102]

Another benefit derived from winglets was the increase in the aspect ratio of wing without compromising its structural integrity. The aspect

---

101. Ibid., pp. 1–2, 5, 13–14. Whitcomb also suggested consultation of the following two references regarding winglet design: John E. Lamar, "A Vortex-Lattice Method for the Mean Camber Shapes of Trimmed Noncoplanar Platforms with Minimum Vortex Drag," NASA TN-D-8090 (1976) and M.I. Goldhammer, "A Lifting Surface Theory for the Analysis of Nonplanar Lifting Systems," AIAA Paper No. 76-16 (Jan. 1976).

102. Stuart G. Flechner, Peter F. Jacobs, and Richard T. Whitcomb, "A High Subsonic Wind Tunnel Investigation of Winglets on a Representative Second-Generation Jet Transport Wing," NASA TN-8264 (July 1976), pp. 1, 13.

ratio of a wing is the relationship between span—the distance from tip to tip—and chord—the distance between the leading and trailing edge. A long, thin wing has a high aspect ratio, which produces longer range at a certain cruise speed because it does not suffer from wingtip vortexes and the corresponding energy losses as badly as a short and wide chord low aspect ratio wing. The drawback to a high aspect ratio wing is that its long, thin structure flexes easily under aerodynamic loads. Making this type of wing structurally stable required strengthening that added weight. Winglets offered increased aspect ratio with no increase in wingspan. For every 1-foot increase in wingspan, meaning aspect ratio, there was an increase in wing-bending force. Wings structurally strong enough to support a 2-foot span increase would also support 3-foot winglets while producing the same gain in aspect ratio.[103]

NASA made sure the American aviation industry was aware of the results of Whitcomb's winglet studies and its part in the ACEE program. Langley organized a meeting focusing on advanced technologies developed by NASA for Conventional Take-Off and Landing (CTOL) aircraft, primarily airliners, business jets, and personal aircraft, from February 28 to March 3, 1978. During the session dedicated to advanced aero-dynamic controls, Flechner and Jacobs summarized the results of wind tunnel results on winglets applied to a Boeing KC-135 aerial tanker, Lockheed L-1011 and McDonnell-Douglas DC-10 airliners, and a generic model with high aspect ratio wings.[104] Presentations from McDonnell-Douglas and Boeing representatives revealed ongoing industry work done under contract with NASA. Interest in winglets was widespread at the conference and after as manufacturers across the United States began to consider their use and current and future designs.[105]

Whitcomb's winglets first found use on general aviation aircraft at the same time he and his colleagues at Langley began testing them on air transport models and a good 4 years before the pivotal CTOL conference. Another visionary aeronautical engineer, Burt Rutan, adopted them for his revolutionary designs. The homebuilt Vari-Eze of 1974 incorporated

---

103. Larson, "Winglets."
104. See also Stuart G. Flechner and Peter F. Jacobs, "Experimental Results of Winglets on First, Second, and Third Generation Jet Transports," NASA TM-72674 (1978).
105. For these articles, see *Conventional Take-off and Landing (CTOL) Transport Technology 1978: Proceedings of a Conference Held at Langley Research Center, Hampton, VA, Feb. 28–Mar. 3, 1978*, NASA CP-2036, Parts I and II (Washington, DC: NASA, 1978); Chambers, *Concept to Reality*, p. 38.

winglets combined with vertical control surfaces. The airplane was an overall innovative aerodynamic configuration with its forward canard, high aspect ratio wings, low-weight composite materials, a lightweight engine, and pusher propeller, Whitcomb's winglets on Rutan's Vari-Eze offered private pilots a stunning alternative to conventional airplanes. His nonstop world-circling Voyager and the Beechcraft Starship of 1986 also featured winglets.[106]

The business jet community was the first to embrace winglets and incorporate them into production aircraft. The first jet-powered airplane to enter production with winglets was the Learjet Model 28 in 1977. Learjet was in the process of developing a new business jet, the Model 55, and built the Model 28 as a testbed to evaluate its new proprietary high aspect ratio wing and winglet system, called the Longhorn. The manufacturer developed the system on its own initiative without assistance from Whitcomb or NASA, but it was clear where the winglets came from. The comparison flight tests of the Model 28 with and without winglets showed that the former increased its range by 6.5 percent. An additional benefit was improved directional stability. Learjet exhibited the Model 28 at the National Business Aircraft Association convention and put it into production because of its impressive performance and included winglets on its successive business jets.[107] Learjet's competitor, Gulfstream, also investigated the value of winglets to its aircraft in the late 1970s. The Gulfstream III, IV, and V aircraft included winglets in their designs. The Gulfstream V, able to cruise at Mach 0.8 for a distance of 6,500 nautical miles, captured over 70 national and world flight records and received the 1997 Collier Trophy. Records aside, the ability to fly business travelers nonstop from New York to Tokyo was unprecedented after the introduction of the Gulfstream V in 1995.[108]

Actual acceptance on the part of the airline industry was mixed in the beginning. Boeing, Lockheed, and Douglas each investigated the possibility of incorporating winglets into current aircraft as part of the ACEE program. Winglets were a fundamental design technology, and each manufacturer

---

106. Chambers, *Concept to Reality*, p. 41. During the takeoff for the world flight, one of Voyager's winglets broke off, and pilot Dick Rutan had to severely maneuver the aircraft to break the other one off before the journey could continue.
107. Ibid., pp. 41–43.
108. Gulfstream, "The History of Gulfstream: 1958–2008," 2009, *http://www.gulfstream.com/history* (Accessed Feb. 15, 2009).

The KC-135 winglet test vehicle in flight over Dryden. NASA.

had to design them for the specific airframe. NASA awarded contracts to manufacturers to experiment with incorporating them into existing and new designs. Boeing concluded in May 1977 that the economic benefits of winglets did not justify the cost of fabrication for the 747. Lockheed chose to extend the wingtips for the L-1011 and install flight controls to alleviate the increased structural loads. McDonnell-Douglas immediately embraced winglets as an alternative to increasing the span of a wing and modified a DC-10 for flight tests.[109]

The next steps for Whitcomb and NASA were flight tests to demonstrate the viability of winglets for first and second transport and airliner generations. Whitcomb and his team chose the Air Force's Boeing KC-135 aerial tanker as the first test airframe. The KC-135 shared with its civilian version, the pioneering 707, and other early airliners and transports an outer wing that exhibited elliptical span loading with high loading at the outer panels. This wingtip loading was ideal for winglets. Additionally, the Air Force wanted to improve the performance and fuel efficiency of the aging aerial tanker. Whitcomb and this team designed the winglet, and Boeing handled the structural design and fabrication of winglets for an Air Force KC-135. NASA and the Air Force performed the flights tests at Dryden Flight Research Center in 1979 and 1980. The tests revealed a 20-percent reduction in drag because of lift, with a

---

109. Chambers, *Concept to Reality*, p. 38.

7-percent gain in the lift-to-drag ratio at cruise, which confirmed Whitcomb's findings at Langley.[110]

McDonnell-Douglas conducted a winglet flight evaluation program with a DC-10 airliner as part of NASA's Energy Efficient Transport (EET) program within the larger ACEE program in 1981. The DC-10 represented a second-generation airliner with a wing designed to produce nonelliptic loading to avoid wingtip pitch-up characteristics. As a result, the wing bending moments and structural requirements were not as dramatic as those found on a first-generation airliner, such as the 707. Whitcomb and his team conducted a preliminary wind tunnel examination of a DC-10 model in the 8-foot TPT. McDonnell-Douglas engineers designed the aerodynamic and structural shape of the winglets and manufacturing personnel fabricated them. The company performed flights tests over 16 months, which included 61 comparison flights with a DC-10 leased from Continental Airlines. These industry flight tests revealed that the addition of winglets to a DC-10, combined with a drooping of the outboard ailerons, produced a 3-percent reduction in fuel consumption at passenger-carrying distances, which met the bottom line for airline operators.[111]

The DC-10 did not receive winglets because of the prohibitive cost of Federal Aviation Administration (FAA) recertification. Nevertheless, McDonnell-Douglas was a zealous convert and used the experience and design data for the advanced derivative of the DC-10, the MD-11, when that program began in 1986. The first flight in January 1990 and the grueling 10-month FAA certification process that followed validated the use of winglets on the MD-11. The extended range version could carry almost 300 passengers at distances over 8,200 miles, which made it one of the farther flying aircraft in history and ideal for expanding Pacific air routes.[112]

Despite its initial reluctance, Boeing justified the incorporation of winglets into the new 747-400 in 1985, making it the first large U.S. commercial transport to incorporate winglets. The technology increased

---

110. *KC-1935 Winglet Program Review: Proceedings of a Symposium Held At Dryden Flight Research Center, Sept. 16, 1981,* NASA CP-2211 (Washington, DC: NASA, 1982), pp. 1, 11–12; Chambers, *Concept to Reality,* pp. 38–39. In the end, the Air Force chose not to equip its KC-135 aerial tankers with winglets, opting for new engines instead.

111. Staff of Douglas Aircraft Company, *DC-10 Winglet Flight Evaluation,* NASA CR-3704 (June 1983), pp. v, 115–116; Chambers, *Concept to Reality,* pp. 38, 39, 41, 43.

112. Chambers, *Concept to Reality,* p. 43; "Winglets for the Airlines," n.p., n.d.; The Boeing Company, "Commercial Airplanes: MD-11 Family," 2009, *http://www.boeing.com/commercial/md-11family/index.html* (Accessed Mar. 1, 2009).

the new airplane's range by 3 percent, enabling it to fly farther and with more passengers or cargo. The Boeing winglet differed from the McDonnell-Douglas design in that it did not have a smaller fin below the wingtip. Boeing engineers felt the low orientation of the 747 wing, combined with the practical presence of airport ground-handling equipment, made the deletion necessary.[113]

It was clear that Boeing included winglets on the 747-400 for improved performance. Boeing also offered winglets as a customer option for its 737 series aircraft and adopted blended winglets for its 737 and the 737-derivative Business Jet provided by Aviation Partners, Inc., of Seattle in the early 1990s. The specialty manufacturer introduced its proprietary "blended winglet" technology—the winglet is joined to the wing via a characteristic curve—and started retrofitting them to Gulfstream II business jets. The performance accessory increased fuel efficiency by 7 percent. That work lead to commercial airliner accounts. Winglets for the 737 offered fuel savings and reduced noise pollution. The relationship with Boeing lead to a joint venture called Aviation Partners Boeing, which now produces winglets for the 757 and 767 airliners. By 2003, there were over 2,500 Boeing jets flying with blended winglets. The going rate for a set of the 8-foot winglets in 2006 was $600,000.[114]

Whitcomb's winglets found use on transport, airliner, and business jet applications in the United States and Europe. Airbus installed them on production A319, A320, A330, and A340 airliners. It was apparent that regardless of national origin, airlines chose a pair of winglets for their aircraft because they offered a savings of 5 percent in fuel costs. Rather than fly at the higher speeds made possible by winglets, most airline operators simply cruised at their pre-winglet speeds to save on fuel.[115]

Whitcomb's aerodynamic winglets also found a place outside aeronautics, as they met the hydrodynamic needs of the international yacht racing community. In preparation for the America's Cup yacht race in 1983, Australian entrepreneur Alan Bond embraced Whitcomb's work on

---

113. Chambers, *Concept to Reality*, pp. 38, 43.
114. Aviation Partners Boeing, "Winglets," 2006, http://www.aviationpartnersboeing.com (Accessed Mar. 27, 2009); Stephen O. Andersen and Durwood Zaelke, *Industry Genius: Inventions and People Protecting the Climate and Fragile Ozone Layer* (Sheffield, UK: Greenleaf Publishing, 2003), pp. 32–52; Aviation Partners Boeing, "Winglets Save Airlines Money: An Interview with Joe Clark and Jason Paur," 2006, http://www.aviationpartnersboeing.com/interview.html (Mar. 27, 2009).
115. Welch, "Whitcomb," p. 5; Larson, "Winglets."

spiraling vortex drag and believed it could be applied to racing yachts. He assembled an international team that designed a winged keel, essentially a winglet tacked onto the bottom of the keel, for Australia II. Stunned by Australia II's upsetting the American 130-year winning streak, the international yachting community heralded the innovation as the key to winning the race. Bond argued that the 1983 America's Cup race was instrumental to the airline industry's adoption of the winglet and erroneously believed that McDonnell-Douglas engineers began experimenting with winglets during the summer of 1984.[116]

Of the three triumphant innovations pioneered by Whitcomb, the area rule fuselage, the supercritical wing, and the winglet, perhaps it is the last that is the most easily recognizable for everyday air travelers and aviation observers. Engineer and historian Joseph R. Chambers remarked that: "no single NASA concept has seen such widespread use on an international level as Whitcomb's winglets." The application to commercial, military, and general aviation aircraft continues.[117]

Whitcomb and History

Aircraft manufacturers tried repeatedly to lure Whitcomb away from NASA Langley with the promise of a substantial salary. At the height of his success during the supercritical wing program, Whitcomb remarked: "What you have here is what most researchers like—independence. In private industry, there is very little chance to think ahead. You have to worry about getting that contract in 5 or 6 months."[118] Whitcomb's independent streak was key to his and the Agency's success. His relationship with his immediate boss, Laurence K. Loftin, the Chief of Aerodynamic Research at Langley, facilitated that autonomy until the late 1970s. When ordered to test a laminar flow concept that he felt was impractical in the 8-foot TPT, which was widely known as "Whitcomb's tunnel," he retired as head of the Transonic Aerodynamics Branch in February 1980. He had worked in that organization since coming to Hampton from Worcester 37 years earlier, in 1943.[119]

Whitcomb's resignation was partly due to the outside threat to his independence, but it was also an expression of his practical belief that

---

116. David Devoss, "The Race to Recover the Cup," *Los Angeles Times*, Aug. 31, 1986, p. X9.
117. Chambers, *Concept to Reality*, p. 44.
118. Grubisich, "Fuel-Saver in Wings."
119. Bert, "Winged Victory"; Welch, "Whitcomb," p. 4.

his work in aeronautics was finished. He was an individual in touch with major national challenges and having the willingness and ability to devise solutions to help. When he made the famous quote "We've done all the easy things—let's do the *hard* [emphasis Whitcomb's] ones," he made the simple statement that his purpose was to make a difference.[120] In the early days of his career, it was national security, when an innovation such as the area rule was a crucial element of the Cold War tensions between the United States and the Soviet Union. The supercritical wing and winglets were Whitcomb's expression of making commercial aviation and, by extension, NASA, viable in an environment shaped by world fuel shortages and a new search for economy in aviation. He was a lifelong workaholic bachelor almost singularly dedicated to subsonic aerodynamics. While Whitcomb exhibited a reserved personality outside the laboratory, it was in the wind tunnel laboratory that he was unrestrained in his pursuit of solutions that resulted from his highly intuitive and individualistic research methods.

With his major work accomplished, Whitcomb remained at Langley as a part-time and unpaid distinguished research associate until 1991. With over 30 published technical papers, numerous formal presentations, and his teaching position in the Langley graduate program, he was a valuable resource for consultation and discussion at Langley's numerous technical symposiums. In his personal life, Whitcomb continued his involvement in community arts in Hampton and pursued a new quest: an alternative source of energy to displace fossil fuels.[121]

Whitcomb's legacy is found in the airliners, transports, business jets, and military aircraft flying today that rely upon the area rule fuselage, supercritical wings, and winglets for improved efficiency. The fastest, highest-flying, and most lethal example is the U.S. Air Force's Lockheed Martin F-22 Raptor multirole air superiority fighter. Known widely as the 21st Century Fighter, the F-22 is capable of Mach 2 and features an area rule fuselage for sustained supersonic cruise, or supercruise, performance and a supercritical wing. The Raptor was an outgrowth of the Advanced Tactical Fighter (ATF) program that ran from 1986 to 1991. Lockheed designers benefited greatly from NASA work in fly-by-wire

---

120. Hallion, *On the Frontier*, p. 202.
121. Bert, "Winged Victory"; NASA History Office, "Richard T. Whitcomb," 2008, http://history.nasa.gov/naca/bio.html (Accessed Feb. 27, 2009); "Richard Travis Whitcomb: Distinguished Research Associate," NASA Langley Research Center, Apr. 1983.

control, composite materials, and stealth design to meet the mission of the new aircraft. The Raptor made its first flight in 1997, and production aircraft reached Air Force units beginning in 2005.[122]

Whitcomb's ideal transonic transport also included an area rule fuselage, but because most transports are truly subsonic, there is no need for that design feature for today's aircraft.[123] The Air Force's C-17 Globemaster III transport is the most illustrative example. In the early 1990s, McDonnell-Douglas used the knowledge generated with the YC-15 to develop a system of new innovations—supercritical airfoils, winglets, advanced structures and materials, and four monstrous high-bypass turbofan engines—that resulted in the award of the 1994 Collier Trophy. After becoming operational in 1995, the C-17 is a crucial element in the Air Force's global operations as a heavy-lift, air-refuelable cargo transport.[124] After the C-17 program, McDonnell-Douglas, which was absorbed into the Boeing Company in 1997, combined NASA-derived advanced blended wing body configurations with advanced supercritical airfoils and winglets with rudder control surfaces in the 1990s.[125]

Unfortunately, Whitcomb's tools are in danger of disappearing. Both the 8-foot HST and the 8-foot TPT are located beside each other on Langley's East Side, situated between Langley Air Force Base and the Back River. The National Register of Historic Places designated the Collier-winning 8-foot HST a national historic landmark in October 1985.[126] Shortly after Whitcomb's discovery of the area rule, the NACA suspended active operations at the tunnel in 1956. As of 2006, the Historic Landmarks program designated it as "threatened," and its future

---

122. James Blackwell, "Influence on Today's Aircraft," in *Proceedings of the F-8 Digital Fly-By-Wire and Supercritical Wing First Flight's 20th Anniversary, May 27, 1992*, pp. 96–97, 100; U.S. Air Force, "F-22 Raptor," Mar. 2009, http://www.af.mil/information/factsheets/factsheet.asp?fsID=199 (May 21, 2009).
123. Hallion, *On the Frontier*, p. 206.
124. Langley Research Center, "NASA Contributions to the C-17 Globemaster III," FS-1996-05-06-LaRC (May 1996): p. 2.
125. Chambers, *Innovation in Flight*, p. 79.
126. The National Register also recognized two other important Langley wind tunnels: the Variable-Density Tunnel of 1922 and the Full-Scale Tunnel of 1931. National Park Service, "From Sand Dunes to Sonic Booms: List of Sites," n.d., http://www.nps.gov/nr/travel/aviation/sitelist.htm (Accessed Mar. 15, 2009); NASA, "Langley Research Center National Historic Landmarks," 1992, http://www.nasa.gov/centers/langley/news/factsheets/Landmarks.html#8FT (Accessed Mar. 15, 2009).

The Boeing C-17 Globemaster III. U.S. Air Force.

disposition was unclear.[127] The 8-foot TPT opened in 1953. He validated the area rule concept and conducted his supercritical wing and winglet research through the 1950s, 1960s, and 1970s in this tunnel, which was located right beside the old 8-foot HST. The tunnel ceased operations in 1996 and has been classified as "abandoned" by NASA.[128] In the early 21st century, the need for space has overridden the historical importance of the tunnel, and it is slated for demolition.

Overall, Whitcomb and Langley shared the quest for aerodynamic efficiency, which became a legacy for both. Whitcomb flourished working in his tunnel, limited only by the wide boundaries of his intellect and enthusiasm. One observer considered him to be "flight

---

127. National Park Service, "Eight-Foot High Speed Tunnel," n.d., http://www.nps.gov/nr/travel/aviation/8ft.htm (Accessed Mar. 5, 2009); National Park Service, "National Historic Landmarks Program: Eight-Foot High Speed Tunnel," 2006, http://tps.cr.nps.gov/nhl/detail.cfm?ResourceId=1916&ResourceType=Structure, (Accessed Mar. 5, 2009).

128. Welch, "Whitcomb," p. 4; NASA, "Audit of Wind Tunnel Utilization," 2003, oig.nasa.gov/audits/reports/FY03/pdfs/ig-03-027.pdf (Accessed Mar. 17, 2009).

Case 2 | Richard Whitcomb and the Quest for Aerodynamic Efficiency

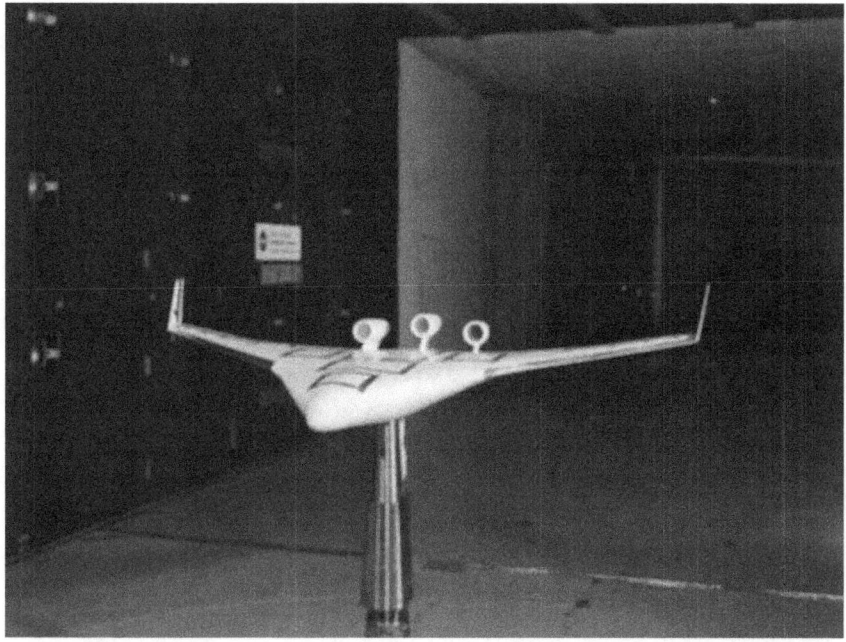

A 3-percent scale model of the Boeing Blended Wing Body 450 passenger subsonic transport in the Langley 14 x 22 Subsonic Tunnel. NASA.

theory personified."[129] More importantly, Whitcomb was the ultimate personification of the importance of the NACA and NASA to American aeronautics during the second aeronautical revolution. The NACA and NASA hired great people, pure and simple, in the quest to serve American aeronautics. These bright minds made up a dynamic community that created innovations and ideas that were greater than the sum of their parts. Whitcomb, as one of those parts, fostered innovations that proved to be of longstanding value to aviation.

129. Welch, "Whitcomb," p. 4.

Recommended Additional Readings
Reports, Papers, Articles, and Presentations:

John D. Anderson, Jr., "Richard Whitcomb and the Area Rule," in *U.S. Air Force: A Complete History*, Dik A. Daso, ed. (New York: Hugh Lauter Levin Associates, 2006).

David A. Anderton, "NACA Formula Eases Supersonic Flight," *Aviation Week* 63 (Sept. 12, 1955).

Clay Blair, Jr., "The Man Who Put the Squeeze on Aircraft Design," *Air Force Magazine* 39 (1956).

Conventional Take-off and Landing (CTOL) Transport Technology 1978: Proceedings of a Conference Held at Langley Research Center, Hampton, VA, Feb. 28–Mar. 3, 1978, NASA CP-2036 (1978).

Douglas Aircraft Company, *DC-10 Winglet Flight Evaluation*, NASA CR-3704 (1983).

Stuart G. Flechner and Peter F. Jacobs, "Experimental Results of Winglets on First, Second, and Third Generation Jet Transports," NASA TM-72674 (1978).

Stuart G. Flechner, Peter F. Jacobs, and Richard T. Whitcomb, "A High Subsonic Wind Tunnel Investigation of Winglets on a Representative Second-Generation Jet Transport Wing," NASA TN-8264 (1976).

M.I. Goldhammer, "A Lifting Surface Theory for the Analysis of Nonplanar Lifting Systems," AIAA Paper No. 76-16 (1976).

Charles D. Harris, "NASA Supercritical Airfoils—A Matrix of Family-Related Airfoils," NASA TP-2969 (1990).

KC-1935 Winglet Program Review: Proceedings of a Symposium Held at Dryden Flight Research Center, Sept. 16, 1981, NASA CP-2211 (1982).

John E. Lamar, "A Vortex-Lattice Method for the Mean Camber Shapes of Trimmed Noncoplanar Platforms with Minimum Vortex Drag," NASA TN-D-8090 (1976).

Langley Research Center, "NASA Contributions to the C-17 Globemaster III," FS-1996-05-06-LaRC (1996).

Lawrence C. Montoya and Richard D. Banner, "F-8 Supercritical Wing Flight Pressure, Boundary-Layer, and Wake Measurements and Comparisons with Wind Tunnel Data," NASA TM-X-3544 (1977).

Proceedings of the F-8 Digital Fly-By-Wire and Supercritical Wing First Flight's 20th Anniversary, May 27, 1992, NASA CP-3256 (1996).

Jon S. Pyle and Louis L. Steers, "Flight-Determined Lift and Drag Characteristics of an F-8 Airplane Modified with a Supercritical Wing with Comparisons to Wind Tunnel Results," NASA TM-X-3250 (1975).

Supercritical Wing Technology—A Progress Report on Flight Evaluations, NASA SP-301 (1972).

Richard T. Whitcomb, "A Design Approach and Selected Wind-Tunnel Results at High Subsonic Speeds for Wing-Tip Mounted Winglets," NASA TN-D-8260 (1976).

Richard T. Whitcomb, "A Study of the Zero-Lift Drag-Rise Characteristics of Wing-Body Combinations Near the Speed of Sound," NACA RM-L52H08 (1952).

Richard T. Whitcomb, "A Study of the Zero-Lift Drag-Rise Characteristics of Wing-Body Combinations Near the Speed of Sound," NACA TR-1273 (1956).

Richard T. Whitcomb, "Research Associated with the Langley 8-Foot Tunnels Branch: Lecture at Ames Research Center, October 21, 1970," NASA TM-108686 (1970).

Richard T. Whitcomb, "Some Considerations Regarding the Application of the Supersonic Area Rule to the Design of Airplane Fuselages," NACA RM-L56E23a (1956).

Richard T. Whitcomb and Larry L. Clark, "An Airfoil Shape for Efficient Flight at Supercritical Mach Numbers," NASA TM-X-1109 (1965).

Richard T. Whitcomb and Thomas L. Fischetti, "Development of a Supersonic Area Rule and an Application to the Design of a Wing-Body Combination Having High Lift-to-Drag Ratios," NACA RM-L53H31A (1953).

Richard T. Whitcomb and Thomas C. Kelly, "A Study of the Flow Over a 45-degree Sweptback Wing-Fuselage Combination at Transonic Mach Numbers," NACA RM-L52DO1 (1952).

Books and Monographs:

John D. Anderson, Jr., *A History of Aerodynamics and its Impact on Flying Machines* (New York: Cambridge University Press, 1997).

Donald D. Baals and William R. Corliss, *Wind Tunnels of NASA* (Washington, DC: NASA, 1981).

John Becker, *The High Speed Frontier: Case Histories of Four NACA Programs 1920–1950*, NASA SP-445 (Washington, DC: NASA, 1980).

Joseph R. Chambers, *Concept to Reality: Contributions of the Langley Research Center to U.S. Civil Aircraft of the 1990s* (Washington, DC: NASA, 2003).

Joseph R. Chambers, *Innovation in Flight: Research of the NASA Langley Research Center on Revolutionary Advanced Concepts for Aeronautics*, NASA SP-2005-4539 (Washington, DC: NASA, 2005).

Eric M. Conway, *High Speed Dreams: NASA and the Technopolitics of Supersonic Transportation, 1945–1999* (Baltimore: Johns Hopkins University Press, 2005).

Eugene S. Ferguson, *Engineering and the Mind's Eye* (Boston: MIT Press, 1994).

Michael Gorn, *Expanding the Envelope: Flight Research at NACA and NASA* (Lexington: University Press of Kentucky, 2001).

Richard P. Hallion, *Designers and Test Pilots* (Alexandria, VA: Time-Life Books, 1983).

Richard P. Hallion, *On the Frontier: Flight Research at Dryden, 1946–1981* (Washington, DC: NASA, 1984).

James R. Hansen, *Engineer in Charge: A History of the Langley Aeronautical Laboratory, 1917–1958*, NASA SP-4305 (Washington, DC: NASA, 1987).

Pam E. Mack, ed., *From Engineering Science to Big Science: The NACA and NASA Collier Trophy Research Project Winners* (Washington, DC: NASA, 1998).

James Schultz, *Crafting Flight: Aircraft Pioneers and the Contributions of the Men and Women of NASA Langley Research Center* (Washington, DC: NASA, 2003).

The Bell XV-1 Convertiplane. NASA.

CASE 3: NACA–NASA and the Rotary Wing Revolution

John F. Ward

The NACA and NASA have always had a strong interest in promoting Vertical/Short Take-Off and Landing (V/STOL) flight, particularly those systems that make use of rotary wings: helicopters, autogiros, and tilt rotors. New structural materials, advanced propulsion concepts, and the advent of fly-by-wire technology influenced emergent rotary wing technology. Work by researchers in various Centers, often in partnership with the military, enabled the United States to achieve dominance in the design and development of advanced military and civilian rotary wing aircraft systems, and continues to address important developments in this field.

IF WORLD WAR I LAUNCHED THE FIXED WING AIRCRAFT INDUSTRY, the Second World War triggered the rotary wing revolution and sowed the seeds of the modern American helicopter industry. The interwar years had witnessed the development of the autogiro, an important short takeoff and landing (STOL) predecessor to the helicopter, but one incapable of true vertical flight, or hovering in flight. The rudimentary helicopter appeared at the end of the interwar era, both in Europe and America. In the United States, the Sikorsky R-4 was the first and only production helicopter used in United States' military operations during the Second World War. R-4 production started in 1943 as a direct outgrowth of the predecessor, VS-300, the first practical American helicopter, which Igor Sikorsky had refined by the end of 1942. That same year, the American Helicopter Society (AHS) was chartered as a professional engineering society representing the rotary wing industry. Also in 1943, the Civil Aeronautics Administration (CAA), forerunner of the Federal Aviation Administration (FAA), issued Aircraft Engineering Division Report No. 32, "Proposed Rotorcraft Airworthiness." Thus was America's rotary wing industry birthed.[1]

---

1. Russell E. Lee, "Famous Firsts in Helicopter History," in Walter J. Boyne and Donald S. Lopez, eds., *Vertical Flight: The Age of the Helicopter* (Washington: Smithsonian Institution Press, 1984), p. 248; Don Fertman, "The Helicopter History of Sikorsky Aircraft," *Vertiflite*, vol. 30, no. 4 (May/June 1984), p. 16; Mike Debraggio, "The American Helicopter Society—A Leader for 40 Years," *Vertiflite*, vol. 30, no. 4 (May/June 1984), p. 56.

Igor Sikorsky flying the experimental VS-300. Sikorsky.

As a result of the industry's growth spurred by continued military demand during the Korean war and the Vietnam conflict, interest in helicopters grew almost exponentially. As a result of the boost in demand for helicopters, Sikorsky Aircraft, Bell Helicopter, Piasecki Helicopter (which evolved into Vertol Aircraft Corporation in 1956, becoming the Vertol Division of the Boeing Company in 1960), Kaman Aircraft, Hughes Helicopter, and Hiller Aircraft entered design evaluations and prototype production contracts with the Department of Defense. Over the past 65 years, the rotary wing industry has become a vital sector of the world aviation system. Types of private, commercial and military utilization abound using aircraft designs of increasing capability, efficiency, reliability, and safety. Helicopters have now been joined by the military V-22, the first operational tilt rotor, and emerging rotary wing unmanned aerial vehicles (UAV), with both successful rotary wing concepts having potential civil applications. Over the past 78 years, the National Advisory Committee for Aeronautics (NACA) and its successor, the National Aeronautics and Space Administration (NASA), have made significant research and technology contributions to the rotary wing revolution, as evidenced by numerous technical publications on rotary wing research testing, database analysis, and theoretical developments published since the 1930s. These technical resources have made significant contributions to the Nation's aircraft industry, military services, and private and commercial enterprises.

The Research Culture

As part of the broad scope of aeronautics research, the rotary wing efforts spanned the full range of research activity, including theoretical study, wind tunnel testing, and ground-based simulation. Flight-test NACA rotary wing research began in the early 1920s with exploratory wind tunnel tests of simple rotor models as the precursor to the basic research undertaken in the 1930s. The Langley Memorial Aeronautical Laboratory, established at Hampton, VA, in 1917, purchased a Pitcairn PCA-2 autogiro in 1931 for research use.[2] The National Advisory Committee for Aeronautics had been formed in 1915 to "supervise and direct scientific study the problems of flight, with a view to their practical solution." Rotary wing research at Langley proceeded under the direction of the Committee with annual inspection meetings by the full Committee to review aeronautical research progress. In the early 1940s, the Ames Aeronautical Laboratory, now known as the Ames Research Center, opened for research at Moffett Field in Sunnyvale, CA. Soon after, the Aircraft Engine Research Laboratory, known for many years as the Lewis Research Center and now known as the Glenn Research Center, opened in Cleveland, OH. Each NACA Center had unique facilities that accommodated rotary wing research needs. Langley Research Center played a major role in NACA–NASA rotary wing research until 1976, when Ames Research Center was assigned the lead role.

The rotary wing research is carried out by a staff of research engineers, scientists, technical support specialists, senior management, and administrative personnel. The rotary wing research staff draws on the expertise of the technical discipline organizations in areas such as aerodynamics, structures and materials, propulsion, dynamics, acoustics, and human factors. Key support functions include such activities as test apparatus design and fabrication, instrumentation research and development (R&D), and research computation support. The constant instrumentation challenge is to adapt the latest technology available to acquiring reliable research data. Over the years, the related challenge for computation tasks is to perform data reduction and analysis for the

---

2. F.B. Gustafson, "A History of NACA Research on Rotating-Wing Aircraft," Journal of the American Helicopter Society, vol. 1, no. 1 (Jan. 1956), p. 16.

increasing sophistication and scope of theoretical investigations and test projects. In the NACA environment, the word "computers" actually referred to a large cadre of female mathematicians. They managed the test measurement recordings, extracted the raw data, analyzed the data using desktop electromechanical calculators, and hand-plotted the results. The NASA era transformed this work from a tedious enterprise into managing the application of the ever-increasing power of modern electronic data recording and computing systems.

The dissemination of the rotary wing research results, which form the basis of NACA–NASA contributions over the years, takes a number of forms. The effectiveness of the contributions depends on making the research results and staff expertise readily available to the Nation's Government and industry users. The primary method has traditionally been the formal publication of technical reports, studies, and compilations that are available for exploitation and use by practitioners. Another method that fosters immediate dialogue with research peers and potential users is the presentation of technical papers at conferences and technical meetings. These papers are published in the conference proceedings and are frequently selected for broader publication as papers or journal articles by technical societies such as the Society of Automotive Engineers (SAE)–Aerospace and the American Institute of Aeronautics and Astronautics (AIAA). Since 1945, NACA–NASA rotary wing research results have been regularly published in the *Proceedings* of the American Helicopter Society Annual Forum and the *Journal* of the AHS. During this time, 30 honorary awards have been presented to NACA and NASA researchers at the Annual Forum Honors Night ceremonies. These awards were given to individual researchers and to technical teams for significant contributions to the advancement of rotary wing technology.

Over the years, the technical expertise of the personnel conducting the ongoing rotary wing research at NACA–NASA has represented a valuable national resource at the disposal of other Government organizations and industry. Until the Second World War, small groups of rotary wing specialists were the prime source of long-term, fundamental research. In the late 1940s, the United States helicopter industry emerged and established technical teams focused on more near-term research in support of their design departments. In turn, the military recognized the need to build an in-house research and development capability to guide their major investments in new rotary wing fleets. The Korean war marked the beginning of the U.S. Army's long-term commitment to the utilization of rotary wing aircraft. In 1962, Gen. Hamilton H. Howze, the first Director of Army Aviation, convened the U.S. Army Tactical Mobility Requirements Board (Howze Board).[3] This milestone launched the emergence of the Air Mobile Airborne Division concept and thereby the steady growth in U.S. military helicopter R&D and production. The working relationship among Government agencies and industry R&D organizations has been close. In particular, the availability of unique facilities and the existence of a pool of experienced rotary wing researchers at NASA led to the United States Army's establishing a "special relationship" with NASA and an initial research presence at the Ames Research Center in 1965. This was followed by the creation of co-located and integrated research organizations at the Ames, Langley, and Glenn Research Centers in the early 1970s. The Army organizations were staffed by specialists in key disciplines such as unsteady aerodynamics, aeroelasticity, acoustics, flight mechanics, and advanced design. In addition, Army civilian and military engineering and support personnel were assigned to work full time in appropriate NASA research facilities and theoretical analysis groups. These assignments included placing active duty military test pilots in the NASA flight research organizations. Over the long term, this teaming arrangement facilitated significant research activity. In addition to Research and Technology Base projects, it made it possible to perform major jointly funded and managed rotary wing Systems Technology and Experimental Aircraft programs. The United States Army partnership was augmented by other research teaming agreements with the United States Navy, FAA, the Defense Advanced Research Projects Agency (DARPA), academia, and industry.

NACA 1930–1958: Establishing Fundamentals

While the helicopter industry did not emerge until the 1950s, the NACA was engaged in significant rotary wing research starting in the 1930s at the Langley Memorial Aeronautical Laboratory (LMAL), now the NASA

---

3. Edgar C. Wood, "The Army Helicopter, Past, Present and Future," Journal of the American Helicopter Society, vol. 1, no. 1 (Jan. 1956), pp 87–92; Lt. Gen. John J. Tolson, *Airmobility, 1961-1971*, a volume in the U.S. Army *Vietnam Studies* series (Washington, DC: Army, 1973), pp. 16–24; and J. A. Stockfisch, *The 1962 Howze Board and Army Combat Developments*, Monograph Report MR-435-A (Santa Monica: The RAND Corporation, 1994).

Pitcairn PCA-2 Autogiro. NASA.

Langley Research Center.[4] The early contributions were the result of studies of the autogiro. The focus was on documenting flight characteristics, performance prediction methods, comparison of flight-test and wind tunnel test results, and theoretical predictions. In addition, fundamental operating problems definition and potential solutions were addressed. In 1931, the NACA made its first direct purchase of a rotary wing aircraft for flight test investigations, a Pitcairn PCA-2 autogiro. (With few exceptions, future test aircraft were acquired as short-term loan or long-term bailment from the military aviation departments.) The Pitcairn was used over the next 5 years in flight-testing and tests of the rotor in the Langley 30- by 60-foot Full-Scale Tunnel. Formal publications of greatest permanent value received "report" status, and the Pitcairn's first study, NACA Technical Report 434, was the first authoritative information on autogiro performance and rotor behavior.[5]

---

4. This case study has drawn upon two major sources covering the period 1930 through 1984 published in Vertiflite, the quarterly magazine of the American Helicopter Society: Frederic B. Gustafson, "History of NACA/NASA Rotating-Wing Aircraft Research, 1915–1970," Vertiflite, Reprint VF-70, (Apr. 1971), pp. 1–27; and John Ward, "An Updated History of NACA/NASA Rotary-Wing Aircraft Research 1915-1984," Vertiflite, vol. 30, No. 4 (May/June 1984), pp. 108–117. The author (who wrote the second of those two) has extended the coverage beyond the original 1984 end date.
5. J.B. Wheatley, "Lift and Drag Characteristics and Gliding Performance of an Autogiro as Determined In Flight," NACA Report No. 434 (1932).

The mid-1930s brought visiting autogiros and manufacturing personnel to Langley Research Center. In addition, analytical and wind tunnel work was carried out on the "Gyroplane," which incorporated a rotor without the usual flapping or lead-lag hinges at the blade root. This was the first systematic research documented and published for what is now called the "rigid" or "hingeless" rotor. This work was the forerunner of the hingeless rotor's reappearance in the 1950s and 1960s with extensive R&D effort by industry and Government. The NACA's early experience with the Gyroplane rotor suggested that "designing toward flexibility rather than toward rigidity would lead to success." In the 1950s, the NACA began to encourage this design approach to those expressing interest in hingeless rotors.

While the NACA worked to provide the fundamentals of rotary wing aerodynamics, the autogiro industry experienced major changes. Approximately 100 autogiros were built in the United States and hundreds more worldwide. Problems in smaller autogiros were readily addressed, but those in larger sizes persisted. They included stick vibration, heavy control forces, vertical bouncing, and destructive out-of-pattern blade behavior known as ground resonance. Private and commercial use underwent a discouraging stage. However, military interest grew in autogiro utility capabilities for safe flight at low airspeed. In an early example of cooperation with the military, the NACA's research effort was linked to the needs of the Army Air Corps (AAC), predecessor of the Army Air Forces (AAF). In quick succession, Langley Laboratory conducted flight and/or wind tunnel tests on a series of Kellett Autogiros, including the KD-1, YG-1, YG-1A, YG-1B, and the Pitcairn YG-2. The NACA provided control force and performance measurements, and pilot assessments of the YG-1. In addition, recommendations were provided on maneuver limitations and redesign for better military serviceability. This led to the NACA providing recommendations and pilot training to enable the Army Air Corps to begin conducting its own rotary wing aircraft experimental and acceptance testing.

In the fall of 1938, international events required that the NACA's emphasis turn to preparedness. The United States required fighters and bombers with superior performance. In the next few years, experimental rotary wing research declined, but important basic groundwork was conducted. Limited effort began on the potentially catastrophic phenomena of ground resonance or coupled rotor-fuselage mechanical instability. Photographs were taken of the rotor-blade out-of-pattern behavior by mounting a camera high on the Langley Field balloon (airship) hangar while an autogiro was operated on the ground. Exploratory flight tests were done using a hub-mounted camera. In these tests blade motion studies were conducted to document the pattern of rotor-blade stalling behavior. In the closing years of the 1930s, analytical progress was also made in the creation of a new theory of rotor aerodynamics that became a classic reference and formed the basis for NACA helicopter experimentation in the 1940s.[6] In these years, the top leadership of the NACA engaged in visible participation in the formal dialogue with the rotating wing community. In 1938, Dr. George W. Lewis, the NACA Headquarters Director of Aeronautical Research, served as Chairman of the Research Programs session of the pioneering Rotating-Wing Aircraft Meeting at the Franklin Institute in Philadelphia. In 1939, Dr. H.J.E. Reid, Director of Langley Laboratory, the NACA's only laboratory, served as Chairman of the session in Dr. Lewis's absence.[7]

The early 1940s continued a period of only modest NACA effort on rotary wing research. However, military interest in the helicopter as a new operational asset started to grow with attention to the need for special missions such as submarine warfare and the rescue of downed pilots. As noted in the introduction to this chapter, the need was met by the Sikorsky R-4 (YR-4B), which was the only production helicopter used in United States military operations during the Second World War. The R-4 production started in 1943 as a direct outgrowth of the Sikorsky VS-300. As the helicopter industry emerged, the NACA rotary wing community enjoyed a productive contact through the interface provided by the NACA Rotating Wing (later renamed Helicopter) Subcommittee. It was in these technical subcommittees that experts from Government, industry, and academia spelled out the research needs and set priorities to be addressed by the NACA rotary wing research specialists. The NACA committee and subcommittee roles were marked by a strong supervisory tone, as called for in the NACA charter. The members lent a definite direction to NACA research based on their technical needs. They also attended annual inspection tours of the three NACA Centers to review the progress on the assigned

---

6. J.B. Wheatley, "A Aerodynamic Analysis of the Autogiro Rotor With Comparison Between Calculated and Experimental Results," NACA Report No. 487 (1934).

7. Anon., "Proceedings of Rotating-Wing Aircraft Meeting of the Franklin Institute, Philadelphia, Pennsylvania, Oct. 20–29, 1938," Philadelphia Section, Institute of the Aeronautical Sciences (IAS); Anon., "Proceedings of the Second Annual Rotating-Wing Aircraft Meeting of the Franklin Institute, Philadelphia, Pennsylvania, Nov. 30–Dec. 1, 1939," Philadelphia Section, Institute of the Aeronautical Sciences (IAS).

Sikorsky YR-4B tested in the Langley 30 x 60 ft. wind tunnel. NASA.

research efforts. In the NASA era, the committees and subcommittees evolved into a more advisory function: commenting upon and ranking the merits of projects proposed by the research teams.

NACA Report 716, published in 1941, constituted a particularly significant contribution to helicopter theory, for it provided simplified methods and charts for determining rotor power required and blade motion.[8] For the first time, design studies could be performed to begin to assess the impacts of blade-section stalling and tip-region compressibility effects. Theoretical work continued throughout the 1940s to extend the simple theory into the region of more extreme operating conditions. Progress began to be made in unraveling the influence of airfoil selection, high blade-section angles of attack, and high tip Mach numbers. The maximum Mach number excursion occurred as the tip passed through the region where the rotor rotational velocity and the forward airspeed combined.

Flight research was begun with the first production helicopter, the Sikorsky YR-4B. This work produced a series of comparisons of flight-test results with theoretical predictions utilizing the new methodology

---

8. F.J. Bailey, Jr., "A Simplified Theoretical Method of Determining the Characteristics of a Lifting Rotor in Forward Flight," NACA Report No. 716 (1941).

for rotor performance and blade motion. The results of the comparisons validated the basic theoretical methods for hover and forward flight in the range of practical steady-state operating conditions. The YR-4B helicopter was also tested in the Langley 30 by 60 tunnel.

This facilitated rotor-off testing to provide fuselage-only lift and drag measurements. This in turn enabled the flight measurements to be adjusted for direct comparison with rotor theory.

With research progressing in flight test, wind tunnel test and theory development, a growing, well-documented open rotary wing database was swiftly established. At the request of industry, Langley airfoil specialists designed and tested airfoils specifically tailored to operating in the challenging unsteady aerodynamic environment of the helicopter rotor. However, the state-of-the-art of airfoil development required that the airfoil be designed on the basis of a single, steady airflow condition. Selecting this artful compromise between rapid excursions into the high angle of attack stall regions and the zero-lift conditions was daunting.[9] Database buildup also included the opportunity offered by the YR-4B 30x60 wind tunnel test setup. This provided the opportunity to document a database from hovering tests on six sets of rotor blades of varying construction and geometry. The testing included single, coaxial, and tandem rotor configurations. Basic single rotor investigations were conducted of rotor-blade pressure distribution, Mach number effects, and extreme operation conditions.

In 1952, Alfred Gessow and Garry Myers published a comprehensive textbook for use by the growing helicopter industry.[10] The authors' training and experience had been gained at Langley Laboratory, and the experimental and theoretical work done by laboratory personnel over the previous 15 years (constituting over 70 published documents) served as the basis of the aerodynamic material developed in the book. The Gessow-Myers textbook remains to this day a classic introduction to helicopter design.

Significant contributions were made in rotor dynamics. The principal contributions addressed the lurking problem of ground resonance, or self-excited mechanical instability—the coupling of in-plane rotor-blade

---

9. F.B. Gustafson, "Effects on Helicopter Performance of Modifications in Profile-Drag Characteristics of Rotor-Blade Airfoil Sections," NACA WR-L-26 [formerly NACA Advanced Confidential Report ACR L4H05] (1944).
10. Alfred Gessow and Garry C. Myers, Jr., *Aerodynamics of the Helicopter* (New York: The Macmillan Company, 1952; reissued by Frederick Ungar Publishing Co., 1967).

oscillations with the rocking motion of the fuselage on its landing gear. First encountered in some autogiro designs, the potential for a catastrophic outcome also existed for the helicopter.[11] Theory developed and disseminated by the NACA enabled the understanding and analysis of ground resonance. This capability was considered essential to the successful design, production, and general use of rotary wing aircraft. Langley pioneered the use of scaled models for the study of dynamic problems such as ground resonance, blade flutter, and control coupling.[12] This contribution to the contemporary state-of-the-art was a forerunner of the all-encompassing development and use of mathematical modeling throughout the modern rotary wing technical community.

As the helicopter flight-testing experience evolved, the research pilots observed problems in holding to steady, precision flight to enable data recording. Frequent control input adjustments were required to prevent diverging into attitudes that were difficult to recover from. Investigation of these flying quality characteristics led to devising standard piloting techniques to produce research-quality data. Deliberate, sharp-step and pulse-control inputs were made, and the resulting aircraft pitch, roll, and yaw responses were recorded for a few seconds. Out of this work came the research specialties of rotary wing flying qualities, stability and control, and handling qualities. Standard criteria for defining required flying qualities specifications gradually emerged from the NACA flight research. The results of this work supported the development of Navy helicopter specifications in the early 1950s and eventually for all military helicopters in 1956. In 1957, research at the NACA Ames Research Center produced a systematic protocol for pilots to assess aircraft handling qualities.[13] The importance of damping of angular velocity and control power, and their interrelation, was investigated in Langley flight-testing. The results provided the basis for a major portion of formal flying-qualities criteria.[14] After modification in 1969 based on exten-

---

11. R.P. Coleman, "Theory of Self-Excited Mechanical Oscillations of Hinged Rotor Blades," NACA WR-L-308 [formerly NACA Advanced Restricted Report 3G29] (1943).

12. G.W. Brooks, "The Application of Models to Helicopter Vibration and Flutter Research," *Proceedings of the ninth annual forum of the American Helicopter Society* (May 1953).

13. George E. Cooper, "Understanding and Interpreting Pilot Opinion," *Aeronautical Engineering Review*, vol. 16, no. 3, (Mar. 1957), p. 47–51.

14. S. Salmirs and R.J. Tapscott, "The Effects of Various Combinations of Damping and Control Power on Helicopter Handling Qualities During Both Instrument and Visual Flight," NASA TN-D-58 (1959).

sive study of in-flight and simulation tasks at Ames, the Cooper-Harper Handling Qualities Rating Scale was published. It remains the standard for evaluating aircraft flying qualities, including rotary wing vehicles.[15]

In the late 1950s, the Army expanded the use of helicopters. The rotary wing industry grew to the point that manufacturers' engineering departments included research and development staff. In addition, the Army established an aviation laboratory (AVLABS), now known as the Aviation Applied Technology Directorate (AATD), at the Army Transportation School, Fort Eustis, VA. This organization was able to sponsor and publish research conducted by the manufacturers. Fort Eustis was situated within 25 miles of the NACA's Langley Research Center in Hampton on the Virginia peninsula. A majority of the key AVLABS personnel were experienced NACA rotary wing researchers. As it turned out, this personnel relocation, amounting to an unplanned "contribution" of expertise to the Army, was the forerunner of significant, long-term, co-located laboratory teaming agreements between the Army and NASA.

NASA 1958–1970: A Time of Transition

The transformation of the NACA into NASA in 1958 was marked by an inevitable subordination of the NACA's aeronautical research charter to NASA's mandated space mission work. The assigned aeronautics staff dropped over 80 percent, from 7,100 to 1,400, as the space program gained momentum in the early 1960s. In the new space-focused environment, aeronautics needed to be product-oriented to attract budget allocation support. In these circumstances, helicopter research lost ground as the focus shifted to new nonrotor Vertical Take-Off and Landing (VTOL) and Short Take-Off and Landing aircraft. In many cases, the rotary wing work formed the base for VTOL investigations. In the case of NACA–NASA rotor-flow studies, experimental and theoretical studies on rotor-time-averaged inflow led to extensive work on establishing wind tunnel jet-boundary layer (wall interference) correction methodology for other VTOL, as well as rotor-borne, lifting systems.[16]

---

15. G.E. Cooper and R.P. Harper, Jr., "The Use of Pilot Rating in the Evaluation of Aircraft Handling Qualities," NASA TN-D-5153 (1969).
16. Harry H. Heyson and S. Katzoff, "Induced Velocities Near a Lifting Rotor with Nonuniform Disk Loading," NACA Report 1319 (1957).

In a sense, it became the U.S. Army's turn to bolster NASA rotary wing endeavors in support of the Army's need for continued helicopter development. In 1965, the Army was granted permission to reactivate, staff, and utilize the Ames 7- by 10-foot Tunnel No. 2. In addition, the Army provided personnel to assist Ames in carrying out projects of interest to the Army. A group of about 45 people was established by the Army and identified as the Army Aeronautical Activity at Ames (AAA–A).[17] In 1970, the working relationship between NASA and the Army was significantly enhanced. Co-located Army research organizations were established at Ames, Langley, and Lewis (now Glenn) Research Centers. They focused on the respective Center's specialty of aeroflight dynamics, structures, and propulsion. This teaming laid the solid groundwork for major rotary wing programs that NASA and the Army jointly planned, executed, and funded in the 1970s and 1980s that influenced both military and civilian rotary wing aircraft development.

One of the unique research facilities authorized in 1939 and operated by the NACA, and then NASA, was the 40- by 80-foot Full-Scale Tunnel at Ames. This research facility also provided the opportunity to work directly with industry on vehicle development programs. In the case of rotary wing aircraft, the tunnel was utilized for investigating new vehicle and rotor system concepts and for thoroughly documenting the basic aerodynamic behavior of prototype and production articles. By the 1960s, numerous in-house and industry full-scale rotary wing hardware were tested. Examples include the Bell XV-1 "convertiplane" in 1953–1954, followed by many other projects, including a modified production rotor incorporating leading edge camber and boundary-layer control; the Bell UH-1 "Huey" helicopter (tested to assist in the development of a high-performance flight-test helicopter); a folded rotor with test data obtained in start-stop and folding conditions at forward speeds; and a four-bladed rotor investigation with extensive rotor-blade pressure measurements taken as a followup to prior flight test measurements made at Langley Research Center.[18]

---

17. Edwin P. Hartman, *Adventures in Research, A History of Ames Research Center, 1940–1965*, NASA SP-4302 (Washington, DC: NASA, 1970), p. 411.
18. William Warmbrodt, Charles Smith, and Wayne Johnson, "Rotorcraft Research Testing in the National Full-Scale Aerodynamics Complex at NASA Ames Research Center," NASA TM-86687 (May 1985); J. Sheiman and L.H. Ludi, "Qualitative Evaluation of Effect of Helicopter Rotor Blade Tip Vortex on Blade Airloads," NASA TN-D-1637 (1963).

The pressure-instrumented blade used in the latter tests had an extremely limited operating life of only 10 hours. This was because of the installation of nearly 50 miniature differential pressure transducers inside the rotor blade. This required that a total of almost 100 small holes be drilled in the upper and lower surface of the primary structure D-spar—normally an absolute "safety of flight" violation. The conservative 10-hour limit was based upon conservative crack-growth-rate limits determined from blade specimen cyclic load tests. The earlier flight test investigation of blade pressure distributions produced a very significant contribution as a primary database for the understanding of basic rotor unsteady aerodynamics. The tabulated pressure data provided time histories of individual differential pressures and simultaneous blade bending moments around the rotor azimuth in a wide assortment of steady and maneuvering flight conditions.[19] This database became the standard experimental data reference source for advancing theoretical comparison work for many years. As an aside, in working with the original flight data to hand-digitize the detailed recordings of differential pressure time-history traces, it became possible, in time, to visually recognize the specific flight-test condition by the periodic pressure trace signature.[20] It was possible to identify the rotor's actual flight condition relative to the surrounding airmass. This still raises the question of the possibility of applying modern signal recognition technology to provide on-board safety-of-flight and noise abatement operating boundary displays for the pilot.

Flying qualities flight investigations emphasized the importance of ample damping of angular velocity and of control power (rotor-generated aircraft pitch and roll control moments) and their interaction. This work at Langley and similar work at Ames provided a significant portion of the helicopter flying qualities criteria. This early work was extended to the use of in-flight simulation using Langley's YHC-1A tandem rotor helicopter with special onboard computing and recording equipment.[21]

---

19. James Sheiman, "A Tabulation of Helicopter Rotor-Blade Differential Pressures, Stresses, and Motions As Measured In Flight," NASA TM-X-952 (1964).

20. John F. Ward, "Helicopter Rotor Periodic Differential Pressures and Structural Response Measured in Transient and Steady-State Maneuvers," *Journal of the American Helicopter Society*, vol. 16, no. 1 (Jan. 1971).

21. F. Garren, J.R. Kelly, and R.W. Summer, "VTOL Flight Investigation to Develop a Decelerating Instrument Approach Capability," Society of Automotive Engineers Paper No. 690693 (1969), presented at the Aeronautics and Space Engineering and Manufacturing Meeting, Los Angeles, CA, Oct. 6–10, 1969.

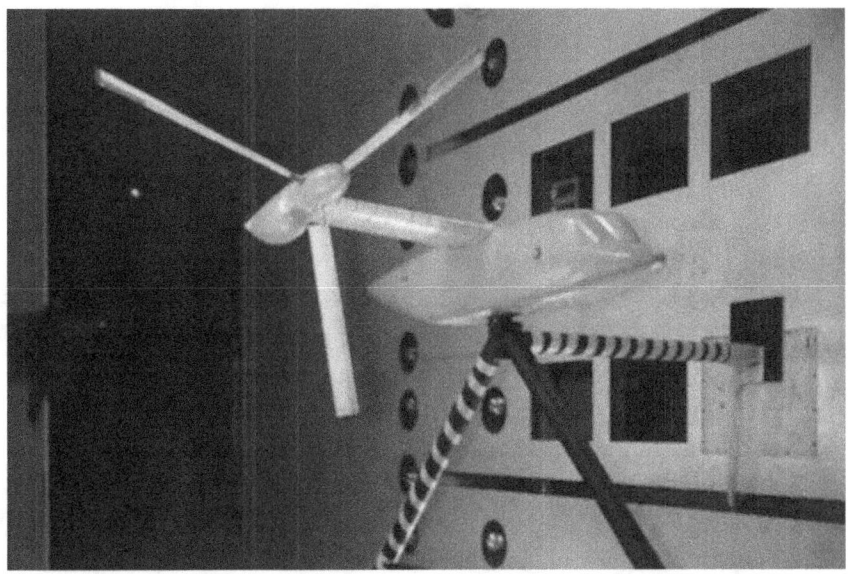

Tilt rotor semi-span dynamic model in the Langley Transonic Dynamics Tunnel. NASA.

The flight operations of most interest were terminal area instrument flight on steep approaches to vertical touchdown landings. The results of this work were initially oriented to nonrotor VTOL operations, but the results were found to be equally applicable to helicopters.

In the area of structural dynamics, investigations addressing the problems of aeroelastic stability of rotor-powered aircraft were conducted utilizing new analytical methods and experimental studies by Langley and Ames researchers. Emphasis was placed on tilt rotor and tilt propeller (i.e., tilt wing) aircraft concepts. Two-degree-of-freedom "air resonance" (akin to rotor-fuselage "ground resonance") and proprotor/propeller whirl instability were among the problems investigated.[22] Rotor-pylon-wing aeroelastic instability problems for tilt rotor designs were explored in the Ames 40 by 80 Full-Scale Tunnel in this period. The aeroelastic stability problems of the tilt rotor and tilt-stopped rotor designs were also investigated at model scale in the unique Freon atmosphere of the Langley Transonic Dynamics Tunnel, which provided full-scale Mach number and Reynolds number scaling.[23] These research

22. Wilmer H. Reed, III, "Review of Propeller-Rotor Whirl Flutter," NASA TR-R-264 (1967).
23. William T. Yeager, Jr., and Raymond G. Kvaternik, "A Historical Overview of Aeroelasticity Branch and Transonic Dynamics Tunnel Contributions to Rotorcraft Technology and Development," NASA TM-2001-211054 / ARL-TR-2564, (Aug. 2001).

investigations resulted in significant contributions to the development of the validated design tools for advanced rotorcraft.

With the increased interest in hingeless rotor concepts, NASA obtained and quickly accomplished flight research with a copy of an experimental Bell Helicopter three-bladed hingeless rotor installed on an H-13 helicopter.[24] Early experience with "rigid" rotors had led the NACA to encourage interest in exploring the possibilities of removing conventional blade-root hinges and substituting instead blade structural flexibility. Another manufacturer, Lockheed Aircraft, made a major commitment to the hingeless rotor concept coupled to a mast-mounted mechanical gyro introduced into the pitch control linkage.[25] The root regions of the blades in this innovative design were "matched stiffness" or "soft in-plane," which meant that the blade chord-wise, or horizontal, structural bending stiffness was matched to the flap-wise, or vertical, bending stiffness. Dynamic model tests of this concept were conducted in the Langley 30 by 60 Full-Scale Tunnel and in the Freon atmosphere of the Langley Transonic Dynamics Tunnel. The use of Freon gas facilitated the testing of the 10-foot-diameter rotor model at full-scale Reynolds number and Mach numbers. This work began the establishment of a documented database for hingeless rotor design. These dynamic model tests were part of a cooperative NASA–Army AVLABS program.

To further explore the problems and practical means for realizing the potential of the hingeless rotor concept, Langley Research Center purchased the Lockheed XH-51N, a high-speed research helicopter. The flight investigation focused on the tendency for hingeless rotors to encounter high in-plane blade loads in roll maneuvers, coupling between the response to longitudinal and lateral control input, ride quality, and pilot handling qualities. In general, it was demonstrated with the flight tests and model tests that the hingeless rotor system was different from the conventional hinged systems. Inherently, the hingeless designs produced increased control moments, quicker response to pilot input and superior handling qualities. It turned out that later rotor designs incorporating elastomeric bearings to replace conventional hinges could provide a practical option to some of the fully hingeless designs.

---

24. R.J. Huston, "An Exploratory Investigation of Factors Affecting the Handling Qualities of a Rudimentary Hingeless Rotor Helicopter," NASA TN-D-3418 (May 1966).
25. I.H. Culver and J.E. Rhodes, "Structural Coupling in the Blades of a Rotating Wing Aircraft," IAS Paper No. 62-33 (1962).

NASA 1970–1990: Joint Program Momentum Peaks

During the early 1970s, the Ames Flight Simulator for Advanced Aircraft (FSAA) became operational and the first tilt rotor simulations were successfully accomplished. By 1975, the Army decided to augment the rotary wing flight dynamics research at Ames as NASA initiated the fabrication of the Vertical Motion Simulator (VMS). This simulator, with very large vertical and horizontal motion capability, was a national asset well suited for rotary wing research.

At Langley, a major instrument flight rules (IFR) investigation was conducted under the VTOL Approach and Landing Technology (VALT) program. The VALT Boeing-Vertol CH-47 Chinook helicopter was the primary research vehicle for exploring the control/display/task relationships. In addition, the Sikorsky SH-3 Sea King helicopter was used as a testbed for exploring the merits and defining the electro-optical parameter requirements associated with advanced "real-world" display concepts. The objective was to identify systems that might be capable of providing a pilot an "out-the-window display" during IFR flight conditions through the use of fog-cutting sensors or advanced computer-generated visual situation displays. The VALT CH-47 flights were conducted at the Wallops Flight Center, where the NASA Aeronautical Research Radar Complex provided omnidirectional tracking coverage. This facility permitted the investigation of a wide variety of approach trajectories and selection of any desired wind direction relative to the final approach heading. Computer-graphic displays were generated on the ground and transmitted via video link to the aircraft for presentation in the pilots' instrument panel. The integrated flight-test system permitted manual, augmented, or completely automatic control for executing flight trajectories that could be optimized from the standpoints of fuel, time, airspace utilization, ride qualities, noise abatement, or air traffic control considerations. Many concepts were explored in the IFR program, including flight director control/display concepts and signal smoothing techniques, which proved valuable in achieving fully automatic approach and landing capability.[26] Extensive flight demonstrations were conducted at Wallops Flight Center with the VALT CH-47 aircraft for Government and industry groups to demonstrate the new progress achieved in IFR approach and landing technology.

---

26. J.R. Kelly, F.R. Niessen, J.J. Thibodeaux, K.R. Yenni, and J.F. Garren, Jr., "Flight Investigation of Manual and Automatic VTOL Decelerating Instrument Approaches and Landings," NASA TN-D-7524 (July 1974).

In structures technology, one of the important outcomes of the space program was the development and implementation of comprehensive computational finite element analyses. State-of-the-art finite element methodology was collected from among the large aerospace companies and unified into the NASA Structural Analysis (NASTRAN) computer program. The basic development contract was managed by NASA's Goddard Space Flight Center and then by Langley for improvements and distribution to approximately 260 installations. During the early 1980s, Langley played a key role in bringing advanced structural design capability into the helicopter industry. The contribution here was the onsite assignment of an experienced structural dynamics specialist at a prime manufacturer's facility to guide the integration of the preliminary static structural design methodology with rotor dynamic analysis methodology.[27] This avoided the tedious process of repeatedly freezing an airframe structural design effort and each time doing a separate dynamic analysis to determine if an acceptable dynamic response criterion was achieved.

During this period, the Army added to its already extensive helicopter crash-test activities by joining with NASA to crash-test the Boeing Vertol CH-47C helicopter in the Impact Dynamics Research Facility at Langley, which accommodated aircraft up to 30,000 pounds.[28] The facility had been converted from a Lunar Landing Research Facility to a center for the study of crash effects on aircraft. A unique feature of this massive gantry structure was the capability to impact full-scale aircraft under free-flight conditions with precise control of attitude and velocity.

The ongoing rotary wing research began to expand in scope with the establishment of the Army co-located research groups at the three NASA Centers. At Ames, full-scale rotor wind tunnel testing continued at an increased pace in the 40- by 80-foot tunnel. In the 1970s, the wind tunnel tests included the Sikorsky Advancing Blade Concept (ABC) rotor. This rotor concept incorporated two counter-rotating coaxial rotors. The hingeless blades were very stiff to allow the advancing blades on both sides of the rotor disk to balance the opposing rolling moments thereby

---

27. R.G. Kvaternik and W.G. Walton, Jr., "A Formulation of Rotor-Airframe Coupling for the Design Analysis of Vibrations of Helicopter Airframes," NASA RP-1089 (June 1982).
28. Karen Jackson, Richard L. Boitnott, Edwin L. Fasanella, Lisa E. Jones, and Karen H. Lyle, "A Summary of DOD-Sponsored Research Performed at NASA Langley's Impact Dynamics Research Facility," Journal of the American Helicopter Society, vol. 51, no. 1 (June 2004).

The Sikorsky XH-59A Advancing Blade Concept helicopter was a joint test program between the Army, Navy, NASA, and Air Force. NASA

maintaining aircraft trim as airspeed is increased. Forward thrust is supplied by auxiliary propulsion rather than by forward tilt of the main rotor as in conventional helicopter designs.

NASA also tested a full-scale semispan wing-pylon-rotor of the Bell Helicopter tilt rotor design.[29] This test was followed by a similar entry of a semispan setup of a Boeing Vertol tilt rotor concept. During this period, improvements were made in the 40- by 80-foot Full-Scale Tunnel to upgrade the research capability. Its online data capability was augmented by introducing a new Dynamic Analysis System for real-time analysis of critical test parameters. A new Rotor Test Apparatus (RTA) was added to facilitate full-scale rotor testing. With this new equipment in place, a Kaman Controllable Twist Rotor (CTR) was first investigated in 1975.

In the early 1970s, the modest in-house research funding level for rotary wing projects led to seeking other sources within the new, more elaborate financial system of NASA. It turned out that contracting out-of-house research had become a staple of the rapidly growing procure-

---

29. H.K. Edenborough, T.M. Gaffey, and J.A. Weiberg, "Analysis and Tests Confirm Design of Proprotor Aircraft," AIAA Paper No. 72-803 (1972).

ment system.³⁰ This offered the opportunity to begin to solicit, select, and fund small supporting research contracts to augment the in-house rotary wing work categorized as Research and Technology Base. In the Flight Research Branch at Langley between 1969 and 1974, over 77 contractor reports (CR) and related technical papers were published. The performing organizations included industry and university research departments. The research topics included analytical and experimental investigations of rotor-blade aeroelastic stability, blade-tip vortex aerodynamics, rotor-blade structural loads prediction, free-wake geometry prediction, nonuniform swash-plate dynamic analysis program, rotor-blade dynamic stall, composite blade structures, and variable geometry rotor concepts, In the mid 1970s, this entry into contracted research to augment in-house work was further augmented by teaming of NASA and Army rotary wing research at the three NASA Centers. Finally, projects between NASA, the Army, and contractors evolved into major joint efforts in Systems Technology and Experimental Aircraft during the following decade.

The mid-1970s brought two major rotary wing experimental aircraft programs, both jointly funded and managed by NASA and the Army. At Langley, the Rotor Systems Research Aircraft (RSRA) program was launched. This was a new approach to conducting flight research on helicopter rotor systems.³¹ Two vehicles were designed and fabricated by Sikorsky Aircraft. The basic airframe, propulsion, and control systems of the two RSRA vehicles were those of the Sikorsky S-61 Sea King helicopter. In addition, the RSRA incorporated a unique rotor force balance system and isolation system, a programmable electronic control system, a variable incidence wing with a force balance system, drag brakes, and two TF34 auxiliary thrust turbofan engines. As a unique safety feature, the three-member-crew ejection system incorporated automatic balanced sequencing of explosive separation of the test rotor-blades as the first step in permitting the rapid ejection of the pilot, copilot, and test engineer. After design and fabrication at Sikorsky, the first of two RSRA vehicles made its first flight in 1976. After initial tests of the helicopter configuration, flight-testing was continued at the NASA Wallops Flight

---

30. James R. Hansen, *Spaceflight Revolution, Langley Research Center From Sputnik to Apollo*, NASA SP-4308 (Washington, DC: NASA 1995), pp. 81–111.
31. A.W. Linden and M.W. Hellyer, "The Rotor Systems Research Aircraft," AIAA Paper No. 74-1277 (1974).

Center with the Langley–Army project team and contractor onsite support. Acceptance testing was completed by the Langley team, which was then joined by Ames flight-test representatives in anticipation of pending transfer of the RSRA program to Ames.

At Ames, a NASA–Army program of equal magnitude was launched to design and fabricate two XV-15 Tilt Rotor Research Aircraft (TRRA). In this case, the program focused on a proof-of-concept flight investigation. This concept, pursued by rotary wing designers since the early 20th century, employs a low-disk-loading rotor at each wingtip that can tilt its axis from vertical, providing lift, to horizontal, providing propulsive thrust in wing-borne forward flight. The TRRA contract was awarded to Bell Helicopter Textron. Late in the program, as the XV-15 reached flight status, the United States Navy added funding for special mission-suitability testing. Eventually, XV-15 testing gave confidence to tilt rotor advocates who successfully pushed for development of an operational system, which emerged as the V-22 Osprey.

The RSRA and TRRA experimental aircraft programs together represented a total initial investment of approximately $90 million, ($337 million in 2009 dollars), shared equally by NASA and the Army. The size and scope of these programs were orders of magnitude beyond previous NACA–NASA rotary wing projects. This represented a new level of

The NASA–Army Sikorsky S-72 Rotor Systems Research Aircraft in flight at NASA's Ames Research Center. NASA.

resources in rotary wing research for NASA and with it came considerably more day-to-day visibility within the NASA aeronautics program.

The bicentennial year of 1976 also marked a year of major organizational change in NASA rotary wing research. As part of an overall Agency reassessment of the roles and missions of each Center, the Ames Research Center was assigned the lead Center responsibility for helicopter research. An objective of the lead Center concept was to consolidate program lead in one Center and, wherever possible, combine research efforts of similar nature. As a result, all rotary wing flight test, guidance, navigation, and terminal area research were consolidated at Ames, which brought these research activities together with the extensive simulation and related flight research facilities. Langley retained supporting research roles in structures, noise, dynamics, and aeroelasticity. The realignment of responsibilities and transfer of flight research aircraft caused unavoidable turbulence in the day-to-day conduct of the rotary wing program from 1976 to 1978. However, the momentum of the program gradually returned, and the program grew to new levels with NASA and Army research teams at Ames, Langley, and Glenn working to carry out their responsibilities in rotary wing research.

At Ames, the testing of full-scale rotor systems continued at an increasing pace in the 40 by 80 Full-Scale Tunnel. In 1976, the Controllable Twist Rotor concept was tested again, this time with multicyclic control. "Two-per-rev" (two control cycles per one rotor revolution), "three-per-rev," and "four-per-rev" cyclic control was added to the CTR's servo flap system to evaluate the effectiveness in reducing blade stresses and vibration of the fuselage module. Both favorable effects were achieved with only minor effect on the rotor power requirements. The Sikorsky S-76 rotor system was tested in 1977 in a joint NASA–Sikorsky investigation of tip shapes. This was followed by a joint NASA–Bell investigation of the Bell Model 222 fuselage drag characteristics. In 1978, the NASA–Army XV-15 Tilt Rotor Research Aircraft arrived from Bell Helicopter for full-scale wind tunnel tests prior to initiation of its own flight tests. The wind tunnel tests revealed a potential tail structural vibration problem that would be further explored in flight following the strengthening of the empennage attachment structure. The next rotor test was the Kaman Circulation Control Rotor (CCR) in 1978.[32]

---

32. Jack N. Nielsen and James C. Biggers, "Recent Progress in Circulation Control Aerodynamics," AIAA Paper No. 87-0001 (1987).

A new concept was introduced based on technology developed at the David Taylor Ship Research and Development Center (since 1992 the Carderock Division of the Naval Surface Weapons Center). The Kaman rotor utilized elliptical-shaped airfoils with trailing edge slots. Lift was augmented by blowing compressed air from these slots. The need for mechanical cyclic blade feathering to provide rotor control was eliminated replaced by cyclic blowing. The wind tunnel testing investigated the amount of blowing control necessary to maintain forward flight. In 1979, the Lockheed X-Wing Stoppable Rotor was tested in the 40 by 80 Full-Scale Tunnel. This concept, funded by the Defense Advanced Research Projects Agency, also incorporated a circulation control concept. The X-Wing rotor was designed to be stoppable (and startable) at high forward flight speed while still carrying lift. Since two of the four blade trailing edges become leading edges when stopped, provisions were made to provide for separate blowing systems for the leading and trailing edges of the blades. When operating as a fixed X-Wing aircraft, aircraft roll and pitch control were provided by differential blowing from the aft edges of opposing, nonrotating blades serving as swept forward and aft wings. The wind tunnel tests of the 25-foot-diameter rotor successfully demonstrated the ability to start and stop the rotor at speeds of approximately 180 knots (maximum tunnel speed).

The Boeing Vertol Bearingless Main Rotor (BMR) was tested in 1980.[33] The BMR used elastic materials in the construction of the rotor hub rather than mechanical bearings for articulation. Such designs have aeroelastic stability characteristics different from conventional mechanical systems. Therefore, the wind tunnel tests investigated the degree of stability present and established appropriate boundaries for safe flight. In addition, in 1980, the Sikorsky Advancing Blade Concept (ABC) coaxial rotor was again tested in the 40 by 80 Full-Scale Tunnel.[34] In this entry, the full-scale rotor was tested atop a configuration replica of the actual XH-59A flight-test aircraft. This testing focused on an investigation of the drag characteristics of the rotor shaft and hubs of the coaxial rotors. In an effort to reduce the drag, tests were made with the actual fuselage modeled and the actual flight demonstrator hardware compo-

---

33. W. Warmbrodt and J.L. McCloud, II, "A Full-Scale Wind Tunnel Investigation of a Helicopter Bearingless Main Rotor," NASA TM-81321 (1981).
34. M. Mosher and R.L. Peterson, "Acoustic Measurements of a Full-Scale Coaxial Helicopter," AIAA Paper No. 83-0722 (1983).

nents utilized to explore several inter-rotor fairing configurations. (In 2008, Sikorsky Aircraft unveiled a new technology demonstrator aircraft incorporating the advancing blade concept identified as the X2. In this design forward thrust is provided by a pusher propeller installation.)

In 1984, Ames shut down the 40- by 80-foot facility for tunnel modification to upgrade the 40- by 80-foot section to a speed capability of 250 knots and add a new 80 by 120 leg to the tunnel facility capable of speeds to 80 knots. The upgraded facility, known as the National Full-Scale Aerodynamics Complex (NFAC), reopened in 1987 and would have been operated by NASA until 2010. However, budgetary pressures forced its closure in 2003. Four years later, in 2007, the United States Air Force's Arnold Engineering Development Center (AEDC) upgraded key operating systems and reopened the facility under a 25-year lease with NASA. The anticipated majority customer for this national asset was seen to be the United States Army, in collaboration with NASA, in support of rotary wing research.

A Helicopter Transmission Technology program was initiated at the Glenn Research Center to foster the application of an extensive technology base in bearings, seals, gears, and new concepts specifically to helicopter propulsion systems.[35] Research continued at a growing pace. In order to upgrade the analytical methods for large spiral bevel gears, NASA supported the development and validation testing of finite element method computer programs by Boeing Vertol. The opportunity was taken to utilize the available aft transmission hardware assets, available from the canceled XCH-62 Heavy Lift Helicopter Program, for analytical methods validation data. Another program at Glenn was the joint NASA–DARPA Convertible Engine Systems Technology (CEST) program. This program involved the modification of a TF34 turbofan engine to a fan/shaft engine configuration for use as a research test engine to investigate the performance, control, noise, and transient characteristics. The potential application of CEST was to the X-Wing vehicle concept by using a single-core engine to provide shaft power to a rotor in hover and low speed, and conversion capability to provide fan thrust for high speed, stopped rotor mode, and flight propulsion.

Ongoing research in helicopter handling qualities continued and expanded at the Ames Research Center. In 1978, one of these programs

---

35. Robert C. Ball, "Summary Highlights of the Advanced Rotor Transmission (ART) Program," AIAA Paper No. 92-3362 (1992).

provided essential simulation data on the effects of large variations in rotor design parameters on handling qualities and agility in helicopter nap-of-the-Earth (NOE) flight. The parameters investigated including flapping hinge offset, flapping hinge restraint, rotor blade inertia, and blade pitch-flap coupling. Experiments were carried out on the Ames piloted simulators to systematically study stability and control augmentation systems designed to improve NOE flying and handling qualities characteristics.

New efforts in computational analysis to increase rotor efficiency began at Ames. An analytical procedure was developed to predict rotor performance trends in relation to changes in the shape of the blade tips. The analytical procedure utilized two full potential flow-field computer programs developed for computation of the transonic flow field about fixed wings and airfoils. The analytical procedure rapidly became a useful tool for predicting aerodynamic performance improvements that may be achieved by modifying blade geometry. The procedure was guided by design studies and reduced the experimental testing required to select blade configurations. NASA continued the long-established tradition of furnishing excellent references for technical practice when, in 1980, research scientist Wayne Johnson, a member of the Army–NASA research team at Ames, published his book *Helicopter Theory*, a comprehensive state-of-the-art coverage of the fundamentals of helicopter theory and engineering analysis. The extensive bibliography of cited literature included an extensive listing of rotary wing technical publications authored by researchers from the NACA, NASA, the Army, industry, and academia.[36]

Research accelerated on advancing the ability of a helicopter to execute a radar approach. Civil weather/mapping radar could be used to provide approach guidance under instrument meteorological conditions (IMC) to select safe landing environments. Onboard radar systems were widely used by helicopter operators to provide approach guidance to offshore oil rigs without the need for electronic navigation aids at the landing site. For use over the water, the radar provided guidance and ensures obstacle awareness and avoidance, but involved very high pilot workload and limited guidance accuracy. For use over land, the ground clutter return made these approaches infeasible without more advanced radar systems. Two programs at Ames resulted from major advances in radar approaches. One program involved the

---

36. Wayne Johnson, *Helicopter Theory* (Princeton: Princeton University Press, 1980).

The NASA/Army/Bell XV-15 Tilt Rotor Research Aircraft in flight. NASA.

use of a video data processor in conjunction with the weather radar for overwater approaches. This system automatically tracked a designated radar target and displayed a pilot-selected approach course. The second radar program involved the development of an innovative technique to suppress ground clutter radar returns in order to locate simple, low-cost radar reflectors near the landing site. This program was extended to provide the pilot with precision localizer and glide-slope information using airborne weather radar and a ground-based beacon or reflector array.

The 1980s brought several major accomplishments in the tilt rotor program.[37] The second XV-15 aircraft was brought to flight status and accepted by the Government after check flights and acceptance ceremonies at NASA's Dryden Flight Research Center on October 28, 1980. It was then used for flight tests aimed at verifying aeroelastic stability, evaluating fatigue load reduction modifications, and expanding the maneuver envelope. Subsequently, this aircraft was ferried to Ames, where tests continued in the areas of handling qualities, flight control, and expansion of the landing approach envelope. The first XV-15 aircraft was brought to flight status in late 1980, and initial work was

---

37. D.C. Dugan, R.G. Erhart, and L.G. Schroers, "The XV-15 Tilt Rotor Research Aircraft," NASA TM-81244 / AVRADCOM Technical Report 80-A-15 (1980).

done on a ground tiedown rig to measure the downwash field and noise environment. Meanwhile, the second XV-15 participated in the Paris Air Show. The aircraft performed daily, on schedule, and received wide acclaim as a demonstration of new aeronautical technological achievement. The XV-15 crew concluded each daily performance with a courteous "bow," the hovering tilt rotor ceremoniously dipping its nose to the audience. After the flight demonstration in France and subsequent flights in Farnborough, England, the aircraft was returned to Ames for continued flight demonstration and proof-of-concept testing. The two vehicles achieved a high level of operational reliability, not the usual attribute of highly specialized research aircraft. One of the vehicles was returned to Bell Helicopter under a cooperative arrangement that made the aircraft available to the contractor at no cost in exchange for doing a number of program flight-test tasks, particularly in the mission suitability category. The overall success of the NASA–Army XV-15 (with a rotor diameter of 25 feet and a gross weight of 13,428 pounds) proof-of-concept program contribution is reflected in the application of the proven technology to the design and production of the new joint-service V-22 Osprey, (rotor diameter: 38 feet; gross weight: 52,000 pounds). The classic claim of research results having to endure a 20-year shelf life before actual engineering design application begins did not apply. It took only 5 years to move from achieving proof-of-concept with the XV-15 research aircraft to initiation of preliminary design of the operational V-22 Osprey.

There has been over a half century of an unbroken series of NACA–NASA research contributions to tilt rotors since early XV-3 flight assessments and wind tunnel testing in the mid-1950s.[38] Since that beginning, NACA–NASA researchers have pursued many subject areas, including tilt rotor analytical investigations to solve a rotor/pylon aeroelastic stability problem, dynamic model aeroelastic testing in the Langley Transonic Dynamics Tunnel, analytical method development and verification, wind tunnel tests of full-scale rotor/wing/pylon assembles, XV-15 vehicle wind tunnel tests and flight tests, and detailed investigation of many other potential problem areas. This sustained effort and the robust demonstration and advocacy of the technology's potential resulted in the XV-15 program being cited in 1993 as "the program that wouldn't

---

38. Martin D. Maisel, Demo J. Giulianetti, and Daniel C. Dugan, *The History of The XV-15 Tilt Rotor Research Aircraft From Concept to Flight*, NASA SP-2000-4517 (Washington, DC: NASA, 2000).

die" in a University of California at Berkeley School of Engineering case study in a course on "The Political Process in Systems Architecture."[39]

During the early 1980s, the rotary wing activity at Glenn Research Center increased with the addition of new transmission test facilities rated at 500 and 3,000 horsepower. Research progressed on traction drive, hybrid drive, and other advanced technology concepts. The problem of efficient engine operation at partial power settings was addressed with initial studies indicating turbine bypass engine concepts offered potential solutions. Similar studies on contingency power for one-engine-inoperative (OEI) emergency operation focused on water injection and cooling flow modulation. Renewed efforts in aircraft icing included rotary wing icing research. A broad scope program was launched to study the icing environment, develop basic ice accretion prediction methods, acquiring in-flight icing data for comparison with wind tunnel data from airfoil icing tests to verify rotor performance prediction methods. In addition, flight tests of a pneumatic deicing boot system were conducted using the Ottawa spray rig and the United States Army CH-47 in-flight icing spray system. In 1983, research testing began on the NASA–DARPA Convertible Engine System Technology program.[40] TF34 fan/shaft engine hardware with variable fan inlet guide vanes for thrust modulation was used to evaluate fan hub design and map the steady-state and transient performance and stability of the concept. New rotary wing efforts were also started in the areas of transmission noise, and flight/propulsion control integration technology.

Langley Research Center activity in rotary wing research increased substantially within the Structures Directorate, with focused programs in acoustics, dynamics, structural materials, and crashworthiness. This research was carried out in close association with the Army Structures Laboratory, now known as the Vehicle Technology Directorate (VTD). NASA and Army joint use of the Langley 4- by 7-meter tunnel for aerodynamic and acoustic model testing became an important feature of the rotary wing program. Confirmed progress was achieved in airframe dynamic analysis methodology addressing the engineering management and execution of the efficient use of finite element methods for

---

39. Brenda Forman, "The V-22 Tiltrotor 'Osprey:' The Program That Wouldn't Die," *Vertiflite*, vol. 39, no. 6, (Nov./Dec. 1993), pp. 20–23.
40. Jack G. McArdle, "Outdoor Test Stand Performance of a Convertible Engine with Variable Inlet Guide Vanes for Advanced Rotorcraft Propulsion," NASA TM-88939 (1986).

simultaneous tasks of static and dynamic airframe preliminary design.[41] These techniques were demonstrated, publicly documented, and verified by comparison with shake test data for the CH-47 helicopter airframe. Other research related to helicopter dynamics included participation with the Army in a program to demonstrate the use of closed-loop multicyclic control of rotor-blade pitch motion for vibration reduction. The program involved flight-testing of an Army OH-6 helicopter by Hughes Helicopters.[42]

One of the more innovative approaches to research teaming was developed in the area of rotary wing noise. In 1982, discussions between the American Helicopter Society and NASA addressed the industry concern that the proposed rulemaking by Federal Aviation Administration would place the helicopter industry at a considerable disadvantage. The issue was based on the point that the state-of-the-art noise prediction did not allow the prediction of noise for new designs with acceptable confidence levels. As a result, NASA and the Society, joined by the FAA and the Helicopter Association International (HAI)—an organization of helicopter commercial operators—embarked on a joint program in noise research. Through the AHS, American helicopter manufacturers pooled their research with that of NASA under a 5-year plan leading to improved noise prediction capability. All research results were shared among the Government and industry participants in periodic technical exchanges. Langley managed the program with full participation by Ames and Glenn Research Centers in their areas of expertise.

After delivery of the two RSRA vehicles to the Ames Research Center in the late 1970s, the helicopter and compound (with wing and TF34 turbofan engines installed) configurations were involved in an extended period of ground- and flight-testing to document the characteristics of the basic vehicles. This included extensive calibrations of the onboard load measurement systems for the rotor forces and moments; wing lift, drag, and pitching moment; and TF34 engine thrust. This work was followed by the initiation the research flight program utilizing the delivered S-61 rotor system. In 1983, NASA and DARPA launched a major research program to design, fabricate and flight-test an X-Wing rotor on the new RSRA. The RSRA was ideally suited to the testing of new rotor

---

41. Raymond G. Kvaternik, "The NASA/Industry Design Analysis Methods for Vibration (DAMVIBS) Program—A Government Overview," AIAA Paper No. 92-2200 (1992).
42. B.P. Gupta, A.H. Logan, and E.R. Wood, "Higher Harmonic Control for Rotary Wing Aircraft," AIAA Paper No. 84-2484 (1984).

concepts, being specifically design for the purpose. One RSRA vehicle was returned to Sikorsky Aircraft for installation of an X-Wing rotor. This aircraft was eventually moved to NASA Dryden Flight Research Center at Edwards Air Force Base, CA, where final preparations were made for flight-testing. The second vehicle embarked on fixed-wing flight testing at the Dryden Center to expand and document the flight envelope of the RSRA beyond 200 knots, the speed range of interest in the start-stop conversion testing for the X-Wing rotor.

Contributions were beginning to emerge from the NASA–American Helicopter Society Rotorcraft Noise Prediction Program, the joint Government-industry effort initiated in 1983.[43] The four major thrusts were: noise prediction, database development, noise reduction, and criteria development. Fundamental experimental and analytical studies were started in-house and under grants to universities. In order to obtain high-quality noise data for comparison with evolving prediction capability, a wind tunnel testing program was initiated. This NASA-sponsored program was performed in 1986 in the Dutch-German wind tunnel (Duits-Nederlandse wind tunnel, DNW) using a model-scale Bo 105 main rotor. This program was performed with the support of the Federal Aviation Administration and the collaboration of the German aerospace research establishment. In these tests and in subsequent tests of the model in the DNW tunnel in 1988, researchers gained valuable insight into the aeroacoustic mechanism of blade vortex interaction (BVI) noise.

In regard to rotor external noise reduction, Langley researchers investigated the possibility of BVI noise reduction using active control of blade pitch. A model-scale wind tunnel test was conducted in the Langley Transonic Dynamics Tunnel (TDT) using the Aeroelastic Rotor Experimental System (ARES).[44] Results were encouraging and demonstrated noise level reductions up to 5 decibels (dB) for low and moderate forward speeds. A major contribution of the NASA–AHS program was the development of a comprehensive rotorcraft system noise prediction capability. The primary objective of this capability, the computer code named ROTONET, was to provide industry with a reliable predictor for

---

43. Ruth M. Martin, "NASA/AHS Rotorcraft Noise Reduction Program: NASA Langley Acoustics Division Contributions," *Vertiflite*, vol. 35, no. 4, (May/June 1989), pp. 48–52.
44. W.R. Mantay, W.T. Yeager, Jr., M.N. Hamouda, R.G. Cramer, Jr., and C.W. Langston, "Aeroelastic model Helicopter Testing in the Langley TDT," NASA TM-86440 / USAAVSCOM TM-85-8-5 (1985).

use in design evaluation and noise certification efforts. ROTONET was developed in several phases, with each phase released to Noise Reduction Program participants for testing and evaluation. Validation data from flight test of production and experimental rotorcraft constituted a vital element of the program. The first was of the McDonnell-Douglas 500E helicopter, tested at NASA's Wallops Flight Facility. The second flight-test effort at Wallops, a joint NASA–Army program, was performed in 1987 using an Aerospatiale Dauphine helicopter, which had a relatively advanced blade design and a Fenestron-type (ducted) tail rotor. The year 1988 saw a joint NASA–Bell Helicopter effort in flight investigation of the noise characteristics the NASA–Army XV-15 Tilt Rotor Research Aircraft. The results indicated that while the aircraft seemed very quiet in the airplane mode, significant blade-vortex interaction noise was evident in the helicopter mode of flight. NASA benefited from the interaction with and participation in the variety of industry noise programs, which helped set the groundwork for subsequent joint participation in rotary wing research.[45]

NASA 1990–2007: Coping with Institutional and Resource Challenges

Over the next decade and a half, the NASA rotary wing program's available organizational and financial resources were significantly impacted by NASA and supporting Agency organizational, mission, and budget management decisions. These decisions were driven by changes in program priorities in the face of severe budget pressures and reorganization mandates seeking to improve operational efficiency. NASA leaders were being tasked with more ambitious space missions and with recovering from two Shuttle losses. In the face of these challenges, the rotary wing program, among others, was adjusted in the effort to continue to make notable research contributions. Examples of the array of real impacts on the rotary wing program over this period were: (1) termination of the NASA–DARPA RSRA–X-Wing program; (2) stopping the NASA–Army flight operations of the only XV-15 TRRA aircraft and the two RSRA vehicles; (3) transfer of all active NASA research aircraft to Dryden Flight Research Center, which essentially closed NASA rotary wing flight operations; (4) elimination of vehicle program offices at NASA Headquarters; (5) closing the National Full-Scale Aerodynamic Complex wind tunnel at

---

45. Robert J. Huston, Robert A. Golub, and James C. Yu, "Noise Considerations for Tilt Rotor," AIAA Paper 89-2359 (1989).

Ames in 2003 (reopened under a lease to the United States Air Force in 2007); (6) converting to full-cost accounting, which represented a new burden on vehicle research funding allocations; and (7) the imposition of a steady and severe decline in aeronautics budget requests, staring in the late 1990s. Overshadowing this retrenching activity in the 1990s was the total reorientation, and hence complete transformation, of the Ames Research Center from an Aeronautics Research Mission Center to a Science Mission Center with the new lead in information technology (IT).[46] Responsibility for Ames's aerodynamics and wind tunnel management was assigned to Langley Research Center. The persistent turbulence in the NASA rotary wing research community presented a growing challenge to the ability to generate research contributions. Here is where the established partnership with the United States Army and co-located laboratories at Ames, Langley, and Glenn Research Centers made it possible to maximize effectiveness by strengthening the combined efforts. In the case of Ames, this was done by creating a new combined Army–NASA Rotorcraft Division. The center of gravity of NASA rotary wing research thus gradually shifted to the Army.

The decision to ground and place in storage the only remaining XV-15 TRRA in 1994 was fortunately turned from a real setback to an unplanned contribution. Bell Helicopter, having lost the other XV-15, N702NA, in an accident in 1992, requested bailment of the Ames aircraft, N703NA, in 1994 to continue its own tilt rotor research, demonstrations, and applications evaluations in support of the ongoing (and troubled) V-22 Osprey program. The NASA and Army management agreed. As part of the extended use, on April 21, 1995, the XV-15 became the first tilt rotor to land at the world's first operational civil vertiport at the Dallas Convention Center Heliport/Vertiport. After its long and successful operation and its retirement in 2003, this aircraft is on permanent display at the Smithsonian Institution's Udvar-Hazy Center at Washington Dulles International Airport, Chantilly, VA.

With the military application of proven tilt rotor technology well underway with the procurement of the V-22 Osprey by the Marine Corps and Air Force, the potential for parallel application of tilt rotor technology to civil transportation was also addressed by NASA. Early studies, funded by the FAA and NASA, indicated that the concept had potential

---

46. Glenn E. Bugos, *Atmosphere of Freedom, Sixty Years at the Ames Research Center*, NASA SP-4314 (Washington, DC: NASA 2000), pp. 211–246.

for worldwide application and could be economically viable.[47] In late 1992, Congress directed the Secretary of Transportation to establish a Civil Tilt Rotor Development Advisory Committee (CTRDAC) to examine the technical, operational, and economic issues associated with integrating the civil tilt rotor (CTR) into the Nation's transportation system. The Committee was also charged with determining the required additional research and development, the regulatory changes required, and the estimated cost of the aircraft and related infrastructure development. In 1995, the Committee issued the findings. The CTR was determined to be technically feasible and could be developed by the United States' industry. It appeared that the CTR could be economically viable in heavily traveled corridors. Additional research and development and infrastructure planning were needed before industry could make a production decision. In response to this finding, elements of work suggested by the CTRDAC were included in the NASA rotorcraft program plans.

Significant advances in several technological areas would be required to enable the tilt rotor concept to be introduced into the transportation system. In 1994, researchers at Ames, Langley, and Glenn Research Centers launched the Advanced Tiltrotor Transport Technology (ATTT) program to develop the new technologies. Because of existing funding limitations, initial research activity was focused on the primary concerns of noise and safety. The noise research activity included the development of refined acoustic analyses, the acquisition of wind tunnel prop-rotor noise data to validate the analytical method, and flight tests to determine the effect of different landing approach profiles on terminal area and community noise. The safety effort was related to the need to execute approaches and departures at confined urban vertiports. For these situations the capability to operate safely with one-engine-inoperative in adverse weather conditions was required. This area was addressed by conducting engine design studies to enable generating high levels of emergency power in OEI situations without adversely impacting weight, reliability, maintenance, or normal fuel economy. Additional operational safety investigations were carried out on the Ames Vertical Motion Simulator to assess crew station issues, control law variations, and assign advanced configurations such as the variable diameter tilt rotor. The principal American rotary wing airframe

---

47. Maisel, et al., *History of the XV-15 Tilt-Rotor,* pp. 110–114.

and engine manufacturers participated in the noise and safety investigations, which assured that proper attention was given to the practical application of the new technology.[48] An initial step in civil tilt rotor aircraft development was taken by Bell Helicopter in September 1998, by teaming with Agusta Helicopter Company of Italy, to design, manufacture, and certify a commercial version of the XV-15 aircraft design designated the BA 609.

Despite the institutional and resource turbulence overshadowing rotary wing activity, the NASA and Army researchers persisted in conducting base research. They continued to make contributions to advance the state of rotary wing technology applicable to civil and military needs, a typical example being the analysis of the influence of the vortex ring state (VRS) flight in rapid, steep descents, brought to the forefront by initial operating problems experienced by the V-22 Osprey.[49] The current NASA Technical Report Server (NTRS) Web site has posted over 2,200 NASA rotary wing technical reports. Of these, approximately 800 entries have been posted since 1991—the peak year, with 143 entries. These postings facilitate public access to the formal documentation of NASA contributions to rotary wing technology. The annual postings gradually declined after 1991. In what may be a mirror image of the state of NASA's realigned rotary wing program, since 2001 the annual totals of posted rotary wing reports are in the 20–40 range, with an increasing percentage reflecting contributions by Army coauthors.

As the Army and NASA rotary wing research was increasingly linked in mutually supporting roles at the co-located centers, outsourcing, cooperation, and partnerships with industry and academia also grew. In 1995, the Army and NASA agreed to form the National Rotorcraft Technology Center (NRTC) occupying a dedicated facility at Ames Research Center. This jointly funded and managed organization was created to provide central coordination of rotary wing research activities of the Government, academia, and industry. Government participation included Army, NASA, Navy, and the FAA. The academic laboratories' participation was accomplished by NRTC having acquired the responsibility to manage the Rotorcraft Centers of Excellence (RCOE) program

---

48. William J. Snyder, John Zuk, and Hans Mark, "Tilt Rotor Technology Takes Off," AIAA Paper 89-2359 (1989).

49. Wayne Johnson, "Model for Vortex Ring State Influence on Rotorcraft Flight Dynamics," NASA TP-2005-213477 (2005).

that had been in existence since 1982 under the Army Research Office. In 1996, the periodic national competition resulted in establishing Georgia Institute of Technology, the University of Maryland at College Park, and Pennsylvania State University as the three RCOE sites.

The Rotorcraft Industry Technology Association (RITA), Inc., was also established in 1996. Principal members of RITA included the United States helicopter manufacturers Bell Helicopter Textron, the Boeing Company, Sikorsky Aircraft Corporation, and Kaman Aerospace Corporation. Supporting members included rotorcraft subsystem manufacturers and other industry entities. Associate Members included a growing number of American universities and nonprofit organizations. RITA was governed by a Board of Directors supported by a Technical Advisory Committee that guided and coordinated the performance of the research projects. This industry-led organization and NRTC signed a unique agreement to be partners in rotary wing research. The Government would share the cost of annual research projects proposed by RITA and approved by NRTC evaluation teams. NASA and the Army each contributed funds for 25 percent of the cost of each project—together they matched the industry-member share of 50 percent. Over the first 5 years of the Government-industry agreement, the total annual investment averaged $20 million. The RITA projects favored mid- and near-term research efforts that complemented mid- and long-term research missions of the Army and NASA. Originally, there was concern that the research staff of industry competitors would be reluctant to share project proposal information and pool results under the RITA banner. This concern quickly turned out to be unfounded as the research teams embarked on work addressing common technical problems faced by all participants.

NRTC was not immune to the challenges posed by limited NASA budgets, which eventually caused some cutbacks in NRTC support of RITA and the RCOE program. In 2005, the name of the RITA enterprise was changed to the Center for Rotorcraft Innovation (CRI), and the principal office was relocated from Connecticut to the Philadelphia area.[50] Accomplishments posted by RITA–CRI include cost-effective integrated helicopter design tools and improved design and manufacturing practices for increased damage tolerance. The area of rotorcraft

---

50. The Center for Rotorcraft Innovation (CRI) Web site is: *http://www.irotor.org*.

operations accomplishments included incorporating developments in synthetic vision and cognitive decision-making systems to enhance the routine performance of critical piloting tasks and enabling changes in the air traffic management system that will help rotorcraft become a more-significant participant in the civil transportation system. The American Helicopter Society International recognized RITA for one of its principal areas of research effort by awarding the Health and Usage Monitoring Project Team the AHS 1998 Grover E. Bell Award for "fostering and encouraging research and experimentation in the important field of helicopters."

As previously noted, in the mid-1990s, NASA Ames's entire aircraft fleet was transferred some 300 miles south to Dryden Flight Research Center at Edwards Air Force Base, CA. This inventory included a number of NASA rotary wing research aircraft that had been actively engaged since the 1970s.[51] However, the U.S. Army Aeroflightdynamics Directorate, co-located at Ames since 1970, chose to retain their research aircraft. In 1997, after several years of negotiation, NASA Headquarters signed a directive that Ames would continue to support the Army's rotorcraft airworthiness research using three military helicopters outfitted for special flight research investigations. The AH-1 Cobra had been configured as the Flying Laboratory for Integrated Test and Evaluation (FLITE). One UH-60 Blackhawk was configured as the Rotorcraft Aircrew Systems Concepts Airborne Laboratory (RASCAL) and remained as the focus for advanced controls and was utilized by the NASA–Army Rotorcraft Division to develop programmable, fly-by-wire controls for nap-of-the-Earth maneuvering studies. This aircraft was also used for investigating noise-abatement, segmented approaches using local differential Global Positioning System (GPS) guidance. The third aircraft, another UH-60 Blackhawk, had been extensively instrumented for the conduct of the UH-60 Airloads Program. The principal focus of the program was the acquisition of detailed rotor-blade pressure distributions in a wide array of flight conditions to improve and validate advanced analytical methodology. The last NACA–NASA rotor air-loads flight program of this nature had been conducted over three decades earlier, before the advent of the modern digital data acquisition and processing revolu-

---

51. David D. Few, "A Perspective on 15 Years of Proof-of-Concept Aircraft Development and Flight Research at Ames—Moffett by the Rotorcraft and Powered-Lift Flight Projects Division, 1970-1985," NASA RP-1187 (1987).

tion.[52] Again, the persistence of the NASA–Army researchers met the institutional and resource challenges and pressed on with fundamental research to advance rotary wing technology.

On December 20, 2006, the White House issued Executive Order 13419 establishing the first National Aeronautics Research and Development Policy. The Executive order was accompanied by the policy statement prepared by the National Science and Technology Council's Committee on Technology. This 13-page document included recommendations to clarify, focus, and coordinate Federal Government aeronautics R&D activities. Of particular note for NASA's rotary wing community was Section V of the policy statement: "Stable and Long-Term Foundational Research Guidelines." The roles and responsibilities of the executive departments and agencies were addressed, noting that several executive organizations should take responsibility for specific parts of the national foundational (i.e., fundamental) aeronautical research program. Specifically, "NASA should maintain a broad foundational research effort aimed at preserving the intellectual stewardship and mastery of aeronautics core competencies." In addition, "NASA should conduct research in key areas related to the development of advanced aircraft technologies and systems that support DOD, FAA, the Joint Planning and Development Office (JPDO) and other executive departments and agencies.[53] NASA may also conduct such research to benefit the broad aeronautics community in its pursuit of advanced aircraft technologies and systems. . . . " In supporting research benefiting the broad aeronautics community, care is to be taken "to ensure that the government is not stepping beyond its legitimate purpose by competing with or unfairly subsidizing commercial ventures." There is a strong implication that the new policy may lead NASA's aeronautics role in a return to the more modest, but successful, ways of NASA's predecessor, the National Advisory Committee for Aeronautics, with a primary focus on fundamental research, with the participation of

---

52. Edwin W. Aiken, Robert A. Jacobson, Michelle M. Eshow, William S. Hindson, and Douglas H. Doane, "Preliminary Design Features of the RASCAL—A NASA/Army Rotorcraft In-Flight Simulator," AIAA Paper 92-4175 (1992); Robert T.N. Chen, William S. Hindson, and Arnold W. Mueller, "Acoustic Flight Tests of Rotorcraft Noise-Abatement Approaches Using Local Differential GPS Guidance," NASA TM-110370 (1995); Robert M. Kufeld and Paul C. Loschke, "UH-60 Airloads Program—Status and Plans," AIAA Paper 91-3142 (1991).

53. In 2003, Congress authorized the Joint Planning and Development Office (JPDO) coordinating the activities of multiple Federal agencies in planning Next Generation Air Transportation System to implement the transformation of the national airspace system.

academia, and the cooperative research support for systems technology and experimental aircraft program investments by the DOD, the FAA, and industry. In the case of rotary wing research, since the 1990s, NASA management decisions had moved the residual effort in this direction under the pressure of limited resources.

As charged, 1 year after the Executive order and policy statement were issued, the National Science and Technology Council issued the "National Plan For Aeronautics Research and Development and Related Infrastructure." Rotary wing R&D is specifically identified as being among the aviation elements vital to national security and homeland defense with a goal of "Developing improved lift, range, and mission capability for rotorcraft." Future NASA rotary wing foundational research contributions may also contribute to other goals and objective of the plan. For example, under Energy Efficiency and Environment Protection, is Goal 2: Advance development of technologies and operations to enable significant increases in energy efficiency of the aviation system, and Goal 3: Advance development of technologies and operational procedures to decrease the significant environmental impacts of the aviation system.

Perhaps the most important long-term challenge for the rotary wing segment of aviation is the need for focused attention on improved safety. In this regard, Goal 2 under the plan section titled "Aviation Safety is Paramount" appears to embrace the rotary wing need in calling for developing technologies to reduce accidents and incidents through enhanced aerospace vehicle operations on the ground and in the air. The opportunity for making significant contributions in this arena may exist through enhanced teaming of NASA and the rotary wing community under the International Helicopter Study Team (IHST).[54] The goal of the ambitious IHST is to work to reduce helicopter accident rates by 80 percent in 10 years. The participating members of the organization include technical societies, helicopter and engine manufacturers, commercial operator and public service organizations, the FAA, and NASA. Past performance suggests that the timely application of NASA rotary wing fundamental research expertise and unique facilities to this international endeavor would spawn significant contributions and accomplishments.

---

54. Mark Liptak, "International Helicopter Study Team (IHST) Overview Briefing," presented at Helicopter Association International HELI EXPO Meeting, Houston, TX, Feb. 21–23, 2009 (see http://www.ihst.org).

Recommended Additional Readings

Reports, Papers, Articles, and Presentations:

Edwin W. Aiken, Robert A. Jacobson, Michelle M. Eshow, William S. Hindson, and Douglas H. Doane, "Preliminary Design Features of the RASCAL—A NASA/Army Rotorcraft In-Flight Simulator," AIAA Paper 92-4175 (1992).

F.J. Bailey, Jr., "A Simplified Theoretical Method of Determining the Characteristics of a Lifting Rotor in Forward Flight," NACA Report 716 (1941).

Robert C. Ball, "Summary Highlights of the Advanced Rotor Transmission (ART) Program," AIAA Paper 92-3362 (1992).

G.W. Brooks, "The Application of Models to Helicopter Vibration and Flutter Research," *Proceedings* of the ninth annual forum of the American Helicopter Society (May 1953).

Robert T.N. Chen, William S. Hindson, and Arnold W. Mueller, "Acoustic Flight Tests of Rotorcraft Noise-Abatement Approaches Using Local Differential GPS Guidance," NASA TM-110370 (1995).

R.P. Coleman, "Theory of Self-Excited Mechanical Oscillations of Hinged Rotor Blades," NACA WR-L-308 [formerly NACA Advanced Restricted Report 3G29] (1943).

G.E. Cooper, "Understanding and Interpreting Pilot Opinion," *Aeronautical Engineering Review*, vol. 16, no. 3, (Mar. 1957), p. 47–51.

G.E. Cooper and R.P. Harper, Jr., "The Use of Pilot Rating in the Evaluation of Aircraft Handling Qualities," NASA TN-D-5153 (1969).

I.H. Culver and J.E. Rhodes, "Structural Coupling in the Blades of a Rotating Wing Aircraft," IAS Paper 62-33 (1962).

Mike Debraggio, "The American Helicopter Society—A Leader for 40 Years," *Vertiflite*, vol. 30, no. 4 (May–June 1984).

D.C. Dugan, R.G. Erhart, and L.G. Schroers, "The XV-15 Tilt Rotor Research Aircraft," NASA TM-81244 / AVRADCOM Technical Report 80-A-15 (1980).

H.K. Edenborough, T.M. Gaffey, and J.A. Weiberg, "Analysis and Tests Confirm Design of Proprotor Aircraft," AIAA Paper 72-803 (1972).

Don Fertman, "The Helicopter History of Sikorsky Aircraft," *Vertiflite*, vol. 30, no. 4 (May–June 1984).

David D. Few, "A Perspective on 15 Years of Proof-of-Concept Aircraft Development and Flight Research at Ames–Moffett by the Rotorcraft and Powered-Lift Flight Projects Division, 1970–1985," NASA RP-1187 (1987).

Brenda Forman, "The V-22 Tiltrotor 'Osprey:' The Program That Wouldn't Die," *Vertiflite*, vol. 39, no. 6, (Nov.–Dec. 1993), pp. 20–23.

F. Garren, J.R. Kelly, and R.W. Summer, "VTOL Flight Investigation to Develop a Decelerating Instrument Approach Capability," Society of Automotive Engineers Paper No. 690693 (1969).

B.P. Gupta, A.H. Logan, and E.R. Wood, "Higher Harmonic Control for Rotary Wing Aircraft," AIAA Paper 84-2484 (1984).

F.B. Gustafson, "Effects on Helicopter Performance of Modifications in Profile-Drag Characteristics of Rotor-Blade Airfoil Sections," NACA WR-L-26 [Formerly NACA Advanced Confidential Report ACR L4H05] (1944).

F.B. Gustafson, "A History of NACA Research on Rotating-Wing Aircraft," *Journal of the American Helicopter Society*, vol. 1, no. 1 (Jan. 1956), p. 16.

F.B. Gustafson, "History of NACA/NASA Rotating-Wing Aircraft Research, 1915–1970," *Vertiflite*, Reprint VF-70 (Apr. 1971), pp. 1–27.

Harry H. Heyson and S. Katzoff, "Induced Velocities Near a Lifting Rotor with Nonuniform Disk Loading," NACA Report 1319 (1957).

R.J. Huston, "An Exploratory Investigation of Factors Affecting the Handling Qualities of a Rudimentary Hingeless Rotor Helicopter," NASA TN-D-3418 (1966).

Robert J. Huston, Robert A. Golub, and James C. Yu, "Noise Considerations for Tilt Rotor," AIAA Paper 89-2359 (1989).

Karen Jackson, Richard L. Boitnott, Edwin L. Fasanella, Lisa E. Jones, and Karen H. Lyle, "A Summary of DOD-Sponsored Research Performed at NASA Langley's Impact Dynamics Research Facility," *Journal of the American Helicopter Society*, vol. 51, no. 1 (June 2004).

Wayne Johnson, "Model for Vortex Ring State Influence on Rotorcraft Flight Dynamics," NASA TP-2005-213477 (2005).

J.R. Kelly, F.R. Niessen, J.J. Thibodeaux, K.R. Yenni, and J.F. Garren, Jr., "Flight Investigation of Manual and Automatic VTOL Decelerating Instrument Approaches and Landings," NASA TN-D-7524 (1974).

Robert M. Kufeld and Paul C. Loschke, "UH-60 Airloads Program—Status and Plans," AIAA Paper 91-3142 (1991).

R.G. Kvaternik and W.G. Walton, Jr., "A Formulation of Rotor-Airframe Coupling for the Design Analysis of Vibrations of Helicopter Airframes," NASA RP-1089 (1982).

R.G. Kvaternik, "The NASA/Industry Design Analysis Methods for Vibration (DAMVIBS) Program—A Government Overview," AIAA Paper 92-2200 (1992).

A.W. Linden and M.W. Hellyer, "The Rotor Systems Research Aircraft," AIAA Paper No. 74-1277 (1974).

Mark Liptak, "International Helicopter Study Team (IHST) Overview Briefing," Helicopter Association International HELI EXPO Meeting, Houston, TX, Feb. 21–23, 2009.

W.R. Mantay, W.T. Yeager, Jr., M.N. Hamouda, R.G. Cramer, Jr., and C.W. Langston, "Aeroelastic model Helicopter Testing in the Langley TDT," NASA TM-86440 / USAAVSCOM TM-85-8-5 (1985).

Ruth M. Martin, "NASA/AHS Rotorcraft Noise Reduction Program: NASA Langley Acoustics Division Contributions," *Vertiflite*, vol. 35, no. 4 (May–June 1989), pp. 48–52.

J.G. McArdle, "Outdoor Test Stand Performance of a Convertible Engine with Variable Inlet Guide Vanes for Advanced Rotorcraft Propulsion," NASA TM-88939 (1986).

M. Mosher and R.L. Peterson, "Acoustic Measurements of a Full-Scale Coaxial Helicopter," AIAA Paper 83-0722 (1983).

Jack N. Nielsen and James C. Biggers, "Recent Progress in Circulation Control Aerodynamics," AIAA Paper 87-0001 (1987).

Wilmer H. Reed, III, "Review of Propeller-Rotor Whirl Flutter," NASA TR-R-264 (1967).

S. Salmirs and R.J. Tapscott, "The Effects of Various Combinations of Damping and Control Power on Helicopter Handling Qualities During Both Instrument and Visual Flight," NASA TN-D-58 (1959).

James Sheiman, "A Tabulation of Helicopter Rotor-Blade Differential Pressures, Stresses, and Motions As Measured In Flight," NASA TM-X-952 (1964).

J. Sheiman and L.H. Ludi, "Qualitative Evaluation of Effect of Helicopter Rotor Blade Tip Vortex on Blade Airloads," NASA TN-D-1637 (1963).

William J. Snyder, John Zuk, and Hans Mark, "Tilt Rotor Technology Takes Off," AIAA Paper 89-2359 (1989).

John F. Ward, "Helicopter Rotor Periodic Differential Pressures and Structural Response Measured in Transient and Steady-State Maneuvers," *Journal of the American Helicopter Society*, vol. 16, no. 1 (Jan. 1971).

John F. Ward, "An Updated History of NACA/NASA Rotary-Wing Aircraft Research 1915-1984," *Vertiflite*, vol. 30, no. 4 (May–June 1984), pp. 108–117.

W. Warmbrodt and J.L. McCloud, II, "A Full-Scale Wind Tunnel Investigation of a Helicopter Bearingless Main Rotor," NASA TM-81321 (1981).

William Warmbrodt, Charles Smith, and Wayne Johnson, "Rotorcraft Research Testing in the National Full-Scale Aerodynamics Complex at NASA Ames Research Center," NASA TM-86687 (1985).

J.B. Wheatley, "Lift and Drag Characteristics and Gliding Performance of an Autogiro as Determined In Flight," NACA Report 434 (1932).

J.B. Wheatley, "A Aerodynamic Analysis of the Autogiro Rotor With Comparison Between Calculated and Experimental Results," NACA Report 487 (1934).

Edgar C. Wood, "The Army Helicopter, Past, Present and Future," *Journal of the American Helicopter Society*, vol. 1, no. 1 (Jan. 1956), pp 87–92.

William T. Yeager, Jr., and Raymond G. Kvaternik, "A Historical Overview of Aeroelasticity Branch and Transonic Dynamics Tunnel Contributions to Rotorcraft Technology and Development," NASA TM-2001-211054, ARL-TR-2564, (2001).

Books and Monographs:

Walter J. Boyne and Donald S. Lopez, eds., *Vertical Flight: The Age of the Helicopter* (Washington: Smithsonian Institution Press, 1984).

Glenn E. Bugos, *Atmosphere of Freedom, Sixty Years at the Ames Research Center*, NASA SP-4314 (Washington, DC: NASA, 2000).

Alfred Gessow and Garry C. Myers, Jr., *Aerodynamics of the Helicopter* (New York: The Macmillan Company, 1952; reissued by Frederick Ungar Publishing Co., 1967).

Michael H. Gorn, *Expanding the Envelope: Flight Research at NACA and NASA* (Lexington: The University Press of Kentucky, 2001).

Richard P. Hallion and Michael H. Gorn, *On the Frontier: Experimental Flight at NASA Dryden* (Washington, DC: Smithsonian Books, 2003).

James R. Hansen, *Engineer in Charge: A History of the Langley Aeronautical Laboratory, 1917–1958*, NASA SP-4305 (Washington, DC: NASA, 1987).

James R. Hansen, *Spaceflight Revolution: Langley Research Center From Sputnik to Apollo*, NASA SP-4308 (Washington, DC: NASA, 1995).

Edwin P. Hartman, *Adventures in Research, A History of Ames Research Center, 1940–1965*, NASA SP-4302 (Washington, DC: NASA, 1970).

Michael J. Hirschberg, *The American Helicopter: An Overview of Helicopter Developments in America, 1908–1999* (Arlington, VA: ANSER, 1999).

Wayne Johnson, *Helicopter Theory* (Princeton: Princeton University Press, 1980).

Martin D. Maisel, Demo J. Giulianetti, and Daniel C. Dugan, *The History of The XV-15 Tilt Rotor Research Aircraft From Concept to Flight*, NASA SP-2000-4517 (Washington, DC: NASA, 2000).

J.A. Stockfisch, *The 1962 Howze Board and Army Combat Developments*, Monograph Report MR-435-A (Santa Monica: The RAND Corporation, 1994).

Lt. Gen. John J. Tolson, *Airmobility, 1961–1971*, a volume in the U.S. Army *Vietnam Studies* series (Washington, DC: U.S. Army, 1973).

THIS PAGE INTENTIONALLY BLANK

Aerodynamic model of NASA's SCAT-15F supersonic transport design attached for a subsonic wind tunnel test in 1969. NASA.

CASE 4

Softening the Sonic Boom: 50 Years of NASA Research

Lawrence R. Benson

The advent of practical supersonic flight brought with it the shattering shock of the sonic boom. From the onset of the supersonic age in 1947, NACA–NASA researchers recognized that the sonic boom would work against acceptance of routine overland supersonic aircraft operation. In concert with researchers from other Federal and military organizations, they developed flight-test programs and innovative design approaches to reshape aircraft to minimize boom effects while retaining desirable high-speed behavior and efficient flight performance.

AFTER ITS FORMATION IN 1958, the National Aeronautics and Space Administration (NASA) began devoting most of its resources to the Nation's new civilian space programs. Yet 1958 also marked the start of a program in the time-honored aviation mission that the Agency inherited from the National Advisory Committee for Aeronautics (NACA). This task was to help foster an advanced passenger plane that would fly at least twice the speed of sound.

Because of economic and political factors, developing such an aircraft became more than a purely technological challenge. One of the major barriers to producing a supersonic transport involved a phenomenon of atmospheric physics barely understood in the late 1950s: the shock waves generated by supersonic flight. Studying these "sonic booms" and learning how to control them became a specialized and enduring field of NASA research for the next five decades. During the first decade of the 21st century, all the study, testing, and experimentation of the past finally began to reap tangible benefits in the same California airspace where supersonic flight began.[1]

---

1. The author is grateful to Karl Bender of NASA's Dryden Research Library for helping to gather source materials. For a concise introduction to sonic boom theory, see Kenneth J. Plotkin and Domenic J. Maglieri, "Sonic Boom Research: History and Future," American Institute of Aeronautics and Astronautics (AIAA), Paper 2003-3575, June 23, 2003.

From Curiosity to Controversy

In 1947, Muroc Army Airfield, CA, was a small collection of aircraft hangars and other austere buildings adjoining the vast Rogers Dry Lake in the high desert of the Antelope Valley, across the San Gabriel Mountains from the Los Angeles basin. Because of the airfield's remoteness and clear skies, a small team of Air Force, the NACA, and contractor personnel was using Muroc for a secret project to explore the still unknown territory of supersonic flight. On October 14, more than 40,000 feet over the little desert town of Boron, visible only by its contrail, Capt. Chuck Yeager's 31-foot-long rocket-propelled Bell XS-1 successfully "broke" the fabled sound barrier.[2] The sonic boom from his little experimental airplane—the first to fly supersonic in level flight—probably did not reach the ground on that historic day.[3] Before long, however, the acoustical signature of the shock waves generated by XS-1s and other supersonic aircraft became a familiar sound at and around the isolated airbase.

In the previous century, an Austrian physicist-philosopher, Ernst Mach, was the first to explain the phenomenon of supersonic shock waves, which he displayed visually in 1887 with a cleverly made photograph showing those formed by a high-velocity projectile, in this case a bullet. The speed of sound, he also determined, varied in relation to the density of the medium though which it passed, such as air molecules. (At sea level, the speed of sound is 760 mph.) In 1929, Jakob Ackeret, a Swiss fluid dynamicist, named this variable "Mach number" in his honor. This guaranteed that Ernst would be remembered by future generations, especially after it became known that the 700 mph speed of Yeager's XS-1, flying at 43,000 feet, was measured as Mach 1.06.[4]

Humans have long been familiar with and often frightened by natural sonic booms in the form of thunder, i.e., sudden surges of air

---

2. For its development and testing, see Richard P. Hallion, *Supersonic Flight: Breaking the Sound Barrier and Beyond: The Story of the Bell X-1 and Douglas D-558* (New York: Macmillan, 1977).
3. Some of the personnel stationed at Muroc when Yeager broke the sound barrier later recalled hearing a sonic boom, but these may have been memories of subsequent flights at higher speeds. One of NASA's top sonic boom experts has calculated that at Mach 1.06 and 41,000 feet above ground level, atmospheric refraction and absorption of the shock waves would almost certainly have dissipated the XS-1's sonic boom before it could reach the surface. E-mail, Edward A. Haering, Dryden Flight Research Center, to Lawrence R. Benson, Apr. 8, 2009.
4. "Ernst Mach," Stanford Encyclopedia of Philosophy, Mar. 21, 2008, http://plato.stanford.edu/entries/Ernst-mach; Jeff Scott, "Ernst Mach and Mach Number," Nov. 9, 2003, http://www.aerospaceweb.org/question/history/q0149.shmtl.

Case 4 Softening the Sonic Boom: 50 Years of NASA Research

Bell XS-1—the first aircraft to exceed Mach 1 in level flight, October 14, 1947. U.S. Air Force.

pressure caused when strokes of lightning instantaneously heat contiguous columns of air molecules. Perhaps the most awesome of sonic booms—heard only rarely—have been produced by large meteoroid fireballs speeding through the atmosphere. On a much smaller scale, the first acoustical shock waves produced by human invention were the modest cracking noises from the snapping of a whip. The high-power explosives perfected in the latter half of the 19th century were able—as Mach explained—to propel projectiles faster than the speed of sound. Their acoustical shock waves would be among the cacophony of fearsome sounds heard by millions of soldiers during the two World Wars.[5]

On a Friday evening, September 8, 1944, an explosion blew out a large crater in Stavely Road, west of London. The first German V-2 ballistic missile aimed at England had announced its arrival. "After the explosion came a double thunderclap caused by the sonic boom catch-

---

5. By the end of World War II, ballistic waves were well understood, e.g., J.W.M. Dumond, et al., "A Determination of the Wave Forms and Laws of Propagation and Dissipation of Ballistic Shock Waves," *Journal of the Acoustical Society of America* (hereinafter cited as *JASA*), vol. 18, no. 1 (Jan. 1946), pp. 97–118.

ing up with the fallen rocket."⁶ For the next 7 months, millions of people would hear these sounds, which would become known as "sonic bangs" in Britain, from more than 3,000 V-2s launched at England as well as liberated portions of France, Belgium, and the Netherlands. Their sound waves would always arrive too late to warn any of those unfortunate enough to be near the missiles' points of impact.⁷ After World War II, these strange noises faded into memory for several years—until the arrival of new jet fighter planes.

In November 1949, the NACA designated its growing detachment at Muroc as the High-Speed Flight Research Station (HSFRS), 1 month before the Air Force renamed the installation Edwards Air Force Base (AFB).⁸ By the early 1950s, the desert and mountains around Edwards reverberated with the occasional sonic booms of experimental and prototype aircraft, as did other flight-test locations in the United States and United Kingdom. Scientists and engineers had been familiar with the "axisymmetric" ballistic shock waves of projectiles such as artillery shells (referred to scientifically as bodies of revolution).⁹ This was one reason the fuselage of the XS-1 was shaped like a 50-caliber bullet. But these new acoustic phenomena—many of which featured a double-boom sound—hinted that they were more complex. In late 1952, the editors of the world's oldest aeronautical weekly stated with some hyperbole that "the 'supersonic bang' phenomenon, if only by reason of its sudden incidence and the enormous public interest it has aroused, is probably the most spectacular and puzzling occurrence in the history of aerodynamics."¹⁰

---

6. David Darling: *The Complete Book of Spaceflight: From Apollo 1 to Zero Gravity* (Hoboken, NJ: John Wiley and Sons, 2003), p. 457. See also "Airpower: Missiles and Rockets in Warfare," http://www.centennialofflight.gov/essay/Air_Power/Missiles/AP29.htm; and Bob Ward, *Dr. Space: The Life of Wernher von Braun* (Annapolis: Naval Institute, 2005), p. 43.

7. The definitive biography, *Van Braun: Dreamer of Space, Engineer of War*, by Michael J. Neufeld (New York: Alfred A. Knopf, 2007), pp. 133–136, leaves open the question of whether the Germans at Peenemünde heard the first manmade sonic booms in 1942 when their A-4 test rockets exceeded Mach 1 about 25 seconds after launch.

8. For the authoritative history of the NACA/NASA mission at Edwards AFB, see Richard P. Hallion and Michael H. Gorn, *On the Frontier: Experimental Flight at NASA Dryden* (Washington, DC: Smithsonian, 2003).

9. Plotkin and Maglieri, "Sonic Boom Research," pp. 1–2.

10. Introduction to "The Battle of the Bangs," *Flight and Aircraft Engineer*, vol. 61, no. 2289 (Dec. 5, 1952), p. 696, http://www.flightglobal.com/pdfarchive/view/1952/%203457.

A young British graduate student, Gerald B. Whitham, was the first to analyze thoroughly the abrupt rise in air pressure upon arrival of a supersonic vehicle's "bow wave," followed by a more gradual but deeper fall in pressure for a fraction of a second, and then a recompression with the passing of the vehicle's tail wave. As shown in a simplified fashion by Figure 1, this can be illustrated graphically by an elongated capital "N" (the solid line) transecting a horizontal axis (the dashed line) representing ambient air pressure during a second or less of elapsed time. For Americans, the pressure change is usually expressed in pounds per square foot (psf—also abbreviated as lb/ft^2).

Because a jet fighter (or a V-2 missile) is much longer than an artillery shell is, the human ear could detect a double boom if its tail shock wave arrived a tenth of a second or more after its bow shock wave. Whitham was first to systematically examine the more complex shock waves, which he called the F-function, generated by "nonaxisymmetrical" (i.e., asymmetrical) configurations, such as airplanes.[11]

The number of these double booms multiplied in the mid-1950s as the Air Force Flight Test Center (AFFTC) at Edwards (assisted by the HSFRS) began putting a new generation of Air Force jet fighters and interceptors, known as the Century Series, through their paces. The remarkably rapid advance in aviation technology (and priorities of the Cold War "arms race") is evident in the sequence of their first flights at Edwards: YF-100 Super Sabre, May 1953; YF-102 Delta Dagger, October 1953; XF-104 Starfighter, February 1954; F-101 Voodoo, September 1954; YF-105 Thunderchief, October 1955; and F-106 Delta Dart, December 1956.[12]

With the sparse population living in California's Mojave Desert region during the 1950s, disturbances caused by the flight tests of new jet aircraft were not a serious issue. But even in the early 1950s, the United

---

11. G.B. Whitham, "The Flow Pattern of a Supersonic Projectile," *Communications on Pure and Applied Mathematics*, vol. 5, no. 3 (1952), pp. 301–348 (available at http://www3.interscience.wiley.com/journal/113395160/issue) and "On the Propagation of Weak Shock Waves," *Journal of Fluid Dynamics*, vol. 1, No. 3 (Sept. 1956), pp. 290–318 (available at http://journals.cambridge.org/action/displayJournal?jid=JFM), and described in Larry J. Runyan, et al., *Sonic Boom Literature Survey, vol. II, Capsule Summaries*, (Seattle: Boeing Commercial Airplane Co. for the FAA), Sept. 1973, pp. 6–8, 59–60. Whitham later taught at both the Massachusetts and California Institutes of Technology.

12. Air Force Flight Test Center History Office, *Ad Inexplorata: The Evolution of Flight Testing at Edwards Air Force Base* (Edwards AFB: AFFTC, 1996), Appendix B, p. 55.

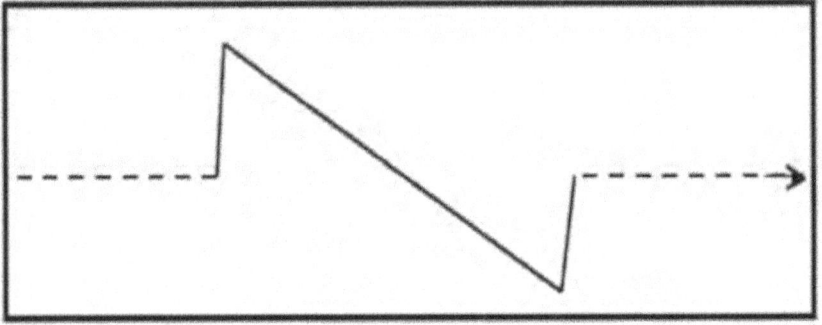

Figure 1. Simplified N-shaped sonic boom signature. NASA.

States Air Force (USAF) became concerned about their future impact. In November 1954, for example, its Aeronautical Research Laboratory at Wright-Patterson AFB, OH, submitted a study to the Air Force Board of top generals on early findings regarding the still somewhat mysterious nature of sonic booms. Although concluding that low-flying aircraft flying at supersonic speeds could cause considerable damage, the report optimistically predicted the possibility of supersonic flight without booms at altitudes over 35,000 feet.[13]

As the latest Air Force and Navy fighters went into full production and began flying from bases throughout the Nation, much of the American public was exposed to jet noise for the first time. This included the thunderclap-like thuds characteristic of sonic booms—often accompanied by rattling windowpanes. Under certain conditions, as the U.S. armed services and British Royal Air Force (RAF) had learned, even maneuvers below Mach 1 (e.g., accelerations, dives, and turns) could generate and focus transonic shock waves in such a manner as to cause strong sonic booms.[14] Indeed, residents of Southern California began hearing such booms in the late 1940s, when North American Aviation was flight-testing its new F-86 Sabre. The first civilian claim against the

---

13. John G. Norris, "AF Says 'Sonic Boom' Can Peril Civilians," *Washington Post and Times Herald* (hereinafter cited as *Washington Post*), Nov. 9, 1954, pp. 1, 12.

14. One of the first studies on focused booms was G.M. Lilley, et al., "Some Aspects of Noise from Supersonic Aircraft," *Journal of the Royal Aeronautical Society*, vol. 57 (June 1953), pp. 396–414, as described in Runyan, *Sonic Boom Capsule Summaries*, p. 54. AFFTC used F-100s to conduct the first in-flight boom measurements: Marshall E. Mullens, "A Flight Test Investigation of the Sonic Boom," AFFTC TN-56-20, May 1956.

USAF for sonic boom damage was apparently filed at Eglin AFB, FL, in 1951, when only subsonic jet fighters were assigned there.[15] Additionally, as shown in 1958 by Frank Walkden, another English mathematician, the lift effect of airplane wings could magnify the strength of sonic booms more than previously estimated.[16]

Sonic boom claims against the Air Force first became statistically significant in 1957, reflecting its growing inventory of Century fighters and the type of maneuvers they sometimes performed, which could focus acoustical rays into what became called "super booms." (It was found that these powerful but localized booms had a U-shaped signature, with the tail shock wave as well as that from the nose of the airplane being above ambient air pressure.) Most claims involved broken windows or cracked plaster, but some were truly bizarre, such as the death of pets or the insanity of livestock. In addition to these formal claims, Air Force bases, local police switchboards, and other agencies received an uncounted number of phone calls about booms, ranging from merely inquisitive to seriously irate.[17] Complaints from constituents also became an issue for the U.S. Congress.[18] Between 1956 and 1968, some 38,831 claims were submitted to the Air Force, which approved 14,006 in whole or in part—65 percent for broken glass, 21 percent for cracked plaster (usually already weakened), 8 percent for fallen objects, and 6 percent for other reasons.[19]

The military's problem with sonic boom complaints seems to have peaked in the 1960s. One reason was the sheer number of fighter-type aircraft stationed around the Nation (over three times as many as today). Secondly, many of these aircraft's missions were air defense. This often meant flying at high speed over populated areas for training in

---

15. History of the 3201 Air Base Group, Eglin AFB, Jul.–Sept. 1951, Abstract from Information Retrieval and Indexing System (IRIS) No. 438908, Air Force Historical Research Center, Maxwell AFB, AL.
16. F. Walkden, "The Shock Pattern of a Wing-Body Combination Far from the Flight Path," *Aeronautical Quarterly*, vol. 9, pt. 2 (May 1958), pp. 164–194; described in Runyan, *Sonic Boom Capsule Summaries*, 8–9. Both Walkden and Whitman did their pioneering studies at the University of Manchester.
17. Fred Keefe and Grover Amen, "Boom," *The New Yorker*, May 16, 1962, pp. 33–34.
18. Albion B. Hailey, "AF Expert Dodges Efforts to Detail 'Sonic Boom' Loss," *Washington Post*, Aug. 25, 1960, p. A15.
19. J.P. and E.G.R Taylor, "A Brief Legal History of the Sonic Boom in America," *Aircraft Engine Noise and Sonic Boom* (Neuilly Sur Seine, France: NATO Advisory Group for Aerospace Research and Development [AGARD], 1969), Conference Proceedings (CP) No. 42, Paris, May 1969, pp. 2-1–2-11.

defending cities and other key targets from aerial attack, sometimes in practice against Strategic Air Command (SAC) bombers. The North American Air Defense Command (NORAD) conducted two of the largest such exercises, Skyshield I and Skyshield II, in 1960 and 1961. The Federal Aviation Agency (FAA) shut down all civilian air traffic while NORAD's interceptors and SAC bombers (augmented by some from the RAF) battled overhead—accompanied by a sporadic drumbeat of sonic booms reaching the surface.[20]

Although most fighters and interceptors deployed in the 1960s could readily fly faster than sound, they could only do so for a short distance because of the rapid fuel consumption of jet engine afterburners. Thus, their sonic boom "carpets" were relatively short. However, one supersonic American warplane that became operational in 1960 was designed to fly faster than Mach 2 for more than 1,000 miles.

This innovative but troublesome aircraft was the SAC's new Convair-built B-58 Hustler medium bomber. On March 5, 1962, the Air Force showed off the long-range speed of the B-58 by flying one from Los Angles to New York in just over 2 hours at an average pace of 1,215 mph (despite having to slow down for an aerial refueling over Kansas). After another refueling over the Atlantic, the same Hustler "outraced the sun" (i.e., flew faster than Earth's rotation) back to Los Angles with one more refueling, completing the record-breaking round trip at an average speed of 1,044 mph.[21]

Capable of sustained Mach 2+ speeds, the four-engine delta-winged Hustler (weighing up to 163,000 pounds) helped demonstrate the feasibility of a supersonic transport. But the B-58's performance revealed at least one troubling omen. Almost wherever it flew supersonic over populated areas, the bomber left sonic boom complaints and claims in its wake. Indeed, on its record-shattering flight of March 1962, flown mostly at an altitude of 50,000 feet (except when coming down to 30,000 feet for refueling), "the jet dragged a sonic boom 20 to 40 miles wide back and forth across the country—frightening residents, breaking windows, crack-

---

20. "Warplanes Fill Skies Over U.S. and Canada," *Los Angeles Times*, Sept. 10, 1960, p. 4; Albion B. Halley and Warren Kornberg, "U.S. Tests Air Defenses in 3000-Plane 'Battle,'" *Washington Post*, Oct. 15, 1961, pp. A1, B1; Richard Witkin, "Civilian Planes Halted 12 Hours in Defense Test," *New York Times*, Oct. 15, 1961, pp. 1, 46.

21. Marcelle S. Knaack, *Post-World War II Bombers, 1945–1973* (Washington, DC: Government Printing Office (hereinafter cited as GPO) for Office of Air Force History, 1988), pp. 394–395 (vol. 2, *Encyclopedia of U.S. Air Force Aircraft and Missile Systems*).

Convair B-58 Hustler, the first airplane capable of sustained supersonic flight and a major contributor to early sonic boom research. USAF.

ing plaster, and setting dogs to barking."[22] As indicated by Figure 2, the B-58 became a symbol for sonic boom complaints (despite its small numbers).

Most Americans, especially during times of increased Cold War tensions, tolerated occasional disruptions justified by national defense. But how would they react to constantly repeated sonic booms generated by civilian jet airliners? Could a practical passenger-carrying supersonic airplane be designed to minimize its sonic signature enough to be acceptable to people below? NASA's attempts to resolve these two questions occupy the remainder of this history.

A Painful Lesson: Sonic Booms and the Supersonic Transport

By the late 1950s, the rapid pace of aeronautical progress—with new turbojet-powered airliners flying twice as fast and high as the propeller-driven transports they were replacing—promised even higher speeds in coming years. At the same time, the perceived challenge to America's technological superiority implied by the Soviet Union's early space triumphs inspired a willingness to pursue ambitious new aerospace ventures. One of these was the Supersonic Commercial Air Transport (SCAT). This program was further motivated by competition from

---

22. "Jet Breaks 3 Records—and Many Windows," *Los Angeles Times*, Mar. 6, 1962, p. 1. In reality, most of the damage was done while accelerating after the refuelings.

Figure 2. Cover of an Air Force pamphlet for sonic boom claim investigators. USAF.

Britain and France to build an airliner that was expected to dominate the future of mid- and long-range commercial aviation.[23]

---

23. For the definitive account of political and economic aspects of the SST and subsequent programs (as well as many technical details), see Erik M. Conway, *High-Speed Dreams: NASA and the Technopolitics of Supersonic Transportation, 1945–1999* (Baltimore: Johns Hopkins, 2005), pp. 27–45 cited here. For an earlier study by an insider, see F. Edward McLean, "Supersonic Cruise Technology," NASA Special Publication (SP) 472 (Washington, DC: GPO, 1985). For an account focused on its political aspects, see Mel Howitch, *Clipped Wings: The American SST Conflict* (Cambridge: MIT, 1982).

From SCAT Research to SST Development

The recently established FAA became the major advocate within the U.S. Government for a supersonic transport, with key personnel at three of the NACA's former laboratories eager to help with this challenging new program. The Langley Research Center in Hampton, VA, (the NACA's oldest and largest lab) and the Ames Research Center at Moffett Field in Sunnyvale, CA, both had airframe design expertise and facilities, while the Lewis Research Center in Cleveland, OH, specialized in the kind of advanced propulsion technologies needed for supersonic cruise.

The strategy for developing the SCAT depended heavily on leveraging technologies being developed for another Air Force bomber—one much larger, faster, and more advanced than the B-58. This would be the revolutionary B-70, designed to cruise several thousand miles at speeds of Mach 3. NACA experts had been helping the Air Force plan this giant intercontinental bomber since the mid-1950s (with aerodynamicist Alfred Eggers of the Ames Laboratory conceiving the innovative design for it to ride partially on compression lift created by its own supersonic shock waves). North American Aviation won the B-70 contract in 1958, but the projected expense of the program and advances in missile technology led President Dwight Eisenhower to cancel all but one prototype in 1959. The administration of President John Kennedy eventually approved production of two XB-70As. Their main purpose would be to serve as Mach 3 testbeds for what had become known simply as the Supersonic Transport (SST). NASA continued to refer to design concepts for the SST using the older acronym for Supersonic Commercial Air Transport. By 1962, these concepts had been narrowed down to three Langley designs (SCAT-4, SCAT-15, and SCAT-16) and one from Ames (SCAT-17). These became the baselines for industry studies and SST proposals.[24]

Even though Department of Defense resources (especially the Air Force's) would be important in supporting the SST program, the aerospace industry made it clear that direct federal funding and assistance would be essential. Thus research and development (R&D) of the SST became a split responsibility between the Federal Aviation Agency and the National Aeronautics and Space Administration—with NASA conducting and sponsoring the supersonic research and the FAA in charge

---

24. McLean, *Supersonic Cruise Technology*, pp 35–46; Joseph R. Chambers, *Innovation in Flight*; "Research of the NASA Langley Research Center on Revolutionary Concepts for Aeronautics," NASA SP-2005-4539, pp. 25–28.

of the SST's overall development. The first two leaders of the FAA, retired Lt. Gen. Elwood R. "Pete" Quesada (1958–1961) and Najeeb E. Halaby (1961–1965), were both staunch proponents of producing an SST, as to a slightly lesser degree was retired Gen. William F. "Bozo" McKee (1965–1968). As heads of an independent agency that reported directly to the president, they were at the same level as NASA Administrators T. Keith Glennan (1958–1961) and James Beggs (1961–1968). The FAA and NASA administrators, together with Secretary of Defense Robert McNamara (somewhat of a skeptic on the SST program), provided interagency oversight and comprised the Presidential Advisory Committee (PAC) for the SST established in April 1964. This arrangement lasted until 1967, when the Federal Aviation Agency became the Federal Aviation Administration under the new Department of Transportation, whose secretary became responsible for the program.[25]

Much of NASA's SST-related research involved advancing the state-of-the-art in such technologies as propulsion, fuels, materials, and aerodynamics. The latter included designing airframe configurations for sustained supersonic cruise at high altitudes, suitable subsonic maneuvering in civilian air traffic patterns at lower altitudes, safe takeoffs and landings at commercial airports, and acceptable noise levels—to include the still-puzzling matter of sonic booms.

Dealing with the sonic boom entailed a multifaceted approach: (1) performing flight tests to better quantify the fluid dynamics and atmospheric physics involved in generating and propagating shock waves, as well as their effects on structures and people; (2) conducting community surveys to gather public opinion data on sample populations exposed to booms; (3) building and using acoustic simulators to further evaluate human and structural responses in controlled settings; (4) performing field studies of possible effects on animals; (5) evaluating various aerodynamic configurations in wind tunnel experiments; and (6) analyzing flight test and wind tunnel data to refine theoretical constructs and mathematical models for lower-boom aircraft designs. Within NASA, the Langley Research Center was a focal point for sonic

---

25. *FAA Historical Chronology, 1926–1996*, http://ww.faa.gov/about/media/b-chron.pdf. For Quesada's role, see Stuart I. Rochester, *Takeoff at Mid-Century: Federal Civil Aviation Policy in the Eisenhower Years, 1953–1961* (Washington, DC: GPO for FAA, 1976). For the activism of Halaby and the demise of the SST after his departure, see Richard J. Kent, Jr., *Safe, Separated, and Soaring: A History of Civil Aviation Policy, 1961–1972* (Washington, DC: GPO for FAA, 1980).

boom studies, with the Flight Research Center (FRC) at Edwards AFB conducting many of the supersonic tests.[26]

Although the NACA, especially at Langley and Ames, had been doing research on supersonic flight since World War II, none of its technical reports (and only one conference paper) published through 1957 dealt directly with sonic booms.[27] That situation began to change when Langley's long-time manager and advocate of supersonic programs, John P. Stack, formalized the SCAT venture in 1958. During the next year, three Langley employees whose names would become well known in the field of sonic boom research began publishing NASA's first scientific papers on the subject. These were Harry W. Carlson, a versatile supersonic aerodynamicist, Harvey H. Hubbard, chief of the Acoustics and Noise Control Division, and Domenic J. Maglieri, a young engineer who became Hubbard's top sonic boom specialist. Carlson would tend to focus on wind tunnel experiments and sonic boom theory, while the other two men specialized in planning and monitoring field tests, then analyzing the data collected.[28] These research activities began to expand under the new pro-SST Kennedy Administration in 1961. After the president formally approved development of the supersonic transport in June 1963, sonic boom research took off. Langley's experts, augmented by NASA contractors and grantees, published 26 papers on sonic booms just 3 years later.[29]

Supersonic Flight Tests and Surveys

The systematic sonic boom testing that NASA began in 1958 would exponentially expand the heretofore largely theoretical and anecdotal

---

26. NASA's HSFRC became the FRC in 1959. For an overall summary of Langley's supersonic activities, see Chambers, *Innovations in Flight*, ch. 1, "Supersonic Civil Aircraft: The Need for Speed," pp. 7–70.
27. Based on author's review of Section 7.4, "Noise, Aircraft" in volumes of the *Index of NACA Technical Publications* (Washington DC: NACA Division of Research Information) covering the years 1915–1957.
28. Telephone interview, Domenic Maglieri by Lawrence Benson, Feb. 6, 2009.
29. A.B. Fryer, et al., "Publications in Acoustics and Noise Control from the NASA Langley Research Center during 1940–1976," NASA TM-X-74042, July 1977. The following abbreviations are used for NASA publications cited in the notes: Conference Publication (CP), Contractor Report (CR), Reference Publication (RP), Special Publication (SP), Technical Memorandum (TM), formerly classified Tech Memo (TM-X), Technical Note (TN), Technical Paper (TP), and Technical Report (TR). Bibliographic data and often full text copies can be accessed through the NASA Technical Reports Server (NTRS), http://ntrs.nasa.gov/search.jsp.

knowledge about sonic booms with a vast amount of "real world" data. The new information would make possible increasingly sophisticated experiments and provide feedback for checking and refining theories and mathematical models. Because of the priority bestowed on sonic boom research by the SST program and the numerous types of aircraft then available for creating booms (including some faster than anything flying today), the data and findings from the tests conducted in the 1960s are still of significant value in the 21st century.[30]

The Langley Research Center (often referred to as NASA Langley) served as the Agency's "team leader" for supersonic research. Langley's acoustics specialists conducted NASA's initial sonic boom tests in 1958 and 1959 at the Wallops Island Station on Virginia's isolated Delmarva Peninsula. During the first year, they used six sorties by NASA F-100 and F-101 fighters, flying at speeds between Mach 1.1 and 1.4 and altitudes from 25,000 to 45,000 feet, to make the first good ground recordings and measurements of sonic booms for steady, level flights (the kind of profile a future airliner would fly). Observers judged some of the booms above 1.0 psf to be objectionable, likening them to nearby thunder, and a sample plate glass window was cracked by one plane flying at 25,000 feet. The 1959 test measured shock waves from 26 flights of a Chance Vought F8U-3 (a highly advanced prototype based on the Navy's Crusader fighter) at speeds up to Mach 2 and altitudes up to 60,000 feet. A B-58 from Edwards AFB also made two supersonic passes at 41,000 feet. Boom intensities from these higher altitudes seemed to be tolerable to observers, with negligible increases in measured overpressures between Mach 1.4 and 2.0. These results were, however, very preliminary.[31]

In July 1960, NASA and the Air Force conducted Project Little Boom at a bombing range north of Nellis AFB, NV, to measure the effects on structures and people of extremely powerful sonic booms. F-104 and F-105 fighters flew slightly over the speed of sound (Mach 1.09 to 1.2) at altitudes

---

30. For a chronological summary of selected projects during first decade, see Johnny M. Sands, "Sonic Boom Research (1958–1968)," FAA, Nov. 1968, Defense Technical Information Center (DTIC) document AD 684806.

31. Domenic J. Maglieri, Harvey H. Hubbard, and Donald L. Lansing, "Ground Measurements of the Shock-Wave Noise from Airplanes in Level Flight at Mach Numbers to 1.4 and Altitudes to 45,000 Feet," NASA TN-D-48, Sept. 1959; Lindsay J. Lina and Domenic J. Maglieri, "Ground Measurements of Airplane Shock-Wave Noise at Mach Numbers to 2.0 and at Altitudes to 60,000 Feet," NASA TN-D-235, Mar. 1960.

as low as 50 feet above ground level. There were more than 50 incidents of sample windows being broken at 20 to 100 psf, but only a few possible breakages below 20 psf, and no physical or psychological harm to volunteers exposed to overpressures as high as 120 psf.[32] At Indian Springs, Air Force fighters flew supersonically over an instrumented C-47 transport from Edwards, both in the process of landing and on the ground. Despite 120 psf overpressures, there was only very minor damage when on the ground and no problems in flight.[33] Air Force fighters once again would test powerful sonic booms in 1965 in support of Joint Task Force 2 at Tonopah, NV. The strongest sonic boom ever recorded, 144 psf, was generated by an Air Force F-4E Phantom II flying Mach 1.26 at 95 feet.[34]

In late 1960 and early 1961, NASA and AFFTC followed up on Little Boom with Project Big Boom. B-58 bombers made 16 passes flying Mach 1.5 at altitudes of 30,000 to 50,000 feet over arrays of sensors, which measured a maximum overpressure of 2.1 psf. Varying the bomber's weight from 82,000 to 120,000 pounds provided the first hard data on how an aircraft's weight and related lift produced higher over-pressures than existing theories based on volume alone would indicate.[35]

Throughout the 1960s, Edwards Air Force Base—with its unequaled combination of Air Force and NASA expertise, facilities, instrumentation, airspace, emergency landing space, and types of aircraft—hosted the largest number of sonic boom tests. NASA researchers from Langley's Acoustics Division spent much of their time there working with the Flight Research Center in a wide variety of flight experiments. The Air Force Flight Test Center usually participated as well.

In an early test in 1961, Gareth Jordan of the FRC led an effort to collect measurements from F-104s and B-58s flying at speeds of Mach 1.2 to 2.0 over sensors located along Edward AFB's supersonic corridor

---

32. Maglieri, Vera Huckel, and Tony L. Parrott, "Ground Measurements of Shock-Wave Pressure for Fighter Airplanes Flying at Very Low Altitudes . . .," NASA TN-D-3443, July 1966 (superseded classified TMX-611, 1961).
33. Gareth H. Jordan, "Flight Measurements of Sonic Booms and Effects of Shock Waves on Aircraft," in *Society of Experimental Test Pilots Quarterly Review*, vol. 5, No. 1 (1961), pp. 117–131, presented at SETP Supersonic Symposium, Sept. 29, 1961.
34. John O. Powers, J.M. Sands, and Maglieri, "Survey of United States Sonic Boom Overflight Experimentation," NASA TM-X-66339, May 1969, p. 5; USAF Fact Sheet, "Sonic Boom," Oct. 2005, http://www.af.mil/factsheets/fsID=184; Telephone interview, Maglieri by Benson, Mar. 19, 2009.
35. Maglieri and Hubbard, "Ground Measurements of the Shock-Wave Noise from Supersonic Bomber Airplanes in the Altitude Range from 30,000 to 50,000 Feet," NASA TN-D-880, July 1961.

and at Air Force Plant 42 in Palmdale, about 20 miles south. Most of the Palmdale measurements were under 1.0 psf, which the vast majority of people surveyed there and in Lancaster (where overpressures tended to be somewhat higher) considered no worse than distant thunder. But there were some exceptions.[36]

Other experiments at Edwards in 1961 conducted by Langley personnel with support from the FRC and AFFTC contributed a variety of new information. With help from a tethered balloon, they made the first good measurements of atmospheric effects, showing that air turbulence in the lower atmosphere (known as the boundary layer) significantly affected wave shape and overpressure. They also gathered the first data on booms from very high altitudes. Using an aggressive flight profile, AFFTC's B-58 crew managed to zoom up to 75,000 feet—25,000 feet higher than the bomber's normal cruising altitude and 15,000 feet over its design limit! The overpressures measured from this high altitude proved stronger than predicted (not a promising result for the planned SST). Much lower down, fighter aircraft performed accelerating and turning maneuvers to generate the kind of acoustical rays that amplified shock waves and produced multiple booms and super booms. The various experiments showed that a combination of atmospheric conditions, altitude, speed, flight path, aircraft configuration, and sensor location determined the shape of the pressure signatures.[37]

Of major significance for future boom minimization efforts, NASA also began making in-flight shock wave measurements. The first of these, at Edwards in 1960, had used an F-100 with a sensor probe to measure supersonic shock waves from the sides of an F-100, F-104, and B-58, as well as from F-100s speeding past with only 100 feet of separation. The data confirmed Whitham's overall theory, with some discrepancies. In early 1963, an F-106 equipped with a sophisticated new sensor probe designed at Langley flew seven sorties both above and below a B-58 at speeds of Mach 1.42 to 1.69 and altitudes of approximately 40,000 to 50,000 feet. The data gathered confirmed Walkden's theory that lift as well as volume increases peak shock wave pressures. As indicated by

---

36. Jordan, "Flight Measurements of Sonic Booms."
37. Ibid.; Maglieri and Donald L. Lansing, "Sonic Booms from Aircraft in Maneuvers," NASA TN-D-2370, July 1964; Hubbard, et al., "Ground Measurements of Sonic-Boom Measurements for the Altitude Range of 10,000 to 75,000 Feet," NASA TR-R-198, July 1964. (Both reports were based on the tests in 1961.)

Figure 3, analysis of the readings also found that the bow and tail shock waves spread farther apart as they flowed from the B-58 and showed how the multiple or "saw tooth" shock waves produced by the rest of an airplane's structure (e.g., fuselage, canopy, wings, engines, nacelles, etc.) merged with the stronger bow and tail waves until—at a distance of between 50 and 90 body lengths—they began to coalesce into the classic N-shaped signature.[38] This marked a major milestone in sonic boom research.

One of the most publicized and extended flight test programs at Edwards had begun in 1959 with the first launch from a B-52 of the fastest aircraft ever flown: the rocket-propelled X-15. Three of these legendary aerospace vehicles expanded the envelope and gathered data on supersonic and hypersonic flight for the next 8 years. Although the X-15 was not specifically dedicated to sonic boom tests, the Flight Research Center did begin placing microphones and tape recorders under the X-15s' flight tracks in the fall of 1961 to gather boom data. FRC researchers much later reported on the measurements of these sonic booms, made at speeds of Mach 3.5 and Mach 4.8.[39]

For the first few years, NASA's sonic boom tests occurred in relative isolation within military airspace in the desert Southwest or over Virginia's rural Eastern Shore. A future SST, however, would have to fly over heavily populated areas. Thus, from July 1961 through January 1962, NASA, the FAA, and the Air Force carried out the Community and Structural Response Program at St. Louis, MO. In Operation Bongo, the Air Force sent B-58 bombers on 76 supersonic training flights over the city at altitudes from 31,000 to 41,000 feet, announcing them as routine SAC radar bomb-scoring missions. F-106 interceptors flew 11 additional flights at 41,000 feet. Langley personnel installed sensors on the ground, which measured overpressures up to 3.1 psf. Investigators from Scott AFB, IL, or for a short time, a NASA-contracted engineering firm, responded to dam-

---

38. Harriet J. Smith, "Experimental and Calculated Flow Fields Produced by Airplanes Flying at Supersonic Speeds," NASA TN-D-621, Nov. 1960; J.F. Bryant, Maglieri, and V.S. Richie, "In-Flight Shock-Wave Measurements Above and Below a Bomber Airplane at Mach Numbers from 1.42 to 1.69," NASA TN-D-1968, Oct. 1963.

39. NASA Flight Research Center, "X-15 Program" [monthly report], Sept. 1961, Dryden archive, File LI-6-10A-13 (Peter Merlin assisted the author in finding this and other archival documents.); Karen S. Green and Terrill W. Putnam, "Measurements of Sonic Booms Generated by an Airplane Flying at Mach 3.5 and 4.8," NASA TM-X-3126, Oct. 1974. (Since hypersonic speeds were not directly relevant for the SST, a formal report was delayed until NASA began planning reentry flights for the Space Shuttle.) For a history of the X-15 program, see Hallion and Gorn, *On the Frontier*, pp. 101–125.

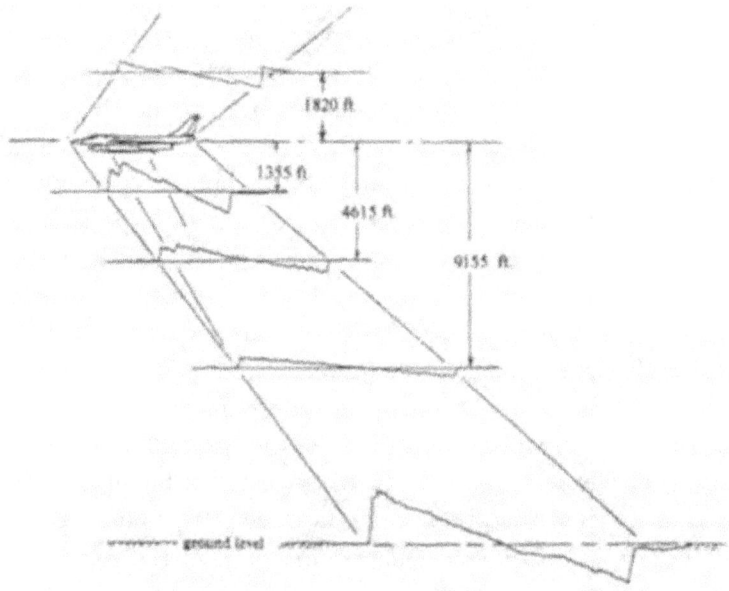

Figure 3. In-flight sonic boom signatures of B-58 at Mach 1.6. USAF.

age claims, finding some possibly legitimate minor damage in about 20 percent of the cases. Repeated interviews with more than 1,000 residents found 90 percent were at least somewhat affected by the booms and about 35 percent were annoyed. Scott AFB (a long-distance phone call from St. Louis) received about 3,000 complaints during the test and another 2,000 in response to 74 sonic booms in the following 3 months. The Air Force eventually approved 825 claims for $58,648. These results served as a warning that repeated sonic booms could pose an issue for SST operations.[40]

To obtain more definitive data on structural damage, NASA in December 1962 resumed tests at Wallops Island using various sample buildings. Air Force F-104s and B-58s and Navy F4H Phantom IIs flew at altitudes from 32,000 to 62,000 feet, creating overpressures up to 3 psf. Results indicated that cracks to plaster, tile, and other brittle materials triggered by sonic booms occurred in spots where the materials were already under stress (a finding that would be repeated in later more comprehensive tests).[41]

---

40. Charles W. Nixon and Hubbard, "Results of the USAF–NASA–FAA Flight Program to Study Community Response to Sonic Booms in the Greater St. Louis Area," NASA TN-D-2705, May 1965; Clark, et al., "Studies of Sonic Boom Damage," NASA CR-227, May 1965.
41. Sands, "Sonic Boom Research (1958–1968)," p. 3.

In February 1963, NASA, the FAA, and the USAF conducted Project Littleman at Edwards AFB to see what happened when two specially instrumented light aircraft were subjected to sonic booms. F-104s made 23 supersonic passes at distances as near as 560 feet from a little Piper Colt and a 2-engine Beech C-45, creating overpressures up to 16 psf. Their responses were "so small as to be insignificant"—dismissing one possible concern about SST operations.[42]

The St. Louis survey had left many unanswered questions on public opinion. To learn more, the FAA's Supersonic Transport Development Office, with support from NASA Langley and the USAF (including Tinker AFB), next conducted the Oklahoma City Public Reaction Study from February through July 1964. This was a much more intensive and systematic test. In an operation named Bongo II, B-58s, F-104s, F-101s, and F-106s were called upon to deliver sonic booms between 1.0 and 2.0 psf, 8 times per day, 7 days a week, for 26 weeks, with another 13 weeks of followup activities. The aircraft flew a total of 1,253 supersonic flights at Mach 1.2 to 2.0 and altitudes between 21,000 and 50,000 feet.

The FAA (which had a major field organization in Oklahoma City) instrumented nine control houses scattered throughout the metropolitan area with various sensors to measure structural effects, while experts from Langley instrumented three houses and set up additional sensors throughout the area to record overpressures, wave patterns, and meteorological conditions. The National Opinion Research Center at the University of Chicago interviewed a sample of 3,000 adults three times during the study. By the end of the test, 73 percent of those surveyed felt that they could live with the number and strength of the booms experienced, and 27 percent would not accept indefinite booms at the level tested. Forty percent believed that they caused some structural damage (even though the control houses showed no significant effects). Analysis of the shock wave patterns by NASA Langley showed that a small number of overpressure measurements were significantly higher than expected, indicating probable atmospheric influences, including heat rising from urban landscapes. One possible result was the breakage of almost 150 windows in the city's two tallest buildings early in the test.[43]

---

42. Maglieri and Garland J. Morris, "Measurement of Response of Two Light Airplanes to Sonic Booms," NASA TN-D-1941, Aug. 1963.
43. D.A. Hilton, Maglieri, and R. Steiner, "Sonic-Boom Exposures during FAA Community Response Studies over a 6-Month Period in the Oklahoma City Area," NASA TN-D-2539, Dec. 1964.

The Oklahoma City study added to the growing knowledge about sonic booms and their acceptance by the public at the cost of negative publicity for the FAA. In view of the reactions to the St Louis and Oklahoma City tests by much of the public and some politicians, plans for another extended sonic boom test over a different city, including flights at night, never materialized.[44]

The FAA and Air Force conducted the next series of tests from November 1964 to February 1965 in a much less populated place: the remote Oscura camp in the vast White Sands Missile Range of New Mexico, where 21 structures of various types and ages with a variety of plaster, windows, and furnishings were studied for possible damage. F-104s from nearby Holloman AFB and B-58s from Edwards generated 1,494 booms producing overpressures from 1.6 to 19 psf. The 680 sonic booms at 5.0 psf caused no real problems, but those above 7.9 psf revealed varying degrees of damage to glass, plaster, tile, and stucco already in vulnerable condition. A parallel study of several thousand incubated chicken eggs showed no reduction in hatchability, and audiology tests on 20 personnel subjected daily to the booms showed no hearing impairment.[45]

Before the White Sands test ended, NASA Langley personnel began collecting boom data from a highly urbanized setting in winter weather. During February and March 1965, they recorded data at five ground stations as B-58 bombers flew 22 training missions in a corridor over downtown Chicago at speeds of Mach 1.2 to 1.66 and altitudes from 38,000 to 48,000 feet. The results showed how amplitude and wave shape varied widely depending upon atmospheric conditions. These 22 flights and 27 others resulted in the Air Force approving 1,442 of 2,964 damage claims for $114,763.[46]

Also in March 1965, the FAA and NASA, in cooperation with the U.S. Forest Service, studied the effect on hazardous mountain snow packs in the Colorado Rockies of Air Force fighters creating boom overpressures up to 5.0 psf. Because of stable snow conditions, none of these created an avalanche. Interestingly enough, in the early 1960s the

---

44. Conway, *High-Speed Dreams*, pp. 121–122.
45. Thomas H. Higgins, "Sonic Boom Research and Design Considerations in the Development of a Commercial Supersonic Transport," JASA, vol. 39, no. 5, pt. 2 (Nov. 1966), pp. 526–531.
46. David. A. Hilton, Vera Huckel, and Maglieri, "Sonic Boom Measurements during Bomber Training Operations in the Chicago Area," NASA TN-D-3655, Oct. 1966.

National Park Service had tried to use newly deployed F-106s at Geiger Field, WA, to create controlled avalanches in Glacier National Park (Project "Safe Slide"), but presumably found traditional artillery fire more suitable.[47]

From the beginning of the SST program, the aircraft most desired for experiments was, of course, the North American XB-70 Valkyrie. The first of the giant testbeds (XB-70-1) arrived at Edwards AFB in September 1964, and the better performing and better instrumented second aircraft (XB-70-2) arrived in July 1965. With a length of 186 feet, a wingspan of 105 feet, and a gross weight of about 500,000 pounds, the six-engine giant was less than two-thirds as long as some of the later SST concepts, but it was the best real-life surrogate available.

Even during the initial flight envelope expansion by contractor and AFFTC test pilots, the Flight Research Center began gathering sonic boom data, including direct comparisons of its shock waves with those of a B-58 flying only 800 feet behind.[48] Using an array of microphones and recording equipment at several ground stations, NASA researchers eventually built a database of boom signatures from 39 flights made by the XB-70s (10 with B-58 chase planes), from March 1965 through May 1966.[49] Because "the XB-70 is capable of duplicating the SST flight profiles and environment in almost every respect," the FRC was looking forward to beginning its own experimental research program using the second Valkyrie on June 15, 1966, with sonic boom testing listed as the first priority.[50]

On June 8, however, XB-70-2 crashed on its 47th flight as the result of an infamous midair collision that killed two pilots and gravely injured

---

47. Histories of the 4700 Air Defense Wing, Jan.–Mar. and Apr.–June 1960, IRIS abstracts; History of the 84th Fighter Group, Jan.–Dec. 1961, IRIS abstract; Telephone interview, Maglieri by Benson, Mar. 13, 2009.
48. William H. Andrews, "Summary of Preliminary Data Derived from the XB-70 Airplanes," NASA TM-X-1240, June 1966, pp. 11–12. Despite being 3.5 times heavier than the B-58, the XB-70's bow wave proved to be only slightly stronger.
49. Maglieri, et al., "A Summary of XB-70 Sonic Boom Signature Data, Final Report," NASA CR-189630, Apr. 1992. Until this report, the 1965–1966 findings were filed away unpublished. The original oscillographs were also scanned and digitized at this time for use in the High-Speed Research Program.
50. FRC, "NASA XB-70 Flight Research Program," Apr. 1966, Dryden archive, File L2-4-4D-3, p. 10 quoted. See also C.M. Plattner, "XB-70A Flight Research: Phase 2 to Emphasize Operational Data," *Aviation Week*, June 13, 1966, pp. 60–62.

An XB-70 taking off at Edwards AFB accompanied by a B-58 chase plane in the mid-1960s, when both were used for sonic boom research. North American.

a third. Despite this tragic setback to the test program, the less capable XB-70-1 (which underwent modifications until November) eventually proved useful for many purposes. After 6 months of joint AFFTC/FRC operations, including the boom testing described below, the plane was turned over full time to NASA in April 1967 after 60 Air Force flights. The FRC, with a more limited budget, then used the Valkyrie for 23 more test missions until February 1969, when the unique aircraft was retired to the USAF Museum in Dayton, OH.[51] All told, NASA acquired sonic boom measurements from 51 of the 129 total flights made by the XB-70s, using two ground stations on Edwards AFB, one at nearby Boron, and two in Nevada.[52] This data proved to be of great value in the future.

The loss of one XB-70 and retirement of the other from supersonic testing was made somewhat less painful by the availability of another smaller but even faster product of advanced aviation technology: the Lockheed YF-12 and its cousin, the SR-71—both nicknamed Blackbirds. On May 1, 1965, shortly after arriving at Edwards, a YF-12A set nine world records, including a closed-course speed of 2,070 mph (Mach 3.14)

---

51. NASA Dryden Fact Sheet, "XB-70," http://www.nasa.gov/centers/dryden/new/FactSheets/FS-084-DFRC_prt.htm; Hallion and Gorn, *On the Frontier*, pp. 176–185, 421.
52. Maglieri, "Summary of XB-70 Sonic Boom," pp. 4–5.

and a sustained altitude of 80,257 feet. Four of that day's five flights also yielded sonic boom measurements. At speeds of Mach 2.6 to 3.1 and altitudes of 60,000 to 76,500 feet above ground level, overpressures varied from 1.2 to 1.7 psf depending on distance from the flight path. During another series of flight tests at slower speeds and lower altitudes, overpressures up to 5.0 psf were measured during accelerations after having slowed to refuel. These early results proved consistent with previous B-58 data.[53] Data gathered over the years from ground arrays measuring the sonic signatures from YF-12s, XB-70s, B-58s, and smaller aircraft flying at various altitudes also showed that the lateral spread of a boom carpet (without the influence of atmospheric variables) could be roughly equated to 1 mile for every 1,000 feet of altitude, with the N-signatures become more rounded with distance until degenerating into the approximate shape of a sine wave.[54]

Although grateful to benefit from the flights of AFFTC's Blackbirds, the FRC wanted its own YF-12 or SR-71 for supersonic research. It finally gained the use of two YF-12s through a NASA–USAF memorandum of understanding signed in June 1969, paying for operations with funding left over from termination of the X-15 and XB-70 programs.[55]

In the fall of 1965, with public acceptance of sonic booms becoming a significant public and political issue, the White House Office of Science and Technology established the National Sonic Boom Evaluation Office (NSBEO) under an interagency Coordinating Committee on Sonic Boom Studies. The new organization, which was attached to Air Force Headquarters for administrative purposes, planned a comprehensive series of tests known as the National Sonic Boom Evaluation Program, to be conducted primarily at Edwards AFB. NASA (in particular the Flight Research and Langley Centers) would be responsible for test operations and data collection, with the Stanford Research Institute hired to help analyze the findings.[56] After careful preparations (including specially built structures

---

53. R.T. Klinger, "YF-12A Flight Test Sonic Boom Measurements," Lockheed Advanced Development Projects Report SP-815, June 1, 1965, Dryden archive, File Ll-4-10A-1.
54. John O. Powers, J.M. Sands, and Maglieri, "Survey of United States Sonic Boom Overflight Experimentation," NASA TM-X-66339, May 1969, pp. 9, 12–13.
55. Peter W. Merlin, *From Archangel to Senior Crown: Design and Development of the Blackbird* (Reston, VA: AIAA, 2008), pp. 106–107, 116–118, 179; Hallion and Gorn, *On the Frontier*, p. 187.
56. NSBEO, "Sonic Boom Experiments at Edwards Air Force Base; Interim Report" (prepared under contract by Stanford Research Institute), pp. 1–2, (hereinafter cited as SRI, "Edwards AFB Report"). For political and bureaucratic background on the NSBEO, see Conway, *High-Speed Dreams*, pp. 122–123.

and extensive sensor and recording arrays), the National Sonic Boom Evaluation began in June 1966. Its main objectives were to address the many issues left unresolved from previous tests. Unfortunately, the loss of XB-70-2 on June 8 forced a 4-month break in the test schedule, with the limited events completed in June designated Phase I. The second phase began in November, when XB-70-1 returned to flight status, and lasted into January 1967. A total of 367 supersonic missions were flown by XB-70s, B-58s, YF-12s, SR-71s, F-104s, and F-106s during the two phases. These were supplemented by 256 subsonic flights by KC-135s, WC-135Bs, C-131Bs, and Cessna 150s. In addition, the Goodyear blimp Mayflower was used in the June phase to measure sonic booms at 2,000 feet.[57]

By the end of testing, the National Sonic Boom Evaluation had obtained new and highly detailed acoustic and seismic signatures from all the different supersonic aircraft in various flight profiles during a variety of atmospheric conditions. The data from 20 XB-70 flights at speeds from Mach 1.38 to 2.94 were to be of particular long-term interest. For example, Langley's sophisticated nose probe used for the pioneering in-flight flow-field measurements of the B-58 in 1963 was installed on one of the FRC's F-104s to do the same for the XB-70. Comparison of data between blimp and ground sensors and variations between the summer and winter tests confirmed the significant influence that atmospheric conditions, such as turbulence and convective heating near the surface, have on boom propagation.[58] Also, the evaluation provided an opportunity to gather data on more than 1,500 sonic boom signatures created during 35 flights by the recently available SR-71s and YF-12s at speeds up to Mach 3.0 and altitudes up to 80,000 feet.[59]

Some of the findings portended serious problems for planned SST operations. The program obtained responses from several hundred volunteers, both outdoors and in houses, to sonic booms of different intensities produced by each of the supersonic aircraft. The time between

---

57. SRI, "Edwards AFB Report," p. 9.
58. Maglieri, et al., "Summary of Variations of Sonic Boom Signatures Resulting from Atmospheric Effects," Feb. 1967, and "Preliminary Results of XB-70 Sonic Boom Field Tests During National Sonic Boom Evaluation Program," Mar. 1967, Annex C-1 and C-2, in SRI, "Edwards AFB Report;" H.H. Hubbard and D.J. Maglieri, "Sonic Boom Signature Data from Cruciform Microphone Array Experiments during the 1966–1967 EAFB National Sonic Boom Evaluation Program," NASA TN-D-6823, May 1972.
59. SRI, "Edwards AFB Report," pp. 17–20, Annexes C-F; Maglieri, et al., "Sonic Boom Measurements for SR-71 Aircraft Operating at Mach Numbers to 3.0 and Altitudes to 24834 Meters," NASA TN-D-6823, Sept. 1972.

the peak overpressure of the bow and tail waves for aircraft at high altitudes ranged from about 0.1 second for the F-104, 0.2 second for the B-58, and 0.3 second for the XB-70. The respondents also compared sonic booms to the jet engine noise of subsonic aircraft. Although data varied for each of the criteria measured, significant minorities tended to find the booms either "just acceptable" or unacceptable, with the "sharper" N-wave signature from the lower flying F-104 more annoying outdoors than the more rounded signatures from the larger aircraft, which had to fly at higher altitudes to create the same overpressure. Other factors included the frequency, time of day or night, and type of boom signature. Correlating how the subjects responded to jet noise (measured in decibels) and sonic booms (normally measured in psf), the SRI researchers used the perceived noise decibel (PNdB) level to assess how loud booms seem to human ears.[60]

Employing sophisticated sensors, civil engineers measured the physical effects on houses and a building with a large interior space (the base bowling alley) to varying degrees of booms created by F-104s, B-58s, and the XB-70. Of special concern for the SST, they found the XB-70's elongated N-wave created more of the low frequencies that cause indoor vibrations, such as rattling windows (although less bothersome to observers outdoors). And although no significant harm was detected to the instrumented structures, 57 complaints of damage were received from residents in the surrounding area, and three windows were broken on base. Finally, monitoring by the Department of Agriculture detected no ill effects on farm animals in the area, although avian species (chickens, turkeys, etc.) reacted more than livestock did.[61] The National Sonic Boom Evaluation remains the most comprehensive such test program yet conducted.

Later, in 1967, the opportunity for collecting additional survey data presented itself when the FAA and NASA learned that SAC was starting an extensive training program for its growing fleet of SR-71s. TRACOR,

---

60. SRI, "Edwards AFB Report," pp. 11–16, Annex B; K.D. Kryter, "Psychological Experiments on Sonic Booms Conducted at Edwards Air Force Base, Final Report," (Menlo Park: SRI, 1968), summarized by Richard M. Roberds, "Sonic Boom and the Supersonic Transport," *Air University Review*, vol. 22, No. 7 (July–Aug. 1971), pp. 25–33.

61. SRI, "Edwards AFB Report," pp. 20–23, Annexes G and H; David Hoffman, "Sonic Boom Tests Fail to Win Any Boosters," *Washington Post*, Aug. 3, 1967, p. A3; A.J. Bloom, et al. (SRI), "Response of Structures to Sonic Booms Produced by XB-70, B-58, and F-104 Aircraft . . . at Edwards Air Force Base, Final Report," NSBEO 2-67, Oct. 1967; D.S. Findley, et al., "Vibration Responses of Test Structure No. 1 during the . . . National Sonic Boom Program," NASA TM-X-72706, June 1975, and "Vibration Responses of Test Structure No. 2 . . . ," NASA TM-X-72704, June 1975.

Inc., of Austin, TX, which was already under contract to NASA doing surveys on airport noise, had its contract expanded in May 1967 to include public responses to the SR-71s' sonic booms in Dallas, Los Angeles, Denver, Atlanta, Chicago, and Minneapolis. Between July 3 and October 2, Air Force SR-71s made 220 flights over these cities at high altitude, ranging from 5 over Atlanta to 60 over Dallas. The minority of sonic booms that were measured were almost all N-waves with overpressures from slightly less than 1.0 psf to 2.0 psf. Although the data from this impromptu test program were less than definitive, the overall findings (based on 6,375 interviews) were fairly consistent with the previous human response surveys. For example, after an initial dropoff, the level of annoyance with the booms tended to increase over time, and almost all those who complained were worried about damage. Among 15 different adjectives supplied to describe the booms (e.g., disturbing, annoying, irritating), the word "startling" was chosen much more than any other.[62]

Although the FRC and AFFTC continued their missions of supersonic flight-testing and experimentation at Edwards, what might be called the heroic era of sonic boom testing was drawing to a close. The FAA and the Environmental Science Services Administration (a precursor of the Environmental Protection Agency) did some sophisticated testing of meteorological effects at Pendleton, OR, from September 1968 until May 1970, using a dense grid of recently invented unattended recording equipment to measure random booms from SR-71s. On the other side of the continent, NASA and the Navy studied sonic booms during Apollo missions in 1970 and 1971.[63]

The most significant NASA testing in 1970 took place from August through October at the Atomic Energy Commission's Jackass Flats test site in Nevada. In conjunction with the FAA and the National Oceanic and Atmospheric Administration (NOAA), NASA took advantage of the 1,527-foot-tall BREN Tower (named for its original purpose, the "Bare Reactor Experiment Nevada" in 1962) to install a vertical array of 15 microphones as well as meteorological sensors. (Until then, a 250-foot tower at Wallops Island had been the highest used in sonic boom testing.) During the summer and fall of 1970, the FRC's F-104s from Edwards made 121 boom-generating flights to provide measurements of several

---

62. TRACOR, Inc., "Public Reactions to Sonic Booms," NASA CR-1665, Sept. 1970.
63. Hilton and Herbert R. Henderson documented the sonic boom measurements from the Apollo 15, 16, and 17 missions in NASA TNs D-6950, D-7606, and D-7806, published from 1972 to 1974.

still poorly understood aspects of the sonic boom, especially the places, known mathematically as caustics, where nonlinear focusing of acoustical rays occurs. Frequently caused by speeds very near Mach 1 or by acceleration, they can result in U-shaped signatures with bow and tail wave overpressures strong enough to create super booms. The BREN Tower allowed such measurements to be made in the vertical dimension for the first time. This test resulted in definitive data on the formation and nature of caustics, information that would be valuable in helping pilots to avoid making focused booms.[64]

For all intents and purposes, the results of earlier testing and surveys had already helped to seal the fate of the SST before the reports on this latest test began coming in. Yet the data gathered from 1958 through 1970 during the SST program contributed tremendously to the international aeronautical and scientific communities' understanding of one of the most baffling and complicated aspects of supersonic flight. As Harry Carlson told the Nation's top sonic boom scientists and engineers on the very same day of the last F-104 mission over Jackass Flats: "The importance of flight-test programs cannot be overemphasized. These tests have provided an impressive amount of high-quality data, which has been of great value in the verification of theoretical methods for the prediction of nominal overpressures and in the estimation from a statistical standpoint of the modifying influence of unpredictable atmospheric nonuniformities."[65]

Laboratory Experiments and Sonic Boom Theory

The rapid progress made in understanding the nature and significance of sonic booms during the 1960s resulted from the synergy among flight testing, wind tunnel experiments, psychoacoustical studies, theoretical refinements, and new computing capabilities. Vital to this process was the largely free exchange of information by NASA, the FAA, the USAF, the airplane manufacturers, academia, and professional organizations such as the American Institute of Aeronautics and Astronautics (AIAA) and the Acoustical Society of America (ASA). The sharing of information

---

64. George T. Haglund and Edward J. Kane, "Flight Test Measurements and Analysis of Sonic Boom Phenomena Near the Shock Wave Extremity," NASA CR-2167, Feb. 1973; Telephone interview, Maglieri by Benson, Mar. 13, 2009.

65. Harry W. Carlson, "Some Notes on the Present Status of Sonic Boom Prediction and Minimization Research," *Third Conference on Sonic Boom Research . . . Washington, DC, Oct. 29–30, 1970*, NASA SP-255, 1971, p. 395.

even extended to potential rivals in Europe, where the Anglo-French Concorde supersonic airliner got off to a headstart on the more ambitious American program.

Designing commercial aircraft has always required tradeoffs between speed, range, capacity, weight, durability, safety, and, of course, costs—both for manufacturing and operations. Balancing such factors was especially challenging with an aircraft as revolutionary as the SST. Unlike with the supersonic military aircraft in the 1950s, NASA's scientists and engineers and their partners in industry also had to increasingly consider the environmental impacts of their designs. At the Agency's aeronautical Centers, especially Langley, this meant that aerodynamicists incorporated the growing knowledge about sonic booms in their equations, models, and wind tunnel experiments.

Harry Carlson of the Langley Center had conducted the first wind tunnel experiment on sonic boom generation in 1959. As reported in December, he tested seven models of various geometrical and airplane-like shapes at differing angles of attack in Langley's original 4 by 4 supersonic wind tunnel at a speed of Mach 2.01. The tunnel's relatively limited interior space mandated the use of very small models to obtain sonic boom signatures: about 2 inches in length for measuring shock waves at 8 body lengths distance and only about three-quarters inch for trying to measure them at 32 body lengths (as close as possible to the "far field," a distance where multiple shock waves coalesce into the typical N-wave signature). Although far-field data were problematic, the overall results correlated with existing theory, such as Whitham's formulas on volume-induced overpressures and Walkden's on those caused by lift.[66] Carlson's attempt to design one of the models to alleviate the strength of the bow shock was unsuccessful, but this might be considered NASA's first attempt at boom minimization.

The small size and extreme precision needed for the models, the disruptive effects of the assemblies needed to hold them, and the extra sensitivity required of pressure-sensing devices all limited a wind tunnel's ability to measure the type of shock waves that would reach the ground from a full-sized aircraft. Even so, substantial progress continued, and the data served as a useful cross-check on flight test data and

---

66. Carlson, "An Investigation of Some Aspects of the Sonic Boom by Means of Wind-Tunnel Measurements of Pressures about Several Bodies at a Mach Number of 2.01," NASA TN-D-161, Dec. 1959. Carlson used Langley's 4 by 4 Supersonic Pressure Wind Tunnel, completed in 1948, for most of his experiments.

mathematical formulas.[67] For example, in 1962 Carlson used a 1-inch model of a B-58 to make the first correlation of flight test data with wind tunnel data and sonic boom theory. Results proved that wind tunnel readings, with appropriate extrapolations, could be used with some confidence to estimate sonic boom signatures.[68]

Exactly 5 years after publishing results of the first wind tunnel sonic boom experiment, Harry Carlson was able to report, "In recent years, intensive research efforts treating all phases of the problem have served to provide a basic understanding of this phenomenon. The theoretical studies [of Whitham and Walkden] have resulted in the correlations with the wind tunnel data…and with the flight data."[69] As for minimization, wind tunnel tests of SCAT models had revealed that some configurations (e.g., the "arrow wing") produced lower overpressures.[70] Such possibilities were soon being explored by aerodynamicists in industry, academia, and NASA. They included Langley's long-time supersonic specialist, F. Edward McLean, who had discovered extended near-field effects that might permit designing airframes for lower overpressures.[71] Of major significance (and even more potential in the future), improved data reduction methods and numerical evaluations of sonic boom theory were being adapted for high-speed processing with new computer codes and hardware, such as Langley's massive IBM 704. Using these new capabilities, Carlson, McLean, and others eventually designed the SCAT-15F, an improved SST concept optimized for highly efficient cruise.[72]

In addition to reports and articles, NASA researchers presented findings from the growing knowledge about sonic booms in various

---

67. For examples of these wind tunnel experiments, see Runyan, "Sonic Boom Capsule Summaries," as well as the NTRS bibliographical database.
68. Carlson, "Wind Tunnel Measurements of the Sonic-Boom Characteristics of a Supersonic Bomber Model and a Correlation with Flight-Test Ground Measurements," NASA TM-X-700, July 1962.
69. Carlson, "Correlation of Sonic-Boom Theory with Wind Tunnel and Flight Measurements," NASA TR-R-213, Dec. 1964. p. 1.
70. Evert Clark, "Reduced Sonic Boom Foreseen for New High-Speed Airliner," *New York Times*, Jan. 1965, pp. 7, 12 (based on visit to NASA Langley).
71. F. Edward McLean, "Some Nonasymptotic Effects of the Sonic Boom of Large Airplanes," NASA TN-D-2877, June 1965.
72. Carlson, "Correlation of Sonic-Boom Theory," pp. 2–23. For an earlier status report on supersonic work at Langley and some at Ames, see William J. Alford and Cornelius Driver, "Recent Supersonic Transport Research," *Astronautics & Aeronautics*, vol. 2, No. 9 (Sept. 1964), pp. 26–37; Chambers, *Innovation in Flight*, pp. 32–34.

meetings and professional symposia. One of the earliest took place September 17–19, 1963, when NASA Headquarters sponsored an SST feasibility studies review at the Langley Center—attended by Government, contractor, and airline personnel—that examined every aspect of the planned airplane. In a session on noise, Harry Carlson warned that "sonic boom considerations alone may dictate allowable minimum altitudes along most of the flight path and have indicated that in many cases the airframe sizing and engine selection depend directly on sonic boom."[73] On top of that, Harvey Hubbard and Domenic Maglieri discussed how atmospheric effects and community response to building vibrations might pose problems with the current SST sonic boom objectives (2 psf during acceleration and 1.5 psf during cruise).[74]

The conferees discussed various other technological challenges for the planned American SST, some indirectly related to the sonic boom issue. For example, because of frictional heating, an airframe covered largely with stainless steel (such as the XB-70) or titanium (such as the then-top secret A-12/YF-12) would be needed to cruise at Mach 2.7+ and over 60,000 feet, an altitude that many hoped would allow the sonic boom to weaken by the time it reached the surface. Manufacturing such a plane, however, would be much more expensive than building a Mach 2.2 SST with aluminum skin, such as the Concorde.

Despite such concerns, the FAA had already released the SST request for proposals (RFP) on August 15, 1963. Thereafter, as explained by Ed McLean, "NASA's role changed from one of having its own concepts evaluated by the airplane industry to one of evaluating the SST concepts of the airplane industry."[75] By January 1964, Boeing, Lockheed, North American, and their jet engine partners had submitted initial proposals. In retrospect, advocates of the SST were obviously hoping that technology would catch up with requirements before it went into production.

Although the SST program was now well underway, a growing awareness of the public response to booms became one factor in many that tri-agency (FAA–NASA–DOD) groups in the mid-1960s, including the PAC

---

73. Carlson, "Configuration Effects on Sonic Boom," *Proceedings of NASA Conference on Supersonic-Transport Feasibility Studies and Supporting Research, Sept. 17–19, 1963 . . . Hampton, VA*, NASA TM-X-905, Dec. 1963, p. 381.
74. Hubbard and Maglieri, "Factors Affecting Community Acceptance of the Sonic Boom," ibid., pp. 399–412.
75. McLean, *Supersonic Transport Technology*, p. 46.

chaired by Robert McNamara, considered in evaluating the proposed SST designs. The sonic boom issue also became the focus of a special committee of the National Academy of Sciences and attracted increasing attention from the academic and scientific community at large.

The Acoustical Society of America, made up of professionals of all fields involving sound (ranging from music to noise to vibrations), sponsored the first Sonic Boom Symposium on November 3, 1965, during its 70th meeting in—appropriately enough—St. Louis. McLean, Hubbard, Carlson, Maglieri, and other Langley experts presented papers on the background of sonic boom research and their latest findings.[76] The paper by McLean and Barrett L. Shrout included details on a breakthrough in using near-field shock waves to evaluate wind tunnel models for boom minimization, in this case a reduction in maximum overpressure in a climb profile from 2.2 to 1.1 psf. This technique also allowed the use of 4-inch models, which were easier to fabricate to the close tolerances required for accurate measurements.[77]

In addition to the scientists and engineers employed by the aircraft manufactures, eminent researchers in academia took on the challenge of discovering ways to minimize the sonic boom, usually with support from NASA. These included the team of Albert George and A. Richard Seebass of Cornell University, which had one of the Nation's premier aeronautical laboratories. Seebass edited the proceedings of NASA's first sonic boom research conference, held on April 12, 1967. The meeting was chaired by another pioneer of minimization, Wallace D. Hayes of Princeton University, and attended by more than 60 other Government, industry, and university experts. Boeing had been selected as the SST contractor less than 4 months earlier, but the sonic boom was becoming recognized far and wide as a possibly fatal flaw for its future production, or at least for allowing it to fly supersonically over land.[78] The two most obvious theoretical ways to reduce sonic booms during supersonic cruise—flying much higher with no increase in weight or building

---

76. *JASA*, vol. 39, no. 5, pt. 2 (Nov. 1966), pp. 519–572.
77. F. Edward McLean and Barrett L. Shrout, "Design Methods for Minimization of Sonic Boom Pressure-Field Disturbances," ibid., 519–525. For an updated report, see Carlson, McLean, and Shrout, "A Wind Tunnel Study of Sonic-Boom Characteristics for Basic and Modified Models of a Supersonic Transport Configuration," NASA TM-X-1236, May 1966.
78. Evert Clark, "Sonic Boom to Limit Speed of Superjets Across U.S.," *New York Times*, Oct. 31, 1966, pp. 1, 71; "George Gardner, "Overland Flights by SST Still in Doubt," *Washington Post*, July 10, 1967, p. A7.

an airframe 50 percent longer at half the weight—were not considered practical.[79] Furthermore, as apparent from a presentation by Domenic Maglieri on flight test findings, such an airplane would still have to deal with the problem of booms caused by maneuvering and accelerating, and from atmospheric conditions.[80]

The stated purpose of this conference was "to determine whether or not all possible aerodynamic means of reducing sonic boom overpressure were being explored."[81] In that regard, Harry Carlson showed how various computer programs then being used at Langley for aerodynamic analyses (e.g., lift and drag) were also proving to be a useful tool for bow wave predictions, complementing improved wind tunnel experiments for examining boom minimization concepts.[82] After presentations by representatives from NASA, Boeing, and Princeton, and follow-on discussions by other experts, some of the attendees thought more avenues of research could be explored. But many were still concerned whether low enough sonic booms were possible using contemporary technologies. Accordingly, NASA's Office of Advanced Research and Technology, which hosted the conference, established specialized research programs on seven aspects of sonic boom theory and applications at five American universities and the Aeronautical Research Institute of Sweden.[83] This mobilization of aeronautical brainpower almost immediately began to pay dividends.

Seebass and Hayes cochaired NASA's second sonic boom conference on May 9–10, 1968. It included 19 papers on the latest boom-related testing, research, experimentation, and theory by specialists from NASA and the universities. The advances made in one year were impressive. In the area of theory, for example, the straightforward linear technique for predicting the propagation of sonic booms from slender airplanes such as the SST had proven reliable, even for calculating some nonlinear (mathematically complex and highly erratic) aspects of their signatures.

---

79. A.R. Seebass, ed., *Sonic Boom Research: Proceedings of a Conference . . . Washington, DC, Apr. 12, 1967*, NASA SP-147, 1967.
80. Maglieri, "Sonic Boom Flight Research—Some Effects of Airplane Operations and the Atmosphere on Sonic Boom Signatures," ibid., pp. 25–48.
81. A.R. Seebass, "Preface," ibid., p. iii.
82. Carlson, "Experimental and Analytical Research on Sonic Boom Generation at NASA," ibid., pp. 9–23.
83. Ira R. Schwartz, ed., *Sonic Boom Research, Second Conference, Washington, DC, May 9–10, 1968*, NASA SP-180, 1968, pp. iv–v.

Additional field testing had improved understanding of the geometrical acoustics caused by atmospheric conditions. Computational capabilities needed to deal with such complexities continued to accelerate. Aeronautical Research Associates of Princeton (ARAP), under a NASA contract, had developed a computer program to calculate overpressure signatures for supersonic aircraft in a horizontally stratified atmosphere. Offering another preview of the digital future, researchers at Ames had begun using a computer with graphic displays to perform flow-field analyses and to experiment with a dozen diverse aircraft configurations for lower boom signatures. Several other papers by academic experts, such as Antonio Ferri of New York University (a notable prewar Italian aerodynamicist who had worked at the NACA's Langley Laboratory after escaping to the United States in 1944), dealt with progress in the aerodynamic techniques to reduce sonic booms.[84]

Nevertheless, several important theoretical problems remained, such as the prediction of sonic boom signatures near a caustic (an objective of the previously described Jackass Flats testing in 1970), the diffraction of shock waves into "shadow zones" beyond the primary sonic boom carpet, nonlinear shock wave behavior near an aircraft, and the still mystifying effects of turbulence. Ira R. Schwartz of NASA's Office of Advanced Research and Technology summed up the state of sonic boom minimization as follows: "It is yet too early to predict whether any of these design techniques will lead the way to development of a domestic SST that will be allowed to fly supersonic over land as well as over water."[85]

Rather than conduct another meeting the following year, NASA deferred to a conference by NATO's Advisory Group for Aerospace Research & Development (AGARD) on aircraft engine noise and sonic boom, held in Paris during May 1969. Experts from the United States and five other nations attended this forum, which consisted of seven sessions. Three of the sessions, plus a roundtable, dealt with the status of boom research and the challenges ahead.[86] As reflected by these conferences, the three-way partnership between NASA, Boeing, and the academic aeronautical community during the late 1960s continued to yield new knowledge about sonic booms as well as technological advance in

---

84. Ibid., 1–193; For more on the ARAP computer program, see Wallace D. Hayes, et al., "Sonic Boom Propagation in a Stratified Atmosphere with Computer Program," NASA CR-1299, Apr. 1969.
85. *Second Conference on Sonic Boom Research*, p. vii.
86. AGARD, *Aircraft Engine Noise and Sonic Boom* (see note 19 for bibliographical data).

exploring ways to deal with them. In addition to more flight test data and improved theoretical constructs, much of this progress was the result of various experimental apparatuses.

The use of wind tunnels (especially Langley's 4 by 4 supersonic wind tunnels and the 9 by 7 and 8 by 7 supersonic sections of Ames's Unitary Wind Tunnel complex) continued to advance the understanding of shock wave generation and aircraft configurations that could minimize the sonic boom.[87] As two of Langley's sonic boom experts reported in 1970, the many challenges caused by nonuniform tunnel conditions, model and probe vibrations, boundary layer effects, and the precision needed for small models "have been met with general success."[88]

Also during the latter half of the 1960s, NASA and its contractors developed several new types of simulators that proved useful in studying the physical and psychoacoustic effects of sonic booms. The smallest (and least expensive) was a spark discharge system. The Langley Center and other laboratories used these "bench-type" devices for basic research into the physics of pressure waves. Langley's system created miniature sonic booms by using parabolic or two-dimensional mirrors to focus the shock waves caused by discharging high voltage bolts of electricity between tungsten electrodes toward precisely placed microphones. Such experiments were used to verify laws of geometrical acoustics. The system's ability to produce shock waves that spread out spherically proved useful for investigating how the cone-shaped waves generated by aircraft interact with buildings.[89]

For studying the effect of temperature gradients on boom propagation, Langley used a ballistic range consisting of a helium gas launcher that shot miniature projectiles at constant Mach numbers through a partially enclosed chamber. The inside could be heated to ensure a stable atmosphere for accuracy in boom measurements. Innovative NASA-sponsored simulators included Ling-Temco-Vought's shock-expansion tube, basically a mobile 13-foot-diameter conical horn mounted on a

---

87. For a survey, see Daniel D. Baals and William R. Corliss, "Wind Tunnels of NASA," SP-440 (Washington, DC: NASA, 1981).
88. Phillip M. Edge and Harvey H. Hubbard, "Review of Sonic-Boom Simulation Devices and Techniques," Dec. 1970, *JASA Journal*, vol. 51, No. 2, pt. 2 (Feb. 1972), p. 723.
89. W.D. Beasly, J.D. Brooks, and R.L. Barger, "A Laboratory Investigation of N-Wave Focusing," NASA TN-D-5306, July 1969; J.D. Brooks, et al., "Laboratory Investigation of Diffraction and Reflection of Sonic Booms by Buildings," NASA TN-D-5830, June 1970.

trailer, and General American Research Division's explosive gas-filled envelopes suspended above sensors at Langley's sonic boom simulation range.[90] NASA also contracted with Stanford Research Institute for simulator experiments that showed how sonic booms could interfere with sleep, especially for older people.[91]

Other simulators were devised to handle both human and structural response to sonic booms. (The need to better understand effects on people was called for in a report released in June 1968 by the National Academy of Sciences.)[92] Unlike the previously described studies using actual sonic booms created by aircraft, these devices had the advantages of a controlled laboratory environment. They allowed researchers to produce multiple boom signatures of varying shapes, pressures, and durations as often as needed at a relatively low cost.[93] The Langley Center's Low-Frequency Noise Facility—built earlier in the 1960s to generate the intense chest-pounding sounds of giant Saturn boosters during Apollo launches—also performed informative sonic boom simulation experiments. Consisting of a cylindrical test chamber 24 feet in diameter and 21 feet long, it could accommodate people, small structures, and materials for testing. Its electrohydraulically operated 14-foot piston was capable of producing sound waves from 1–50 hertz (sort of a super subwoofer) and sonic boom N-waves from 0.5 to 20 psf at durations from 100 to 500 milliseconds.[94]

To provide an even more versatile system designed specifically for sonic boom research, NASA contracted with General Applied Science Laboratories (GASL) of Long Island, NY, to develop an ideal simulator using a quick action valve and shock tube design. (Antonio Ferri was

---

90. Edge and Hubbard, "Review of Sonic Boom Simulation," pp. 724–728; Hugo E. Dahlke, et al., "The Shock-Expansion Tube and Its Application as a Sonic Boom Simulator," NASA CR-1055, June 1968; R.T Sturgielski, et al., "The Development of a Sonic Boom Simulator with Detonable Gases," NASA CR-1844, Nov. 1971.
91. Jerome Lukas and Karl D. Kryler, "A Preliminary Study of the Awakening and Startle Effects of Simulated Sonic Booms," NASA CR-1193, Sept. 1968, "Awakening Effects of Simulated Sonic Booms and Subsonic Aircraft Noise . . . ," NASA CR-1599, May 1970.
92. David Hoffman, "Report Sees Need for Study on Sonic Boom Tolerance," *Washington Post*, June 26, 1968, p. A3.
93. Ira R. Schwartz, Sonic Boom Simulation Facilities," AGARD, *Aircraft Engine Noise and Sonic Boom*, p. 29-1.
94. Philip M. Edge and William H. Mayes, "Description of Langley Low-Frequency Noise Facility and Study of Human Response to Noise Frequencies below 50 cps," NASA TN-D-3204, Jan. 1966.

the president of GASL, which he had cofounded with the illustrious aeronautical scientist Theodore von Kármán in 1956). Completed in 1969, this new simulator consisted of a high-speed flow valve that sent pressure wave bursts through a heavily reinforced 100-foot-long conical duct that expanded into an 8 by 8 test section with an instrumentation and model room. It could generate overpressures up to 10 psf with durations from 50 to 500 milliseconds. Able to operate at less than a 1-minute interval between bursts, its sonic boom signatures proved very accurate and easy to control.[95] In the opinion of Ira Schwartz, "the GASL/NASA facility represents the most advanced state of the art in sonic boom simulation."[96]

While NASA and its partners were learning more about the nature of sonic booms, the SST was becoming mired in controversy. Many in the public, the press, and the political arena were concerned about the noise SSTs would create, with a growing number expressing hostility to the entire SST program. As one of the more reputable critics wrote in 1966, with a map showing a dense network of future boom carpets crossing the United States, "the introduction of supersonic flight, as it is at present conceived, would mean that hundreds of millions of people would not only be seriously disturbed by the sonic booms . . . they would also have to pay out of their own pockets (through subsidies) to keep the noise-creating activity alive."[97]

Opposition to the SST grew rapidly in the late 1960s, becoming a cause celebre for the burgeoning environmental movement as well as target for small-Government conservatives opposed to Federal subsidies.[98] Typical of the growing trend among opinion makers, the *New York Times* published its first strongly anti–sonic-boom editorial in June 1968, linking the SST's potential sounds with an embarrassing incident the week before when an F-105 flyover shattered 200 windows at the Air Force Academy, injuring a dozen people.[99] The next 2 years brought a

---

95. Roger Tomboulian, Research and Development of a Sonic Boom Simulation Device, NASA CR-1378, July 1969; Stacy V. Jones, "Sonic Boom Researchers Use Simulator," *New York Times*, May 10, 1969, pp. 37, 41.
96. Ira R. Schwartz, "Sonic Boom Simulation Facilities," p. 29-6.
97. B.K. O. Lundberg, "Aviation Safety and the SST," *Astronautics & Aeronautics*, vol. 3, No. 1 (Jan. 1966), p. 28. Lundberg was a Swedish scientist very critical of SSTs.
98. See Conway, *High-Speed Dreams*, pp. 118–156.
99. "The Shattering Boom," *New York Times*, June 8, 1968, p. 30.

Case 4 | Softening the Sonic Boom: 50 Years of NASA Research

growing crescendo of complaints about the supersonic transport, both for its expense and the problems it could cause—even as research on controlling sonic booms began to bear some fruit.

By the time 150 scientists and engineers gathered in Washington, DC, for NASA's third sonic boom research conference on October 29–30, 1970, the American supersonic transport program was less than 6 months away from cancellation. Thus the 29 papers presented at the conference and others at the ASA's second sonic boom symposium in Houston the following month might be considered, in their entirety, as a final status report on sonic boom research during the SST decade.[100] Of future if not near-term significance, considerable progress was being made in understanding how to design airplanes that could fly faster than the speed of sound while leaving behind a gentler sonic footprint.

As summarized by Ira Schwartz: "In the area of boom minimization, the NASA program has utilized the combined talents of Messrs. E. McLean, H.L. Runyan, and H.R. Henderson at NASA Langley Research Center, Dr. W.D. Hayes at Princeton University, Drs. R. Seebass and A.R. George at Cornell University, and Dr. A. Ferri at New York University to determine the optimum equivalent bodies of rotation [a technique for relating airframe shapes to standard aerodynamic rules governing simple projectiles with round cross sections] that minimize the overpressure, shock pressure rise, and impulse for given aircraft weight, length, Mach number, and altitude of operation. Simultaneously, research efforts of NASA and those of Dr. A. Ferri at New York University have provided indications of how real aircraft can be designed to provide values approaching these optimums. . . . This research must be continued or even expanded if practical supersonic transports with minimum and acceptable sonic boom characteristics are to be built."[101]

Any consensus among the attendees about the progress they were making was no doubt tempered by their awareness of the financial problems now plaguing the Boeing Company and the political difficulties facing the administration of President Richard Nixon in continuing to subsidize the American SST. From a technological standpoint, many of them also seemed resigned that Boeing's final 2707-300

---

100. Ira R. Schwartz, ed., *Third Conference on Sonic Boom Research . . . Washington, DC, Oct. 29–30, 1970,* NASA SP-255, 1971. The papers from the ASA's Houston symposium were published in *JASA,* vol. 51, No. 2 (Feb. 1972), pt. 2.
101. *Third Conference on Sonic Boom Research,* Preface by Ira Schwartz, p. iv.

design (despite its 306-foot length and 64,000-foot cruising altitude) would not pass the overland sonic boom test. Richard Seebass, who was in the vanguard of minimization research, admitted that "the first few generations of supersonic transport (SST) aircraft, if they are built at all, will be limited to supersonic flight over oceanic and polar regions."[102] In view of such concerns, some of the attendees were even looking toward hypersonic aerospace vehicles, in case they might cruise high enough to leave an acceptable boom carpet.

As for the more immediate prospects of a domestic supersonic transport, Lynn Hunton of the Ames Research Center warned that "with regard to experimental problems in sonic boom research, it is essential that the techniques and assumptions used be continuously questioned as a requisite for assuring the maximum in reliability."[103] Harry Carlson probably expressed the general opinion of Langley's aerodynamicists when he cautioned that "the problem of sonic boom minimization through airplane shaping is inseparable from the problems of optimization of aerodynamic efficiency, propulsion efficiency, and structural weight. . . . In fact, if great care is not taken in the application of sonic boom design principles, the whole purpose can be defeated by performance degradation, weight penalties, and a myriad of other practical considerations."[104]

After both the House and Senate voted in March 1971 to eliminate SST funding, a joint conference committee confirmed its termination in May.[105] This and related cuts in supersonic research inevitably slowed momentum in dealing with sonic booms. Even so, researchers in NASA, as well as in academia and the aerospace industry, would keep alive the possibility of civilian supersonic flight in a more constrained and less technologically ambitious era. Fortunately for them, the ill-fated SST program left behind a wealth of data and discoveries about sonic booms. As evidence, the Langley Research Center produced or sponsored more than 200 technical publications on the subject over 19 years, most related to the SST program. (Many of those published in the early 1970s were based on previous research and testing.) This

---

102. R. Seebass, "Comments on Sonic Boom Research," ibid., p. 411.
103. Lynn W. Hunton, "Comments on Low Sonic Boom Configuration Research, ibid., p. 417.
104. Carlson, "Sonic Boom Prediction and Minimization Research," ibid., p. 397.
105. For a detailed postmortem, see Edward Wenk, "SST—Implications of a Political Decision," *Astronautics & Aeronautics*, vol. 9, No. 10 (Oct. 1971), pp. 40–49.

literature, depicted in Figure 4, would be a legacy of enduring value in the future.[106]

Keeping Hopes Alive: Supersonic Cruise Research

"The number one technological tragedy of our time." That was how President Nixon characterized the votes by the Congress to stop funding an American supersonic transport.[107] Despite its cancellation, the White House, the Department of Transportation (DOT), and NASA—as well as some in Congress—did not allow the progress in supersonic technologies the SST had engendered to completely dissipate. During 1971 and 1972, the DOT and NASA allocated funds for completing some of the tests and experiments that were underway when the program was terminated. The administration then added line-item funding to NASA's fiscal year (FY) 1973 budget for scaled-down supersonic research, especially as related to environmental problems. In response, NASA established the Advanced Supersonic Technology (AST) program in July 1972.

To more clearly indicate the exploratory nature of this effort and allay fears that it might be a potential follow-on to the SST, the AST program was renamed Supersonic Cruise Aircraft Research (SCAR) in 1974. When the term aircraft in its title continued to raise suspicion in some quarters that the goal might be some sort of prototype, NASA shortened the program's name to Supersonic Cruise Research (SCR) in 1979.[108] For the sake of simplicity, the latter name is often applied to all 9 years of the program's existence. For NASA, the principal purpose of AST, SCAR, and SCR was to conduct and support focused research into the problems of supersonic flight while advancing related technologies. NASA's aeronautical Centers, most of the major airframe manufactures, and many research organizations and universities participated. From Washington, NASA's Office of Aeronautics and Space Technology (OAST)

---

106. Compiled by screening B.A. Fryer, et al., "Publications in Acoustics and Noise Control from the NASA Langley Research Center During 1940–1976," NASA TM-X-7402, July 1977. Five reports for 1967 that Maglieri (in reviewing the draft of this chapter) found missing from Fryer's compilation have been added to that column.

107. Stephen D. Ambrose, *Nixon: Triumph of a Politician*, vol. 2 (New York: Simon and Schuster, 1989), p. 433, cited by Conway, *High-Speed Dreams*, p. 153. For the political and bureaucratic background of the AST program, see Conway, pp. 153–158.

108. F. Edward McLean, "SCAR Program Overview," *Proceedings of the SCAR Conference . . . Hampton, VA, Nov. 9–12, 1976*, pt. 1, NASA CP-001, 1976, pp. 1–3; McLean, *Supersonic Cruise Technology*, pp. 101–102.

provided overall supervision but delegated day-to-day management to the Langley Research Center, which established an AST Project Office in its Directorate of Aeronautics, soon placed under a new Aeronautics System Division. The AST program was organized into four major elements—propulsion, structure and materials, stability and control, and aerodynamic performance—plus airframe-propulsion integration. (NASA spun off the propulsion work on a variable cycle engine [VCE] as a separate program in 1976.) Sonic boom research was one of 16 subelements.[109]

At the Aeronautical Systems Division, Cornelius "Neil" Driver, who headed the Vehicle Integration Branch, and Ed McLean, as chief of the AST Project Office, were key officials in planning and managing the AST/SCAR effort. After McLean retired in 1978, the AST Project Office passed to a fellow aerodynamicist, Vincent R. Mascitti, while Driver took over the Aeronautical Systems Division. One year later, Domenic Maglieri replaced Mascitti in the AST Project Office.[110] Despite Maglieri's sonic boom expertise, the goal of minimizing the AST's sonic boom for overland cruise had long since ceased being an SCR objective. As later explained by McLean: "The basic approach of the SCR program . . . was to search for the solution of supersonic problems through disciplinary research. Most of these problems were well known, but no satisfactory solution had been found. When the new SCR research suggested a potential solution . . . the applicability of the suggested solution was assessed by determining if it could be integrated into a practical commercial supersonic airplane and mission. . . . If the potential solution could not be integrated, it was discarded."[111]

To meet the practicality standard for integration into a supersonic airplane, solving the sonic boom problem had to clear a new and almost insurmountable hurdle. In April 1973, responding to years of political pressure, the FAA announced a new rule that banned commercial or civil aircraft from supersonic flight over the land mass or territorial waters of the United States if measurable overpressure would reach the surface.[112] One of the initial objectives of the AST's sonic boom research had been to

---

109. McLean, *Supersonic Cruise Technology*, pp. 104–108; Sherwood Hoffman, "Bibliography of Supersonic Cruise Aircraft Research (SCAR)," NASA RP-1003, Nov. 1977, pp. 1–5.
110. Chambers, *Innovation in Flight*, pp. 39–40.
111. McLean, *Supersonic Cruise Technology*, p. 103.
112. FAA Chronology, Apr. 27, 1973. The rule was included as Federal Aviation Regulation (FAR) Section 91.817, Civil Aircraft Sonic Boom, effective Sept. 30, 1963.

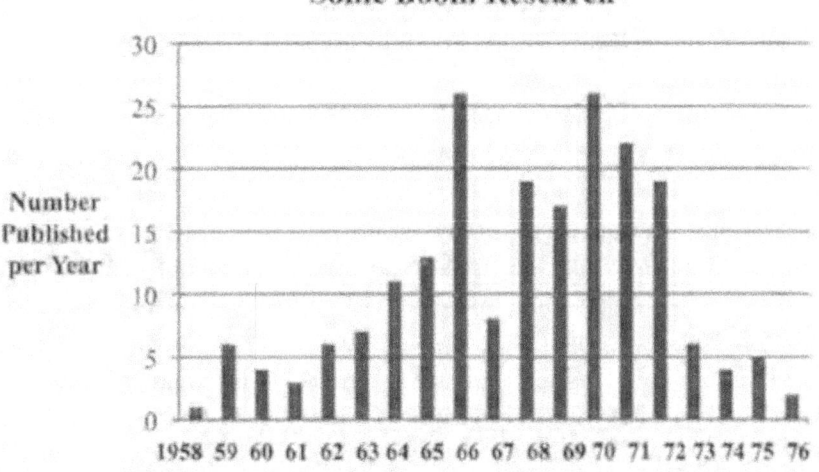

Figure 4. Reports produced or sponsored by NASA Langley, 1958–1976. NASA.

establish a metric on public acceptability of sonic boom signatures for use in the aerodynamic design process. The FAA's stringent new regulation seemed to rule out any such flexibility.

As a result, when Congress cut FY 1974 funding for the AST program from $40 million to about $10 million, the subelement for sonic boom research went on NASA's chopping block. The design criteria for the SCAR/SCR program became a 300-foot-long, 270-passenger airplane that could fly as effectively as possible over land at subsonic speeds yet still cruise efficiently at 60,000 feet and Mach 2.2 over water. To meet these criteria, Langley aerodynamicists modified their SCAT-15F design from the late 1960s into a notional concept with better low-speed performance (but higher sonic boom potential) called the ATF-100. This served as a baseline for three industry teams in coming up with their own designs.[113]

When the AST program began, however, prospects for a significant quieting of its sonic footprint appeared possible. Sonic boom theory had advanced significantly during the 1960s, and some promising if not yet practical ideas for reducing boom signatures had begun to emerge. As

---

113. Marvin Miles, "Hopes for SST Are Dim but R&D Continues—Just in Case," *Los Angeles Times*, Nov. 25, 1973, pp. G1, 11; McLean, *Supersonic Cruise Technology*, pp. 117-118; Conway, *High-Speed Dreams*, pp. 176–180.

indicated by Figure 4, some findings based on that research continued to come out in print during the early 1970s.

As far back as 1965, NASA's Ed McLean had discovered that the sonic boom signature from a very long supersonic aircraft flying at the proper altitude could be nonasymptotic (i.e., not reach the ground in the form of an N-wave). This confirmed the possibility of tailoring an airplane's shape into something more acceptable.[114] Some subsequent theoretical suggestions, such as various ways of projecting heat fields to create a longer "phantom" fuselage, are still decidedly futuristic, while others, such as adding a long spike to the nose of an SST to slow the rise of the bow shock wave, would (as described later) eventually prove more realistic.[115] Meanwhile, researchers under contract to NASA kept advancing the state of the art in more conventional directions. For example, Antonio Ferri of New York University in partnership with Hans Sorensen of the Aeronautical Research Institute of Sweden used new 3-D measuring techniques in Sweden's trisonic wind tunnel to more accurately correlate near-field effects with linear theory. Testing NYU's model of a 300-foot-long SST cruising at Mach 2.7 at 60,000 feet, it showed the opportunity for sonic booms of less than 1.0 psf.[116] Ferri's early death in 1975 left a void in supersonic aerodynamics, not least in sonic boom research.[117]

By the end of the SST program, Albert George and Richard Seebass had formulated a mathematical foundation for many of the previous theories. They also devised a near-field boom-minimization theory, applicable in an isothermal atmosphere, for reducing the overpressures of flattop and ramp-type signatures. It was applicable to both front and rear shock waves along with their parametric correlation to airframe lift and area. In a number of seminal papers and articles in the early 1970s, they explained the theory along with some ideas on possible aerodynamic

---

114. F. Edward McLean, "Some Non-asymptotic Effects on the Sonic Boom of Large Airplanes," NASA TN-D-2877, June 1965, as interpreted by Plotkin and Maglieri, "Sonic Boom Research," p. 5.
115. Miles, "Sonic Boom Not Insoluble, Scientist Says," *Los Angeles Times*, Dec. 10, 1970, pp. E4–E5; David S. Miller and Carlson, "Sonic Boom Minimization by Application of Heat or Force Fields to Airplane Airflow," NASA TN-D-5582, Dec. 1969; Rudolph J. Swigart, "An Experimental Study in the Validity of the Heat-Field Concept for Sonic Boom Alleviation," NASA CR-2381, Mar. 1974.
116. Antonio Ferri, Huai-Chu Wang, and Hans Sorensen, "Experimental Verification of Low Sonic Boom Configuration," NASA CR-2070, June 1973.
117. For a retrospective, see Percy J. Bobbit and Maglieri, "Dr. Antonio Ferri's Contribution to Supersonic Transport Sonic-Boom Technology," *Journal of Spacecraft and Rockets*, vol. 40, no. 4 (July–Aug. 2003), pp.459–466.

shaping (e.g., slightly blunting an aircraft's nose) and the optimum cruise altitude (lower than previously thought) for reducing boom signatures.[118]

Theoretical refinements and new computer modeling techniques continued to appear in the early 1970s. For example, in June 1972, Charles Thomas of the Ames Research Center explained a mathematical procedure using new algorithms for waveform parameters to extrapolate the formation of far-field N-waves. This was an alternative to using F-function effects (the pattern of near-field shock waves emanating from an airframe), which were the basis of the previously discussed program developed by Wallace Hayes and colleagues at ARAP. Although both methods accounted for acoustical ray tracing and could arrive at similar results, Thomas's program allowed easier input of flight information (speed, altitude, atmospheric conditions, etc.) for automated data processing.[119]

In June 1973, at the end of the AST program's first year, NASA Langley's Harry Carlson, Raymond Barger, and Robert Mack published a study on the applicability of sonic boom minimization concepts for overland supersonic transport designs. They examined four reduced-boom concepts for a commercially viable Mach 2.7 SST with a range of 2,500 nautical miles (i.e., coast to coast in the United States). Using experimentally verified minimization concepts of George, Seebass, Hayes, Ferri, Barger, and the English researcher L.B. Jones, along with computational techniques developed at Langley during the SST program, Carlson's team examined ways to manipulate the F-function to project a flatter far-field sonic boom signature. In doing this, the team was handicapped by the continuing lack of established signature characteristics (the combinations of initial peak overpressure, maximum shock strength, rise time, and duration) that people would best tolerate, both outdoors and especially indoors. Also, the complexity of aft aircraft geometry made measuring effects on tail shocks difficult.[120]

Even so, their study confirmed the advantages of designs with highly swept wings toward the rear of the fuselage with twist and camber for

---

118. A.R. George and R. Seebass, "Sonic-Boom Minimization," Nov. 1970, in *JASA*, vol. 51, no. 2, pt. 3 (Feb. 1972), pp. 686–694; A.R. Seebass and A.R George, "Sonic Boom Minimization through Aircraft Design and Operation," AIAA Paper 73-241, Jan. 1973; A.R. Seebass and A.R. George, "The Design and Operation of Aircraft to Minimize Their Sonic Boom," *Journal of Aircraft*, vol. 11, no. 9 (Sept. 1974), pp. 509–517. (Quote is from p. 516.)
119. Charles L. Thomas, "Extrapolation of Sonic Boom Pressure Signatures by the Waveform Parameter Method," NASA TN-D-6823, June 1972.
120. Carlson, Raymond L. Barger, and Robert J. Mack, "Application of Sonic-Boom Minimization Concepts in Supersonic Transport Design," NASA TN-D-7218, June 1973.

sonic boom shaping. It also found the use of canards (small airfoils used as horizontal stabilizers near the nose of rear-winged aircraft) could optimize lift distribution for sonic boom benefits. Although two designs showed bow shocks of less than 1.0 psf, their report noted "that there can be no assurance at this time that [their] shock-strength values . . . if attainable, would permit unrestricted overland operations of supersonic transports."[121] Ironically, these words were written just before the new FAA rule rendered them largely irrelevant.

In October 1973, Edward J. Kane of Boeing, who had been a key sonic boom expert during the SST program, released the results of a similar NASA-sponsored study on the feasibility of a commercially viable low-boom transport using technologies projected to be available in 1985. Based on the latest theories, Boeing explored two longer-range concepts: a high-speed (Mach 2.7) design that would produce a sonic boom of 1.0 psf or less, and a medium-speed (Mach 1.5) design with a signature of 0.5 psf or less.[122] In retrospect, this study, which reported mixed results, represented industry's perspective on the prospects for boom minimization just as the AST program dropped plans for supersonic cruise over land.

Obviously, the virtual ban on civilian supersonic flight in the United States dampened any enthusiasm by private industry to continue investing very much capital in sonic boom research. Within NASA, some of those with experience in sonic boom research also redirected their efforts into other areas of expertise. Of the approximately 1,000 technical reports, conference papers, and articles by NASA and its contractors listed in bibliographies of the SCR program from 1972 to 1980, only 8 dealt directly with the sonic boom.[123]

Even so, progress in understanding sonic booms did not come to a complete standstill. In 1972, Christine M. Darden, a Langley mathematician in an engineering position, had developed a computer code to adapt Seebass and George's minimization theory, which was based on an isothermal (uniform) atmosphere, into a program that applied to a standard (stratified) atmosphere. It also allowed more design flexibility than

---

121. Ibid., p. 28.
122. Edward J. Kane, "A Study to Determine the Feasibility of a Low Sonic Boom Supersonic Transport," AIAA Paper 73-1035, Oct. 1973. See also NASA CR-2332, Dec. 1973.
123. Sherwood Hoffman, "Bibliography of Supersonic Cruise Aircraft Research (SCAR)" [1972–1977], NASA RP-1003, Nov. 1977, and "Bibliography of Supersonic Cruise Research (SCR) Program from 1977 to Mid-1980," NASA RP-1063, Dec. 1980.

previous low-boom configuration theory did, such as better aerodynamics in the nose area.[124]

Using this new computer program, Darden and Robert Mack followed up on the previously described study by Carlson's team by designing wing-body models with low-boom characteristics: one for cruise at Mach 1.5 and two for cruise at Mach 2.7. At 6 inches in length, these were the largest yet tested for sonic boom propagation in a 4 by 4 supersonic wind tunnel—an improvement made possible by continued progress in measuring and extrapolating near-field effects to signatures in the far field. The specially shaped models (all arrow-wing configurations, which distributed lift effects to the rear) showed significantly lower overpressures and flatter signatures than standard designs did, especially at Mach 1.5, at which both the bow and tail shocks were softened. Because of funding limitations, this promising research could not be sustained long enough to develop definitive boom minimization techniques.[125] It was apparently the last significant experimentation on sonic boom minimization for more than a decade.

While this work was underway, Darden and Mack presented a paper on current sonic boom research at the first SCAR conference, held at Langley on November 9–12, 1976 (the only paper on that subject among the 47 presentations). "Contrary to earlier beliefs," they explained, "it has been found that improved efficiency and lower sonic boom characteristics do not always go hand in hand." As for the acceptability of sonic booms, they reported that the only research in North America was being done at the University of Toronto.[126] Another NASA contribution to understanding sonic booms came in early 1978 with the publication by Harry Carlson

---

124. Christine M. Darden, "Minimization of Sonic-Boom Parameters in Real and Isothermal Atmospheres," NASA TN D-7842, Mar. 1975; Darden, "Sonic Boom Minimization with Nose-Bluntness Relaxation," NASA TP-1348, Jan. 1979; Darden, "Affordable/Acceptable Supersonic Flight: Is It Near?" 40th Aircraft Symposium, Japan Society for Aeronautical and Space Sciences (JSASS), Yokohama, Oct. 9–11, 2002.
125. Robert J. Mack and Darden, "Wind-Tunnel Investigation of the Validity of a Sonic-Boom-Minimization Concept," NASA TP-1421, Oct. 1979. They had previously presented their findings at an AIAA conference in Seattle on Mar. 12–14, 1979 as "Some Effects of Applying Sonic Boom Minimization to Supersonic Cruise Aircraft Design," AIAA Paper 79-0652, also published in *Journal of Aircraft*, vol. 17, no. 3 (Mar. 1980), pp. 182–186.
126. Darden and Mack, "Current Research in Sonic-Boom Minimization," *Proceedings of the SCAR Conference* [1976], pt. 1, pp. 525–541 (quote from p. 526). Darden had discussed some of these topics in "Sonic Boom Theory – Its Status in Prediction and Minimization," AIAA Paper 76-1, presented at the AIAA Aerospace Sciences Meeting, Washington, DC, Jan. 26–28, 1976.

of "Simplified Sonic-Boom Prediction," a how-to guide on a relatively quick and easy method to determine sonic boom characteristics. It could be applied to a wide variety of supersonic aircraft configurations as well as spacecraft at altitudes up to 76 kilometers. Although his clever series of graphs and equations would not provide the accuracy needed to predict booms from maneuvering aircraft or in designing airframe configurations, Carlson explained that "for many purposes (including the conduct of preliminary engineering studies or environmental impact statements), sonic-boom predictions of sufficient accuracy can be obtained by using a simplified method that does not require a wind tunnel or elaborate computing equipment. Computational requirements can in fact be met by hand-held scientific calculators, or even slide rules."[127]

The month after publication of this study, NASA released its final environmental impact statement (EIS) for the Space Shuttle program, which benefited greatly from the Agency's previous research on sonic booms, including that with the X-15 and Apollo missions, and adaptations of Charles Thomas's waveform-based computer program.[128] While ascending, the EIS estimated maximum overpressures of 6 psf (possibly up to 30 psf with focusing effects) about 40 miles downrange over open water, caused by both its long exhaust plume and its curving flight profile while accelerating toward orbit. During reentry of the manned vehicle, the sonic boom was estimated at a more modest 2.1 psf, which would affect about 500,000 people as it crossed the Florida peninsula or 50,000 when landing at Edwards.[129] In following decades, as populations in those areas boomed, millions more would be hearing the sonic signatures of returning Shuttles, more than 120 of which would be monitored for their sonic booms.[130]

Some other limited experimental and theoretical work on sonic booms continued in the late 1970s. Richard Seebass at Cornell and Kenneth Plotkin of Wyle Research, for example, delved deeper into the

---

127. Carlson, "Simplified Sonic-Boom Prediction," NASA TP-1122, Mar. 1978, p. 1.
128. Paul Holloway of Langley and colleagues from the Ames, Marshall, and Johnson centers presented an early analysis, "Shuttle Sonic Boom—Technology and Predictions," in AIAA Paper 73-1039, Oct. 1973.
129. NASA HQ (Myron S. Malkin), *Environmental Impact Statement: Space Shuttle Program (Final)*, Apr. 1978, pp. 106–116.
130. Including measurements in Hawaii, with the Shuttle at 253,000 feet and moving at Mach 23. Telephone interview, Maglieri by Benson, Mar. 18, 2009.

challenging phenomena of caustics and focused booms.[131] At the end of the decade, Langley's Raymond Barger published a study on the relationship of caustics to the shape and curvature of acoustical wave fronts caused by actual aircraft maneuvers. To graphically display these effects, he programmed a computer to draw simulated three-dimensional line plots of the acoustical rays in the wave fronts. Figure 5 shows how even a simple decelerating turn, in this case from Mach 2.4 to Mach 1.5 in a radius of 23 kilometers (14.3 miles), can focus the kind of caustic that might result in a super boom.[132]

Unlike in the 1960s, there was little if any NASA sonic boom flight testing during the 1970s. As a case in point, NASA's YF-12 Blackbirds at Edwards (where the Flight Research Center was renamed the Dryden Flight Research Center in 1976) flew numerous supersonic missions in support of the AST/SCAR/SCR program, but none of them were dedicated to sonic boom issues.[133] On the other hand, operations of the Concorde began providing a good deal of empirical data on sonic booms.

One discovery about secondary booms came after British Airways and Air France began Concorde service to the United Sates in May 1976. Although the Concordes slowed to subsonic speeds while well offshore, residents along the Atlantic seaboard began hearing what were called the "East Coast Mystery Booms." These were detected all the way from Nova Scotia to South Carolina, some measurable on seismographs.[134] Although a significant number of the sounds defied explanation, studies by the Naval Research Laboratory, the Federation of American Scientists, a committee of the Jason DOD scientific advisory group, and the FAA eventually determined that most of the low rumbles heard in Nova Scotia and New England were secondary booms from the Concorde. They were reaching land after being bent or reflected by temperature varia-

---

131. Plotkin and Maglieri, "Sonic Boom Research," pp. 5–6, 10.
132. Raymond L. Barger, "Sonic-Boom Wave-Front Shapes and Curvatures Associated with Maneuvering Flight," NASA TP-1611, Dec. 1979. Fig. 5 is from p. 23.
133. James and Associates, ed., *YF-12 Experiments Symposium: A conference held at Dryden Flight Research Center . . . Sept. 13–15, 1978*, NASA CP-2054, 1978; Hallion and Gorn, *On the Frontier*, Appendix P [YF-12 Flight Chronology, 1969–1978], pp. 423–429. The Dryden Center tested an oblique wing aircraft, the AD-1, from 1979 to 1982. Although this configuration might have sonic boom benefits at mid-Mach speeds, it was not a consideration in this experimental program.
134. "Second Concorde Noise Report for Dulles Shows Consistency," *Aviation Week*, July 19, 1976, p. 235; William Claiborne, "Those Mystery Booms Defy Expert Explanation," *Washington Post*, Dec. 24, 1977, p. A1.

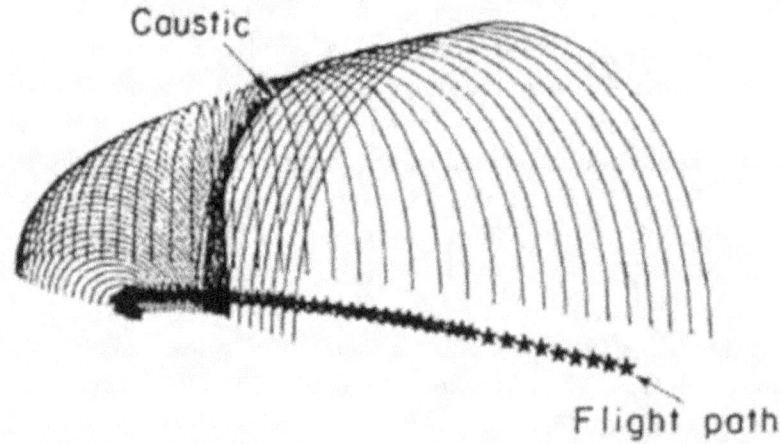

Figure 5. Acoustic wave front above a maneuvering aircraft. NASA.

tions high up in the thermosphere from Concordes still about 75 to 150 miles offshore. In July 1978, the FAA issued new rules prohibiting the Concorde from creating sonic booms that could be heard in the United States. The new FAA rules did not address the issue of secondary booms because of their low intensity; nevertheless, after Concorde secondary booms were heard by coastal communities, the Agency became even more sensitive to the sonic boom potential inherent in AST designs.[135]

The second conference on Supersonic Cruise Research, held at NASA Langley in November 1979, was the first and last under its new name. More than 140 people from NASA, other Government agencies, and the aerospace industry attended. This time there were no presentations on the sonic boom, but a representative from North American Rockwell did describe the concept of a Mach 2.7 business jet for 8–10 passengers that would generate a sonic boom of only 0.5 psf.[136] It would take another

---

135. Deborah Shapely, "East Coast Mystery Booms: A Scientific Suspense Tale," *Science*, vol. 199, no. 4336 (Mar. 31, 1978), pp. 1416–1417; "Concordes Exempted from Noise Rules," *Aviation Week*, July 3, 1978, p. 33; G.J. MacDonald, et al., "Jason 1978 Sonic Boom Report," JSR-78-09 (Arlington, VA: SRI International, Nov. 1978); Richard Kerr, "East Coast Mystery Booms: Mystery Gone but Booms Linger On," *Science*, vol. 203, no. 4337 (Jan. 19, 1979), p. 256; John H. Gardner and Peter H. Rogers, "Thermospheric Propagation of Sonic Booms from the Concorde Supersonic Transport," Naval Research Laboratory Memo Report 3904, Feb. 14, 1979 (DTIC AD A067201).

136. Robert Kelly, "Supersonic Cruise Vehicle Research/Business Jet," *Supersonic Cruise Research '79: Proceedings of a Conference . . . Hampton, VA, Nov. 13–16, 1979*, NASA CP-2108, pt. 2, pp. 935–944.

20 years for ideas about low-boom supersonic business jets to result in more than just paper studies.

Despite SCR's relatively modest cost versus its significant technological accomplishments, the program suffered a premature death in 1981. Reasons for this included the Concorde's economic woes, opposition to civilian R&D spending by key officials in the new administration of President Ronald Reagan, and a growing federal deficit. These factors, combined with cost overruns for the Space Shuttle, forced NASA to abruptly cancel Supersonic Cruise Research without even funding completion of many final reports.[137] As regards sonic boom research, an exception to this was a compilation of charts for estimating minimum sonic boom levels published by Christine Darden in May 1981. She and Robert Mack also published results of their previous experimentation that would be influential when efforts to soften the sonic boom resumed.[138]

SST Reincarnated: Birth of the High-Speed Civil Transport

For much of the next decade, the most active sonic boom research took place as part of the Air Force's Noise and Sonic Boom Impact Technology (NSBIT) program. This was a comprehensive effort started in 1981 to study the noises resulting from military training and operations, especially those involving environmental impact statements and similar assessments. Although NASA was not intimately involved with NSBIT, Domenic Maglieri (just before his retirement from the Langley Center) and the recently retired Harvey Hubbard compiled a comprehensive annotated bibliography of sonic boom research, organized into 10 major areas, to help inform NSBIT participants of the most relevant sources of information.[139]

One of the noteworthy achievements of the NSBIT program was to continue building a detailed sonic boom database (known as Boomfile) on all U.S. supersonic aircraft by flying them over a large array of newly developed sensors at Edwards AFB in the summer of 1987. Called the Boom Event Analyzer Recorder (BEAR), these unmanned devices

---

137. Conway, *High-Speed Dreams*, pp. 180–188; Chambers, *Innovations in Flight*, p. 48.
138. Darden, "Charts for Determining Potential Minimum Sonic-Boom Overpressures for Supersonic Cruise Aircraft," NASA TP-1820, May 1981; Darden and R.J. Mack, "Some Effects of Applying Sonic Boom Minimization to Supersonic Aircraft," Journal of Aircraft, vol. 17, no. 3 (Mar. 1980), pp. 182–186.
139. Hubbard, Maglieri, and David G. Stephens, "Sonic-Boom Research—Selected Bibliography with Annotation," NASA TM-87685, Sept. 1986.

recorded the full sonic boom waveform in digital format.[140] Other contributions of NSBIT were long-term sonic boom monitoring of combat training areas, continued assessment of structures exposed to sonic booms, studies of the effects of sonic booms on livestock and wildlife, and intensified research on focused booms (long an issue with maneuvering fighter aircraft). The latter included a specialized computer program (derived from that originated by NASA's Thomas) called PCBoom to predict these events.[141] In a separate project, fighter pilots were successfully trained to lay down super booms at specified locations (an idea first broached in the early 1950s).[142]

By the mid-1980s, the growing economic importance of nations in Asia was drawing attention to the long flight times required to cross the Pacific Ocean or the ability to reach most of Asia from Europe. The White House Office of Science and Technology (OST), reversing the administration's initial opposition to civilian aeronautical research, took various steps to gain support for such activities. In March 1985, the OST released a report, "National Aeronautical R&D Goals: Technology for America's Future," which included a long-range supersonic transport.[143] Then, in his State of the Union Address in January 1986, President Reagan ignited interest in the possibility of a hypersonic transport—the National Aero-Space Plane (NASP)—dubbed the "Orient Express." The Battelle Memorial Institute, which established the Center for High-Speed Commercial Flight in April 1986, became a focal point and influential advocate for these proposals.[144]

NASA had been working with the Defense Advanced Research Projects Agency (DARPA) on hypersonic technology for what became the NASP since the early 1980s. In February 1987, the OST issued an updated National Aeronautical R&D Goals, subtitled "Agenda for Achievement."

---

140. J. Micah Downing, "Lateral Spread of Sonic Boom Measurements from US Air Force Boomfile Flight Tests," *High Speed Research: Sonic Boom; Proceedings of a Conference . . . Hampton, VA, Feb. 25–27, 1992*, vol. 1, NASA CR-3172, pp. 117–129. For a description, see Robert E. Lee and Downing, "*Boom Event Analyzer Recorder: the USAF Unmanned Sonic Boom Monitor*," AIAA Paper 93-4431, Oct. 1993.
141. Plotkin and Maglieri, "Sonic Boom Research," p. 6.
142. John G. Norris, "AF Says 'Sonic Boom' Can Peril Civilians—Might be Used as Weapon," *Washington Post*, Nov. 9, 1954, pp. 1, 12; Downing, et al., "Measurement of Controlled Focused Sonic Booms from Maneuvering Aircraft," *JASA*, vol. 104, no. 1 (July 1998), pp. 112–121.
143. Judy A. Rumerman, *NASA Historical Data Book, vol. 6, NASA Space Applications, Aeronautics . . . and Resources, 1979–1988*, NASA SP-2000-4012, 2000, pp. 177–178.
144. Conway, *High-Speed Flight*, pp. 201–215; Paul Proctor, "Conference [sponsored by Battelle] Cites Potential Demand for Mach 5 Transports by Year 2000," *Aviation Week*, Nov. 10, 1986, pp. 42–46.

It called for both aggressively pursuing the NASP and developing the "fundamental technology, design, and business foundation for a long-range supersonic transport."[145] In response, NASA accelerated its hypersonic research and began a new quest to develop commercially viable supersonic technology. This started with contracts to Boeing and Douglas aircraft companies in October 1986 for market and feasibility studies on what was now named the High-Speed Civil Transport (HSCT), accompanied by several internal NASA assessments. These studies soon ruled out hypersonic speeds (above Mach 5) as being impractical for passenger service. Eventually, NASA and its industry partners settled on a cruise speed of Mach 2.4.[146] Although only marginally faster than the Concorde, the HSCT was expected to double its range and carry three times as many passengers. Meanwhile, the NASP survived as a NASA–DOD program (the X-30) until 1994, with its sonic boom potential studied by current and former NASA specialists.[147]

The contractual studies on the HSCT emphasized the need to resolve environmental issues, including the restrictions on cruising over land because of sonic booms, before it could meet the goal of efficient long-distance supersonic flight. On January 19–20, 1988, the Langley Center hosted a workshop on the status of sonic boom methodology and understanding. Sixty representatives from Government, academia, and industry attended—including many of those involved in the SST and SCR efforts and several from the Air Force's NSBIT program. Working groups on sonic boom theory, minimization, atmospheric effects, and human response determined that the following areas most needed more research: boom carpets, focused booms, high-Mach predictions, atmospheric effects, acceptability metrics, signature prediction, and low-boom airframe designs.

The report from this workshop served as a baseline on the latest knowledge about sonic booms and some of the challenges that lay ahead. One of these was the disconnect between aerodynamic efficiency and lowering shock strength that had long plagued efforts at boom minimization. Simply stated, near-field shockwaves from a streamlined

---

145. Rumerman, *NASA Historical Data Book*, vol. 6, p. 178.
146. Conway, *High-Speed Dreams*, pp. 218–228; Chambers, *Innovations in Flight*, p. 50.
147. Maglieri, Victor E. Sothcroft, and John Hicks, "Influence of Vehicle Configurations and Flight Profile on X-30 Sonic Booms," AIAA Paper 90-5224, Oct. 29, 1990; Maglieri, "A Brief Review of the National Aero-Space Plane Sonic Booms Final Report," USAF Aeronautical Systems Center TR-94-9344, Dec. 1992.

airframe coalesce more readily into strong front and tail shocks, while the near-field shock waves from a higher-drag airframe are less likely to join together, thus allowing a more relaxed N-wave signature. This paradox (illustrated by Figure 6) would have to be solved before a low-boom supersonic transport would be both permissible and practical.[148]

Trying Once More: The High-Speed Research Program

While Boeing and Douglas were reporting on early phases of their HSCT studies, the U.S. Congress approved an ambitious new program for High-Speed Research (HSR) in NASA's budget for FY 1990. This effort envisioned Government and industry sharing the cost, with NASA taking the lead for the first several years and industry expanding its role as research progressed. (Because of the intermingling of sensitive and proprietary information, much of the work done during the HSR program was protected by a limited distribution system, and some has yet to enter the public domain.) Although the aircraft companies made some early progress on lower-boom concepts for the HSCT, they identified the need for more sonic boom research by NASA, especially on public acceptability and minimization techniques, before they could design a practical HSCT able to cruise over land.[149]

Because solving environmental issues would be a prerequisite to developing the HSCT, NASA structured the HSR program into two phases. Phase I—focusing on engine emissions, noise around airports, and sonic booms, as well as preliminary design work—was scheduled for 1990–1995. Among the objectives of Phase I were predicting HCST sonic boom signatures, determining feasible reduction levels, and finding a scientific basis on which to set acceptability criteria. After hopefully making sufficient progress on the environmental problems, Phase II would begin in 1994. With more industry participation and greater funding, it would focus on economically realistic airframe and propulsion technologies and was hoped to have extended until 2001.[150]

---

148. Darden, et al., *Status of Sonic Boom Methodology and Understanding; Proceedings . . . Langley Research Center . . . Jan. 19–20, 1988*, NASA CP-3027, June 1989. Fig. 6 is copied from p. 6.
149. Boeing Commercial Airplanes, "High-Speed Civil Transport Study; Final Report," NASA CR-4234, Sept. 1989; Douglas Aircraft Company, "1989 High-Speed Civil Transport Studies," NASA CR-4375, May 1991 (published late with an extension). For a summary of Boeing's early design process, see George T. Haglund, "HSCT Designs for Reduced Sonic Boom," AIAA Paper 91-3103, Sept. 1991.
150. Allen H. Whitehead, ed., *First Annual High-Speed Research Workshop; Proceedings . . . Williamsburg, VA, May 14–16, 1991*, NASA CP-10087, Apr. 1992, pt. 1, pp. 5–22, 202.

When NASA convened its first workshop for the entire High-Speed Research program in Williamsburg, VA, from May 14–16, 1991, the headquarters status report on sonic boom technology warned that "the importance of reducing sonic boom cannot be overstated." One of the Douglas studies had projected that even by 2010, overwater-only routes would account for only 28 percent of long-range air traffic, but with overland cruise, the proposed HSCT could capture up to 70 percent of all such traffic. Based on past experience, the study admitted that research on low boom designs "is viewed with some skepticism as to its practical application. Therefore an early assessment is warranted."[151]

NASA, its contractors, academic grantees, and the manufactures were already busy conducting a wide range of sonic boom research projects. The main goals were to demonstrate a waveform shape that could be acceptable to the public, to prove that a viable airplane could be built to generate such a waveform, to determine that such a shape would not be too badly disrupted during its propagation through the atmosphere, and to estimate that the economic benefit of overland supersonic flight would make up for any performance penalties imposed by a low-boom design.[152]

During the next 3 years, NASA and its partners went into a full-court press against the sonic boom. They began several dozen major experiments and studies, the results of which were published in reports and presented at conferences and workshops dealing solely with the sonic boom. These were held at the Langley Research Center in February 1992,[153] the Ames Research Center in May 1993,[154] the Langley Center in June 1994,[155] and again at Langley in September 1995.[156] The workshops, like the sonic boom effort itself, were organized into three major

---

151. Ibid., pp. 272 and 275 quoted.
152. Ibid., Table of Contents, pp. iv-v.
153. Darden, ed., *High-Speed Research: Sonic Boom; Proceeding . . . Langley Research Center . . . Feb. 25–27, 1992*, NASA CR-3172, Oct. 1992, vols. 1, 2.
154. Thomas A. Edwards, ed., *High-Speed Research: Sonic Boom; Proceedings . . . Ames Research Center . . . May 12–14, 1993*, NASA CP-10132, vol. 1.
155. David A. McCurdy, ed., *High-Speed Research: 1994 Sonic Boom Workshop, Atmospheric Propagation and Acceptability Studies; Proceedings . . . Hampton, VA, June 1–3, 1994*, NASA CP-3209; *High-Speed Research: 1994 Sonic Boom Workshop: Configuration, Design, Analysis, and Testing . . . Hampton, VA, June 1–3, 1994*, NASA CP-209669, Dec. 1999.
156. Daniel G. Baize, *1995 NASA High-Speed Research Program Sonic Boom Workshop: Proceedings . . . Langley Research Center . . . Sept. 12–13, 1995*, NASA CP-3335, vol. 1, July 1966.

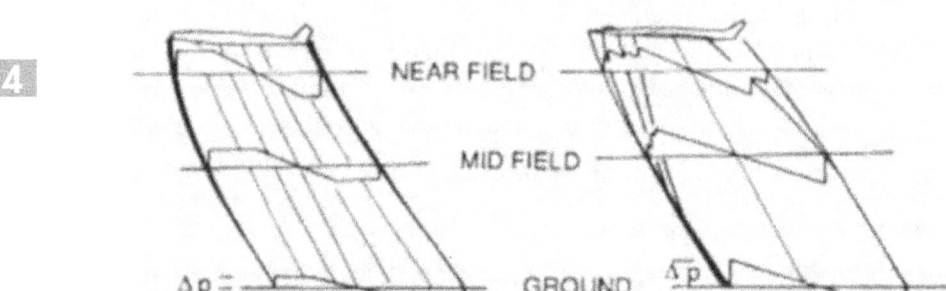

Figure 6. Low-boom/high-drag paradox. NASA.

areas of research: (1) configuration design and analysis (managed by Langley's Advanced Vehicles Division), (2) atmospheric propagation, and (3) human acceptability (both managed by Langley's Acoustics Division). The reports from these workshops were each well over 500 pages long and included dozens of papers on the progress or completion of various projects.[157]

The HSR program precipitated major advances in the design of supersonic configurations for reduced sonic boom signatures. Many of these were made possible by the new field of computational fluid dynamics (CFD). Researchers were now able to use complex computational algorithms processed by supercomputers to calculate the nonlinear aspects of near-field shock waves, even at high Mach numbers and angles of attack. Results could be graphically displayed in mesh and grid formats that emulated three dimensions. (In simple terms: before CFD, the nonlinear characteristics of shock waves generated by a realistic airframe had involved too many variables and permutations to calculate by conventional means.)

The Ames Research Center, with its location in the rapidly growing Silicon Valley area, was a pioneer in applying CFD capabilities to aerodynamics. At the 1991 HSR workshop, an Ames team led by Thomas Edwards and including modeling expert Samsun Cheung predicted that "in many ways, CFD paves the way to much more rapid progress

---

157. For help in deciding which of the many research projects to cover, the author referred to Darden, "Progress in Sonic-Boom Understanding: Lessons Learned and Next Steps," *1994 Sonic Boom Workshop*, pp. 269–292, for guidance.

in boom minimization. . . . Furthermore, CFD offers fast turnaround and low cost, so high-risk concepts and perturbations to existing geometries can be investigated quickly."[158]

At the same time, Christine Darden and a team that included Robert Mack and Peter G. Coen, who had recently devised a computer program for predicting sonic booms, used very realistic 12-inch wind tunnel models (the largest yet to measure for sonic boom). Although the model was more realistic than previous ones and validating much about the designs, including such details as engine nacelles, signature measurements in Langley's 4 by 4 Unitary Wind Tunnel and even Ames 9 by 7 Unitary Wind Tunnel still left much to be desired.[159] During subsequent workshops and at other venues, experts from Ames, Langley and their local contractors reported optimistically on the potential of new CFD computer codes—with names like UPS3D, OVERFLOW, AIRPLANE, and TEAM—to help design configurations optimized for both constrained sonic booms and aerodynamic efficiency. In addition to promoting the use of CFD, former Langley employee Percy "Bud" Bobbitt of Eagle Aeronautics pointed out the potential of hybrid laminar flow control (HLFC) for both aerodynamic and low-boom purposes.[160] At the 1992 workshop, Darden and Mack admitted how recent experiments at Langley had revealed limitations in using near-field wind tunnel data for extrapolating sonic boom signatures.[161]

Even the numbers-crunching capability of supercomputers was not yet powerful enough for CFD codes and the grids they produced to accurately depict effects beyond the near field, but the use of parallel computing held the promise of eventually doing so. It was becoming apparent that, for most aerodynamic purposes, CFD was the design tool of the future, with wind tunnel models becoming more a means of verification. As just one example of the value of CFD methods, Ames researchers were able to design an airframe that generated a type of multishock signature that

---

158. Thomas A. Edwards, et al., "Sonic Boom Prediction and Minimization using Computational Fluid Dynamics," *First Annual High Speed Workshop [1991]*, pt. 2, p. 732.
159. Darden, et al., "Design and Analysis of Low Boom Concepts at Langley Research Center," ibid., pp. 675–699); Peter G. Coen, "Development of a Computer Technique for Prediction of Transport Aircraft Flight Profile Sonic Boom Signatures," NASA CR-188117, Mar. 1991 (M.S. Thesis, George Washington University).
160. Percy J. Bobbit, "Application of Computational Fluid Dynamics and Laminar Flow Technology for Improved Performance and Sonic Boom Reduction," *1992 Sonic Boom Workshop*, vol. 2, pp. 137–144.
161. Mack and Darden, "Limitations on Wind-Tunnel Pressure Signature Extrapolation," *1992 Sonic Boom Workshop*, vol. 2, pp. 201–220.

might reach the ground with a quieter sonic boom than either the ramp or flattop wave forms that were a goal of traditional minimization theories.[162] Although not part of the HSCT effort, Ames and its contractors also used CFD to continue exploring the possible advantages of oblique-wing aircraft, including sonic boom minimization.[163]

Since neither wind tunnels nor CFD could as yet prove the persistence of waveforms for more than a small fraction of the 200 to 300 body lengths needed to represent the distance from an HSCT to the surface, Domenic Maglieri of Eagle Aeronautics led a feasibility study in 1992 on the most cost effective ways to verify design concepts with realistic testing. After exploring a wide range of alternatives, the team selected the Teledyne-Ryan BQM-34 Firebee II remotely piloted vehicle (RPV), which the Air Force and Navy had used as a supersonic target drone. Four of these 28-feet-long RPVs, which could sustain a speed of Mach 1.3 at 9,000 feet (300 body lengths from the surface) were still available as surplus. Modifying them with low-boom design features such as specially configured 40-inch nose extensions (shown in Figure 7 with projected waveforms from 20,000 feet), could provide far-field measurements needed to verify the waveform shaping projected by CFD and wind tunnel models.[164] Meanwhile, a complementary plan at the Dryden Flight Research Center led to NASA's first significant sonic boom testing there since the 1960s. SR-71 program manager David Lux, atmospheric specialist L.J. Ehernberger, aerodynamicist Timothy R. Moes, and principal investigator Edward A. Haering came up with a proposal to demonstrate CFD design concepts by having one of Dryden's SR-71s modified with a low-boom configuration. As well as being much larger, faster, and higher-flying than the little Firebee, an SR-71 would also allow easier acquisition of near-field measurements for direct comparison with CFD predic-

---

162. Susan E. Cliff, et al., "Design and Computational/Experimental Analysis of Low Sonic Boom Configurations," *1994 Sonic Boom Workshop*, vol. 2, pp. 33–57. For a review of CFD work at Ames from 1989–1994, see Samsun Cheung, "Supersonic Civil Airplane Study and Design: Performance and Sonic Boom," NASA CR-197745, Jan. 1995.

163. Christopher A. Lee, "Design and Testing of Low Sonic Boom Configurations and an Oblique All-Wing Supersonic Transport," NASA CR-197744, Feb. 1995.

164. Maglieri, et al., "Feasibility Study on Conducting Overflight Measurements of Shaped Sonic Boom Signatures Using the Firebee BQM-34E RPV," NASA CR-189715, Feb. 1993. Fig. 7 is copied from p. 52, with waveforms based on a speed of Mach 1.3 at 20,000 feet rather than the 9,000 feet of planned flight tests.

tions.[165] To lay the groundwork for this modification, the Dryden Center obtained baseline data from a standard SR-71 using one of its distinctive F-16XL aircraft (built by General Dynamics in the early 1980s for evaluation by the Air Force as a long-range strike version of the F-16 fighter). In tests at Edwards during July 1993, the F-16XL flew as close as 40 feet below and behind an SR-71 cruising at Mach 1.8 to collect near-field pressure measurements. Both the Langley Center and McDonnell-Douglas analyzed this data, which had been gathered by a standard flight-test nose boom. Both reached generally favorable conclusions about the ability of CFD and McDonnell-Douglas's proprietary MDBOOM program (derived from PCBoom) to serve as design tools. Based on these results, McDonnell-Douglas Aerospace West designed modifications to reduce the bow and middle shock waves of the SR-71 by reshaping the front of the airframe with a "nose glove" and adding to the midfuselage cross-section. An assessment of these modifications by Lockheed Engineering & Sciences found them feasible.[166] The next step would be to obtain the considerable funding that would be needed for the modifications and testing.

In May 1994, the Dryden Center used two of its fleet of F-18 Hornets to measure how near-field shockwaves merged to assess the feasibility of a similar low-cost experiment in waveform shaping using two SR-71s. Flying at Mach 1.2 with one aircraft below and slightly behind the other, the first experiment positioned the canopy of the lower F-18 in the tail shock extending down from the upper F-18 (called a tail-canopy formation). The second experiment had the lower F-18 fly with its canopy in the inlet shock of the upper F-18 (inlet-canopy). Ground sensor recordings revealed that the tail-canopy formation caused two separate N-wave signatures, but the inlet-canopy formation yielded a single modified signature, which two of the recorders measured as a flat-top waveform. Even with the excellent visibility from the F-18's bubble canopy (one pilot used the inlet shock wave as a visual cue for positioning

---

165. David Lux, et al., "Low-Boom SR-71 Modified Signature Demonstration Program," *1994 Sonic Boom Workshop: Configuration, Design, Analysis and Testing*, pp. 237–248.
166. Edward H. Haering, et al., "Measurement of the Basic SR-71 Airplane Near-Field Signature," *1994 Sonic Boom Workshop: Configuration, Design, Analysis, and Testing*, pp. 171–197; John M. Morgenstern, et al., "SR-71A Reduced Sonic Boom Modification Design," ibid., pp. 199–217; Kamran Fouladi, "CFD Predictions of Sonic-Boom Characteristics for Unmodified and Modified SR-71 Configurations," ibid., pp. 219–235. Fig. 8 is copied from p. 222.

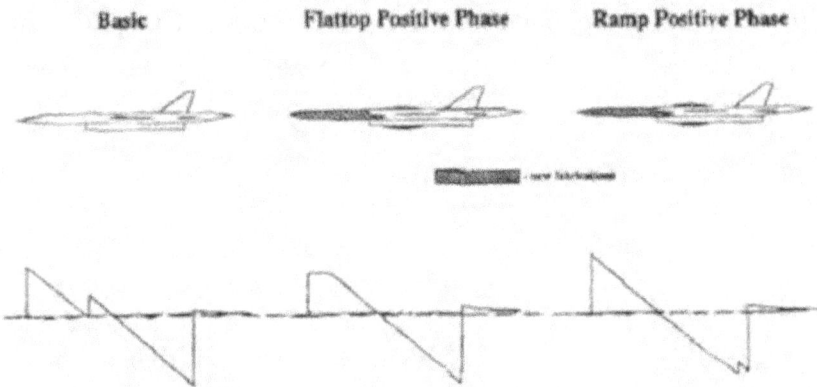

Figure 7. Proposed modifications to BQM-34 Firebee II. NASA.

his aircraft) and its responsive flight controls, maintaining such precise positions was still not easy, and the pilots recommended against trying to do the same with the SR-71, considering its larger size, slower response, and limited visibility.[167]

Atmospheric effects had long posed many uncertainties in understanding sonic booms, but advances in acoustics and atmospheric science since the SCR program promised better results. Scientists needed a better understanding not only of the way air molecules absorb sound waves, but also old issue of turbulence. In addition to using the Air Force's Boomfile and other available material, Langley's Acoustic Division had Eagle Aeronautics, in a project led by Domenic Maglieri, restore and digitize data from the irreplaceable XB-70 records.[168]

The division also took advantage of the NATO Joint Acoustic Propagation Experiment (JAPE) at the White Sands Missile Range in August 1991 to do some new testing. The researchers arranged for F-15, F-111, T-38 aircraft, and one of Dryden's SR-71s to make 59 supersonic passes over an extensive array of BEAR, other recording systems, and meteorological sensors—both early in the morning (when the air was still) and during the afternoon (when there was usually more turbulence). Although meteorological data were incomplete, results later showed

---

167. Catherine M. Bahm and Edward A. Haering, "Ground-Recorded Sonic Boom Signatures of F-18 Aircraft in Formation Flight," *1995 Sonic Boom Workshop*, vol. 1, pp. 220–243.
168. J. Micah Downing, "Lateral Spread of Sonic Boom Measurements from US Air Force Boomfile Flight Tests," *1992 Sonic Boom Workshop*, vol. 1, pp. 117–136; Maglieri, et al., "A Summary of XB-70 Sonic Boom Signature Data, Final Report," NASA CR-189630, Apr. 1992.

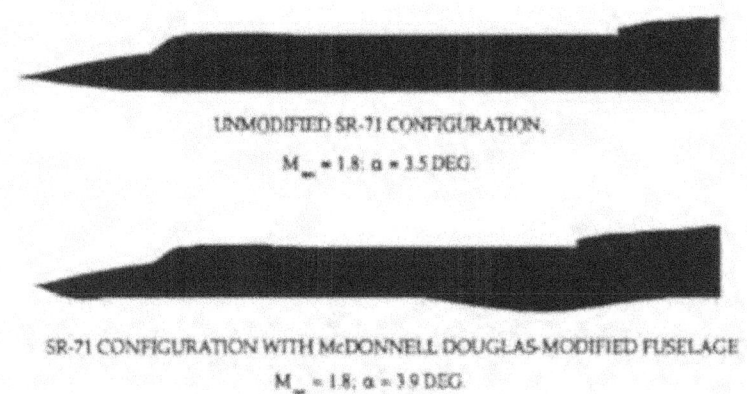

Figure 8. Proposed SR-71 low-boom modification. NASA.

the effects of molecular relaxation and turbulence on both the rise time and overpressure of bow shocks.[169] Additional atmospheric information came from experiments on waveform freezing (persistence), measuring diffraction and distortion of sound waves, and trying to discover the actual relationship among molecular relaxation, turbulence, humidity, and other weather conditions.[170]

Leonard Weinstein of the Langley Center even developed a way to capture images of shock waves in the real atmosphere. He did this using a ground-based schlieren system (a specially masked and filtered tracking camera with the Sun providing backlighting). As shown in the accompanying photo, this was first demonstrated in December 1993 with a T-38 flying just over Mach 1 at Wallops Island.[171] All of the research into the theoretical, aerodynamic, and atmospheric aspects of sonic boom—no matter how successful—would not protect the Achilles' heel of previous programs: the subjective response of human beings.

---

169. William L. Willshire and David W. DeVilbiss, "Preliminary Results from the White Sands Missile Range Sonic Boom Propagation Experiment," *1992 Sonic Boom Workshop*, vol. 1, pp. 137–144.
170. Gerry L. McAnich, "Atmospheric Effects on Sonic Boom—A Program Review," *First Annual HSR Workshop*, pp. 1201–1207; Allan D. Pierce and Victor W. Sparrow, "Relaxation and Turbulence Effects on Sonic Boom Signatures," ibid., pp. 1211–1234; Kenneth J. Plotkin, "The Effect of Turbulence and Molecular Relaxation on Sonic Boom Signatures," ibid., pp. 1241–1261; Lixin Yao, et al., "Statistical and Numerical Study of the Relation Between Weather and Sonic Boom," ibid., pp. 1263–1284.
171. Leonard M. Weinstein, "An Optical Technique for Examining Aircraft Shock Wave Structures in Flight," *1994 Sonic Boom Workshop, Atmospheric Propagation*, pp. 1–18. The following year Weinstein demonstrated improved results using a digital camera: "An Electronic Schlieren Camera for Aircraft Shock Wave Visualization," *1995 Sonic Boom Workshop*, vol. 1, pp. 244–258.

As a result, the Langley Center, led by Kevin Shepherd of the Acoustics Division, had begun a systematic effort to measure human responses to different strengths and shapes of sonic booms and hopefully determine a tolerable level for community acceptance. As an early step, the division built an airtight foam-lined sonic boom simulator booth (known as the "boom box") based on one at the University of Toronto. Using the latest in computer-generated digital amplification and loudspeaker technology, it was capable of generating shaped waveforms up to 4 psf (140 decibels). Based on responses from subjects, researchers selected the perceived-level decibel (PLdB) as the preferred metric. For responses outside a laboratory setting, Langley planned several additional acceptance studies.[172]

By 1994, early results had become available from two of these projects. The Langley Center and Wyle Laboratories had developed mobile boom simulator equipment called the In-Home Noise Generation/Response System (IHONORS). It consisted of computerized sound systems installed in 33 houses for 8 weeks at a time in a network connected by modems to a monitor at Langley. From February to December 1993, these households were subjected to almost 58,500 randomly timed sonic booms of various signatures for 14 hours a day. Although definitive analyses were not available until the following year, the initial results confirmed how the level of annoyance increased whenever subjects were startled or trying to rest.[173]

Preliminary results were also in from the first phase of the Western USA Sonic Boom Survey of civilians who had been exposed to such sounds for many years. This part of the survey took place in remote desert towns around the Air Force's vast Nellis combat training range complex in Nevada. Unlike previous community surveys, it correlated citizen responses to accurately measured sonic boom signatures (using BEAR devices) in places where booms were a regular occurrence, yet where the subjects did not live on or near a military installation (i.e., where

---

172. Kevin P. Shepherd, "Overview of NASA Human Response to Sonic Boom Program," *First Annual HSR Workshop*, pt. 3, pp. 1287–1291; Shepherd, et al., "Sonic Boom Acceptability Studies," ibid., pp. 1295–1311.
173. David A. McCurdy, et al., "An In-Home Study of Subjective Response to Simulated Sonic Booms," *1994 Sonic Boom Workshop: Atmospheric Propagation and Acceptability*, pp. 193–207; McCurdy and Sherilyn A. Brown, "Subjective Response to Simulated Sonic Boom in Homes," *1995 Sonic Boom Workshop*, pp. 278–297.

Case 4 | Softening the Sonic Boom: 50 Years of NASA Research

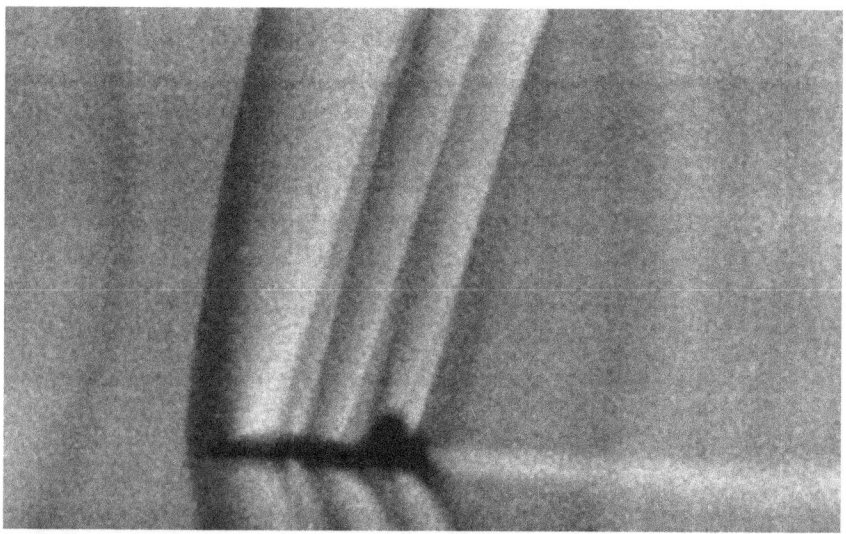

Leonard Weinstein's innovative schlieren photograph showing shock waves emanating from a T-38 flying Mach 1.1 at 13,000 feet, December 1993. NASA.

the economic benefits of the base for the local economy might influence their opinions). Although findings were not yet definitive, these 1,042 interviews proved more decisive than any of the many other research projects in determining the future direction of the HSCT effort. Based on a metric called day-night average noise level, the respondents found the booms much more annoying than previous studies on other types of aircraft noise had, even at the levels projected for low-boom designs. Their negative responses, in effect, dashed hopes that the HSR program might lead to an overland supersonic transport.[174]

Well before the paper on this survey was presented at the 1994 Sonic Boom Workshop, its early findings had prompted NASA Headquarters to reorient High-Speed Research toward an HSCT design that would only fly supersonic over water. Just as with the AST program 20 years earlier, this became the goal of Phase II of the HSR program (which began using FY 1994 funding left over from the canceled NASP).[175]

At the end of the 1994 workshop, Christine Darden discussed lessons learned so far and future directions. While the design efforts had shown outstanding progress, dispersal of the work among two NASA Centers

---

174. James M. Fields, et al., "Residents' Reactions to Long-Term Sonic Boom Exposure: Preliminary Results," *1994 Sonic Boom Workshop: Atmospheric Propagation and Acceptability,* vol. 1, pp. 193–217.
175. Conway, *High-Speed Dreams,* p. 253.

and two major aircraft manufacturers had resulted in communication problems as well as a certain amount of unhelpful competition. The milestone-driven HSR effort required concurrent progress in various different areas, which is inherently difficult to coordinate and manage. And even if low-boom airplane designs had been perfected to meet acoustic criteria, they would have been heavier and suffer from less acceptable low-speed performance than unconstrained designs. Under the new HSR strategy, any continued minimization research would now aim at lowering the sonic boom of the "baseline" overwater design, while propagation studies would concentrate on predicting boom carpets, focused booms, secondary booms, and ground disturbances. In view of the HSCT's overwater mission, new environmental studies would devote more attention to the penetration of shock waves into water and the effect of sonic booms on marine mammals and birds.[176]

Although the preliminary results of the first phase of the Western USA Survey had already had a decisive impact, Wyle Laboratories completed Phase II with a similar polling of civilians in Mojave Desert communities exposed regularly to sonic booms, mostly from Edwards AFB. Surprisingly, this phase of the survey found the people there much more amenable to sonic booms than the desert dwellers in Nevada were, but they were still more annoyed by booms than by other aircraft noise of comparable perceived loudness.[177]

With the decision to end work on a low-boom HSCT, the proposed modifications of the Firebee RPVs and SR-71 had of course been canceled (postponing for another decade any full-scale demonstrations of boom shaping). Nevertheless, some testing continued that would prove of future value. From February through April 1995, the Dryden Center conducted more SR-71 and F-16XL sonic boom flight tests. Led by Ed Haering, this experiment included an instrumented YO-3A light aircraft from the Ames Center, an extensive array of various ground sensors, a network of new differential Global Positioning System (GPS) receivers accurate to within 12 inches, and installation of a sophisticated new nose boom with four pressure sensors on the F-16XL. On eight long missions, one of Dryden's SR-71s flew at speeds between Mach 1.25 and Mach

---

176. Darden, "Progress in Sonic-Boom Understanding: Lessons Learned and Next Steps," *1994 Sonic Boom Workshop, Atmospheric Propagation and Acceptability*, pp. 269–290.
177. James M. Fields, "Reactions of Residents to Long-Term Sonic Boom Noise Environments," NASA CR-201704, June 1997.

1.6 at 31,000–48,000 feet, while the F-16XL (kept aloft by in-flight refuelings) made numerous near- and mid-field measurements at distances from 80 to 8,000 feet. Some of these showed that the canopy shock waves were still distinct from the bow shock after 4,000–6,000 feet. Comparisons of far-field measurements obtained by the YO-3A flying at 10,000 feet above ground level and the recording devices on the surface revealed effects of atmospheric turbulence. Analysis of the data validated two existing sonic boom propagation codes and clearly showed how variations in the SR-71's gross weight, speed, and altitude caused differences in shock wave patterns and their coalescence into N-shaped waveforms.[178]

This successful experiment marked the end of dedicated sonic boom flight-testing during the HSR program.

By late 1998, a combination of economic, technological, political, and budgetary problems (including NASA's cost overruns for the International Space Station) convinced Boeing to cut its support and the Administration of President Bill Clinton to terminate the HSR program at the end of FY 1999. Ironically, NASA's success in helping the aircraft industry develop quieter subsonic aircraft, which had the effect of moving the goalpost for acceptable airport noise, was one of the factors convincing Boeing to drop plans for the HSCT. Nevertheless, the HSR program was responsible for significant advances in technologies, techniques, and scientific knowledge, including a better understand of the sonic boom and ways to diminish it.[179]

Proof at Last: The Shaped Sonic Boom Demonstration

After the HSR program dropped plans for an overland supersonic airliner, Domenic Maglieri compiled a NASA study of all known proposals

---

178. Edward A. Haering, L.J. Ehernberger, and Stephen A. Whitmore, "Preliminary Airborne Experiments for the SR-71 Sonic Boom Propagation Experiment," *1995 Sonic Boom Workshop*, vol. 1, pp. 176–198; Stephen R. Norris, Haering, and James E. Murray, "Ground-Bases Sensors for the SR-71 Sonic Boom Propagation Experiment," ibid., pp. 199–218; Hugh W. Poling, "Sonic Boom Propagation Codes Validated by Flight Test," NASA CR-201634, Oct. 1996.

179. Conway, *High-Speed Dreams*, pp. 286–300; James Schultz, "HSR Leaves Legacy of Spinoffs," *Aerospace America*, vol. 37, no. 9 (Sept. 1999), pp. 28–32. The Acoustical Society held its third sonic boom symposium in Norfolk from Oct. 15–16, 1998. Because of HSR distribution limitations, many of the presentations could be oral only, but a few years later, the ASA was able to publish some of them in a special edition of its journal. For a status report as of the end of the HSR, see Kenneth J. Plotkin, "State of the Art of Sonic Boom Modeling," *JASA*, vol. 111, No. 1, pt. 3 (Jan. 2002), pp. 530–536.

for smaller supersonic aircraft intended for business customers.[180] In 1998, one year after the drafting of this report, Richard Seebass (now with the University of Colorado) gave some lectures at NATO's von Kármán Institute in Belgium. He reflected on NASA's conclusion that a practical, commercial-sized supersonic transport would have a sonic boom that was not acceptable to enough people. On the other hand, he believed the recent high-speed research "leads us to conclude that a small, appropriately designed supersonic business jet's sonic boom may be nearly inaudible outdoors and hardly discernible indoors." Such an airplane, he stated, "appears to have a significant market . . . if . . . certifiable over most land areas."[181]

At the start of the new century, the prospects for a small supersonic aircraft received a shot in the arm from the Defense Advanced Research Projects Agency, well known for encouraging innovative technologies. DARPA received $7 million in funding starting in FY 2001 to explore design concepts for a Quiet Supersonic Platform (QSP)—an airplane that could have both military and civilian potential. Richard W. Wlezien, a NASA official on loan to DARPA as QSP program manager, wanted ideas that might lead to a Mach 2.4, 100,000-pound aircraft that "won't rattle your windows or shake the china in your cabinet." It was hoped that a shaped sonic boom signature of no more than 0.3 psf would allow unrestricted operations over land. By the end of 2000, 16 companies and laboratories had been selected to participate in the QSP project, with the University of Colorado and Stanford University to work on sonic boom propagation and minimization.[182] Support from NASA would include modeling expertise, wind tunnel facilities, and flight-test operations.

Although the later phase of the QSP program emphasized military requirements, its most publicized achievement was the Shaped Sonic Boom Demonstration (SSBD). This was not one of its original components.

---

180. The study, originally completed in 1997, was in the process of being formally published by NASA as "A Compilation and Review of Supersonic Business Jet Studies from 1960–1995" as this history was being written.

181. Richard Seebass, "History and Economics of, and Prospects for, Commercial Supersonic Transport," (Paper 1) and "Sonic Boom Minimization" (Paper 6), *NATO Research and Technology Organization, Fluid Dynamics Research on Supersonic Aircraft [proceedings . . . Rhode Saint-Genèse, Belgium, May 25–29, 1998]*, RTO-EN-4, Nov. 1998 (pp. 1–6 of Paper 1 and abstract of Paper 2 quoted). Sadly, Seebass would not live to see a low boom airplane configuration finally demonstrated in 2003.

182. Robert Wall, "Darpa Envisions New Supersonic Designs," *Aviation Week*, Aug. 28, 2000, p. 47; and "Novel Technologies in Quest for Quiet Flight," *Aviation Week*, Jan. 8, 2001, p. 61.

In 1995, the Dryden Flight Research Center used an F-16XL to make detailed in-flight supersonic shock wave measurements as near as 80 feet from an SR-71. NASA.

Resurrecting an idea from the HSR program, Domenic Maglieri and colleagues at Eagle Aeronautics recommended that DARPA include a flight-test program using the BQM-34E Firebee II as a proof-of-concept for the QSP's sonic boom objectives. Liking this idea, Northrop Grumman Corporation (NGC) wasted no time in acquiring the last remaining Firebee IIs from the Naval Air Weapons Station at Point Mugu, CA, but later determined that they were now too old for test purposes. As an alternative, NGC aerodynamicist David Graham recommended using different versions of the Northrop F-5 (which had been modified into larger training and reconnaissance models) for sonic boom comparisons. Maglieri then suggested modifications to an F-5E that could flatten its sonic boom signature. Based largely on NGC's proposal for an F-5E Shaped Sonic Boom Demonstration, DARPA in July 2001 selected it over QSP proposals from the other two system integrators, Boeing Phantom Works and Lockheed Martin's Skunk Works.[183]

In designing the modifications, a Northrop Grumman team in El Segundo, CA, led by David Graham, benefited from its partnership with

---

183. Joseph W. Pawlowski, David H. Graham, Charles H. Boccadoro (NGC), Peter G. Coen (LaRC), and Domenic J. Maglieri (Eagle Aero.), "Origins and Overview of the Shaped Sonic Boom Demonstration Program," AIAA Paper 2005-5, presented at the *43rd Aerospace Sciences Meeting, Reno, NV, Jan. 10–13, 2005*, pp. 3–7 (also published with briefing slides by the Air Force Research Laboratory as AFRL-VA-WP-2005-300, Jan. 2005).

a multitalented working group. This team included Kenneth Plotkin of Wyle Laboratories, Domenic Maglieri and Percy Bobbitt of Eagle Aeronautics, Peter G. Coen and colleagues at the Langley Center, John Morgenstern of Lockheed Martin, and other experts from Boeing, Gulfstream, and Raytheon. They applied knowledge gained from the HSR with the latest in CFD technology to begin design of a nose extension and other modifications to reshape the F-5E's sonic boom. The moderate size and flexibility of the basic F-5E design, which had allowed different configurations in the past, made it the perfect choice for the SSBD. The shaped-signature modifications (which harked back to the stillborn SR-71 proposal of the HSR program) were tested in a supersonic wind tunnel at NASA's Glenn Research Center with favorable results.[184]

In further preparation for the SSBD, the Dryden Center conducted the Inlet Spillage Shock Measurement (ISSM) experiment in February 2002. One of its F-15Bs equipped with an instrumented nose boom gathered pressure data from a standard F-5E flying at about Mach 1.4 and 32,000 feet. The F-15 did these probes at separation distances ranging from 60 to 1,355 feet. In addition to serving as a helpful "dry run" for the planned demonstration, the ISSM experiment proved to be of great value in validating and refining Grumman's proprietary GCNSfv CFD code (based on the Ames Center's ARC3D code), which was being used to design the SSBD configuration. Application of the flight test measurements nearly doubled the size of the CFD grid, to approximately 14 million points.[185]

For use in the Shaped Sonic Boom Demonstration, the Navy loaned Northrop Grumman one of its standard F-5Es, which the company began to modify at its depot facility in St. Augustine, FL, in January 2003. Under supervision of the company's QSP program manager, Charles Boccadoro, NGC technicians installed a nose glove and 35-foot fairing under the fuselage (resulting in a "pelican-shaped" profile). The modifications, which extended the plane's length from 46 to approximately 50 feet, were designed to strengthen the bow shock but weaken and stretch out the shock waves from the cockpit, inlets, and wings—keep-

---

184. Ibid., p. 8; Edward D. Flinn, "Lowering the Boom on Supersonic Flight Noise," *Aerospace America*, vol. 40, No. 2 (Feb. 2002), pp. 20–21; Wyle Laboratories, "Wyle Engineers Play Significant Role in Northrop Grumman Sonic Boom Test Program," News Release 09-11, Sept. 11, 2003.
185. Keith H. Meredith, et al., "Computational Fluid Dynamics Comparison and Flight Test Measurement of F-5E Off-Body Pressures," AIAA Paper 2005-6, presented at *43rd Aerospace Sciences Meeting*, Reno, NV, Jan. 10–13, 2005.

ing them from coalescing to form the sharp initial peak of the N-wave signature.[186] After checkout flights in Florida starting on July 25, 2003, the modified F-5E, now called the SSBD F-5E, arrived in early August at Palmdale, CA, for more functional check flights.

On August 27, 2003, on repeated runs through an Edwards supersonic corridor, the SSBD F-5E, piloted by NGC's Roy Martin, proved for the first time that—as theorized since the 1960s—a shaped sonic boom signature from a supersonic aircraft could persist through the real atmosphere to the ground. Flying at Mach 1.36 and 32,000 feet on an early-morning run, the SSBD F-5E was followed 45 seconds later by an unmodified F-5E from the Navy's aggressor training squadron at Fallon, NV. They flew over a high-tech ground array of various sensors manned by personnel from Dryden, Langley, and almost all the organizations in the SSBD working group. Figure 9 shows the subtle but significant difference between the flattened waveform from the SSBD F-5E (blue) and the peaked N-wave from its unmodified counterpart (red) as recorded by a Base Amplitude and Direction Sensor (BADS) on this historic occasion. As a bonus, the initial rise in pressure of the shaped signature was only about 0.83 psf as compared with the 1.2 psf from the standard F-5E—resulting in a much quieter sonic boom.[187]

During the last week of August, the two F-5Es flew three missions to provide many more comparative sonic boom recordings. On two other missions, using the technique developed for the SR-71 during HSR, a Dryden F-15B with a specially instrumented nose boom followed the SSBD-modified F-5E to gather near-field measurements. The data from the F-15B probing missions showed how the F-5E's modifications changed its normal shock wave signature, which data from the ground sensors revealed as persisting down through the atmosphere to consistently produce the quieter flat-topped sonic boom signatures. The SSBD met expectations, but unusually high temperatures (even for the Antelope Valley in August) limited the top speed and endurance of the F-5Es. Because of this and a desire to gather more data on maneuvers and different atmospheric conditions,

---

186. Graham Warwick, "F-5E Shapes Up to Change Sonic Boom," *Flight International*, Aug. 5, 2003, p. 30; T.A. Heppenheimer, "The Boom Stops Here," *Air and Space Magazine*, Nov. 2005, http://www.airspacemag.com/fight-today/boom.html.

187. Pawlowski, et al., "Origins and Overview of the SSBD," pp. 10–12; Peter G. Coen and Roy Martin, "Fixing the Sound Barrier: The DARPA/NASA/Northrop-Grumman Shaped Sonic Boom Flight Demonstration," Briefing at EAA AirVenture, Oshkosh, WI, July 2004. (Fig. 9 is taken from slide 21.)

Peter Coen, Langley's manager for supersonic vehicles technology, and researchers at Dryden led by SSBD project manager David Richwine and principal investigator Ed Haering, began planning a NASA-funded Shaped Super Boom Experiment (SSBE) to follow up on the SSBD.[188]

NASA successfully conducted the SSBE with 21 more flights during 11 days in January 2004. These met or exceeded all test objectives. Eight of these flights were again accompanied by an unmodified Navy F-5E from Fallon, while Dryden's F-15B flew four more probing flights to acquire additional near-field measurements. An instrumented L-23 sailplane from the USAF Test Pilot School obtained boom measurements from 8,000 feet (well above the ground turbulence layer) on 13 flights. All events were precisely tracked by differential GPS receivers and Edwards AFB's extensive telemetry system. In all, the SSBE yielded over 1,300 sonic boom signature recordings and 45 probe datasets—obtaining more information about the effects of turbulence, helping to confirm CFD predictions and wind tunnel validations, and bequeathing a wealth of data for future engineers and designers.[189] In addition to a series of scientific papers, the SSBD–SSBE accomplishments were the subject of numerous articles in the trade and popular press, and participants presented well-received briefings at various aeronautics and aviation venues.

Breaking Up Shock Waves with "Quiet Spike"

In June 2003, the FAA—citing a finding by the National Research Council that there were no insurmountable obstacles to building a quiet supersonic aircraft—began seeking comments on its noise standards in advance of a technical workshop on the issue. In response, the Aerospace Industries Association, the General Aviation Manufactures Association, and most aircraft companies felt that the FAA's sonic boom restriction

---

188. Pawlowski, et al., "Origins and Overview of the SSBD," pp. 11–12; NASA Press Release 03-50, "NASA Opens New Chapter in Supersonic Flight," Sept. 4, 2003; Gary Creech, "NASA, Northrop Study Sonic Boom Reduction," *Dryden X-Press*, vol. 46, issue 2 (Mar. 2004). p. 1.
189. Pawlowski, et al., "Origins and Overview of the SSBD," pp 12–13; David H. Graham, et al., "Wind Tunnel Validation of Shaped Sonic Boom Demonstration Aircraft Design," AIAA Paper 2005-7; Haering, et al., "Airborne Shaped Sonic Boom Demonstration Pressure Measurements with Computational Fluid Dynamics Comparisons," AIAA Paper 2005-9; Plotkin, et al., "Ground Data Collection of Shaped Sonic Boom Experiment Aircraft Pressure Signatures," AIAA Paper 2005-10; John M. Morgenstern, et al., "F-5 Shaped Sonic Boom Demonstrator's Persistence of Boom Shaping Reduction through Turbulence," AIAA 2005-12; all papers presented at *43rd Aerospace Sciences Meeting*, Reno, NV, Jan. 10–13, 2005.

was the still the most serious impediment to creating the market for a supersonic business jet (SSBJ), which would be severely handicapped if unable to fly faster than sound over land.[190]

By the time the FAA workshop was held in mid-November, Peter Coen of the Langley Center and a Gulfstream vice president were able to report on the success of the SSBD. Coen also outlined future initiatives in NASA's Supersonic Vehicles Technology program. In addition to leveraging the results of DARPA's QSP research, NASA hoped to engage industry partners for follow-on projects on the sonic boom, and was also working with Eagle Aeronautics on new three-dimensional CFD boom propagation models. For additional psychoacoustical studies, Langley had reconditioned its boom simulator booth. And as a possible followup to the SSBD, NASA was considering a shaped low-boom demonstrator that could fly over populated areas, allowing definitive surveys on public acceptance of minimized boom signatures.[191]

The Concorde made its final transatlantic flights just a week after the FAA's workshop. Its demise marked the first time in modern history that a mode of transportation had retreated back to slower speeds. This did, however, leave the future supersonic market entirely open to business jets. Although the success of the SSBD hinted at the feasibility of such an aircraft, designing one—as explained in a new study by Langley's Robert Mack—would still not be at all easy.[192]

During the next several years, a few individual investors and a number of American and European aircraft companies—including Gulfstream, Boeing, Lockheed, Cessna, Raytheon, Dassault, Sukhoi, and the privately held Aerion Corporation—pursued assorted SSBJ concepts with varying degrees of cooperation, competition, and commitment. Some of these and other aviation-related companies also worked together on supersonic strategies through three consortiums: Supersonic Aerospace International

---

190. James R. Asker, "FAA Seeks Information on Sonic Boom Research," *Aviation Week*, June 2, 2003, p. 21; David Bond, "The Time is Right," *Aviation Week*, Oct. 20, 2003, pp. 57–58.
191. Peter G. Coen, "Supersonic Vehicles Technology: Sonic Boom Technology Development and Demonstration" and Preston A. Henne, "A Gulfstream Perspective on the DARPA QSP Program and Future Supersonic Initiatives," Briefing Slides, FAA Civil Supersonic Aircraft Workshop, Washington, DC, Nov. 13, 2003; Aimee Cunningham, "Sonic Booms and Human Ears: How Much Can the Public Tolerate," *Popular Science*, July 30, 2004, http://www.popsci.com/military-aviation-space/article/2004-07/sonic-booms-and-human-ears.
192. Robert J. Mack, "A Supersonic Business Jet Concept Designed for Low Sonic Boom," NASA TM-2003-212435, Oct. 2003.

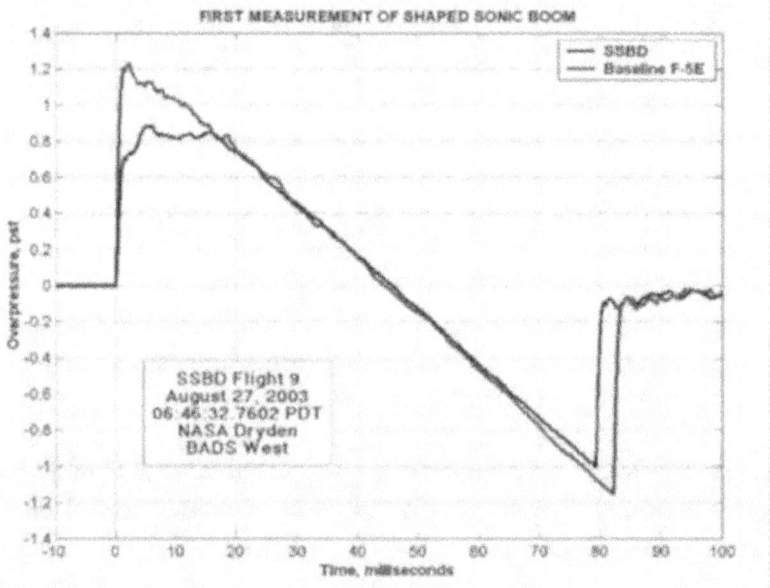

Figure 9. Normal and flattened F-5E sonic boom signatures. NASA.

(SAI), which had support from Lockheed-Martin; the 10-member Supersonic Cruise Industry Alliance (SCIA); and Europe's High-Speed Aircraft Industrial Project (HISAC), comprising more than 30 companies, universities, and other members. Meanwhile, the FAA began the lengthy process for considering a new metric on acceptable sonic booms and, in the interest of global consistency, prompted the International Civil Aviation Organization (ICAO) to also put the issue on its agenda. It was in this environment of both renewed enthusiasm and ongoing uncertainty about commercial supersonic flight that NASA continued to study and experiment on ways to make the sonicboom more acceptable to the public.[193]

Richard Wlezien (back from DARPA as NASA's vehicle systems manager) hoped to follow up on the SSBD with a truly low-boom super-

---

193. For examples, see Graham Warwick, "Quiet Progress: Aircraft Designers Believe They Can Take the Loud Boom out of Supersonic Travel," *Flight International*, Oct. 20, 2004, pp. 32–33; Edward H. Phillips, "Boom Could Doom: Debate over Hybrid SSBJ Versus Pure Supersonic Is Heating Up," *Aviation Week*, June 13, 2005, pp. 84–85; Francis Fiorino, "Lowering the Boom," *Aviation Week*, Nov. 7, 2005, p. 72; "Supersonic Private Jets in Development," *Business Travel News Online*, Oct. 23, 2006; John Wiley, "The Super-Slow Emergence of Supersonic," *Business and Commercial Aviation*, Sept. 1, 2007, pp. 48–50; Edward H. Phillips, "Shock Wave: Flying Faster than Sound Is the Holy Grail of Business Aviation," *Aviation Week*, Oct. 8, 2007, pp. 50–51; Mark Huber, "Mach 1 for Millionaires," *Air and Space Magazine*, Mar.–Apr. 2006, http://www.airspacemag.com/flight-today/millionaire.html

sonic demonstrator, possibly by 2010. In July 2005, NASA announced the Sonic Boom Mitigation Project, which began with concept explorations by major aerospace companies on the feasibility of either modifying another existing aircraft or designing a new demonstrator.[194] As explained by Peter Coen, "these studies will determine whether a low sonic boom demonstrator can be built at an affordable cost in a reasonable amount of time."[195] Although numerous options for using existing aircraft were under investigation, most of the studies were leaning toward the need to build a new experimental airplane as the most effective solution. On August 30, 2005, however, NASA Headquarters announced the end of the short-lived Sonic Boom Mitigation Project because of changing priorities.[196]

Despite this setback, there was still one significant boom lowering experiment in the making. Gulfstream Aerospace Corporation, which had been teamed with Northrop Grumman in one of the canceled studies, had already patented a new sonic boom mitigation technique.[197] Testing this invention—a retractable lance-shaped device to extend the length of an aircraft—would become the next major sonic boom flight experiment.

In the meantime, NASA continued some relatively modest sonic boom testing at the Dryden Center, mainly to help improve simulation capabilities. In a joint project with the FAA and Transport Canada in the summer of 2005, researchers from Pennsylvania State University strung an array of advanced microphones at Edwards AFB to record sonic booms created by Dryden F-18s passing overhead. Eighteen volunteers, who sat on lawn chairs alongside the row of microphones during the flyovers to experience the real thing, later gauged the fidelity of the played-back recordings. These were then used to help improve the accuracy of the booms replicated in simulators.[198]

"*Quiet Spike*" was the name that Gulfstream gave to its nose boom concept. Based on CFD models and results from Langley's 4 by 4 super-

---

194. Graham Warwick, "NASA Narrows R&D Agenda," *Flight International*, Feb. 15, 2006, p. 28; Ellen H. Thompson, et al., "NASA Funds Studies for Quieter Supersonic Boom," NASA News Release 05-176, July 8, 2005.
195. David Collogan, "Manufacturers, NASA working on Bizjet Sonic Boom Project," *The Weekly of Business Aviation*, July 18, 2005, p. 21.
196. Michael A. Dornheim, "Will Low Boom Fly? NASA Cutbacks Delay Flight Test of Shaped Demonstrator . . . ," *Aviation Week*, Nov. 7, 2005, pp. 68–69.
197. U.S. Patent No. 6,698,684, "Supersonic Aircraft with Spike for Controlling and Reducing Sonic Boom," Mar. 2, 2004, http://www.patentstorm.us/patents/6698684/description.html.
198. Jay Levine, "Lowering the Boom," July 29, 2005, http://www.nasa.gov/centers/dryden/news/X-Press/stories/2005/072905.

Close-up view of the SSBD F-5E, showing its enlarged "pelican" nose and lower fuselage designed to shape the shock waves from the front of the airframe. NASA.

sonic wind tunnel, Gulfstream was convinced that the Quiet Spike device could greatly mitigate a sonic boom by breaking up the typical nose shock into three less-powerful waves that would propagate in parallel to the ground.[199] However, the company needed to test the structural and aerodynamic suitability of the device and also obtain supersonic in-flight data on its shock scattering ability. NASA's Dryden Flight Research Center had the capabilities needed to accomplish these tasks. Under this latest public-private partnership, Gulfstream fabricated a telescoping 30-foot-long nose boom (made of molded graphite epoxy over an aluminum frame) to attach to the radar bulkhead of Dryden's frequently modified F-15B No. 836. A motorized cable and pulley system could extend the spike up to 24 feet and retract it back to 14 feet. After extensive static testing at its Savannah, GA, facility, Gulfstream and NASA technicians at Dryden attached the specially instrumented spike to the F-15's radar bulkhead in April 2006 and began conducting further ground tests, such as for vibration.[200]

---

199. Donald C. Howe, et al., "Development of the Gulfstream Quiet Spike for Sonic Boom Minimization," AIAA Paper 2008-124, presented at *46th Aerospace Sciences Meeting, Reno, NV, Jan. 7–10, 2008*; Natalie D. Spivey, et al., "Quiet Spike Build-up Ground Vibration Testing Approach," NASA TN-2007-214625, Nov. 2007.
200. James W. Smolka, et al., "Flight Testing of the Gulfstream Quiet Spike on a NASA F-15B," paper presented to the Society of Experimental Test Pilots, Anaheim, CA, Sept. 27, 2007, 1-24; Stephen B. Cumming, et al., "Aerodynamic Effects of a 24-foot Multi-segmented Telescoping Nose Boom on an F-15B Airplane," NASA TM-2008-214634, Apr. 2008.

After various safety checks, aerodynamic assessments and checkout flights, Dryden conducted Quiet Spike flight tests from August 10, 2006 until February 14, 2007. Key engineers on the project included Dryden's Leslie Molzahn and Thomas Grindle, and Gulfstream's Robbie Cowart. Veteran NASA test pilot Jim Smolka gradually expanded the F-15B's flight envelope up to Mach 1.8 and performed sonic boom experiments with the telescoping nose boom at speeds up to Mach 1.4 at 40,000 feet. Aerial refueling by AFFTC's KC-135 allowed extended missions with multiple test points. Because it was known that the weak shock waves from the spike would rather quickly coalesce with the more powerful shock waves generated by the rest of the F-15's unmodified high-boom airframe, data were collected from distances of no more than 1,000 feet. These measurements, made by a chase plane using similar probing techniques to those of the SR-71 and SSBD tests, confirmed CFD models on the spike's ability to generate a sawtooth wave pattern that, if reaching the surface, would cause only a muffled sonic boom. Analysis of the data appeared to confirm that shocks of equal strength would not coalesce into a single strong shock. In February 2007, with all major test objectives having been accomplished, the Quiet Spike F-15B was flown to Savannah for Gulfstream to restore to its normal configuration.[201]

For this successful test of an innovative design concept for a future SSBJ, James Smolka, Leslie Molzahn, and three Gulfstream employees subsequently received *Aviation Week and Space Technology*'s Laureate Award in Aeronautics and Propulsion. One month later, however, both the Gulfstream Corporation and the Dryden Center were saddened by the death in an airshow accident of Gerard Schkolnik, Gulfstream's Director of Supersonic Technology Programs, who had been a Dryden employee for 15 years.[202]

Focusing on Fundamentals: The Supersonics Project

In January 2006, NASA Headquarters announced its restructured aeronautics mission. As explained by Associate Administrator for Aeronautics Lisa J. Porter, "NASA is returning to long-term investments

---

201. Smolka, et al., "Flight Testing of the Gulfstream Quiet Spike," pp. 28–38; Haering, et al., "Preliminary Results from the Quiet Spike Flight Test," briefing presented at the Fundamental Aeronautics Program meeting, New Orleans, Oct. 30–Nov. 1, 2007.
202. "Aeronautics/Propulsion Laureate," *Aviation Week*, Mar. 17, 2008, p. 40; "Obituary: Gerard Schkolnik," ibid., Apr. 21, 2008, p. 22.

in cutting-edge fundamental research in traditional aeronautical disciplines . . . appropriate to NASA's unique capabilities." One of the four new program areas announced was Fundamental Aeronautics (which included supersonic research), with Rich Wlezien as acting director.[203]

During May, NASA released more details on Fundamental Aeronautics, including plans for what was called the Supersonics Project, managed by Mary Jo Long-Davis with Peter Coen as its principal investigator. One of the project's major technical challenges was to accurately model the propagation of sonic booms from aircraft to the ground incorporating all relevant physical phenomena. These included realistic atmospheric conditions and the effects of vibrations on structures and the people inside (for which most existing research involved military firing ranges and explosives). "The research goal is to model sonic boom impact as perceived both indoors and outdoors." Developing the propagation models would involve exploitation of existing databases and additional flight tests as necessary to validate the effects of molecular relaxation, rise time, and turbulence on the loudness of sonic booms.[204]

As the Supersonics Project evolved, it added aircraft concepts more challenging than an SSBJ to serve as longer-range targets on which to focus advanced research and technologies. These were a medium-sized (100–200 passenger) Mach 1.6–1.8 supersonic airliner that could have an acceptable sonic boom by about 2020 and an efficient multi-Mach aircraft that might have an acceptably low boom when flying at a speed somewhat below Mach 2 by the years 2030–2035. NASA awarded advanced concept studies for these in October 2008.[205] NASA

---

203. Michael Braukus/Doc Mirleson, "NASA Restructures Aeronautics Research," NASA News Release 06-008, Jan. 12, 2006; Lisa Porter, "Reshaping NASA's Aeronautics Program, Briefing," Jan. 12, 2006, http://www.nasa.gov/home/hqnews/2006/jan/ HQ_06008_ARMD_ Restructuring.html.
204. Peter Coen, Mary Jo Long-Davis, and Louis Povinelli, "Fundamental Aeronautics Program Supersonics Project, Reference Document," May 26, 2006, pp. 36–37, http://www.aeronautics. nasa.gov/fap/documents.html (quote from p. 36).
205. Jefferson Morris, "Quiet, Please: With More Emphasis on Partnering, NASA Continues Pursuit of Quieter Aircraft," Aviation Week, June 25, 2007, p. 57; Lisa J. Porter, "NASA's Aeronautics Program," Fundamental Aeronautics Annual Meeting, New Orleans, Oct. 30, 2007, slide on "Supersonics System Level Metrics," http://www.aeronautics.nasa.gov/fap/PowerPoints/ARMD&FA_Intro.pdf; Beth Hickey, "NASA Awards Future Aircraft Research Contracts," Contract Release C08-060, Oct. 6, 2008; Graham Warwick, "Forward Pitch," Aviation Week, Oct. 20, 2008, p. 22.

also began working with Japan's Aerospace Exploration Agency (JAXA) on supersonic research, including sonic boom modeling.[206] Although NASA was not ready as yet to develop a new low-boom supersonic research airplane, it supported an application by Gulfstream to the Air Force that reserved the designation X-54A just in case this would be done in the future.[207]

Meanwhile, existing aircraft had continued to prove their value for sonic boom research. During 2005, the Dryden Center began applying a creative new flight technique called low-boom/no-boom to produce controlled booms. Ed Haering used PCBoom4 modeling in developing this concept, which Jim Smolka then refined into a flyable maneuver with flight tests over an extensive array of pressure sensors and microphones. The new technique allowed F-18s to generate shaped ("low boom") signatures as well as the evanescent sound waves ("no-boom") that remain after the refraction and absorption of shock waves generated allow Mach speeds (known as the Mach cutoff) before they reach the surface.

The basic low-boom/no-boom technique requires cruising just below Mach 1 at about 50,000 feet, rolling into an inverted position, diving at a 53-degree angle, keeping the aircraft's speed at Mach 1.1 during a portion of the dive, and pulling out to recover at about 32,000 feet. This flight profile took advantage of four attributes that contribute to reduced overpressures: a long propagation distance (the relatively high altitude of the dive), the weaker shock waves generated from the top of an aircraft (by diving while upside down), low airframe weight and volume (the relatively small size of an F-18), and a low Mach number. This technique allowed Dryden's F-18s, which normally generate overpressures of 1.5 psf in level flight, to produce overpressures under 0.1 psf. Using these maneuvers, Dryden's skilled test pilots could precisely place these focused quiet booms on specific locations, such as those with observers and sensors. Not only were the overpressures low, they had a slower rise time than the typical N-shaped sonic

---

206. Hans Greimel, "Japan to Talk with NASA on Supersonic Jet," *Washington Post*, May 8, 2006, http://www.washingtonpost.com/wp-dyn/content/article/2006/05/09/AR2226050800267; Beth Dickey, "NASA and JAXA to Conduct Joint Research on Sonic Boom Modeling," NASA News Release 09-117, May 8, 2008.
207. "X-54A Designation Issued as Placeholder for Future Boom Research Aircraft," *Aerospace Daily*, July 21, 2008, p. 1.

NASA F-15B No. 836 in flight with Quiet Spike, September 2006. NASA.

signature. The technique also resulted in systematic recordings of evanescent waves—the kind that sound merely like distant thunder.[208]

Dryden researchers used this technique in July 2007 during a test called House Variable Intensity Boom Effect on Structures (House VIBES). Following up on a similar test from the year before with an old (early 1960s) Edwards AFB house slated for demolition,[209] Langley engineers installed 112 sensors (a mix of accelerometers and microphones) inside the unoccupied half of a modern (late 1990s) duplex house. Other sensors were placed outside the house and on a nearby 35-foot tower. These measured pressures and vibrations from 12 normal intensity N-shaped booms (up to 2.2 psf) created by F-18s in steady and level flight at Mach 1.25 and 32,000 feet as well as 31 shaped booms (registering only 0.1 to 0.7 psf) from F-18s using the Low Boom/No Boom flight profile. The latter booms were similar to those that would

---

208. Haering, Smolka, James E. Murray, and Plotkin, "Flight Demonstration of Low Overpressure N-Wave Sonic Booms and Evanescent Waves," *Innovations in Non-Linear Acoustics: 17th International Symposium on Nonlinear Acoustics, International Sonic Boom Forum, State College, PA, July 21–22, 2005*, American Institute of Physics Conference Proceedings, vol. 838 (May 2006), pp. 647–650.
209. Jacob Klos and R.D. Bruel, "Vibro-Acoustical Response of Buildings Due to Sonic Boom Exposure: June 2006 Field Test," NASA TM-2007-214900, Sept. 2007.

be expected from an acceptable supersonic business jet. The specially instrumented F-15B No. 852 performed six flights, and an F-18A did one flight. Above the surface boundary layer, an instrumented L-23 sailplane from the Air Force Test Pilot School recorded shock waves at precise locations in the path of the focused booms to account for atmospheric effects. The data from the house sensors confirmed fewer vibrations and noise levels in the modern house than had been the case with the older house. At the same time, data gathered by the outdoor sensors added greatly to NASA's variable intensity sonic boom database, which was expected to help program and validate sonic boom propagation codes for years to come, including more advanced three-dimensional versions of PCBoom.[210]

With the awakening of interest in an SSBJ, NASA Langley acoustics specialists including Brenda Sullivan and Kevin Shepherd had resumed an active program of studies and experiments on human and structural response to sonic booms. They upgraded the HSR-era simulator booth with an improved computer-controlled playback system, new loudspeakers, and other equipment to more accurately replicate the sound of various boom signatures, such as those recorded at Edwards. In 2005, they also added predicted boom shapes from several low-boom aircraft designs.[211] At the same time, Gulfstream created a new mobile sonic boom simulator to help demonstrate the difference between traditional and shaped sonic booms to a wider audience. Although Gulfstream's folded horn design could not reproduce the very low frequencies of Langley's simulator booth, it created a "traveling" pressure wave that moved past the listener and resonated with postboom noises, features that were judged more realistic than other simulators.

Under the aegis of the Supersonics Project, plans for additional simulation capabilities accelerated. Based on multiple studies that had long cited the more bothersome effects of booms experienced indoors, the

---

210. Creech, "Sonic Boom Tests Scheduled," Dryden News Release 07-38, July 5, 2007; Guy Norris, "Sonic Spike," *Aviation Week*, Oct. 8, 2007, p. 52; Haering, et al., "Initial Results from the Variable Intensity Sonic Boom Propagation Database," AIAA Paper 2008-3034, presented at the 14th AIAA/CEAS Aeroacoustics Conference, Vancouver, BC, Canada, May 5–7, 2008; Jacob Klos, "Vibro-Acoustic Response of Buildings Due to Sonic Boom Exposure: July 2007 Field Test," NASA TM-2008-215349, Sept. 2008.
211. Brenda M. Sullivan, "Research on Subjective Response to Simulated Sonic Booms at NASA Langley Research Center," paper presented at International Sonic Boom Forum, State College, PA, July 21–22, 2005.

Langley Center began in the summer of 2008 to build one of the most sophisticated sonic boom simulation systems yet. Scheduled for completion in early 2009, it would consist of a carefully constructed 12- by 14-foot room with sound and pressure systems that would replicate all the noises and vibrations caused by various levels and types of sonic booms.[212] Such studies would be vital if most concepts for supersonic business jets were ever to be realized. When the FAA updated its policy on supersonic noise certification in October 2008, it acknowledged the promising results of recent experiments but cautioned that any future changes in the rules against supersonic flight would still depend on public acceptance.[213]

NASA's Supersonics Project also put a new flight test on its agenda: the Lift and Nozzle Change Effects on Tail Shocks (LaNCETS). Both the SSBD and Quiet Spike experiments had only involved shock waves from the front of an aircraft. Yet shocks from the rear of an aircraft as well as jet engine exhaust plumes also contribute to sonic booms—especially the recompression phase of the typical N-wave signature—but have long been more difficult to control. NASA initiated the LaNCETS experiment to address this issue. As described in the Supersonic Project's original planning document, one of the metrics for LaNCETS was to "investigate control of aft shock structure using nozzle and/or lift tailoring with the goal of a 20% reduction in near-field tail shock strength."[214]

NASA Dryden had just the airplane with which to do this: F-15B No. 837. Originally built in 1973 as the Air Force's first preproduction TF-15A two-seat trainer (soon redesignated as the F-15B), it had been extensively modified for various experiments over its long lifespan. These included the Short Takeoff and Landing Maneuvering Technology Demonstration, the High-Stability Engine Control project, the Advanced Control Technology for Integrated Vehicles Experiment (ACTIVE),

---

212. Brenda M. Sullivan, "Design of an Indoor Sonic Boom Simulator at NASA Langley Research Center," July 28, 2008, paper presented at *Noise-Con 2008, Baltimore, July 12–14, 2008*, and "Research at NASA on Human Response to Sonic Booms," Nov. 17, 2008, at *5th International Conference on Flow Dynamics, Sendai, Japan, Nov. 17–20, 2008*; Coen, Lou Povinelli, and Kaz Civinskas, "Supersonics Project Overview," Fundamental Aeronautics Annual Meeting, Atlanta, Oct. 7, 2008, http://www.aeronautics.nasa.gov/fap/ PowerPoints/SUP_ATL_Overview.pdf.

213. "FAA Updates Policy on SST Noise Certification," *The Weekly of Business Aviation*, Oct. 27, 2008, p. 195.

214. "Supersonics Project Reference Document," p. 43. The acronym "LaNCETS" was devised by Haering: Discussion with Benson, at the Dryden Flight Research Center, Dec. 12, 2008.

and Intelligent Flight Control Systems (IFCS). F-15B No. 837 had the following special features: digital fly-by-wire controls, canards ahead of the wings for changing longitudinal lift distribution, and thrust-vectoring variable area ratio nozzles on its twin jet engines that could (1) constrict and expand to change the shape the exhaust plumes and (2) change the pitch and yaw of the exhaust flow.[215] It was planned to use these capabilities for validating computational tools developed at Langley, Ames, and Dryden to predict the interactions between shocks from the tail and exhaust under various lift and plume conditions.

Tim Moes, one of the Supersonics Project's associate managers, was the LaNCETS project manager at the Dryden Center. Jim Smolka, who had flown most of F-15B No. 837's previous missions at Dryden, was its test pilot. He and Nils Larson in F-15B No. 836 conducted Phase I of the test program with three missions from June 17–19, 2008. They gathered baseline measurements with 29 probes, all at 40,000 feet and speeds of Mach 1.2, 1.4, and 1.6.[216]

Several months before Phase II of LaNCETS, NASA specialists and affiliated researchers in the Supersonics Project announced significant progress in near-field simulation tools using the latest in computational fluid dynamics. They even reported having success as far out as 10 body lengths (a mid-field distance). As seven of these researchers claimed in August 2008, "[It] is reasonable to expect the expeditious development of an efficient sonic boom prediction methodology that will eventually become compatible with an optimization environment."[217] Of course, more data from flight-testing would increase the likelihood of this prediction.

LaNCETS Phase II began on November 24, 2008, with nine missions flown by December 11. After being interrupted by a freak snowstorm during the third week of December and then having to break for the holiday season, the LaNCETS team completed the project with flight

---

215. Dryden Flight Research Center, "F-15B #837," http://www.nasa.gov/centers/dryden/aircraft/F-15B-837/index.html, accessed Feb. 11, 2009.

216. Larry Cliatt, et al., "Overview of the LaNCETS Flight Experiment and CFD Analysis," Briefing, *Fundamental Aeronautics Annual Meeting*, Atlanta, Oct. 2, 2008.

217. J.H. Casper, et al., "Assessment of Near-Field Sonic Boom Simulation Tools," AIAA Paper 2008-6592 (p. 8 quoted) and Richard L. Campbell, et al., "Efficient Unstructured Grid Adaptation Methods for Sonic Boom Prediction," AIAA Paper 2008-7327, both presented at *26th Applied Aerodynamics Conference*, Honolulu, Aug. 18–21, 2008.

tests on January 12, 15, and 30, 2009. In all, Jim Smolka flew 13 missions in F-15B No. 837, 11 of which included in-flight shock wave measurements by No. 836 from distances of 100 to 500 feet. Nils Larson piloted the probing flights, with Jason Cudnik or Carrie Rhoades in the back seat. The aircrews tested the effects of both positive and negative canard trim at Mach 1.2, 1.4, and 1.6 as well as thrust vectoring at Mach 1.2 and 1.4. They also gathered supersonic data on plume effects with different nozzle areas and exit pressure ratios. Once again, GPS equipment recorded the exact locations of the two aircraft for each of the datasets. On January 30, 2009, with Jim Smolka at the controls for the last time, No. 837 made a final flight before its well-earned retirement.[218]

The large amount of data collected will be made available to industry and academia, in addition to NASA researchers at Langley, Ames, and Dryden. For the first time, analysts and engineers will be able to use actual flight test results to validate and improve CFD models on tail shocks and exhaust plumes—taking another step toward the design of a truly low-boom supersonic airplane.[219]

Perspectives on the Past, Prospects for the Future

Unfortunately for the immediate future of civilian supersonic flight, the successful LaNCETS project coincided almost exactly with the spread of the global financial crisis and the start of a severe recession. These negative economic developments hit almost all major industries, not the least being air carriers and aircraft manufacturers. The impact on those recently thriving companies making business jets was aggravated even more by populist and political backlash at executives of troubled corporations, some now being subsidized by the Federal Government, for continuing to fly in corporate jets. Lamenting this unsought negative publicity, *Aviation Week and Space Technology* examined the plight

---

218. Norris, "Sonic Solutions: NASA Uses Unique F-15B to Complete Design Tools for Quiet Supersonic Aircraft," *Aviation Week*, Jan. 5, 2009, p. 53; Creech and Beth Dickey, "Lancets Flights Probe Supersonic Shockwaves," Dryden News Release 09-04, Jan. 22, 2009; Tim Moes, "Sonic Boom Research at NASA Dryden: Objectives and Flight Results of the Lift and Nozzle Change Effects on Tail Shock (LaNCETS) Project," Partial Briefing Slides, International Test & Evaluation Association, Antelope Valley Chapter, Feb. 24, 2009; E-mail, Tim Moes to Benson, "Re: More Details on LaNCETS," Mar. 11, 2009.

219. For an early analysis, see Trong T. Bui, "CFD Analysis of Nozzle Jet Plume Effects on Sonic Boom Signature," AIAA Paper 2009-1054, presented at *47th Aerospace Sciences Meeting, Orlando, FL, Jan. 5–8, 2009*.

of the small-jet manufacturers in a story with following subheading: "As if the economy were not enough, business aviation becomes a scapegoat for executive excess."[220] Nevertheless, NASA was continuing to invest in supersonic technologies and sonic boom research, and the aircraft industry was not ready to abandon the ultimate goal of supersonic civilian flight. For example, Boeing—under a Supersonics Project contract—was studying low-boom modifications for one of NASA's F-16XL aircraft as one way to seek the holy grail for practical supersonic commercial flight: acceptance by the public. This relatively low-cost idea for a shaped sonic boom demonstrator had been one of the options being considered during NASA's short-lived Sonic Boom Mitigation Project in 2005. Since then, findings from the Quiet Spike and LaNCETS experiments, along with continued progress in computational fluid dynamics, were helping to confirm and refine the aerodynamic and propulsion attributes needed to mitigate the strength of sonic booms.

In the case of the F-16XL, the modifications proposed by Boeing included an extended nose glove (reminiscent of the SSBD), lateral chines that blend into the wings (as with the SR-71), a sharpened V-shaped front canopy (like those of the F-106 and SR-71), an expanded nozzle for its jet engine (similar to those of F-15B No. 837), and a dorsal extension (called a "stinger") to lengthen the rear of the airplane. Although such add-ons would preclude the low-drag characteristics also desired in a demonstrator, Boeing felt that its "initial design studies have been encouraging with respect to shock mitigation of the forebody, canopy, inlet, wing leading edge, and aft lift/volume distribution features." Positive results from more detailed designs and successful wind tunnel testing would be the next requirements for continuing consideration of the proposed modifications.[221]

It was clear that NASA's discoveries about sonic booms and how to control them were beginning to pay dividends. Whatever the fate of Boeing's idea or any other proposals yet to come, NASA was committed to finding the best way to demonstrate fully shaped sonic booms. As another encouraging sign, the FAA was working with NASA on a roadmap for studying community reactions to sonic booms, one that would soon be presented to the ICAO.[222]

---

220. Graham Warwick, et al., "Open Season," *Aviation Week*, Mar. 2, 2009, pp. 20–21.
221. Warwick, "Beyond the N-Wave: Modifying NASA's Arrow-Wing F-16XL Could Help Pave the Way for Low-Boom Supersonic Transports," *Aviation Week*, Mar. 23, 2009, p. 52.
222. E-mail, Coen, Langley Research Center, to Benson, Apr. 10, 2009.

As shown in this study, past expectations for a quiet civilian supersonic transport had repeatedly run up against scientific, technical, economic, and political hurdles too high to overcome. That is why such an airplane has yet to fly. Yet the knowledge gained and lessons learned from each attempt attest to the value of persistence in pursuing both basic and applied research. Recent progress in shaping sonic booms builds upon the work of dedicated NASA civil servants over more than half a century, the data and documentation preserved through NASA's scientific and technical information program, the special facilities and test resources maintained and operated by NASA's research Centers, and NASA's support of and partnership with contractors and universities.

Since the dawn of civilization, conquering the twin tyrannies of time and distance has been a powerful human aspiration, one that has served as a catalyst for many technological innovations. It seems reasonable to assume that this need for speed will eventually break down the barriers in the way of practical supersonic transportation, to include solving the problem of the sonic boom. When that time finally does come, it will have been made possible by NASA's many years of meticulous research, careful testing, and inventive experimentation on ways to soften the sonic footprint.

Recommended Additional Readings
Reports, Papers, Articles, and Presentations:

William J. Alford and Cornelius Driver, "Recent Supersonic Transport Research," *Astronautics & Aeronautics*, vol. 2, no. 9 (Sept. 1964), pp. 26–37.

William H. Andrews, "Summary of Preliminary Data Derived from the XB-70 Airplanes," NASA TM-X-1240 (1966).

Daniel G. Baize, "1995 NASA High-Speed Research Program Sonic Boom Workshop," NASA CP-3335, vol. 1 (1996).

Raymond L. Barger, "Sonic-Boom Wave-Front Shapes and Curvatures Associated with Maneuvering Flight," NASA TP-1611 (1979).

W.D. Beasly, J.D. Brooks, and R.L. Barger, "A Laboratory Investigation of N-Wave Focusing," NASA TN-D-5306 (1969).

Percy J. Bobbit and Domenic J. Maglieri, "Dr. Antonio Ferri's Contribution to Supersonic Transport Sonic-Boom Technology," *Journal of Spacecraft and Rockets*, vol. 40, no. 4 (July–Aug. 2003), pp. 459–466.

J.D. Brooks, et al., "Laboratory Investigation of Diffraction and Reflection of Sonic Booms by Buildings," NASA TN-D-5830 (1970).

J.F. Bryant, D.J. Maglieri, and V.S. Richie, "In-Flight Shock-Wave Measurements Above and Below a Bomber Airplane at Mach Numbers from 1.42 to 1.69," NASA TN-D-1968 (1963).

Trong T. Bui, "CFD Analysis of Nozzle Jet Plume Effects on Sonic Boom Signature," AIAA Paper 2009-1054 (2009).

Richard L. Campbell, et al., "Efficient Unstructured Grid Adaptation Methods for Sonic Boom Prediction," AIAA Paper 2008-7327 (2008).

Harry W. Carlson, "An Investigation of Some Aspects of the Sonic Boom by Means of Wind-Tunnel Measurements of Pressures about Several Bodies at a Mach Number of 2.01," NASA TN-D-161 (1959).

Harry W. Carlson, "Configuration Effects on Sonic Boom," *Proceedings of NASA Conference on Supersonic-Transport Feasibility Studies and Supporting Research, Sept. 17–19, 1963 . . . Hampton, VA,* NASA TM-X-905 (1963).

Harry W. Carlson, "Correlation of Sonic-Boom Theory with Wind Tunnel and Flight Measurements," NASA TR-R-213 (1964).

Harry W. Carlson, "Some Notes on the Present Status of Sonic Boom Prediction and Minimization Research," *Third Conference on Sonic Boom Research . . . Washington, DC, Oct. 29–30, 1970,* NASA SP-255 (1971).

Harry W. Carlson, "Simplified Sonic-Boom Prediction," NASA TP-1122 (1978).

Harry W. Carlson, et al., "A Wind Tunnel Study of Sonic-Boom Characteristics for Basic and Modified Models of a Supersonic Transport Configuration," NASA TM-X-1236 (1966).

Harry W. Carlson, et al., "Application of Sonic-Boom Minimization Concepts in Supersonic Transport Design," NASA TN-D-7218 (1973).

J.H. Casper, et al., "Assessment of Near-Field Sonic Boom Simulation Tools," AIAA Paper 2008-6592 (2008).

Samsun Cheung, "Supersonic Civil Airplane Study and Design: Performance and Sonic Boom," NASA CR-197745 (1995).

Clark, Buhr, & Nexen [firm]. "Studies of Sonic Boom Damage," NASA CR-227 (1965).

Peter G. Coen, "Development of a Computer Technique for Prediction of Transport Aircraft Flight Profile Sonic Boom Signatures," NASA CR-188117 (1991).

Hugo E. Dahlke, et al., "The Shock-Expansion Tube and Its Application as a Sonic Boom Simulator," NASA CR-1055 (1968).

Christine M. Darden, "Minimization of Sonic-Boom Parameters in Real and Isothermal Atmospheres," NASA TN-D-7842 (1975).

Christine M. Darden, "Sonic Boom Theory—Its Status in Prediction and Minimization," AIAA Paper 76-1 (1976).

Christine M. Darden, "Sonic Boom Minimization with Nose-Bluntness Relaxation," NASA TP-1348 (1979).

Christine M. Darden, "Charts for Determining Potential Minimum Sonic-Boom Overpressures for Supersonic Cruise Aircraft," NASA TP-1820 (1981).

Christine M. Darden, ed., *High Speed Research: Sonic Boom; Proceedings of a Conference at the Langley Research Center, Feb. 25–27, 1992*, NASA CR-3172 (1992), vols. 1, 2.

Christine M. Darden and R.J. Mack, "Some Effects of Applying Sonic Boom Minimization to Supersonic Aircraft," *Journal of Aircraft*, vol. 17, no. 3 (Mar. 1980), pp. 182–186.

Christine M. Darden, et al., "Status of Sonic Boom Methodology and Understanding," NASA CP-3027 (1989).

J.W.M. Dumond, et al., "A Determination of the Wave Forms and Laws of Propagation and Dissipation of Ballistic Shock Waves," *Journal of the Acoustical Society of America*, vol. 18, no. 1 (Jan. 1946), pp. 97–118.

Philip M. Edge and William H. Mayes, "Description of Langley Low-Frequency Noise Facility and Study of Human Response to Noise Frequencies below 50 cps," NASA TN-D-3204 (1966).

Phillip M. Edge and Harvey H. Hubbard, "Review of Sonic-Boom Simulation Devices and Techniques," Dec. 1970, *Journal of the Acoustical Society of America*, vol. 51, no. 2, pt. 2 (Feb. 1972), p. 723.

Thomas A. Edwards, ed., *High-Speed Research: Sonic Boom; Proceedings of a Conference at the Ames Research Center, May 12–14, 1993*, NASA CP-10132 (1993), vol. 1.

Antonio Ferri, Huai-Chu Wang, and Hans Sorensen, "Experimental Verification of Low Sonic Boom Configuration," NASA CR-2070 (1973).

James M. Fields, "Reactions of Residents to Long-Term Sonic Boom Noise Environments," NASA CR-201704 (1997).

A.B. Fryer, et al., "Publications in Acoustics and Noise Control from the NASA Langley Research Center during 1940–1976," NASA TM-X-74042 (1977).

John H. Gardner and Peter H. Rogers, "Thermospheric Propagation of Sonic Booms from the Concorde Supersonic Transport," Naval Research Laboratory Memo Report 3904, Feb. 14, 1979 (DTIC AD A067201).

A.R. George and A.R. Seebass, "Sonic-Boom Minimization," Nov. 1970, *Journal of the Acoustical Society of America*, vol. 51, no. 2, pt. 3 (Feb. 1972), pp. 686–694.

David H. Graham, et al., "Wind Tunnel Validation of Shaped Sonic Boom Demonstration Aircraft Design," AIAA Paper 2005-7 (2005).

Karen S. Green and Terrill W. Putnam, "Measurements of Sonic Booms Generated by an Airplane Flying at Mach 3.5 and 4.8," NASA TM-X-3126 (1974).

Edward A. Haering, et al., "Airborne Shaped Sonic Boom Demonstration Pressure Measurements with Computational Fluid Dynamics Comparisons," AIAA Paper 2005-9 (2005).

Edward A. Haering, James W. Smolka, James E. Murray, and Kenneth J. Plotkin, "Flight Demonstration of Low Overpressure N-Wave Sonic Booms and Evanescent Waves," *Innovations in Non-Linear Acoustics: 17th International Symposium on Nonlinear Acoustics, International Sonic Boom Forum, State College, PA, July 21–22, 2005*, American Institute of Physics Conference Proceedings, vol. 838 (May 2006), pp. 647–650.

Edward A. Haering, et al., "Initial Results from the Variable Intensity Sonic Boom Propagation Database," AIAA Paper 2008-3034 (2008).

George T. Haglund and Edward J. Kane, "Flight Test Measurements and Analysis of Sonic Boom Phenomena Near the Shock Wave Extremity," NASA CR-2167 (1973).

Thomas H. Higgins, "Sonic Boom Research and Design Considerations in the Development of a Commercial Supersonic Transport," *Journal of the Acoustical Society of America*, vol. 39, no. 5, pt. 2 (Nov. 1966), pp. 526–531.

David. A. Hilton, Vera Huckel, and Domenic J. Maglieri, "Sonic Boom Measurements during Bomber Training Operations in the Chicago Area," NASA TN-D-3655 (1966).

Harvey H. Hubbard, et al., "Ground Measurements of Sonic-Boom Measurements for the Altitude Range of 10,000 to 75,000 Feet," NASA TR-R-198 (1964).

D.A. Hilton, D.J. Maglieri, and R. Steiner, "Sonic-Boom Exposures during FAA Community Response Studies over a 6-Month Period in the Oklahoma City Area," NASA TN-D-2539 (1964).

Sherwood Hoffman, "Bibliography of Supersonic Cruise Aircraft Research (SCAR)" [1972–1977], NASA RP-1003 (1977).

Sherwood Hoffman, "Bibliography of Supersonic Cruise Research (SCR) Program from 1977 to Mid-1980," NASA RP-1063 (1980).

Donald C. Howe, et al., "Development of the Gulfstream Quiet Spike for Sonic Boom Minimization," AIAA Paper 2008-124 (2008).

Harvey H. Hubbard and Domenic J. Maglieri, "Sonic Boom Signature Data from Cruciform Microphone Array Experiments during the 1966–67 EAFB National Sonic Boom Evaluation Program," NASA TN-D-6823 (1972).

Harvey H. Hubbard, Domenic J. Maglieri, and David G. Stephens, "Sonic-Boom Research—Selected Bibliography with Annotation," NASA TM-87685 (1986).

James and Associates, ed., *YF-12 Experiments Symposium: A conference held at Dryden Flight Research Center . . . Sept. 13–15, 1978*, NASA CP-2054 (1978).

Gareth H. Jordan, "Flight Measurements of Sonic Booms and Effects of Shock Waves on Aircraft," in *Society of Experimental Test Pilots Quarterly Review*, vol. 5, no. 1 (1961), pp. 117–131.

Jacob Klos and R.D. Bruel, "Vibro-Acoustical Response of Buildings Due to Sonic Boom Exposure: June 2006 Field Test," NASA TM-2007-214900 (2007).

Jacob Klos, "Vibro-Acoustic Response of Buildings Due to Sonic Boom Exposure: July 2007 Field Test," NASA TM-2008-215349 (2008).

Christopher A. Lee, "Design and Testing of Low Sonic Boom Configurations and an Oblique All-Wing Supersonic Transport," NASA CR-197744 (1995).

G.M. Lilley, et al., "Some Aspects of Noise from Supersonic Aircraft," *Journal of the Royal Aeronautical Society*, vol. 57 (June 1953), pp. 396–414.

Lindsay J. Lina and Domenic J. Maglieri, "Ground Measurements of Airplane Shock-Wave Noise at Mach Numbers to 2.0 and at Altitudes to 60,000 Feet," NASA TN-D-235, Mar. 1960.

Jerome Lukas and Karl D. Kryler, "A Preliminary Study of the Awakening and Startle Effects of Simulated Sonic Booms," NASA CR-1193 (1968).

Jerome Lukas and K.D. Kryler, "Awakening Effects of Simulated Sonic Booms and Subsonic Aircraft Noise," NASA CR-1599 (1970).

B.K.O. Lundberg, "Aviation Safety and the SST," *Astronautics & Aeronautics*, vol. 3, no. 1 (Jan. 1966), p. 28.

G.J. MacDonald, et al., "Jason 1978 Sonic Boom Report," JSR-78-09 (Arlington, VA: SRI International, Nov. 1978).

Robert J. Mack, "A Supersonic Business Jet Concept Designed for Low Sonic Boom," NASA TM-2003-212435 (2003).

Robert J. Mack and Christine M. Darden, "Wind-Tunnel Investigation of the Validity of a Sonic-Boom-Minimization Concept," NASA TP-1421 (1979).

Robert J. Mack and Christine M. Darden, "Some Effects of Applying Sonic Boom Minimization to Supersonic Cruise Aircraft Design," *Journal of Aircraft*, vol. 17, no. 3 (Mar. 1980), pp. 182–186.

Domenic J. Maglieri, "A Brief Review of the National Aero-Space Plane Sonic Booms Final Report," USAF Aeronautical Systems Center Report TR-94-9344 (1992).

Domenic J. Maglieri, Vera Huckel, and Tony L. Parrott, "Ground Measurements of Shock-Wave Pressure for Fighter Airplanes Flying at Very Low Altitude," NASA TN-D-3443 (1966) (superseded classified TM-X-611, 1961).

Domenic J. Maglieri, Harvey H. Hubbard, and Donald L. Lansing, "Ground Measurements of the Shock-Wave Noise from Airplanes in Level Flight at Mach Numbers to 1.4 and Altitudes to 45,000 Feet," NASA TN-D-48 (1959).

Domenic J. Maglieri and Harvey H. Hubbard, "Ground Measurements of the Shock-Wave Noise from Supersonic Bomber Airplanes in the Altitude Range from 30,000 to 50,000 Feet," NASA TN-D-880 (1961).

Domenic J. Maglieri and Donald L. Lansing, "Sonic Booms from Aircraft in Maneuvers," NASA TN-D-2370 (1964).

Domenic J. Maglieri and Garland J. Morris, "Measurement of Response of Two Light Airplanes to Sonic Booms," NASA TN-D-1941 (1963).

Domenic J. Maglieri, et al., "Sonic Boom Measurements for SR-71 Aircraft Operating at Mach Numbers to 3.0 and Altitudes to 24834 Meters," NASA TN-D-6823 (1972).

Domenic J. Maglieri, Victor E. Sothcroft, and John Hicks, "Influence of Vehicle Configurations and Flight Profile on X-30 Sonic Booms," AIAA Paper 90-5224 (1990).

Domenic J. Maglieri, et al., "A Summary of XB-70 Sonic Boom Signature Data, Final Report," NASA CR-189630 (1992).

Domenic J. Maglieri, et al., "Feasibility Study on Conducting Overflight Measurements of Shaped Sonic Boom Signatures Using the Firebee BQM-34E RPV," NASA CR-189715 (1993).

David A. McCurdy, ed., "High-Speed Research: 1994 Sonic Boom Workshop, Atmospheric Propagation and Acceptability Studies," NASA CP-3209, and " Configuration, Design, Analysis, and Testing," NASA CP-209669 (1999).

F. Edward McLean, "Some Nonasymptotic Effects of the Sonic Boom of Large Airplanes," NASA TN-D-2877 (1965).

Keith H. Meredith, et al., "Computational Fluid Dynamics Comparison and Flight Test Measurement of F-5E Off-Body Pressures," AIAA Paper 2005-6 (2005).

David S. Miller and Harry W. Carlson, "Sonic Boom Minimization by Application of Heat or Force Fields to Airplane Airflow," NASA TN-D-5582 (1969).

John M. Morgenstern, et al., "F-5 Shaped Sonic Boom Demonstrator's Persistence of Boom Shaping Reduction through Turbulence," AIAA 2005-12 (2005).

Marshall E. Mullens, "A Flight Test Investigation of the Sonic Boom," AFFTC TN-56-20 (1956).

Charles W. Nixon and Harvey H. Hubbard, "Results of the USAF–NASA–FAA Flight Program to Study Community Response to Sonic Booms in the Greater St. Louis Area," NASA TN-D-2705 (1965).

Joseph W. Pawlowski, David H. Graham, Charles H. Boccadoro (NGC), Peter G. Coen (LaRC), and Domenic J. Maglieri (Eagle Aero.), "Origins and Overview of the Shaped Sonic Boom Demonstration Program," AIAA Paper 2005-5 (2005), pp. 3–7.

Kenneth J. Plotkin, "State of the Art of Sonic Boom Modeling," *Journal of the Acoustical Society of America*, vol. 111, No. 1, pt. 3 (Jan. 2002), pp. 530–536.

Kenneth J. Plotkin and Domenic J. Maglieri, "Sonic Boom Research: History and Future," AIAA Paper 2003-3575 (2003).

Kenneth J. Plotkin, et al., "Ground Data Collection of Shaped Sonic Boom Experiment Aircraft Pressure Signatures," AIAA Paper 2005-10 (2005).

Hugh W. Poling, "Sonic Boom Propagation Codes Validated by Flight Test," NASA CR-201634 (1996).

John O. Powers, J.M. Sands, and Domenic Maglieri, "Survey of United States Sonic Boom Overflight Experimentation," NASA TM-X-66339 (1969).

Johnny M. Sands, "Sonic Boom Research (1958-1968)," FAA, Nov. 1968, Defense Technical Information Center (DTIC) document AD-684806.

Ira R. Schwartz, ed., *Sonic Boom Research, Second Conference, Washington, DC, May 9–10, 1968*, NASA SP-180 (1968).

Ira R. Schwartz, ed., *Third Conference on Sonic Boom Research. . . Washington, DC, Oct. 29–30, 1970*, NASA SP-255 (1971).

A.R. Seebass, ed., *Sonic Boom Research: Proceedings of a Conference. . . Washington, DC, Apr. 12, 1967*, NASA SP-147 (1967).

A.R. Seebass and A.R. George, "Sonic Boom Minimization through Aircraft Design and Operation," AIAA Paper 73-241 (1973).

A.R. Seebass and A.R. George, "The Design and Operation of Aircraft to Minimize Their Sonic Boom," *Journal of Aircraft*, vol. 11, no. 9 (Sept. 1974), pp. 509–517.

Harriet J. Smith, "Experimental and Calculated Flow Fields Produced by Airplanes Flying at Supersonic Speeds," NASA TN-D-621 (1960).

James W. Smolka, et al., "Flight Testing of the Gulfstream Quiet Spike on a NASA F-15B," in The Society of Experimental Test Pilots, *2007 Report to the Aerospace Profession* (Lancaster, CA: SETP, 2007).

R.T. Sturgielski, et al., "The Development of a Sonic Boom Simulator with Detonable Gases," NASA CR-1844 (1971).

Rudolph J. Swigart, "An Experimental Study in the Validity of the Heat-Field Concept for Sonic Boom Alleviation," NASA CR-2381 (1974).

J.P. Taylor and E.R. Taylor, "A Brief Legal History of the Sonic Boom in America," *Aircraft Engine Noise and Sonic Boom* (Neuilly Sur Seine, France: NATO Advisory Group for Aerospace Research and Development [AGARD], 1969), Conference Proceedings (CP) No. 42, Paris, May 1969, pp. 2-1–2-11.

Charles L. Thomas, "Extrapolation of Sonic Boom Pressure Signatures by the Waveform Parameter Method," NASA TN-D-6823 (1972).

Roger Tomboulian, "Research and Development of a Sonic Boom Simulation Device," NASA CR-1378 (1969).

F. Walkden, "The Shock Pattern of a Wing-Body Combination Far from the Flight Path," *Aeronautical Quarterly*, vol. 9, pt. 2 (May 1958), pp. 164–194.

G.B. Whitham, "The Flow Pattern of a Supersonic Projectile," *Communications on Pure and Applied Mathematics*, vol. 5, No. 3 (1952), pp. 301–348.

G.B. Whitham, "On the Propagation of Weak Shock Waves," *Journal of Fluid Dynamics*, vol. 1, no. 3 (Sept. 1956), pp. 290–318.

Books and Monographs:

Donald D. Baals and William R. Corliss, *Wind Tunnels of NASA*, SP-440 (Washington, DC: NASA, 1981).

Joseph R. Chambers, *Innovation in Flight; Research of the NASA Langley Research Center on Revolutionary Concepts for Aeronautics*, NASA SP-2005-4539 (Washington, DC: NASA, 2005).

Erik M. Conway, *High Speed Dreams: NASA and the Technopolitics of Supersonic Transportation, 1945–1999* (Baltimore: Johns Hopkins, 2005).

Richard P. Hallion, *Supersonic Flight: Breaking the Sound Barrier and Beyond—The Story of the Bell X-1 and Douglas D-558* (New York: The Macmillan Co., 1977).

Richard P. Hallion and Michael H. Gorn, *On the Frontier: Experimental Flight at NASA Dryden* (Washington, DC: Smithsonian, 2003).

Mel Horwitch, *Clipped Wings: The American SST Conflict* (Cambridge: MIT, 1982).

Richard J. Kent, Jr., *Safe, Separated, and Soaring: A History of Civil Aviation Policy, 1961–1972* (Washington, DC: FAA, 1980).

F. Edward McLean, *Supersonic Cruise Technology*, NASA SP-472 (Washington, DC: NASA, 1985).

Peter W. Merlin, *From Archangel to Senior Crown: Design and Development of the Blackbird* (Reston, VA: AIAA, 2008).

Kenneth Owen, *Concorde: Story of a Supersonic Pioneer* (London: Science Museum, 2001).

Stuart I. Rochester, *Takeoff at Mid-Century: Federal Civil Aviation Policy in the Eisenhower Years, 1953–1961* (Washington, DC: FAA, 1976).

THIS PAGE INTENTIONALLY BLANK

X-15 research pilot (and, subsequently, Gemini and Apollo astronaut) Neil A. Armstrong, wearing the X-15's Clark MC-2 full-pressure suit, 1960. NASA.

CASE 5

Toward Transatmospheric Flight: From V-2 to the X-51

T.A. Heppenheimer

The expansion of high-speed aerothermodynamic knowledge enabled the attainment of hypersonic speeds, that is, flight at speeds of Mach 5 and above. Blending the challenge of space flight and flight within the atmosphere, this led to the emergence of the field of transatmospherics: systems that would operated in the upper atmosphere, transitioning from lifting flight to ballistic flight, and back again. NACA–NASA research proved essential to mastery of this field, from the earliest days of blunt body reentry theory to the advent of increasingly sophisticated transatmospheric concepts, such as the X-15, the Shuttle, the X-43A, and the X-51.

ON DECEMBER 7, 1995, the entry probe of the Galileo spacecraft plunged downward into the atmosphere of Jupiter. It sliced into the planet's hydrogen-rich envelope at a gentle angle and entered at Mach 50, with its speed of 29.5 miles per second being four times that of a return to Earth from the Moon. The deceleration peaked at 228 g's, equivalent to slamming from 5,000 mph to a standstill in a single second. Yet the probe survived. It deployed a parachute and transmitted data from its onboard instruments for nearly an hour, until overwhelmed by the increasing pressures it encountered within the depths of the Jovian atmosphere.[1]

The Galileo probe offered dramatic proof of how well the National Aeronautics and Space Administration (NASA) had mastered the field of hypersonics, particularly the aerothermodynamic challenges of double-digit high-Mach atmospheric entries. That level of performance was impressive, a performance foreshadowed by the equally impressive (certainly for their time) earlier programs such as Mercury, Gemini, Apollo, Pioneer, and Viking. But NASA had, arguably, an even greater challenge before it: developing the technology of transatmospheric flight—the ability to transit, routinely, from flight within the atmosphere to flight out

---

1. Richard E. Young, Martha A. Smith, and Charles K. Sobeck, "Galileo Probe: In Situ Observations of Jupiter's Atmosphere," *Science*, no. 272 (May 10, 1996), pp. 837–838.

into space, and to return again. It was a field where challenge and contradiction readily mixed: a world of missiles, aircraft, spacecraft, rockets, ramjets, and combinations of all of these, some crewed by human operators, some not.

Transatmospheric flight requires mastery of hypersonics, flight at speeds of Mach 5 and higher in which aerodynamic heating predominates over other concerns. Since its inception after the Second World War, three problems have largely driven its development.

First, the advent of the nuclear-armed intercontinental ballistic missile (ICBM), during the 1950s, brought the science of reentry physics and took the problem of thermal protection to the forefront. Missile nose cones had to be protected against the enormous heat of their atmosphere entry. This challenge was resolved by 1960.

Associated derivative problems were dealt with as well, including that of protecting astronauts during demanding entries from the Moon. Maneuvering hypersonic entry became a practical reality with the Martin SV-5D Precision Recovery Including Maneuvering Entry (PRIME) in 1967. In 1981, the Space Shuttle introduced reusable thermal protection—the "tiles"—that enabled its design as a "cool" aluminum airplane rather than one with an exotic hot structure. Then in 1995, the Galileo mission met demands considerably greater than those of a return from the Moon.

A second and contemporary problem, during the 1950s, involved the expectation that flight speeds would increase essentially without limit. This hope lay behind the unpiloted air-launched Lockheed X-7, which used a ramjet engine and ultimately reached Mach 4.31. There also was the rocket-powered and air-launched North American X-15, the first transatmospheric aircraft. One X-15 achieved Mach 6.70 (4,520 mph) in October 1967. This set a record for winged hypersonic flight that stood until the flight of the Space Shuttle Columbia in 1981. The X-15 introduced reaction thrusters for aircraft attitude, and they subsequently became standard on spacecraft, beginning with Project Mercury. But the X-15 also used a "rolling tail" with elevons (combined elevators and ailerons) in the atmosphere and had to transition to and from space flight. The flight control system that did this later flew aboard the Space Shuttle. The X-15 also brought the first spacesuit that was flexible when pressurized rather than being rigid like an inflated balloon. It too became standard. In aviation, the X-15 was first to use a simulator as a basic tool for development, which became a critical instrument for pilot training. Since then, simulators have entered general use and today are employed with all aircraft.[2]

A third problem, emphasized during the era of President Ronald Reagan's Strategic Defense Initiative (SDI) in the 1980s, involved the prospect that hypersonic single-stage-to-orbit (SSTO) air-breathing vehicles would shortly replace the Shuttle and other multistage rocket-boosted systems. This concept depended upon the scramjet, a variant of the ramjet engine that sustained a supersonic internal airflow to run cool. But while scramjets indeed outperformed conventional ramjets and rockets, their immaturity and higher drag made their early application as space access systems impossible. The abortive National Aero-Space Plane (NASP) program consumed roughly a decade of development time. It ballooned enormously in size, weight, complexity, and cost as time progressed and still lacked, in the final stages, the ability to reach orbit. Yet while NASP faltered, it gave a major boost to computational fluid dynamics, which use supercomputers to study airflows in aviation. This represents another form of simulation that also is entering general use. NASP also supported the introduction of rapid-solidification techniques in metallurgy. These enhance alloys' temperature resistance, resulting in such achievements as the advent of a new type of titanium that can withstand 1,500 degrees Fahrenheit (°F).[3] Out of it have come more practical and achievable concepts, as evidenced by the NASA X-43 program and the multiparty X-51A program of the present.

Applications of practical hypersonics to the present era have been almost exclusively within reentry and thermal protection. Military hypersonics, while attracting great interest across a range of mission areas, such as surveillance, reconnaissance, and global strike, has remained the stuff of warhead and reentry shape research. Ambitious concepts for transatmospheric aircraft have received little support outside the laboratory environment. Concepts for global-ranging hypersonic "cruisers" withered in the face of the cheaper and more easily achievable rocket.

---

2. Mark Wolverton, "The Airplane That Flew Into Space," *American Heritage of Invention and Technology*, (summer 2001), pp. 12–20.

3. J. Sorensen, "Titanium Matrix Composites—NASP Materials and Structures Augmentation Program," AIAA Paper 90-5207 (1990); Stanley W. Kandebo, "Boeing *777* to Incorporate New Alloy Developed for NASP," *Aviation Week*, May 3, 1993, p. 36; "NASP Materials and Structures Program, Titanium Matrix Composites," McDonnell-Douglas, Dec. 31, 1991, DTIC ADB-192559, Defense Technical Information Center.

Moving Beyond the V-2: John Becker Births American Hypersonics

During the Second World War, Germany held global leadership in high-speed aerodynamics. The most impressive expression of its technical interest and competence in high-speed aircraft and missile design was the V-2 terror weapon, which introduced the age of the long-range rocket. It had a range of over 200 miles at a speed of approximately Mach 5.[4] A longer-range experimental variant tested in 1945, the A-4b, sported swept wings and flew at 2,700 mph, reentering and leveling off in the upper atmosphere for a supersonic glide to its target. In its one semi-successful flight, it completed a launch and reentry, though one wing broke off during its terminal Mach 4+ glide.[5] One appreciates the ambitious nature and technical magnitude of the German achievement given that the far wealthier and more technically advantaged United States pursued a vigorous program in piloted rocket planes all through the 1950s without matching the basic performance sought with the A-4b.

Key to the German success was a strong academic-industry partnership and, particularly, a highly advanced complex of supersonic wind tunnels. The noted tunnel designer Carl Wieselsberger (who died of cancer during the war) introduced a blow-down design that initially operated at Mach 3.3 and later reached Mach 4.4. The latter instrument supported supersonic aerodynamic and dynamic stability studies of various craft, including the A-4b. German researchers had ambitious plans for even more advanced tunnels, including an Alpine complex capable of attaining Mach 10. This tunnel work inspired American emulation after the war and, in particular, stimulated establishment of the Air Force's Arnold Engineering Development Center at Tullahoma, TN.[6]

---

4. Walter Dornberger, *V-2* (New York: The Viking Press, 1958 ed.), relates its history from the point of view of the German military commander of V-2 development and its principal research facility.

5. Michael J. Neufeld, *The Rocket and the Reich: Peenemünde and the Coming of the Ballistic Missile Era* (Cambridge: Harvard University Press, 1995), pp. 250–251.

6. Ronald Smelt, "A Critical Review of German Research on High-Speed Airflow," *Journal of the Royal Aeronautical Society*, vol. 50, No. 432 (Dec. 1946), pp. 899–934; Theodore von Kármán, "Where We Stand: First Report to General of the Army H. H. Arnold on Long Range Research Problems of the AIR FORCES with a Review of German Plans and Developments," Aug. 22, 1945, vol. II-1, Copy No. 13, including Hsue-shen Tsien, "Reports on the Recent Aeronautical Developments of Several Selected Fields in Germany and Switzerland," July 1945; Hsue-shen Tsien, "High Speed Aerodynamics," Dec. 1945; and F.L. Wattendorf, "Reports on Selected Topics of German and Swiss Aeronautical Developments," June 1945; Peter P. Wegener, *The Peenemünde Wind Tunnels: A Memoir* (New Haven: Yale University Press, 1996), pp. 22–24, 70.

Case 5 | Toward Transatmospheric Flight: From V-2 to the X-51

The German A-4b, being readied for a test flight, January 1945. USAF.

At war's end, America had nothing comparable to the investment Germany had made in high-speed flight, either in rockets or in wind tunnels and other specialized research facilities. The best American wartime tunnel only reached Mach 2.5. As a stopgap, the Navy seized a German facility, transported it to the United States, and ran it at Mach 5.18, but

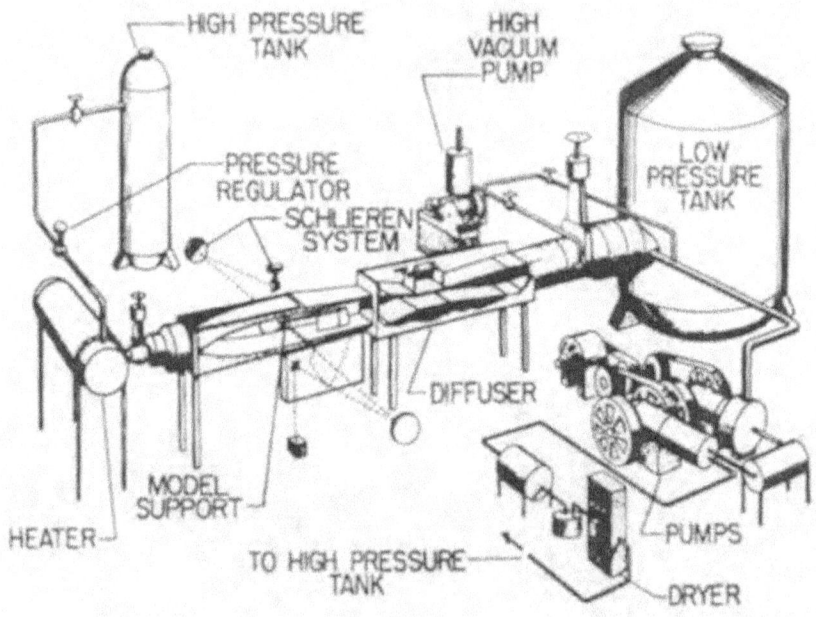

The layout of the Langley 11-inch hypersonic tunnel advocated by John V. Becker. NASA.

it did this only beginning in 1948.[7] Even so, aerodynamicist John Becker, a young and gifted engineer working at the National Advisory Committee for Aeronautics (NACA) Langley Laboratory, took the initiative in introducing Agency research in hypersonics. He used the V-2 as his rationale. In an August 1945 memo to Langley's chief of research, written 3 days before the United States atom-bombed Hiroshima, he noted that planned NACA facilities were to reach no higher than Mach 3. With the V-2 having already flown at Mach 5, he declared, this capability was clearly inadequate.

He outlined an alternative design concept for "a supersonic tunnel having a test section four-foot square and a maximum test Mach number of 7.0."[8] A preliminary estimate indicated a cost of $350,000. This was no mean sum. It was equivalent six decades later to approximately $4.2 million. Becker sweetened his proposal's appeal by suggesting that Langley

---

7. William B. Anspacher, Betty Gay, Donald Marlowe, Paul Morgan, and Samuel Raff, *The Legacy of the White Oak Laboratory* (Dahlgren, VA: Naval Surface Warfare Center, 2000), pp. 209–210; Donald D. Baals and William R. Corliss, *Wind Tunnels of NASA*, SP-440 (Washington, DC: NASA, 1981), pp. 51–52; James R. Hansen, *Engineer in Charge: A History of the Langley Aeronautical Laboratory, 1917–1958*, SP-4305 (Washington, DC: NASA, 1987), p. 467.
8. Quoted in Hansen, *Engineer in Charge*, pp. 344–345.

begin modestly with a small demonstration wind tunnel. It could be built for roughly one-tenth of this sum and would operate in the blow-down mode, passing flow through a 1-foot-square test section. If it proved successful and useful, a larger tunnel could follow. His reasoned idea received approval from the NACA's Washington office later in 1945, and out of this emerged the Langley 11-Inch Hypersonic Tunnel. Slightly later, Alfred J. Eggers began designing a hypersonic tunnel at the NACA's West Coast Ames Aeronautical Laboratory, though this tunnel, with a 10-inch by 14-inch test section, used continuous, not blow-down, flow. Langley's was first. When the 11-inch tunnel first demonstrated successful operation (to Mach 6.9) on November 26, 1947, American aeronautical science entered the hypersonic era. This was slightly over a month after Air Force test pilot Capt. Charles E. Yeager first flew faster than sound in the Bell XS-1 rocket plane.[9]

Though ostensibly a simple demonstration model for a larger tunnel, the 11-inch tunnel itself became an important training and research tool that served to study a wide range of topics, including nozzle development and hypersonic flow visualization. It made practical contributions to aircraft development as well. Research with the 11-inch tunnel led to a key discovery incorporated on the X-15, namely that a wedge-shaped vertical tail markedly increased directional stability, eliminating the need for very large stabilizing surfaces. So useful was it that it remained in service until 1973, staying active even with a successor, the larger Continuous Flow Hypersonic Tunnel (CFHT), which entered service in 1962. The CFHT had a 31-inch test section and reached Mach 10 but took a long time to become operational. Even after entering service, it operated much of the time in a blow-down mode rather than in continuous flow.[10]

---

9. John V. Becker, "Results of Recent Hypersonic and Unsteady Flow Research at the Langley Aeronautical Laboratory," *Journal of Applied Physics*, vol. 21 (July 1950), pp. 619–628; Patrick J. Johnston and Wallace C. Sawyer, "An Historical Perspective on Hypersonic Aerodynamic Research at the Langley Research Center," AIAA Paper 88-0230 (1988). For examples of its research, see Charles H. McLellan, Thomas W. Williams, and Mitchel H. Bertram, "Investigation of a Two-Step Nozzle in the Langley 11-inch Hypersonic Tunnel," NACA TN-2171 (1950); Charles H. McLellan and Thomas W. Williams, "Liquefaction of Air in the Langley 11-inch Hypersonic Tunnel," NACA TN-3302 (1954).

10. John V. Becker, "The X-15 Project: Part I—Origins and Research Background," *Astronautics & Aeronautics*, vol. 2, No. 2 (Feb. 1964), pp. 52–61; Charles H. McLellan, "A Method for Increasing the Effectiveness of Stabilizing Surfaces at High Supersonic Mach Numbers," NACA RM-L54F21 (1954); Baals and Corliss, *Wind Tunnels of NASA*, pp. 56–57, 94–95; William T. Schaefer, Jr., "Characteristics of Major Active Wind Tunnels at the Langley Research Center," NASA TM-X-1130 (1965), pp. 12, 27.

Emergent Hypersonic Technology and the Onset of the Missile Era

The ballistic missile and atomic bomb became realities within a year of each other. At a stroke, the expectation arose that one might increase the range of the former to intercontinental distance and, by installing an atomic tip, generate a weapon—and a threat—of almost incomprehensible destructive power. But such visions ran afoul of perplexing technical issues involving rocket propulsion, guidance, and reentry. Engineers knew they could do something about propulsion, but guidance posed a formidable challenge. MIT's Charles Stark Draper was seeking inertial guidance, but he couldn't approach the Air Force requirement, which set an allowed miss distance of only 1,500 feet at a range of 5,000 miles for a ballistic missile warhead.[11]

Reentry posed an even more daunting prospect. A reentering 5,000-mile-range missile would reach 9,000 kelvins, hotter than the solar surface, while its kinetic energy would vaporize five times its weight in iron.[12] Rand Corporation studies encouraged Air Force and industry missile studies. Convair engineers, working under Karel J. "Charlie" Bossart, began development of the Atlas ICBM in 1951. Even with this seemingly rapid implementation of the ballistic missile idea, time scales remained long term. As late as October 1953, the Air Force declared that it would not complete research and development until "sometime after 1964."[13]

Matters changed dramatically immediately after the Castle Bravo nuclear test on March 1, 1954, a weaponizable 15-megaton H-bomb, fully 1,000 times more powerful than the atomic bomb that devastated Hiroshima less than a decade previously. The "Teapot Committee," chaired by the Hungarian emigree mathematician John von Neumann, had anticipated success with Bravo and with similar tests. Echoing Bruno Augenstein of the Rand Corporation, the Teapot group recom-

---

11. Jacob Neufeld, *The Development of Ballistic Missiles in the United States Air Force, 1945–1960* (Washington, DC: USAF, 1990), p. 293; Col. Edward N. Hall, USAF, "Air Force Missile Experience," in Lt. Col. Kenneth F. Gantz, ed., *The United States Air Force Report on the Ballistic Missile: Its Technology, Logistics, and Strategy* (Garden City, NY: Doubleday & Co., Inc., 1958), pp. 47–59; Donald MacKenzie, *Inventing Accuracy* (Cambridge: MIT Press, 1990).
12. P.H. Rose and W.I. Stark, "Stagnation Point Heat-Transfer Measurements in Dissociated Air," *Journal of the Aeronautical Sciences*, vol. 25, no. 2 (Feb. 1958), pp. 86–97.
13. John L. Chapman, *Atlas: the Story of a Missile* (New York: Harper & Brothers, 1960), pp. 28–34, 74; Neufeld, *Development of Ballistic Missiles*, pp. 78, 44–50, 68–77; G. Harry Stine, *ICBM: The Making of the Weapon that Changed the World* (New York: Orion Books, 1996), pp. 140–146, 162–174, 186–188.

> It is well known that for any truly blunt body, the bow shock wave is detached and there exists a stagnation point at the nose. Consider conditions at this point and assume that the local radius of curvature of the body is σ (see sketch).

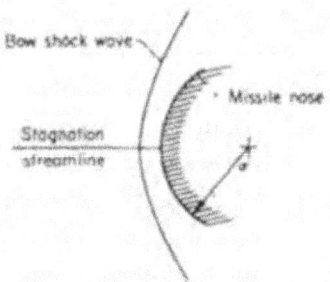

> The bow shock wave is normal to the stagnation streamline and converts the supersonic flow ahead of the shock to a low subsonic speed flow at high static temperature downstream of the shock. Thus, it is suggested that conditions near the stagnation point may be investigated by treating the nose section as if it were a segment of a sphere in a subsonic flow field.

Extract of text from NACA Report 1381 (1953), in which H. Julian Allen and Alfred J. Eggers postulated using a blunt-body reentry shape to reduce surface heating of a reentry body. NASA.

mended that the Atlas miss distance should be relaxed "from the present 1,500 feet to at least two, and probably three, nautical miles."[14] This was feasible because the new H-bomb had such destructive power that such a "miss" distance seemed irrelevant. The Air Force leadership concurred, and only weeks after the Castle Bravo shot, in May 1954, Vice Chief of Staff Gen. Thomas D. White granted Atlas the service's highest developmental priority.

But there remained the thorny problem of reentry. Only recently, most people had expected an ICBM nose cone to possess the needle-nose sharpness of futurist and science fiction imagination. The realities of aerothermodynamic heating at near-orbital speeds dictated otherwise. In 1953, NACA Ames aerodynamicists H. Julian Allen and Alfred

---

14. Bruno Augenstein, "Rand and North American Aviation's Aerophysics Laboratory: An Early Interaction in Missiles and Space," International Astronautical Federation, Paper IAA-98-IAA.2.2.06 (1998); Neufeld, *Development of Ballistic Missiles*, pp. 259, 102–106, 117; Robert L. Perry, "The Atlas, Thor, Titan, and Minuteman," in Eugene M. Emme, ed., *The History of Rocket Technology: Essays on Research, Development, and Utility* (Detroit: Wayne State University Press, 1964), pp. 142–161.

Eggers concluded that an ideal reentry shape should be bluntly rounded, not sharply streamlined. A sharp nose produced a very strong attached shock wave, resulting in high surface heating. In contrast, a blunt nose generated a detached shock standing much further off the nose surface, allowing the airflow to carry away most of the heat. What heating remained could be alleviated via radiative cooling or by using hot structures and high-temperature coatings.[15]

There was need for experimental verification of blunt body theory, but the hypersonic wind tunnel, previously so useful, was suddenly inadequate, much as the conventional wind tunnel a decade earlier had been inadequate to obtaining the fullest understanding of transonic flows. As the slotted throat tunnel had replaced it, so now a new research tool, the shock tube, emerged for hypersonic studies. Conceived by Arthur Kantrowitz, a Langley veteran working at Cornell, the shock tube enabled far closer simulation of hypersonic pressures and temperatures. From the outset, Kantrowitz aimed at orbital velocity, writing in 1952 that: "it is possible to obtain shock Mach numbers in the neighborhood of 25 with reasonable pressures and shock tube sizes."[16]

Despite the advantages of blunt body design, the hypersonic environment remained so extreme that it was still necessary to furnish thermal protection to the nose cone. The answer was ablation: covering the nose with a lightweight coating that melts and flakes off to carry away the heat. Wernher von Braun's U.S. Army team invented ablation while working on the Jupiter intermediate-range ballistic missile (IRBM), though General Electric scientist George Sutton made particularly notable contributions. He worked for the Air Force, which built and successfully protected a succession of ICBMs: Atlas, Titan, and Minuteman.[17]

---

15. H. Julian Allen and A.J. Eggers, Jr., "A Study of the Motion and Aerodynamic Heating of Ballistic Missiles Entering the Earth's Atmosphere at High Supersonic Speeds," NACA TR-1381 (1953); H. Julian Allen, "The Aerodynamic Heating of Atmospheric Entry Vehicles," in J. Gordon Hall, ed., *Fundamental Phenomena in Hypersonic Flow: Proceedings of the International Symposium Sponsored by Cornell Aeronautical Laboratory* (Ithaca, NY: Cornell University Press, 1966), pp. 6–10; Edwin P. Hartman, *Adventures in Research: A History of the Ames Research Center, 1940–1965*, NASA SP-4302 (Washington, DC: NASA, 1970), pp. 215–218.
16. E.L. Resler, Shao-Chi Lin, and Arthur Kantrowitz, "The Production of High Temperature Gases in Shock Tubes," *Journal of Applied Physics*, vol. 23 (Dec. 1952), p. 1397.
17. Frank Kreith, *Principles of Heat Transfer* (Scranton, PA: International Textbook Co., 1965), pp. 538–545; George W. Sutton, "The Initial Development of Ablation Heat Protection: an Historical Perspective," *Journal of Spacecraft and Rockets*, vol. 19 (1982), pp. 3–11.

Case 5 | Toward Transatmospheric Flight: From V-2 to the X-51

A Jupiter IRBM launches from Cape Canaveral on May 18, 1958, on an ablation reentry test. U.S. Army.

Flight tests were critical for successful nose cone development, and they began in 1956 with launches of the multistage Lockheed X-17. It rose high into the atmosphere before firing its final test stage back at Earth, ensuring the achievement of a high-heat load, as the test nose cone would typically attain velocities of at least Mach 12 at only 40,000 feet. This was half the speed of a satellite, at an altitude typically traversed by today's subsonic airliners. In the pre-ablation era, the warheads typically burned up in the atmosphere, making the X-17 effectively a flying shock tube whose nose cones only lived long enough to return data by telemetry. Yet out of such limited beginnings (analogous to the rudimentary test methodologies of the early transonic and supersonic era just a decade previously) came a technical base that swiftly resolved the reentry challenge.[18]

Tests followed with various Army and Air Force ballistic missiles. In August 1957, a Jupiter-C (an uprated Redstone) returned a nose cone after a flight of 1,343 miles. President Dwight D. Eisenhower subsequently showed it to the public during a TV appearance that sought to bolster American morale a month after Sputnik had shocked the world. Two Thor-Able flights went to 5,500 miles in July 1958, though their nose cones both were lost at sea. But the agenda also included Atlas, which first reached its full range of 6,300 miles in November 1958. Two nose cones built by GE, the RVX-1 and –2, flew subsequently as payloads. An RVX-2 flew 5,000 miles in July 1959 and was recovered, thereby becoming the largest object yet to be brought back. Attention now turned to a weaponized nose cone shape, GE's Mark 3. Flight tests began in October, with this nose cone entering operational service the following April.[19]

Success in reentry now was a reality, yet there was much more for the future. The early nose cones were symmetric, which gave good ballistic characteristics but made no provision for significant aerodynamic maneuver and cross-range. The military sought both as a means of achieving greater operational flexibility. An Air Force experimental uncrewed lifting body design, the Martin SV-5D (X-23) PRIME, flew three flights between December 1966 and April 1967, lofted over the Pacific Test Range by modified Atlas boosters. The first flew 4,300 miles, maneuvering in pitch (but not in cross-range), and missed its target aim point by only 900 feet. The third mission demonstrated a turning cross-range of 800 miles, the SV-5D impacting within 4 miles of its aim point and subsequently was recovered.[20]

Other challenges remained. These included piloted return from the Moon, reusable thermal protection for the Shuttle, and planetary entry into the Jovian atmosphere, which was the most demanding of all. Even

---

18. "Re-Entry Research: The Lockheed X-17," *Flight* (Feb. 6, 1959), p. 181.
19. James M. Grimwood and Francis Strowd, *History of the Jupiter Missile System* (Huntsville, AL: U.S. Army Ordnance Missile Command, July 27, 1962), pp. 18–20; *Time* (Nov. 18, 1957, pp. 19–20, and Dec. 8, 1958, p. 15); Joel W. Powell, "Thor-Able and Atlas Able," *Journal of the British Interplanetary Society*, vol. 37, No. 5 (May 1984), pp. 219–225; General Electric, "Thermal Flight Test Summary Report for Mark 3 Mod 1 Re-Entry Vehicles" (1960), Defense Technical Information Center [DTIC] Report AD-362539; Convair, "Flight Test Evaluation Report, Missile 7D" (1959), DTIC AD-832686.
20. Martin Marietta, "SV-5 PRIME Final Flight Test Summary," Report ER 14465 (1967).

so, by the time of PRIME in 1967, the reentry problem had been resolved, manifested by the success of both ballistic missile nose cone development and the crewed spacecraft effort. The latter was arguably the most significant expression of hypersonic competency until the return to Earth from orbit by the Space Shuttle Columbia in 1981.

Transitioning from the Supersonic to the Hypersonic: X-7 to X-15

During the 1950s and early 1960s, aviation advanced from flight at high altitude and Mach 1 to flight in orbit at Mach 25. Within the atmosphere, a number of these advances stemmed from the use of the ramjet, at a time when turbojets could barely pass Mach 1 but ramjets could aim at Mach 3 and above. Ramjets needed an auxiliary rocket stage as a booster, which brought their general demise after high-performance afterburning turbojets succeeded in catching up. But in the heady days of the 1950s, the ramjet stood on the threshold of becoming a mainstream engine. Many plans and proposals existed to take advantage of their power for a variety of aircraft and missile applications.

The burgeoning ramjet industry included Marquardt and Wright Aeronautical, though other firms such as Bendix developed them as well. There were also numerous hardware projects. One was the Air Force-Lockheed X-7, an air-launched high-speed propulsion, aerodynamic, and structures testbed. Two were surface-to-air ramjet-powered missiles: the Navy's ship-based Mach 2.5+ Talos and the Air Force's Mach 3+ Bomarc. Both went on to years of service, with the Talos flying "in anger" as a MiG-killer and antiradiation SAM-killer in Vietnam. The Air Force also was developing a 6,300-mile-range Mach 3+ cruise missile—the North American SM-64 Navaho—and a Mach 3+ interceptor fighter—the Republic XF-103. Neither entered the operational inventory. The Air Force canceled the troublesome Navaho in July 1957, weeks after the first flight of its rival, Atlas, but some flight hardware remained, and Navaho flew in test for as far as 1,237 miles, though this was a rare success. The XF-103 was to fly at Mach 3.7 using a combined turbojet-ramjet engine. It was to be built largely of titanium, at a time when this metal was little understood; it thus lived for 6 years without approaching flight test. Still, its engine was built and underwent test in December 1956.[21]

---

21. Marcelle Size Knaack, *Post-World War II Fighters*, vol. 1 of *Encyclopedia of U.S. Air Force Aircraft and Missile Systems* (Washington, DC: Office of Air Force History, 1978), p. 329; Richard A. DeMeis, "The Trisonic Titanium Republic," *Air Enthusiast*, vol. 7 (1978), pp. 198–213.

The steel-structured X-7 proved surprisingly and consistently productive. The initial concept of the X-7 dated to December 1946 and constituted a three-stage vehicle. A B-29 (later a B-50) served as a "first stage" launch aircraft; a solid rocket booster functioned as a "second stage" accelerating it to Mach 2, at which the ramjet would take over. First flying in April 1951, the X-7 family completed 100 missions between 1955 and program termination in 1960. After achieving its Mach 3 design goal, the program kept going. In August 1957, an X-7 reached Mach 3.95 with a 28-inch diameter Marquardt ramjet. The following April, the X-7 attained Mach 4.31—2,881 mph—with a more-powerful 36-inch Marquardt ramjet. This established an air-breathing propulsion record that remains unsurpassed for a conventional subsonic combustion ramjet.[22]

At the same time that the X-7 was edging toward the hypersonic frontier, the NACA, Air Force, Navy, and North American Aviation had a far more ambitious project underway: the hypersonic X-15. This was Round Two, following the earlier Round One research airplanes that had taken flight faster than sound. The concept of the X-15 was first proposed by Robert Woods, a cofounder and chief engineer of Bell Aircraft (manufacturer of the X-1 and X-2), at three successive meetings of the NACA's influential Committee on Aerodynamics between October 1951 and June 1952. It was a time when speed was king, when ambitious technology-pushing projects were flying off the drawing board. These included the Navaho, X-2, and XF-103, and the first supersonic operational fighters—the Century series of the F-100, F-101, F-102, F-104, and F-105.[23]

Some contemplated even faster speeds. Walter Dornberger, former commander of the Nazi research center at Peenemünde turned senior Bell Aircraft Corporation executive, was advocating BoMi, a proposed skip-gliding "Bomber-Missile" intended for Mach 12. Dornberger supported Woods in his recommendations, which were adopted by the NACA's Executive Committee in July 1952. This gave them the status of policy, while the Air Force added its own support. This was significant because

---

22. Lee L. Peterson, "Evaluation Report on X-7A," AFMDC [Holloman AFB], ADJ 57-8184 (1957); and William A. Ritchie, "Evaluation Report on X-7A (System 601B)," AFMDC DAS-58-8129 (1959).
23. Robert S. Houston, Richard P. Hallion, and Ronald G. Boston, "Transiting from Air to Space: The North American X-15," and John V. Becker, "The Development of Winged Reentry Vehicles: An Essay from the NACA-NASA Perspective, 1952–1963," in Richard P. Hallion, ed., *The Hypersonic Revolution: Eight Case Studies in the History of Hypersonic Technology*, vol. 1: *From Max Valier to Project PRIME, 1924–1967* (Wright-Patterson AFB: Aeronautical Systems Division, 1987), pp. I–xii, No. 1, 383–386.

its budget was 300 times larger than that of the NACA.[24] The NACA alone lacked funds to build the X-15, but the Air Force could do this easily. It also covered the program's massive cost overruns. These took the airframe from $38.7 million to $74.5 million and the large engine from $10 million to $68.4 million, which was nearly as much as the airframe.[25]

The Air Force had its own test equipment at its Arnold Engineering Development Center (AEDC) at Tullahoma, TN, an outgrowth of the Theodore von Kármán technical intelligence mission that Army Air Forces Gen. Henry H. "Hap" Arnold had sent into Germany at the end of the Second World War. The AEDC, with brand-new ground test and research facilities, took care to complement, not duplicate, the NACA's research facilities. It specialized in air-breathing and rocket-engine testing. Its largest installation accommodated full-size engines and provided continuous flow at Mach 4.75. But the X-15 was to fly well above this, to over Mach 6, highlighting the national facilities shortfall in hypersonic test capabilities existing at the time of its creation.[26]

While the Air Force had the deep pockets, the NACA—specifically Langley—conducted the research that furnished the basis for a design. This took the form of a 1954 feasibility study conducted by John Becker, assisted by structures expert Norris Dow, rocket expert Maxime Faget, configuration and controls specialist Thomas Toll, and test pilot James Whitten. They began by considering that during reentry, the vehicle should point its nose in the direction of flight. This proved impossible, as the heating was too high. He considered that the vehicle might alleviate this problem by using lift, which he was to obtain by raising the nose. He found that the thermal environment became far more manageable. He concluded that the craft should enter with its nose high, presenting its flat undersurface to the atmosphere. The Allen-Eggers paper was in print, and he later wrote that: "it was obvious to us that what we were seeing here was a new manifestation of H.J. Allen's 'blunt-body' principle."[27]

---

24. Harry Hansen, *Engineer in Charge*, NASA SP-4305, p. 428; Hansen, ed., *The World Almanac and Book of Facts for 1956* (New York: New York World-Telegram Corp., 1956), p. 757.
25. Dennis Jenkins, *X-15: Extending the Frontiers of Flight*, NASA SP-2007-562 (Washington, DC: NASA, 2007), pp. 336–337.
26. U.S. Air Force Systems Command, *History of the Arnold Engineering Development Center* (Arnold Air Force Station, TN: AEDC, n.d.); Julius Lukasiewicz, *Experimental Methods of Hypersonics* (New York: Marcel Dekker, Inc., 1973), p. 247.
27. Becker, "Development of Winged Reentry Vehicles," in Hallion, *Hypersonic Revolution*, vol. 1, p. 386.

To address the rigors of the daunting aerothermodynamic environment, Norris Dow selected Inconel X (a nickel alloy from International Nickel) as the temperature-resistant superalloy that was to serve for the aircraft structure. Dow began by ignoring heating and calculated the skin gauges needed only from considerations of strength and stiffness. Then he determined the thicknesses needed to serve as a heat sink. He found that the thicknesses that would suffice for the latter were nearly the same as those that would serve merely for structural strength. This meant that he could design his airplane and include heat sink as a bonus, with little or no additional weight. Inconel X was a wise choice; with a density of 0.30 pounds per cubic inch, a tensile strength of over 200,000 pounds per square inch (psi), and yield strength of 160,000 psi, it was robust, and its melting temperature of over 2,500 °F ensured that the rigors of the anticipated 1,200 °F surface temperatures would not weaken it.[28]

Work at Langley also addressed the important issue of stability. Just then, in 1954, this topic was in the forefront because it had nearly cost the life of the test pilot Chuck Yeager. On the previous December 12, he had flown the X-1A to Mach 2.44 (approximately 1,650 mph). This exceeded the plane's stability limits; it went out of control and plunged out of the sky. Only Yeager's skill as a pilot had saved him and his airplane. The problem of stability would be far more severe at higher speeds.[29]

Analysis, confirmed by experiments in the 11-inch wind tunnel, had shown that most of the stability imparted by an aircraft's tail surfaces was produced by its wedge-shaped forward portion. The aft portion contributed little to the effectiveness because it experienced lower air pressure. Charles McLellan, another Langley aerodynamicist, now proposed to address the problem of hypersonic stability by using tail surfaces that would be wedge-shaped along their entire length. Subsequent tests in the 11-inch tunnel, as mentioned previously, confirmed that this solution worked. As a consequence, the size of the tail surfaces shrank from being almost as large as the wings to a more nearly conventional appearance.[30]

---

28. Becker, "The X-15 Project," pp. 52–61. Technical characteristics of Inconel X are from "Inconel X-750 Technical Data" (Sylmar, CA: High Temp Metals, Inc., 2009).
29. Richard P. Hallion, On the Frontier: Flight Research at Dryden, 1946–1982, SP-4303 (Washington, DC: NASA, 1984), pp. 70–71.
30. McLellan, "A Method for Increasing the Effectiveness of Stabilizing Surfaces," NACA RM-L54F21 (1954).

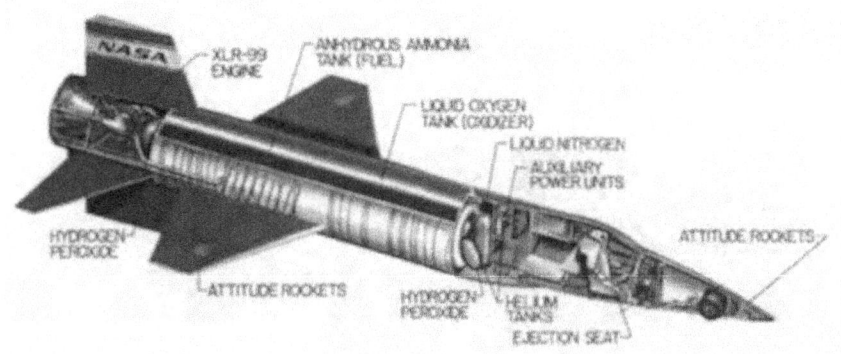

A schematic drawing of the X-15's internal layout. NASA.

This study made it possible to proceed toward program approval and the award of contracts both for the X-15 airframe and its powerplant, a 57,000-pound-thrust rocket engine burning a mix of liquid oxygen and anhydrous ammonia. But while the X-15 promised to advance the research airplane concept to over Mach 6, it demanded something more than the conventional aluminum and stainless steel structures of earlier craft such as the X-1 and X-2. Titanium was only beginning to enter use, primarily for reducing heating effects around jet engine exhausts and afterburners. Magnesium, which Douglas favored for its own high-speed designs, was flammable and lost strength at temperatures higher than 600 °F. Inconel X was heat-resistant, reasonably well known, and relatively easily worked. Accordingly, it was swiftly selected as the structural material of choice when Becker's Langley team assessed the possibility of designing and fabricating a rocket-boosted air-launched hypersonic research airplane. The Becker study, completed in April 1954, chose Mach 6 as the goal and proposed to fly to altitudes as great as 350,000 feet. Both marks proved remarkably prescient: the X-15 eventually flew to 354,200 feet in 1963 and Mach 6.70 in 1967. This was above 100 kilometers and well above the sensible atmosphere. Hence, at that early date, more than 3 years before Sputnik, Becker and his colleagues already were contemplating piloted flight into space.[31]

The X-15: Pioneering Piloted Hypersonics

North American Aviation won the contract to build the X-15. It first flew under power in September 1959, by which time an Atlas had hurled an

---

31. John V. Becker, Norris F. Dow, Maxime A. Faget, Thomas A. Toll, and J.B. Whitten, "Research Airplane Study," NACA Langley (April 1954).

The North American X-15 at NASA's Flight Research Center (now the Dryden Flight Research Center) in 1961. NASA.

RVX-2 nose cone to its fullest range. However, as a hypersonic experiment, the X-15 was a complete airplane. It thus was far more complex than a simple reentry body, and it took several years of cautious flight-testing before it reached peak speed of above Mach 6, and peak altitude as well.

Testing began with two so-called "Little Engines," a pair of vintage Reaction Motors XLR11s that had earlier served in the X-1 series and the Douglas D-558-2 Skyrocket. Using these, the X-15 topped the records of the earlier X-2, reaching Mach 3.50 and 136,500 feet. Starting in 1961, using the "Big Engine"—the Thiokol XLR99 with its 57,000 pounds of thrust—the X-15 flew to its Mach 6 design speed and 50+ mile design altitude, with test pilot Maj. Robert White reaching Mach 6.04 and NASA pilot Joseph Walker an altitude of 354,200 feet. After a landing accident, the second X-15 was modified with external tanks and an ablative coating, with Air Force Maj. William "Pete" Knight subsequently flying this variant to Mach 6.70 (4,520 mph) in 1967. However, it sustained severe thermal damage, partly as a result of inadequate understanding of the interactions of impinging hypersonic shock-on-shock flows. It never flew again.[32]

The X-15's cautious buildup proved a wise approach, for this gave leeway when problems arose. Unexpected thermal expansion leading to localized buckling and deformation showed up during early high-Mach flights. The skin behind the wing leading edge exhibited localized buckling after the first flight to Mach 5.3, but modifications to the wings eliminated hot

---

32. Johnny G. Armstrong, "Flight Planning and Conduct of the X-15A-2 Envelope Expansion Program," AFFTC TD-69-4 (1969).

spots and prevented subsequent problems, enabling the airplane to reach beyond Mach 6. In addition, a flight to Mach 6.04 caused a windshield to crack because of thermal expansion. This forced redesign of its frame to incorporate titanium, which has a much lower coefficient of expansion. The problem—a rare case in which Inconel caused rather than resolved a heating problem—was fixed by this simple substitution.[33]

Altitude flights brought their own problems, involving potentially dangerous auxiliary power unit (APU) failures. These issues arose in 1962 as flights began to reach well above 100,000 feet; the APUs began to experience gear failure after lubricating oil foamed and lost its lubricating properties. A different oil had much less tendency to foam; it now became standard. Designers also enclosed the APU gearbox within a pressurized enclosure. The gear failures ceased.[34]

The X-15 substantially expanded the use of flight simulators. These had been in use since the famed Link Trainer of Second World War and now included analog computers, but now they also took on a new role as they supported the development of control systems and flight equipment. Analog computers had been used in flight simulation since 1949. Still, in 1955, when the X-15 program began, it was not at all customary to use flight simulators to support aircraft design and development. But program managers turned to such simulators because they offered effective means to study new issues in cockpit displays, control systems, and aircraft handling qualities. A 1956 paper stated that simulation had "heretofore been considered somewhat of a luxury for high-speed aircraft," but now "has been demonstrated as almost a necessity," in all three axes, "to insure [sic] consistent and successful entries into the atmosphere." Indeed, pilots spent much more time practicing in simulators than they did in actual flight, as much as an hour per minute of actual flying time.[35]

---

33. William H. Dana, "The X-15 Airplane—Lessons Learned," AIAA Paper 93-0309 (1993); Joseph Weil, "Review of the X-15 Program," NASA TN-D-1278 (1962).
34. Perry V. Row and Jack Fischel, "X-15 Flight Test Experience," *Astronautics and Aerospace Engineering*, vol. 1 (June 1963), pp. 25–32.
35. Quotes from "Research Airplane Committee Report on Conference on the Progress of the X-15 Project," NACA Langley Aeronautical Laboratory, 1956, p. 84; James I. Kilgore, "The Planes that Never Leave the Ground," *American Heritage of Invention and Technology* (winter 1989), pp. 60–62; John P. Smith, Lawrence J. Schilling, and Charles A. Wagner, "Simulation at Dryden Flight Research Facility from 1957 to 1982," NASA TM-101695 (1989), p. 4; Milton O. Thompson, *At the Edge of Space: The X-15 Flight Program* (Washington, DC: Smithsonian Institution Press, 1992), pp. 70–71.

The most important flight simulator was built by North American. Located originally in Los Angeles, Paul Bikle, the Director of NASA's Flight Research Center, moved it to that Center in 1961. It replicated the X-15 cockpit and included actual hydraulic and control-system hardware. Three analog computers implemented equations of motion that governed translation and rotation of the X-15 about all three axes, transforming pilot inputs into instrument displays.[36]

The North American simulator became critical in training X-15 pilots as they prepared to execute specific planned flights. A particular mission might take little more than 10 minutes, from ignition of the main engine to touchdown on the lakebed, but a test pilot could easily spend 10 hours making practice runs in this facility. Training began with repeated trials of the normal flight profile with the pilot in the simulator cockpit and a ground controller close at hand. The pilot was welcome to recommend changes, which often went into the flight plan. Next came rehearsals of off-design missions: too much thrust from the main engine, too high a pitch angle when leaving the stratosphere.

Much time was spent practicing for emergencies. The X-15 had an inertial reference unit that used analog circuitry to display attitude, altitude, velocity, and rate of climb. Pilots dealt with simulated failures in this unit as they worked to complete the normal mission or, at least, to execute a safe return. Similar exercises addressed failures in the stability augmentation system. When the flight plan raised issues of possible flight instability, tests in the simulator used highly pessimistic assumptions concerning stability of the vehicle. Other simulations introduced in-flight failures of the radio or Q-ball multifunction sensor. Premature engine shutdown imposed a requirement for safe landing on an alternate lakebed that was available for emergency use.[37]

The simulations indeed had realistic cockpit displays, but they left out an essential feature: the g-loads, produced both by rocket thrust and by deceleration during reentry. In addition, a failure of the stability augmentation system, during reentry, could allow the airplane to oscillate

---

36. NASA FRC, "Experience with the X-15 Adaptive Flight Control System," NASA TN-D-6208 (1971); Perry V. Row and Jack Fischel, "Operational Flight-test Experience with the X-15 Airplane," AIAA Paper 63-075 (1963).
37. Wendell H. Stillwell, *X-15 Research Results*, NASA SP-60 (Washington, DC: NASA, 1965), pp. 37–38; Robert G. Hoey and Richard E. Day, "Mission Planning and Operational Procedures for the X-15 Airplane," NASA TN-D-1158 (1962), NTRS Document ID 19710070140.

in pitch and yaw. This changed the drag characteristics and imposed a substantial cyclical force.

To address such issues, investigators installed a flight simulator within the gondola of an existing centrifuge at the Naval Air Development Center in Johnsville, PA. The gondola could rotate on two axes while the centrifuge as a whole was turning. It not only produced g-forces; its g-forces increased during the simulated rocket burn. The centrifuge imposed such forces anew during reentry while adding a cyclical component to give the effect of an oscillation in yaw or pitch.[38]

There also were advances in pressure suits, under development since the 1930s. Already an early pressure suit had saved the life of Maj. Frank K. Everest during a high-altitude flight in the X-1, when it had suffered cabin decompression from a cracked canopy. Marine test pilot Lt. Col. Marion Carl had worn another during a flight to 83,235 feet in the D-558-2 Skyrocket in 1953, as had Capt. Iven Kincheloe during his record flight to 126,200 feet in the Bell X-2 in 1956. But these early suits, while effective in protecting pilots, were almost rigid when inflated, nearly immobilizing them. In contrast, the David G. Clark Company, a girdle manufacturer, introduced a fabric that contracted in circumference while it stretched in length. An exchange between these effects created a balance that maintained a constant volume, preserving a pilot's freedom of movement. The result was the Clark MC-2 suit, which, in addition to the X-15, formed the basis for American spacesuit development from Project Mercury forward. Refined as the A/P22S-2, the X-15's suit became the standard high-altitude pressure suit for NASA and the Air Force. It formed the basis for the Gemini suit and, after 1972, was adopted by the U.S. Navy as well, subsequently being employed by pilots and aircrew in the SR-71, U-2, and Space Shuttle.[39]

---

38. C.C. Clark and C.H. Woodling, "Centrifuge Simulation of the X-15 Research Aircraft," NADC MA-5916 (1959); Jenkins, *X-15*, p. 279; NASA, "Research Airplane Committee Report on Conference on the Progress of the X-15 Project," (1958), pp. 107–116.
39. For the Clark suit's development, see A. Scott Crossfield with Clay Blair, Jr., *Always Another Dawn: The Story of a Rocket Test Pilot* (Cleveland: The World Publishing Co., 1960), pp. 253–261; Paul Crickmore, Lockheed SR-71 Blackbird (London: Osprey Publishing Ltd., 1986), pp. 100–102; T.A. Heppenheimer, *History of the Space Shuttle*, vol. 2, *Development of the Shuttle, 1972–81* (Washington, DC: Smithsonian Institution Press, 2002), pp. 274–277; Jenkins, *X-15*, pp. 131–146; Loyd S. Swenson, James M. Grimwood, and Charles Alexander, *This New Ocean: A History of Project Mercury*, NASA SP-4201 (Washington, DC: NASA, 1998), pp. 225–231.

The X-15 also accelerated development of specialized instrumentation, including a unique gimbaled nose sensor developed by Northrop. It furnished precise speed and positioning data by evaluation of dynamic pressure ("q" in aero engineering shorthand), and thus was known as the Q-ball. The Q-ball took the form of a movable sphere set in the nose of the craft, giving it the appearance of the enlarged tip of a ballpoint pen. "The Q-ball is a go-no go item," NASA test pilot Joseph Walker told *Time* magazine reporters in 1961, adding: "Only if she checks okay do we go."[40] The X-15 also incorporated "cold jet" hydrogen peroxide reaction controls for maintaining vehicle attitude in the tenuous upper atmosphere, when dynamic air pressure alone would be insufficient to permit adequate flight control functionality. When Iven Kincheloe reached 126,200 feet, his X-2 was essentially a free ballistic object, uncontrollable in pitch, roll, and yaw as it reached peak altitude and then began its descent. This situation made reaction controls imperative for the new research airplane, and the NACA (later NASA) had evaluated them on a so-called "Iron Cross" simulator on the ground and then in flight on the Bell X-1B and on a modified Lockheed F-104 Starfighter. They then proved their worth on the X-15 and, as with the Clark pressure suit, were incorporated on Mercury and subsequent American spacecraft.

The X-15 introduced a side stick flight controller that the pilot would utilize during acceleration (when under loads of approximately 3 g's), relying on a fighter-type conventional control column for approach and landing. The third X-15 had a very different flight control system than the other two, differing greatly from the now-standard stability-augmented hydromechanical system carried by operational military and civilian aircraft. The third aircraft introduced a so-called "adaptive" flight control system, the MH-96. Built by Minneapolis Honeywell, the MH-96 relied on rate gyros, which sensed rates of motion in pitch, roll, and yaw. It also incorporated "gain," defined as the proportion between sensed rates of angular motion and a deflection of the ailerons or other controls. This variable gain, which changed automatically in response to flight conditions, functioned to maintain desired handling qualities across the spectrum of X-15 performance. This arrangement made it possible to introduce blended reaction and aerodynamic controls on the same stick, with this blending occurring automatically in response to

---

40. *Time*, Oct. 27, 1961, p. 89.

the values determined for gain as the X-15 flew out of the atmosphere and back again. Experience, alas, would reveal the MH-96 as an immature, troublesome system, one that, for all its ambition, posed significant headaches. It played an ultimately fatal role in the loss of X-15 pilot Maj. Michael Adams in 1967.[41]

The three X-15s accumulated a total of 199 flights from 1959 through 1968. As airborne instruments of hypersonic research, they accumulated nearly 9 hours above Mach 3, close to 6 hours above Mach 4, and 87 minutes above Mach 5. Many concepts existed for X-15 derivatives and spinoffs, including using it as a second stage to launch small satellite-lofting boosters, to be modified with a delta wing and scramjet, and even to form the basis itself for some sort of orbital spacecraft; for a variety of reasons, NASA did not proceed with any of these. More significantly, however, was the strong influence the X-15 exerted upon subsequent hypersonic projects, particularly the National Hypersonic Flight Research Facility (NHFRF, pronounced "nerf"), intended to reach Mach 8.

A derivative of the Air Force Flight Dynamics Laboratory's X-24C study effort, NHFRF was also to cruise at Mach 6 for 40 seconds. A joint Air Force-NASA committee approved a proposal in July 1976 with an estimated program cost of $200 million, and NHFRF had strong support from NASA's hypersonic partisans in the Langley and Dryden Centers. Unfortunately, its rising costs, at a time when the Shuttle demanded an ever-increasing proportion of the Agency's budget and effort, doomed it, and it was canceled in September 1977. Overall, the X-15 set speed and altitude records that were not surpassed until the advent of the Space Shuttle.[42]

---

41. Dana, "The X-15 Airplane—Lessons Learned," AIAA Paper 93-0309 (1993); Thompson, *At the Edge of Space*, pp. 200–202; Lawrence W. Taylor and George B. Merrick, "X-15 Airplane Stability Augmentation Systems," NASA TN-D-1157 (1962); Robert A. Tremant, "Operational Experience and Characteristics of the X-15 Flight Control System," NASA TN-D-1402 (Dec. 1962), Donald R. Bellman, et al., *Investigation of the Crash of the X-15-3 Aircraft on November 15, 1967* (Edwards: NASA Flight Research Center, Jan. 1968), pp. 8–15.

42. Kenneth E. Hodge, et al., *Proceedings of the X-15 First Flight 30th Anniversary Celebration*, CP 3105 (Edwards: NASA, June 8, 1989); Hallion, *On the Frontier*, pp. 170–172; Donald P. Hearth and Albert E. Preyss, "Hypersonic Technology: Approach to an Expanded Program," *Astronautics and Aeronautics*, (Dec. 1976), pp. 20–37; "NASA to End Hypersonic Effort," *Aviation Week and Space Technology* (Sept. 26, 1977).

The X-20 Dyna-Soar

During the 1950s, as the X-15 was taking shape, a parallel set of initiatives sought to define a follow-on hypersonic program that could actually achieve orbit. They were inspired in large measure by the 1938–1944 Silbervögel ("Silver Bird") proposal of Austrian space flight advocate Eugen Sänger and his wife, mathematician Irene Sänger-Bredt, which greatly influenced postwar Soviet, American, and European thinking about hypersonics and long-range "antipodal" flight. Influenced by Sänger's work and urged onward by the advocacy of Walter Dornberger, Bell Aircraft Corporation in 1952 proposed the BoMi, intended to fly 3,500 miles. Bell officials gained funding from the Air Force's Wright Air Development Center (WADC) to study longer-range 4,000-mile and 6,000-mile systems under the aegis of Air Force project MX-2276.

Support took a giant step forward in February 1956, when Gen. Thomas Power, Chief of the Air Research and Development Command (ARDC, predecessor of Air Force Systems Command) and a future Air Force Chief of Staff, stated that the service should stop merely considering such radical craft and instead start building them. With this level of interest, events naturally moved rapidly. A month later, Bell received a study contract for Brass Bell, a follow-on Mach 15 rocket-lofted boost-glider for strategic reconnaissance. Power preferred another orbital glider concept, RoBo (for Rocket Bomber), which was to serve as a global strike system. To accelerate transition of hypersonics from the research to the operational community, the ARDC proposed its own concept, Hypersonic Weapons Research and Development Supporting System (HYWARDS). With so many cooks in the kitchen, the Air Force needed a coordinated plan. An initial step came in December 1956, as Bell raised the velocity of Brass Bell to Mach 18. A month later, a group headed by John Becker, at Langley, recommended the same design goal for HYWARDS. RoBo still remained separate, but it emerged as a long-term project that could be operational by the mid-1970s.[43]

NACA researchers split along centerlines over the issue of what kind of wing design to employ for HYWARDS. At NACA Ames, Alfred Eggers and Clarence Syvertson emphasized achieving maximum lift. They proposed a high-wing configuration with a flat top, calculating its hypersonic

---

43. Clarence J. Geiger, "Strangled Infant: The Boeing X-20A Dyna-Soar," in Hallion, *Hypersonic Revolution*, vol. 1, pp. 189–201; Capt. Roy F. Houchin, "The Rise and Fall of Dyna-Soar: A History of Air Force Hypersonic R&D, 1944–1963," Air Force Institute of Technology (1995), DTIC ADA-303832.

life-to-drag (L/D) as 6.85 and measuring a value of 6.65 during hypersonic tunnel tests. Langley researchers John Becker and Peter Korycinski argued that Ames had the configuration "upside down." Emphasizing lighter weight, they showed that a flat-bottom Mach 18 shape gave a weight of 21,400 pounds, which rose only modestly at higher speeds. By contrast, the Ames "flat-top" weight was 27,600 pounds and rising steeply. NASA officials diplomatically described the Ames and Langley HYWARDS concepts respectively as "high L/D" and "low heating," but while the imbroglio persisted, there still was no acceptable design. It fell to Becker and Korycinski to break the impasse in August 1957, and they did so by considering heating. It was generally expected that such craft required active cooling. But Becker and his Langley colleagues found that a glider of global range achieved peak uncooled skin temperatures of 2,000 °F, which was survivable by using improved materials. Accordingly, the flat-bottom design needed no coolant, dramatically reducing both its weight and complexity.[44]

This was a seminal conclusion that reshaped hypersonic thinking and influenced all future development down to the Space Shuttle. In October 1957, coincident with the Soviet success with Sputnik, the ARDC issued a coordinated plan that anticipated building HYWARDS for research at 18,000 feet per second, following it with Brass Bell for reconnaissance at the same speed and then RoBo, which was to carry nuclear bombs into orbit. HYWARDS now took on the new name of Dyna-Soar, for "Dynamic Soaring," an allusion to the Sänger-legacy skip-gliding hypersonic reentry. (It was later designated X-20.) To the NACA, it constituted a Round Three following the Round One X-1, X-2, and Skyrocket, and the Round Two X-15.

The flat-bottom configuration quickly showed that it was robust enough to accommodate flight at much higher speeds. In 1959, Herbert York, the Defense Director of Research and Engineering, stated that Dyna-Soar was to fly at 15,000 mph, lofted by the Martin Company's Titan I missile, though this was significantly below orbital speed. But

---

44. John V. Becker, "The Development of Winged Reentry Vehicles: An Essay from the NACA-NASA Perspective, 1952–1963," in Hallion, *Hypersonic Revolution*, vol. 1, pp. 391–407; Alvin Seiff and H. Julian Allen, "Some Aspects of the Design of Hypersonic Boost-Glide Aircraft," NACA RM-A55E26 (1955); Alfred J. Eggers and Clarence Syvertson, "Aircraft Configurations Developing High Lift-Drag Ratios at High Supersonic Speeds," NACA RM-A55L05 (1956); Hansen, *Engineer in Charge*, pp. 467–473.

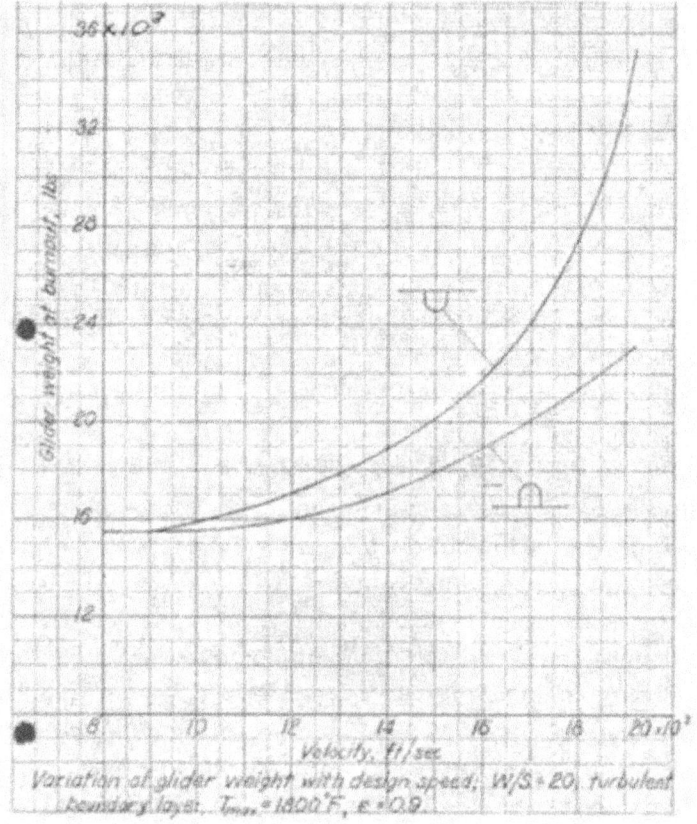

This 1957 Langley trade-study shows weight advantage of flat-bottom reentry vehicles at higher Mach numbers. This led to abandonment of high-wing designs in favor of flat-bottom ones such as the X-20 Dyna-Soar and the Space Shuttle. NASA.

during subsequent years it changed to the more-capable Titan II and then to the powerful Titan III-C. With two solid-fuel boosters augmenting its liquid hypergolic main stage, it could easily boost Dyna-Soar to the 18,000 mph necessary for it to achieve orbit. A new plan of December 1961 dropped suborbital missions and called for "the early attainment of orbital flight."[45]

By then, though, Dyna-Soar was in deep political trouble. It had been conceived initially as a prelude to the boost-glider Brass Bell for

---

45. Capt. Roy Houchin, "Hypersonic Technology and Aerospace Doctrine," *Air Power History*, vol. 46, no. 3 (fall 1999), pp. 4–17; Terry L. Sunday and John R. London, "The X-20 Space Plane: Past Innovation, Future Vision," in John Becklake, ed., *History of Rocketry and Astronautics*, vol. 17 (San Diego: American Astronautical Society/Univelt, 1995), pp. 253–284.

Case 5 | Toward Transatmospheric Flight: From V-2 to the X-51

This full-size mockup of the X-20 gives an indication of its small, compact design. USAF.

reconnaissance and to the orbital RoBo for bombardment. But Brass Bell gave way to a purpose-built concept for a small-piloted station, the Manned Orbiting Laboratory (MOL), which could carry more sophisticated reconnaissance equipment. (Ironically, though a team of MOL astronauts was selected, MOL itself likewise was eventually canceled.) RoBo, a strategic weapon, fell out of the picture completely, for the success of the solid-propellant Minuteman ICBM established the silo-launched ICBM as the Nation's prime strategic force, augmented by the Navy's fleet of Polaris-launching ballistic missile submarines.[46]

In mid-1961, Secretary of Defense Robert S. McNamara directed the Air Force to justify Dyna-Soar on military grounds. Service advocates responded by proposing a host of applications, including orbital reconnaissance, rescue, inspection of Soviet spacecraft, orbital bombardment,

---

46. Wyndham D. Miles, "The Polaris," in Emme, *History of Rocket Technology*, pp. 162–175.

and use of the craft as a ferry vehicle. McNamara found these rationalizations unconvincing but was willing to allow the program to proceed as a research effort, at least for the time being. In an October 1961 memo to President John F. Kennedy, he proposed to "re-orient the program to solve the difficult technical problems involved in boosting a body of high lift into orbit, sustaining man in it and recovering the vehicle at a designated place."[47] This reorientation gave the project 2 more years of life.

Then in 1963, he asked what the Air Force intended to do with it after using it to demonstrate maneuvering entry. He insisted he could not justify continuing the program if it was a dead-end effort with no ultimate purpose. But it had little potential utility, for it was not a cargo rocket, nor could it carry substantial payloads, nor could it conduct long-duration missions. And so, in December McNamara canceled it, after 6 years of development time, a Government contract investment of $410 million, the expenditure of 16 million man-hours by nearly 8,000 contractor personnel, 14,000 hours of wind tunnel testing, 9,000 hours of simulator runs, and the preparation of 3,035 detailed technical reports.[48]

Ironically, by time of its cancellation, the X-20 was so far advanced that the Air Force had already set aside a block of serial numbers for the 10 production aircraft. Its construction was well underway, Boeing having completed an estimated 42 percent of design and fabrication tasks.[49] Though the X-20 never flew, portions of its principal purposes were fulfilled by other programs. Even before cancellation, the Air Force launched the first of several McDonnell Aerothermodynamic/elastic Structural Systems Environmental Test (ASSET) hot-structure radiative-cooled flat-bottom cone-cylinder shapes sharing important configuration similarities to the Dyna-Soar vehicle. Slightly later, its Project PRIME demonstrated cross-range maneuver after atmospheric entry. This used the Martin SV-5D lifting body, a vehicle differing significantly from the X-20 but which complemented it nonetheless. In this fashion, the Air Force succeeded at least partially in obtaining lifting reentry data from winged vehicles and lifting bodies that widened the future prospects for reentry.

---

47. Curtis Peebles, "The Origin of the U.S. Space Shuttle—1," *Spaceflight*, vol. 21, no. 11 (Nov. 1979), pp. 435–442.
48. Geiger, "Strangled Infant," in Hallion, *Hypersonic Revolution*, vol. 1, pp. 313, 319–320.
49. Ibid., pp. 294–310, 313.

Hot Structures and Return from Space: X-20's Legacy and ASSET

Dyna-Soar never flew, but it sharply extended both the technology and the temperature limits of hot structures and associated aircraft elements, at a time when the American space program was in its infancy.[50] The United States successfully returned a satellite from orbit in April 1959, while ICBM nose cones were still under test, when the Discoverer II test vehicle supporting development of the National Reconnaissance Office's secret Corona spy satellite returned from orbit. Unfortunately, it came down in Russian-occupied territory far removed from its intended recovery area near Hawaii. Still, it offered proof that practical hypersonic reentry and recovery were at hand.

An ICBM nose cone quickly transited the atmosphere, whereas recoverable satellite reentry took place over a number of minutes. Hence a satellite encountered milder aerothermodynamic conditions that imposed strong heat but brought little or no ablation. For a satellite, the heat of ablation, measured in British thermal units (BTU) per pound of protective material, was usually irrelevant. Instead, insulative properties were more significant: Teflon, for example, had poor ablative properties but was an excellent insulator.[51]

Production Dyna-Soar vehicles would have had a four-flight service life before retirement or scrapping, depending upon a hot structure comprised of various materials, each with different but complementary properties. A hot structure typically used a strong material capable of withstanding intermediate temperatures to bear flights loads. Set off from it were outer panels of a temperature-resistant material that did not have to support loads but that could withstand greatly elevated temperatures as high as 3,000 °F. In between was a lightweight insulator (in Dyna-Soar's case, Q-felt, a silica fiber from the firm of Johns Manville). It had a tendency to shrink, thus risking dangerous gaps where high heat could bypass it. But it exhibited little shrinkage above 2,000 °F

---

50. Boeing, "Summary of Technical Advances: X-20 Program," Report D2-23418 (July 1964).
51. Robert L. Perry, *Management of the National Reconnaissance Program, 1960–1965* (Chantilly, VA: NRO, 2001 edition of a Jan. 1969 work), p. 9; Jeffrey Richelson, *American Espionage and the Soviet Target* (New York: William Morrow, 1987), p. 184; F.R. Riddell and J.D. Teare, "The Differences Between Satellite and Ballistic Missile Re-Entry Problems," in Morton Alperin and Hollingsworth F. Gregory, eds., *Vistas in Aeronautics*, vol. 2 (New York: Pergamon Press, 1959), pp. 174–190; Leo Steg, "Materials for Re-Entry Heat Protection of Satellites," *American Rocket Society Journal* (Sept. 1960), pp. 815–822.

and could withstand 3,000 °F. By "preshrinking" this material, it qualified for operational use.[52]

For its primary structure, Dyna-Soar used René 41, a nickel alloy that included chromium, cobalt, and molybdenum. Its use was pioneered by General Electric for hot-section applications in its jet engines. The alloy had room temperature yield strength of 130,000 psi, declining slightly at 1,200 °F, and was still strong at 1,800 °F. Some of the X-20's panels were molybdenum alloy, which offered clear advantages for such hot areas as the wing leading edges. D-36 columbium alloy covered most other areas of the vehicle, including the flat underside of the wings.

These panels had to resist flutter, which brought a risk of cracking because of fatigue, as well as permitting the entry of superheated hypersonic flows that could destroy the internal structure within seconds. Because of the risks to wind tunnels from hasty and ill-considered flutter testing (where a test model for example can disintegrate, damaging the interior of the tunnel), X-20 flutter testing consumed 18 months of Boeing's time. Its people started testing at modest stress levels and reached levels that exceeded the vehicle's anticipated design requirements.[53]

The X-20's nose cap had to function in a thermal and dynamic pressure environment more extreme even than that experienced by the X-15's Q-ball. It was a critical item that faced temperatures of 3,680 °F, accompanied by a daunting peak heat flux of 143 BTU per square foot per second. Both Boeing and its subcontractor Chance Vought pursued independent approaches to development, resulting in two different designs. Vought built its cap of siliconized graphite with an insulating layer of a temperature-resistant zirconium oxide ceramic tiles. Their melting point was above 4,500 °F, and they covered its forward area, being held in place by thick zirconium oxide pins. The Boeing design was simpler, using a solid zirconium oxide nose cap reinforced against cracking with two screens of platinum-rhodium wire. Like the airframe, the nose caps were rated through four orbital flights and reentries.[54]

---

52. Geiger, "Strangled Infant," in Hallion, *Hypersonic Revolution*, vol. 1, pp. 347–370.
53. Aeronautical Systems Division, *Proceedings of 1962 X-20A (Dyna-Soar) Symposium*, vol. 3: *Structures and Materials* (Wright-Patterson AFB, OH: USAF, Mar. 1963), DTIC AD-346192; Howard J. Middendorf, "Materials and Processes for X-20A (Dyna-Soar)," Air Force Systems Command (June 1964), DTIC AD-449685; and William Cowie, "Utilization of Refractory Metals on the X-20A (Dyna-Soar)," Air Force Systems Command (June 1964), DTIC AD-609169.
54. ASD, *X-20A Proceedings*, vol. 3, DTIC AD-346192.

Generally, the design of the X-20 reflected the thinking of Langley's John Becker and Peter Korycinski. It relied on insulation and radiation of the accumulated thermal load for primary thermal protection. But portions of the vehicle demanded other approaches, with specialized areas and equipment demanding specialized solutions. Ball bearings, facing a 1,600 °F thermal environment, were fabricated as small spheres of René 41 nickel alloy covered with gold. Antifriction bearings used titanium carbide with nickel as a binder. Antenna windows had to survive hot hypersonic flows yet be transparent to radio waves. A mix of oxides of cobalt, aluminum, and nickel gave a coating that showed a suitable emittance while furnishing requisite temperature protection.

The pilot looked through five clear panes: three that faced forward and two on the sides. The three forward panes were protected by a jettisonable protective shield and could only be used below Mach 5 after reentry, but the side ones faced a less severe aerothermodynamic environment and were left unshielded. But could the X-20 be landed if the protective shield failed to jettison after reentry? NASA test pilot Neil Armstrong, later the first human to set foot upon the Moon, flew approaches using a modified Douglas F5D Skylancer. He showed it was possible to land the Dyna-Soar using only visual cues obtained through the side windows.

The cockpit, equipment bay, and a power bay were thermally isolated and cooled via a "water wall" using lightweight panels filled with a jelled water mix. The hydraulic system was cooled as well. To avoid overheating and bursting problems with conventional inflated rubber tires, Boeing designed the X-20 to incorporate tricycle-landing skids with wire brush landing pads.[55] Dyna-Soar, then, despite never having flown, significantly advanced the technology of hypersonic aerospace vehicle design. Its contributions were many and can be illustrated by examining the confidence with which engineers could approach the design of critical technical elements of a hypersonic craft, in 1958 (the year North American began fabricating the X-15) and 1963 (the year Boeing began fabricating the X-20):[56] In short, within the 5 years that took the X-20 from a paper study to a project well underway, the "art of the possible"

---

55. Ibid.; Geiger, "Strangled Infant," in Hallion, *Hypersonic Revolution*, vol. 1, pp. 347–349, 361–370.

56. Geiger, "Strangled Infant," in Hallion, *Hypersonic Revolution*, vol. 1, pp. 344–346; R.L. Schleicher, "Structural Design of the X-15," *Journal of the Royal Aeronautical Society* (Oct. 1963), pp. 618–636.

TABLE 1
INDUSTRY HYPERSONIC "DESIGN CONFIDENCE"
AS MEASURED BY ACHIEVABLE DESIGN TEMPERATURE CRITERIA, °F

ELEMENT	X-15	X-20
Nose cap	3,200	4,300
Surface panels	1,200	2,750
Primary structure	1,200	1,800
Leading edges	1,200	3,000
Control surfaces	1,200	1,800
Bearings	1,200	1,800

in hypersonics witnessed a one-third increase in possible nose cap temperatures, a more than double increase in the acceptable temperatures of surface panels and leading edges, and a one-third increase in the acceptable temperatures of primary structures, control surfaces, and bearings.

The winddown and cancellation of Dyna-Soar coincided with the first flight tests of the much smaller but nevertheless still very technically ambitious McDonnell ASSET hypersonic lifting reentry test vehicle. Lofted down the Atlantic Test Range on modified Thor and Thor-Delta boosters, they demonstrated reentry at over Mach 18. ASSET dated to 1959, when Air Force hypersonic advocates advanced it as a means of assessing the accuracy of existing hypersonic theory and predictive techniques. In 1961, McDonnell Aircraft, a manufacturer of fighter aircraft and also the Project Mercury spacecraft, began design and fabrication of ASSET's small sharply swept delta wing flat-bottom boost-gliders. They had a length of 69 inches and a span of 55 inches.

Though in many respects they resembled the soon-to-be-canceled X-20, unlike that larger, crewed transatmospheric vehicle, the ASSET gliders were more akin to lifting nose cone shapes. Instead of the X-20's primary reliance upon René 41, the ASSET gliders largely used columbium alloys, with molybdenum alloy on their forward lower heat shield, graphite wing leading edges, various insulative materials, and columbium, molybdenum, and graphite coatings as needed. There were also three nose caps: one fabricated from zirconium oxide rods, another from tungsten coated with thorium, and a third of siliconized graphite coated with zirconium oxide. Though all six ASSETs looked alike, they were built in two differing variants: four Aerothermodynamic Structural Vehicles (ASV) and two Aerothermodynamic Elastic Vehicles (AEV). The former reentered from higher velocities (between 16,000 and 19,500 feet per second) and altitudes (from 202,000 to 212,000 feet), necessitating use of two-stage Thor-Delta boosters. The latter (only one of which flew successfully) used a single-stage Thor booster and reentered at 13,000 feet per second from an altitude of 173,000 feet. It was a hypersonic flutter research vehicle, analyzing as well the behavior of a trailing-edge flap representing a hypersonic control surface. Both the ASV and AEV flew with a variety of experimental panels installed at various locations and fabricated by Boeing, Bell, and Martin.[57] The ASSET program conducted six flights between September 1963 and February 1965, all successful save for one AEV launch in March 1964. Though intended for recovery from the Atlantic, only one survived the rigors of parachute deployment, descent, and being plunged into the ocean. But that survivor, the ASV-3, proved to be in excellent condition, with the builder, International Harvester, rightly concluding it "could have been used again."[58] ASV-4, the best flight flown, was also the last one, with the final flight-test report declaring that it returned "the highest quality data of the ASSET program." It flew at a peak speed of Mach 18.4, including a hypersonic glide that covered 2,300 nautical miles.[59]

Overall, the ASSET program scored a host of successes. It was all the more impressive for the modest investment made in its development: just $21.2 million. It furnished the first proof of the magnitude and seriousness of upper-surface leeside heating and the dangers of hypersonic flow impingement into interior structures. It dealt with practical issues of fabrication, including fasteners and coatings. It contributed to understanding of hypersonic flutter and of the use of movable control surfaces. It also demonstrated successful use of an attitude-adjusting reaction control system, in near vacuum and at speeds much higher than those of the X-15. It complemented Dyna-Soar and left the aerospace industry believing that hot structure design technology would be the normative technical approach taken on future launch vehicles and orbital spacecraft.[60]

---

57. Richard P. Hallion, "ASSET: Pioneer of Lifting Reentry," in Hallion, ed., *Hypersonic Revolution*, vol. 1, pp. 451, 461–465, 501–505; USAF Flight Dynamics Laboratory, "Advanced Technology Program: Technical Development Plan for Aerothermodynamic/Elastic Structural Systems Environmental Tests (ASSET)" (Sept. 1963), pp. 1–5.
58. *Aviation Week* (May 24, 1965), p. 62; McDonnell, "ASSET ASV-3 Flight Test Report," Report B251 (65FD-234) (Jan. 4, 1965).
59. McDonnell, "ASSET ASV-4 Flight Test Report," Report B707 (65FD-938) (June 25, 1965), p. 156; Hallion, "ASSET," in Hallion, ed., *Hypersonic Revolution*, vol. 1, p. 519.
60. USAF Flight Dynamics Laboratory, "ASSET Final Briefing," Report 65FD-850 (Oct. 5, 1965).

TABLE 2					
MCDONNELL ASSET FLIGHT TEST PROGRAM					
DATE	VEHICLE	BOOSTER	VELOCITY (FEET/ SECOND)	ALTITUDE (FEET)	RANGE (NAUTICAL MILES)
Sept. 18, 1963	ASV-1	Thor	16,000	205,000	987
Mar. 24, 1964	ASV-2	Thor-Delta	18,000	195,000	1,800
July 22, 1964	ASV-3	Thor-Delta	19,500	225,000	1,830
Oct. 27, 1964	AEV-1	Thor	13,000	168,000	830
Dec. 8, 1964	AEV-2	Thor	13,000	187,000	620
Feb. 23, 1965	ASV-4	Thor-Delta	19,500	206,000	2,300

Hypersonic Aerothermodynamic Protection and the Space Shuttle

Certainly over much of the Shuttle's early conceptual period, advocates thought such logistical transatmospheric aerospace craft would employ hot structure thermal protection. But undertaking such structures on large airliner-size vehicles proved troublesome and thus premature. Then, as though given a gift, NASA learned that Lockheed had built a pilot plant and could mass-produce silica "tiles" that could be attached to a conventional aluminum structure, an approach far more appealing than designing a hot structure. Accordingly, when the Agency undertook development of the Space Shuttle in the 1970s, it selected this approach, meaning that the new Shuttle was, in effect, a simple aluminum airplane. Not surprisingly, Lockheed received a NASA subcontract in 1973 for the Shuttle's thermal-protection system.

Lockheed had begun its work more than a decade earlier, when investigators at Lockheed Missiles and Space began studying ceramic fiber mats, filing a patent on the technology in December 1960. Key people included R.M. Beasley, Ronald Banas, Douglas Izu, and Wilson Schramm. By 1965, subsequent Lockheed work had led to LI-1500, a material that was 89 percent porous and weighed 15 pounds per cubic foot (lb/ft^3). Thicknesses of no more than an inch protected test surfaces during simulations of reentry heating. LI-1500 used methyl methacrylate (Plexiglas), which volatilized when hot, producing an outward flow of cool gas that protected the heat shield, though also compromising its reusability.[61]

Lockheed's work coincided with NASA plans in 1965 to build a space station as is main post-Apollo venture and, consequently, the first great wave of interest in designing practical logistical Shuttle-like spacecraft to fly between Earth and the orbital stations. These typically were conceived as large winged two-stage-to-orbit systems with fly-back boosters and orbital spaceplanes. Lockheed's Maxwell Hunter devised an influential design, the Star Clipper, with two expendable propellant tanks and LI-1500 thermal protection.[62] The Star Clipper also was large enough to benefit from the Allen-Eggers blunt-body principle, which lowered its temperatures and heating rates during reentry. This made it possible to dispense with the outgassing impregnant, permitting use—and, more importantly, reuse—of unfilled LI-1500. Lockheed also introduced LI-900, a variant of LI-1500 with a porosity of 93 percent and a weight of only 9 pounds per cubic foot. As insulation, both LI-900 and LI-1500 were astonishing. Laboratory personnel found that they could heat a tile in a furnace until it was white hot, remove it, allow its surface to cool for a couple of minutes, and pick it up at its edges with their fingers, with its interior still glowing at white heat.[63]

---

61. Paul Cooper and Paul F. Holloway, "The Shuttle Tile Story," *Astronautics & Aeronautics*, vol. 19, no. 1 (Jan. 1981), pp. 24–34; Wilson B. Schramm, Ronald P. Banas, and Y. Douglas Izu, "Space Shuttle Tile—The Early Lockheed Years," *Lockheed Horizons*, Issue 13 (1983), pp. 2–15; T.A. Heppenheimer, *The Space Shuttle Decision*, SP-4221 (Washington, DC: NASA, 1999).
62. Lockheed Missiles and Space Corporation (LMSC), "Space Transport and Recovery System (Space Shuttle)," LMSC A946332 (Mar. 1969), Shuttle Historical Documents Collection, N. SHHDC-0048, NASA Marshall Space Flight Center; LMSC, "Final Report: Integral Launch and Re-Entry Vehicle," LMSC A959837 (Dec. 1969), Center for Aerospace Information 70N-31831.
63. Richard C. Thuss, Harry G. Thibault, and Arnold Hiltz, "The Utilization of Silica Based Surface Insulation for the Space Shuttle Thermal Protection System," SAMPE National Technical Conference on Space Shuttle Materials, Huntsville, AL (Oct. 1971), pp. 453–464, Center for Aerospace Information 72A-10764; Schramm, et al., "Space Shuttle Tile"; L.J. Korb, C.A. Morant, R.M. Calland, and C.S. Thatcher, "The Shuttle Orbiter Thermal Protection System," and Wilson Schramm, "HRSI and LRSI—The Early Years," both in *American Ceramic Society Bulletin*, vol. 60 (1981), pp. 1188–1195; L.J. Graham, F.E. Sugg, and W. Gonzalez, "Nondestructive Evaluation of Space Shuttle tiles," *Ceramic Engineering and Science Proceedings*, vol. 3 (1982), pp. 680–697; Robert L. Dotts, Donald M. Curry, and Donald L. Tillian, "Orbiter Thermal Protection System," and William C. Schneider and Glenn J. Miller, "The Changing 'Scales of the Bird' (Shuttle Tile Structural Integrity)," in Norman Chaffee, ed., "Space Shuttle Technical Conference," NASA Conference Publication 2343 (1983).

Previous company work had amounted to general materials research. But Lockheed now understood in 1971 that NASA wished to build the Shuttle without simultaneously proceeding with the station, opening a strong possibility that the company could participate. The program had started with a Phase A preliminary study effort, advancing then to Phase B, which was much more detailed. Hot structures were initially ascendant but posed serious challenges, as NASA Langley researchers found when they tried to build a columbium heat shield suitable for the Shuttle. The exercise showed that despite the promise of reusability and long life, coatings were fragile and damaged easily, leading to rapid oxygen-induced embrittlement at high temperatures. Unprotected columbium oxidized particularly readily and, when hot, could burst into flame. Other refractory metals were available, but they were little understood because they had been used mostly in turbine blades.

Even titanium amounted literally to a black art. Only one firm, Lockheed, had significant experience with a titanium hot structure. That experience came from the Central Intelligence Agency-sponsored Blackbird strategic reconnaissance program, so most of the pertinent shop-floor experience was classified. The aerospace community knew that Lockheed had experienced serious difficulties in learning how to work with titanium, which for the Shuttle amounted to an open invitation to difficulties, delays, and cost overruns.

The complexity of a hot structure—with large numbers of clips, brackets, standoffs, frames, beams, and fasteners—also militated against its use. Each of the many panel geometries needed their own structural analysis that was to show with confidence that the panel could withstand creep, buckling, flutter, or stress under load, and in the early computer era, this posed daunting analytical challenges. Hot structures were also known generally to have little tolerance for "overtemps," in which temperatures exceeded the structure's design point.[64]

Thus, having taken a long look at hot structures, NASA embraced the new Lockheed pilot plant and gave close examination to Shuttle designs that used tiles, which were formally called Reusable Surface Installation (RSI). Again, the choice of hot structures versus RSI reflected the deep pockets of the Air Force, for hot structures were

---

64. Korb, et al., "Shuttle Orbiter TPS"; L. J. Korb and H. M. Clancy, "The Shuttle Thermal Protection System—A Material and Structural Overview," SAMPE 26th National Symposium, Los Angeles, CA (Apr. 1981), pp. 232–249 (Center for Aerospace Information 81A-44344).

costly and complex. But RSI was inexpensive, flexible, and simple. It suited NASA's budget while hot structures did not, so the Agency chose it.

In January 1972, President Richard M. Nixon approved the Shuttle as a program, thereby raising it to the level of a Presidential initiative. Within days, Dale Myers, a senior official, announced that NASA had made the basic decision to use RSI. The North American Rockwell concept that won the $2.6 billion prime contract in July therefore specified RSI as well—but not Lockheed's. North American Rockwell's version came from General Electric and was made from mullite.[65]

Which was better, the version from GE or the one from Lockheed? Only tests would tell—and exposure to temperature cycles of 2,300 °F gave Lockheed a clear advantage. NASA then added acoustic tests that simulated the loud roars of rocket flight. This led to a "sudden-death shootout," in which competing tiles went into single arrays at NASA Johnson. After 20 cycles, only Lockheed's entrants remained intact. In separate tests, Lockheed's LI-1500 withstood 100 cycles to 2,500 °F and survived a thermal overshoot to 3,000 °F as well as an acoustic overshoot to 174 decibels (dB).

Lockheed won the thermal-protection subcontract in 1973, with NASA specifying LI-900 as the baseline RSI. The firm responded by preparing to move beyond the pilot-plant level and to construct a full-scale production facility in Sunnyvale, CA. With this, tiles entered the mainstream of thermal protection systems available for spacecraft design, in much the same way that blunt bodies and ablative approaches had before them, first flying into space aboard the Space Shuttle Columbia in April 1981. But getting them operational and into space was far from easy.[66]

---

65. "NASA Space Shuttle Technology Conference," vol. 2: "Structures and Materials," NASA TM-X-2273 (1971); Heppenheimer, *The Space Shuttle Decision*, pp. 341–346.

66. Schramm, et al., "Space Shuttle Tile." See also Donald H. Humes, "Hypervelocity Impact Tests on Space Shuttle Orbiter Thermal Protection Material," NASA TM-X-74039 (1977); M.J. Suppans and C.J. Schroeder, "Space Shuttle Orbiter Thermal Protection Development and Verification Test Program," AIAA Paper 78-485 (1978); R. Jeffrey Smith, "Shuttle Problems Compromise Space Program," *Science* (Nov. 23, 1979), pp. 910–912, 914; Mitch Waldrop, "Space Shuttle Tiles: A Question of Bonding," *Chemical and Engineering News*, vol. 58 (May 12, 1980), pp. 27–29; W.C. Rochelle, et al., "Orbiter TPS Development and Certification testing at the NASA/JSC 10 MW Atmospheric Reentry Materials and Structures Evaluation Facility," AIAA Paper 83-0147 (1983).

The Tiles Become Operational

Manufacture of the silica tiles was straightforward, at least in its basic steps. The raw material consisted of short lengths of silica fiber of l.0-micron diameter. A measured quantity of fibers, mixed with water, formed a slurry. The water was drained away, and workers added a binder of colloidal silica, then pressed the material into rectangular blocks that were 10 to 20 inches in diameter and more than 6 inches thick. These blocks were the crudest form of LI-900, the basic choice of RSI for the entire Shuttle. They sat for 3 hours to allow the binder to jell, then were dried thoroughly in a microwave oven. The blocks moved through sintering kilns that baked them at 2,375 °F for 2 hours, fusing binder and fibers together. Band saws trimmed distortions from the blocks, which were cut into cubes and then carved into individual tiles using milling machines driven by computer. The programs contained data from Rockwell International on the desired tile dimensions.

Next, the tiles were given a spray-on coating. After being oven-dried, they returned to the kilns for glazing at temperatures of 2,200 °F for 90 minutes. To verify that the tiles had received the proper amount of coating, technicians weighed samples before and after the coating and glazing. The glazed tiles then were made waterproof by vacuum deposition of a silicon compound from Dow Corning while being held in a furnace at 350 °F. These tiles were given finishing touches before being loaded into arrays for final milling.[67]

Although the basic LI-900 material showed its merits during 1972, it was another matter to produce it in quantity, to manufacture tiles that were suitable for operational use, and to provide effective coatings. To avoid having to purify raw fibers from Johns Manville, Lockheed asked that company to find a natural source of silica sand with the necessary purity. The amount needed was small, about 20 truckloads, and was not of great interest to quarry operators. Nevertheless, Johns Manville found a suitable source in Minnesota.

Problems arose when shaping the finished tiles. Initial plans called for a large number of identical flat tiles, varying only in thickness and trimmed to fit at the time of installation. But flat tiles on the curved surface of the Shuttle produced a faceted surface that promoted the onset of turbulence in the airflow, resulting in higher rates of heating. The tiles

---

67. Richard G. O'Lone, "Thermal Tile Production Ready to Roll," *Aviation Week* (Nov. 8, 1976), pp. 51–54; L.J. Korb, C.A. Morant, R.M. Calland, and C.S. Thatcher, "The Shuttle Orbiter Thermal Protection System," *American Ceramic Society Bulletin*, vol. 60, 1981, pp. 1188–1193.

then would have had to be thicker, which threatened to add weight. The alternative was an external RSI contour closely matching that of the orbiter's outer surface. Lockheed expected to produce 34,000 tiles for each orbiter, grouping most of them in arrays of two dozen or so and machining their back faces, away from the glazed coating, to curves matching the contours of the Shuttle's aluminum skin. Each of the many thousands of tiles was to be individually numbered, and none had precisely the same dimensions. Instead, each was defined by its own set of dimensions. This cost money, but it saved weight.

Difficulties also arose in the development of coatings. The first good one, LI-0042, was a borosilicate glass that used silicon carbide to enhance its high-temperature thermal emissivity. It dated to the late 1960s; a variant, LI-0050, initially was the choice for operational use. This coating easily withstood the rated temperature of 2,300 °F, but in tests, it persistently developed hairline cracks after 20 to 60 thermal cycles. This was unacceptable; it had to stand up to 100 such cycles. The cracks were too small to see with the unaided eye and did not grow large or cause tile failure. But they would have allowed rainstorms to penetrate the tiles during the weeks that an orbiter was on the ground between missions, with the rain adding to the launch weight. Help came from NASA Ames, where researchers were close to Lockheed, both in their shared interests and in their facilities being only a few miles apart. Howard Goldstein at Ames, a colleague of the branch chief, Howard Larson, set up a task group and brought in a consultant from Stanford University, which also was just up the road. They spent less than $100,000 in direct costs and came up with a new and superior coating called reaction-cured glass. Like LI-0050, it was a borosilicate, consisting of more than 90 percent silica along with boria or boron oxide along with an emittance agent. The agent in LI-0050 had been silicon carbide; the new one was silicon tetraboride, SiB4. During glazing, it reacted with silica in a way that increased the level of boria, which played a critical role in controlling the coating's thermal expansion. This coating could be glazed at lower temperature than LI-0050 could, reducing the residual stress that led to the cracking. SiB4 oxidized during reentry, but in doing so, it produced boria and silica, the ingredients of the glass coating itself.[68]

---

68. L.J. Korb and H.M. Clancy, "Symposium on Reusable Surface Insulation for Space Shuttle," vol. 1, NASA TM-X-2719 (1973), pp. 14–15; "The Shuttle Thermal Protection System—A Material and Structural Overview," Apr. 1981, pp. 232–249, CASI 81A-44344.

The Shuttle's distinctive mix of black-and-white tiles was all designed as standard LI-900 with its borosilicate coating, but the black ones had SiB4 and the white ones did not. Still, they all lacked structural strength and were brittle. They could not be bonded directly to the orbiter's aluminum skin, for they would fracture and break because of their inability to follow the flexing of this skin under its loads. Designers therefore placed an intermediate layer between tiles and skin, called a strain isolator pad (SIP). It was a felt made of Nomex nylon from DuPont, which would neither melt nor burn. It had useful elasticity and could stretch in response to Shuttle skin flexing without transmitting excessive strain to the tiles.[69]

Testing of tiles and other thermal-protection components continued through the 1970s, with NASA Ames being particularly active. A particular challenge lay in creating turbulent flows, which demanded close study because they increased the heat-transfer rates many times over. During reentry, hypersonic flow over a wing is laminar near the leading edge, transitioning to turbulence at some distance to the rear. No hypersonic wind tunnel could accommodate anything resembling a full-scale wing, and it took considerable power as well as a strong airflow to produce turbulence in the available facilities. Ames had a 60-megawatt arc-jet, but even that facility could not accomplish this. Ames succeeded in producing such flows by using a 20-megawatt arc-jet that fed its flow into a duct that was 9 inches across and 2 inches deep. The narrow depth gave a compressed flow that readily produced turbulence, while the test chamber was large enough to accommodate panels with size of 8 by 20 inches. This facility supported the study of coatings that led to the use of reaction-cured glass. Tiles of LI-900, 6 inches square and treated with this coating, survived 100 simulated reentries at 2,300 °F in turbulent flow.[70]

The Ames 20-megawatt arc-jet facility made its own contribution in a separate program that improved the basic silica tile. Excessive temperatures caused these tiles to fail by shrinking and becoming denser.

---

69. David H. Greenshields, "Orbiter Thermal Protection System Development" (Apr. 1977), pp. 1-28–1-42, CASI 77A-35304; North American-Rockwell, "Space Shuttle System Summary Briefing," Report SV 72-19 (July 8, 1972); Korb, et al., "Shuttle"; Robert M. Powers, *Shuttle: The World's First Spaceship* (Harrisburg, PA: Stackpole Books, 1979), p. 241.

70. Frank Kreith, *Principles of Heat Transfer* (Scranton, PA: International Textbook Co., 1965), pp. 534–538; Benjamin M. Elson, "New Unit to Test Shuttle Thermal Guard," *Aviation Week* (Mar. 31, 1975), pp. 52–53; H.K. Larson and H.E. Goldstein, "Space Shuttle Orbiter Thermal Protection Material Development and Testing," (Mar. 1978), pp. 189–194, CASI 79A-17673.

Investigators succeeded in reducing the shrinkage by raising the tile density and adding silicon carbide to the silica, rendering it opaque and reducing internal heat transfer. This led to a new grade of silica RSI with density of 22 lb/ft^3 that had greater strength as well as improved thermal performance.[71]

The Ames researchers carried through with this work during 1974 and 1975, with Lockheed taking this material and putting it into production as LI-2200. Its method of manufacture largely followed that of standard LI-900, but whereas that material relied on sintered colloidal silica to bind the fibers together, LI-2200 dispensed with this and depended entirely on fiber-to-fiber sintering. LI-2200 was adopted in 1977 for operational use on the Shuttle, where it found application in specialized areas. These included regions of high concentrated heat near penetrations such as landing-gear doors as well as near interfaces with the carbon-carbon nose cap, where surface temperatures could reach 2,600 °F.[72]

Testing proceeded in four overlapping phases. Material selection ran through 1973 and 1974 into 1975; the work that led to LI-2200 was an example. Material characterization proceeded concurrently and extended midway through 1976. Design development tests covered 1974 through 1977; design verification activity began in 1977 and ran through subsequent years. Materials characterization called for some 10,000 test specimens, with investigators using statistical methods to determine basic material properties. These were not the well-defined properties that engineers find listed in handbooks; they showed ranges of values that often formed a Gaussian distribution, with its bell-shaped curve. This activity addressed such issues as the lifetime of a given material, the effects of changes in processing, or the residual strength after a given number of flights. A related topic was simple but far-reaching: to be able to calculate the minimum tile thickness, at a given location, that would hold the skin temperature below the maximum allowable.[73]

Design development tests used only 350 articles but spanned 4 years, because each of them required close attention. An important goal involved validating the specific engineering solutions to a number

---

71. CASI 79A-17673; CASI 81A-44344.
72. Korb, et al., "Shuttle"; Elizabeth A. Muenger, *Searching the Horizon: A History of Ames Research Center, 1940–1976*, NASA SP-4304 (Washington, DC: NASA, 1985).
73. *Aviation Week* (Mar. 31, 1975), pp. 52–53; CASI 81A-44344; Gregory P. McIntosh and Thomas P. Larkin, "The Space Shuttle's Testing Gauntlet," *Astronautics and Aeronautics* (Jan. 1976), pp. 60, 62–64.

of individual thermal-protection problems. Thus the nose cap and wing leading edges were made of carbon-carbon, in anticipation of their being subjected to the highest temperatures. Their attachments were exercised in structural tests that simulated flight loads up to design limits, with design temperature gradients.

Design development testing also addressed basic questions of the tiles themselves. There were narrow gaps between them, and while Rockwell had ways to fill them, these gap-fillers required their own trials by fire. A related question was frequently asked: What happens if a tile falls off? A test program addressed this and found that in some areas of intense heating, the aluminum skin indeed would burn through. The only way to prevent this was to be sure that the tiles were firmly bonded in place, and this meant all those located in critical areas.[74]

Design verification tests used fewer than 50 articles, but these represented substantial portions of the vehicle. An important test article, evaluated at NASA Johnson, reproduced a wing leading edge and measured 5 by 8 feet. It had two leading-edge panels of carbon-carbon set side by side, a section of wing structure that included its principal spars, and aluminum skin covered with RSI. It could not have been fabricated earlier in the program, for its detailed design drew on lessons from previous tests. It withstood simulated air loads, launch acoustics, and mission-temperature-pressure environments, not once, but many times.[75]

The testing ranged beyond the principal concerns of aerodynamics, heating, and acoustics. There also was concern that meteoroids might not only put craters in the carbon-carbon but also cause it to crack. At NASA Langley, the researcher Donald Humes studied this by shooting small glass and nylon spheres at target samples using a light-gas gun driven by compressed helium. Helium is better than gunpowder, as it can expand at much higher velocities. Humes wrote that carbon-carbon: "does not have the penetration resistance of the metals on a thickness basis, but on a weight basis, that is, mass per unit area required to stop projectiles, it is superior to steel."[76]

---

74. McIntosh and Larkin, "Space Shuttle's Testing Gauntlet," pp. 60, 62–64; *Aviation Week* (Mar. 31, 1975), p. 52; M.J. Suppans and C.J. Schroeder, "Space Shuttle Orbiter Thermal Protection Material," AIAA Paper 78-485 (1978).
75. McIntosh and Larkin, "Space Shuttle's Testing Gauntlet," pp. 60, 62–64.
76. Donald H. Humes, "Hypervelocity Impact Tests on Space Shuttle Orbiter Thermal Protection Material," NASA TM X-74039 (1977), p. 12.

Yet amid the advanced technology of arc-jets, light-gas guns, and hypersonic wind tunnels, one of the most important tests was also one of the simplest. It involved nothing more than taking tiles that were bonded with adhesive to the SIP and the underlying aluminum skin and physically pulling them off.

It was no new thing for people to show concern that the tiles might not stick. In 1974, a researcher at Ames noted that aerodynamic noise was potentially destructive, telling a reporter for Aviation Week that: "We'd hate to shake them all off when we're leaving." At NASA Johnson, a 10-MW arc-jet saw extensive use in lost-tile investigations. Tests indicated there was reason to believe that the forces acting to pull off a tile would be as low as 2 psi, just some 70 pounds for a tile measuring 6 by 6 inches square. This was low indeed; the adhesive, SIP, and RSI material all were considerably stronger. The thermal-protection testing therefore had given priority to thermal rather than to mechanical work, essentially taking it for granted that the tiles would stay on. Thus, attachment of the tiles to the Shuttle lacked adequate structural analysis, failing to take into account the peculiarities in the components. For example, the SIP had some fibers oriented perpendicular to the cemented tile undersurface. The tile was made of ceramic fibers, with these fibers concentrating the loads. This meant that the actual stresses they faced were substantially greater than anticipated.[77]

Columbia orbiter OV-102 was the first to receive working tiles. Columbia was also slated to be first into space. It underwent final assembly at the Rockwell plant in Palmdale, CA, during 1978. Checkout of onboard systems began in September, and installation of tiles proceeded concurrently, with Columbia to be rolled out in February 1979. But mounting the tiles was not at all like laying bricks. Measured gaps were to separate them; near the front of the orbiter, they had to be positioned to within 0.17 inches of vertical tolerance to form a smooth surface that

---

77. W.C. Rochelle, et al., "Orbiter TPS Development and Certification Testing at the NASA/JSC 10 MW Atmospheric Reentry Materials and Structures Evaluation Facility," AIAA Paper 83-0147 (1983); Richard G. O'Lone, "Shuttle Test Pace Intensifies at Ames," Aviation Week (June 24, 1974), p. 71; Mitch Waldrop, "Space Shuttle Tiles: A Question of Bonding," Chemical and Engineering News, vol. 58 (May 12, 1980), pp. 27–29; Paul A. Cooper and Paul F. Holloway, "Shuttle Tile Story," pp. 27, 29; William C. Schneider and Glenn J. Miller, "The Challenging 'Scales of the Bird' (Shuttle Tile Structural Integrity)" in Norman Chaffee, ed., "Space Shuttle Technical Conference," NASA Conference Publication CP-2342 (1983), pp. 403–404.

would not trip the airflow into turbulence. This would not have been difficult if the tiles had rested directly on the aluminum skin, but they were separated from that skin by the spongy SIP. The tiles were also fragile. An accidental tap with a wrench, a hard hat, even a key chain could crack the glassy coating. When that happened, the damaged tile had to be removed and the process of installation had to start again with a new one.[78]

The tiles came in arrays, each array numbering about three-dozen tiles. It took 1,092 arrays to cover this orbiter, and NASA reached a high mark when technicians installed 41 of them in a single week. But unfortunate news came midway through 1979 as detailed studies showed that in many areas the combined loads due to aerodynamic pressure, vibration, and acoustics would produce excessively large forces on the tiles. Work to date had treated a 2-psi level as part of normal testing, but now it was clear that only a small proportion of the tiles already installed faced stresses that low. Over 5,000 tiles faced force levels of 8.5 to 13 psi, with 3,000 being in the range of 2 to 6.5 psi. The usefulness of tiles as thermal protection was suddenly in doubt.[79]

What caused this? The fault lay in the nylon felt SIP, which had been modified by "needling" to increase its through-the-thickness tensile strength and elasticity. This was accomplished by punching a barbed needle through the felt fabric, some 1,000 times per square inch, which oriented fiber bundles transversely to the SIP pad. Tensile loads applied across the SIP pad, acting to pull off a tile, were transmitted into the SIP at discrete regions along these transverse fibers. This created localized stress concentrations, where the stresses approached twice the mean value. These local areas failed readily under load, causing the glued bond to break.[80]

There also was a clear need to increase the strength of the tiles' adhesive bonds. The solution came during October and involved modifying a thin layer at the bottom of each tile to make it denser. The process was called, quite logically, "densification." It used DuPont's Ludox

---

78. "First Shuttle Launch Vehicle Being Assembled at Palmdale," *Aviation Week* (Nov. 27, 1978), p. 64; Waldrop, "Tiles"; Craig Covault, "Thermal Tile Application Accelerated," *Aviation Week* (May 21, 1979), pp. 59–63.
79. *Aviation Week* (Nov. 27, 1978), p. 64; Waldrop, "Tiles"; Craig Covault, "Administration Backs Shuttle Fund Rise," *Aviation Week* (Sept. 17, 1979), pp. 22–23.
80. L.J. Graham, F.E. Sugg, and W. Gonzalez, "Nondestructive Evaluation of Space Shuttle Tiles," *Ceramic Engineering and Science Proceedings*, vol. 3, (1982), pp. 681–683; Schneider and Miller, "Challenging"; *Astronautics and Aeronautics* (Jan. 1981), p. 29.

with a silica "slip." Ludox was colloidal silica stirred into water and stabilized with ammonia; the slip had fine silica particles dispersed in water. The Ludox acted like cement; the slip provided reinforcement, in the manner of sand in concrete. It worked: the densification process clearly restored the lost strength.[81]

By then, Columbia had been moved to the Kennedy Space Center. The work nevertheless went badly during 1979, for as people continued to install new tiles, they found more and more that needed to be removed and replaced. Orderly installation procedures broke down. Rockwell had received the tiles from Lockheed in arrays and had attached them in well-defined sequences. Even so, that work had gone slowly, with 550 tiles in a week being a good job. But now Columbia showed a patchwork of good ones, bad ones, and open areas with no tiles. Each individual tile had been shaped to a predetermined pattern at Lockheed using that firm's numerically controlled milling machines. But the haphazardness of the layout made it likely that any precut tile would fail to fit into its assigned cavity, leaving too wide a gap with the adjacent ones.

Many tiles therefore were installed one by one, in a time-consuming process that fitted two into place and then carefully measured space for a third, designing it to fill the space between them. The measurements went to Sunnyvale, CA, where Lockheed carved that tile to its unique specification and shipped it to the Kennedy Space Center (KSC). Hence, each person took as long as 3 weeks to install just 4 tiles. Densification also took time; a tile removed from Columbia for rework needed 2 weeks until it was ready for reinstallation.[82]

How could these problems have been avoided? They all stemmed from the fact that the tile work was well advanced before NASA learned that the tile-SIP-adhesive bonds had less strength than the Agency needed. The analysis that disclosed the strength requirements was neither costly nor demanding; it might readily have been in hand during 1976 or 1977. Had this happened, Lockheed could have begun shipping densified tiles at an early date. Their development and installation would have occurred within the normal flow of the Shuttle program, with the change amounting perhaps to little more than an engineering detail.

---

81. "Densification Process Applied to Shuttle Tiles," *Aviation Week* (Feb. 25, 1980), p. 22; *Astronautics and Aeronautics* (Jan. 1981), pp. 29–30.
82. *Aviation Week* (Feb. 25, 1980), pp. 22–24; Craig Covault, "Mated Shuttle Reaches Pad 39," *Aviation Week* (May 7, 1979), p. 14.

The Space Shuttle Columbia descends to land at Edwards following its hypersonic reentry from orbit in April 1981. NASA.

The reason this did not happen was far-reaching, for it stemmed from the basic nature of the program. The Shuttle effort followed "concurrent development," with design, manufacture, and testing proceeding in parallel rather than in sequence. This approach carried risk, but the Air Force had used it with success during the 1960s. It allowed new technologies to enter service at the earliest possible date. But within the Shuttle program, funds were tight. Managers had to allocate their budgets adroitly, setting priorities and deferring what they could put off. To do this properly was a high art, calling for much experience and judgment, for program executives had to be able to conclude that the low-priority action items would contain no unpleasant surprises. The calculation of tile strength requirements was low on the action list because it appeared unnecessary; there was good reason to believe that the tiles would face nothing worse than 2 psi. Had this been true, and had the main engines been ready, Columbia might have flown by mid-1980. It did not fly until April 1981, and, in this sense, tile problems brought a delay of close to 1 year.

The delay in carrying through the tile-strength computation was not mandatory. Had there been good reason to upgrade its priority, it could readily have been done earlier. The budget stringency that brought this deferral (along with many others) thus was false economy par excellence, for the program did not halt during that year of launch delay. It kept writing checks for its contractors and employees. The missing tile-strength analysis thus ramified in its consequences, contributing substantially to a cost overrun in the Shuttle program.[83]

During 1979, NASA gave the same intense level of attention to the tiles' mechanical problems that it had previously reserved for their thermal development. The effort nevertheless continued to follow the pattern of three steps forward and two steps back, and, for a while, more tiles were removed than were put on in a given week. Even so, by the fall of 1980, the end was in sight.[84]

During the spring of 1979, before the main tile problems had come to light, the schedule had called for the complete assembly of Columbia, with its external tank and solid boosters, to take place on November 24, 1979. Exactly 1 year later, a tow vehicle pulled Columbia into the Vehicle Assembly Building as a large crowd watched and cheered. Within 2 days, Columbia was mounted to its tank, forming a live Shuttle in flight configuration. Kenneth Kleinknecht, an X-series and space flight veteran and now Shuttle manager at NASA Johnson, put it succinctly: "The vehicle is ready to launch."[85]

Shuttle Aerodynamics and Structures

The Shuttle was one of the last major aircraft to rely almost entirely on wind tunnels for studies of its aerodynamics. There was much interest in an alternative: the use of supercomputers to derive aerodynamic data through solution of the governing equations of airflow, known as the Navier-Stokes equations. Solution of the complete equations was out of the question, for they carried the complete physics of turbulence, with turbulent eddies that spanned a range of sizes covering several orders of magnitude. But during the 1970s, investigators made headway by dropping the terms within these equations that contained viscosity, thereby suppressing turbulence.[86]

---

83. Waldrop, "Tiles."
84. "Shuttle Engine, Tile Work Proceeding on Schedule," *Aviation Week* (Sept. 15, 1980), p. 26.
85. "NASA Finishes Shuttle Mating," *Aviation Week* (Dec. 1, 1980), pp. 18–19; *Aviation Week* (Sept. 17, 1979), p. 22; "NASA Presses to Hold Tight Shuttle Schedule," *Aviation Week* (Aug. 4, 1980), p. 24.
86. John D. Anderson, *A History of Aerodynamics* (New York: Cambridge University Press, 1997), pp. 441–443.

People pursued numerical simulation because it offered hope of overcoming the limitations of wind tunnels. Such facilities usually tested small models that failed to capture important details of the aerodynamics of full-scale aircraft. Other errors arose from tunnel walls and model supports. Hypersonic flight brought its own restrictions. No installation had the power to accommodate a large model, realistic in size, at the velocity and temperatures of reentry.[87]

By piecing together results from specialized facilities, it was possible to gain insights into flows at near-orbital speeds. The Shuttle reentered at Mach 27. NASA Langley had a pair of wind tunnels that used helium, which expands to very high flow velocities. These attained Mach 20, Mach 26, and even Mach 50. But their test models were only a few inches in size, and their flows were very cold and could not duplicate the high temperatures of atmosphere entry. Shock tunnels, which heated and compressed air using shock waves, gave true temperature up to Mach 17 while accommodating somewhat larger models. Yet their flow durations were measured in milliseconds.[88]

During the 1970s, the largest commercially available mainframe computers included the Control Data 7600 and the IBM 370-195.[89] These sufficed to treat complete aircraft—but only at the lowest level of approximation, which used linearized equations and treated the airflow over an airplane as a small disturbance within a uniform free stream. The full Navier-Stokes equations contained 60 partial derivatives; the linearized approximation retained only 3 of these terms. It nevertheless gave good accuracy in computing lift, successfully treating such complex configurations as a Shuttle orbiter mated to its 747. The next level of approximation restored the most important nonlinear terms and treated transonic and hypersonic flows, which were particularly difficult to simulate in wind tunnels. The inadequacies of wind tunnel work had brought such errors as faulty predictions of the location of shock waves along the wings of the C-141, an Air Force transport. In flight test, this plane tended to nose downward, and its design had to be modified at considerable expense.

---

87. Dean R. Chapman, Hans Mark, and Melvin W. Pirtle, "Computers vs. Wind Tunnels for Aerodynamic Flow Simulations," *Astronautics and Aeronautics* (Apr. 1975), p. 26.
88. T.A. Heppenheimer, *Hypersonic Technologies and the National Aerospace Plane* (Arlington, VA: Pasha Publications, 1990), pp. 128–134.
89. William D. Metz, "Midwest Computer Architect Struggles with the Speed of Light," *Science* (Jan. 27, 1978), pp. 404–405.

Computers such as the 7600 could not treat complete aircraft in transonic flow, for the equations were more complex and the computation requirements more severe. HiMAT, a highly maneuverable NASA experimental aircraft, flew at Dryden and showed excessive drag at Mach 0.9. Redesign of its wing used a transonic-flow computational code and approached the design point. The same program, used to reshape the wing of the Grumman Gulfstream, gave considerable increases in range and fuel economy while reducing the takeoff distance and landing speed.[90]

During the 1970s, NASA's most powerful computer was the Illiac IV, at Ames Research Center. It used parallel processing and had 64 processing units, achieving speeds up to 25 million operations per second. Built by Burroughs Corporation with support from the Pentagon, this machine was one of a kind. It entered service at Ames in 1973 and soon showed that it could run flow-simulation codes an order of magnitude more rapidly than a 7600. Indeed, its performance foreshadowed the Cray-1, a true supercomputer that became commercially available only after 1976.

The Illiac IV was a research tool, not an instrument of mainstream Shuttle development. It extended the reach of flow codes, treating three-dimensional inviscid problems while supporting simulations of viscous flows that used approximate equations to model the turbulence.[91] In the realm of Space Shuttle studies, Ames's Walter Reinhardt used it to run a three-dimensional inviscid code that included equations of atmospheric chemistry. Near-peak-entry heating of the Shuttle would be surrounded by dissociated air that was chemically reacting and not in chemical equilibrium. Reinhardt's code treated the full-scale orbiter during entry and gave a fine example of the computational simulation of flows that were impossible to reproduce in ground facilities.[92]

---

90. Gina Bari Kolata, "Who Will Build the Next Supercomputer?" *Science*, Jan. 16, 1981, pp. 268–269; Randolph A. Graves, Jr., "Computational Fluid Dynamics: The Coming Revolution," *Astronautics and Aeronautics* (Mar. 1982), pp. 20–28; *Astronautics and Aeronautics* (Apr. 1975), pp. 22–30. For numbers of partial derivatives, see Dean R. Chapman, "Computational Aerodynamics Development and Outlook," *AIAA Journal*, vol. 17, No. 12 (1979), p. 1294.

91. *Science*, Jan. 16, 1981, pp. 268–269; Benjamin B. Elson, "Computer Seen Assuming Shuttle Tasks." *Aviation Week* (Sept. 3, 1973), pp. 14–16; D.L. Slotnick, "The Fastest Computer," *Scientific American* (Feb. 1971), pp. 76–87.

92. Walter A. Reinhardt, "Parallel Computation of Unsteady, Three-Dimensional, Chemically Reacting, Nonequilibrium Flow Using a Time-Split Finite Volume Method on the Illiac IV," *Journal of Physical Chemistry*, vol. 81, no. 25 (1977), pp. 2427–2435.

Such exercises gave tantalizing hints of what would be done with computers of the next generation. Still, the Shuttle program was at least a decade too early to use computational simulations both routinely and effectively. NASA therefore used its wind tunnels. The wind tunnel program gave close attention to low-speed flight, which included approach and landing as well as separation from the 747 during the 1977 flight tests of Enterprise.

In 1975, Rockwell built a $1 million model of the orbiter at 0.36 scale, lemon yellow in color and marked with the blue NASA logo. It went into the 40- by 80-foot test section of Ames's largest tunnel, which was easily visible from the adjacent freeway. It gave parameters for the astronauts' flight simulators, which previously had used data from models at 3-percent scale. The big one had grooves in its surface that simulated the gaps between thermal protection tiles, permitting assessment of the consequences of the resulting roughness of the skin. It calibrated and tested systems for making aerodynamic measurements during flight test and verified the design of the elevons and other flight control surfaces as well as of their actuators.[93]

Other wind tunnel work strongly influenced design changes that occurred early in development. The most important was the introduction of the lightweight delta wing late in 1972, which reduced the size of the solid boosters and chopped 1 million pounds from the overall weight. Additional results changed the front of the external tank from a cone to an ogive and moved the solid boosters rearward, placing their nozzles farther from the orbiter. The modifications reduced drag, minimized aerodynamic interference on the orbiter, and increased stability by moving the aerodynamic center aft.

The activity disclosed and addressed problems that initially had not been known to exist. Because both the liquid main engines and the solids had nozzles that gimbaled, it was clear that they had enough power to provide control during ascent. Aerodynamic control would not be necessary, and managers believed that the orbiter could set its elevons in a single position through the entire flight to orbit. But work in wind tunnels subsequently showed that aerodynamic forces during ascent would impose excessive loads on the wings. This required elevons to move while in powered flight to relieve these loads. Uncertainties in the

---

93. Richard G. O'Lone, "Tunnel Tests Yield New Orbiter Data," *Aviation Week* (June 30, 1975), pp. 43–44.

wind tunnel data then broadened this requirement to incorporate an active system that prevented overloading the elevon actuators. This system also helped the Shuttle to fly a variety of ascent trajectories, which imposed different elevon loads from one flight to the next.[94]

Much wind tunnel work involved issues of separation: Enterprise from its carrier aircraft, solid boosters from the external tank after burnout. At NASA Ames, a 14-foot transonic tunnel investigated problems of Enterprise and its 747. Using the same equipment, engineers addressed the separation of an orbiter from its external tank. This was supposed to occur in near-vacuum, but it posed aerodynamic problems during an abort.

The solid boosters brought their own special issues and nuances. They had to separate cleanly; under no circumstances could a heavy steel casing strike a wing. Small solid rocket motors, mounted fore and aft on each booster, were to push them away safely. It then was necessary to understand the behavior of their exhaust plumes, for these small motors were to blast into onrushing airflow that could blow their plumes against the orbiter's sensitive tiles or the delicate aluminum skin of the external tank. Wind tunnel tests helped to define appropriate angles of fire while also showing that a short, sharp burst from the motors was best.[95]

Prior to the first orbital flight in 1981, the program racked up 46,000 wind tunnel hours. This consisted of 24,900 hours for the orbiter, 17,200 for the mated launch configuration, and 3,900 for the carrier aircraft program. During the 9 years from contract award to first flight, this was equivalent to operating a facility 16 hours a day, 6 days a week. Specialized projects demanded unusual effort, such as an ongoing attempt to minimize model-to-model and tunnel-to-tunnel discrepancies. This work alone conducted 28 test series and used 14 wind tunnels.[96]

Structural tests complemented the work in aerodynamics. The mathematics of structural analysis was well developed, with computer

---

94. Charlie C. Dill, et al., "The Space Shuttle Ascent Vehicle Aerodynamic Challenges: Configuration Design and Data Base Development," in Norman Chaffee, ed., "Space Shuttle Technical Conference," NASA CP-2342, (1983), pp. 151–152, 161.
95. Richard G. O'Lone, "Shuttle Task Pace Intensifies at Ames," *Aviation Week* (June 24, 1974), p. 71; Craig Covault, "Thermal, Weight Concerns Force Changes to Shuttle," *Aviation Week* (Dec. 9, 1974), p. 19; Dill, et al., "Ascent," pp. 154, 165.
96. James C. Young, et al., "The Aerodynamic Challenges of the Design and Development of the Space Shuttle Orbiter," in Chaffee, "Conference," pp. 217–220.

programs called NASTRAN that dealt with strength under load while addressing issues of vibration, bending, and flexing. The equations of NASTRAN were linear and algebraic, which meant that in principle they were easy to solve. The problem was that there were too many of them, for the most detailed mathematical model of the orbiter's structure had some 50,000 degrees of freedom. Analysts introduced abridged versions that cut this number to 1,000 and then relied on experimental tests for data that could be compared with the predictions of the computers.[97]

There were numerous modes of vibration, with frequencies that changed as the Shuttle burned its propellants. Knowledge of these frequencies was essential, particularly in dealing with "pogo." This involved a longitudinal oscillation like that of a pogo stick, with propellant flowing in periodic surges within its main feed line. Such surges arose when their frequency matched that of one of the structural modes, producing resonance. The consequent variations in propellant-flow rate then caused the engine thrust to oscillate at that same rate. This turned the engines into sledgehammers, striking the vehicle structure at its resonant frequency, and made the pogo stronger. It weakened only when consumption of propellant brought a further change in the structural frequency that broke the resonance, allowing the surges to die out.

Pogo was common; it had been present on earlier launch vehicles. It had brought vibrations with acceleration of 9 g's in a Titan II, which was unacceptably severe. Engineering changes cut this to below 0.25 g, which enabled this rocket to launch the manned Gemini spacecraft. Pogo reappeared in Apollo during the flight of a test Saturn V in 1968. For the Shuttle, the cure was relatively simple, calling for installation of a gas-filled accumulator within the main oxygen line. This damped the pogo oscillations, though design of this accumulator called for close understanding of the pertinent frequencies.[98]

---

97. C. Thomas Modlin, Jr., and George A. Zupp, Jr., "Shuttle Structural Dynamics Characteristics—The Analysis and Verification," in Chaffee, "Conference," p. 326; "COSMIC Software Catalog," NASA CR-191005 (1993).

98. Dennis Jenkins, *Space Shuttle: The History of the National Space Transportation System* (Stillwater, MN: Voyageur Press), 2001, p. 416; James M. Grimwood, Barton C. Hacker, and Peter J. Vorzimmer, *Project Gemini Technology and Operations: A Chronology*, NASA SP-4002 (Washington, DC: NASA, 1969), pp. 68, 76, 121; Roger E. Bilstein, *Stages to Saturn: A Technological History of the Apollo/Saturn Launch Vehicles*, NASA SP-4206 (Washington, DC: NASA, 1980), pp. 360, 362–363.

The most important structural tests used actual flight hardware, including the orbiter Enterprise and STA-099, a full-size test article that later became the Challenger. In 1978, Enterprise went to NASA Marshall, where the work now included studies on the external tank. For vibrational tests, engineers assembled a complete Shuttle by mating Enterprise to such a tank and to a pair of dummy solid boosters. One problem that these models addressed came at lift-off. The ignition of the three main engines imposes a sudden load of more than 1 million pounds of thrust. This force bends the solid boosters, placing considerable stress at their forward attachments to the tank. If the solid boosters were to ignite at that moment, their thrust would add to the stress.

To reduce the force on the attachment, analysts took advantage of the fact that the solid boosters would not only bend but would sway back and forth somewhat slowly, like an upright fishing rod. The strain on the attachment would increase and decrease with the sway, and it was possible to have the solid boosters ignite at an instant of minimum load. This called for delaying their ignition by 2.7 seconds, which cut the total load by 25 percent. The main engines fired during this interval, which consumed propellant, cutting the payload by 600 pounds. Still, this was acceptable.[99]

While Enterprise underwent vibration tests, STA-099 showed the orbiter's structural strength by standing up to applied forces. Like a newborn baby that lacks hair, this nascent form of Challenger had no thermal-protection tiles. Built of aluminum, it looked like a large fighter plane. For the structural tests, tiles were not only unnecessary; they were counterproductive. The tiles had no structural strength of their own that had to be taken into account, and they would have received severe damage from the hydraulic jacks that applied the loads and forces.

STA-099 and Columbia had both been designed to accommodate a set of loads defined by a database designated 5.1. In 1978, there was a new database, 5.4, and STA-099 had to withstand its loads without acquiring strains or deformations that would render it unfit for flight. Yet in an important respect, this vehicle was untestable; it was not possible to validate the strength of its structural design merely by applying loads with those jacks. The Shuttle structure had evolved under such strong emphasis on saving weight that it was necessary to take full account

---

99. Modlin and Zupp, "Structural," p. 326; Alden C. Mackey and Ralph E. Gotto, "Structural Load Challenges During Space Shuttle Development," in Chaffee, "Conference," pp. 335–339.

of thermal stresses that resulted from temperature differences across structural elements during reentry. No facility existed that could impose thermal stresses on so large an object as STA-099, for that would have required heating the entire vehicle.

STA-099 and Columbia had both been designed to withstand ultimate loads 140 percent greater than those of the 5.1 database. The structural tests on STA-099 now had to validate this safety factor for the new 5.4 database. Unfortunately, a test to 140 percent of the 5.4 loads threatened to produce permanent deformations in the structure. This was unacceptable, for STA-099 was slated for refurbishment into Challenger. Moreover, because thermal stresses could not be reproduced over the entire vehicle, a test to 140 percent would sacrifice the prospect of building Challenger while still leaving questions as to whether an orbiter could meet the safety factor of 140 percent.

NASA managers shaped the tests accordingly. For the entire vehicle, they used the jacks to apply stresses only up to 120 percent of the 5.4 loads. When the observed strains proved to match closely the values predicted by stress analysis, the 140 percent safety factor was deemed to be validated. In addition, the forward fuselage underwent the most severe aerodynamic heating, yet it was relatively small. It was subjected to a combination of thermal and mechanical loads that simulated the complete reentry stress environment in at least this limited region. STA-099 then was given a detailed and well-documented posttest inspection. After these tests, STA-099 was readied as the flight vehicle Challenger, joining Columbia as part of NASA's growing Shuttle fleet.[100]

Aerospaceplane to NASP: The Lure of Air-Breathing Hypersonics

The Space Shuttle represented a rocket-lofted approach to hypersonic space access. But rockets were not the only means of propulsion contemplated for hypersonic vehicles. One of the most important aspects of hypersonic evolution since the 1950s has been the development of the supersonic combustion ramjet, popularly known as a scramjet. The ramjet in its simplest form is a tube and nozzle, into which air is introduced, mixed with fuel, and ignited, the combustion products passing

---

100. Craig Covault, "NASA Evaluating Major Shuttle Orbiter Changes," *Aviation Week* (Oct. 10, 1977), p. 26; Photo of Structural Test Article, *Aviation Week*, Mar. 6, 1978, p. 13; Philip C. Glynn and Thomas L. Moser, "Orbital Structural Design and Verification," in Chaffee, "Conference," pp. 353–356.

through a classic nozzle and propelling the engine forward. Unlike a conventional gas turbine, the ramjet does not have a compressor wheel or staged compressor blades, cannot typically function at speeds less than Mach 0.5, and does not come into its own until the inlet velocity is near or greater than the speed of sound. Then it functions remarkably well as an accelerator, to speeds well in excess of Mach 3.

Conventional subsonic-combustion ramjets, as employed by the Mach 4.31 X-7, held promise as hypersonic accelerators for a time, but they could not approach higher hypersonic speeds because their subsonic internal airflow heated excessively at high Mach. If a ramjet could be designed that had a supersonic internal flow, it would run much cooler and at the same time be able to accelerate a vehicle to double-digit hypersonic Mach numbers, perhaps reaching the magic Mach 25, signifying orbital velocity. Such an engine would be a scramjet. Such engines have only recently made their first flights, but they nevertheless are important in hypersonics and point the way toward future practical air-breathing hypersonics.

An important concern explored at the NACA's Lewis Flight Propulsion Laboratory during the 1950s was whether it was possible to achieve supersonic combustion without producing attendant shock waves that slow internal flow and heat it. Investigators Irving Pinkel and John Serafini proposed experiments in supersonic combustion under a supersonic wing, postulating that this might afford a means of furnishing additional lift. Lewis researchers also studied supersonic combustion testing in wind tunnels. Supersonic tunnels produced very low air pressure, but it was known that aluminum borohydride could promote the ignition of pentane fuel even at pressures as low as 0.03 atmospheres. In 1955, Robert Dorsch and Edward Fletcher successfully demonstrated such tunnel combustion, and subsequent research indicated that combustion more than doubled lift at Mach 3.

Though encouraging, this work involved flow near a wing, not in a ramjet-like duct. Even so, NACA aerodynamicists Richard Weber and John MacKay posited that shock-free flow in a supersonic duct could be attained, publishing the first open-literature discussion of theoretical scramjet performance in 1958, which concluded: "the trends developed herein indicate that the [scramjet] will provide superior performance at higher hypersonic flight speeds."[101] The Weber-MacKay study came a year after Marquardt researchers had demonstrated supersonic combustion of a hydrogen and air mix. Other investigators working contemporaneously were the manager William Avery and the experimentalist Frederick Billig, who independently achieved supersonic combustion at the Johns Hopkins University Applied Physics Laboratory (APL), and J. Arthur Nicholls at the University of Michigan.[102]

The most influential of all scramjet advocates was the colorful Italian aerodynamicist, partisan leader, and wartime emigree, Antonio Ferri. Before the war, as a young military engineer, he had directed supersonic wind tunnel studies at Guidonia, Benito Mussolini's showcase aeronautical research establishment outside Rome. In 1943, after the collapse of the Fascist regime and the Nazi assumption of power, he left Guidonia, leading a notably successful band of anti-Nazi, anti-Fascist partisans. Brought to America by Moe Berg, a baseball player turned intelligence agent, Ferri joined NACA Langley, becoming Director of its Gas Dynamics Branch. Turning to the academic world, he secured a professorship at Brooklyn Polytechnic Institute. He formed a close association with

---

101. Richard J. Weber and John S. MacKay, "An Analysis of Ramjet Engines Using Supersonic Combustion," NACA TN-4386 (1958), p. 22; Irving Pinkel and John S. Serafini, "Graphical Method for Obtaining Flow Field in Two-Dimensional Supersonic Stream to Which Heat is Added," NACA TN-2206 (1950); Irving Pinkel, John S. Serafini, and John L. Gregg, "Pressure Distribution and Aerodynamic Coefficients Associated with Heat Addition to Supersonic Air Stream Adjacent to Two-Dimensional Supersonic Wing," NACA RM-E51K26 (1952).

102. Alan Newman, "Speed Ahead of its Time," Johns Hopkins Magazine, vol. 40, no. 4 (Dec. 1988), pp. 26–31; Harold E. Gilreath, "The Beginning of Hypersonic Ramjet Research at APL," Johns Hopkins Applied Physics Laboratory Technical Digest, vol. 11, no. 3-4 (1990), pp. 319–335; G.L. Dugger, F.S. Billig, and W.H. Avery, "Hypersonic Propulsion Studies at the Applied Physics Laboratory, The Johns Hopkins University," Johns Hopkins University Applied Physics Laboratory Report TG 405 (June 14, 1961), esp. pp. 1–3; Frank D. Stull, Robert A. Jones, and William P. Zima, "Propulsion Concepts for High Speed Aircraft," Paper 75-1092, Society of Automotive Engineers (1975); Paul J. Waltrup, Griffin Y. Anderson, and Frank D. Stull, "Supersonic Combustion Ramjet (Scramjet) Engine Development in the United States," Johns Hopkins University Applied Physics Laboratory Paper 76-042 (1976); Paul J. Waltrup, "Liquid Fueled Supersonic Combustion Ramjets: A Research Perspective of the Past, Present and Future," AIAA Paper 86-0158 (1986); Paul J. Waltrup, "Hypersonic Airbreathing Propulsion: Evolution and Opportunities," in Advisory Group for Aeronautical Research and Development, Conference Proceedings on the Aerodynamics of Hypersonic Lifting Vehicles (Neuilly sur Seine, France: NATO-AGARD, 1987), pp. 1–29; and in Thomas C. Adamson, Jr., "Aeronautical and Aerospace Engineering Education at the University of Michigan," in Barnes McCormick, et al., Aerospace Engineering Education During the First Century of Flight (Reston, VA: AIAA, 2004), p. 54.

Alexander Kartveli, chief designer at Republic Aviation, and designer of the P-47, F-84, XF-103, and F-105. Indeed, Kartveli's XF-103 (which, alas, never was completed or flown) employed a Ferri engine concept. In 1956, he established General Applied Science Laboratories (GASL), with financial backing from the Rockefellers.[103]

Ferri emphasized that scramjets could offer sustained performance far higher than rockets could, and his strong reputation ensured that people listened to him. At a time when shock-free flow in a duct still loomed as a major problem, Ferri did not flinch from it but instead took it as a point of departure. He declared in September 1958 that he had achieved it, thus taking a position midway between the demonstrations at Marquardt and APL. Because he was well known, he therefore turned the scramjet from a wish into an invention, which might be made practical.

He presented his thoughts publicly at a technical colloquium in Milan in 1960 ("Many of the older men present," John Becker wrote subsequently, "were politely skeptical") and went on to give a far more detailed discussion in May 1964, at the Royal Aeronautical Society in London. This was the first extensive public presentation on hypersonic propulsion, and the attendees responded with enthusiasm. One declared that whereas investigators "had been thinking of how high in flight speed they could stretch conventional subsonic burning engines, it was now clear that they should be thinking of how far down they could stretch supersonic burning engines," and another added that Ferri now was "assailing the field which until recently was regarded as the undisputed regime of the rocket."[104]

---

103. Edward T. Curran, "Scramjet Engines: The First Forty Years," *Journal of Propulsion and Power*, vol. 17, No. 6 (Nov.–Dec. 2001), pp. 1138–1148; Antonio Ferri, "Review of Scramjet Technology," *Journal of Aircraft*, vol. 5, no. 1 (Jan. 1968), pp. 3–10; T.A. Heppenheimer, *Facing the Heat Barrier: A History of Hypersonics*, SP-2007-4232 (Washington, DC: NASA, 2007), p. 103; R.R. Jamison, "Hypersonic Air Breathing Engines," in A.R. Collar and J. Tinkler, *Hypersonic Flow: Proceedings of the Eleventh Symposium of the Colston Research Society held in the University of Bristol, Apr. 6–8, 1959* (London: Butterworths Scientific Publications, 1960), pp. 391–408; S.W. Greenwood, "Spaceplane Propulsion," *The Aeroplane and Astronautics* (May 25, 1961), pp. 597–599.
104. John V. Becker, "Confronting Scramjet: The NASA Hypersonic Ramjet Experiment," in Richard P. Hallion, ed., *The Hypersonic Revolution: Eight Case Studies in the History of Hypersonic Technology*, vol. 2: *From Scramjet to the National Aero-Space Plane, 1964–1986* (Wright-Patterson AFB: Aeronautical Systems Division, 1987), p. 752; Antonio Ferri, "Review of Problems in Application of Supersonic Combustion," 7th Lanchester Memorial Lecture, *Journal of the Royal Aeronautical Society*, vol. 68, no. 645 (Sept. 1964), pp. 595, 597; Antonio Ferri, "Supersonic Combustion Progress," *Astronautics & Aeronautics*, vol. 2, no. 8 (Aug. 1964), pp. 32–37; Heppenheimer, *Facing the Heat Barrier*, pp. 104–105.

Scramjet advocates were offered their first opportunity to actually build such propulsion systems with the Air Force's abortive Aerospaceplane program of the late 1950s–mid-1960s. A contemporary to Dyna-Soar but far less practical, Aerospaceplane was a bold yet premature effort to produce a logistical transatmospheric vehicle and possible orbital strike system. Conceived in 1957 and initially known as the Recoverable Orbital Launch System (ROLS), Aerospaceplane attracted surprising interest from industry. Seventeen aerospace companies submitted contract proposals and related studies; Convair, Lockheed, and Republic submitted detailed designs. The Republic concept had the greatest degree of engine-airframe integration, a legacy of Ferri's partnership with Kartveli.

By the early 1960s, Aerospaceplane not surprisingly was beset with numerous developmental problems, along with a continued debate over whether it should be a single- or two-stage system, and what proportion of its propulsion should be turbine, scramjet, and pure rocket. Though it briefly outlasted Dyna-Soar, it met the same harsh fate. In the fall of 1963, the Air Force Scientific Advisory Board damned the program in no uncertain terms, noting: "Aerospaceplane has had such an erratic history, has involved so many clearly infeasible factors, and has been subjected to so much ridicule that from now on this name should be dropped. It is recommended that the Air Force increase [its] vigilance [so] that no new program achieves such a difficult position."[105] The next year, Congress slashed its remaining funding, and Aerospaceplane was at last consigned to a merciful oblivion.

In the wake of Aerospaceplane's cancellation, both the Air Force and NASA maintained an interest in advancing scramjet propulsion for transatmospheric aircraft. The Navy's scramjet interest, though great, was primarily in smaller engines for missile applications. But Air Force and NASA partisans formed an Ad-Hoc Working Group on Hypersonic Scramjet Aircraft Technology.

---

105. Air Force Scientific Advisory Board, "Memo-Report of the USAF Scientific Advisory Board Aerospace Vehicles/Propulsion Panels on Aerospaceplane, VTOL, and Strategic Manned Aircraft" (Oct. 24, 1963), pp. 1, 3, SAB Office files, USAF HQ, Pentagon, Washington, DC; see also F.E. Jariett and G. Karel, "Aerospaceplane: The Payload Capabilities of Various Recoverable Systems All Using Hydrogen Fuel," General Dynamics Astronautics Report AE62-0892 (Oct. 25, 1962), esp. pp. 8–23; "Aerospaceplane May be a Two-Stage Vehicle," *Aviation Week & Space Technology* (July 22, 1963), pp. 245–249; Heppenheimer, *Facing the Heat Barrier*, pp. 112–128.

Both agencies pursued development programs that sought to build and test small scramjet modules. The Air Force Aero-Propulsion Laboratory sponsored development of an Incremental Scramjet flight-test program at Marquardt. This proposed test vehicle underwent extensive analysis and study, though without actually flying as a functioning scramjet testbed. The first manifestation of Langley work was the so-called Hypersonic Research Engine (HRE), an axisymmetric scramjet of circular cross section with a simple Oswatitsch spike inlet, designed by Anthony duPont. Garrett AiResearch built this engine, planned for a derivative of the X-15. The HRE never actually flew as a "hot" functioning engine, though the X-15A-2 flew repeatedly with a boilerplate test article mounted on the stub ventral fin (during its record flight to Mach 6.70 on October 3, 1967, searing hypersonic shock interactions melted it off the plane). Subsequent tunnel tests revealed that the HRE was, unfortunately, the wrong design. A podded and axisymmetric design, like an airliner's fanjet, it could only capture a small fraction of the air that flowed past a vehicle, resulting in greatly reduced thrust. Integrating the scramjet with the airframe, so that it used the forebody to assist inlet performance and the afterbody as a nozzle enhancement, would more than double its thrust.[106]

Investigation of such concepts began at Langley in 1968, with pioneering studies by researchers John Henry, Shimer Pinckney, and others. Their work expanded upon a largely Ferri-inspired base, defining what emerged as common basic elements of subsequent Langley scramjet research. It included a strong emphasis upon airframe integration, use of fixed geometry, a swept inlet that could readily spill excess airflow, and the use of struts for fuel injection. Early observations, published in 1970, showed that struts were practical for a large supersonic combustor at Mach 8. The program went on to construct test scramjets and conducted almost 1,000 wind tunnel test runs of engines at Mach 4

---

106. M.L. Brown and R.L. Maxwell, Marquardt Corporation, "Scramjet Incremental Flight Test Program," Summary, Marquardt-AF Aero-Propulsion Laboratory Report AFAPLTR-67-112 (1968), pp. 3–4; Becker, "Confronting Scramjet," in Hallion, ed., *Hypersonic Revolution*, vol. 2, pp. 747–861, examines the HRE in detail; John R. Henry and Griffin Y. Anderson, "Design Considerations for the Airframe-Integrated Scramjet," NASA TM-X-2895 (1973); Robert A. Jones and Paul W. Huber, "Toward Scramjet Aircraft," *Astronautics and Aeronautics*, vol. 16, no. 2 (Feb. 1978), pp. 38–48; G. Burton Northam and G.Y. Anderson, "Supersonic Combustion Ramjet Research at Langley," AIAA Paper 86-0159 (1986).

and Mach 7. Inlets at Mach 4 proved sensitive to "unstarts," a condition where the shock wave is displaced, disrupting airflow and essentially starving the engine of its oxygen. Flight at Mach 7 raised the question of whether fuel could mix and burn in the short available combustor length.[107]

Langley test engines, like engines at GASL, Marquardt, and other scramjet research organizations, encountered numerous difficulties. Large disparities existed between predicted performance and that actually achieved in the laboratory. Indeed, the scramjet, advanced so boldly in the mid-1950s, would not be ready for serious pursuit as a propulsive element until 1986. Then, on the eve of the National Aerospace Plane development program, Langley researchers Burton Northam and Griffin Anderson announced that NASA had succeeded at last in developing a practical scramjet. They proclaimed triumphantly: "At both Mach 4 and Mach 7 flight conditions, there is ample thrust both for acceleration and cruise."[108]

Out of such optimism sprang the National Aero-Space Plane program, which became a central feature of the presidency of Ronald Reagan. It was linked to other Reagan-era defense initiatives, particularly his Strategic Defense Initiative, a ballistic missile shield intended to reduce the threat of nuclear war, which critics caustically belittled as "Star Wars." SDI called for the large-scale deployment of defensive arms in space, and it became clear that the Space Shuttle would not be their carrier. Experience since the Shuttle's first launch in April 1981 had shown that it was costly and took a long time to prepare for relaunch. The Air Force was unwilling to place the national eggs in such a basket. In February 1984, Defense Secretary Caspar Weinberger approved a document stating that total reliance upon the Shuttle represented an unacceptable risk.

---

107. NASA Langley completed 963 successful runs of three strut, parametric, and step-strut scramjets between 1976–1987; from Edward G. Ruf, "Airframe-Integrated Scramjet Engine tests in NASA Langley Scramjet Engine Test Facilities," at *http://hapb-www.larc.nasa.gov/Public/Engines/engine_tests.html*, accessed on May 1, 2009.

108. Northam and Anderson, "Supersonic Combustion Ramjet Research," AIAA Paper 86-0159, p. 7; for NASA and other work, see J. Menzler and T.W. Mertz, "Large Scale Supersonic Combustor Testing at Conditions Simulating Mach 8 Flight," AIAA Paper 70-715 (1970); Carl A. Trexler, "Inlet Performance of the Integrated Langley Scramjet Module (Mach 2.3 to 7.6)," AIAA Paper 75-1212 (1975); Robert W. Guy and Ernest A. Mackley, "Initial Wind Tunnel Tests at Mach 4 and 7 of a Hydrogen-Burning, Airframe Integrated Scramjet," AIAA Paper 79-7045 (1979); and R.C. Rogers, D.P. Capriotti, and R.W. Guy, "Experimental Supersonic Combustion Research at NASA Langley," AIAA Paper 2506 (1998).

An Air Force initiative was under way at the time that looked toward an alternative. Gen. Lawrence A. Skantze, Chief of Air Force Systems Command (AFSC), had sponsored studies of Trans Atmospheric Vehicles (TAVs) by Air Force Aeronautical Systems Division (ASD). These reflected concepts advanced by ASD's chief planner, Stanley A. Tremaine, as well as interest from Air Force Space Division (SD), the Defense Advanced Research Projects Agency (DARPA), and Boeing and other companies. TAVs were SSTO craft intended to use the Space Shuttle Main Engine (SSME) and possibly would be air-launched from derivatives of the Boeing 747 or Lockheed C-5. In August 1982, ASD had hosted a 3-day conference on TAVs, attended by representatives from AFSC's Space Division and DARPA. In December 1984, ASD went further. It established a TAV Program Office to "streamline activities related to long-term, preconceptual design studies."[109]

DARPA's participation was not surprising, for Robert Cooper, head of this research agency, had elected to put new money into ramjet research. His decision opened a timely opportunity for Anthony duPont, who had designed the HRE for NASA. DuPont held a strong interest in "combined-cycle engines" that might function as a turbine air breather, translate to ram/scram, and then perhaps use some sophisticated air collection and liquefaction process to enable them to boost as rockets into orbit. There are several types of these engines, and duPont had patented such a design as early as 1972. A decade later, he still believed in it, and he learned that Anthony Tether was the DARPA representative who had been attending TAV meetings.

Tether sent him to Cooper, who introduced him to DARPA aerodynamicist Robert Williams, who brought in Arthur Thomas, who had been studying scramjet-powered spaceplanes as early as Sputnik. Out of this climate of growing interest came a $5.5 million DARPA study program, Copper Canyon. Its results were so encouraging that DARPA took the notion of an air-breathing single-stage-to-orbit vehicle to Presidential science adviser George Keyworth and other senior officials, including Air Force Systems Command's Gen. Skantze. As Thomas recalled: "The people were amazed at the component efficiencies that had been

---

109. USAF Aeronautical Systems Division news release 84-211; Richard P. Hallion, "Yesterday, Today, and Tomorrow: From Shuttle to the National Aero-Space Plane," in Hallion, ed., *Hypersonic Revolution*, vol. 2, pp. 1336–1337, 1361; S.A. Tremaine and Jerry B. Arnett, "Transatmospheric Vehicles—A Challenge for the Next Century," AIAA Paper 84-2414 (1984).

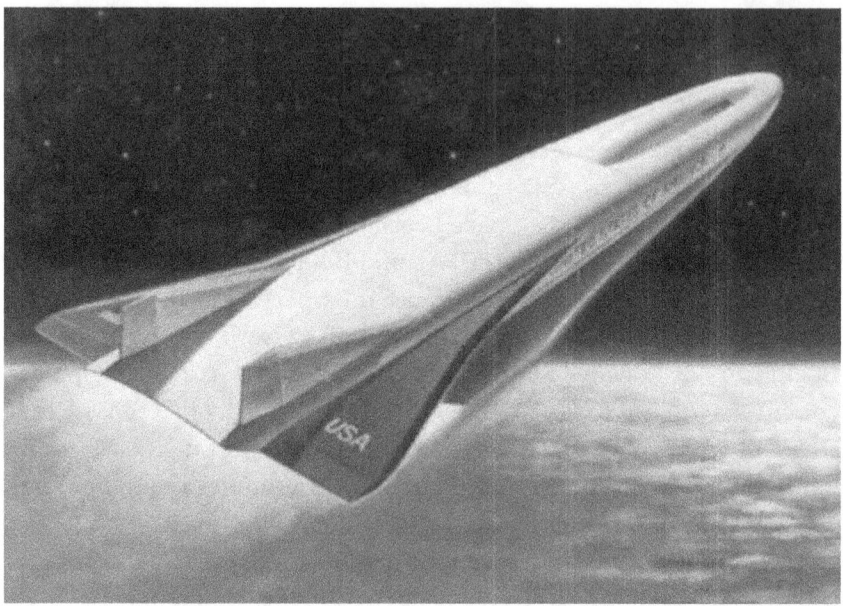

The National Aero-Space Plane concept in final form, showing its modified lifting body design approach. NASA.

assumed in the study. They got me aside and asked if I really believed it. Were these things achievable? Tony [duPont] was optimistic everywhere: on mass fraction, on drag of the vehicle, on inlet performance, on nozzle performance, on combustor performance. The whole thing, across the board. But what salved our consciences was that even if these things weren't all achieved, we still could have something worthwhile. Whatever we got would still be exciting."[110]

Gen. Skantze realized that SDI needed something better than the Shuttle—and Copper Canyon could possibly be it. Briefings were encouraging, but he needed to see technical proof. That evidence came when he visited GASL and witnessed a subscale duPont engine in operation. Afterward, as DARPA's Bob Williams recalled subsequently: "the Air Force system began to move with the speed of a spaceplane."[111] Secretary of Defense Caspar Weinberger received a briefing and put his support behind the effort. In January 1986, the Air Force established a joint-service Air Force-Navy-NASA National Aero-Space Plane Joint Program Office at Aeronautical Systems Division, transferring into it all the personnel

---

110. Quote from Heppenheimer interview with Arthur Thomas, Sept. 24, 1987.
111. Quote from Heppenheimer interview with Robert Williams, May 1, 1986.

previously assigned to the TAV Program Office established previously. (The program soon received an X-series designation, as the X-30.) Then came the clincher. President Ronald Reagan announced his support for what he now called the "Orient Express" in his State of the Union Address to the Nation on February 4, 1986. President Reagan's support was not the product of some casual whim: the previous spring, he had ordered a joint Department of Defense-NASA space launch study of future space needs and, additionally, established a national space commission. Both strongly endorsed "aerospace plane development," the space commission recommending it be given "the highest national priority."[112]

Though advocates of NASP attempted to sharply differentiate their effort from that of the discredited Aerospaceplane of the 1960s, the NASP effort shared some distressing commonality with its predecessor, particularly an exuberant and increasingly unwarranted optimism that afforded ample opportunity for the program to run into difficulties. In 1984, with optimism at its height, DARPA's Cooper declared that the X-30 could be ready in 3 years. DuPont, closer to the technology, estimated that the Government could build a 50,000-pound fighter-size vehicle in 5 years for $5 billion. Such predictions proved wildly off the mark. As early as 1986, the "Government baseline" estimate of the aircraft rose to 80,000 pounds. Six years later, in 1992, its gross weight had risen eightfold, to 400,000 pounds. It also had a "velocity deficit" of 3,000 feet per second, meaning that it could not possibly attain orbit. By the next year, NASP "lay on life support."[113]

It had evolved from a small, seductively streamlined speedster to a fatter and far less appealing shape more akin to a wooden shoe, entering a death spiral along the way. It lacked performance, so it needed greater power and fuel, which made it bigger, which meant it lacked performance so that it needed greater power and fuel, which made it bigger . . . and bigger . . . and bigger. X-30 could never attain the "design closure" permitting it to reach orbit. NASP's support continuously softened,

---

112. National Commission on Space, *Pioneering the Space Frontier* (New York: Bantam Books, 1986), p. 184; Joint DOD–NASA Task Team Report, *National Space Transportation and Support Study* (Washington, DC: GPO, 1986); President Ronald Reagan, State of the Union Address, Feb. 4, 1986.

113. Larry Schweikart, *The Quest for the Orbital Jet: The National Aero-Space Plane Program (1983–1995)*, vol. 3 of Hallion, ed., *The Hypersonic Revolution* (Washington, DC: USAF, 1998), pp. 349, 279–351.

particularly as technical challenges rose, performance estimates fell, and other national issues grew in prominence. It finally withered in the mid-1990s, leaving unresolved what, if anything, scramjets might achieve.[114]

Transatmospherics after NASP

Two developments have paced work in hypersonics since NASP died in 1995. Continuing advances in computers, aided markedly by advancements in wind tunnels, have brought forth computational fluid dynamics (CFD). Today, CFD simulates the aerodynamics of flight vehicles with increasing (though not perfect) fidelity. In addition, NASA and the Air Force have pursued a sequence of projects that now aim clearly at developing operational scramjet-powered military systems.

Early in the NASP effort, in 1984, Robert Whitehead of the Office of Naval Research spoke on CFD to its people. Robert Williams recalls that Williams presented the equations of fluid dynamics "so the computer could solve them, then showed that the computer technology was also there. We realized that we could compute our way to Mach 25 with high confidence."[115] Unfortunately, in reality, DARPA could not do that. In 1987, the trade journal *Aerospace America* reported: "almost nothing is known about the effects of heat transfer, pressure gradient, three-dimensionality, chemical reactions, shock waves, and other influences on hypersonic transition."[116] (This transition causes a flow to change from laminar to turbulent, a matter of fundamental importance.)

---

114. Stuart O. Schmitt, Theodore J. Wierzbanowski, and Johnny Johnson, "The Challenge of X-30 Flight Test," 31st Symposium, Society of Experimental Test Pilots, Beverly Hills, CA, Sept. 26, 1987; United States General Accounting Office, "National Aero-Space Plane: A Technology Development and Demonstration Program to Build the X-30," Report GAO/NSIAD-88-122 (Apr. 1988); Alan W. Wilhite, et al., "Concepts Leading to the National Aero-Space Plane Program," AIAA Paper 90-0294 (1990); Robert B. Barthelemy, "The National Aero Space Plane Program: A Revolutionary Concept," *Johns Hopkins Applied Physics Laboratory Technical Digest*, vol. 11, no. 2 & 3 (1990), pp. 312–318; United States General Accounting Office, "National Aero-Space Plane: Key Issues Facing the Program," Report GAO/T-NSIAD-92-26 (Mar. 1992), pp. 4–15; Joseph F. Shea, et al., "Report of the Defense Science Board Task Force on National Aero-Space Plane (NASP) Program" (1992); United States General Accounting Office, *National Aero-Space Plane: Restructuring Future Research and Development Efforts*, Report GAO/NSIAD-93-71 (Dec. 1992), p. 4; and Ray L. Chase and Ming H. Tang, "A History of the NASP Program from the Formation of the Joint Program Office to the Termination of the HySTP Scramjet Performance Demonstration Program," AIAA Paper 95-6031 (1995).

115. Author interview with Robert Williams, May 1, 1986.

116. Quote from Douglas L. Dwoyer, Paul Kutler, and Louis A. Povinelli, "Retooling CFD for Hypersonic Aircraft," *Aerospace America* (Oct. 1987), p. 35.

Code development did mature so that it could adequately support the next hypersonic system, NASA's X-43A program. In supporting the X-43A effort, NASA's most important code was GASP. NASP had used version 2.0; the X-43A used 3.0.[117] Like any flow code, it could not calculate the turbulence directly but had to model it. GASP 3.0 used the Baldwin-Lomax algebraic model that Princeton's Antony Jameson, a leading writer of flow codes, describes as: "the most popular model in the industry, primarily because it's easy to program."[118] GASP 3.0 also uses "eddy-viscosity" models, which Stanford's Peter Bradshaw rejects out of hand: "Eddy viscosity does not even deserve to be described as a 'theory' of turbulence!" More broadly, he adds, "Even the most sophisticated turbulence models are based on brutal simplifications" of the pertinent nonlinear partial differential equations.[119]

Can increasing computer power make up for this? Calculations of the NASP era had been rated in gigaflops, billions of floating point operations per second (FLOPS).[120] An IBM computer has recently cracked the petaflop mark—at a quadrillion operations per second, and even greater performance is being contemplated.[121] At Stanford University's Center for Turbulence Research, analyst Krishnan Mahesh studied flow within a commercial turbojet and found a mean pressure drop that differs from the observed value by only 2 percent. An earlier computation had given an error of 26 percent, an order of magnitude higher.[122] He used Large Eddy Simulation, which calculates the larger turbulent eddies and models the small ones that have a more universal character. But John Anderson, a historian of fluid dynamics, notes that LES

---

117. Charles E. Cockrell, Jr., Walter C. Engelund, Robert D. Bittner, Tom N. Jentinck, Arthur D. Dilley, and Abdelkader Frendi, "Integrative Propulsive Computational Fluid Dynamics Methodology for the Hyper-X Flight Experiment." *Journal of Spacecraft and Rockets* (Nov.–Dec. 2001), pp. 838, 843; S. Srinivasan, R.D. Bittner, and B.J. Bobskill, "Summary of the GASP Code Application and Evaluation Effort for Scramjet Combustor Flow-fields," AIAA Paper 93-1973 (1993).
118. T.A. Heppenheimer, "Some Tractable Mathematics for Some Intractable Physics," *Mosaic* (National Science Foundation), spring 1991, p. 30.
119. Quote in P. Bradshaw, "Progress in Turbulence Research," AIAA Paper 90-1480 (1990), p. 3.
120. Jack J. Dongarra, "Performance of Various Computers Using Standard Linear Equations Software in a Fortran Environment," Argonne National Laboratory, Technical Memorandum 23 (Sept. 30, 1988).
121. "Super Supercomputers," *Aviation Week* (Feb. 16, 2009).
122. Krishnan Mahesh, et al., "Large-Eddy Simulation of Gas Turbine Combustors." *Annual Research Briefs* (Stanford: Center for Turbulence Research, Stanford University, 2001), pp. 3–17.

"is not viewed as an industry standard." He sees no prospect other than direct numerical simulation (DNS), which directly calculates all scales of turbulence. "It's clear-cut," he adds. "The best way to calculate turbulence is to use DNS. Put in a fine enough grid and calculate the entire flow field, including the turbulence. You don't need any kind of model and the turbulence comes out in the wash as part of the solution." But in seeking to apply DNS, even petaflops aren't enough. Use of DNS for practical problems in industry is "many decades down the road. Nobody to my knowledge has used DNS to deal with flow through a scramjet. That type of application is decades away."[123] With the limitations as well as benefits of CFD more readily apparent, it thus is significant that more traditional hypersonic test facilities are also improving. As just one example, NASA Langley's largest hypersonic facility, the 8-foot High Temperature Tunnel (HTT), has been refitted to burn methane and use its combustion products, with oxygen replenishment, as the test gas. This heats the gas. As reviewed by the *Journal of Spacecraft and Rockets*: "the oxygen content of the freestream gas is representative of flight conditions as is the Mach number, total enthalpy, dynamic pressure, and Reynolds number."[124]

One fruitful area with NASP had been its aggressive research on scramjets, which benefited substantially because of NASA's increasing investment in high-temperature hypersonic test facilities.[125]

Table 3 enumerates the range of hypersonic test facilities for scramjet and aerothermodynamic research available to researchers at the NASA Langley Research Center. Between 1987 and the end of 1994, Langley researchers ran over 1,500 tests on 10 NASP engine modules, over 1,200 in a single 3-year period, from the end of 1987 to 1990. After NASP wound down, Agency researchers ran nearly 700 tests on four other configurations between 1994 and 1996. These tests, ranging from Mach 4 to Mach

---

123. Author interview, John D. Anderson, Jr., Nov. 19, 2008.
124. Quote: Charles E. Cockrell, Jr., et al., *Journal of Spacecraft and Rockets* (2001), p. 841; L.D. Huebner, K.E. Rock, R.T. Voland, and A.R. Wieting, "Calibration of the Langley 8-Foot High Temperature Tunnel for Hypersonic Propulsion Airbreathing Testing," AIAA Paper 96-2197 (1996).
125. R.W. Guy, et al., "Operating Characteristics of the Langley Mach 7 Scramjet Test Facility," NASA TM-81929 (1981); S.R. Thomas and R.W. Guy, "Expanded Operational Capabilities of the Langley Mach 7 Scramjet Test Facility, NASA TP-2186 (1983); E.H. Andrews, Jr., et al., "Langley Mach 4 Scramjet Test Facility," NASA TM-86277 (1985); D.E. Reubush and R.L. Puster, "Modification to the Langley 8-Ft. High Temperature Tunnel for Hypersonic Propulsion Testing," AIAA Paper 1987-1887 (1987); D.W. Witte, et al., "1998 Calibration of the Mach 4.7 and Mach 6 Arc-Heated Scramjet Test Facility Nozzles," NASA TM-2004-213250 (2004).

TABLE 3			
NASA LRC SCRAMJET PROPULSION AND AEROTHERMODYNAMIC TEST FACILITIES			
FACILITY NAME	MACH	REYNOLDS NUMBER	SIZE
8-foot High Temperature Tunnel	4, 5, 7	$0.3–5.1 \times 10^6$ / ft.	8-ft. dia.
Arc-Heated Scramjet Test Facility	4.7–8.0	$0.04–2.2 \times 10^6$ / ft.	4-ft. dia.
Combustion-Heated Scramjet Test Facility	3.5–6.0	$1.0–6.8 \times 10^6$ / ft.	42" x 30"
Direct Connect Supersonic Combustion Test Facility	4.0–7.5	$1.8–31.0 \times 10^6$ / ft.	[Note (a)]
HYPULSE Shock Tunnel [Note (b)]	5.0–25	$0.5–2.5 \times 10^6$ / ft.	7-ft dia.
15-inch Mach 6 High Temperature Tunnel	6	$0.5–8.0 \times 10^6$ / ft.	15" dia.
20-inch Mach 6 CF4 Tunnel	6	$0.05–0.7 \times 10^6$ / ft.	20" dia.
20-inch Mach 6 Tunnel	6	$0.5–8.0 \times 10^6$ / ft.	20" x 20"
31-inch Mach 10 Tunnel	10	$0.2–2.2 \times 10^6$ / ft.	31" x 31"

Source: Data from NASA LRC Facility brochures.
(a) DCSCTF section varies: 1.52" x 3.46" with a M = 2 nozzle and 1.50" x 6.69" with a M = 2.7 nozzle.
(b) LRC's HYPULSE shock tunnel is at the GASL Division of Allied Aerospace Industries, Ronkonkoma, NY.

8, so encouraged scramjet proponents that they went ahead with plans for a much-scaled-back effort, the Hyper-X (later designated X-43A), which compared in some respects with the ASSET program undertaken after cancellation of the X-20 Dyna-Soar three decades earlier.[126]

The X-43, managed at Langley Research Center by Vincent Rausch, a veteran of the earlier TAV and NASP efforts, began in 1995 as Hyper-X, coincident with the winddown of NASP. It combined a GASL scramjet engine with a 100-inch-long by 60-inch-span slender lifting body and an Orbital Sciences Pegasus booster, this combination being carried to a launch altitude of 40,000 feet by NASA Dryden's NB-52B Stratofortress. After launch, the Pegasus took the X-43 to approximately 100,000 feet,

---

126. The engines and their total successful test runs were Gov't Baseline (114); Engine A (69); Engine A-1 (55); Engine A-2 (321); Engine A-2+ (72); Engine C (233); Engine B-1 (359); NASP SX-20 (160); NASP SXPE (142); and NASP CDE (24), a total of 1,549 successful test runs. See the previously cited Ruf, "Airframe-Integrated Scramjet Engine Tests in NASA Langley Scramjet Engine Test Facilities."

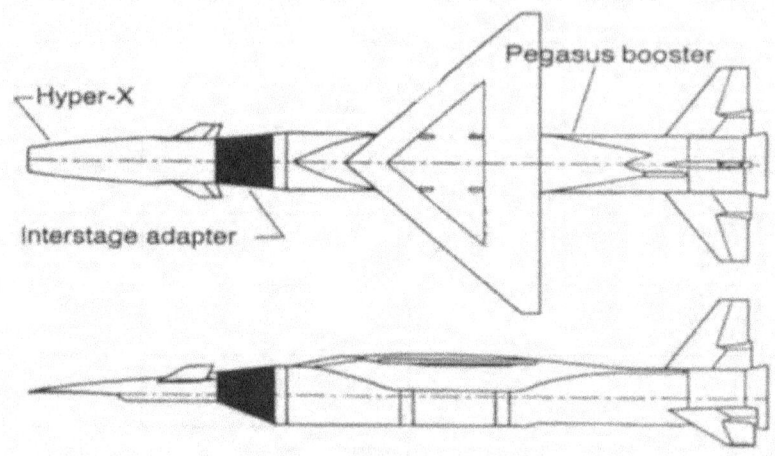

Schematic layout of the Hyper-X (subsequently X-43A) scramjet test vehicle and its Orbital Sciences Pegasus winged booster, itself a hypersonic vehicle. NASA.

where it would separate, demonstrating scramjet ignition (using silane and then adding gaseous hydrogen) and operation at velocities as high as Mach 10.

The X-43 program cost $230 million and consumed not quite a decade of development time. Built by Microcraft, Inc., of Tullahoma, TN, the X-43 used the shape of a Boeing study for a Mach 10 global reconnaissance and space access vehicle, conceived by a team under the leadership of George Orton. Langley Research Center furnished vital support, executing nearly 900 test runs of 4 engine configurations between 1996 and 2003.[127]

Microcraft completed three X-43A flight-test vehicles for testing by NASA Dryden Flight Research Center. Unfortunately, the first flight attempt failed in 2001, when the Pegasus booster shed a control fin after launch. A 3-year reexamination and review of the program led to a successful flight on March 27, 2004, the first successful hypersonic flight of a scramjet-powered airplane. The Pegasus boosted the X-43A to Mach 6.8. After separation, the X-43A burned silane, which ignites on contact

---

127. Engine tests totaled 876, at Mach 5, 7, 10, and 15. The engine configurations, tunnels, and test runs were: DFX, 467 runs (97 in CHSTF and 370 in the AHSTF); HXEM, 146 runs (130 in AHSTF and 16 in the 8-ft. HTT); HXFE, 54 runs (all in 8-ft. HTT); and HSM, 209 runs (all in HYPULSE); see the previously cited Ruf, "Airframe-Integrated Scramjet Engine Tests in NASA Langley Scramjet Engine Test Facilities."

with the air, for 3 seconds. Then it ramped down the silane and began injecting gaseous hydrogen, burning this gas for 8 seconds. This was the world's first flight test of such a scramjet.[128]

That November, NASA did it again with its third X-43A. On November 16, it separated from its booster at 110,000 feet and Mach 9.7 and its engine burned for 10 to 12 seconds with silane off. On its face, this looked like the fastest air-breathing flight in history, but this speed (approximately 6,500 mph) resulted from its use of Pegasus, a rocket. The key point was that the scramjet worked, however briefly. During the flight, the X-43A experienced airframe temperatures as high as 3,600 °F.[129]

Meanwhile, the Air Force was preparing to take the next step with its HyTech program. Within it, Pratt & Whitney, now merged with Rocketdyne, has been a major participant. In January 2001, it demonstrated the Performance Test Engine (PTE), an airframe-integrated scramjet that operated at hypersonic speeds using the hydrocarbon JP-7. Like the X-43A engine, though, the PTE was heavy. Its successor, the Ground Demonstrator Engine (GDE), was flight-weight. It also used fuel to cool the engine structure. One GDE went to Langley for testing in the HTT in 2005. It made the important demonstration that the cooling could be achieved using no more fuel than was to be employed for propulsion.

Next on transatmospheric agenda is a new X-test vehicle, the X-51A, built by Boeing, with a scramjet by Pratt & Whitney Rocketdyne. These firms are also participants in a consortium that includes support from NASA, DARPA, and the Air Force. The X-51A scramjet is fuel-cooled, with the cooling allowing it to be built of Inconel 625 nickel alloy rather than an exotic superalloy. Lofted to Mach 4.7 by a modified Army Tactical Missile System (ATACMS) artillery rocket booster, the X-51A is intended to fly at Mach 7 for minutes at a time, burning JP-7, a hydrocarbon fuel

---

128. Bruce A. Smith, "Elevon Failure Precedes Loss of First X-43A," *Aviation Week*, June 11, 2001, pp. 50–51; Michael Dornheim: "X-43 to Fly in Fall," July 28, 2003, pp. 36–37; "A Breath of Fast Air," April 5, 2004, pp. 28–29.

129. Thomas J. Bogar, Edwards A. Eiswirth, Lana M. Couch, James L. Hunt, and Charles R. McClinton, "Conceptual Design of a Mach 10, Global Reach Reconnaissance Aircraft," AIAA Paper 96-2894 (1996); Charles R. McClinton, Vincent L. Rausch, Joel Sitz, and Paul Reukauf, "Hyper-X Program Status," AIAA Paper 01-0828, 39th Aerospace Sciences Meeting, Reno, NV, Nov. 8–11, 2001; David E. Reubush, Luat T. Nguyen, and Vincent L. Rausch, "Review of X-43A Return to Flight Activities and Current Status," AIAA Paper 03-7085 (2003); Jay Levine, "Exploring the Hypersonic Realm," *The X-Press*, vol. 46, no. 10 (Nov. 26, 2004), pp. 1, 8; Michael Dornheim, "But Now What?" *Aviation Week*, Nov. 22, 2004, pp. 24–26.

The first Boeing X-51 WaveRider undergoing final preparations for flight, Edwards AFB, California, 2010. USAF

used previously on the Lockheed SR-71. The X-51A uses ethylene to start the combustion. Then the flight continues on JP-7. Following checkout trials beginning in late 2009, the X-51 made its first powered flight on May 26, 2010. After being air-launched from a B-52, it demonstrated successful hydrocarbon scramjet ignition and acceleration. Further tests will hopefully advance the era of practical scramjet-powered flight, likely beginning with long-range missiles. As this review indicates, the story of transatmospherics illustrates the complexity of hypersonics; the tenacity and dedication of NASA's aerodynamics, structures, and propulsion community; and the Agency's commitment to take on challenges, no matter how difficult, if the end promises to be the advancement of flight and humanity's ability to utilize the air and space medium.[130]

---

130. W. J. Hennigan, "Test Flight Shatters Records," *Los Angeles Times*, May 27, 2010; Matthew Shaer, "Scramjet-Powered X-51A WaveRider Missile Breaks Mach 6 Record," *Christian Science Monitor*, May 27, 2010; "WaveRider Sets Record for Hypersonic Flight," Associated Press Release, May 27, 2010.

Recommended Additional Readings

Note: The following list represents significant research by NASA and its predecessor, the NACA, in the field of transatmospheric flight. These references are readily available through the NASA Technical Reports Server. Its complete holdings include over half a million citations, of which some 90,000 show full text. Users can access it via *http://ntrs.nasa.gov/search.jsp*.

Contributions of Hypersonic Leaders:

H. Julian Allen, "Twenty-first Wright Brothers lecture: Hypersonic flight and the re-entry problem," NASA TM-108690, 1957.

H. Julian Allen and Stanford E. Neice, "Problems of performance and heating of hypersonic vehicles," NASA RM-A55L15, 1956.

H. Julian Allen and Murray Toback, "Dynamic stability of vehicles traversing ascending or descending paths through the atmosphere," NASA TN-4275, 1958.

John V. Becker and Peter F. Korycinski, "Heat transfer and pressure distribution at a Mach number of 6.8 on bodies with conical flares and extensive flow separation," NASA RM-L56F22, 1956.

Alfred J. Eggers, H. Julian Allen, and Stanford E. Neice, "A comparative analysis of the performance of long-range hypervelocity vehicles." NASA TR-1382, 1958.

Maxime A. Faget and H. Rudolph Dettwyler, "Initial flight investigations of a twin-engine supersonic ramjet," NASA RM-L50H10, 1950.

Maxime A. Faget and O.G. Smith, "Potential improvements to the Shuttle through evolution," AIAA Paper 93-2417, 1993.

Reentry:

R.A. Allen, J.C. Camm, and P.H. Rose, "Nonequilibrium and equilibrium radiation at super-satellite re-entry velocities," NASA CR-51743, 1962.

Kim S. Bey, Robert C. Scott, Robert E. Bartels, William A. Waters, and Roger Chen, "Analysis of the Shuttle Tile Overlay repair concept," NASA TM-2007-214857, 2007.

G. Bedjai, R.C. Brumfield, S. Demetriades, P.D. Lenn, and D.L. Ward, "Design and performance study on a 500 kilowatt linear crossed-field air-plasma accelerator for a re-entry facility," NASA CR-58543, 1964.

H.R. Bredfeldt and W.E. Scharfman, "Use of the Langmuir probe to determine the electron density and temperature surrounding re-entry vehicles," NASA CR-66275, 1966.

A.E. Bryson, Jr., and J.L. Speyer, "A neighboring optimum feedback control scheme based on estimated time-to-go with application to re-entry flight paths," NASA CR-87501, 1967.

D.L. Compton, B.J. Short, and S.C. Sommer, "Free-flight measurements of static and dynamic stability models of the Project Mercury re-entry capsule at Mach numbers 3 and 9.5," NASA TM-X-373, 1960.

R. Hermann, "Hypersonic aerodynamic problems at re-entry of space vehicles," NASA CR-69486, 1965.

E.J. Hopkins and A.D. Levin, "Reentry glide maneuvers for recovery of a winged first-stage rocket booster," NASA TN-D-1295, 1962.

P.O. Jarvinen, "On the use of magnetohydrodynamics during high speed re-entry," NASA CR-206, 1965.

E. Kaplan and F.D. Linzer, "Considerations in design of calorimeters for the Project Fire superorbital re-entry test vehicle," NASA CR-52196, 1963.

G.C. Kenyon, "The lateral and directional aerodynamic characteristics of a re-entry configuration based on a blunt 13 deg half-cone at Mach numbers to 0.90," NASA TM-X-583, 1961.

G.C. Kenyon and F.B. Sutton, "The longitudinal aerodynamic characteristics of a re-entry configuration based on a blunt 13 deg half-cone at Mach numbers to 0.92," NASA TM-X-571, 1961.

A.L. Laganelli, "Analysis of flight test transition and turbulent heating data. Part 2: Turbulent heating results," NASA CR-130251, 1972.

R. Lehnert and B. Rosenbaum, "Plasma effects on Apollo re-entry communication," NASA TN-D-2732, 1965.

A. Martellucci, B.L. Maguire, and R.S. Neff, "Analysis of flight test transition and turbulent heating data. Part l: Boundary layer transition results," NASA CR-129045, 1972.

R.B. Miles, D.A. Santavicca, and M. Zimmermann, "Evaluation of nonintrusive flow measurement techniques for a re-entry flight experiment," NASA CR-172142, 1983.

A.B. Miller, "Pilot re-entry guidance and control," NASA CR-331, 1965.

C.T. Swift, "Radiation from slotted-cylinder antennas in a re-entry plasma environment," NASA TN-D-2187, 1964.

F.J. Tischer, "A rough estimate of the 'blackout' time in re-entry communications," NASA TM-X-55059, 1962.

F.J. Tischer, "Attenuation in re-entry communications plans office," TM-X-51027, NASA 1963.

X-15:

E.H. Andrews, Jr., and R.C. Rogers, "Study of underexpanded exhaust jets of an X-15 airplane model and attached ramjet engine simulator at Mach 6.86," NASA TM-X-1571, 1968.

R.G. Bailey, "High total temperature sensing probe for the X-15 hypersonic aircraft," NASA CR-116772, 1968.

R.D. Banner, A.E. Kuhl, and R.D. Quinn, "Preliminary results of aerodynamic heating studies on the X-15 airplane," NASA TM-X-638, 1962.

C. Barret, "Review of our national heritage of launch vehicles using aerodynamic surfaces and current use of these by other nations," NASA TP-3615, 1996.

S.R. Bland and L.S. Young, "Transonic flutter investigation of models of the X-15 airplane horizontal tail," NASA TM-X-447, 1961.

R.K. Bogue and L.D. Webb, "Advanced air data sensing techniques," NASA TM-X-61115, 1968.

W.G. Cockayne, "Description of an energy management system for the X-15," NASA CR-96006, 1968.

W.R. Deazley, "A study of two proposed stabilization techniques for the X-15 horizontal surface control system," NASA CR-95955, 1961.

R.W. Dunning, "The control characteristics of two preliminary models of the X-15 research airplane at Mach numbers of 2.98 and 4.01," NASA TM-X-212, 1960.

D.E. Fetterman, Jr., and J.A. Penland, "Static longitudinal, directional, and lateral stability and control data at a Mach number of 6.83 on the final configuration of the X-15 research airplane," NASA TM-X-236, 1960.

G.M. Goranson, "Test of a North American X-15 research-vehicle model in the JPL 21-inch hypersonic wind tunnel," NASA CR-53710, 1963.

K.S. Green and T.W. Putnam, "Measurements of sonic booms generated by an airplane flying at Mach 3.5 and 4.8," NASA TM-X-3126, 1974.

E.C. Holleman. M.O. Thompson, and J. Weil, "An assessment of lifting reentry flight control requirements during abort, terminal glide, and approach and landing situations," NASA TM-X-59119, 1967.

E.J. Montoya and M. Palitz, "Wind-tunnel investigation of the flow field beneath the fuselage of the X-15 airplane at Mach numbers from 4 to 8," NASA TM-X-1469, 1967.

A.B. Price, "Full scale test report, X-15A-2 ablative thermal protection system," NASA CR-82004, 1968.

A.B. Price, "Thermal protection system X-15A-2. Design report," NASA CR-82003, 1968.

"Progress of the X-15 research airplane program," NASA SP-90, 1965.

W.J. Sefic, "Friction characteristic of steel skids equipped with skegs on a lakebed surface," NASA TM-81347, 1979.

M.O. Thompson, "General review of piloting problems encountered during simulation and flights of the X-15," NASA TM-X-56884, 1964.

L.S. Young, "Transonic flutter investigation of models of proposed horizontal tails for the X-15 airplane," NASA TM-X-442, 1961.

Hot Structures, X-20 Dyna-Soar:

Bianca Trujillo Anderson, Robert R. Meyer, Jr., and Harry R. Chiles, "Techniques used in the F-14 variable-sweep transition flight experiment," NASA TM-100444, 1988.

R.P. Bielat, "Transonic aerodynamic characteristics of the Dyna-Soar glider and Titan 3 launch vehicle configuration with various fin arrangements," NASA TM-X-809, 1963.

C.W. Boquist, "Graphite-metal composites," NASA CR-74308, 1965.

D.A. Buell and G.B. McCullough, "The wind-induced loads on a dynamically scaled model of the Dyna Soar glider and Titan 2 booster in launch position," NASA TM-X-659, 1962.

Victor A. Canacci and Jose C. Gonsalez, "Flow quality measurements in an aerodynamic model of NASA Lewis' Icing Research Tunnel," NASA CR-1999-202353, 1999.

A.J. Chellman, "Development of powder metallurgy 2XXX series Al alloy plate and sheet materials for high temperature aircraft structural applications," NASA CR-172521, 1985.

Kim K. deGroh, Donald A. Jaworske, and Daniela C. Smith, "Optical property enhancement and durability evaluation of heat receiver aperture shield materials," NASA TM-1998-206623, 1998.

H.L. Giles and J.W. Thomas, "Analysis of hypersonic pressure and heat transfer tests on a flat plate with a flap and a delta wing with body, elevons, fins, and rudders," NASA CR-536, 1966.

A.M. Hall, A.F. Hoenie, and C.J. Slunder, "Thermal and mechanical treatment for precipitation-hardening stainless steels," NASA SP-5089, 1967.

A.N. Hammer, G.C. Aigret, T.P. Psichogios, and C. Rodgers, "Fabrication of cooled radial turbine rotor," NASA CR-179503, 1986.

A.K. Hepler and A.R. Swegle, "Design and fabrication of brazed Rene 41 honeycomb sandwich structural panels for advanced space transportation systems," NASA CR-165801, 1981.

L.S. Jernell and C.D. Babb, "Effect of booster fins on static stability characteristics of a 0.02 scale model of the Titan 3 launch vehicle with the Dyna-Soar glider and a bulbous nose at Mach numbers from 1.60 to 3.50," NASA TM-X-885, 1963.

M.H. Leipold and C.M. Kapadia, "The role of anions in mechanical failure," NASA CR-121937, 1971.

D.L. McDaniels and R.A. Signorelli, "Evaluation of low-cost aluminum composites for aircraft engine structural applications," NASA TM-83357, 1983.

M.O. McKinney and J. Scheiman, "Evaluation of turbulence reduction devices for the Langley 8-foot Transonic Pressure Tunnel," NASA TM-81792, 1981.

Shouichi Ochiai, Nanabu Ueno, and Samu Noguchi, "Martensitic transformation and microstructures in sintered NiAl alloys," NASA TT-20192, 1988.

J. Scheiman, "Considerations for the installation of honeycomb and screens to reduce wind-tunnel turbulence," NASA TM-81868, 1981.

R.E. Shanks and G.N. Ware, "Investigation of the flight characteristics of a 1/5-scale model of a Dyna-Soar glider configuration at low subsonic speeds," NASA TM-X-683, 1962.

A. Tsuge, "Grain boundary engineering of mechanical strength of silicon nitride (Si3N4)," NASA TM-77433, 1984.

S. Yajima, M. Omori, J. Hayashi, H. Kayano, and M. Hamano, "Process for the production of metal nitride sintered bodies and resultant silicon nitride and aluminum nitride sintered bodies," NASA TM-77253, 1983.

Shuttle Tiles:

P.J. Bobbitt, C.L.W. Edwards, and R.W. Barnwell, "Simulation of time-varying loads on arrays of Shuttle tiles in a large transonic tunnel," NASA TM-84529, 1982.

O.L. Flowers and D.A. Stewart, "Catalytic surface effects on contaminated Space Shuttle tile in a dissociated nitrogen stream," NASA TM-86770, 1985.

G.L. Giles, "Substructure procedure for including the flexibility in stress analysis of Shuttle thermal protection system," NASA TM-81864, 1980.

G.L. Giles and M. Wallas, "Computer program for nonlinear static stress analysis of Shuttle thermal protection system: User's manual," NASA TM-81856, 1981.

G.L. Giles and M. Vallas, "Use of an engineering data management system in the analysis of Space Shuttle orbiter tiles," NASA TM-83215, 1981.

L.R. Hunt, "Aerodynamic heating in large cavities in an array of RSI tiles," NASA TN-D-8400, 1977.

Gregory N. Katnick, "Debris/ice/TPS assessment and integrated photographic analysis on Shuttle Mission STS-89," NASA TM-1998-207684, 1998.

Anita Macdonald and Paul Friederich, "Millimeter wave dielectric measurements of Space Shuttle tiles," NASA CR-187473, 1990.

J. Marroquin and R.R. Burrows, "Results of wind tunnel test OA253 in the AEDC 16-T propulsion wind tunnel using a 0.035 scale Space Shuttle launch vehicle model 84-OTS and entry vehicle model 84-O," NASA CR-167369, 1982.

B.A. Marshall, "Space Shuttle AFRSI full-scale credibility test in the NASA Ames Research Center 11 x 11 foot wind tunnel using model 124-0 installed in the 96-0 test fixture," NASA CR-167651, 1982.

A.B. Mattson and C.J. Schwindt, "FTIR instrument to monitor vapors from Shuttle tile waterproofing materials," NASA CR-199959, 1995.

R. Miserentino, L.D. Pinson, and S.A. Leadbetter, "Some Space Shuttle tile/strain-isolator-pad sinusoidal vibration tests," NASA TM-81853, 1980.

Timothy R. Moes and Robert R. Meyer, Jr., "In-flight investigation of Shuttle tile pressure orifice installations," NASA TM-4219, 1990.

G.J. Neuner and C.B. Delano, "Development of an improved coating for polybenzimidazole foam," NASA CR-2697, 1976.

A.P. Shore and R. Garcia, "Effects of substate deformation and SIP thickness on tile/SIP interface stresses for Shuttle thermal protection," NASA TM-81855, 1980.

A.A. Stewart, M. Cuellar, and O. Flowers, "Performance of an ablator for Space Shuttle in-orbit repair in an arc-plasma airstream," NASA TP-2150, 1983.

S.S. Tompkins, W.D. Brewer, R.K. Clark, C.M. Pittman, and K.L. Brinkley, "An assessment of the readiness of ablative materials for preflight application to the Shuttle orbiter," NASA TM-81823, 1980.

W.L. Wells and J. Hudgins, "Experimental assessment of a computer program used in Space Shuttle orbiter entry heating analysis," NASA TM-84572, 1983.

A.J. Zuckerwar and D.R. Sprinkle, "Proposed dynamic phase difference method for the detection of tile debonding from the Space Shuttle orbiter," NASA TM-83140, 1981.

X-30 National Aerospace Plane, Scramjet:

A. Abtahi and P. Dean, "Heat flux sensor research and development. The cool film calorimeter," NASA CR-189789, 1990.

E.H. Andrews, Jr., et al., "Langley Mach 4 Scramjet Test Facility," NASA TM-86277, 1985

Steven L. Baughcum and Stephen C. Henderson, "Aircraft emission scenarios projected in the year 2015 for the NASA Technology Concept Aircraft (TCA) High Speed Civil Transport," NASA CR-1998-207635, 1998.

Edwin C. Cady, "Slush hydrogen technology program," NASA CR-195353, 1994.

J.A. Cerro, et al., "A study of facilities and fixtures for testing of a high speed civil transport wing component," NASA CR-198352, 1996.

R. Edelman, "Diffusion controlled combustion for scramjet applications," NASA CR-66363, 1965.

J.C. Evvard, "The scramjet," NASA TM-X-56755, 1965.

Roger A. Fields, W. Lance Richards, and Michael V. DeAngelis, "Combined loads test fixture for thermal-structural testing aerospace vehicle panel concepts," NASA TM-2005-212039, 2004.

R.W. Guy, et al., "Operating Characteristics of the Langley Mach 7 Scramjet Test Facility," NASA TM-81929, 1981.

Terry L. Hardy, "FLUSH: a tool for the design of slush hydrogen flow systems," NASA TM-102467, 1990.

Scott D. Holland, "Computational parametric study of sidewall-compression scramjet inlet performance at Mach 10," NASA TM-4411, 1993.

Marvin Kussoy, George Huang, and Florian Menter, ""Hypersonic flows as related to the National Aerospace Plane," NASA CR-199365, 1995.

Unmeel B. Mehta, "The Aerospace Plane design challenge: Credible computational fluid dynamic results," NASA TM-102887, 1990.

"National Aero-Space Plane," NASA TM-109450, 1990.

"National Aerospace Plane thermal development. (Latest citations from the Aerospace Database)," NASA TM-97-113072, 1997.

Surya N. Patniak, James D. Guptill, Dale A. Hopkins, and Thomas M. Lavelle, "Neural network and regression approximations in high speed civil transport aircraft design optimization," NASA TM-1998-206316, 1998.

Terrill W. Putnam and Theodore G. Ayers, "Flight research and testing," NASA TM-100439, 1988.

Rodney H. Ricketts, Thomas E. Noll, and Lawrence J. Huttsell, "An overview of aeroelasticity studies for the National Aerospace Plane," NASA TM-107728, 1993.

William C. Rose, "Numerical investigations in three-dimensional internal flows," NASA CR-183108, 1988.

"Shuttle to Space Station. Heart assist implant. Hubble update. X-30 mock-up," NASA TM-110837, 1992.

J.C. Tannehill and G. Wadawadigi, "Development of a 3-D upwind PNS code for chemically reacting hypersonic flowfields," NASA CR-190182, 1992.

S.R. Thomas and R.W. Guy, "Expanded Operational Capabilities of the Langley Mach 7 Scramjet Test Facility," NASA TP-2186, 1983.

M.E. Tuttle and J. Rogacki, "Thermoviscoplastic response of Ti-15-3 under various loading conditions," NASA CR-187621, 1991.

D.W. Witte, et al., "1998 Calibration of the Mach 4.7 and Mach 6 Arc-Heated Scramjet Test Facility Nozzles," NASA TM-2004-213250, 2004.

X-43 Hyper-X:

"2000–2001 Research Engineering Annual Report," NASA TM-2004-212025, 2004.

Ethan Baumann, Catherine Bahm, Brian Strovers, Roger Beck, and Michael Richard, "The X-43A six degree of freedom Monte Carlo analysis," NASA TM-2007-214630, 2007.

Scott A. Berry, Michael DiFulvio, and Matthew K. Kowalkowski, "Forced Boundary-Layer Transition on X-43 (Hyper-X) in NASA LaRC 20-Inch Mach 6 Air Tunnel," NASA TM-2000-210316.

Scott A. Berry, Michael DiFulvio, and Matthew K. Kowalkowski, "Forced Boundary-Layer Transition on X-43 (Hyper-X) in NASA LaRC 31-Inch Mach 6 Air Tunnel," NASA TM-2000-210315.

Edward H. Glaessgen, et al., "X-43A rudder spindle fatigue life estimate and testing," NASA TM-2005-213525, 2005.

William L. Ko and Leslie Gong, "Thermoelastic analysis of Hyper-X camera windows suddenly exposed to Mach 7 stagnation aerothermal shock," NASA TP-2000-209030, 2000.

J.A. Lee and P.S. Chen, "Aluminum-scandium alloys: material characterization, friction stir welding, and compatibility with hydrogen peroxide," NASA TM-2004-213604, 2004.

Jessica Lux-Baumann, Ray Dees, and David Fratello, "Control room training for the Hyper-X project utilizing aircraft simulation," NASA TM-2006-213685, 2006.

Chan-gi Pak, "Aeroservoelastic stability analysis of the X-43A stack," NASA TM-2008-214365, 2008.

Matthew Redif, Yohan Lin, Courtney Amos Besssent, and Carole Barklow, "The Hyper-X flight systems validation program," NASA TM-2007-214620, 2007.

Karla S. Shy, Jacob J. Hageman, and Jeanette H. Le, "The aircraft simulation role in improving safety through control room training," NASA TM-2002-210731, 2002.

THIS PAGE INTENTIONALLY BLANK

The Northrop HL-10 turning onto final approach for a lakebed landing. NASA.

CASE 6

Physical Problems, Challenges, and Pragmatic Solutions

Robert G. Hoey

The advent of the supersonic and hypersonic era introduced a wide range of operational challenges that required creative insight by the flight research community. Among these were phenomena such as inertial (roll) coupling, transonic pitch-up, panel flutter, structural resonances, pilot-induced oscillations, and aerothermodynamic heating. Researchers had to incorporate a variety of solutions and refine simulation techniques to better predict the realities of flight. The efforts of the NACA and NASA, in partnership with other organizations, including the military, enabled development and refinement of reliable aerospace vehicle systems.

THE HISTORY OF AVIATION is replete with challenges and difficulties overcome by creative scientists and engineers whose insight, coupled with often-pragmatic solutions, broke through what had appeared to be barriers to future flight. At the dawn of aviation, the problems were largely evident to all: for example, simply developing a winged vehicle that could take off, sustain itself in the air, fly in a controlled fashion, and then land. As aviation progressed, the problems and challenges became more subtle but no less demanding. The National Advisory Committee on Aeronautics (NACA) had been created in 1915 to pursue the "scientific study of the problems of flight, with a view to their practical solution," and that spirit carried over into the aeronautics programs of the National Aeronautics and Space Administration (NASA), which succeeded the NACA on October 1, 1958, not quite a year after Sputnik had electrified the world. The role of the NACA, and later NASA, is mentioned often in the following discussion. Both have been instrumental in the discovery and solution to many of these problems.

As aircraft flight speeds moved from the firmly subsonic through the transonic and into the supersonic and even hypersonic regimes, the continuing challenge of addressing unexpected interactions and problems followed right along. Since an airplane is an integrated system, many of these problems crossed multiple discipline areas and affected multiple aspects of an aircraft's performance, or flight safety. Numerous examples could be selected, but the author has chosen to examine a representative sampling from several areas: experience with flight control systems and their design, structures, and their aeroelastic manifestations; flight simulation; flight dynamics (the motions and experience of the airplane in flight); and aerothermodynamics, the demanding environment of aerodynamic heating that affects a vehicle and its structure at higher velocities.

Flight Control Systems and Their Design

During the Second World War, there were multiple documented incidents and several fatalities that occurred when fighter pilots dove their propeller-driven airplanes at speeds approaching the speed of sound. Pilots reported increasing levels of buffet and loss of control at these speeds. Wind tunnels at that time were incapable of producing reliable meaningful data in the transonic speed range because the local shock waves were reflected off the wind tunnel walls, thus invalidating the data measurements. The NACA and the Department of Defense (DOD) created a new research airplane program to obtain a better understanding of transonic phenomena through flight-testing. The first of the resulting aircraft was the Bell XS-1 (later X-1) rocket-powered research airplane.

On NACA advice, Bell had designed the X-1 with a horizontal tail configuration consisting of an adjustable horizontal stabilizer with a hinged elevator at the rear for pitch control, at a time when a fixed horizontal tail and hinged elevator constituted the standard pitch control configuration for that period.[1] The X-1 incorporated this as an emergency means to increase its longitudinal (pitch) control authority at transonic speeds. It proved a wise precaution because, during the early buildup flights, the X-1 encountered similar buffet and loss of control as reported by earlier fighters. Analysis showed that local shock waves were forming on the tail surface, eventually migrating to the elevator hinge line. When they reached the hinge line, the effectiveness of the elevator was significantly reduced, thus causing the loss of control. The X-1 NACA–U.S. Air Force (USAF) test team determined to go ahead, thanks to the

---

1. R.M. Stanley and R.J. Sandstrom, "Development of the XS-1 Airplane," HQ Air Materiel Command, *Air Force Supersonic Research Airplane XS-1, Report No. 1* (Wright Field: Air Materiel Command, Jan. 9, 1948), p. 7.

X-1 having an adjustable horizontal tail. They subsequently validated that the airplane could be controlled in the transonic region by moving the horizontal stabilizer and the elevator together as a single unit. This discovery allowed Capt. Charles E. Yeager to exceed the speed of sound in controlled flight with the X-1 on October 14, 1947.[2]

An extensive program of transonic testing was undertaken at the NACA High-Speed Flight Station (HSFS; subsequently the Dryden Flight Research Center) on evaluating aircraft handling qualities using the conventional elevator and then the elevator with adjustable stabilizer.[3] As a result, subsequent transonic airplanes were all designed to use a one-piece, all-flying horizontal stabilizer, which solved the control problem and was incorporated on the prototypes of the first supersonic American jet fighters, the North American YF-100A, and the Vought XF8U-1 Crusader, flown in 1953 and 1954. Today, the all-moving tail is a standard design element of virtually all high-speed aircraft developed around the globe.[4]

Resolving the Challenge of Aerodynamic Damping
Researchers in the early supersonic era also faced the challenges posed by the lack of aerodynamic damping. Aerodynamic damping is the natural resistance of an airplane to rotational movement about its center of gravity while flying in the atmosphere. In its simplest form, it consists of forces created on aerodynamic surfaces that are some distance from the center of gravity (cg). For example, when an airplane rotates about the cg in the pitch axis, the horizontal tail, being some distance aft of the cg, will translate up or down. This translational motion produces a vertical lift force on the tail surface and a moment (force times distance) that tends to resist the rotational motion. This lift force opposes the rotation regardless of the direction of the motion. The resisting force will be proportional to the rate of rotation, or pitch rate. The faster the rotational rate, the larger will be the resisting force. The magnitude of

---

2. Walter C. Williams, "Instrumentation, Airspeed Calibration, Tests, Results and Conclusions," HQ AMC, *Air Force Supersonic Research Airplane XS-1*, p. 24.
3. W.C. Williams and A.S. Crossfield, "Handling Qualities of High-Speed Airplanes," RM L52A08 (Jan. 28, 1952).
4. It should be noted, of course, that the all-moving tail was essentially a "rediscovery" of earlier design practice. All-moving tails, for very different reasons, had been a feature of early airplanes, typified by the Wright Flyer and numerous European examples such as the Blériot and the Fokker Eindecker.

the resisting tail lift force is dependent on the change in angle of attack created by the rotation. This change in angle of attack is the vector sum of the rotational velocity and the forward velocity of the airplane. For low forward velocities, the angle of attack change is quite large and the natural damping is also large. The high aerodynamic damping associated with the low speeds of the Wright brothers flights contributed a great deal to the brothers' ability to control the static longitudinal instability of their early vehicles.

At very high forward speed, the same pitch rate will produce a much smaller change in angle of attack and thus lower damping. For practical purposes, all aerodynamic damping can be considered to be inversely proportional to true velocity. The significance of this is that an airplane's natural resistance to oscillatory motion, in all axes, disappears as the true speed increases. At hypersonic speeds (above Mach 5), any rotational disturbance will create an oscillation that will essentially not damp out by itself.

As airplanes flew ever faster, this lightly damped, oscillatory tendency became more obvious and was a hindrance to accurate weapons delivery for military aircraft, and pilot and passenger comfort for commercial aircraft. Evaluating the seriousness of the damping challenge in an era when aircraft design was changing markedly (from the straight-wing propeller-driven airplane to the swept and delta wing jet and beyond). It occupied a great amount of attention from the NACA and early NASA researchers, who recognized that it would pose a continuing hindrance to the exploitation of the transonic and supersonic region, and the hypersonic beyond.[5]

In general, aerodynamic damping has a positive influence on handling qualities, because it tends to suppress the oscillatory tendencies of a naturally stable airplane. Unfortunately, it gradually disappears as the speed increases, indicating the need for some artificial method of suppressing these oscillations during high-speed flight. In the preelectronic flight control era, the solution was the modification of flight control systems to incorporate electronic damper systems, often referred to as Stability Augmentation Systems (SAS). A damper system for one axis consisted of a rate gyro measuring rotational rate in that axis, a gain-changing circuit that adjusted the size of the needed control command,

---

5. For cases, see Edwin J. Saltzman and Theodore G. Ayers, *Selected Examples of NACA/NASA Supersonic Flight Research*, SP-513 (Edwards, CA: NASA Dryden Flight Research Center, 1995).

and a servo mechanism that added additional control surface commands to the commands from the pilot's stick. Control surface commands were generated that were proportional to the measured rotational rate (feedback) but opposite in sign, thus driving the rotational rate toward zero.

Damper systems were installed in at least one axis of all of the Century-series fighters (F-100 through F-107), and all were successful in stabilizing the aircraft in high-speed flight.[6] Development of stability augmentation systems—and their refinement through contractor, Air Force–Navy, and NACA–NASA testing—was crucial to meeting the challenge of developing Cold War airpower forces, made yet more demanding because the United States and the larger NATO alliance chose a conscious strategy of using advanced technology to generate high-leverage aircraft systems that could offset larger numbers of less-individually capable Soviet-bloc designs.[7]

Early, simple damper systems were so-called single-string systems and were designed to be "fail-safe." A single gyro, servo, and wiring system were installed for each axis. The feedback gains were quite low, tailored to the damping requirements at high speed, at which very little control surface travel was necessary. The servo travel was limited to a very small value, usually less than 2 degrees of control surface movement. A failure in the system could drive the servo to its maximum travel, but the transient motion was small and easily compensated by the pilot. Loss of a damper at high speed thus reduced the comfort level, or weapons delivery accuracy, but was tolerable, and, at lower speeds associated with takeoff and landing, the natural aerodynamic damping was adequate.

One of the first airplanes to utilize electronic redundancy in the design of its flight control system was the X-15 rocket-powered research airplane, which, at the time of its design, faced numerous unknowns. Because of the extreme flight conditions (Mach 6 and 250,000-foot altitude), the servo travel needed for damping was quite large, and the pilot could not compensate if the servo received a hard-over signal.

---

6. Robert. G. Hoey and Capt. Iven C. Kincheloe, "F-104A Stability and Control," AFFTC TR-56-14, April 1958.
7. See Joseph R. Chambers, *Partners in Freedom: Contributions of the Langley Research Center to U.S. Military Aircraft of the 1990s*, SP-2000-4519 (Washington, DC: NASA, 2000), passim; Robert K. Geiger, et al., *The AGARD History, 1952–1987* (Neuilly sur Seine: NATO Advisory Group for Aeronautical Research and Development, 1988 ed.), pp. ix–xxv; and Thomas C. Lassman, *Sources of Weapon Systems Innovation in the Department of Defense: The Role of In-House Research and Development, 1945–2000* (Washington, DC: Center for Military History, 2008), pp. 93–97.

The solution was the incorporation of an independent, but identical, feedback "monitoring" channel in addition to the "working" channel in each axis. The servo commands from the monitor and working channel were continuously compared, and when a disagreement was detected, the system was automatically disengaged and the servo centered. This provided the equivalent level of protection to the limited-authority fail-safe damper systems incorporated in the Century series fighters. Two of the three X-15s retained this fail-safe damper system throughout the 9-year NASA–Air Force–Navy test program, although a backup roll rate gyro was added to provide fail-operational, fail-safe capability in the roll axis.[8] Refining the X-15's SAS system necessitated a great amount of analysis and simulator work before the pilots deemed it acceptable, particularly as the X-15's stability deteriorated markedly at higher angles of attack above Mach 2. Indeed, one of the major aspects of the X-15's research program was refining understanding of the complexities of hypersonic stability and control, particularly during reentry at high angles of attack.[9]

The electronic revolution dramatically reshaped design approaches to damping and stability. Once it was recognized that electronic assistance was beneficial to a pilot's ability to control an airplane, the concept evolved rapidly. By adding a third independent channel, and some electronic voting logic, a failed channel could be identified and its signal "voted out," while retaining the remaining two channels active. If a second failure occurred (that is, the two remaining channels did not agree), the system would be disconnected and the damper would become inoperable. Damper systems of this type were referred to as fail-operational, fail-safe (FOFS) systems. Further enhancement was provided by comparing the pilot's stick commands with the measured airplane response and using analog computer circuits to tailor servo commands so that the airplane response was nearly the same for all flight conditions. These systems were referred to as Command Augmentation Systems (CAS). The next step in the evolution was the incorporation of a mathematical model of the desired aircraft response into the analog computer circuitry. An error signal was generated by comparing the instantaneous

---

8. Personal Experience as an Air Force Flight Planner during the X-15 envelope expansion flight-testing.
9. Robert A. Tremant, "Operational Experience and Characteristics of the X-15 Flight Control System," NASA Technical Note D-1402 (Dec. 1962), and Wendell H. Stillwell, *X-15 Research Results*, SP-60 (Washington, DC: NASA, 1965), pp. 51–52.

measured aircraft response with the desired mathematical-model response, and the servo commands forced the airplane to fly per the mathematical model, regardless of the airplane's inherent aerodynamic tendencies. These systems were called "model-following."

Even higher levels of redundancy were necessary for safe operation of these advanced control concepts after multiple failures, and the failure logic became increasingly more complex. Establishing the proper "trip" levels, where an erroneous comparison would result in the exclusion of one channel, was an especially challenging task. If the trip levels were too tight, a small difference between the outputs of two perfectly good gyros would result in nuisance trips, while trip levels that were too loose could result in a failed gyro not being recognized in a timely manner. Trip levels were usually adjusted during flight test to provide the safest settings.

NASA's Space Shuttle orbiter utilized five independent control system computers. Four had identical software. This provided fail-operational, fail-operational, fail-safe (FOFOFS) capability. The fifth computer used a different software program with a "get-me-home" capability as a last resort (often referred to as the "freeze-dried" control system computer).

The Challenge of Limit-Cycles

The success of the new electronic control system concepts was based on the use of electrical signals from sensors (primarily rate gyros and accelerometers) that could be fed into the flight control system to control aircraft motion. As these electronic elements began to play a larger role, a different dynamic phenomenon came into play. "Limit-cycles" are a common characteristic of nearly all mechanical-electrical closed-loop systems and are related to the total gain of the feedback loop. For an aircraft flight control system, total loop gain is the product of two variables: (1) the magnitude of the aerodynamic effectiveness of the control surface for creating rotational motion (aerodynamic gain) and (2) the magnitude of the artificially created control surface command to the control surface (electrical gain). When the aerodynamic gain is low, such as at very low airspeeds, the electrical gain will be correspondingly high to command large surface deflections and rapid aircraft response. Conversely, when the aerodynamic gain is high, such as at high airspeed, low electrical gains and small surface deflections are needed for rapid airplane response.

These systems all have small dead bands, lags, and rate limits (nonlinearities) inherent in their final, real-world construction. When the total feedback gain is increased, the closed-loop system will eventually exhibit a small oscillation (limit-cycle) within this nonlinear region. The resultant total loop gain, which causes a continuous, undamped limit-cycle to begin, represents the practical upper limit for the system gain since a further increase in gain will cause the system to become unstable and diverge rapidly, a condition which could result in structural failure of the system. Typically the limit-cycle frequency for an aircraft control system is between two and four cycles per second.

Notice that the limit-cycle characteristics, or boundaries, are dependent upon an accurate knowledge of control surface effectiveness. Ground tests for limit-cycle boundaries were first devised by NASA Dryden Flight Research Center (DFRC) for the X-15 program and were accomplished by using a portable analog computer, positioned next to the airplane, to generate the predicted aerodynamic control effectiveness portion of the feedback path.[10] The control system rate gyro on the airplane was bypassed, and the analog computer was used to generate the predicted aircraft response that would have been generated had the airplane been actually flying. This equivalent rate gyro output was then inserted into the control system. The total loop gain was then gradually increased at the analog computer until a sustained limit-cycle was observed at the control surface. Small stick raps were used to introduce a disturbance in the closed-loop system in order to observe the damping characteristics. Once the limit-cycle total loop gain boundaries were determined, the predicted aerodynamic gains for various flight conditions were used to establish electrical gain limits over the flight envelope. These ground tests became routine at NASA Dryden and at the Air Force Flight Test Center (AFFTC) for all new aircraft.[11] For subsequent production aircraft, the resulting gain schedules were programmed within the flight control system computer. Real-time, direct measurements of airspeed, altitude, Mach number, and angle of attack were used to access and adjust the electrical gain schedules while in flight to provide the highest safe feedback gain while avoiding limit-cycle boundaries.

Although the limit-cycle ground tests described above had been performed, the NASA–Northrop HL-10 lifting body encountered limit-cycle

---

10. L.W. Taylor, Jr., and J.W. Smith, "An Analysis of the Limit-Cycle and Structural-Resonance Characteristics of the X-15 Stability Augmentation System," NASA TN-D-4287 (Dec. 1967).
11. Weneth D. Painter and George J. Sitterle, "Ground and Flight Test Methods for Determining Limit Cycle and Structural Resonance Characteristics of Aircraft Stability Augmentation Systems," NASA TN-D-6867 (1972).

oscillations on its maiden flight. After launch from the NB-52, the telemetry data showed a large limit-cycle oscillation of the elevons. The oscillations were large enough that the pilot could feel the aircraft motion in the cockpit. NASA pilot Bruce Peterson manually lowered the pitch gain, which reduced the severity of the limit-cycle. Additional aerodynamic problems were present during the short flight requiring that the final landing approach be performed at a higher-than-normal airspeed. This caused the limit-cycle oscillations to begin again, and the pitch gain was reduced even further by Peterson, who then capped his already impressive performance by landing the craft safely at well over 300 mph. NASA engineer Weneth Painter insisted the flight be thoroughly analyzed before the test team made another flight attempt, and subsequent analysis by Robert Kempel and a team of engineers concluded that the wind tunnel predictions of elevon control effectiveness were considerably lower than the effectiveness experienced in flight.[12] This resulted in a higher aerodynamic gain than expected in the total loop feedback path and required a reassessment of the maximum electrical gain that could be tolerated.[13]

Flight Control Systems and Pilot-Induced Oscillations

Pilot-induced oscillations (PIO) occur when the pilot commands the control surfaces to move at a frequency and/or magnitude beyond the capability of the surface actuators. When a hydraulic actuator is commanded to move beyond its design rate limit, it will lag behind the commanded deflection. If the command is oscillatory in nature, then the resulting surface movement will be smaller, and at a lower rate, than commanded. The pilot senses a lack of responsiveness and commands even larger surface deflections. This is the same instability that can be generated by a high-gain limit-cycle, except that the feedback path is through the pilot's stick, rather than through a sensor and an electronic servo. The instability will continue until the pilot reduces his gain (ceases to command large rapid surface movement), thus allowing the actuator to return to its normal operating range.

---

12. R.W. Kempel and J.A. Manke, "Flight Evaluation of HL-10 Lifting Body Handling Qualities at Mach Numbers from 0.30 to 1.86", NASA TN-D-7537, (Jan. 1974).
13. Milton O. Thompson with J.D. Hunley, *Flight Research: Problems Encountered and What They Should Teach Us*, SP-2000-4522 (Washington, DC: NASA, 2000), pp. 19–20; see also R. Dale Reed with Darlene Lister, *Wingless Flight: The Lifting Body Story*, SP-4220 (Washington, DC: NASA, 1997), pp. 96–102.

The prototype General Dynamics YF-16 Lightweight Fighter (LWF) unexpectedly encountered a serious PIO problem on a high-speed taxi test in 1974. The airplane began to oscillate in roll near the end of the test. The pilot, Philip Oestricher, applied large, corrective stick inputs, which saturated the control actuators and produced a pilot-induced oscillation. When the airplane began heading toward the side of the runway, the pilot elected to add power and fly the airplane rather than veer into the dirt along side of the runway. Shortly after the airplane became airborne, his large stick inputs ceased, and the PIO and limit-cycle stopped. Oestricher then flew a normal pattern and landed the airplane safely. Several days later, after suitable modifications to its flight control system, it completed its "official" first flight.

The cause of this problem was primarily related to the "force stick" used in the prototype YF-16. The control stick was rigidly attached to the airplane, and strain gages on the stick measured the force being applied by the pilot. This electrical signal was transmitted to the flight control system as the pilot's command. There was no motion of the stick, thus no feedback to the pilot of how much control deflection he was commanding. During the taxi test, the pilot was unaware that he was commanding full deflection in roll, thus saturating the actuators. The solution was a reduction in the gain of the pilot's command signal, as well as a geometry change to the stick that allowed a small amount of stick movement. This gave the pilot some tactile feedback as to the amount of control deflection being commanded, and a hard stop when the stick was commanding full deflection.[14] The incident offered lessons in both control system design and in human factors engineering, particularly on the importance of ensuring that pilots receive indications of the magnitude of their control inputs via movable sticks. Subsequent fly-by-wire (FBW) aircraft have incorporated this feature, as opposed to the "fixed" stick concept tried on the YF-16. As for the YF-16, it won the Lightweight Fighter design competition, was placed in service in more developed form as the F-16 Fighting Falcon, and subsequently became a widely produced Western jet fighter.

Another PIO occurred during the first runway landing of the NASA–Rockwell Space Shuttle orbiter during its approach and landing tests in 1978. After the flare, and just before touchdown, astronaut pilot Fred Haise commanded a fairly large pitch control input that saturated the

---

14. Personal recollections from serving as a member of the YF-16 Taxi Test Incident review team.

The General Dynamics YF-16 prototype Lightweight Fighter (LWF) in flight over the Edwards range. USAF.

control actuators. At touchdown, the orbiter bounced slightly and the rate-limiting saturation transferred to the roll axis. In an effort to keep the wings level, the pilot made additional roll inputs that created a momentary pilot-induced oscillation that continued until the final touchdown. At one point, it seemed the orbiter might veer toward spectators, one of whom was Britain's Prince Charles, then on a VIP tour of the United States. (Ironically, days earlier, the Prince of Wales had "flown" the Shuttle simulator at the NASA Johnson Space Center, encountering the same kind of lateral PIO that Haise did on touchdown.) Again, the cause was related to the high sensitivity of the stick in comparison with the Shuttle's slow-moving elevon actuators. The incident sparked a long and detailed study of the orbiter's control system in simulators and on the actual vehicle. Several changes were made to the control system, including a reduced sensitivity of the stick and an increase in the maximum actuator rates.[15]

---

15. Robert G. Hoey, et al., "Flight Test Results from the Entry and Landing of the Space Shuttle Orbiter for the First Twelve Orbital Flights," AFFTC TR-85-11 (1985), p. 104. Robert G. Hoey, et al., "AFFTC Evaluation of the Space Shuttle Orbiter and Carrier Aircraft—NASA Approach and Landing Test," AFFTC TR-78-14, May 1978, pp. 104, 114, 117. See also Richard P. Hallion, *On the Frontier: Flight Research at Dryden, 1946–1981*, SP-4303 (Washington, DC: NASA, 1984), pp. 249–250.

The above discussion of electronic control system evolution has sequentially addressed the increasing complexity of the systems. This was not necessarily the actual chronological sequence. The North American F-107, an experimental nuclear strike fighter derived from the earlier F-100 Super Sabre, utilized one of the first fly-by-wire control systems—Augmented Longitudinal Control System (ALCS)—in 1956. One of the three prototypes was used by NASA, thus providing the Agency with its first exposure to fly-by-wire technology. Difficult maintenance of the one-of-kind subsystems in the F-107 forced NASA to abandon its use as a research airplane after about 1 year of flying.

Self-Adaptive Flight Control Systems

One of the more sophisticated electronic control system concepts was funded by the AF Flight Dynamics Lab and created by Minneapolis Honeywell in the late 1950s for use in the Air Force-NASA-Boeing X-20 Dyna-Soar reentry glider. The extreme environment associated with a reentry from space (across a large range of dynamic pressures and Mach numbers) caused engineers to seek a better way of adjusting the feedback gains than stored programs and direct measurements of the atmospheric variables. The concept was based on increasing the electrical gain until a small limit-cycle was measured at the control surface, then alternately lowering and raising the electrical gain to maintain a small continuous, but controlled, limit-cycle throughout the flight. This allowed the total loop gains to remain at their highest safe value but avoided the need to accurately predict (or measure) the aerodynamic gains (control surface effectiveness).

This system, the MH-96 Adaptive Flight Control System (AFCS), was installed in a McDonnell F-101 Voodoo testbed and flown successfully by Minneapolis Honeywell in 1959–1960. It proved to be fairly robust in flight, and further system development occurred after the cancellation of the X-20 Dyna-Soar program in 1963. After a ground-test explosion during an engine run with the third X-15 in June 1960, NASA and the Air Force decided to install the MH-96 in the hypersonic research aircraft when it was rebuilt. The system was expanded to include several autopilot features, as well as a blending of the aerodynamic and reaction controls for the entry environment. The system was triply redundant, thus providing fail-operational, fail-safe capability. This was an improvement over the other two X-15s, which had only fail-safe features. Because of the added features of the MH-96, and the additional redundancy it provided, NASA and the Air Force used the third X-15 for all planned high-altitude flights (above 250,000 feet) after an initial envelope expansion program to validate the aircraft's basic performance.[16]

Unfortunately, on November 15, 1967, the third X-15 crashed, killing its pilot, Major Michael J. Adams. The loss of X-15 No. 3 was related to the MH-96 Adaptive Flight Control System design, along with several other factors. The aircraft began a drift off its heading and then entered a spin at high altitude (where dynamic pressure—"q" in engineering shorthand—is very low). The flight control system gain was at its maximum when the spin started. The control surfaces were all deflected to their respective stops attempting to counter the spin, thus no limit-cycle motion—4 hertz (Hz) for this airplane—was being detected by the gain changer. Thus, it remained at maximum gain, even though the dynamic pressure (and hence the structural loading) was increasing rapidly during entry. When the spin finally broke and the airplane returned to a normal angle of attack, the gain was well above normal, and the system commanded maximum pitch rate response from the all-moving elevon surface actuators. With the surface actuators operating at their maximum rate, there was still no 4-Hz limit-cycle being sensed by the gain changer, and the gain remained at the maximum value, driving the airplane into structural failure at approximately 60,000 feet and at a velocity of Mach 3.93.[17]

As the accident to the third X-15 indicated, the self-adaptive control system concept, although used successfully for several years, had some subtle yet profound difficulties that resulted in it being used in only one subsequent production aircraft, the General Dynamics F-111 multipurpose strike aircraft. One characteristic common to most of the model-following systems was a disturbing tendency to mask deteriorating handling qualities. The system was capable of providing good handling qualities to the pilot right up until the system became saturated, resulting in an instantaneous loss of control without the typical warning a pilot would receive from any of the traditional signs of impending loss of control, such as lightening of control forces and the beginning

---

16. Dennis R. Jenkins, *X-15: Extending the Frontiers of Flight*, SP-2007-562 (Washington, DC: NASA, 2007), p. 402.
17. Donald R. Bellman, et al., *Investigation of the Crash of the X-15-3 Aircraft on November 15, 1967* (Edwards: NASA Flight Research Center, Jan. 1968), pp. 8–15.

of control reversal.[18] A second serious drawback that affected the F-111 was the relative ease with which the self-adaptive system's gain changer could be "fooled," as with the accident to the third X-15. During early testing of the self-adaptive flight control system on the F-111, testers discovered that, while the plane was flying in very still air, the gain changer in the flight control system could drive the gain to quite high values before the limit-cycle was observed. Then a divergent limit-cycle would occur for several seconds while the gain changer stepped the gain back to the proper levels. The solution was to install a "thumper" in the system that periodically introduced a small bump in the control system to start an oscillation that the gain changer could recognize. These oscillations were small and not detectable by the pilot, and thus, by inducing a little "acceptable" perturbation, the danger of encountering an unexpected larger one was avoided.

For most current airplane applications, flight control systems use stored gain schedules as a function of measured flight conditions (altitude, airspeed, etc.). The air data measurement systems are already installed on the airplane for pilot displays and navigational purposes, so the additional complication of a self-adaptive feature is considered unnecessary. As the third X-15's accident indicated, even a well-designed adaptive flight control system can be fooled, resulting in tragic consequences.[19] The "lesson learned," of course (or, more properly, the "lesson relearned") is that the more complex the system, the harder it is to identify the potential hazards. It is a lesson that engineers and designers might profitably take to heart, no matter what their specialty.

Induced Structural Resonances

Overall, electronic enhancements introduced significant challenges with respect to their practical incorporation in an airplane. Model-following systems required highly responsive servos and high gain levels for the feedback from the motion sensors (gyros and accelerometers) to the control surfaces. These high-feedback-gain requirements introduced serious issues regarding the aircraft structure. An aircraft structure is surprisingly vulnerable to induced frequencies, which, like a struck musical tuning fork, can result in resonant motions that may reach the naturally

---

18. For more on its strengths and weaknesses, see L.W. Taylor, Jr., and E.J. Adkins, "Adaptive Flight Control Systems—Pro and Con," NASA TM-X-56008 (1964).

19. Personal experience as an X-15 flight planner and X-20 stability and control flight test engineer.

Case 6 | Physical Problems, Challenges, and Pragmatic Solutions

The General Dynamics F-111A was the first production aircraft to use a self-adaptive flight control system. NASA.

destructive frequency of the structure, breaking it apart. Rapid movement of a control surface could trigger a lightly damped oscillation of one of the structural modes of the airplane (first mode tail bending, for example). This structural oscillation could be detected by the flight control system sensors, resulting in further rapid movement of the control surface. The resulting structural/control surface oscillation could thus be sustained, or even amplified. These additive vibrations were typically at higher frequencies (5–30 Hz) than the limit-cycle described earlier (2–4 Hz), although some of the landing gear modes and wing bending modes for larger aircraft are typically below 5 Hz. If seemingly esoteric, this phenomenon, called structural resonance, is profoundly serious.

Even the stiff and dense X-15 encountered serious structural resonance effects. Ground tests had uncovered a potential resonance between the pitch control system and a landing gear structural mode. Initially, researchers concluded that the effect was related to the ground-test equipment and its setup, and thus would not occur in flight. However, after several successful landings, the X-15 did experience a high-frequency vibration upon one touchdown. Additionally, a second and more severe structural resonance occurred at 13 Hz (coincident with the horizontal tail bending mode) during one entry from high altitude by the third

X-15 outfitted with the MH-96 adaptive flight control system.[20] The pilot would first note a rumbling vibration that swiftly became louder. As the structure resonated, the vibrations were transmitted to the gyros in the flight control system, which attempted to "correct" for them but actually fed them instead. They were so severe that the pilot could not read the cockpit instruments and had to disengage the pitch damper in order to stop them. As a consequence, a 13 Hz "notch" filter was installed in the electrical feedback path to reduce the gain at the observed structural frequency. Thereafter, the third X-15 flew far more predictably[21]

Structural resonance problems are further complicated by the fact that the predicted structural frequencies are often in error, thus the flight control designers cannot accurately anticipate the proper filters for the sensors. Further, structural resonance is related to a structural feedback path, not an aerodynamic one as described for limit-cycles. As a precaution, ground vibration tests (GVT) are usually conducted on a new airplane to accurately determine the actual structural mode frequencies of the airplane.[22] Researchers attach electrically driven and controlled actuators to various locations on the airplane and perform a small amplitude frequency "sweep" of the structure, essentially a "shake test." Accelerometers at strategic locations on the airplane detect and record the structural response. Though this results in a more accurate determination of the actual structural frequencies for the control system designer, it still does not identify the structural path to the control system sensors.

The flight control resonance characteristics can be duplicated on the ground by placing the airplane on a soft mounting structure, (airbags, or deflated tires and struts) then artificially raising the electrical gain in each flight control closed loop until a vibration is observed. Based on its experience with ground- and flight-testing of research airplanes, NASA DFRC established a ground rule that the flight gains could only be allowed to reach one-half of the gain that triggered a resonance (a gain margin of

---

20. G.B. Merrick and L.W. Taylor, Jr., "X-15 Stability Augmentation System," NASA Report H-271 (Jan. 1961); L.W. Taylor, Jr., and J.W. Smith, "An Analysis of the Limit Cycle and Structural Resonance Characteristics of the X-15 Stability Augmentation System," NASA TN-D-4287 (Dec. 1967).
21. John P. Smith, Lawrence J. Schilling, and Charles A. Wagner, "Simulation at Dryden Flight Research Facility from 1957 to 1982," NASA TM-101695 (1989), p. 4; Stillwell, *X-15 Research Results*, pp. 61–69.
22. Weneth D. Painter and George J. Sitterle, "Ground and Flight Test Methods for Determining Limit Cycle and Structural Resonance Characteristics of Aircraft Stability Augmentation Systems," NASA TN-D-6867 (June 1972).

2.0). This rule-of-thumb ground rule has been generally accepted within the aircraft industry, and ground tests to establish resonance gain margins are performed prior to first flights. If insufficient gain margin is present, the solution is sometimes a relocation of a sensor, or a stiffening of the sensor mounting structure. For most cases, the solution is the placement of an electronic notch filter within the control loop to reduce the system gain at the identified structural frequency. Many times the followup ground test identifies a second resonant frequency for a different structural mode that was masked during the first test. A typical notch filter will lower the gain at the selected notch frequency as desired but will also introduce additional lag at nearby frequencies. The additional lag will result in a lowering of the limit-cycle boundaries. The control system designer is thus faced with the task of reducing the gain at one structural frequency while minimizing any increase in the lag characteristics at the limit-cycle frequency (typically 2–4 Hz). This challenge resulted in the creation of lead-lag filters to minimize the additional lag in the system when notch filters were required to avoid structural resonance.[23]

Fighter aircraft usually are designed for 7–9 g load factors and, as a consequence, their structures are quite stiff, exhibiting high natural frequencies. Larger transport and reconnaissance airplanes are designed for much lower load factors, and the structures are more limber. Since structural frequencies are often only slightly above the natural aerodynamic frequencies—as well as the limit-cycle frequencies—of the airplane, this poses a challenge for the flight control system designer who is trying to aggressively control the aerodynamic characteristics, avoid limit-cycles, and avoid any control system response at the structural mode frequencies.

Structural mode interactions can occur across a range of flight activities. For example, Rockwell and Air Force testers detected a resonant vibration of the horizontal stabilizer during early taxi tests of the B-1 bomber. It was traced to a landing gear structural mode, and a notch filter was installed to reduce the flight control gain at that frequency. The ground test for resonance is fairly simple, but the structural modes that need to be tested can produce a fairly large matrix of test conditions. External stores and fuel loadings can alter the structural frequencies of the airplane and thus change the control system feedback

---

23. Ibid.

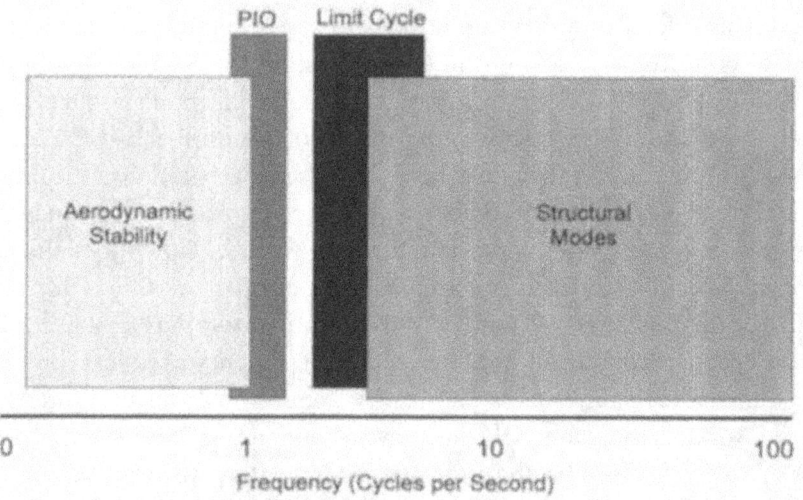

Aircraft frequency spectrum for flight control system design. USAF.

characteristics.[24] The frequency of the wing torsion mode of the General Dynamics YF-16 Lightweight Fighter (the prototype of the F-16A Fighting Falcon) was dramatically altered when AIM-9 Sidewinder air-to-air missiles were mounted at the wingtip.

The transformed dynamics of the installed missiles induced a serious aileron/wing-twist vibration at 6 Hz (coincident with the wing torsion mode), a motion that could also be classified as flutter, but in this case was obviously driven by the flight control system. Again, the solution was the installation of a notch filter to reduce the aileron response at 6 Hz.[25]

NASA researchers at the Dryden Flight Research Center had an unpleasant encounter with structural mode resonance during the Northrop–NASA HL-10 lifting body flight-test program. After an aborted launch attempt on the HL-10 lifting body, the NB-52B mother ship was returning with the HL-10 still mounted under the wing pylon. When the HL-10 pilot initiated propellant jettison, the launch airplane immediately experienced a violent vibration of the launch pylon attaching the lifting body to the NB-52B. The pilot stopped jettisoning and turned the flight control system off, whereupon the vibration stopped. The solution to this problem was strictly a change in operational procedure—in

---

24. Personal experience as a member of the B-1 Flight Readiness Review Team.
25. Personal experience as a member of the Light Weight Fighter Joint Test Force.

the future, the control system was to be disengaged before jettisoning while in captive flight.[26]

Fly-By-Wire: Fulfilling Promise and Navigating Around Nuance

As designers and flightcrews became more comfortable with electronic flight control systems and the systems became more reliable, the idea of removing the extra weight of the pilot's mechanical control system began to emerge. Pilots resisted the idea because electrical systems do fail, and the pilots (especially military pilots) wanted a "get-me-home" capability. One flight-test program received little attention but contributed a great deal to the acceptance of fly-by-wire technology. The Air Force initiated a program to demonstrate that a properly designed fly-by-wire control system could be more reliable and survivable than a mechanical system. The F-4 Survivable Flight Control System (SFCS) program was initiated in the early 1970s. Many of the then-current accepted practices for flight control installations were revised to improve survivability. Four independent analog computer systems provided fail-op, fail-op (FOFO) redundancy. A self-adaptive gain changer was also included in the control logic (similar to the MH-96 in the X-15). Redundant computers, gyros, and accelerometers were eventually mounted in separate locations in the airplane, as were power supplies. Flight control system wire bundles for redundant channels were separated and routed through different parts of the airplane. Individual surface actuators (one aileron for example) could be operated to continue to maintain control when the opposite control surface was inoperative. The result was a flight control system that was lighter yet more robust than a mechanical system (which could be disabled by a single failure of a pushrod or cable). After development flight-testing of the SFCS airplane was completed, the standard F-4 mechanical backup system was removed, and the airplane was flown in a completely fly-by-wire configuration.[27]

The first production fly-by-wire airplane was the YF-16. It used four redundant analog computers with FOFO capability. The airplane was not only the first production aircraft to use FBW control, it was also the first airplane intentionally designed to be unstable in the pitch axis while

---

26. Interview with John Manke, HL-10 test pilot.
27. Maj. Robert Ettinger, Capt. Robert Majoros, and Lt. Col. Cecil W. Powell, "Air Force Evaluation of the Fly-By-Wire Portion of the Surviveable Flight Control System Advanced Development Program," AFFTC TR-73-32 (Aug. 1973).

flying at subsonic speeds ("relaxed static stability"). The YF-16 prototype test program allowed the Air Force and General Dynamics to iron out the quirks of the FBW control system as well as the airplane aerodynamics before entering the full scale development of the F-16A/B. The high gains required for flying the unstable airplane resulted in some structural resonance and limit-cycle problems. The addition of external stores (tanks, bombs, and rockets) altered the structural mode frequencies and required fine-tuning of the control laws. Researchers and designers learned that flight control system design and aircraft interactions in the emergent FBW era were clearly far more complex and nuanced than control system design in the era of direct mechanical feedback and the augmented hydromechanical era that had followed.[28]

The Advent of Digital Flight Control Systems

Digital flight control systems were more nuanced still.[29] Analog computers calculate solutions simultaneously, thus producing an instantaneous output for any input. Digital computers, although more precise than analog, calculate solutions in sequence, thus introducing a time delay between the input and the output, often referred to as "transport delay." Early digital computers were far too slow to function in a real-time, flight control feedback system and could not compute the required servo commands fast enough to control the aircraft motions. As digital computation become faster and faster, control system designers gave serious attention to using them in aircraft flight control systems. NASA Dryden undertook the modification and flight-testing of a Vought F-8C Crusader Navy fighter to incorporate a digital fly-by-wire (DFBW) control system, based on the Apollo Guidance Computer used in the Apollo space capsule. The F-8 DFBW's first flight was in 1972, and the test program completed 248 DFBW flights before its retirement at the end of 1985.

It constituted a very bold and aggressive research program. The F-8 used redundant digital computers and was the first airplane relying solely on fly-by-wire technology for all of its flights. (Earlier FBW efforts, such as the AF F-4 Survivable Flight Control System, used a mechanical backup system for the first few flights.) NASA's F-8 DFBW program

---

28. Maj. James A. Eggers and Maj. William Bryant, Jr., "Flying Qualities Evaluation of the YF-16 Prototype Light Weight Fighter," AFFTC TR-75-15 (1975).

29. The early advent of digital fly-by-wire is the subject of another case study in this volume (Piccirillo) and so is not examined in great detail here.

Case 6 | Physical Problems, Challenges, and Pragmatic Solutions

not only set the stage for future military and civil digital flight control systems and fly-by-wire concepts, it also established the precedent for the operational procedures and built-in-test (BIT) requirements for this family of flight control systems.[30] The ground-testing and general operating methods that were established by NASA DFRC in order to ensure safety of their F-8 DFBW airplane are still being used by most modern military and civilian airplanes.

After the completion of the basic digital FBW demonstration program, the F8 DFBW airplane was used for additional research testing, such as identifying the maximum allowable transport delay for a digital system to avoid pilot-induced oscillations. This is a key measurement in determining whether digital computations are fast enough to be used successfully in a control system. (The number turned out to be quite small, on the order of only 100 to 120 milliseconds.) The stimulus for this research was the PIO experienced by Shuttle pilot-astronaut Fred Haise during the fifth and last of the approach and landing tests flown at Edwards by the Space Shuttle orbiter Enterprise on October 26, 1977. Afterward, the Shuttle test team asked the DFBW test team if they could run in-flight simulations of the Shuttle using the F-8 DFBW testbed, to determine the effect of transport delays upon control response. During this follow-on research-testing phase, NASA Dryden Flight Research Center pilot John Manke experienced a dramatic, and very scary, landing. As he touched down, he added power to execute a "touch and go" to fly another landing pattern. But instead of climbing smoothly away, the F-8 began a series of violent pitching motions that Manke could not control. He disengaged the test system (which then reverted to a digital FBW version of the basic F-8 control system) just seconds before hitting the ground. The airplane returned to normal control, and the pilot landed safely. The culprit was an old set of control laws resident in the computer memory that had never been tested or removed. A momentary high pitch rate during the short ground roll had caused the airplane to automatically switch to these old control laws, which were later

---

30. Dwain A. Deets and Kenneth J. Szalai, "Design and Flight Experience with a Digital Fly-By-Wire Control System in an F-8 Airplane," NATO Advisory Group for Aeronautical Research and Development Conference Paper AGARD-CP-137 (1974); see also James E. Tomayko, *Computers Take Flight: A History of NASA's Pioneering Digital Fly-By-Wire Project*, SP-2000-4224 (Washington, DC: NASA, 2000).

The Ling-Temco-Vought A-7D DIGITAC testbed was the first U.S. Air Force airplane with a digital flight control system. USAF.

determined to be unflyable.[31] This event further reinforced the need for extensive validation and verification tests of all software used in digital flight control systems, no matter how expensive or time-consuming.

In 1975, the Air Force began its own flight-testing of a digital flight control system, using a Ling-Temco-Vought A-7D Corsair II attack aircraft modified with a digital flight control system (dubbed DIGITAC) to duplicate the handling qualities of the analog Command Augmentation System of the baseline A-7D aircraft. As well, testers intended to evaluate several multimode features.

The model-following system was enhanced to allow several models to be selected in flight. The objective was to determine if the pilot might desire a different model response during takeoff and landing, for example, than during air-to-air or air-to-ground gunnery maneuvers. The program was completed successfully in only 1 year of testing, primarily because the airplane was equipped with the standard A-7D mechanical backup system. The airplane used two digital computers that were continuously compared. If a disagreement occurred, the entire system would disengage, and the backup mechanical system was used to safely recover the airplane. The pilot also had a paddle switch on the stick that

---

31. Interview with Manke; see also Tomayko, *Computers Take Flight*, pp. 111–114.

immediately disconnected the digital system. This allowed software changes to be made quickly and safely and avoided most of the necessary, but time-consuming, preflight safety procedures that were associated with NASA's F-8 DFBW program.[32]

One of the more challenging flight control system designs was associated with the Grumman X-29 research airplane. The X-29 was designed to demonstrate the advantages of a forward-swept wing (FSW), along with other new technologies.

The airplane would fly with an unusually large level of pitch instability. The F-16, while flying at subsonic speeds, had a negative static margin of about 6 percent. The X-29 static margin was 35 percent unstable. (In practical terms, this meant that the divergence time to double amplitude was about half a second, effectively meaning that the airplane would destroy itself if it went out of control before the pilot could even recognize the problem!) This level of instability required extremely fast control surface actuators and state-of-the-art computer software. The primary system was a triplex of digital computers, each of which was backed up by an analog computer. A failure of one digital channel did not prevent the remaining two digital computers from continuing to function. After two digital channel failures, the system reverted to the three all-analog computers, thus maintaining fail-op, fail-op, fail-safe capability.

After completing the limit-cycle and resonance ground tests mentioned earlier, plus a lengthy software validation and verification effort, the flight-testing began in 1984 at NASA's Dryden Flight Research Center.[33] The control system handled the high level of instability quite well, and the test program on two airplanes was very successful, ending in 1992. Although the forward-swept wing concept has not been incorporated in any modern airplanes, the successful completion of the X-29 program further boosted the confidence in digital FBW control systems.[34]

In recent years, the digital FBW systems have become the norm in military aircraft. The later models of the F-15, F-16, and F/A-18 were

---

32. Capt. Lawrence Damman, Capt. Ronald Grabe, Robert Kennington, and Paul W. Kirsten, "Flight Test Development of a Multimode Digital Flight Control System Implemented in an A-7D (DIGITAC)," AFFTC TR-76-15 (June 1976).
33. Personal experience as a member of the X-29 Flight Readiness Review Team.
34. Paul Pellicano, Joseph Krumenacker, and David Van Hoy, "X-29 High Angle-of-Attack Flight Test Procedures, Results, and Lessons Learned," Society of Flight Test Engineers 21st Annual Symposium, Aug. 1990.

equipped with digital FBW flight control systems. The C-17 Globemaster III airlifter and F-117 Nighthawk stealth fighter performed their first flights with digital FBW systems. The Lockheed Martin F-22 Raptor and F-35 Lightning II Joint Strike Fighter exploit later digital FBW technology. Each has three digital computers and, for added safety, three of each critical component within its control systems. (Such "cross-strapping" of the various components allows FOFOFS redundancy.) There are dual-air data systems providing the various state variables that are backed up by an inertial system. The various "survivability" features first examined and demonstrated decades previously with the F-4 SFCS program (wire-routing, separate component locations, etc.) were also included in their basic design.

Enhanced Electrical Actuators: Critical Enablers for FBW/DFBW

Nearly all high-speed airplanes use hydraulic actuators to operate the control surfaces. This provides a significant boost to the pilot's ability to move a large control surface, which is experiencing very high aerodynamic loads. The computers and other electronic devices mentioned above merely provided signals to servos, which in turn commanded movement of hydraulic actuators. The hydraulic system provided the real muscle to move the surfaces. When Lockheed Martin's "Skunk Works" was designing the planned X-33 Research Vehicle (intended to explore one possible design for a single-stage-to-orbit logistical spacecraft), keeping gross lift-off weight (GLOW) as low as possible was a crucial design goal. Because the hydraulic system would have been employed only during boost and entry, the entire hydraulic system would have been dead weight while the vehicle was in the space environment. Thus, control system designers elected to use electro-mechanical actuators to move the control surfaces, eliminating any need for a hydraulic system. Though X-33 was canceled for a variety of other reasons, its provision for electrical actuators clearly pointed toward future design practice.

Following up on this were a series of three flight-test projects during 1997–1998 as part of the Electrically Powered Actuator Design (EPAD) program sponsored by the Naval Air Warfare Center and Air Force Research Laboratory. Each project tested a different advanced flight control actuator for the left aileron of NASA Dryden's F/A-18 Systems Research Aircraft (SRA). The first was the "smart actuator" that used fiber optics instead of the normal fly-by-wire system to control an otherwise conventional hydraulic actuator.[35] The next project flew an electro-hydrostatic actuator that used an electric motor to drive a small hydraulic pump that actuated the left aileron; the actuator was independent of the normal aircraft hydraulic system.[36] The third project used an electro-mechanical actuator (EMA) that used electrical power generated by the F/A-18's engines to power the left aileron actuator. A fiber-optic controller, self-contained control-surface actuator promises a significant reduction in weight and complexity over conventional actuation systems for future advanced air and space vehicles.[37]

Load Feedback for Flight Controls: Imitating the Birds

Among their many distinctive attributes, birds possess a particularly unique characteristic not experienced by humans: they are continuously aware of the loads their wings and control feathers bear, and they can adjust the wing shape to alleviate or redistribute these loads in real time. This allows a bird to optimize its wing shape across its entire range of flight; for example, a different wing shape for low-speed soaring than for high-speed cruising. Humans are not so fortunate. In the earliest days of flight, most aircraft designers consciously emulated the design of birds for both the planform and airfoil cross section of wings. Indeed, the frail fabric and wood structure of thin wings used by pioneers such as the Wright brothers, Louis Blériot, the Morane brothers, and Anthony Fokker permitted use of aeroelastic wing-warping (twisting) of a wing to bank an airplane, until superseded by the invention of the pivoted aileron. Naturally, when thicker wings appeared, the option of wing-warping became a thing of the past, not revived until the far later jet age and the era of thin composite structures.

For human-created flight, structural loads can be measured via strain gages, and, indeed, the YF-16 utilized strain gages on the main wing spar to adjust the g limiter in the control laws for various fuel loadings and external store configurations. Though the system worked

---

35. Eddie Zavala, "Fiber Optic Experience with the Smart Actuation System on the F-18 Systems Research Aircraft," NASA TM-97-206223, Oct. 1997.
36. Robert Navarro, "Performance of an Electro-Hydrostatic Actuator on the F-18 Systems Research Aircraft," NASA TM-97-206224, Oct. 1997.
37. Joel R. Sitz, "F-18 Systems Research Aircraft," NASA TM-4433 (1992); Lane E. Wallace, *Flights of Discovery: 50 Years at the NASA Dryden Flight Research Center*, SP-4309 (Washington, DC: NASA, 1996), pp. 124–125.

and showed great promise, General Dynamics and the Air Force abandoned this approach for the production F-16 out of concern over the relatively low reliability of the strain gages. The technology still has not yet evolved to the point where designers are willing to put the strain gage outputs directly into the flight control system in a load-feedback manner.[38] Certainly this technology will continue, and changing wing shapes based on load measurements will evolve.

The NASA–Air Force Transonic Aircraft Technology (TACT) program, a joint cooperative effort from 1969 to 1988 between the Langley, Ames, and Dryden Centers, and the Air Force Flight Dynamics Laboratory, led to the first significant test of a so-called mission-adaptive wing (MAW), one blending a Langley-designed flexible supercritical wing planform joined to complex hydraulic mechanisms that could vary its shape in flight. Installed on an F-111A testbed, the MAW could "recontour" itself from a thick supercritical low-speed airfoil section suitable for transonic performance to a thinner symmetrical section ideal for supersonic flight.[39] The MAW, a "first generation" approach to flexible skin and wing approaches, inspired follow-on work including tests by NASA Dryden on its Systems Research Aircraft, a McDonnell-Douglas (now Boeing) F/A-18B Hornet attack fighter using wing deformation as a means of achieving transonic and supersonic roll control.[40]

NASA DFRC is continuing its research on adaptive wing shapes and airfoils to improve efficiency in various flight environments. Thus, over a century after the Wrights first flew a bird-imitative wing-warping airplane at Kitty Hawk, wing-warping has returned to aeronautics, in a "back to the future—back to nature" technique used by the Wright brothers (and birds) to bank, and to perform turns. This cutting-edge technology is not yet in use on any operational airplanes, but it is only a matter of time before these performance enhancement features will increase the efficiency of future military and civilian aircraft.

---

38. Eggers and Bryant, "Flying Qualities Evaluation of the YF-16," AFFTC TR-75-15 (1975).
39. Theodore G. Ayers and James B. Hallissy, "Historical Background and Design Evolution of the Transonic Aircraft Technology Supercritical Wing," NASA TM-81356 (1981); Paul W. Phillips and Stephen B. Smith, "AFTI/F-111 Mission Adaptive Wing (MAW) Automatic Flight Control System Modes Lift and Drag Characteristics," AFFTC TR-89-03 (1989).
40. Andrew M. Lizotte and Michael J. Allen, "Twist Model Development and Results From the Active Aeroelastic Wing F/A-18 Aircraft," NASA TM-2005-212861 (2005); see also Chambers, *Partners in Freedom*, pp. 78–81.

Case 6 | Physical Problems, Challenges, and Pragmatic Solutions

Structures and their Aeroelastic Manifestations

Though an airplane looks rigidly solid, in fact it is a surprisingly flexible machine. The loadings it experiences in flight can manifest themselves in a variety of ways that affect and "move" the structure, and, as discussed previously, the flight control system itself can adversely affect the structure. The convoluted field in which aerodynamics and structures collide both statically and dynamically has led to some of the most complex and challenging problems that engineers, researchers, and designers have faced in the history of aeronautics.

The safety factor for a railroad bridge is usually "10," meaning that the structural members are sized to carry 10 times the design load without failing. Since weight is so crucial to the performance of an airplane, however, its structural safety factor is typically "1.5," that is, the structure can fail if the loads are only 50 percent higher than the design value. As a result of the low aircraft design safety factor, aircraft structures receive far more attention during the design than do bridge structures and are subject to much larger deformations when loaded. This structural deformation can also interact with the aerodynamics of an airplane, both dynamically and statically, independently from the control system interaction mentioned earlier.

Flutter: The Insidious Threat

The most dramatic interaction of airplane structure with aerodynamics is "flutter": a dynamic, high-frequency oscillation of some part of the structure. Aeroelastic flutter is a rapid, self-excited motion, potentially destructive to aircraft structures and control surfaces. It has been a particularly persistent problem since invention of the cantilever monoplane at the end of the First World War. The monoplane lacked the "bridge truss" rigidity found in the redundant structure of the externally braced biplane and, as it consisted of a single surface unsupported except at the wing root, was prone to aerodynamic induced flutter. The simplest example of flutter is a free-floating, hinged control surface at the trailing edge of a wing, such as an aileron. The control surface will begin to oscillate (flap, like the trailing edge of a flag) as the speed increases. Eventually the motion will feed back through the hinge, into the structure, and the entire wing will vibrate and eventually self-destruct. A similar situation can develop on a single fixed aerodynamic surface, like a wing or tail surface. When aerodynamic forces and moments are applied to the surface, the structure will respond by twisting or bending about its elastic axis. Depending on the relationship between the elastic axis of the structure and the axis of the applied forces and moments, the motion can become self-energizing and a divergent vibration—one increasing in both frequency and amplitude—can follow. The high frequency and very rapid divergence of flutter causes it to be one of the most feared, and potentially catastrophic, events that can occur on an aircraft. Accordingly, extensive detailed flutter analyses are performed during the design of most modern aircraft using mathematical models of the structure and the aerodynamics. Flight tests are usually performed by temporarily fitting the aircraft with a flutter generator. This consists of an oscillating mass, or small vane, which can be controlled and driven at different frequencies and amplitudes to force an aerodynamic surface to vibrate. Instrumentation monitors and measures the natural damping characteristics of the structure when the flutter generator is suddenly turned off. In this way, the flutter mathematical model (frequency and damping) can be validated at flight conditions below the point of critical divergence.

Traditionally, if flight tests show that flutter margins are insufficient, operational limits are imposed, or structural beef-ups might be accomplished for extreme cases. But as electronic flight control technology advances, the prospect exists for so-called "active" suppression of flutter by using rapid, computer-directed control surface deflections. In the 1970s, NASA Langley undertook the first tests of such a system, on a one-seventeenth scale model of a proposed Boeing Supersonic Transport (SST) design, in the Langley Transonic Dynamics Tunnel (TDT). Encouraged, Center researchers followed this with TDT tests of a stores flutter suppression system on the model of the Northrop YF-17, in concert with the Air Force Flight Dynamics Laboratory (AFFDL, now the Air Force Research Laboratory's Air Vehicles Directorate), later implementing a similar program on the General Dynamics YF-16. Then, NASA DFRC researchers modified a Ryan Firebee drone with such a system. This program, Drones for Aerodynamic and Structural Testing (DAST), used a Ryan BQM-34 Firebee II, an uncrewed aerial vehicle, rather than an inhabited system, because of the obvious risk to the pilot for such an experiment.

The modified Firebee made two successful flights but then, in June 1980, crashed on its third flight. Postflight analysis showed that one of the software gains had been inadvertently set three times higher than planned, causing the airplane wing to flutter explosively right after launch

Case 6 | Physical Problems, Challenges, and Pragmatic Solutions

A Drones for Aerodynamic and Structural Testing (DAST) unpiloted structural test vehicle, derived from the Ryan Firebee, during a 1980 flight test. NASA.

from the B-52 mother ship. In spite of the accident, progress was made in the definition of various control laws that could be used in the future for control and suppression of flutter.[41] Overall, NASA research on active flutter suppression has been generally so encouraging that the fruits of it were applied to new aircraft designs, most notably in the "growth" version of the YF-17, the McDonnell-Douglas (now Boeing) F/A-18 Hornet strike fighter. It used an Active Oscillation Suppression (AOS) system to suppress flutter tendencies induced by its wing-mounted stores and wingtip Sidewinder missiles, inspired to a significant degree by earlier YF-17 and YF-16 Transonic Dynamics Tunnel testing.[42]

---

41. E. Nissim, "Design of Control Laws for Flutter Suppression Based on the Aerodynamic Energy Concept and Comparisons With Other Design Methods," Technical Report TP-3056, Research Engineering, NASA Dryden Flight Research Center (1990) [given also as American Institute of Aeronautics and Astronautics Conference Paper 89-1212 (1989)].

42. J.T. Foughner, Jr., and C.T. Bensinger, "F-16 Flutter Model Studies With External Wing Stores," NASA TM-74078 (1977); C. Hwang, E. Jonson, G. Mills, T. Noll, and M. Farmer, "Wind Tunnel Test of a Fighter Aircraft Wing/Store Flutter Suppression System: An International Effort," AGARD R-689 (1980); R.P. Peloubet, Jr., and R.L. Haller, "Wind-Tunnel Demonstration of Actrive Flutter Suppression Using F-16 Model with Stores," AFWAL TR-83-3046, vol. 1 (1983); Joseph R. Chambers, *Innovation in Flight: Research of the NASA Langley Research Center on Revolutionary Advanced Concepts for Aeronautics*, SP-2005-4539 (Washington, DC: NASA, 2005), pp. 196–203, 212–215.

Elastic Aerostructural Effects

The distortion of the shape of an airplane structure because of applied loads also creates a static aerodynamic interaction. When air loads are applied to an aerodynamic surface, it will bend or twist proportional to the applied load, just like a spring. Depending on the surface configuration, the distorted shape can produce different aerodynamic properties when compared with the rigid shape. A swept wing, for example, will bend upward at the tip and may also twist as it is loaded.

This new shape may exhibit higher dihedral effect and altered spanwise lift distribution when compared with a rigid shape, impacting the performance of the aircraft. Because virtually all fighter aircraft have short wings and can withstand 7 to 9 g, their aeroelastic deformation is relatively small. In contrast, bomber, cargo, or high-altitude reconnaissance airplanes are typically designed for lower g levels, and the resulting structure, particularly its long, high aspect ratio wings, is often quite limber.

Notice that this is not a dynamic, oscillatory event, but a static condition that alters the steady-state handling qualities of the airplane. The prediction of these aeroelastic effects is a complex and not altogether accurate process, though the trends are usually correct. Because the effect is a static condition, the boundaries for safe flight can usually be determined during the buildup flight-test program, and, if necessary, placards, can be applied to avoid serious incidents once the aircraft enters operational service.

The six-engine Boeing B-47 Stratojet was the first airplane designed with a highly swept, relatively thin, high aspect ratio wing. At higher transonic Mach numbers, deflection of the ailerons would cause the wing to twist sufficiently to cancel, and eventually exceed, the rolling moment produced by the aileron, thus producing an aileron reversal. (In effect, the aileron was acting like a big trim tab, twisting the wing and causing the exact opposite of what the pilot intended.) Aerodynamic loads are proportional to dynamic pressure, so the aeroelastic effects are usually more pronounced at high airspeed and low altitude, and this combination caused several fatal accidents with the B-47 during its flight-testing and early deployment. After flight-testing determined the magnitude and region of reduced roll effectiveness, the airplane was placarded to 425 knots to avoid roll reversal. In sum, then, an aeroelastic problem forced the limiting of the maximum performance achievable by the airplane, rendering it more vulnerable to enemy defenses. The B-47's successor, the B-52, had a much thicker wing root and more robust structure to avoid the problems its predecessor had encountered.

The Mach 3.2+ Lockheed SR-71 Blackbird, designed to cruise at supersonic speeds at very high altitude, was another aircraft that exhibited significant aeroelastic structural deformation.[43] The Blackbird's structure was quite limber, and the aeroelastic predictions for its behavior at cruise conditions were in error for the pitch axis. The SR-71 was a blended wing-body design with chines running along the forward sides of the fuselage and the engine nacelles, then blending smoothly into the rounded delta wing. These chines added lift to the airplane, and because they were well forward of the center of gravity, added a significant amount of pitching moment (much like a canard surface on an airplane such as the Wright Flyer or the Saab AJ-37 Viggen). Flight-testing revealed that the airplane required more "nose-up" elevon deflection at cruise than predicted, adding a substantial amount of trim drag. This reduced the range the Blackbird could attain, degrading its operational performance. To correct the problem, a small shim was added to the production fuselage break just forward of the cockpit. The shim tilted the forebody nose cone and its attached chine surfaces slightly upward, producing a nose-up pitching moment. This allowed the elevons to be returned to their trim faired position at cruise flight conditions, thus regaining the lost range capability.

Sadly, the missed prediction of the aeroelastic effects also contributed to the loss of one of the early SR-71s. While the nose cone forebody shim was being designed and manufactured, the contractor desired to demonstrate that the airplane could attain its desired range if the elevons were faired. To achieve this, Lockheed technicians added trim-altering ballast to the third production SR-71, then being used for systems and sensor testing. The ballast shifted the center of gravity about 2 percent aft from its normal position and at the aft design limit for the airplane. The engineers calculated that this would permit the elevons to be set in their faired position at cruise conditions for this one flight so that the SR-71 could meet its desired range performance. Instead, the aft cg, combined with the nonlinear aerodynamics

---

43. For perspectives on the various members of the Blackbird family, see Peter W. Merlin, *From Archangel to Senior Crown: Design and Development of the Blackbird*, (Reston, VA: American Institute for Aeronautics and Astronautics, 2008); and also his *Mach 3+: NASA/USAF YF-12 Flight Research, 1969–1979*, SP-2001-4525 (Washington, DC: NASA, 2001).

and aeroelastic bending of the fuselage, resulted in the airplane going out of control at the start of a turn at a cruise Mach number. The airplane broke in half, catapulting the pilot, who survived, from the cockpit. Unfortunately, his flight-test engineer/navigator perished.[44] Shim installation, together with other minor changes to the control system and engine inlets, subsequently enabled the SR-71 to meet its performance goals, and it became a mainstay of America's national reconnaissance fleet until its retirement in early 1990.

Lockheed, the Air Force, and NASA continued to study Blackbird aeroelastic dynamics. In 1970, Lockheed proposed installation of a Loads Alleviation and Mode Suppression (LAMS) system on the YF-12A, installing very small canards called "exciter-" or "shaker-vanes" on the forebody to induce in-flight motions and subsequent suppression techniques that could be compared with analytical models, particularly NASA's NASTRAN and Boeing's FLEXSTAB computerized load prediction and response tools. The LAMS testing complemented Air Force-NASA research on other canard-configured aircraft such as the Mach 3+ North American XB-70A Valkyrie, a surrogate for large transport-sized supersonic cruise aircraft. The fruits of this research could be found on the trim canards used on the Rockwell B-1A and B-1B strategic bombers, which entered service in the late 1980s and notably improved their high-speed "on the deck" ride qualities, compared with their three low-altitude predecessors, the Boeing B-52 Stratofortress, Convair B-58 Hustler, and General Dynamics FB-111.[45]

The Advent of Fixed-Base Simulation

Simulating flight has been an important part of aviation research since even before the Wright brothers. The wind tunnel, invented in the 1870s, represented one means of simulating flight conditions. The rudimentary Link trainer of the Second World War, although it did not attempt to represent any particular airplane, was used to train pilots on the proper navigation techniques to use while flying in the clouds. Toward the end of the Second World War, researchers within Government, the military services, academia, and private industry began experimenting with

---

44. Personal experience during SR-71 accident investigation; Ben R. Rich and Leo Janos, *Skunk Works: A Personal Memoir of My Years of Lockheed* (Boston: Little, Brown, and Co., 1994), pp. 192–237.

45. Merlin, *Mach 3+*, pp. 39–42.

Case 6 | Physical Problems, Challenges, and Pragmatic Solutions

analog computers to solve differential equations in real time. Electronic components, such as amplifiers, resistors, capacitors, and servos, were linked together to perform mathematical operations, such as arithmetic and integration. By patching many of these components together, it was possible to continuously solve the equations of motion for a moving object. There are six differential equations that can be used to describe the motion of an object. Three rotational equations identify pitching, rolling, and yawing motions, and three translational equations identify linear motion in fore-and-aft, sideways, and up-and-down directions. Each of these equations requires two independent integration processes to solve for the vehicle velocities and positions. Prior to the advent of analog computers, the integration process was a very tedious, manual operation and not amenable to real-time solutions. Analog computers allowed the integration to be accomplished in real time, opening the door to pilot-in-the-loop simulation. The next step was the addition of controlling inputs from an operator (stick and rudder pedals) and output displays (dials and oscilloscopes) to permit continuous, real-time control of a simulated moving object. Early simulations only solved three of the equations of motion, usually pitch rotation and the horizontal and vertical translational equations, neglecting some of the minor coupling terms that linked all six equations. As analog computers became more available and affordable, the simulation capabilities expanded to include five and eventually all six of the equations of motion (commonly referred to as "six degrees of freedom" or 6DOF).

By the mid-1950s, the Air Force, on NACA advice, had acquired a Goodyear Electronic Differential Analyzer (GEDA) to predict aircraft handling qualities based on the extrapolation of data acquired from previous test flights. One of the first practical applications of simulation was the analysis of the F-100A roll-coupling accident that killed North American Aviation (NAA) test pilot George "Wheaties" Welch on October 12, 1954, one of six similar accidents that triggered an emergency grounding of the Super Sabre. By programming the pilot's inputs into a set of equations of motion representing the F-100A, researchers duplicated the circumstances of the accident. The combination of simulation and flight-testing on another F-100A at the NACA High-Speed Flight Station (now the Dryden Center) forced redesign of the aircraft. North American increased the size of the vertical fin by 10 percent and, when even this proved insufficient, increased it again by nearly 30 percent, modifying existing and new production Super Sabres with the larger tail. Thus modified, the F-100 went on to a very successful career as a mainstay Air Force fighter-bomber.[46]

Another early application of computerized simulation analysis occurred during the Air Force-NACA X-2 research airplane program in 1956. NACA engineer Richard E. Day established a simulation of the X-2 on the Air Force's GEDA analog computer. He used a B-17 bombardier's stick as an input control and a simple oscilloscope with a line representing the horizon as a display along with some voltmeters for airspeed, angle of attack, etc. Although the controls and display were crude, the simulation did accurately duplicate the motions of the airplane. Day learned that lateral control inputs near Mach 3 could result in a roll reversal and loss of control. He showed these characteristics to Capt. Iven Kincheloe on the simulator before his flight to 126,200 feet on September 7, 1956. When the rocket engine quit near Mach 3, the airplane was climbing steeply but was in a 45-degree bank. Kincheloe remembered the simulation results and did not attempt to right the airplane with lateral controls until well into the entry at a lower Mach number, thus avoiding the potentially disastrous coupled motion observed on the simulator.[47]

Kincheloe's successor as X-2 project pilot, Capt. Milburn Apt, also flew the simulator before his ill-fated high-speed flight in the X-2 on September 27, 1956. When the engine exhausted its propellants, Apt was at Mach 3.2 and over 65,000 feet, heading away from Edwards and apparently concerned that the speeding plane would be unable to turn and glide home to its planned landing on Rodgers Dry Lake. When he used the lateral controls to begin a gradual turn back toward the base,

---

46. Marcelle Size Knaack, *Post-World War II Fighters*, vol. 1 of *Encyclopedia of U.S. Air Force Aircraft and Missile Systems* (Washington, DC: Office of Air Force History, 1978), pp. 114–116; Bill Gunston, *Early Supersonic Fighters of the West* (New York: Charles Scribner's Sons, 1975), pp. 153–157; HSFS, "Flight Experience With Two High-Speed Airplanes Having Violent Lateral-Longitudinal Coupling in Aileron Rolls," RM H55A13 (1955); Hubert M. Drake and Wendell H. Stillwell, "Behavior of the Bell X-1A Research Airplane During Exploratory Flights at Mach Numbers Near 2.0 and at Extreme Altitudes," RM H55G25 (1955); Hubert M. Drake, Thomas W. Finch, and James R. Peele, "Flight Measurements of Directional Stability to a Mach Number of 1.48 for an Airplane Tested with Three Different Vertical Tail Configurations," RM H55G26 (1955).

47. Hubert M. Drake and Wendell H. Stillwell, "Behavior of the Bell X-1A Research Airplane During Exploratory Flights at Mach Numbers Near 2.0 and at Extreme Altitudes," RM H55G25 (1955); Capt. Iven C. Kincheloe, USAF, "Flight Research at High Altitude, Part II," in *Proceedings of the Seventh AGARD General Assembly, Nov. 18–26, 1957* (Washington, DC: NATO Advisory Group for Aeronautical Research and Development, 1958).

Case 6 | Physical Problems, Challenges, and Pragmatic Solutions

the X-2 went out of control. Apt was badly battered in the violent motions that ensued, was unable to use his personal parachute, and was killed.[48]

The loss of the X-2 and Apt shocked the Edwards community. The accident could be duplicated on the simulator, solidifying the value of simulation in the field of aviation and particularly flight-testing.[49] The X-2 experience convinced the NACA (later NASA) that simulation must play a significant role in the forthcoming X-15 hypersonic research aircraft program. The industry responded to the need with larger and more capable analog computer equipment.[50]

The X-15 simulator constituted a significant step in both simulator design and flight-test practice. It consisted of several analog computers connected to a fixed-base cockpit replicating that of the aircraft, and an "iron bird" duplication of all control system hardware (hydraulic actuators, cable runs, control surface mass balances, etc.). Computer output parameters were displayed on the normal cockpit instruments, though there were no visual displays outside the cockpit. This simulator was first used at the North American plant in Inglewood, CA, during the design and manufacture of the airplane. It was later transferred to NASA DFRC at Edwards AFB and became the primary tool used by the X-15 test team for mission planning, pilot training, and emergency procedure definition.

The high g environment and the high pilot workload during the 10-minute X-15 flights required that the pilot and the operational support team in the control room be intimately familiar with each flight plan. There was no time to communicate emergency procedures if an emergency occurred—they had to be already imbedded in the memories of the pilot and team members. That necessity highlighted another issue underscored by the X-15's simulator experience: the necessity of replicating with great fidelity the actual cockpit layout and instrumentation in the simulator. On at least two occasions, X-15 pilots nearly misread their instrumentation or reached for the wrong switch because of seemingly minor differences between the simulator and the instrumentation layout of the X-15 aircraft.[51]

---

48. Bell X-2 No. 1 Accident Report, copy in History Office archives, Air Force Flight Test Center, Edwards AFB, CA.
49. Ronald Bel Stiffler, *The Bell X-2 Rocket Research Aircraft: The Flight Test Program* (Edwards AFB: Air Force Flight Test Center, Aug. 12, 1957), p. 87; Richard E. Day, "Coupling Dynamics in Aircraft: A Historical Perspective," SP-532 (1997).
50. Smith, Schilling, and Wagner, "Simulation at Dryden," p. 1.
51. Ibid., p. 3.

Overall, test pilots and flight-test engineers uniformly agreed that the X-15 program could not have been accomplished safely or productively without the use of the simulator. Once the X-15 began flying, engineers updated the simulator using data extracted from actual flight experience, steadily refining and increasing its fidelity. An X-15 pilot "flew" the simulator an average of 15 hours for every flight, roughly 1 hour of simulation for every minute of flying time. The X-15 experience emphasized the profound value of simulation, and soon, nearly all new airplanes and spacecraft were accompanied by fixed-base simulators for engineering analysis and pilot/astronaut training.

Updating Simulator Prediction with Flight-Test Experience

Test pilots who "flew" the early simulators were skeptical of the results that they observed, because there was usually some aspect of the simulation that did not match the real airplane. Stick forces and control surface hinge moments were often not properly matched on the simulator, and thus the apparent effectiveness of the ailerons or elevators was often higher or lower than experienced with the airplane. For procedural trainers (used for checking out pilots in new airplanes) mathematical models were often changed erroneously based strictly on pilot comments, such as "the airplane rolls faster than the simulator." Since these early simulators were based strictly on wind tunnel or theoretical aerodynamic predictions and calculated moments of inertia, the flight-test community began to explore methods for measuring and validating the mathematical models to improve the acceptance of simulators as valid tools for analysis and training. Ground procedures and support equipment were devised by NASA to measure the moments of inertia of small aircraft and were used for many of the research airplanes flown at DFRC.[52]

A large inertia table was constructed in the Air Force Flight Test Center Weight and Balance facility at Edwards AFB for the purpose of measuring the inertia of large airplanes. Unfortunately, the system was never able to provide accurate results, as fluctuations in temperature and humidity adversely affected the performance of the table's sensitive bearings, so the concept was discarded.

During the X-15 flight-test program, NASA researchers at Edwards developed several methods for extracting the aerodynamic stability

---

52. Capt. John Retelle, "Measured Weight, Balance, and Moments of Inertia of the X-24A Lifting Body," AFFTC TD-71-6 (1971).

derivatives from specific flight-test maneuvers. Researchers then compared these results with wind tunnel or theoretical predictions and, where necessary, revised the simulator mathematical models to reflect the flight-test-derived information. For the X-15, the predictions were quite good, and only minor simulator corrections were needed to allow flight maneuvers to be replicated quite accurately on the simulator. The most useful of these methods was an automatic computer analysis of pulse-type maneuvers, originally referred to as Newton-Raphson Parameter Identification.[53,54] This system evolved into a very useful tool subsequently used as an industry standard for identifying the real-world stability and control derivatives during early testing of new aircraft.[55] The resulting updates are usually also transplanted into the final training simulators to provide the pilots with the best possible duplication of the airplanes' handling qualities. Bookkeeping methods for determining moments of inertia of a new aircraft (i.e., tracking the weight and location of each individual component or structural member during aircraft manufacture) have also been given more attention.

Characteristically, the predicted aerodynamics for a new airplane are often in error for at least a few of the derivatives. These errors are usually a result of either a discrepancy between the wind tunnel model that was tested and the actual airplane that was manufactured, or a result of a misinterpretation or poor interpolation of the wind tunnel data. In some cases, these discrepancies have been significant and have led to major incidents (such as the HL-10 first flight described earlier). Another source of prediction errors for simulation is the prediction of the aeroelastic effects from applied air loads to the structure. These aeroelastic effects are quite complex and difficult to predict for a limber airplane. They usually require flight-test maneuvers to identify or validate the actual handling quality effects of structural deformation. There have been several small, business aircraft that have been built, developed, and sold commercially wherein calculated predictions of the aerodynamics were the primary data source, and very little if any wind tunnel tests were ever accomplished. Accurate simulators for pilot

---

53. K.W. Iliff, B.G. Powers, and L.W. Taylor, Jr., "A Comparison of Newton-Raphson and Other Methods for Determining Stability Derivatives from Flight Data," NASA Report H-544 (Mar. 1969).
54. K.W. Iliff and L.W. Taylor, Jr., "Determination of Stability Derivatives from Flight Data Using a Newton-Raphson Minimization Technique," NASA TN-D-6579 (Mar. 1972).
55. Kenneth W. Iliff, "Aircraft Parameter Estimation," AIAA Meeting Paper 1987-0623 (1987).

training have been created by conducting a brief flight test of each airplane, performing required test maneuvers, then applying the flight-test parameter estimation methods developed by NASA. With a little bit of attention during the flight-test program, a highly accurate mathematical model of a new airplane can be assembled and used to produce excellent simulators, even without wind tunnel data.[56]

Moving Base Cockpit and Centrifuge Simulators

As the computational capability to accurately model the handling qualities of an airplane improved, there was recognition that the lack of motion cues was a distraction to the realism of the simulation. An early attempt to simulate motion for the pilot consisted of mounting the entire simulator cockpit on a set of large hydraulic actuators. These actuators would generate a small positive or negative bump to simulate g onset, while any steady-state acceleration was washed out over time (i.e., back to 1 g). The actuators could also tilt the simulator cockpit to produce a side force, or fore and aft force, on the crew. When correlated with a horizon on a visual screen, the result was a quite realistic sensation of motion. These moving-base cockpit systems were rather expensive and difficult to maintain compared with a simple fixed-base cockpit. Since both the magnitude of the g vector and the rotational motion required were false, the systems were not widely accepted in the flight-testing community, where the goal is to evaluate the pilot's response and capabilities in a true flight environment. They found ready acceptance as airline procedures trainers when the maneuvers are slow and g forces are typically small and proved a source of entertainment in amusement parks, aerospace museums, and science centers.

In the 1950s, the Naval Air Development Center (NADC) at Johnsville, PA, developed a large, powerful centrifuge to explore human tolerance to high g forces. The centrifuge consisted of a 182-ton electric DC motor turning a 50-foot arm with a gondola at the end of the arm. The motor could generate g forces at the gondola as high as 40 g's. The gondola was mounted with two controllable gimbals that allowed the g vector to be oriented in different directions for the gondola occupant.[57]

---

56. David L Kohlman, William G. Schweikhard, and Donald R.L Renz, "Advances in Flight Test Instrumentation and Analysis" SAE Doc. No. 871802, Oct. 1987.
57. C.C. Clark and C.H. Woodling, "Centrifuge Simulation of the X-15 Research Aircraft," NADC MA-5916 (1959).

Case 6 | Physical Problems, Challenges, and Pragmatic Solutions

Test pilot entering the centrifuge gondola at the Naval Air Development Center (NADC) in Johnsville, PA. NASA.

Many detailed studies defining human tolerance to g forces were performed on the centrifuge using programmed g profiles. NADC devised a method for installing a cockpit in the gondola, connecting it to a large analog computer, and allowing the pilot to control the computer simulation, which in turn controlled the centrifuge rotation rate and gimbal angles. This allowed the pilot in the gondola to not only see the pilot displays of the simulated flight, but also to feel the associated translational g levels in all three axes. Although the translational g forces were correctly simulated, the gimbal rotations necessary to properly align the total g vector with the cockpit were artificial and were not representative of a flight environment.

One of the first applications of this closed-loop, moving base simulation was in support of the X-15 program in 1958. There were two prime objectives of the X-15 centrifuge program associated with the high g exit and entry: assessment and validation of the crew station (side arm controller, head and arm restraints, displays, etc.), and evaluation of the handling qualities with and without the Stability Augmentation System. The g environment during exit consisted of a forward acceleration (eyeballs-in) increasing from 2 to 4 g, combined with a 2 g pullup (eyeballs-down). The entry g environment was more severe, consisting of a deceleration (eyeballs-out) of 3 g combined with a simultaneous pullout acceleration of 6 g (eyeballs-down).

The results of the X-15 centrifuge program were quite useful to the X-15's overall development; however, the pilots felt that the centrifuge did not provide a very realistic simulation of an aircraft flight environment. The false rotational movement of the gondola was apparent to the pilots and was a distraction to the piloting task during entry. The exit phase of an X-15 flight was a fairly steady acceleration with little rotational motion, and the pilots judged the simulation a good representation of that environment.[58]

The NADC centrifuge was also used in support of the launch phase of the Mercury, Gemini, and Apollo space programs. These provided valuable information regarding the physiological condition of the astronauts and the crew station design but generally did not include closed-loop piloting tasks with the pilot controlling the simulated vehicle and trajectory.

A second closed-loop centrifuge simulation was performed in support of the Boeing X-20 Dyna-Soar program. Dyna-Soar constituted an ambitious but feasible Air Force effort to develop a hypersonic lofted boost-glider capable of an orbital flight. Unfortunately, it was prematurely canceled in 1963 by then-Secretary of Defense Robert S. McNamara. The Dyna-Soar centrifuge study effort was similar to the X-15 centrifuge program, but the acceleration lasted considerably longer and peaked at 6 g (eyeballs-in) at burnout of the Titan III booster. The pilots were "flying" the vehicle in all three axes during these centrifuge runs, and valuable data were obtained relative to the pilot's ability to function effectively during long periods of acceleration. Some of the piloting demonstrations included alleviating wind spikes during the early ascent phase and successfully guiding the booster to an accurate orbital insertion using simple backup guidance concepts in the event of a booster guidance failure.[59] The Mercury and Gemini programs used automatic guidance during the ascent phase, and the only piloting task during boost was to initiate an abort by firing the escape rockets. The Apollo program included a backup piloting mode during the boost based on the results of the X-20 and other centrifuge programs.

---

58. Personal recollections as a flight planning engineer participating in the X-15 centrifuge program. Also see Dennis Jenkins, *X-15: Extending the Frontiers of Flight*.
59. Robert G. Hoey, Lt. Col. Harry R. Bratt, and Maj. Russell L. Rogers, "A Dynamic Simulation of Pilot Controlled Boost for the X-20A Air Vehicle," AFFTC TDR-63-21 (1964).

Variable Stability Airplanes

Although the centrifuge was effective in simulating relatively steady high g accelerations, it lacked realism with respect to normal aircraft motions. There was even concern that some amount of negative training might be occurring in a centrifuge. One possible method of improving the fidelity of motion simulation was to install the entire simulation (computational mathematical model, cockpit displays, and controls) in an airplane, then forcing the airplane to reproduce the flight motions of the simulated airplane, thus exposing the simulator pilot to the correct motion environment. An airplane so equipped is usually referred to as a "variable stability aircraft."

Since their invention, variable stability aircraft have played a significant role in advancing flight technology. Beginning in 1948, the Cornell Aeronautical Laboratory (now Calspan) undertook pioneering work on variable stability using conventional aircraft modified in such a fashion that their dynamic characteristics reasonably approximated those of different kinds of designs. Waldemar Breuhaus supervised modification of a Vought F4U-5 Corsair fighter as a variable stability testbed. From this sprang a wide range of subsequent "v-stab" testbeds. NACA Ames researchers modified another Navy fighter, a Grumman F6F-5 Hellcat, so that it could fly as if its wing were set at a variety of dihedral angles; this research, and that of a later North American F-86 Sabre jet fighter likewise modified for v-stab research, was applied to design of early Century series fighters, among them the Lockheed F-104 Starfighter, a design with pronounced anhedral (negative wing dihedral).[60]

As the analog simulation capability was evolving, Cornell researchers developed a concept of installing a simulator in one cockpit of a

---

60. Edwin P. Hartman, *Adventures in Research: A History of Ames Research Center, 1940–1965*, SP-4302 (Washington, DC: NASA, 1970), pp. 164–166; 257–258; Paul F. Borchers, James A. Franklin, Jay W. Fletcher, *Flight Research at Ames: Fifty-Seven Years of Development and Validation of Aeronautical Technology*, SP-3300 (Washington, DC: NASA, 1998), passim; William M. Kauffman, Charles J. Liddell, Jr., G. Allan Smith, and Rudolph D. Van Dyke, Jr., "An Apparatus for Varying Effective Dihedral in Flight with Application to a Study of Tolerable Dihedral on a Conventional Fighter Airplane," NACA Report 948 (1949); Walter E. McNeill and Brent Y. Creer, "A Summary of Results Obtained during Flight Simulation of Several Aircraft Prototypes with Variable Stability Airplanes," NACA RM-A56C08 (1956); Richard F. Vomaske, Melvin Sadoff, and Fred J. Drinkwater, III, "The Effect of Lateral-Directional Control Coupling on Pilot Control of an Airplane as Determined in Flight and a Fixed-Base Flight Simulator," NASA TN-D-1141 (1961); William M. Kauffman and Fred J. Drinkwater, III, "Variable Stability Airplanes in Lateral Stability Research," *Aeronautical Engineering Review*, vol. 14, No. 8 (Aug. 1955), pp. 29–30.

two-seat Lockheed NT-33A Shooting Star aircraft. By carefully measuring the stability and controllability characteristics of the "T-Bird" and then subtracting those characteristics from the simulated mathematical model, the researchers could program the airplane with a completely different dataset that would effectively represent a different airplane.[61] Initially the variable stability feature was used to perform general research tests by changing various controlled variables and evaluating their effect on pilot performance. Eventually mathematical models were introduced that represented the complete predicted aerodynamic and control system characteristics of new designs. The NT-33A became the most-recognized variable-stability testbed in the world, having "modeled" aircraft as diverse as the X-15, the B-1 bomber, and the Rockwell Space Shuttle orbiter, and flying from the early 1950s until retirement after the end of the Cold War. Thanks to its contributions and those of other v-stab testbeds developed subsequently,[62] engineers and pilots have had a greater understanding of anticipated flying qualities and performance of new aircraft before the crucial first flight.[63] In particular, the variable stability aircraft did not exhibit the false rotations associated with the centrifuge simulation and were thus more realistic in simulating rapid aircraft-like maneuvers. Several YF-22 control law variations were tested using the **CALSPAN NT-33** prior to the first flight. Before the first flight of the F-22, the control laws were tested on the **CALSPAN VISTA**. Today it is inconceivable that a new aircraft would fly before researchers had first evaluated its anticipated handling qualities via variable-stability research.

Low L/D Approach and Landing Trainers

In addition to the need to simulate the handling qualities of a new airplane, a need to accurately duplicate the approach and landing performance also evolved. The air-launched, rocket-powered research airplane concept, pioneered by the X-1, allowed quick access to high-speed flight

---

61. G. Warren Hall, "Research and Development History of USAF Stability T-33," *Journal of the American Aviation Historical Society*, vol. 19, no. 4 (winter 1974).
62. Mostly notably of these were a North American JF-100C Super Sabre (another Ames project), a Martin-Air Force v-stab Convair F-106 Delta Dart; the NASA FRC General Purpose Airborne Simulator (a modified Lockheed Jetstar executive jet transport); the CALSPAN–Air Force Convair NC-131H Total In-Flight Simulator (TIFS), retired in late 2008; the CALSPAN variable stability Douglas B-26 Invader; its successor, the CALSPAN v-stab Learjet; and the most recent, the CALSPAN VISTA Lockheed Martin NF-16.
63. Shafer, "In-Flight Simulation Studies at the NASA Dryden Flight Research Facility."

for research purposes. It also brought with it unpowered, gliding landings, after the rocket fuel was expended. For the X-1 series of airplanes, the landings were not particular stressful because most landings were on the 7-mile dry lakebed at Edwards AFB and the approach glide angles were 8 degrees or less (lift-to-drag (L/D) ratios of about 8). As the rocket-powered airplanes reached toward higher speeds and altitudes, the landing approach angles increased rather dramatically. The approach glide angle for the X-15 was predicted to be between 15 and 20 degrees (lift-to-drag ratios between 2.8 and 4.25) primarily because of the larger base area at the rear of the fuselage. The L/D was further reduced to about 2.5 after landing gear and flap deployment. These steep unpowered approaches prompted a reassessment of the piloting technique to be used. Higher-than-normal approach speeds were suggested as well as a delay of the landing gear and flap deployment until after completion of the landing flare. These new landing methods also indicated a need for a training "simulator" that could duplicate the landing performance of the X-15 in order to explore different landing techniques and train test pilots.

Out-of-the-cockpit, simulated visual displays available at that time were of very poor quality and were not even considered for the X-15 fixed-base simulator. Simulated missions on the X-15 fixed-base simulator were flown to a high-key location over the lakebed using the cockpit instruments, but the simulation was not considered valid for the landing pattern or the actual landing, which was to be done using visual, out-of-the-window references.

North American added a small drag chute to one of its F-100s to allow its pilots to fly landing approaches simulating the X-15. Additionally, both the Air Force and NASA began to survey available jet aircraft that could match the expected X-15 landing maneuver so that the Government pilots could develop a consistent landing method and identify what external cues were necessary to perform accurate landings. The F-104 had just entered the inventory at the AFFTC and NASA. Flight-testing showed that it was an excellent candidate for duplicating the X-15 landing pattern.[64]

---

64. Gene J. Matranga and Neil A. Armstrong, "Approach and Landing Investigation at Lift-Drag Ratios of 2 to 4 Utilizing a Straight-Wing Fighter Airplane," NASA TM-X-31 (1959); Gene J. Matranga and Neil A. Armstrong, "Approach and Landing Investigation at Lift-Drag Ratios of 2 to 4 Utilizing a Delta-Wing Fighter Airplane," NASA TM-X-125 (1959); Stillwell, *X-15 Research Results*, pp. 38–39; Milton O. Thompson, *At the Edge of Space: The X-15 Flight Program* (Washington: Smithsonian Institution Press, 1992).

Various combinations of landing gear and flap settings, plus partial power on the engine, could be used to simulate the entire X-15 landing trajectory from high key to touchdown. F-104s were used throughout the program for chase, for training new X-15 pilots, for practicing approaches prior to each flight, and also for practicing approaches into uprange emergency lakebeds. The combination of the X-15 fixed-base simulator and the F-104 in-flight landing simulation worked very well for pilot training and emergency planning over the entire X-15 test program, and the F-104 did yeoman work supporting the subsequent lifting body research effort as well, through the X-24B.

In the late 1960s, engineers at the Air Force Flight Dynamics Laboratory had evolved a family of reentry shapes (particularly the AFFDL 5, 7, and 8) that blended a lifting body approach with an extensible variable-sweep wing for terminal approach and landing. In support of these studies, in 1969, the Air Force Flight Test Center undertook a series of low L/D approach tests using a General Dynamics F-111A as a surrogate for a variable-sweep Space Shuttle-like craft returning from orbit. The supersonic variable-sweep F-111 could emulate the track of such a design from Mach 2 and 50,000 feet down to landing, and its sophisticated navigation system and two-crew-member layout enabled a flight-test engineer/navigator to undertake terminal area navigation. The result of these tests demonstrated conclusively that a trained crew could fly unpowered instrument approaches from Mach 2 and 50,000 feet down to a precise runway landing, even at night, an important confidence-building milestone on the path to the development of practical lifting reentry logistical spacecraft such as the Shuttle.[65]

Notice that the landing-pattern simulators discussed above did not duplicate the handling qualities of the simulated airplane, only the performance and landing trajectory. Early in the Space Shuttle program, management decided to create a Shuttle Training Aircraft (STA). A Grumman G II was selected as the host airplane. Modifications were made to this unique airplane to not only duplicate the orbiter's handling qualities (a variable-stability airplane), but also to duplicate the landing trajectory and the out-of-the-window visibility from the orbiter cockpit. This NASA training device represents the ultimate in a complete electronic and

---

65. B.L. Schofield, D.F. Richardson, and P.C. Hoag, "Terminal Area Energy Management, Approach, and Landing Investigation for Maneuvering Reentry Vehicles using F-111A and NB-52B Aircraft," AFFTC TD-70-2 (1970).

Case 6 | Physical Problems, Challenges, and Pragmatic Solutions

A Lockheed F-104 flying chase for an X-15 lakebed landing. NASA.

motion-based training simulator. The success of the gliding entries and landings of the Space Shuttle orbiter confirm the value of this trainer.

Digital Computer Simulation

The computational mathematical models for the early simulators mentioned previously were performed on analog computers. Analog computers were capable of solving complex differential equations in real time. The digital computers available in the 1950s were mechanical units that were extremely slow and not capable of the rapid integration that was required for simulation. One difficulty with analog computers was the existence of electronic noise within the equipment, which caused the solutions to drift and become inaccurate after several minutes of operation. For short simulation exercises (such as a 10-minute X-15 flight) the results were quite acceptable. A second difficulty was storing data, such as aerodynamic functions.

The X-20 Dyna-Soar program mentioned previously posed a challenge to the field of simulation. The shortest flight was to be a once-around orbital flight with a flight time of over 90 minutes. A large volume

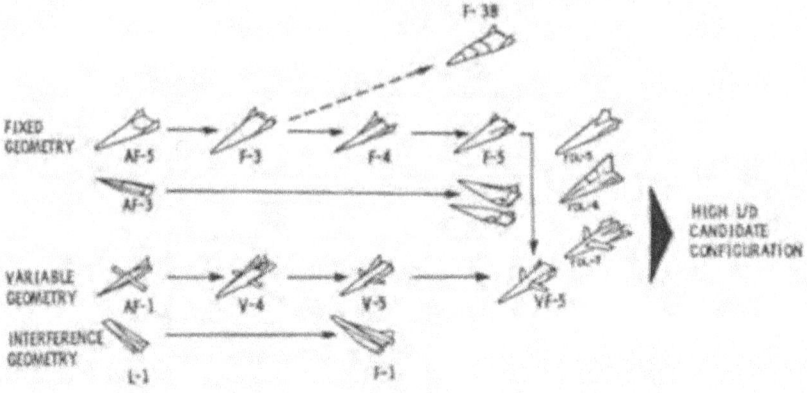

The family of 1960s–1970s reentry shapes developed by the Air Force Flight Dynamics Laboratory. USAF.

of aerodynamic data needed to be stored covering a very large range of Mach numbers and angles of attack. The analog inaccuracy problem was tackled by University of Michigan researchers, who revised the standard equations of motion so that the reference point for integration was a 300-mile circular orbit, rather than the starting Earth coordinates at takeoff. These equations greatly improved the accuracy of analog simulations of orbiting vehicles. As the AFFTC and NASA began to prepare for testing of the X-20, an analog simulation was created at Edwards that was used to develop test techniques and to train pilots. Comparing the real-time simulation solutions with non-real-time digital solutions showed that the closure after 90 minutes was within about 20,000 feet—probably adequate for training, but they still dictated that the mission be broken into segments for accurate results. The solution was the creation of a hybrid computer simulation that solved the three rotational equations using analog computers but solved the three translational equations at a slower rate using digital computers. The hybrid computer equipment was purchased for installation at the AFFTC before the X-20 program was canceled in 1963. When the system was delivered, it was reprogrammed to represent the X-15A-2, a rebuilt variant of the second X-15 intended for possible flight to Mach 7, carrying a scramjet aerodynamic test article on a stub ventral fin.[66] Although quite complex (it necessitated a myriad of analog-to-digital and digital-to-analog conversions), this hybrid system was subsequently

---

66. Capt. Austin J. Lyons, "AFFTC Experiences with Hybrid Computation in a Real-Time Simulation of the X-15A-2," AFFTC TR-66-44 (1967).

used in the AFFTC simulation lab to successfully simulate several other airplanes, including the C-5, F-15, and SR-71, as well as the M2-F2 and X-24A/B Lifting Bodies and Space Shuttle orbiter.

The speed of digital computers increased rapidly in the 1970s, and soon all real-time simulation was being done with digital equipment. Out-of-the-window visual displays also improved dramatically and began to be used in conjunction with the cockpit instruments to provide very realistic training for flight crews. One of the last features to be developed in the field of visual displays was the accurate representation of the terrain surface during the last few feet of descent before touchdown.

Simulation has now become a primary tool for designers, flight-test engineers, and pilots during the design, development, and flight-testing of new aircraft and spacecraft.

Dynamic Instabilities

There are dangerous situations that can occur because of either a coupling of the aerodynamics in different axes or a coupling of the aerodynamics with the inertial characteristics of an airplane. Several of these—Chuck Yeager's close call with the X-1A in December 1953 and Milburn Apt's fatal encounter in September 1956—have been mentioned previously.

Inertial Roll Coupling

Inertial roll coupling is the dynamic loss of control of an airplane occurring during a rapid roll maneuver. The phenomenon of inertial roll coupling is directly related to the evolution of aircraft design. At the time of the Wrights through much of the interwar years, wingspan greatly exceeded fuselage length. As aircraft flight speeds rose, the aspect ratio of wings decreased, and the fineness ratio of fuselages rose, so that by the end of the Second World War, wingspan and fuselage length were roughly equal. In the supersonic era that followed, wingspan reduced dramatically, and fuselage length grew appreciably (think, for example, of an aircraft such as the Lockheed F-104). Such aircraft were highly vulnerable to pitch/yaw/roll-coupling when a rapid rolling maneuver was initiated.

The late NACA–NASA engineer and roll-coupling expert Dick Day described inertial roll coupling as "a resonant divergence in pitch or yaw when roll rate equals the lower of the pitch or yaw natural frequencies."[67]

---

67. Richard E. Day, "Coupling Dynamics in Aircraft: A Historical Perspective," SP-532 (1997), p. 1.

The existence of inertial roll coupling was first revealed by NACA Langley engineer William H. Phillips in 1948, 5 years before it became a dangerous phenomenon.[68] Phillips not only described the reason for the potential loss of control but also defined the criteria for identifying the boundaries of loss of control for different aircraft. During the 1950s, several research airplanes and the Century series fighters encountered fairly severe inertial coupling problems exactly as predicted by Phillips. These airplanes differed from the earlier prop-driven airplanes by having thin, short wings and the mass of the jet engine and fuel concentrated along the fuselage longitudinal axis. This resulted in a higher moment of inertia in the pitch and yaw axis but a significantly lower inertia in the roll axis. The low roll inertia also allowed these airplanes to achieve higher roll rates than their predecessors had. The combination allowed the mass along the fuselage to be slung outward when the airplane was rolled rapidly, producing an unexpected increase in pitching and yawing motion. This divergence in pitch or yaw was related to the magnitude of the roll rate and the duration of the roll. If the roll were sustained long enough, the pitch or yaw angles would become quite large, and the airplane would tumble out of control. In most cases, the yaw axis had the lowest level of static stability, so the divergence was observed as a steady increase in sideslip.[69]

In 1954, after North American Aviation had encountered roll instability with its F-100 aircraft, the Air Force and NAA transferred an F-100A to NACA FRC to allow the NACA to explore the problem through flight-testing and identify a fix. The NACA X-3 research airplane was of a configuration much like the modern fighters and was also used by NACA FRC to explore the inertial coupling problem. These results essentially confirmed Phillips's earlier predictions and determined that increasing the directional stability via larger vertical fin area would mitigate the

---

68. William H. Phillips, "Effect of Steady Rolling on Longitudinal and Directional Stability, NACA TN-627 (1948).
69. Joseph Weil, Ordway B. Gates, Jr., Richard D. Banner, and Albert E. Kuhl, "Flight Experience of Inertia Coupling in Rolling Maneuvers," RM H55WEIL (1955); HSFS, "Flight Experience With Two High-Speed Airplanes Having Violent Lateral-Longitudinal Coupling in Aileron Rolls," RM H55A13 (1955); Hubert M. Drake and Wendell H. Stillwell, "Behavior of the Bell X-1A Research Airplane During Exploratory Flights at Mach Numbers Near 2.0 and at Extreme Altitudes," RM H55G25 (1955); Hubert M. Drake, Thomas W. Finch, and James R. Peele, "Flight Measurements of Directional Stability to a Mach Number of 1.48 for an Airplane Tested with Three Different Vertical Tail Configurations," RM H55G26 (1955); Walter C. Williams and William H. Phillips, "Some Recent Research on the Handling Qualities of Airplanes," RM H55L29a (1956).

problem. The Century series fighters were all reconfigured to reduce their susceptibility to inertial coupling. The vertical tail size was increased for the F-100C and D airplanes.[70] All F-104s were retrofitted with a ventral fin on the lower aft fuselage, which increased their directional stability by 10 to 15 percent. The F-104B, and later models, also had a larger vertical fin and rudder. The F-102 and F-105 received a larger vertical tails than their predecessors (the YF-102 and YF-105) did, and the Mach 2+ F-106 had a larger vertical tail than the F-102 had. Control limiting and placards against continuous rolls (more than 720 degrees of bank) were instituted to ensure safe operation. The X-15 was also susceptible to inertial coupling, and its roll divergence tendencies could be demonstrated on the X-15 simulator. Since high roll rates were not necessary for the high-speed, high-altitude mission of the airplane, the pilots were instructed to avoid high roll rates, and, fortunately, no inertial coupling problems occurred during its flight-testing.

Flight Control Coupling

Flight control coupling is a slow loss of control of an airplane because of a unique combination of static stability and control effectiveness. Day described control coupling—the second mode of dynamic coupling—as " a coupling of static yaw and roll stability and control moments which can produce untrimmability, control reversal, or pilot-induced oscillation (PIO)."[71] So-called "adverse yaw" is a common phenomenon associated with control of an aircraft equipped with ailerons. The down-going aileron creates an increase in lift and drag for one wing, while the up-going aileron creates a decrease in lift and drag for the opposite wing. The change in lift causes the airplane to roll toward the up-going aileron. The change in drag, however, results in the nose of the airplane swinging away from the direction of the roll (adverse yaw). If the airplane exhibits strong dihedral effect (roll produced by sideslip, a quality more pronounced in a swept wing design), the sideslip produced by the aileron deflections will tend to detract from the commanded roll. In the extreme case, with high dihedral effect and strong adverse yaw, the roll can actually reverse, and the airplane will roll in the opposite direction to that commanded by the pilot—as sometimes happened with the Boeing

---

70. Robert G. Hoey and Capt. Milburn G. Apt, "F-100C Phase IV Stability and Control Test" AFFTC TR-56-25, Oct. 1956, pp 8, 144, 145.
71. Day, "Coupling Dynamics in Aircraft," p. 1.

B-47, though by aeroelastic twisting of a wing because of air loads. If the pilot responds by adding more aileron deflection, the roll reversal and sideslip will increase, and the airplane could go out of control.

As discussed previously, the most dramatic incident of control coupling occurred during the last flight of the X-2 rocket-powered research airplane in September 1956. The dihedral effect for the X-2 was quite strong because of the influence of wing sweep rather than the existence of actual wing dihedral. Dihedral effect because of wing sweep is nonexistent at zero-lift but increases proportionally as the angle of attack of the wing increases. After the rocket burned out, which occurred at the end of a ballistic, zero-lift trajectory, the pilot started a gradual turn by applying aileron. He also increased the angle of attack slightly to facilitate the turn, and the airplane entered a region of roll reversal. The sideslip increased until the airplane went out of control, tumbling violently. The data from this accident were fully recovered, and the maneuver was analyzed extensively by the NACA, resulting in a better understanding of the control-coupling phenomenon. The concept of a control parameter was subsequently created by the NACA and introduced to the industry. This was a simple equation that predicted the boundary conditions for aileron reversal based on four stability derivatives. When the yawing moment due to sideslip divided by the yawing moment due to aileron is equal to the rolling moment due to sideslip divided by the rolling moment due to aileron, the airplane remains in balance and aileron deflection will not cause the airplane to roll in either direction.[72]

Dutch Roll Coupling

Dutch roll coupling is another case of a dynamic loss of control of an airplane because of an unusual combination of lateral-directional static stability characteristics. Dutch roll coupling is a more subtle but nevertheless potentially violent motion, one that (again quoting Day) is a "dynamic lateral–directional stability of the stability axis. This coupling of body axis yaw and roll moments with sideslip can produce lateral–directional instability or PIO."[73] A typical airplane design includes "static directional stability" produced by a vertical fin, and a small amount of "dihedral effect" (roll produced by sideslip). Dihedral effect is created by designing the wing with actual dihedral (wingtips higher than the

---

72. In engineering shorthand, $Cn/Cn\alpha = Cl/Cl\alpha$.
73. Day, "Coupling Dynamics in Aircraft," p. 1.

wing root), wing sweep (wingtips aft of the wing root), or some combination of the two. Generally static directional stability and normal dihedral effect are both stabilizing and both contribute to a stable Dutch roll mode (first named for the lateral-directional motions of smooth-bottom Dutch coastal craft, which tend to roll and yaw in disturbed seas). When the interactive effects of other surfaces of an airplane are introduced, there can be potential regions of the flight envelope where these two contributions to Dutch roll stability are not stabilizing (i.e., regions of negative static directional stability or negative dihedral effect). In these regions, if the negative effect is smaller than the positive influence of the other, then the airplane will exhibit an oscillatory roll-yaw motion. (If both effects are negative, the airplane will show a static divergence in both the roll and yaw axes.) All aircraft that are statically stable exhibit some amount of Dutch roll motion. Most are well damped, and the Dutch roll only becomes apparent in turbulent conditions.

The Douglas DC-3 airliner (equivalent to the military C-47 airlifter) had a persistent Dutch roll that could be discerned by passengers watching the wingtips as they described a slow horizontal "figure eight" with respect to the horizon.

Even the dart-like X-15 manifested Dutch roll characteristics. The very large vertical tail configuration of the X-15 was established by the need to control the airplane near engine burnout if the rocket engine was misaligned, a "lesson learned" from tests of earlier rocket-powered aircraft such as the X-1, X-2, and D-558-2. This led to a large symmetrical vertical tail with large rudder surfaces both above and below the airplane centerline. (The rocket engine mechanics and engineers at Edwards later devised a method for accurately aligning the engine, so that the large rudder control surfaces were no longer needed.) The X-15 simulator accurately predicted a strange Dutch roll characteristic in the Mach 3–4 region at angles of attack above 8 degrees with the roll and yaw dampers off. This Dutch roll mode was oscillatory and stable without pilot inputs but would rapidly diverge into an uncontrollable pilot-induced-oscillation when pilot control inputs were introduced.

During wind tunnel tests after the airplane was constructed, it was discovered that the lower segment of the vertical tail, which was operating in a high compression flow field at hypersonic speeds, was highly effective at reentry angles of attack. The resulting rolling motions produced by the lower fin and rudder were contributing a large negative dihedral effect. Fortunately, this destabilizing influence was not enough to overpower the high directional stability produced by the very large vertical tail area, so the Dutch roll mode remained oscillatory and stable. The airplane motions associated with this stable oscillation were completely foreign to the test pilots, however. Whereas a normal Dutch roll is described as "like a marble rolling inside a barrel," NASA test pilot Joe Walker described the X-15 Dutch roll as "like a marble rolling on the outside of the barrel" because the phase relationship between rolling and yawing were reversed. Normal pilot aileron inputs to maintain the wings level were out of phase and actually drove the oscillation to larger magnitudes rather quickly. The roll damper, operating at high gain, was fairly effective at damping the oscillation, thus minimizing the pilot's need to actively control the motion when the roll damper was on.[74]

Because the X-15 roll damper was a single string system (fail-safe), a roll damper failure above about 200,000 feet altitude would have caused the entry to be uncontrollable by the pilot. The X-15 envelope expansion to altitudes above 200,000 feet was delayed until this problem could be resolved. The flight control team proposed installing a backup roll damper, while members of the aerodynamic team proposed removing the lower ventral rudder. Removing the rudder was expected to reduce the directional stability but also would cause the dihedral effect to be stable, thus the overall Dutch roll stability would be more like a normal airplane. The Air Force-NASA team pursued both options. Installation of the backup roll damper allowed the altitude envelope to be expanded to the design value of 250,000 feet. The removal of the lower rudder, however, solved the PIO problem completely, and all subsequent flights, after the initial ventral-off demonstration flights, were conducted without the lower rudder.[75]

The incident described above was unique to the X-15 configuration, but the analysis and resolution of the problem is instructive in that it offers a prudent cautioning to designers and engineers to avoid designs that exhibit negative dihedral effect.[76]

Configuration Influence upon Stall and Departure Behavior

Another maneuver that can lead to loss of control is a stall. An aircraft "stalls" when the wing's angle of attack exceeds a critical angle beyond

---

74. Stillwell, *X-15 Research Results*, pp. 51–52; Thompson, *At the Edge of Space*, pp. 200–202.
75. Personal experience as a flight planning engineer during the X-15 initial envelope expansion tests.
76. Robert G. Hoey, "Correlation of X-15 Simulation Experience with Flight Test Results, AGARD Report 530 (1966).

which the wing can no longer generate the lift necessary to support the airplane. A typical stall consists of some pre-stall warning buffet as the flow over the wing begins to break down, followed by stall onset, usually accompanied by an uncommanded nose-down pitching rotation of the aircraft, as gravity takes over and the airplane naturally tries to regain lost airspeed. The loss of control for a normal stall is quite brief and can usually be overcome, or prevented, by proper control application at the time of pre-stall warning. There are design features of some aircraft that result in quite different stall characteristics. Stalls may be a straightforward wings-level gentle drop (typically leading to a swift and smooth recovery), or sharply abrupt, or an unsymmetrical wing drop leading to a spin entry. The latter can be quite hazardous.

High-performance T-tail aircraft are particularly vulnerable to abnormal stall effects. Lockheed's sleek F-104 Starfighter incorporated a T-tail operating behind a short, stubby, and extremely thin wing. As the wing approached the critical stall angle, the wing tip vortexes impinged on the horizontal tail creating an abrupt nose-up pitching moment, commonly referred to as a "pitch-up." The pitch-up placed the airplane in an uncontrollable flight environment: either a highly oscillatory spin or a deep stall (a stable condition where the airplane remains locked in a high angle of attack vertical descent). To prevent inadvertent pitch-ups, the aircraft was equipped with a "stick shaker," and a "stick kicker." The stick shaker created an artificial vibration of the stick, simulating stall buffet, as the airplane approached a stall. The stick kicker applied a sharp nose-down command to the horizontal tail when the airplane reached the critical condition for an impending pitch-up. A similar situation developed for the McDonnell F-101 Voodoo (also a T-tail behind a short, stubby wing). Stick shakers and kickers were quite successful in allowing these airplanes to operate safely throughout their operational lifespan. Overall, however, the T-tail layout was largely discredited for high-performance fighter and attack aircraft, the most successful postwar fighters being those with low-placed horizontal tails. Such a configuration, typified by the F-100, F-101, F-105, F-5, F-14, F-15, F-16, F/A-18, F-22, F-35, and a host of foreign aircraft, is now a design standard for tailed transonic and supersonic military aircraft. It was a direct outgrowth of the extensive testing the NACA did in the late 1940s and early 1950s on such aircraft as the D-558-2, the North American F-86, and the Bell X-5, all of which, to greater or lesser extents, suffered from pitch-up.

The advent of the swept wing induced its own challenges. In 1935, German aerodynamicist Adolf Busemann discovered that aircraft could operate at higher speeds, and closer to the speed of sound (Mach 1), by using swept wings. By the end of the Second World War, American NACA researcher Robert T. Jones of Langley Memorial Aeronautical Laboratory had independently discovered its benefits as well. The swept wing subsequently transformed postwar military and civil aircraft design, but it was not without its own quite serious problems. The airflow over a swept wing tends to move aft and outboard, toward the tip. This results in the wingtip stalling before the rest of the wing. Because the wingtip is aft of the wing root, the loss of lift at the tip causes an uncommanded nose-rise as the airplane approaches a stall. This nose-rise is similar to a pitch-up but not nearly as abrupt. It can be controlled by the pilot, and for most swept wing airplanes there are no control system features specifically to correct nose-rise problems. Understanding the manifestations of swept wing stall and swept wing pitch-up commanded a great deal of NACA and Air Force interest in the early years of the jet age, for reasons of both safety and combat effectiveness. Much of the NACA's research program on its three swept wing Douglas D-558-2 Skyrockets involved examination of these problems. Research included analysis of a variety of technological "fixes," such as sawtooth leading edge extensions, wing fences, and fixed and retracting slots. Afterward, various combinations of flaps, flow direction fences, wing twist, and other design features have been used to overcome the tip-stall characteristic in modern swept wing airplanes, which, of course, include most commercial airliners.[77]

Aerothermodynamics: Meeting the Heating Challenge

The prediction of structural heating on airplanes flying at hypersonic speeds preceded the actual capability to attain these speeds in controlled flight. There were dire predictions of airplanes burning up when they encountered the "thermal thicket," similar to the dire predictions that preceded flight through the sound barrier. Aerodynamic heating is created by friction of an object moving at very high speed through the atmosphere. Temperatures on the order of 200 degrees Fahrenheit (°F) are generated at Mach 2 (the speed of an F-104 Starfighter of the mid-1950s), 600 °F at Mach 3 (that of a 1960s SR-71 Blackbird), and 1,200

---

77. See, for example, Document 25 in J.D. Hunley, ed., *Toward Mach 2: The Douglas D-558 Program*, SP-4222 (Washington, DC: NASA, 1999), pp. 101–103.

°F at Mach 6 (typical of the X-15). Reentry from orbital speeds (Mach 26—the entry velocity of the Space Shuttle orbiter) will generate temperatures of around 2,400 °F. Airplanes or spacecraft that fly in, or reenter, the atmosphere above Mach 2 must be designed to withstand not only aerodynamic forces associated with high Mach number but also the high temperatures associated with aerodynamic heating. The advent of blunt body reentry theory radically transformed the mental image of the spacecraft, from a "pointy" rocket to one having a far more bluff and rounded body. Conceived by H. Julian Allen with the assistance of Alfred Eggers of the then-NACA Ames Aeronautical Laboratory (now NASA Ames Research Center), blunt-body design postulated using a blunt reentry shape to form a strong "detached" shock wave that could act to relieve up to 90 percent of the thermal load experienced by a body entering Earth's atmosphere from space.[78] Such a technical approach was first applied to missile warhead design and the first crewed spacecraft, both Soviet and American. But blunt bodies, for all their commendable thermodynamic characteristics, likewise have high drag and poor entry down-range and cross-range predictability. Tailored higher L/D lifting body and blended wing-body shapes (such as those pioneered by the Air Force Flight Dynamics Laboratory), while offering far better aerodynamic and cross-range performance and predictability, pose far greater cooling challenges. So, too, do concepts for hypersonic air-breathing vehicles. These diverse requirements have stimulated the design and development of several potential solutions for thermal protection of a vehicle. For purposes of discussion, these concepts are addressed as heat sink structures, ablation, hot structures, active cooling, and advanced ceramic protection.

Heat-Sink Structures

Prior to the X-15 flight-test program, there were several theories predicting the amount of friction heat that would transfer to the surface of a winged aircraft, with substantial differences in the theories. Wind tunnels, ballistic ranges, and high-temperature facilities such as arc-jets were unable to adequately duplicate the flight environment necessitating

---

78. H. Julian Allen and Alfred J. Eggers, Jr., "A Study of the Motion and Aerodynamic Heating of Ballistic Missiles Entering the Earth's Atmosphere at High Supersonic Speeds," NACA Technical Report 1381 (1958) [this widely distributed report was preceded by a more restricted limited-issue classified report for Government and industry earlier]; see also Hartman, *Adventures in Research*, pp. 215–218.

full-scale flight test to determine which theory was correct. The X-15 was that test aircraft. The design needed to be robust in order to survive the worst-case heating predictions if the theories proved to be correct.

A heat-sink structure was selected as the safest and simplest method for providing thermal protection. Inconel X, a nickel alloy normally used for jet engine exhaust pipes, was selected as the primary structural material. It maintained adequate structural strength to about 1,200 °F. The design proceeded by first defining the size of each structural member based on the air loads anticipated during entry, then increasing the size of each member to absorb the expected heat load that would occur during the short exposure time of an X-15 flight.

As with most of the first missile and aircraft explorations, early hypersonic flights in the X-15 showed that none of the prediction methods was completely accurate, although each method showed some validity in a certain Mach range. In general, the measured heat transfer was less than predicted. Thus, one of the most significant flight-test results from the X-15 program was development of more accurate prediction methods based upon real-world data for the thermal protection of future hypersonic and entry vehicles.[79] The majority of aerodynamic heating issues that required attention during the X-15 flight-test program were associated with localized heating: that is, unexpected hot spots that required modification. Some typical examples included loss of cockpit pressurization because of a burned canopy seal (resolved by installing a protective shield in front of the canopy gap), cockpit glass cracked because of deformation of the glass retainer ring (resolved by increasing the clearance around the glass), wing skin buckling behind the slot in the leading edge expansion joint (resolved by installing a thin cover over the expansion joint), thermal expansion of the fuselage triggering nose gear door deployment with resulting damage to internal instrumentation (resolved by increasing the slack in the deployment cable), and buckling of skin on side tunnel fairings because of large temperature difference between outer skin and liquid oxygen (LOX) tank (resolved by adding expansion joints along the side tunnels).

Most of these issues were discovered and resolved fairly easily since the flight envelope was expanded gradually on successive flights with small increases in Mach number on each flight. Had the airplane been exposed to the design entry environment on its very first flight, the

---

79. Jenkins, *X-15: Extending the Frontiers of Flight*, passim.

combined results of these local heating problems would probably have been catastrophic.

Ablation Cooling

Another potential method for disbursing heat during high-speed flight was the application of an "ablation" material to the outer surface of the structure. An ablator is a material that is applied to the outside of a vehicle that burns or chars when exposed to high temperature, thus carrying away much of the associated heat and hot gases. Ablators are quite efficient for short duration, one-time entries such as an intercontinental ballistic missile (ICBM) nose cone. Ablators were also used on the early crewed orbiting capsules (Mercury, Gemini, and Apollo), which used ballistic or semiballistic entry trajectories with relatively short peak heating exposure times. They seemed to offer special promise for lifting bodies, with developers hoping to build classes of aluminum-structured spacecraft that could have a cheap, refurbishable ablative coating re-applied after each flight. Indeed, on April 19, 1967, the Air Force did fly and recover one such subscale experimental vehicle, the Mach 27 Martin SV-5D (X-23) Precision Recovery Including Maneuvering Entry (PRIME) lofted over the Pacific Test Range by a modified Atlas ballistic missile.[80]

But for all their merits, ablators are hardly a panacea. Subsonic and transonic testing of several rocket-powered aluminum lifting bodies at NASA's Flight Research Center showed that this class of vehicle could be landed; however, later analysis indicated that the rough surface of an exposed ablator would probably have reduced the lift and increased the drag so that successful landings would have been questionable.[81]

Flight-test experience with the X-15 confirmed such conclusions. When the decision was made to rebuild the second X-15 after a crash landing, it seemed a perfect opportunity to demonstrate the potential of ablative coatings as a means of furnishing refurbishable thermal protection to hypersonic aircraft and spacecraft. The X-15A-2 was designed to reach Mach 7, absorbing the additional heat load it would experience via MA-25S, a thin Martin-developed silica ablative coating. Coating the aircraft with the MA-25S proved surprisingly time-consuming, as did the refurbishment between flights.

---

80. Joel W. Powell and Ed Hengeveld, "ASSET and PRIME: Gliding Re-Entry Test Vehicles," *Journal of the British Interplanetary Society*, vol. 36 (1983), pp. 369–376.
81. Personal inspection of the SV-5D (X-23) heat shield following vehicle recovery.

During a flight to Mach 6.7 by Maj. William J. "Pete" Knight, unanticipated heating actions severely damaged the aircraft, melting a scramjet boilerplate test module off the airplane and burning holes in the external skin. Though Knight landed safely—a great tribute to his piloting skills—the X-15A-2 was in no condition to fly without major repairs. Although the ablator did provide the added protection needed for most of the airplane, the tedious process of applying it and the operational problems associated with repairing and protecting the soft coating were quite time-consuming and impracticable for an operational military or civilian system.[82] The postentry ablated surface also increased the drag of the airplane by about the same percentage that was observed on the PRIME vehicle. Clearly the X-15A-2's record flight emphasized, as NASA engineer John V. Becker subsequently wrote, "the need for maximum attention to aerothermodynamic detail in design and preflight testing."[83] The "lifting body" concept evolved as a means of using ablative protection for entries of wingless, but landable, vehicles. As a result of the X-15 and lifting body testing by NASA, an ablative coating has not been seriously considered for any subsequent reusable lifting entry vehicle.

Hot Structure Approaches

Another option for thermal protection during entry was the use of exotic, high-temperature materials for the external surface that could reradiate the heat back into space. This concept was proposed for the X-20 Dyna-Soar program, and the vehicle was well under construction at the time of cancellation.[84] In parallel with the X-20 program, the Air Force Flight Dynamics Laboratory developed a small radiative-cooled hot structure vehicle (essentially the first 4 feet of the X-20 Dyna Soar's nose), called the McDonnell Aerothermodynamic/elastic Structural Systems Environmental Tests (ASSET). The ASSET design used the same materials and thermal protection concepts as the X-20 and first flew in September 1963, 3 months before cancellation of the Dyna-Soar. The fourth ASSET vehicle successfully completed a Mach 18.4 entry from 202,000 feet in 1965. Postflight examination indicated it

---

82. Johnny G. Armstrong, "Flight Planning and Conduct of the X-15A-2 Envelope Expansion Program", FTC TD-69-4, July 1969.
83. John V. Becker, "The X-15 Program in Retrospect," *Raumfahrtforschung* (Mar.–Apr. 1969).
84. Robert Godwin, ed., *Dyna-Soar Hypersonic Strategic Weapons System* (Burlington, ON, Canada: Apogee Books, 2003) has an excellent compilation of contemporary Dyna-Soar documents.

survived the entry well, although the operational problems and manufacturing methods for these exotic materials were expensive and time-consuming. Since that time, joint NASA-Air Force-Navy-industry developmental programs such as the X-30 National Aero-Space Plane (NASP) effort of the late 1980s to early 1990s have advanced materials and fabrication technologies that, in due course, may be applied to future hypersonic systems.[85]

Lightweight Ceramic Tiles

Ceramic tiles, of the kind used in a blast furnace or fireplace to insulate the surrounding structure from the extreme temperatures, were far too heavy to be considered for use on a flight vehicle. The concept of a lightweight ceramic tile for thermal protection was conceived by Lockheed and developed into operational use by NASA Ames Research Center, NASA Johnson Space Center, and Rockwell International for use on the Space Shuttle orbiter, first flown into orbit in April 1981. The resulting tiles and ceramic blankets provided exceptionally light and efficient thermal protection for the orbiter without altering the external shape. Although highly efficient for thermal protection, the tiles were—and are—quite fragile and time-consuming to repair and maintain. The Shuttle program experienced considerable delays prior to its first flight because of bonding, breaking, and other installation issues. (Unlike the X-15 gradual envelope expansion program, the Shuttle orbiter was exposed to its full operational flight envelope on its very first orbital flight and entry, thus introducing a great deal of analysis and caution during flight preparation.) Subsequent Shuttle history confirmed the high-maintenance nature of the tiles, and their vulnerability to external damage such as ice or insulation shedding from the super-cold external propellant tank. Even with these limitations, however, they do constitute the most promising technology for future lifting entry vehicles.[86]

---

85. M.H. Shirk, "ASSET: Aerothermoelastic Vehicles (AEV) Results and Conclusions," AFFDL 65FD-1197 (1965); "ASSET Final Briefing," AFFDL 65FD-850 (1965).

86. Paul Cooper and Paul F. Holloway, "The Shuttle Tile Story," *Astronautics & Aeronautics*, vol. 19, no. 1 (Jan. 1981), pp. 24–34; Robert G. Hoey, et al., "Evaluation of the Space Shuttle Orbiter First Orbital Flight: Final Report," AFFTC TR-81-21 (1981); NASA LRC, "Shuttle Performance: Lessons Learned," NASA Conference Publication 2283 (1983); and Robert G. Hoey, et al., "Flight Test Results from the Entry and Landing of the Space Shuttle Orbiter for the First Twelve Orbital Flights: Final Report," AFFTC TR-85-11 (1985).

Active Cooling Approaches

There are other proposed methods for protecting vehicles from high temperature while flying at high speed or during reentry. Several active cooling concepts have been proposed where liquid is circulated through a hot area, then through a radiator to dissipate the heat. These concepts are quite complex and the risk is very high: failure of an active cooling system could result in loss of a hypersonic vehicle within a few seconds. None has been demonstrated in flight. Although work is continuing on active cooling concepts, their application will probably not be realized for many years.

As we look ahead to the future of aviation, it is easy to merely assess the current fleet of successful aircraft or spacecraft, and decide on what improvements we can provide, without considering the history and evolution that produced these vehicles. The danger is that some of the past problems will reappear unless the design and test communities are aware of their history. This paper has attempted to summarize some of the problems that have been encountered, and resolved, during the technology explosion in aviation that has occurred over the last 60 years. The manner in which the problems were discovered, the methods used to determine causes, and the final resolution or correction that was implemented have been presented. Hopefully these brief summaries of historical events will stimulate further research by our younger engineers and historians into the various subjects covered, and to that end, the following works are particularly relevant.

Recommended Additional Readings

Reports, Papers, Articles, and Presentations:

Johnny G. Armstrong, "Flight Planning and Conduct of the X-15A-2 Envelope Expansion Program," AFFTC TD-69-4 (1969).

Theodore G. Ayers and James B. Hallissy, "Historical Background and Design Evolution of the Transonic Aircraft Technology Supercritical Wing," NASA TM-81356 (1981).

John V. Becker, "The X-15 Project," *Astronautics and Aeronautics,* vol. 2, no. 2 (Feb. 1964).

John V. Becker, "The X-15 Program in Retrospect," *Raumfahrtforschung* (Mar.–Apr. 1969).

C.C. Clark and C.H. Woodling, "Centrifuge Simulation of the X-15 Research Aircraft," Naval Air Development Center NADC MA-5916 (1959).

Dwain A. Deets and Kenneth J. Szalai, "Design and Flight Experience with a Digital Fly-By-Wire Control System in an F-8 Airplane," NATO Advisory Group for Aeronautical Research and Development, Conference Paper AGARD-CP-137 (1974).

Robert G. Hoey, "Horizontal Landing Techniques for Hypersonic Vehicles," NATO Advisory Group for Aeronautical Research and Development, AGARD Report No. 428 (1963).

Robert G. Hoey and Richard E. Day, "Mission Planning and Operational Procedures for the X-15 Airplane," NASA TN D-1158 (1962).

Robert G. Hoey, et al., "Flight Test Results from the Entry and Landing of the Space Shuttle Orbiter for the First Twelve Orbital Flights," AFFTC TR-85-11 (1985).

E.C. Holloman, "Piloting Performance During the Boost of the X-15 Airplane to High-Altitude," NASA TN-D-2289 (1964).

C. Hwang, E. Jonson, G. Mills, T. Noll, and M. Farmer, "Wind Tunnel Test of a Fighter Aircraft Wing/Store Flutter Suppression System: An International Effort," NATO Advisory Group for Aeronautical Research and Development, AGARD R-689 (1980).

Kenneth W. Iliff and L.W. Taylor, Jr., "Determination of Stability Derivatives from Flight Data Using a Newton-Raphson Minimization Technique," NASA TN-D-6579 (1972).

Kenneth W. Iliff, "Aircraft Parameter Estimation," AIAA Paper 1987-0623 (1987).

E.E. Kordes and R.B. Noll, "Flight Flutter Results for Flat Rectangular Panels," NASA TN-D-1058 (1962).

E.E. Kordes, R.D. Reed, and A.L. Dawdy, "Structural Heating Experiences on the X-15 Airplane," NASA TM-X-711 (1962).

Andrew M. Lizotte and Michael J. Allen, "Twist Model Development and Results From the Active Aeroelastic Wing F/A-18 Aircraft," NASA TM-2005-212861 (2005).

NASA, "Flight Test Results Pertaining to the Space Shuttlecraft," NASA TM-X-2101 (1970).

R.B. Noll, C.R. Jarvis, C. Pembo, W.P. Lock, and B.J. Scott, "Aerodynamic and Control-System Contributions to the X-15 Airplane Landing-Gear Loads," NASA TN-D-2090 (1963).

Weneth D. Painter and George J. Sitterle, "Ground and Flight Test Methods for Determining Limit Cycle and Structural Resonance Characteristics of Aircraft Stability Augmentation Systems," NASA TN-D-6867 (1972).

NASA, "Shuttle Performance: Lessons Learned," NASA CP-2283, pt. 1–2, (1983).

Paul Pellicano, Joseph Krumenacker, and David Van Hoy, "X-29 High Angle-of-Attack Flight Test Procedures, Results, and Lessons Learned," Society of Flight Test Engineers 21st Annual Symposium (1990).

F.S. Petersen, H.A. Rediess, and J. Weil, "Lateral-Directional Control Characteristics of the X-15 Airplane," NASA TM-X-726 (1962).

Paul W. Phillips and Stephen B. Smith, "AFTI/F-111 Mission Adaptive Wing (MAW) Automatic Flight Control System Modes Lift and Drag Characteristics," AFFTC TR-89-03 (1989).

William H. Phillips, "Effect of Steady Rolling on Longitudinal and Directional Stability," NACA TN-627 (1948).

B.L. Schofield, D.F. Richardson, and P.C. Hoag, "Terminal Area Energy Management, Approach, and Landing Investigation for Maneuvering Reentry Vehicles using F-111A and NB-52B Aircraft," AFFTC TD-70-2 (1970).

L.W. Taylor, Jr., "Analysis of a Pilot-Airplane Lateral Instability Experienced with the X-15 Airplane," NASA TN-D-1059 (1961).

L.W. Taylor, Jr., and E.J. Adkins, "Adaptive Flight Control Systems—Pro and Con," NASA TM-X-56008 (1964).

L.W. Taylor, Jr., and G.B. Merrick, "X-15 Airplane Stability Augmentation System," NASA TN-D-1157 (1962).

L.W. Taylor, Jr., and J.W. Smith, "An Analysis of the Limit-Cycle and Structural-Resonance Characteristics of the X-15 Stability Augmentation System," NASA TN-D-4287 (1967).

Robert A. Tremant, "Operational Experience and Characteristics of the X-15 Flight Control System," NASA TN-D-1402 (1962).

Books and Monographs:

Joseph R. Chambers, *Partners in Freedom: Contributions of the Langley Research Center to U.S. Military Aircraft of the 1990s*, SP-2000-4519 (Washington, DC: NASA, 2000).

Richard E. Day, *Coupling Dynamics in Aircraft: A Historical Perspective*, SP-532 (1997).

Dennis R. Jenkins, *X-15: Extending the Frontiers of Flight*, SP-2007-562 (Washington, DC: NASA, 2007).

Ben R. Rich and Leo Janos, *Skunk Works: A Personal Memoir of My Years of Lockheed* (Boston: Little, Brown, and Co., 1994).

Milton O. Thompson, *At the Edge of Space: The X-15 Flight Program* (Washington, DC: Smithsonian Institution Press, 1992).

Milton O. Thompson with J.D. Hunley, *Flight Research: Problems Encountered and What They Should Teach Us*, SP-2000-4522 (Washington, DC: NASA, 2000).

THIS PAGE INTENTIONALLY BLANK

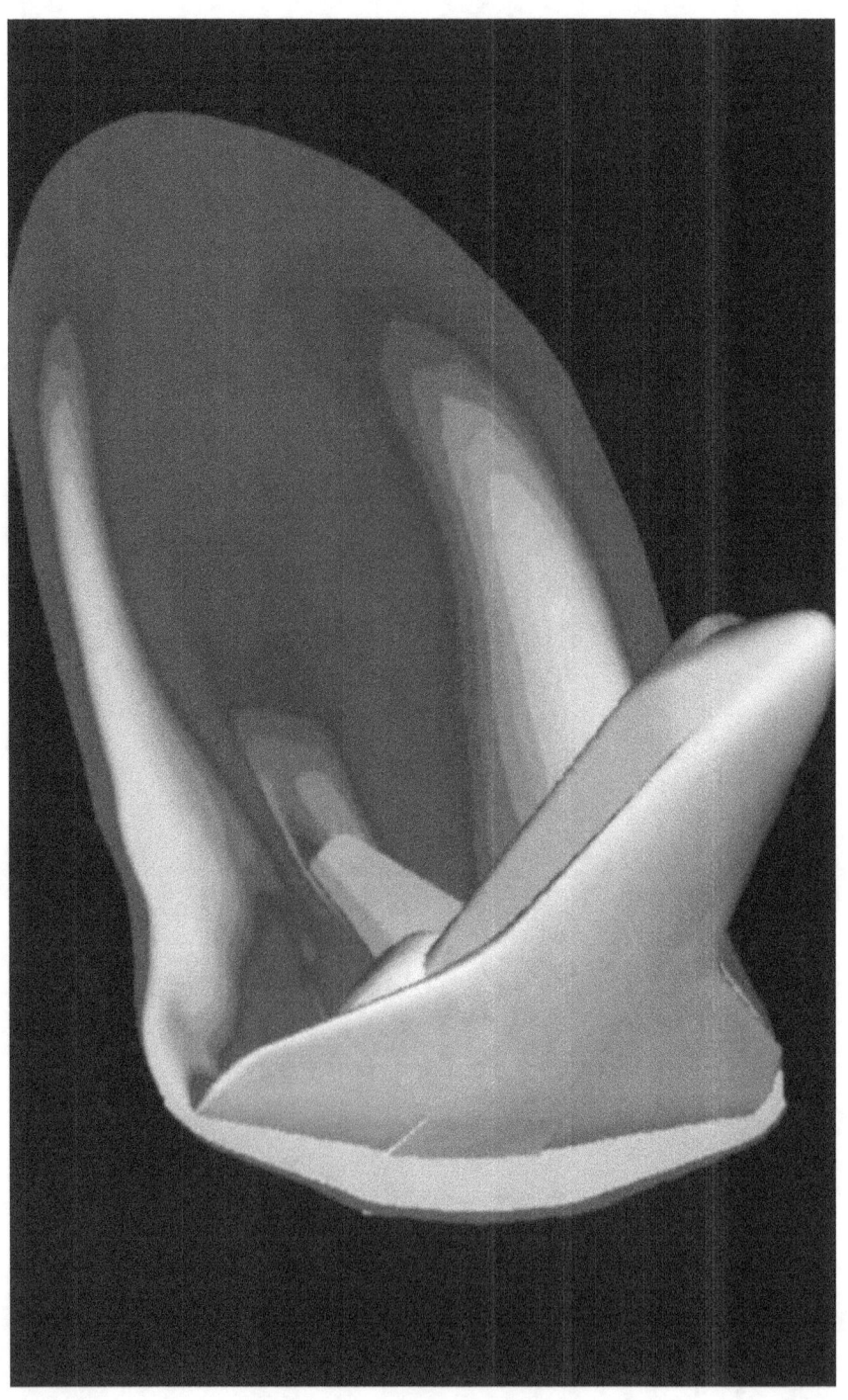

Langley CFD Shuttle study, showing the hypersonic flow pattern as it executes atmospheric entry from orbit. NASA.

CASE 7

NASA and the Evolution of Computational Fluid Dynamics

John D. Anderson, Jr.

The expanding capabilities of the computer readily led to its increasing application to the aerospace sciences. NACA–NASA researchers were quick to realize how the computer could supplement traditional test methodologies, such as the wind tunnel and structural test rig. Out of this came a series of studies leading to the evolution of computer codes used to undertake computational fluid dynamics and structural predictive studies. Those codes, refined over the last quarter century and available to the public, are embodied in many current aircraft and spacecraft systems.

THE VISITOR TO THE SMITHSONIAN INSTITUTION'S National Air and Space Museum (NASM) in Washington, DC, who takes the east escalator to the second floor, turns left into the Beyond the Limits exhibit gallery, and then turns left again into the gallery's main bay is suddenly confronted by three long equations with a bunch of squiggly symbols neatly painted on the wall. These are the Navier-Stokes equations, and the NASM (to this author's knowledge) is the world's only museum displaying them so prominently. These are not some introductory equations drawn for a first course in algebra, with simple symbols like a + b = c. Rather, these are "partial derivatives" strung together from the depths of university-level differential calculus. What are the Navier-Stokes equations, why are they in a gallery devoted to the history of the computer as applied to flight vehicles, and what do they have to do with the National Aeronautics and Space Administration (which, by the way, dominates the artifacts and technical content exhibited in this gallery)?

The answers to all these questions have to do with computational fluid dynamics (CFD) and the pivotal role played by the National Aeronautics and Space Administration (NASA) in the development of CFD over the past 50 years. The role played by CFD in the study and understanding of fluid dynamics in general and in aerospace engineering in particular has grown from a fledgling research activity in the 1960s to a powerful "third" dimension in the profession, an equal partner with pure experiment and pure theory. Today it is used to help design airplanes, study the aerodynamics of automobiles, enhance wind tunnel testing, develop global weather models, and predict the tracts of hurricanes, to name just a few. New jet engines are developed with an extensive use of CFD to model flows and combustion processes, and even the flow field in the reciprocating engine of the average family automobile is laid bare for engineers to examine and study using the techniques of CFD.

The history of the development of computational fluid dynamics is an exciting and provocative story. In the whole spectrum of the history of technology, CFD is still very young, but its importance today and in the future is of the first magnitude. This essay offers a capsule history of the development of theoretical fluid dynamics, tracing how the Navier-Stokes equations came about, discussing just what they are and what they mean, and examining their importance and what they have to do with the evolution of computational fluid dynamics. It then discusses what CFD means to NASA—and what NASA means to CFD. Of course, many other players have been active in CFD, in universities, other Government laboratories, and in industry, and some of their work will be noted here. But NASA has been the major engine that powered the rise of CFD for the solution of what were otherwise unsolvable problems in the fields of fluid dynamics and aerodynamics.

The Evolution of Fluid Dynamics from da Vinci to Navier-Stokes

Fluid flow has fascinated humans since antiquity. The Phoenicians and Greeks built ships that glided over the water, creating bow waves and leaving turbulent wakes behind. Leonardo da Vinci made detailed sketches of the complex flow fields over objects in a flowing stream, showing even the smallest vortexes created in the flow. He observed that the force exerted by the water flow over the bodies was proportional to the cross-sectional area of the bodies. But nobody at that time had a clue about the physical laws that governed such flows. This prompted some substantive experimental fluid dynamics in the 17th and 18th centuries. In the early 1600s, Galileo observed from the falling of bodies through the air that the resistance force (drag) on the body was proportional to the air density. In 1673, the French scientist Edme Mariotte published the first experiments that proved the important fact that the aerodynamic force on an object in a flow varied as the *square* of the flow velocity, not *directly* with the velocity itself as believed by da Vinci and Galileo before him.[1] Seventeen years later, Dutch scientist Christiaan Huygens published the same result from his experiments. Clearly, by this time, fluid dynamics was of intense interest, yet the only way to learn about it was by experiment, that is, empiricism.[2]

This situation began to change with the onset of the scientific revolution in the 17th century, spearheaded by the theoretical work of British polymath Isaac Newton. Newton was interested in the flow of fluids, devoting the whole Book II of his *Principia* to the subject of fluid dynamics. He conjured up a theoretical picture of fluid flow as a stream of particles in straight-line rectilinear motion that, upon impact with an object, instantly changed their motion to follow the surface of the object. This picture of fluid flow proved totally wrong, as Newton himself suspected, and it led to Newton's "sine-squared law" for the force on a object immersed in a flow, which famously misled many early aeronautical pioneers. But if quantitatively incorrect, it was nevertheless the first to theoretically attempt an explanation of why the aerodynamic force varied directly with the square of the flow velocity.[3]

Newton, through his second law contributed indirectly to the breakthroughs in theoretical fluid dynamics that occurred in the 18th century. Newton's second law states that the force exerted on a moving object is directly proportional to the time rate of change of momentum of that object. (It is more commonly known as "force equals mass time acceleration," but this is not found in the *Principia*). Applying Newton's second law to an infinitesimally small fluid element moving as part of a

---

1. John D. Anderson, Jr., *A History of Aerodynamics*, (Cambridge: Cambridge University Press, 1997), pp. 23, 25, 31, 35; Edme Mariotte, *Traite de la percussion ou choc des corps* (Paris: Academy of Sciences, 1673).
2. For further references in the history of fluid dynamics, see R. Giacomelli and E. Pistolesi, "Historical Sketch," in William F. Durand, ed., *Aerodynamic Theory*, vol. 1 (Berlin: Julius Springer Verlag, 1934); Theodore von Kármán, *Aerodynamics* (Cornell: Cornell University Press, 1954); G.A. Tokaty, *A History and Philosophy of Fluid Mechanics* (Henley, UK: G.T. Foulis and Co., Ltd., 1971); Olivier Darrigol, *Worlds of Flow: A History of Hydrodynamics from the Bernoullis to Prandtl* (Oxford: Oxford University Press, 2005); and Michael Eckert, *The Dawn of Fluid Dynamics: A Discipline Between Science and Technology* (Weinheim, Germany: Wiley-VCH Verlag GmbH & Co. KGaA, 2006). All trace the evolution of fluid mechanics and aerodynamics in great detail, with excellent references to primary sources and other works.
3. Isaac Newton, *Mathematical Principles of Natural Philosophy* (Chicago: Encyclopedia Britannica, Inc., 1952 ed. of a 1687 work), pp. 159–267.

fluid flow that is actually a continuum material, Leonhard Euler constructed an equation for the motion of the fluid as dictated by Newton's second law. Euler, arguably the greatest scientist and mathematician of the 18th century, modeled a fluid as a continuous collection of infinitesimally small fluid elements moving with the flow, where each fluid element can continually change its size and shape as it moves with the flow, but, at the same time, all the fluid elements taken as a whole constitute an overall picture of the flow as a continuum. That was somewhat in contrast to the individual and distinct particles in Newton's impact theory model mentioned previously. To his infinitesimally small fluid element, Euler applied Newton's second law in a form that used differential calculus, leading to a differential equation relating the variation of velocity and pressure throughout the flow. This equation, simply labeled the "momentum equation," came to be known simply as Euler's equation. In the 18th century, it constituted a bombshell in launching the field of theoretical fluid dynamics and was to become a pivotal equation in CFD in the 20th century, a testament to Euler's insight and its application.

There is a second fundamental principle that underlies all of fluid dynamics, namely that mass is conserved. Euler applied this principle also to his model of an infinitesimally small moving fluid element, constructing another differential equation labeled the "continuity equation." These two equations, the continuity equation and the momentum equation, were published in 1753, considered one of his finest works. Moreover, these two equations, 200 years later, were to become the physical foundations of the early work in computational fluid dynamics.[4]

After Euler's publication, for the next century all serious efforts to theoretically calculate the details of a fluid flow centered on efforts to solve these Euler equations. There were two problems, however. The first was mathematical: Euler's equations are nonlinear partial differential equations. In general, nonlinear partial differential equations are not easy to solve. (Indeed, to this day there exists no general analytical solution to the Euler equations.) When faced with the need to solve a practical problem, such as the airflow over an airplane wing, in most cases an exact solution of the Euler equations is unachievable. Only by simplifying the fluid dynamic problem and allowing certain terms in the

---

4. As explicated by Leonhard Euler in his *Principles of the Motion of Fluids* (1752), *General Principles of the State of Equilibrium of Fluids* (1753), and *General Principles of the Motion of Fluids* (1755).

equations to be either dropped or modified in such a fashion to make the equations linear rather than nonlinear can these equations be solved in a useful manner. But a penalty usually must be paid for this simplification because in the process at least some of the physical or geometrical accuracy of the flow is lost.

The second problem is physical: when applying Newton's second law to his moving fluid element, Euler did not account for the effects of friction in the flow, that is, the force due to the frictional shear stresses rubbing on the surfaces of the fluid element as it moves in the flow. Some fluid dynamic problems are reasonably characterized by ignoring the effects of friction, but the 18th and 19th century theoretical fluid dynamicists were not sure, and they always worried about what role friction plays in a flow. However, a myriad of other problems are dominated by the effect of friction in the flow, and such problems could not even be addressed by applying the Euler equations. This physical problem was exacerbated by controversy as to what happens to the flow moving along a solid surface. We know today that the effect of friction between a fluid flow and a solid surface (such as the surface of an airplane wing) is to cause the flow velocity right at the surface to be zero (relative to the surface). This is called the no-slip condition in modern terminology, and in aerodynamic theory, it represents a "boundary condition" that must be accounted for in conjunction with the solution of the governing flow equations. The no-slip condition is fully understood in modern fluid dynamics, but it was by no means clear to 19th century scientists. The debate over whether there was a finite relative velocity between a solid surface and the flow immediately adjacent to the surface continued into the 2nd decade of the 20th century.[5] In short, the world of theoretical fluid dynamics in the 18th and 19th centuries was hopelessly cast adrift from many desired practical applications.

The second problem, that of properly accounting for the effects of friction in the flow, was dealt with by two mathematicians in the middle 19th century, France's Claude-Louis-Marie-Henri Navier, and Britain's Sir George Gabriel Stokes. Navier, an instructor at the famed *École nationale des ponts et chaussées*, changed the pedagogical style of teaching civil engineering from one based mainly on cut-and-try empiricism to a program emphasizing physics and mathematical analysis. In 1822, he

---

5. Anderson, *History of Aerodynamics*, p. 89.

gave a paper to the Academy of Sciences that contained the first accurate representation of the effects of friction in the general partial differential momentum equation for fluid flow.[6] Although Navier's equations were in the correct form, his theoretical reasoning was greatly flawed, and it was almost a fluke that he arrived at the correct terms. Moreover, he did not fully understand the physical significance of what he had derived. Later, quite independently from Navier, Stokes, a professor at Cambridge who occupied the Lucasian Chair at Cambridge University (the same chair Newton had occupied a century and a half earlier) took up the derivation of the momentum equation including the effects of friction. He began with the concept of internal shear stress caused by friction in the fluid and derived the governing momentum equation much like it would be derived today in a fluid dynamics class, publishing it in 1845.[7] Working independently, then, Navier and Stokes derived the basic equations that describe fluid flows and contain terms to account for friction. They remain today the fundamental equations that fluid dynamicists employ for analyzing frictional flows.

Finally, in addition to the continuity and momentum equations, a third fundamental physical principle is required for any flow that involves high speeds and in which the density of the flow changes from one point to another. This is the principle of conservation of energy, which holds that energy cannot be created or destroyed; it can only change its form. The origin of this principle in the form of the first law of thermo-dynamics is found in the history of the development of thermodynamics in the late 19th century. When applied to a moving fluid element in Euler's model, and including frictional dissipation and heat transfer by thermal conduction, this principle leads to the energy equation for fluid flow.

So there it is, the origin of the three Navier-Stokes equations exhibited so prominently at the National Air and Space Museum. They are horribly nonlinear partial differential equations. They are also fully coupled together because the variables of pressure, density, and velocity that appear in these equations are all dependent on each other. Obtaining a

---

6. C.L.M.H. Navier, "Mémoire sur les du mouvement des fluids," *Mémoires de l'Academie Royale des Sciences*, No. 6 (1823), pp. 389–416.

7. G.G. Stokes, "On the Theories of the Internal Friction of Fluids in Motion, and of the Equilibrium and Motion of Elastic Solids," *Transactions of the Cambridge Philosophical Society*, vol. 8, no. 22 (1845), pp. 287–342.

general analytical solution of the Navier-Stokes equations is much more daunting than the problem of obtaining a general analytical solution of the Euler equations, for they are far more complex. There is today no general analytical solution of the Navier-Stokes equations (as is likewise true in the case of the Euler equations). Yet almost all of modern computational fluid dynamics is based on the Navier-Stokes equations, and all of the modern solutions of the Navier-Stokes equations are based on computational fluid dynamics.

Computational Fluid Dynamics: What It Is, What It Does

What constitutes computational fluid dynamics? The basic equations of fluid dynamics, the Navier-Stokes equations, are expressions of three fundamental principles: (1) mass is conserved (the continuity equation), (2) Newton's second law (the momentum equation), and (3) the energy equation (the first law of thermodynamics). Moreover, these equations in their most general form are either partial differential equations (as we have discussed) or integral equations (an alternate form we have not discussed involving integrals from calculus).

The partial differential equations are those exhibited at the NASM. Computational fluid dynamics is the art and science of replacing the partial derivatives (or integrals, as the case may be) in these equations with discrete algebraic forms, which in turn are solved to obtain numbers for the flow-field values (pressure, density, velocity, etc.) at discrete points in time and or space.[8] At these selected points in the flow, called grid points, each of the derivatives in each of the equations are simply replaced with numbers that are advanced in time or space to obtain a solution for the flow. In this fashion, the partial differential equations are replaced by a large number of algebraic equations, which can then be solved simultaneously for the flow variables at all the grid points.

The end product of the CFD process is thus a collection of numbers, in contrast to a closed-form analytical solution (equations). However, in the long run, the objective of most engineering analyses, closed-form or otherwise, is a quantitative description of the problem: that is, numbers. Along these lines, in 1856, the famous British scientist James Clerk Maxwell wrote: "All the mathematical sciences are founded on relations between physical laws and laws of numbers, so that the aim of exact

---

8. This discussion is elaborated upon in the author's *Computational Fluid Dynamics: The Basics With Applications* (New York: McGraw-Hill, 1995).

science is to reduce the problems of nature to the determination of quantities by operations with numbers."[9] Well over a century later, it is worth noting how well Maxwell captured the essence of CFD: operations with numbers.

Note that computational fluid dynamics results in solutions for the flow only at the distinct points in the flow called grid points, which were identified earlier. In a CFD solution, grid points are either initially distributed throughout the flow and/or generated during the course of the solution (called an "adaptive grid"). This is in theoretical contrast with a closed-form analytical solution for the flow, where the solution is in the form of equations that allow the calculation of the flow variables at any point of one's choosing, that is, an analytical solution is like a continuous answer spread over the whole flow field. Closed-form analytical solutions may be likened to a traditionalist Dutch master's painting consisting of continuous brush strokes, while a CFD solution is akin to a French pointillist consisting of multicolored dots made with a brush tip.

Generating a grid is an essential part of the art of CFD. The spacing between grid points and the geometric ways in which they are arrayed is critical to obtaining an accurate numerical CFD solution. Poor grids almost always ensure poor CFD solutions. Though good grids do not guarantee good CFD solutions, they are essential for useful solutions. Grid generation is a discipline all by itself, a subspecialty of CFD. And grid generation can become very labor-intensive—for some flows over complex three-dimensional configurations, it may take months to generate a proper grid.

To summarize, the Navier-Stokes equations, the governing equations of fluid dynamics, have been in existence for more than 160 years, their creation a triumph of derivative insight. But few knew how to analytically solve them except for a few simple cases. Because of their complexity, they thus could not serve as a practical widely employed tool in the engineer's arsenal. It took the invention of the computer to make that possible. And because it did so, it likewise permitted the advent of computational fluid dynamics. So how did the idea of numerical solutions to the Navier-Stokes equations evolve?

The Concept of Finite Differences Enters the Mathematical Scene

The earliest concrete idea of how to simulate a partial derivative with an algebraic difference quotient was the brainchild of L.F. Richardson in

---

9. Ibid., p. 3.

1910.[10] He was the first to introduce the numerical solution of partial differential equations by replacing each derivative in the equations with an algebraic expression involving the values of the unknown dependent variables in the immediate neighborhood of a point and then solving simultaneously the resulting massive system of algebraic equations at all grid points. Richardson named this approach a "finite-difference solution," a name that has come down without change since 1910. Richardson did not attempt to solve the Navier-Stokes equations, however. He chose a problem reasonably described by a simpler partial differential equation, Laplace's equation, which in mathematical speak is a linear partial differential equation and which the mathematicians classify as an elliptic partial differential equation.[11] He set up a numerical approach that is still used today for the solution of elliptic partial differential equations called a relaxation method, wherein a sweep is taken throughout the whole grid and new values of the dependent variables are calculated from the old values at neighboring grid points, and then the sweep is repeated over and over until the new values at each grid point converges to the old value from the previous sweep, i.e., the numbers "relax" eventually to the correct solution.

In 1928, Richard Courant, K.O. Friedrichs, and Hans Lewy published "On the Partial Difference Equations of Mathematical Physics," a paper many consider as marking the real beginning of modern finite difference solutions; "Problems involving the classical linear partial differential equations of mathematical physics can be reduced to algebraic ones of a very much simpler structure," they wrote, "by replacing the differentials by difference quotients on some (say rectilinear) mesh."[12] Courant, Friedrichs, and Lewy introduced the idea of "marching solutions," whereby a spatial marching solution starts at one end of the flow and literally marches the finite-difference solution step by step from one

---

10. L.F. Richardson, "The Approximate Arithmetical Solution by Finite Differences of Physical Problems Involving Differential Equations, With Application to the Stresses in a Masonry Dam," *Philosophical Transactions of the Royal Society of London*, ser. A, vol. 210 (1910), pp. 307–357.
11. Many partial differential equations fall within one of the following categories: elliptic equations, parabolic equations, and hyperbolic equations, differentiated by their mathematical (and graphical) behavior. The setting-up of a given numerical solution of these equations depends critically on whether the equation in question is elliptic, parabolic, or hyperbolic.
12. Richard Courant, K.O. Friedrichs, and Hans Lewy, "Über die Partiellen Differenzengleichungen der Mathematischen Physik," *Mathematische Annalen*, vol. 100, (1928), pp. 32–74. All three emigrated to the U.S.

end to the other end of the flow. A time marching solution starts with the all the flow variables at each grid point at some instant in time and marches the finite-difference solution at all the grid points in steps of time to some later value of time. These marching solutions can only be carried out for parabolic or hyperbolic partial differential equations, not for elliptic equations.

Courant, Friedrichs, and Lewy highlighted another important aspect of numerical solutions of partial differential equations. Anyone attempting numerical solutions of this nature quickly finds out that the numbers being calculated begin to look funny, make no sense, oscillate wildly, and finally result in some impossible operation such as dividing by zero or taking the square root of a negative number. When this happens, the solution has blown up, i.e., it becomes no solution at all. This is not a ramification of the physics, but rather, a peculiarity of the numerical processes. Courant, Friedrichs, and Lewy studied the stability aspects of numerical solutions and discovered some essential criteria to maintain stability in the numerical calculations. Today, this stability criterion is referred to as the "CFL criterion" in honor of the three who identified it. Without it, many attempted CFD solutions would end in frustration.

So by 1928, the academic foundations of finite difference solutions of partial differential equations were in place. The Navier-Stokes equations finally stood on the edge of being solved, albeit numerically. But who had the time to carry out the literally millions of calculations that are required to step through the solution? For all practical purposes, it was an impossible task, one beyond human endurance. Then came the electronic revolution and, with it, the digital computer.

The Critical Tool: Emergent High-Speed Electronic Digital Computing

During the Second World War, J. Presper Eckert and John Mauchly at the University of Pennsylvania's Moore School of Electrical Engineering designed and built the ENIAC, an electronic calculator that inaugurated the era of digital computing in the United States. By 1951, they had turned this expensive and fragile instrument into a product that was manufactured and sold, a computer they called the UNIVAC, which stands for Universal Automatic Computer. The National Advisory Committee for Aeronautics (NACA) was quick to realize the potential of a high-speed computer for the calculation of fluid dynamic problems. After all, the NACA was in the business of aerodynamics and after 40 years of trying to solve the equations of motion by simplified analysis, it recognized the breakthrough supplied by the computer to solve these equations numerically on a potentially practical basis. In 1954, Remington Rand delivered an ERA 1103 digital computer intended for scientific and engineering calculations to the NACA Ames Aeronautical Laboratory at Sunnyvale, CA. This was a state-of-the-art computer that was the first to employ a magnetic core in place of vacuum tubes for memory. The ERA 1103 used binary arithmetic, a 36-bit word length, and operated on all the bits of a word at a time. One year later, Ames acquired its first stored-program electronic computer, an IBM 650. In 1958, the 650 was replaced by an IBM 704, which in turn was replaced with an IBM 7090 mainframe in 1961.[13]

The IBM 7090 had enough storage and enough speed to allow the first generation of practical CFD solutions to be carried out. By 1963, four additional index registers were added to the 7090, making it the IBM 7094. This computer became the workhorse for the CFD of the 1960s and early 1970s, not just at Ames, but throughout the aerodynamics community; the author cut his teeth solving dissertation on an IBM 7094 at the Ohio State University in 1966. The calculation speed of a digital computer is measured in its number of floating point operations per second (FLOPS). The IBM 7094 could do 100,000 FLOPS, making it about the fastest computer available in the 1960s. With this number of FLOPS, it was possible to carry out for the first time detailed flow-field calculations around a body moving at hypersonic speeds, one of the major activities within the newly formed NASA that drove both computer and algorithm development for CFD. The IBM 7094 was a "mainframe" computer, a large electronic machine that usually filled a room with equipment. The users would write their programs (usually in the FORTRAN language) as a series of logically constructed line statements that would be punched on cards, and the decks of punched cards (sometimes occupying many boxes for just one program) would be fed into a reader that would read the punches and tell the computer what calculations to make. The output from the calculations would be printed on large sheets and returned to the user. One program at a time was fed into the computer, the so-called "batch" operation. The user would submit his or her batch to the computer desk and then return hours or days later to pick up the printed output. As cumbersome as it

---

13. Paul E. Ceruzzi, *Beyond the Limits: Flight Enters the Computer Age* (Cambridge: MIT Press, 1989), p. 15.

may appear today, the batch operation worked. The field of CFD was launched with such batch operations on mainframe computers like the IBM 7094. And NASA Ames was a spearhead of such activities. Indeed, because of the synergism between CFD and the computers on which it worked, the demands on the central IBM installation at Ames grew at a compounded rate of over 100 percent per year in the 1960s.

With these computers, it became practical to set up CFD solutions of the Euler equations for two-dimensional flows. These solutions could be carried out with a relatively small number of grid points in the flow, typically 10,000 to 100,000 points, and still have computer run times on the order of hours. Users of CFD in the 1960s were happy to have this capability, and the three primary NASA Research Centers—Langley, Ames, and Lewis (now Glenn)—made major strides in the numerical analysis of many types of flows, especially in the transonic and hypersonic regimes. The practical calculation of inviscid (that is, frictionless), three-dimensional flows and especially any type of high Reynolds number flows was beyond the computer capabilities at that time.

This situation changed markedly when the supercomputer came on the scene in the 1970s. NASA Ames acquired the Illiac IV advanced parallel-processing machine. Designed at the University of Illinois, this was an early and controversial supercomputer, one bridging both older and newer computer architectures and processor approaches. Ames quickly followed with the installation of an IBM 360 time-sharing computer. These machines provided the capability to make CFD calculations with over 1 million grid points in the flow field with a computational speed of more than 10^6 FLOPS. NASA installed similar machines at the Langley and Lewis Research Centers. On these machines, NASA researchers made the first meaningful three-dimensional inviscid flow-field calculations and significant two-dimensional high Reynolds number calculations. Supercomputers became the engine that propelled CFD into the forefront of aerospace design as well as research. Bigger and better supercomputers, such as the pioneering Cray-1 and its successor, the Cray X-MP, allowed grids of tens of millions of grid points to be used in a flow-field calculation with speeds beginning to approach the hallowed goal of gigaflops (10^9 floating point operations per second). Such machines made it possible to carry out numerical solutions of the Navier-Stokes equations for three-dimensional fairly high Reynolds number viscous flows. The first three-dimensional Navier-Stokes solutions of the complete flow field around a complete airplane at angle of attack came on the scene in the 1980s, enabled by these supercomputers. Subsonic, transonic, supersonic, and hypersonic flow solutions covered the whole flight regime. Again, the major drivers for these solutions were the aerospace research and development problems tackled by NASA engineers and scientists. This headlong development of supercomputers has continued unabated. The holy grail of CFD researchers in the 1990s was the teraflop machine (10^{12} FLOPS); today, it is the petaflop (10^{15} FLOPS) machine. Indeed, recently the U.S. Energy Department has contracted with IBM to build a 20-petaflop machine in 2012 for calculations involving the safety and reliability of the Nation's aging nuclear arsenal.[14] Such a machine will aid the CFD practitioner's quest for the ultimate flow-field calculations—direct numerical simulation (DNS) of turbulent flows, an area of particularly interest to NASA researchers.

Some Seminal Solutions and Applications

We have discussed the historical evolution of the governing flow equations, the first essential element of CFD. We then discussed the evolution of the high-speed digital computer, the second essential element of CFD. We now come to the crux of this article, the actual CFD flow-field solutions, their evolution, and their importance. Computational fluid dynamics has grown exponentially in the past four decades, rendering any selective examination of applications problematical. This case study examines four applications that have driven the development of CFD to its present place of prominence: the supersonic blunt body problem, transonic airfoils and wings, Navier-Stokes solutions, and hypersonic vehicles.

The Supersonic Blunt Body Problem

On November 1, 1952, the United States detonated a 10.4-megaton hydrogen test device on Eniwetok Atoll in the Marshall Islands, the first implementation of physicist Edward Teller's concept for a "super bomb" and a major milestone toward the development of the American hydrogen bomb. With it came the need for a new entry vehicle beyond the long-range strategic bomber, namely the intercontinental ballistic missile (ICBM). This vehicle would be launched by a rocket booster, go into a suborbital trajectory in space, and then enter Earth's atmosphere

---

14. "Super Supercomputers," *Aviation Week* (Feb. 16, 2009).

at hypersonic speeds near orbital velocity. This was a brand-new flight regime, and the design of the entry vehicle was dominated by an emerging design consideration: aerodynamic heating. Knowledge of the existence of aerodynamic heating was not new. Indeed, in 1876, Lord Rayleigh published a paper in which he noted that the compression process that creates a high stagnation pressure on a high-velocity body also results in a correspondingly large increase in temperature. In particular, he commented on the flow-field characteristic of a meteor entering Earth's atmosphere, noting: "The resistance to a meteor moving at speeds comparable with 20 miles per second must be enormous, as also the rise of temperature due to the compression of the air. In fact it seems quite unnecessary to appeal to friction in order to explain the phenomena of light and heat attending the entrance of a meteor into the earth's atmosphere."[15] We note that 20 miles per second is a Mach number greater than 100. Thus, the concept of aerodynamic heating on very high-speed bodies dates back before the 20th century. However, it was not until the middle of the 20th century that aerodynamic heating suddenly became a showstopper in the design of high-speed vehicles, initiated by the pressing need to design the nose cones of ICBMs.

In 1952, conventional wisdom dictated that the shape of a missile's nose cone should be a slender, sharp-nosed configuration. This was a natural extension of good supersonic design in which the supersonic body should be thin and slender with a sharp nose, all designed to reduce the strength of the shock wave at the nose and therefore reduce the supersonic wave drag. (Among airplanes, the Douglas X-3 Stiletto and the Lockheed F-104A Starfighter constituted perfect exemplars of good supersonic vehicle design, with long slender fuselage, sharp noses, and very thin low aspect ratio [that is, stubby] wings having extremely sharp leading edges. This is all to reduce the strength of the shock waves on the vehicle. The X-3 and F-104 were the first jet airplanes designed for flight at Mach 2, hence their design was driven by the desire to reduce wave drag.) With this tradition in mind, early thinking of ICBM nose cones for hypersonic flight was more of the same, only more so. On the other hand, early calculations showed that the aerodynamic heating to such slender bodies would be enormous. This conventional wisdom was turned on its head in 1951 because of an epiphany by Harry

15. J.W. Strutt (Lord Rayleigh), "On the Resistance of Fluids," *Philosophical Magazine*, ser. 57 (1876), pp. 430–441.

Julian Allen ("Harvey" Allen to his friends because of Allen's delight in the rabbit character named Harvey, played by Jimmy Stewart in the movie of the same name). Allen was at that time the Chief of the High-Speed Research Division at the NACA Ames Research Laboratory. One day, Harvey Allen walked into the office and simply stated that hypersonic bodies should "look like cannonballs."

His reasoning was so fundamental and straightforward that it is worth noting here. Imagine a vehicle coming in from space and entering the atmosphere. At the edge of the atmosphere the vehicle velocity is high, hence it has a lot of kinetic energy (one-half the product of its mass and velocity squared). Also, because it is so far above the surface of Earth (the outer edge of the atmosphere is about 400,000 feet), it has a lot of potential energy (its mass times its distance from Earth times the acceleration of gravity). At the outer edge of the atmosphere, the vehicle simply has a lot of energy. By the time it impacts the surface of Earth, its velocity is zero and its height is zero—no kinetic or potential energy remains. Where has all the energy gone? The answer is the only two places it could: the air itself and the body. To reduce aerodynamic heating to the body, you want more of this energy to go into the air and less into the body. Now imagine two bodies of opposite shapes, a very blunt body (like a cannonball) and a very slender body (like a needle), both coming into the atmosphere at hypersonic speeds. In front of the blunt body, there will be a very strong bow shock wave detached from the surface with a very high gas temperature behind the strong shock (typically about 8,000 kelvins). Hence the air is massively heated by the strong shock wave. A lot of energy goes into the air, and therefore, only a moderate amount of energy goes into the body. In contrast, in front of the slender body there will be a much weaker attached shock wave with more moderate gas temperatures behind the shock. Hence the air is only moderately heated, and a massive amount of energy is left to go into the body. As a result, a blunt body shape will reduce the aerodynamic heating in comparison to a slender body. Indeed, if a slender body would be used, the heating would melt and blunt the nose anyway. This was Allen's thinking. It led to the use of blunt noses on all modern hypersonic vehicles, and it stands as one of the most important aerodynamic contributions of the NACA over its history.

When Allen introduced his blunt body concept in the early 1950s, there were no theoretical solutions of the flow over a blunt body moving at supersonic or hypersonic speeds. In the flow behind the strong curved bow shock wave, the flow behind the almost vertical portion of the shock near the centerline is subsonic, and that behind the weaker, more inclined part of the shock wave further above the centerline is supersonic. There were no pure theoretical solutions to this flow. Numerical solutions of this flow were tried in the 1950s, but all without success. Whatever technique worked in the subsonic region of the flow fell apart in the supersonic region, and whatever technique worked in the supersonic region of the flow fell apart in the subsonic region. This was a potential disaster, because the United States was locked in a grim struggle with the Soviet Union to field and employ intercontinental and intermediate-range ballistic missiles, and the design of new missile nose cones desperately needed solutions of the flow over the body were the United States to ever successfully field a strategic missile arsenal.

On the scene now crept CFD. A small ray of hope came from one of the NACA's and later NASA's most respected theoreticians, Milton O. Van Dyke. Spurred by the importance of solving the supersonic blunt body problem, Van Dyke developed an early numerical solution for the blunt body flow field using an inverse approach: take a curved shock wave of given shape, calculate the flow behind the shock, and solve for the shape of the body that would generate the assumed shock shape. In turn, the flow over a blunt body of given shape could be approached by repetitive applications of this inverse solution, eventually converging to the shape of interest. If critical, it was nevertheless a potentially tedious task that could have consumed thousands of hours by hand calculation, but by using the early IBM computers at Ames, Van Dyke was able to obtain the first reliable numerical solution of the supersonic blunt body flow field, publishing his pioneering work in the first NASA Technical Report issued after the establishment of the Agency.[16] Van Dyke's solution constituted the first important and practical use of CFD but was not without limitations. Although the first major advancement toward the solution of the supersonic blunt body problem, it was only half a loaf. His procedure worked well in the subsonic region of the flow field, but it could penetrate only a small distance into the supersonic region before blowing up. A uniform solution of the whole flow field, including both the subsonic and supersonic regions, was still not obtainable. The supersonic blunt body problem rode into the decade of

---

16. Milton O. Van Dyke, "Supersonic Flow Past a Family of Blunt Axisymmetric Bodies," NASA Technical Report R-1 (1959).

the 1960s as daunting as it ever was. Then came the breakthrough, which was both conceptual and numerical.

First the conceptual breakthrough: at this time the flow was being calculated as a steady flow using the Euler equations, i.e., the flow was assumed to be inviscid (frictionless). For this flow, the governing partial differential equations of continuity, momentum, and energy (the Euler equations) exhibited one type of mathematical behavior (called elliptic behavior) in the subsonic region of the flow and a completely different type of mathematical behavior (called hyperbolic behavior) in the supersonic region of the flow. The equations themselves remain identical in these two regions, but the actual behavior of the mathematical solutions is different. (This is no real surprise because the physical behavior of the flow is certainly different between a subsonic and a supersonic flow.) This change in the mathematical characteristics of the equations was the root cause of all the problems in obtaining a solution to the supersonic blunt body problem. Hence, any numerical solution appropriate for the elliptic (subsonic) region simply was ill-posed in the supersonic region, and any numerical solution appropriate for the hyperbolic (supersonic) region was ill-posed in the subsonic region. Hence, no unified solutions for the whole flow field could be obtained. Then, in the middle 1960s, the following idea surfaced: the Euler equations written for an unsteady flow (carrying along the time derivatives in the equations) were completely hyperbolic with respect to time no matter whether the flow were locally subsonic or supersonic. Why not solve the blunt body flow field by first arbitrarily assuming flow-field properties at all the grid points, calling this the initial flow field at time zero, and then solving the unsteady Euler equations in steps of time, obtaining new flow-field values at each new step in time? The problem is properly posed because the unsteady equations are hyperbolic with respect to time throughout the whole flow field. After continuing this process over a large number of time steps, eventually the changes in the flow properties from one time step to the next grow smaller, and if one goes out to a sufficiently large number of time steps, the flow converges to the steady-state solution. It is this steady-state solution that is desired. The time-marching process is simply a means to the end of obtaining the solution.[17]

---

17. In the old days of obtaining the computer results printed out on paper, I would tell my students to tear off the last page, keep it, and throw the rest out, because the last page contained the answer.

The numerical breakthrough was the implementation of this time-marching approach by means of CFD. Indeed, this process can only be carried out in a practical fashion on a high-speed computer using CFD techniques. The time-marching approach revolutionized CFD. Today, this approach is used for the solution of a whole host of different flow problems, but it got its start with the supersonic blunt body problem. The first practical implementation of the time-marching idea to the supersonic blunt body was carried out by Gino Moretti and Mike Abbett in 1966.[18] Their work transformed the field of CFD. The supersonic blunt body problem in the 1950s and 1960s was worked on by platoons of researchers leading to hundreds of research papers at an untold number of conferences, and it cost millions of dollars. Today, because of the implementation of the time-marching approach by Moretti and Abbett using a finite-difference CFD solution, the blunt body solution is readily carried out in many Government and university aerodynamic laboratories, and is a staple of those aerospace companies concerned with supersonic and hypersonic flight. Indeed, this approach is so straightforward that I have assigned the solution of the supersonic blunt body problem as a homework problem in a graduate course in CFD. What better testimonial of the power of CFD! A problem that used to be unsolvable and for which much time and money was expended to obtain its solution is now reduced to being a "teachable moment" in a graduate engineering course.

CFD and Transonic Airfoils

The analysis of transonic flows suffers from the same problems as those for the supersonic blunt body discussed above. Just considering the flow to be inviscid, the governing Euler equations are highly nonlinear for both transonic and hypersonic flows. From the numerical point of view, both flow fields are mixed regions of locally subsonic and supersonic flows. Thus, the numerical solution of transonic flows originally encountered the same problem as that for the supersonic blunt body problem: whatever worked in the subsonic region did not work in the supersonic region, and vice versa. Ultimately, this problem was solved from two points of view. Historically, the first truly successful CFD solution for the inviscid transonic flow over an airfoil was carried out in

---

18. G. Moretti and M. Abbett, "A Time-Dependent Computational Method for Blunt Body Flows," *AIAA Journal*, vol. 4, no. 12 (1966), pp. 2136–2141.

1971 by Earll Murman and Julian Cole of Boeing Scientific Research Laboratories, whose collaborative research began at the urging of Arnold "Bud" Goldburg, then Chief Scientist of Boeing.[19] They treated a simplified version of the Euler equations called the small-perturbation velocity potential equation. This limited their solutions to the flows over thin airfoils at small angles of attack. Nevertheless, Murman and Cole introduced the concept of writing the finite differences in the equations such that they reached in both the upstream and downstream directions when in the subsonic region, but they reached in only the upwind direction in the supersonic regions. This is motivated by the physical process that in subsonic flow disturbances propagate in all directions but in a supersonic flow disturbances propagate only in the downstream direction. Thus it is proper to form the finite differences in the supersonic region such that they take only information from the upstream side of the grid point.

Today, this approach in modern CFD is called "upwinding" and is part of many modern algorithms in use for all kinds of flows. In 1971, this idea was groundbreaking, and it allowed Murman and Cole to obtain the first successful numerical solutions of the transonic flow over a body. In addition to the restriction of thin airfoils at small angles of attack, however, their use of the small perturbation velocity potential equation also limited their solutions to isentropic flows. This meant that, although their solution captured the semblance of a shock wave in the flow, the location and flow changes across a shock wave were not accurate. Because many transonic flows involve shock waves embedded in the flow, this was definitely a bit of a problem. The solution to this problem involved the numerical treatment of the Euler equations, which, as we have discussed early in this article, accurately pertain to any inviscid flow, not just one with small perturbations and free of shocks.

The finest in such CFD solutions were developed by Antony Jameson, then a professor at Princeton University (and now at Stanford), whose work was heavily sponsored by the NASA Langley Research Laboratory. Using the concept of time marching in combination with a Runge-Kutta time integration of the unsteady equations, Jameson constructed a series of outstanding transonic airfoil codes under the general code name of the FLO codes. These codes entered standard use in many aircraft companies and laboratories. Once again, NASA had been responsible for a

---

19. Earll Murman and Julian D. Cole, "Calculation of Plane Steady Transonic flows," *AIAA Journal*, vol. 9, no. 1 (Jan. 1971), pp 114–121.

major advancement in CFD, helping to develop transonic flow codes that advanced the design of many airfoil shapes used today on modern commercial jet transports.[20]

Navier-Stokes CFD Solutions

As described earlier in this article, the Navier-Stokes equations are the full equations that govern a viscous flow. Solutions of the Navier-Stokes equations are the ultimate in fluid dynamics. To date, no general analytical solutions of these highly nonlinear equations have been obtained. Yet they are the equations that reflect the real world of fluid dynamics. The only way to obtain useful solutions for the Navier-Stokes equations is by means of CFD. And even here such solutions have been slow in coming. The problem has been the very fine grids that are necessary to define certain regions of a viscous flow (in boundary layers, shear layers, separated flows, etc.), thus demanding huge numbers of grid point in the flow field. Practical solutions of the Navier-Stokes equations had to wait for supercomputers such as the Cray X-MP and Cyber 205 to come on the scene. NASA became a recognized and emulated leader in CFD solutions of Navier-Stokes equations, its professionalism evident by its having established the Institute for Computer Applications in Science and Engineering (ICASE) at Langley Research Center, though other Centers as well, particularly Ames, shared this interest in burgeoning CFD.[21]

In particular, NASA researcher Robert MacCormack was responsible for the development of a Navier-Stokes CFD code that, by far, became the most popular and most widely used Navier-Stokes CFD algorithm in the last quarter of the 20th century. MacCormack, an applied mathematician at NASA Ames (and now a professor at Stanford), conceived a straightforward algorithm for the solution of the Navier-Stokes equations, simply identified everywhere as "MacCormack's method."

To understand the significance of MacCormack's method, one must understand the concept of numerical accuracy. Whenever the derivatives in a partial differential equation are replaced by algebraic difference

---

20. A. Jameson, W. Schmidt, and E. Turkel, "Numerical Solution of the Euler Equations by Finite Volume Methods Using Runge-Kutta Time-Stepping Schemes," AIAA Paper 81-1259 (1981); see also A. Jameson, "Successes and Challenges in Computation Aerodynamics," AIAA Paper 87-1184 (1987).
21. Eli Turkel, "Algorithms for the Euler and Navier-Stokes equations for supercomputers," NASA CR-172543 (1985), p. 1.

quotients, there is always a truncation error that introduces a degree of inaccuracy in the numerical calculations. The simplest finite differences, usually involving only two distinct grid points in their formulation, are identified as "first-order" accurate (the least accurate formulation). The next step up, using a more sophisticated finite difference reaching to three grid points, is identified as second-order accurate. For the numerical solution of most fluid flow problems, first-order accuracy is not sufficient; not only is the accuracy compromised, but such algorithms frequently blow up on the computer. (The author's experience, however, has shown that second-order accuracy is usually sufficient for many types of flows.) On the other hand, some of the early second-order algorithms required a large computation effort to obtain this second-order accuracy, requiring many pages of paper to write the algorithm and a lot of computations to execute the solution. MacCormack developed a predictor-corrector two-step scheme that was second-order accurate but required much less effort to program and many fewer calculations to execute. He introduced this scheme in an imaginative paper on hypervelocity impact cratering published in 1969.[22]

MacCormack's method broke open the field of Navier-Stokes solutions, allowing calculation of myriad viscous flow problems, beginning in the 1970s and continuing to the present time, as was as well (in this author's opinion) the most "graduate-student friendly" CFD scheme in existence. Many graduate students have cut their CFD teeth on this method and have been able to solve many viscous flow problems that otherwise could not have attempted. Today, MacCormack's method has been supplanted by several very sophisticated modern CFD algorithms, but even so, MacCormack's method goes down in history as one of NASA's finest contributions to the aeronautical sciences.

Three-Dimensional Flows and Hypersonic Vehicles

Three-dimensional flow-field calculation was, for decades, a frustrating impossibility. I recall colleagues in the 1960s who would have sold their children (at least they said) to be able to calculate three-dimensional flow fields. The number of grid points required for such calculations simply exceeded the capability of any computer at that time. With the advent of supercomputers, however, the practical calculation

---

22. R.W. MacCormack, "The Effect of Viscosity in Hypervelocity Impact Cratering," AIAA Paper 69-354 (1969).

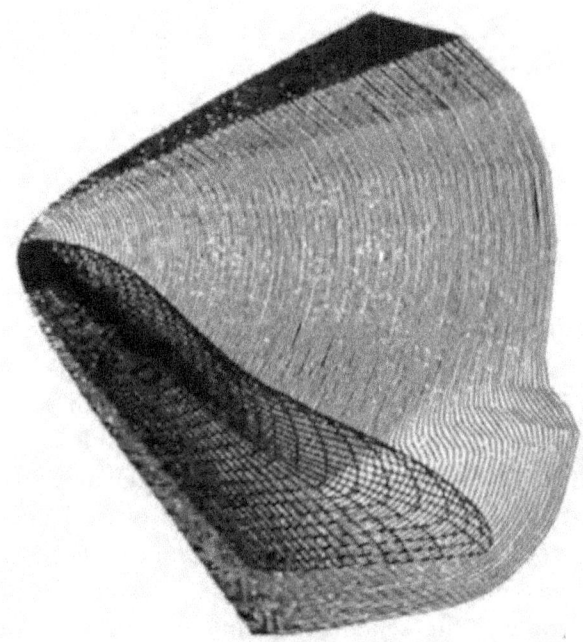

Steady-state three-dimensional shock wave shape, showing the CFD-generated Mach 16.25 flow around a Space Shuttle-like body at a reentry angle of attack of over 39 degrees. From author's collection.

of three-dimensional flow fields became realizable. Once again, NASA researchers led the way. The first truly three-dimensional flow calculation of real importance was carried out by K.J. Weilmuenster in 1983 at the NASA Langley Research Center. He calculated the inviscid flow over a Shuttle-like body at angle of attack, including the shape and location of the three-dimensional bow shock wave. This was no small feat at the time, and it proved to the CFD community that the time had come for such three-dimensional calculations.[23]

This was followed by an even more spectacular success. In 1986, using the predictor-corrector method conceived by NASA Ames Research Center's MacCormack, Joseph S. Shang and S.J. Scherr of the Air Force Flight Dynamics Laboratory (AFFDL) published the first Navier-Stokes calculation of the flow field around a complete airplane. The airplane

---

23. K.J. Weilmuenster, "High Angle of Attack Inviscid Flow Calculations of Shuttle-Like Vehicles with Comparisons to Flight Data," AIAA Paper No. 83-1798 (1983). For Weilmuenster's earlier work, see his "Solution of a large hydrodynamic problem using the STAR-100 computer," NASA TM-X-7394 (1976).

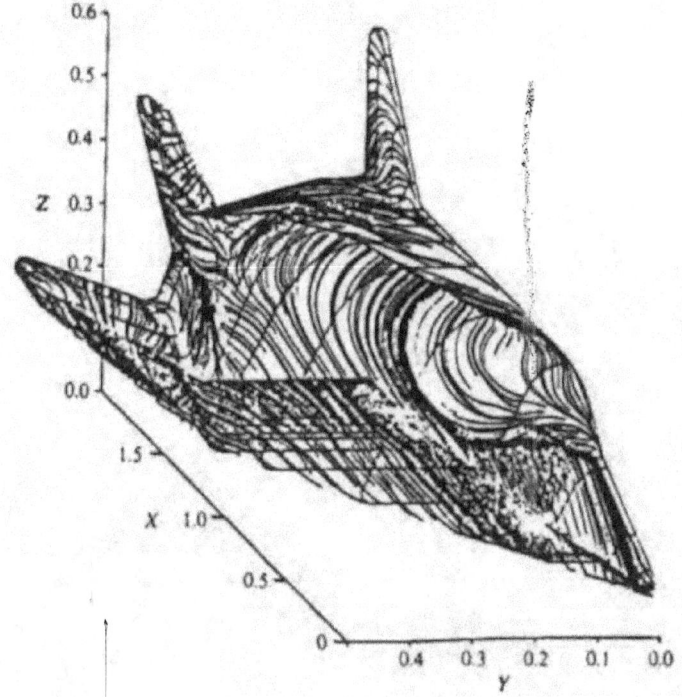

X-24C computed surface streamlines. From author's collection.

was the "X-24C," a proposed (though never completed) rocket-powered Mach 6+ hypersonic test vehicle conceived by the AFFDL, and the calculation was made for flow conditions at Mach 5.95. The mesh system consisted of 475,200 grid points throughout the flow field, and the explicit time-marching procedure took days of computational time on a Cray computer. But it was the first such calculation and a genuine watershed in the advancement of computational fluid dynamics.[24]

Note that both of these pioneering three-dimensional calculations were carried out for hypersonic vehicles, once again underscoring the importance of hypersonic aerodynamics as a major driving force behind the development of computational fluid dynamics and of the leading role played by NASA in driving the whole field of hypersonics.[25]

---

24. J.S. Shang and S.J. Scherr, "Navier-Stokes Solution for a Complete Re-entry Configuration," *Journal of Aircraft*, vol. 23, no. 12 (1986), pp. 881–888.

25. See P. Perrier, "Industrial Methodologies for the Design of Hypersonic Vehicles," and Richard D. Neumann, "Defining the Aerothermodynamic Methodology," in J.J. Bertin, et al., *Hypersonics*, vol. 1: *Defining the Hypersonic Environment* (Boston: Birkhäuser Boston, Inc., 1989), pp. 93–160.

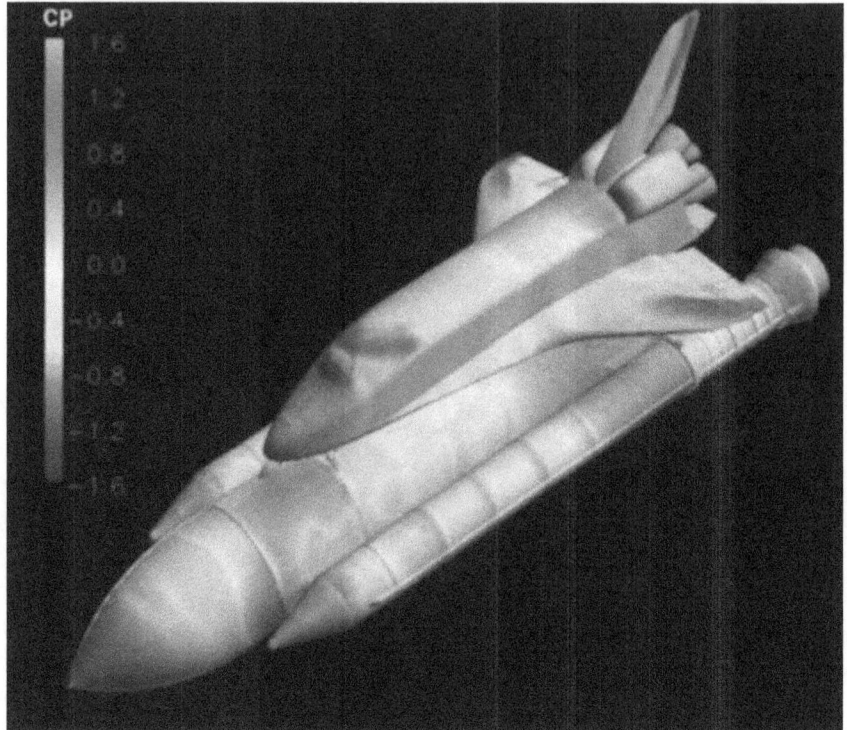

USAF–NASA cooperative CFD Shuttle study. NASA.

Direct Numerical Simulation

What of the future of CFD? Most flows of practical interest are turbulent flows. Turbulence is still one of the few unsolved problems in classical physics. In the calculation of turbulent flows, we therefore have to model the effect of turbulence. Any turbulence model involves some empirical data, and all models are inaccurate to some greater or lesser degree. The uncertainty in turbulence models is the reason for much uncertainty in the calculation of turbulent flows in computational fluid dynamics. This will continue for years to come. There is, however, an approach that requires no turbulence modeling. Nature creates a turbulent flow using the same fundamental principles that are embodied in the Navier-Stokes equations. Indeed, turbulence on its most detailed scale is simply a flow field developed by nature. If one can put enough grid points in the flow, then a Navier-Stokes solution will calculate all the detailed turbulence without the need for any type of model. This is called direct numerical simulation. The key is "enough grid points," which even for the simplest flow over a flat plate requires millions of points.

Once again, NASA researchers have been leading the way. Calculations made at NASA Ames for flow over a flat plate have required over 10 million grid points taking hundreds of hours on supercomputers, an indication of what would be required to calculate the whole flow field around a complete airplane using direct numerical simulation. But this is the future, perhaps, indeed, as far as three decades away. By that time, the computational power of computers will have undoubtedly continued to increase many-fold, and, as well, NASA will be continuing to play a leading role in advancing CFD, even as it is today and has in the past.

Some Important NASA CFD Computer Codes

Not only has NASA played a strong role in the development of new CFD algorithms, it has delivered these contributions to the technical public in the form of highly developed computer codes for the user. In the context of this survey, it would be remiss not to underscore the importance of these codes, three in particular, which this author (and his students) have used as numerical tools for carrying out research: LAURA, OVERFLOW, and CFL3D.

The LAURA code was developed principally by Dr. Peter Gnoffo at the NASA Langley Research Laboratory.[26] This code solves the three-dimensional Euler or Navier-Stokes equations for high-speed supersonic and hypersonic flow fields. It is particularly noteworthy because

---

26. The author has had the privilege of looking over Peter Gnoffo's shoulder for two decades as he progressively developed LAURA. Some examples of work carried out with the LAURA code are given in Robert B. Greendyke and Peter A. Gnoffo "Convective and Radiative Heating for Vehicle Return from the Moon and Mars," NASA TM-110185 (1995); Kenneth Sutton and Peter A. Gnoffo, "Multi-Component Diffusion with Application to Computational Aerothermodynamics," AIAA Paper 98-2575 (1998); William L. Kleb, et al., "Collaborative Software Development in Support of Fast Adaptive AeroSpace Tools (FAAST)," AIAA Paper 2003-3978 (2003); Peter A. Gnoffo and Jeffrey A. White, "Computational Aerothermodynamic Simulation Issues on Unstructured Grids," AIAA Paper 2004-2371 (2004); Peter A. Gnoffo, "Simulation of Stagnation Region Heating in Hypersonic Flow on Tetrahedral Grids," AIAA Paper 2007-3960 (2007); Karen L. Bibb, et al., "Parallel, Gradient-Based Anisotropic Mesh Adaption for Re-Entry Vehicle Configurations," AIAA Paper 2006-3579 (2007); Christopher O. Johnson, et al., "The Influence on Radiative Heating for Earth Entry," NASA Document ID 20080023455 (2008); Richard A. Thompson and Peter A. Gnoffo, "Implementation of a Blowing Boundary Condition in the LAURA Code," NASA Document ID 20080008560 (2008); Peter A. Gnoffo, "Multi-Dimensional, Inviscid Flux Reconstruction for Simulation of Hypersonic Heating on Tetrahedral Grids," AIAA Paper 2009-0599 (2009); and Gnoffo, Peter A., et al., "Implementation of Radiation, Ablation, and Free Energy Minimization Modules for Coupled Simulations of Hypersonic Flow," AIAA Paper 2009-1399 (2009).

it deals with very detailed nonequilibrium and equilibrium chemically reacting flows pertaining to hypersonic reentry vehicles in Earth's and foreign planetary atmospheres. Some applications involve flow-field temperatures so high that radiation becomes a dominant physical feature. The LAURA program readily handles radiative gas dynamics, and, to this author's knowledge, it is the only existing standard code to do so. The LAURA code has been used for the design and analysis of all NASA entry bodies in recent experience and is the most powerful and useful code in existence for high-temperature flow fields.

Of particular use for computing lower speed subsonic and transonic flows is OVERFLOW. This code was developed in the early 1990s by Pieter Burning and Dennis Jesperson as a collaborative effort between NASA Johnson Space Center and the NASA Ames Research Center. It solves the compressible three-dimensional Reynolds-averaged Navier-Stokes equations by means of a time-marching algorithm. OVERFLOW is widely used for the calculation of three-dimensional subsonic and transonic flows, and it proved particularly valuable for computing subsonic viscous flows over airfoils in a recent graduate study of innovative new airfoil shapes for high lift undertaken at the University of Maryland at College Park's Department of Aerospace Engineering.[27]

In the mid-1980s, Dr. Jim Thomas and his colleagues at the NASA Langley Research Center recognized the need for a code that contained the latest advancements in CFD methodology being developed by the applied mathematics community. Out of their interest sprang CFL3D, one of the earliest (yet still most powerful) CFD codes developed by NASA.[28]

---

27. Judy Conlon, "OVERFLOW Code Empowers Computational Fluid Dynamics," *NASA Insights*, vol. 5 (Apr. 1998).

28. For more on CFL3D and its applications, see J.L. Thomas, and R.W. Walters, "Upwind Relaxation Algorithms for the Navier-Stokes Equations," AIAA Paper 85-1501-CP (1985); J.L. Thomas, B. van Leer, and R. W. Walters, "Implicit Flux-Split Schemes for the Euler Equations," AIAA Paper 85-1680 (1985); W. K. Anderson, J. L. Thomas, and B. van Leer, "Comparison of Finite Volume Flux Vector Splitting for the Euler Equations," *AIAA Journal*, vol. 24, no. 9 (Sept. 1986), pp. 1453–1460; J.L. Thomas, et al., "High-Speed Inlet Flows," Symposium on Advances and Applications in CFD, Winter Annual Meeting of ASME, Chicago, IL, Nov. 1988; W.B. Compton, III, J.L. Thomas, W.K. Abeyounis, and M.L. Mason, "Transonic Navier-Stokes Solutions of Three-Dimensional Afterbody Flows," NASA TM-4111 (1989); J.L. Thomas, "An Implicit Multigrid Scheme for Hypersonic Strong-Interaction Flowfields," Comm. *Applied Numerical Methods*, vol. 8 (1992), pp. 683–693; and J.L. Thomas, "Reynolds Number Effects on Supersonic Asymmetrical Flows over a Cone," *Journal of Aircraft*, vol. 30 no. 4, (Apr. 1993), pp. 488–495.

This code is applicable across the whole flight spectrum, from low-speed subsonic flow to hypersonic flow. Not only does it handle steady flows, but it calculates time-accurate unsteady flows as well. Much effort was invested in the development of detailed grids so that it readily handles flows over complex three-dimensional bodies. An appreciation of the power, usefulness, and widespread acceptance of CFL3D can be gained by noting that it is used by over 100 researchers in 22 companies, 13 universities, NASA, and the military services.

LAURA, OVERFLOW, and CFL3D are just three of the CFD codes NASA researchers have generated. Most importantly, because they are the product of taxpayer-supported research, all are readily available, free of charge, to the general public, making NASA unique among other organizations working in the field of CFD. NASA's commitment to making scientific and technical information of the highest quality available to the public—a legacy of its predecessor, the National Advisory Committee for Aeronautics—has influenced its approach to CFD code development and may be counted one of the Agency's most valuable contributions to the whole discipline of computational fluid dynamics. When students and professional practitioners alike need viable computer codes for complex fluid dynamic applications, they have ready access to such codes and the extremely competent individuals who develop them. This is perhaps the highest accolade one can pronounce upon NASA's computational fluid dynamics efforts.

In closing, a proper history of CFD would require a lengthy book and a greater perspective of the past: something yet impossible, for the history of this rather young discipline is still evolving. The challenge is akin to what one might have expected trying to write a history of the balloon in the early 1800s, or a history of flight in 1914. In this case, I have tried to share my perspective in an accessible format, based in part on my own experiences and on my familiarity with the work of many colleagues, especially those within NASA. I have had to leave out so many others and so much great work in CFD just to tell a short story in a limited amount of pages that I feel compelled to apologize to those many others that I have not mentioned. To them I would say that their absence from this case certainly does not mean their contributions were any less important. But this has been an effort to paint a broad-stroke picture, and, like any such picture, it is somewhat subjective. My best wishes go out to all those researchers, present and future, who have and will continue to make computational fluid dynamics a vital, essential, and lasting tool for the study of fluid dynamics.

Recommended Additional Readings
Reports, Papers, Articles, and Presentations:

Anon., "The Fluidity of Thought and Vision: 40 Years of CFD at Iowa State," *Innovate* [Magazine of the Iowa State University College of Engineering] (fall 2007).

K. Anderson, J.L. Thomas, and B. van Leer, "Comparison of Finite Volume Flux Vector Splitting for the Euler Equations," *AIAA Journal*, vol. 24, no. 9 (Sept. 1986), pp. 1453–1460.

Karen L. Bibb, et al., "Parallel, Gradient-Based Anisotropic Mesh Adaption for Re-Entry Vehicle Configurations," AIAA Paper 2006-3579 (2007).

W.B. Compton, III, J.L. Thomas, W.K. Abeyounis, and M.L. Mason, "Transonic Navier-Stokes Solutions of Three-Dimensional Afterbody Flows," NASA TM-4111 (1989).

Judy Conlon, "**OVERFLOW** Code Empowers Computational Fluid Dynamics," *NASA Insights*, vol. 5 (Apr. 1998).

Peter A. Gnoffo, "Simulation of Stagnation Region Heating in Hypersonic Flow on Tetrahedral Grids," AIAA Paper 2007-3960 (2007).

Peter A. Gnoffo, "Multi-Dimensional, Inviscid Flux Reconstruction for Simulation of Hypersonic Heating on Tetrahedral Grids," AIAA Paper 2009-0599 (2009).

Peter A. Gnoffo and Jeffrey A. White, "Computational Aerothermodynamic Simulation Issues on Unstructured Grids," AIAA Paper 2004-2371 (2004).

Peter A. Gnoffo, et al., "Implementation of Radiation, Ablation, and Free Energy Minimization Modules for Coupled Simulations of Hypersonic Flow," AIAA Paper 2009-1399 (2009).

Robert B. Greendyke and Peter A. Gnoffo, "Convective and Radiative Heating for Vehicle Return from the Moon and Mars," NASA TM-110185 (1995).

William L. Kleb, et al., "Collaborative Software Development in Support of Fast Adaptive AeroSpace Tools (FAAST)," AIAA Paper 2003-3978 (2003).

Antony Jameson, "Successes and Challenges in Computation Aerodynamics," AIAA Paper 87-1184 (1987).

Antony Jameson, "Computational Aerodynamics for Aircraft Design," *Science*, vol. 245 (July 28, 1989), pp. 361–371.

A. Jameson, W. Schmidt, and E. Turkel, "Numerical Solution of the Euler Equations by Finite Volume Methods Using Runge-Kutta Time-Stepping Schemes," AIAA Paper 81-1259 (1981).

Christopher O. Johnson, et al., "The Influence on Radiative Heating for Earth Entry," NASA Document ID 20080023455 (2008).

R.W. MacCormack, "The Effect of Viscosity in Hypervelocity Impact Cratering," AIAA Paper 69-354 (1969).

G. Moretti and M. Abbett, "A Time-Dependent Computational Method for Blunt Body Flows," AIAA Journal, vol. 4, no. 12 (1966), pp. 2136–2141.

E. Murman and J.D. Cole, "Calculation of Plane Steady Transonic flows," *AIAA Journal*, vol. 9 no. 1, (Jan. 1971), pp 114–121.

C.L.M.H. Navier, "Mémoire sur les du mouvement des fluids," *Mémoires de l'Academie Royale des Sciences*, no. 6 (1823), pp. 389–416.

J.S. Shang and S.J. Scherr, "Navier-Stokes Solution for a Complete Re-entry Configuration," *Journal of Aircraft*, vol. 23, no. 12 (1986), pp. 881–888.

G.G. Stokes, "On the Theories of the Internal Friction of Fluids in Motion, and of the Equilibrium and Motion of Elastic Solids," *Transactions of the Cambridge Philosophical Society*, vol. 8, no. 22 (1845), p. 287–342.

Kenneth Sutton and Peter A. Gnoffo, "Multi-Component Diffusion with Application to Computational Aerothermodynamics," AIAA Paper 98-2575 (1998).

J.L. Thomas, "An Implicit Multigrid Scheme for Hypersonic Strong-Interaction Flowfields," *Comm. Applied Numerical Methods*, vol. 8 (1992), pp. 683–693.

J.L. Thomas, "Reynolds Number Effects on Supersonic Asymmetrical Flows over a Cone," *Journal of Aircraft*, vol. 30, no. 4 (Apr. 1993), pp. 488–495.

J.L. Thomas and R.W. Walters, "Upwind Relaxation Algorithms for the Navier-Stokes Equations," AIAA Paper 85-1501-CP (1985).

J.L. Thomas, B. van Leer, and R.W. Walters, "Implicit Flux-Split Schemes for the Euler Equations," AIAA Paper 85-1680 (1985).

J.L. Thomas, et al., "High-Speed Inlet Flows," Symposium on Advances and Applications in CFD, Winter Annual Meeting of ASME, Chicago, IL, Nov. 1988.

Richard A. Thompson and Peter A. Gnoffo, "Implementation of a Blowing Boundary Condition in the LAURA Code," NASA Document ID 20080008560 (2008).

Eli Turkel, "Algorithms for the Euler and Navier-Stokes equations for supercomputers," NASA CR-172543 (1985).

Milton O. Van Dyke, "Supersonic Flow Past a Family of Blunt Axisymmetric Bodies," NASA Technical Report R-1 (1959).

K.J. Weilmuenster, "Solution of a large hydrodynamic problem using the STAR-100 computer," NASA TM-X-7394 (1976).

K.J. Weilmuenster, "High Angle of Attack Inviscid Flow Calculations of Shuttle-Like Vehicles with Comparisons to Flight Data," AIAA Paper 83-1798 (1983).

Books and Monographs:

John D. Anderson, Jr., *A History of Aerodynamics* (Cambridge: Cambridge University Press, 1997).

John D. Anderson, Jr., *Computational Fluid Dynamics: The Basics With Applications* (New York: McGraw-Hill, 1995).

J.J. Bertin, et al., *Hypersonics,* vol. 1: *Defining the Hypersonic Environment* (Boston: Birkhäuser Boston, Inc., 1989).

J. Blazek, *Computational Fluid Dynamics: Principles and Applications* (Oxford: Elsevier, 2001 ed.)

Paul E. Ceruzzi, *Beyond the Limits: Flight Enters the Computer Age* (Cambridge: MIT Press, 1989), p. 15.

Olivier Darrigol, *Worlds of Flow: A History of Hydrodynamics from the Bernoullis to Prandtl* (Oxford: Oxford University Press, 2005).

William F. Durand, ed., *Aerodynamic Theory*, vol. 1 (Berlin: Julius Springer Verlag, 1934).

Michael Eckert, *The Dawn of Fluid Dynamics: A Discipline Between Science and Technology* (Weinheim, Germany: Wiley-VCH Verlag GmbH & Co. KGaA, 2006).

Alfred Gessow and Dana J. Morris, *A Survey of Computational Aerodynamics in the United States,* NASA SP-394 (Washington, DC: NASA, 1977).

Charles Hirsch, *Numerical Computation of Internal and External Flows: The Fundamentals of Computational Fluid Dynamics* (Oxford: Elsevier, 2007 ed.).

Theodore von Kármán, *Aerodynamics* (Cornell: Cornell University Press, 1954).

Culbert B. Laney, *Computational Gas Dynamics* (Cambridge: Cambridge University Press, 1998 ed.).

Harvard Lomax, Thomas H. Pulliam, and David W. Zingg, *Fundamentals of Computational Fluid Dynamics* (Berlin: Springer-Verlag, 2001).

John C. Tannehill, Dale A. Anderson, and Richard H. Pletcher, *Computational Fluid Dynamics and Heat Transfer* (London: Taylor and Francis, 1997 ed.).

G.A. Tokaty, *A History and Philosophy of Fluid Mechanics* (Henley, UK: G.T. Foulis and Co., Ltd., 1971).

Jiyuan Tu, Guan Heng Yeoh, and Chaoqun Liu, *Computational Fluid Dynamics: A Practical Approach* (Oxford: Elsevier, 2008).

Z.U.A. Warsi, *Fluid Dynamics: Theoretical and Computational Approaches* (Boca Raton, FL: CRC Press, 1998 ed.).

John Wendt, ed., *Computational Fluid Dynamics: An Introduction* (Berlin: Springer-Verlag GmbH, 2008 ed.).

THIS PAGE INTENTIONALLY BLANK

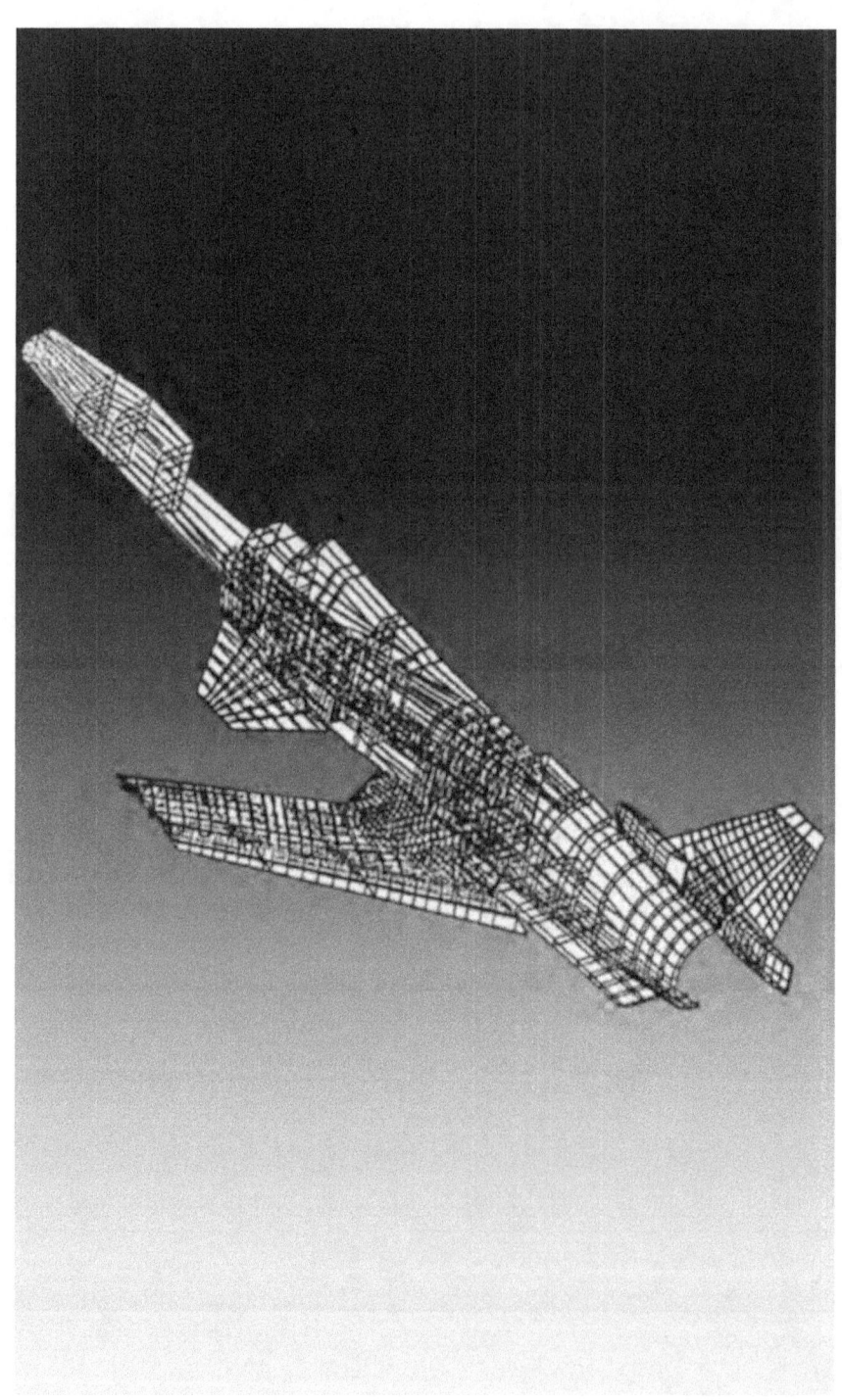

NASTRAN model of the X29A aircraft. NASA.

CASE 8: NASA and Computational Structural Analysis

David C. Aronstein

NASA research has been pivotal in its support of computational analytical methods for structural analysis and design, particularly through the NASTRAN program. NASA Centers have evolved structural analysis programs tailored to their own needs, such as assessing high-temperature aerothermodynamic structural loading for high-performance aircraft. NASA-developed structural tools have been adopted throughout the aerospace industry and are available on the Agency Web site.

THE FIELD OF COMPUTER METHODS in structural analysis, and the contributions of the National Aeronautics and Space Administration (NASA) to it, is wide-ranging. Nearly every NASA Center has a structural analysis group in some form. These groups conduct research and assist industry in grappling with a broad spectrum of problems. This paper is an attempt to show both aspects: the origins, evolution, and application of NASA Structural Analysis System (NASTRAN), and the variety and depth of other NASA activities and contributions to the field of computational structural methods.

In general terms, the goal of structural analysis is to establish that a product has the required strength and stiffness—structural integrity—to perform its function throughout its intended life. Its strength must exceed the loads to which the product is subjected, by some safety margin, the value of which depends on the application.

With aircraft, loads derive from level flight, maneuvering flight, gusts, landings, engine thrust and torque, vibration, temperature and pressure differences, and other sources. Load cases may be specified by regulatory agency, by the customer, and/or by the company practice and experience. Many of the loads depend on the weight of the aircraft, and the weight in turn depends on the design of the structure. This makes the structural design process iterative. Because of this, and also because a large fraction of an aircraft's weight is not actually accounted for by primary structure, initial weight estimates are usually based on experience rather than on a detailed buildup of structural material. A sizing process must be performed to reconcile the predicted empty weight and its relationship to the assumed maximum gross weight, with the required payload, fuel, and mission performance.[1]

After the sizing process has converged, the initial design is documented in the form of a three-view drawing with supporting data. From there, the process is approximately as follows:

- The weights group generates an initial estimate of the weights of the major airframe components.
- The loads group analyzes the vehicle at the defined condition(s) to determine forces, bending moments, etc., in the major components and interfaces.
- The structures group defines the primary load paths and sizes the primary structural members to provide the required strength.
- Secondary load paths, etc., are defined to the required level of detail.

Process details vary between different organizations, but at some point, the structural definition reaches a level of maturity to enable a check of the initial weight estimate. Then the whole design may be iterated, if required. Iteration may also be driven by maturing requirements or by evolution in other aspects of the design, e.g., aerodynamics, propulsion, etc.

Structural Analysis Prior to Computers

Basic principles of structural analysis—static equilibrium, trusses, and beam theory—were known long before computers, or airplanes, existed. Bridges, towers and other buildings, and ships were designed

---

1. There are many topics touching on structural analysis and loads prediction, which the author has not covered. Materials science is excluded, for this paper considers methods for predicting the macroscopic characteristics of a structure, not analyzing the microscopic properties of the materials from which it is built. So too are computer-aided design, computational fluid dynamics, multidisciplinary optimization, structural test and data analysis techniques, and nondestructive inspection/evaluation, aside from passing reference. Though this paper focuses on NASA activities and contributions, there is no intent to minimize the importance of contributions from the aircraft industry, the technical software industry, universities, and other research organizations.

by a combination of experience and some amount of analysis—more so as designs became larger and more ambitious during and after the Industrial Revolution.

With airplanes came much greater emphasis on weight minimization. Massive overdesign was no longer an acceptable means to achieve structural integrity. More rigorous analysis and structural sizing was required. Simplifications allowed the analysis of primary members under simple loading conditions:

- Slender beams: axial load, shear, bending, torsion.
- Trusses: members carry axial load only, joined to other such members at ends.
- Simple shells: pressure loading.
- Semi-monocoque (skin and stringer) structures: shear flow, etc.
- Superposition of loading conditions.

With these simplifications, primary structural members could be sized appropriately to the expected loads. In the days of wood, wire, and fabric, many aircraft structures could be analyzed as trusses: externally braced biplane wings; fuselage structures consisting of longerons, uprights, and cross braces, with diagonal braces or wires carrying torsion; landing gears; and engine mounts. As early as the First World War and in the 1920s, researchers were working to cover every required aspect of the problem: general analysis methods, analysis of wings, horizontal and vertical tails, gust loads, test methods, etc. The National Advisory Committee for Aeronautics (NACA) contributed significantly to the building of this early body of methodology.[2]

Structures with redundancy—multiple structural members capable of sharing one or more loading components—may be desirable for safety,

---

2. A.F. Zahm and L.H. Crook, "Airplane Stress Analysis," Report No. 82, Aerodynamical Laboratory, Bureau of Construction and Repair, U.S. Navy (1918); Roy G. Miller, "Torsion of wing trusses at diving speeds," NACA TR-104, Langley Research Center (1921); F.H. Norton and D.L. Bacon, "The pressure distribution over the horizontal tail surfaces of an airplane II," NACA TR-119, Langley Research Center (1921); Norton and W.G. Brown, "Pressure distribution over the rudder and fin of an airplane in flight," NACA TR-149, Langley Research Center (1923); J.C. Hunsaker and E.B. Wilson, "Report on Behavior of Aeroplanes in Gusts," Massachusetts Institute of Technology (1915), (prepared by MIT and predating the establishment of the NACA but listed and marked as NACA-TR-1); E.P. Warner, "Static Testing and Proposed Standard Specifications," NACA TN-6, Langley Field, VA (July 1920), p. 1.

but they posed new problems for analysis. Redundant structures cannot be analyzed by force equilibrium alone. A conservative simplification, often practiced in the early days of aviation, was to analyze the structure with redundant members missing. A more precise solution would require the consideration of displacements and "compatibility" conditions: members that are connected to one another must deform in such a manner that they move together at the point of connection. Analysis was feasible but time-consuming. Large-scale solutions to redundant ("statically indeterminate") structure problems would become practical with the aid of computers. Until then, more simplifications were made, and specific types of solutions—very useful ones—were developed.

While these analysis methods were being developed, there was a lot of airplane building going on without very much analysis at all. In the "golden age of aviation," many airplanes were built in garages or at small companies that lacked the resources for extensive analysis. "In many cases people who flew the airplanes were the same people who carried out the analysis and design. They also owned the company. There was very little of what we now call structural analysis. Engineers were brought in and paid—not to design the aircraft—but to certify that the aircraft met certain safety requirements."[3]

Through the 1930s, as aircraft structures began to be formed out of aluminum, the semi-monocoque or skin-and-stringer structure became prevalent, and analysis methods were developed to suit. "In the 1930s, '40s, and '50s, techniques were being developed to analyze specific structural components, such as wing boxes and shear panels, with combined bending, torsion, and shear loads and with stiffeners on the skins."[4]

A number of exact solutions to the differential equations for stress and strain in a structural member were known, but these generally exist only for very simple geometric shapes and very limited sets of loading conditions and boundary conditions. Exact solutions were of little practical value to the aircraft designer or stress analyst. Instead, "free body diagrams" were used to analyze structures at selected locations, or "stations." The structure was considered to be cut by a theoretical plane at the station of interest. All loads, applied and inertial, on the portion of the aircraft outboard of the cut had to be borne (reacted) by the structure at the cut.

3. W. Jefferson Stroud, e-mail message to author, Mar. 29, 2009.
4. Stroud, e-mail to author.

In principle, this allowed the stress at any point in the structure to be analyzed—given the time to make an arbitrarily large number of these theoretical cuts through the aircraft. In practice, free body diagrams were used to analyze the structure at key locations—selected fuselage stations, the root, and selected stations of wings and tail surfaces. Structural members were left constant, or tapered appropriately, according to experience and judgment, between the analyzed sections. For major projects such as airliners or bombers, the analysis would be more thorough, and consequently, major design organizations had rooms full of people whose jobs were to perform the required calculations.

The NACA also utilized this brute-force approach to large calculations, and the people who performed the calculations—overwhelmingly women—were called "computers." Annie J. Easley, who worked at the NASA Lewis (now Glenn) Research Center starting in 1955, recalls:

> . . . we were called computers until we started to get the machines, and then we were changed over to either math technicians or mathematicians. . . . The engineers and the scientists are working away in their labs and their test cells, and they come up with problems that need mathematical computation. At that time, they would bring that portion to the computers, and our equipment then were the huge calculators, where you'd put in some numbers and it would clonk, clonk, clonk out some answers, and you would record them by hand. Could add, subtract, multiply, and divide. That was pretty much what those big machines, those big desktop machines, could do. If we needed to find a logarithm or an exponential, we then pulled out the tables.[5]

After World War II, with jet engines pushing aircraft into ever more demanding flight regimes, the analytical community sought to keep up. The NACA continued to improve the methodologies for calculating loads on various parts of an aircraft, and some of the reports generated during that time are still used by industry practitioners today. NACA Technical Report (TR) 1007, for horizontal tail loads in pitch maneuvers, is a good

---

5. Interview of Annie J. Easley by Sandra Johnson, Cleveland, OH, Aug. 21, 2001, pp .2, 8, in *NASA Oral History Project*, on NASA Glenn Research Center History Office Web site at http://grchistory.grc.nasa.gov/index.cfm, accessed Apr. 29, 2009.

example, although it does not cover all of the conditions required by recent airworthiness regulations.[6]

For structural analysis, energy methods and matrix methods began to receive more attention. Energy methods work as follows: one first expresses the deflection of a member as a set of assumed shape functions, each multiplied by an (initially unknown) coefficient; expresses the total strain energy in terms of these unknown coefficients; and finally, finds the values of the coefficients that minimize the strain energy. If the shape functions, from which the solution is built, satisfy the boundary conditions of the problem, then so does the final solution.

Energy methods were not new. The concept of energy minimization was introduced by Lord Rayleigh in the late 19th century and extended by Walter Ritz in two papers of 1908 and 1909.[7] Rayleigh and Ritz were particularly concerned with vibrations. Carlo Alberto Castigliano, an Italian engineer, published a dissertation in 1873 that included two important theorems for applying energy principles to forces and static displacements in structures.[8] However, in the early works, the shape functions were continuous over the domain of interest. The idea of breaking up (discretizing) a complex structure into many simple elements for numerical solution would lead to the concept of finite elements, but for this to be useful, computing technology needed to mature.

The Advent of Direct Analog Computers

The first computers were analog computers. Direct analog computers are networks of physical components (most commonly, electrical components: resistors, capacitors, inductances, and transformers) whose behavior is governed by the same equations as some system of interest that is being modeled. Direct analog computers were used in the 1950s and 1960s to solve problems in structural analysis, heat transfer, fluid flow, and other fields.

The method of analysis and the needs that were driving the move from classical idealizations such as slender-beam theory toward computational

---

6. Henry Pearson, William McGowan, and James Donegan, "Horizontal Tail Loads in Maneuvering Flight," NACA TR-1007, Langley Aeronautical Laboratory, VA (1950); Interview of Gonzalo Mendoza and Zachary Hazen by author, Wichita, KS, Apr. 6, 2009.
7. Arthur W. Leissa, "The Historical Basis of the Rayleigh and Ritz Methods," in *7th International Symposium on Vibrations of Continuous Systems, Zakopane, Poland,* July 2009.
8. "Carlo Alberto Castigliano," at http://www-groups.dcs.st-and.ac.uk/~history/Biographies/Castigliano.html, accessed May 22, 2009.

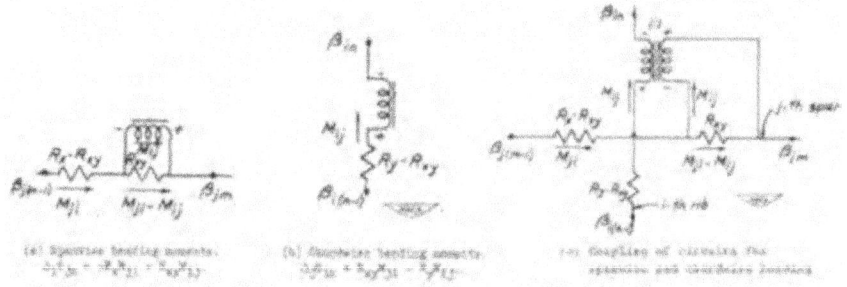

Representation of structural elements by analog circuits. NASA.

methods are well stated in the following passage, from an NACA-sponsored paper by Stanley Benscoter and Richard MacNeal (subsequently a cofounder of the MacNeal Schwendler Corporation [MSC] and member of the NASTRAN development team):

> The theory is expressed entirely in terms of first-order difference equations in order that analogous electrical circuits can be readily designed and solutions obtained on the Caltech analog computer.... In the process of designing thin supersonic wings for minimum weight it is found that a convenient construction with aluminum alloy consists of a rather thick skin with closely spaced spars and no stringers. Such a wing deflects in the manner of a plate rather than as a beam. Internal stress distributions may be considerably different from those given by beam theory.[9]

Their implementation of analog circuitry for bending loads is illustrated here and serves as an example of the direct analog modeling of structures.[10]

Direct analog computing had its advocates well into the 1960s. "For complex problems [direct analog] computers are inherently faster than digital machines since they solve the equations for the several nodes simultaneously, while the digital machines solve them sequentially. Direct analogs have, moreover, the advantage of visualization;

---

9. Stanley U. Benscoter and Richard H. MacNeal, "Equivalent Plate Theory for a Straight Multicell Wing," NACA TN-2786 (1952), p. 1.
10. Benscoter and MacNeal, "Equivalent Plate Theory," NACA TN-2786, p. 28.

computer setups as well as programming are more closely related to the actual problem and are based primarily on physical insight rather than on numerical skills."[11]

The advantages came at a price, however. It could take weeks, in some cases, to set up an analog computer to solve a particular type of problem. And there was no way to store a problem to be revisited at a later date. These drawbacks may not have seemed so important when there was no other recourse available, but they became more and more apparent as the programmable digital computer began to mature.

Hybrid direct-analog/digital computers were hypothesized in the 1960s: essentially a direct analog computer controlled by a digital computer capable of storing and executing program instructions. This would have overcome some of the drawbacks of direct analog computers.[12] However, this possibility was most likely overtaken by the rapid progress of digital computers. At the same time these hybrid analog/digital computers were just being thought about, **NASTRAN** was already in development.

A different type of analog computer—the active-element, or indirect, analog—consisted of operational amplifiers that performed arithmetic operations. These solved programmed mathematical equations, rather than mimicking a physical system. Several NACA locations—including Langley, Ames, and the Flight Research Center (now Dryden Flight Research Center)—used analog computers of this type for flight simulation. Ames installed its first analog computer in 1947.[13] The Flight Research Center flight simulators used analog computers exclusively from 1955 to 1964 and in combination with digital computers until 1975.[14] This type of analog computer can be thought of as simply a less precise, less reliable, and less versatile predecessor to the digital computer.

Digital Computation Triggers Automated Structural Analysis
In 1946, the ENIAC, "commonly accepted as the first successful high-speed electronic digital computer," became operational at the University

---

11. Victor Paschkis and Frederick Ryder, *Direct Analog Computers* (New York: John Wiley & Sons, Inc., 1968), preface.
12. Paschkis and Ryder, p. 383.
13. Glenn Bugos, *Atmosphere of Freedom: Sixty Years at the NASA Ames Research Center*, NASA SP-4314 (Washington, DC: NASA, 2000), pp. 181–182.
14. Gene Waltman, *Black Magic and Gremlins*, pp. 1–4.

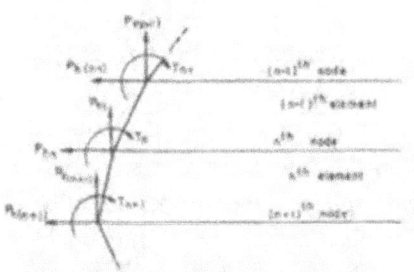
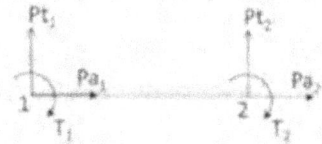

Simple example of discretized structure and single element. NASA.

of Pennsylvania.[15] It took up as much floor space as a medium-sized house and had to be "programmed" by physically rearranging its control connections. Many advances followed rapidly: storing instructions in memory, conditional control transfer, random access memory, magnetic core memory, and the transistor-circuit element. With these and other advances, digital computers progressed from large and ungainly experimental devices to programmable, useful, commercially available (albeit expensive) machines by the mid-1950s.[16]

The FORTRAN programming language was also developed in the mid-1950s and rapidly gained acceptance in technical communities. This was a "high level language," which allowed programming instructions to be written in terms that an engineer or analyst could understand; a compiler handled the translation into "machine language" that the computer could understand. International Business Machines (IBM) developed the original FORTRAN language and also some of the early practical digital computers. Other early digital computers were produced by Control Data Corporation (CDC) and UNIVAC. These developments made it pos-

---

15. Jeremy Meyers, "A Brief History of the Computer (b.c. – 1993a.d.)," on Meyers's Web site at http://www.jeremymeyers.com/comp#EDC, accessed Apr. 29, 2009; see also Paul E. Ceruzzi, *A History of Modern Computing* (Cambridge: MIT Press, 1999), pp. 7, 15, 20–21.
16. Meyers, "A Brief History of the Computer (b.c. – 1993a.d.)."

sible to take the new methods of structural analysis that were emerging and implement them in an automated, repeatable manner.

The essence of these new methods was to treat a structure as a finite number of discrete elastic elements, rather than as a continuum. Reactions (forces and moments) and deflections are only calculated at specific points, called "nodes." Elements connect the nodes. The stress and strain fields in the regions between the nodes do not need to be solved in the global analysis. They only need to be solved when developing the element-level solution, and once this is done for a particular type of element, that element is available as a prepackaged building block. Complex shapes and structures can then be built up from the simple elements. A simple example—using straight beam elements to model a curved beam structure—is illustrated here.

To find, for example, the relationship between the displacements of the nodes and the corresponding reactions, one could do the following (called the unit displacement method). First, a hypothetical unit displacement of one node in one degree of freedom (d.o.f.) only is assumed. This displacement is transposed into the local element coordinate systems of all affected elements. (In the corresponding figure, this would entail the relatively simple transformation between global horizontal and vertical displacements, and element axial and transverse displacements. The angular displacements would require no transformation, except in some cases a sign change.) The predetermined element stiffness matrices are used to find the element-level reactions. The element reactions are then translated back into global coordinates and summed to give the total structure reactions—to the single hypothetical displacement. This set of global reactions, plus zeroes for all forces unaffected by the assumed displacement, constitutes one column in the "stiffness matrix." By repeating the exercise for every degree of freedom of every node, the stiffness matrix can be built. Then the reactions to any set of nodal displacements may be found by multiplying the stiffness matrix by the displacement vector, i.e., the ordered list of displacements. This entails difficult bookkeeping but simple math.

It is more common in engineering, however, to have to find unknown displacements and stresses from known applied forces. This answer is not possible to obtain so directly. (That is, if the process just described seems direct to you. If it does, you are probably an engineer. If it seems too trivial to have even mentioned, then you are probably a mathematician.)

Instead, after the stiffness matrix is found, it must be inverted to obtain the flexibility matrix. The inversion of large matrices is a science in itself. But it can be done, using a computer, if one has time to wait. Most of the science lies in improving the efficiency of the process. Another important output is the stress distribution throughout the structure. But this problem has already been solved at the element level for a hypothetical set of element nodal displacements. Scaling the generic stress distribution by the actual displacements, for all elements, yields the stress state throughout the structure.

There are, of course, many variations on this theme and many complexities that cannot be addressed here. The important point is that we have gone from an insoluble differential equation to a soluble matrix arithmetic problem. This, in turn, has enabled a change from individual analyses by hand of local portions of a structure to a modeling effort followed by an automated calculation of the stresses and deflections of the entire structure.

Pioneering papers on discretization of structures were published by Alexander Hrennikoff in 1941 at the Massachusetts Institute of Technology and by Richard Courant in 1943 at the mathematics institute he founded at New York University that would later bear his name. These papers did not lead to immediate application, in part perhaps because they were ahead of the necessary computational technology and in part because they were still somewhat theoretical and had not yet developed a well-formed practical implementation. The first example of what we now call the finite element method (FEM) is commonly considered to be a paper by M.J. Turner (Boeing), R.W. Clough (University of California at Berkeley, Civil Engineering Department), H.C. Martin (University of Washington, Aeronautical Engineering Department), and L.J. Topp in 1956.[17] This paper presented a method for plane stress problems, using triangular elements. John Argyris at the University of Stuttgart, Germany, also made important early contributions. The term "finite element method" was actually coined by Clough in 1960. The Civil Engineering Department at Berkeley became a major center of early finite element methods development.[18]

---

17. M.J. Turner, R.W. Clough, H.C. Martin, and L.J. Topp , "Stiffness and deflection analysis of complex structures," *Journal of the Aeronautical Sciences*, vol. 23 (1956), pp. 805–823.
18. John R. Brauer, *What Every Engineer Should Know about Finite Element Analysis* (New York: Marcel Dekker, 1993), pp. 2–3; Edward R. Champion, Jr., and J. Michael Ensminger, *Finite Element Analysis with Personal Computers* (New York: Marcel Dekker, Inc., 1988), p. 1.

By the mid-1960s, aircraft companies, computing companies, universities, and Government research centers were beginning to explore the possibilities—although the method allegedly suffered some initial lack of interest in the academic world, because it bypassed elegant mathematical solutions in favor of numerical brute force.[19] However, the practical value could not long be ignored. The following insightful comment, made by a research team at the University of Denver in 1966 (working under NASA sponsorship), sums up the expectation of the period: "It is certain that this concept is going to become one of the most important tools of engineering in the future as structures become more complex and computers more versatile and available."[20]

NASA Spawns NASTRAN, Its Greatest Computational Success

The project to develop a general-purpose finite element structural analysis system was conceived in the midst of this rapid expansion of finite element research in the 1960s. The development, and subsequent management, enhancement, and distribution, of the **NASA Structural Analysis System**, or **NASTRAN**, unquestionably constitutes **NASA's** greatest single contribution to computerized structural analysis—and arguably the single most influential contribution to the field from any source. NASTRAN is the workhorse of structural analysis: there may be more advanced programs in use for certain applications or in certain proprietary or research environments, but **NASTRAN** is the most capable general-purpose, generally available, program for structural analysis in existence today, even more than 40 years after it was introduced.

Origins of NASTRAN

In the early 1960s, structures researchers from the various **NASA** Centers were gathering annually at Headquarters in Washington, DC, to exchange ideas and coordinate their efforts. They began to realize that many organizations—NASA Centers and industry—were independently developing computer programs to solve similar types of structural problems. There were several drawbacks to this situation. Effort was being duplicated needlessly. There was no compatibility of input and output formats, or consistency of naming conventions. The programs

---

19. Thomas J. Butler, "Operating in the Age of NASTRAN ," p. 1.
20. Anita S. West and William F. Hubka, "Matrix Methods and Automation in Structural Engineering," NASA CR-71230 (1966), p. 12.

were only as versatile as the developers cared to make them; the inherent versatility of the finite element method was not being exploited. More benefit might be achieved by pooling resources and developing a truly general-purpose program. Thomas G. Butler of the Goddard Space Flight Center (GSFC), who led the team that developed NASTRAN between 1965 and 1970, recalled in 1982:

> NASA's Office of Advanced Research and Technology (OART) under Dr. Raymond Bisplinghoff sponsored a considerable amount of research in the area of flight structures through its operating centers. Representatives from the centers who managed research in structures convened annually to exchange ideas. I was one of the representatives from Goddard Space Flight Center at the meeting in January 1964. . . . Center after center described research programs to improve analysis of structures. Shells of different kinds were logical for NASA to analyze at the time because rockets are shell-like. Each research concentrated on a different aspect of shells. Some were closed with discontinuous boundaries. Other shells had cutouts. Others were noncircular. Others were partial spans of less than 360°. This all seemed quite worthwhile if the products of the research resulted in exact closed-form solutions. However, all of them were geared toward making some simplifying assumption that made it possible to write a computer program to give numerical solutions for their behavior. . . . Each of these computer programs required data organization different from every other. . . . Each was intended for exploring localized conditions rather than complete shell-like structures, such as a whole rocket. My reaction to these programs was that . . . technology was currently available to give engineering solutions to not just localized shells but to whole, highly varied structures. The method was finite elements.[21]

Doug Michel led the meetings at NASA Headquarters. Butler, Harry Runyan of Langley Research Center, and probably others proposed that NASA develop its own finite element program, if a suitable one could

---

21. Butler, "Operating in the Age of NASTRAN," p. 2.

not be found already existing. "The group thought this was a good idea, and Doug followed up with forming the Ad Hoc Group for Structural Analysis, which was headed by Tom Butler of Goddard," recalled C. Thomas Modlin, Jr., who was one of the representatives from what is now Johnson Space Center.[22] The committee included representatives from all of the NASA Centers that had any significant activity in structural analysis methods at the time, plus an adjunct member from the U.S. Air Force at Wright-Patterson Air Force Base, as listed in the accompanying table.[23]

CENTER	REPRESENTATIVE(S)
Ames	Richard M. Beam and Perry P. Polentz
Flight Research (now Dryden)	Richard J. Rosecrans
Goddard	Thomas G. Butler (Chair) and Peter A. Smidinger
Jet Propulsion Laboratory	Marshall E. Alper and Robert M. Bamford
Langley	Herbert J. Cunningham
Lewis	William C. Scott and James D. McAleese
Manned Spacecraft (now Johnson)	C. Thomas Modlin, Jr., and William W. Renegar
Marshall	Robert L. McComas
Wright-Patterson AFB	James Johnson (adjunct member)

After visiting several aerospace companies, all of whom were "extremely cooperative and candid," and reviewing the existing methods, the committee recommended to Headquarters that NASA sponsor the development of its own finite element program "to update the analytical capability of the whole aerospace community. The program should incorporate the best of the state of the arts, which were currently splintered."[24]

The effort was launched, under the management of Butler at the Goddard Space Flight Center, to define and implement the General Purpose Structural Analysis program. Requirements were collected from the information brought from the various Centers, from the industry visits, and from a conference on "Matrix Methods in Structural Mechanics"

---

22. C. Thomas Modlin, Jr., e-mail message to author, Apr. 23, 2009.
23. Caleb W. McCormick, *The NASTRAN Users' Manual*, NASA SP-222 (1970), p. i.
24. Butler, "Age of NASTRAN," p. 2.

Case 8 | NASA and Computational Structural Analysis

held at Wright-Patterson Air Force Base.[25] Key requirements included the following:[26]

- General-purpose. The system must allow different analysis types—static, transient, thermal, etc.—to be performed on the same structural model without alteration.
- Problem size. At least 2,000 degrees of freedom for static and dynamic analyses alike. (Prior state of the art was approximately 100 d.o.f. for dynamic mode analysis and 100 to 600 d.o.f. for static analysis.)
- Modular. Parts of the program could be changed without disrupting other parts.
- Open-ended. New types of elements, new analysis modules, and new formats could be added.
- Maintainable and capable of being updated.
- Machine-independent. Capable of operating on IBM 360, CDC 6000 Series, and UNIVAC 1108 (the only 3 commercially available computers capable of performing such analysis at the time), and future generations of computers.

After an initial design phase, the implementation contract was awarded to a team led by Computer Sciences Corporation (CSC), with MacNeal Schwendler Corporation and Martin Baltimore as subcontractors. Coding began in July 1966. Dr. Paul R. Peabody was the principal architect of the overall system design. Dr. Richard H. MacNeal (MacNeal Schwendler) designed the solution structure, taking each type of solution from physics, to math, to programming, assisted by David Harting. Keith Redner was the implementation team lead and head programmer, assisted by Steven D. Wall and Richard S. Pyle. Frank J. Douglas coded the element routines and wrote the programmer's manual. Caleb W. McCormick was the author of the user's manual and supervised NASTRAN installation and training. Other members of the development team included Stanley Kaufman (Martin Baltimore), Thomas L. Clark,

---

25. "Matrix Methods in Structural Mechanics," Air Force Flight Dynamics Laboratory, Wright-Patterson Air Force Base, OH, (proceedings published Nov. 1966, conference date unknown); abstract on Defense Technical Information Center Web site at *http://oai.dtic.mil/oai/oai?verb=getRecord&metadataPrefix=html&identifier=AD0646300*, accessed May 25, 2009.
26. Butler, "Age of NASTRAN," pp. 3–4.

David B. Hall, Carl Hennrich, and Howard Dielmann. The project staff at Goddard included Richard D. McConnell, William R. Case, James B. Mason, William L. Cook, and Edward F. Puccinelli.[27]

NASTRAN embodied many technically advanced features that are beyond the scope of this paper (and, admittedly, beyond the scope of this author's understanding), which provided the inherent capability to handle large problems accurately and efficiently. It was referred to as a "system" rather than just a program by its developers, and for good reasons. It had its own internal control language, called Digital Matrix Abstraction Programming (DMAP), which gave flexibility in the use of its different modules. There were 151,000 FORTRAN statements, equating to more than 1 million machine language statements. Twelve prepackaged "rigid formats" permitted multiple types of analysis on the same structural model, including statics, steady-state frequency response, transient response, etc.[28]

The initial development of NASTRAN was not without setbacks and delays, and at introduction it did not have all of the intended capabilities. But the team stayed focused on the essentials, choosing which features to defer until later and which characteristics absolutely had to be maintained to keep NASTRAN true to its intent.[29] According to Butler: "One thing that must be mentioned about the project, that is remarkable, pertains to the spirit that infused it everywhere. Every man thought that he was the key man on the whole project. As it turned out, every man was key because for the whole to mesh no effort was inconsequential. The marvelous thing was that every man felt it inside. There was a feeling of destiny on the project."[30]

That the developers adhered to the original principles to make NASTRAN modular, open-ended, and general-purpose—with common formats and interfaces among its different routines—proved to be more important in the long term than how many elements and analysis

---

27. McCormick, *The NASTRAN Users' Manual*, p. iii; Butler, "Age of NASTRAN," pp. 4–6; Butler also lists many more members and describes their contributions. At time of writing, document is available in the *Tenth NASTRAN Users' Colloquium*, NASA CP-2249, on the NASA Technical Reports Server at http://ntrs.nasa.gov/archive/nasa/casi.ntrs.nasa.gov/198.300.04182_198.300.4182.pdf, accessed Aug. 13, 2009.

28. J.P. Raney, D.J. Weidman, and H.M. Adelman, "NASTRAN: Status, Maintenance, and Future Development," in *First NASTRAN Users' Colloquium*, NASA Langley Research Center (1971), p. 1.

29. McCormick, *NASTRAN User's Manual*, pp. iii–iv.

30. Butler, "Age of NASTRAN," p. 7.

capabilities were available at introduction. Preserving the intended architecture ensured that the details could be filled in later.

Early Use and Continuing Development of NASTRAN

The first components of NASTRAN became operational at Goddard in May 1968. Distribution to other Centers, training, and a debugging period followed through 1969 and into 1970.[31] With completion of the initial development, "the management of NASTRAN was transferred to the Langley Research Center. The NASTRAN Systems Management Office (NSMO) was established in the Structures Division at Langley October 4, 1970."[32] Initial public release followed just 1 month later, in November 1970.

NSMO responsibilities included:[33]

- Centralized program development (advisory committees).
- Coordinating user experiences (bimonthly NASTRAN Bulletin and annual Users' Colloquia).
- System maintenance (error correction and essential improvements).
- Development and addition of new capability.
- NASTRAN-focused research and development (R&D).

The actual distribution of NASTRAN to the public was handled by the Computer Software Management and Information Center (COSMIC), NASA's clearinghouse for software distribution (which is described in a subsequent section of this paper). The price at initial release was $1,700, "which covers reproducing and supplying the necessary system tapes and documentation."[34] Documentation was published in four volumes, each with a distinct purpose: one for users, one for programmers who would be involved in maintenance and subsequent development, a theory manual, and finally a volume of demonstration problems. (The 900-page user's manual could be obtained from COSMIC for $10, if purchased separately from the program itself. The author assumes that the other volumes were similarly priced.)[35]

---

31. Ibid., pp. 2–3.
32. Raney, Weidman, and Adelman, 1971, in *First NASTRAN Users' Colloquium*, p. 1.
33. Ibid., p. 2.
34. Foreword to the *First User's Colloquium*, 1971.
35. McCormick, *The NASTRAN User's Manual*, pp. vii–viii.

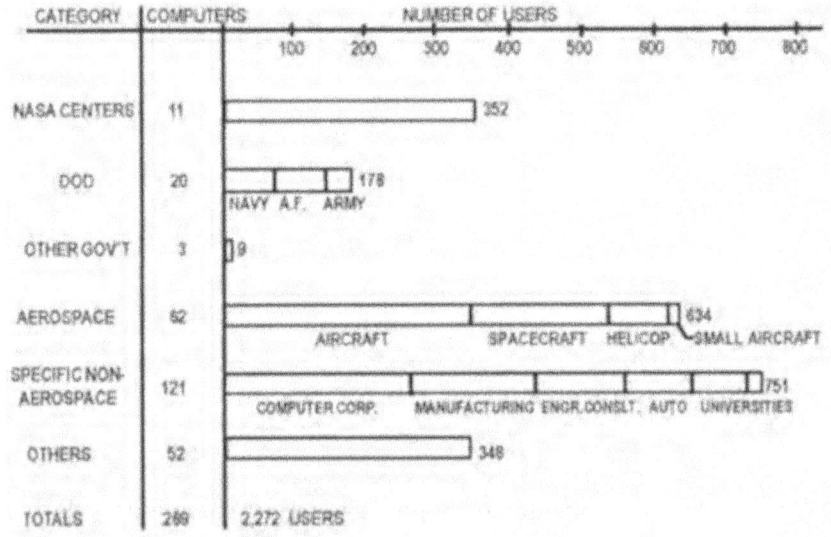

NASTRAN user community profile in 1974. NASA.

Things were happening quickly. Within the first year after public release, NASTRAN was installed on over 60 machines across the United States. There were urgent needs requiring immediate attention. "When NSMO was established in October 1970, there existed a dire need for maintenance of the NASTRAN system. With the cooperation of Goddard Space Flight Center, an interim maintenance contract was negotiated with Computer Sciences Corporation through a contract in effect at GSFC. This contract provided for the essential function of error correction until a contract for full time maintenance could be negotiated through an open competition. The interim maintenance activity was restricted to the correction of over 75 errors reported to the NSMO, together with all associated documentation changes. New thermal bending and hydroelastic elements previously developed by the MacNeal-Schwendler Corporation under contract to GSFC were also installed. Levels 13 and 14 were created for government testing and evaluation. The next version of NASTRAN to be released to the public . . . will be built upon the results of this interim maintenance activity and will be designated Level 15," according to a status report to the user community in 1971.[36]

---

36. Raney, Weidman, and Adelman, "NASTRAN: Status, Maintence, and Future Development," in *First NASTRAN Users' Colloquium, NASA Langley Research Center* (1971).

In June 1971, the contract for full-time maintenance was awarded to MacNeal Schwendler Corporation, which then opened an office near Langley. A bug reporting and correction system was established. Bell Aerospace Company received a contract to develop new elements and a thermal analysis capability. Other efforts were underway to improve efficiency and execution time. A prioritized list of future upgrades was started, with input from all of the NASA Centers. However, for the time being, the pace of adding new capability would be limited by the need to also keep up with essential maintenance.[37]

By 1975, NASTRAN was installed on 269 computers. The estimated composition of the user community (based on a survey taken by the NSMO) is illustrated here.

By this time, the NSMO was feeling the pressure of trying to keep up with maintenance, bug fixes, and requested upgrades from a large and rapidly growing user community. There was also a need to keep up with changing hardware technology. Measures under consideration included improvements to the Error Correction Information System (ECIS); more user involvement in the development of improvements (although this would also require effort to enforce coding standards and interface requirements, and to develop procedures for verification and implementation); and a price increase to help support the NSMO's maintenance costs and also possibly recoup some of the NASTRAN development costs. COSMIC eventually changed its terms for all software distribution to help offset the costs of maintenance.

An annual NASTRAN Users' Colloquium was initiated, the first of which occurred approximately 1 year after initial public release. Each Colloquium usually began with an overview from the NSMO on NASTRAN status, including usage trends, what to expect in the next release, and planned changes in NASTRAN management or terms of distribution. Other papers covered experiences and lessons learned in deployment, installation, and training; technical presentations on new types of elements or new solution capabilities that had recently been, were being, or could be, implemented; evaluation and comparison of NASTRAN with test data or other analysis methods; and user experiences and applications. (The early NASTRAN Users' Colloquia proceedings were available from the National Technical Information Service for $6.)

---

37. Ibid.

The first Colloquia were held at Langley and hosted by the NSMO staff. As the routine became more established and the user community grew, the Colloquia were moved to different Centers and cochaired, usually by the current NSMO Manager and a representative from the hosting Center. There were 21 Users' Colloquia, at which 429 papers were presented. The breakdown of papers by contributing organization is shown here.(Note: collaborative papers are counted under each contributing organization, so the sum of the subtotals exceeds the overall total.)

ORGANIZATIONS PRESENTING PAPERS AT NASTRAN USERS' COLLOQUIA	
TOTAL PAPERS	429
NASA SUBTOTAL:	91
Goddard	33
Langley	35
Other NASA	23
INDUSTRY SUBTOTAL:	274
Computer and software companies	104
Aircraft and spacecraft industry	116
Nonaerospace industry	54
UNIVERSITIES:	26
OTHER GOVERNMENT SUBTOTAL:	91
Air Force	10
Army	15
Navy	61
National Laboratories	5

Computing companies were typically involved in theory, modeling technique, resolution of operational issues, and capability improvements (sometimes on contracts to NASA or other agencies), but also collaborated with "user" organizations assisting with NASTRAN application to problems of interest. All participants were actively involved in the improvement of NASTRAN, as well as its application.

Major aircraft companies—Boeing, General Dynamics, Grumman, Lockheed, McDonnell-Douglas, Northrop, Rockwell, and Vought—were frequent participants, presenting a total of 70 papers. Smaller aerospace companies also began to use NASTRAN. Gates Learjet modeled the Lear 35/36 wing as a test case in 1976 and then used NASTRAN in the design phase of the Lear 28/29 and Lear 55 business jets.[38] Beechcraft used NASTRAN in the design of the Super King Air 200 twin turboprop and the T-34C Mentor military trainer.[39] Dynamic Engineering, Inc., (DEI) began using NASTRAN in the design and analysis of wind tunnel models in the 1980s.[40]

Nonaerospace applications appeared almost immediately. By 1972, NASTRAN was being used in the automotive industry, in architectural engineering, by the Department of Transportation, and by the Atomic Energy Commission. The NSMO had "received expressions of interest in NASTRAN from firms in nearly every West European country, Japan, and Israel."[41] That same year, "NASTRAN was chosen as the principal analytical tool" in the design and construction of the 40-story Illinois Center Plaza Hotel building.[42]

Other nonaerospace applications included:

- Nuclear power plants.
- Automotive industry, including tires as well as primary structure.
- Ships and submarines.
- Munitions.
- Acoustic and electromagnetic applications.
- Chemical processing plants.
- Steam turbines and gas turbines.
- Marine structures.
- Electronic circuitry.

B.F. Goodrich, General Motors, Tennessee Eastman, and Texas Instruments were common presenters at the Colloquia. Frequent Government participants, apart from the NASA Centers, included the David Taylor Naval Ship Research & Development Center, the Naval Underwater Systems Cener, the U.S. Army Armament

---

38. *5th NASTRAN Users' Colloquium*, 1976, pp. 331–352; *8th*, 1979, pp. 11–32; *9th*, 1980, pp. 201–223; *11th*, 1983, pp. 226–248; and *13th*, 1985, pp. 320–340.
39. "Aircraft Design Analysis," in *Spinoff*, NASA, 1979, p. 63.
40. *15th NASTRAN Users' Colloquium*, 1987, pp. 166–183.
41. Rainey and Weidman, "NASTRAN: A Progress Report," *2nd NASTRAN Users' Colloquium* (1972), p. 1.
42. *2nd NASTRAN Users' Colloquium*, p. 421.

Research & Development Command, and several U.S. Army arsenals and laboratories.[43]

Technical improvements, too numerous to describe them all, were continually being made. At introduction (Level 12), **NASTRAN** offered linear static and dynamic analysis. There were two main classes of new capability: analysis routines and structural elements. Developments were often tried out on an experimental basis by users and reported on at the Colloquia before being incorporated into standard **NASTRAN**. Evaluations and further improvements to the capability would typically follow. In addition, of course, there were bug fixes and operational improvements. A few key developments are identified below. Where dates are given, they represent the initial introduction of a capability into standard **NASTRAN**, not to imply full maturity:

- Thermal analysis: initial capability introduced at Level 15 (1973).
- Pre- and post-processing: continuous.
- Performance improvements: continuous.
- Adaptation to new platforms and operating systems: continuous. (The earliest mention the author has found of NASTRAN running on a PC is 1992.[44])
- New elements: continuous. Level 15 included a dummy structural element to facilitate user experimentation.
- Substructuring: the decomposition of a larger model into smaller models that could be constructed, manipulated, and/or analyzed independently. It was identified as an important need when **NASTRAN** was first introduced. Initial substructuring capability was introduced at Level 15 in 1973.
- Aeroelastics and flutter: studies were conducted in the early 1970s. Initial capability was introduced in Level 16 by 1976. **NASTRAN** aeroelastic, flutter, and gust load analysis uses a doublet-lattice aerodynamic method, which approximates the wing as an initially flat surface for the aerodynamic calculation (does not include camber or

---

43. Additional information is provided in the appendixes.
44. Harry G. Schaefer, "Implementation of Mixed Formulation Elements in PC/NASTRAN," in *21st NASTRAN Users' Colloquium* (1992), pp. 1–7.

thickness). The calculation is much simpler than full-fledged computational fluid dynamics (CFD) analysis but neglects many real flow effects as well as configuration geometry details. Accuracy is provided by using correction factors to match the static characteristics of the doublet-lattice model to higher fidelity data from flight test, wind tunnel test, and/or CFD. One of the classic references on correcting lifting surface predictions is a paper by J.P. Giesing, T.P. Kalman, and W.P. Rodden of McDonnell-Douglas, on contract to NASA, in 1976.[45]

- Automated design and analysis: automated fully stressed design was introduced in Level 16 (1976). Design automation is a much broader field than this, however, and most attempts to further automate design, analysis, and/or optimization have taken the form of applications outside of, and interfacing with, **NASTRAN**. In many industries, automated design has become routine; in others, the status of automated design remains largely experimental, primarily because of the inherent complexity of design problems.[46]

- Nonlinear problems: geometric and/or material. Geometric nonlinearity is introduced, for example, when displacements are large enough to change the geometric configuration of the structure in significant ways. Material nonlinearity occurs when local stresses exceed the linear elastic limit. Applications of nonlinear analysis include engine hot section parts experiencing regions of local plasticity, automotive and aircraft crash simulation, and lightweight space structures that may experience large elastic deformations—to name a few. Studies and experimental implementations were made during the 1970s. There are many different classes of nonlinear problems encompassed in this category, requiring a variety of solutions, many of which were added to standard **NASTRAN** through the 1980s.

---

45. J.P. Giesing, T.P. Kalman, and W.P. Rodden, "Correction Factor Techniques for Improving Aerodynamic Prediction Methods," NASA CR-144967 (1976).
46. Interview of Thomas Christy and John Splichal (aircraft industry) by author, Wichita, KS, May 12, 2009.

Moving to Diversify and Commercialize NASTRAN

All of the improvements described above took time to implement. However, many of the using organizations had their own priorities. Several organizations therefore developed their own versions of NASTRAN for internal use, including IBM, Rockwell, and the David Taylor Naval Ship Research & Development Center, not to mention the different NASA Centers. These organizations sometimes contracted with software development companies to make enhancements to their internal versions. Thus, there developed several centers of expertise forging the way forward on somewhat separate paths, but sharing experiences with each other at the Users' Colloquia and other venues. The NSMO did not take responsibility for maintenance of these disparate versions but did consider capabilities developed elsewhere for inclusion in the standard NASTRAN, with appropriate review. This was possible because of the modular structure of NASTRAN to accept new solutions or new elements with little or no disruption to anything else, and it allowed the NSMO's standard NASTRAN to keep up, somewhat, with developments being made by various users.

The first commercial version was announced by MacNeal Schwendler Corporation in 1971.[47] Others followed. SPERRY/NASTRAN was marketed by Sperry Support Services in Europe and by Nippon Univac Kaisha, Ltd., (NUK) in Japan from 1974 to at least 1986. (Sperry was also the UNIVAC parent company—producer of one of the three computers that could run NASTRAN when it was first created.) SPERRY/NASTRAN was developed from COSMIC NASTRAN Level 15.5.[48]

At the 10th NASTRAN Users' Colloquium in 1982, the following commercial versions were identified:[49]

- UAI/NASTRAN (Universal Analytics).
- UNIVAC NASTRAN (Sperry).
- DTNSRDC NASTRAN (David Taylor Naval Ship Research & Development Center).
- MARC NASTRAN (Marc Analysis & Research).
- NKF NASTRAN (NKF Engineering Associates).

---

47. "MSC Software Highlights" on MSC Web site at *http://www.mscsoftware.com/about/history.cfm?Q=135&Z=339&Y=346*, accessed Feb. 12, 2009.
48. *14th User Colloquium*, 1986, p. 1.
49. Butler, "Age of NASTRAN," p. 7.

In spite of this proliferation, common input and output formats were generally maintained. In 1982, Thomas Butler compared COSMIC NASTRAN with the "competition," which included at that time, in addition to the commercial NASTRAN versions, the following programs: ASKA, STRUDL, DET VERITAS NORSKE, STARDYNE, MARC, SPAR, ANSYS, SAP, ATLAS, EASE, and SUPERB. He noted that: "during the period in which NASTRAN maintenance decisions emphasized the intensive debugging of existing capability in preference to adding new capabilities and conveniences, the competitive programs strove hard to excel in something that NASTRAN didn't. They succeeded. They added some elastic elements, e.g.:

- Bending plate with offsets, tapered thickness, and a 10:1 aspect ratio.
- Pipe elbow element.
- Tapered beam element.
- Membrane without excessive stiffness.
- High-order hexagon.

"These new elements make it a bit easier to do some analyses in the category of mechanical structures, listed above."

In addition to new elements, some of the commercial codes added such capabilities as dynamic reduction to assist in condensing a large-order model to a smaller-order one prior to conducting eigenvalue analysis, nonlinear analysis, and packaged tutorials on FEM analysis.

> Viewed from the standpoint of the tools that an analyst needs . . . NASTRAN Level 17.7 can accommodate him with 95 percent of those tools. . . . The effect of delaying the addition of new capability during this 'scrubbing up' period is to temporarily lose the ability to serve the analyst in about 5 percent of his work with the tools that he needs. In the meantime NASTRAN has achieved a good state of health due to the caring efforts of P.R. Pamidi [of CSC] and Bob Brugh [of COSMIC].[50]

The commitment to maintaining the integrity of NASTRAN, rather than adding capability at an unsustainable pace, paid off in the long run.

---

50. Ibid., p. 8.

Computerized Structural Analysis and Research (CSAR) introduced a version of NASTRAN in 1988.[51] However, the trend from the 1980s through the 1990s was toward consolidation of commercial sources for NASTRAN. In 1999, MSC acquired Universal Analytics and CSAR. The Federal Trade Commission (FTC) alleged that by these acquisitions, MSC had "eliminated its only advanced NASTRAN competitors." In 2002, MSC agreed to divest at least one copy of its software, with source code, to restore competition.[52]

At the time of this writing, there are several versions of NASTRAN commercially available. Commercial versions have almost completely superseded NASA's own version, although it is still available through the Open Channel Foundation (as discussed elsewhere in this paper). Even NASA now uses commercial versions of NASTRAN, in addition to other commercial and in-house structural analysis programs, when they meet a particular need.[53]

If one had to sum up the reasons for NASTRAN's extraordinary history, it might be: ripe opportunity, followed by excellent execution. Finite elements were on the cusp. The concepts, and the technology to carry them out, were just emerging. The 1960s were the decade in which the course of the technology would be determined—splintered, or integrated—not that every single activity could possibly be brought under one roof. But, if a single centerpiece of finite element analysis was going to emerge, to serve as a standard and reference point for everything else, it had to happen in that decade, before the technical community took off running in a myriad of different directions.

In execution, the project was characterized by focus, passion, establishment of rules, and adherence to those rules, all coming together under an organization that was dedicated to getting its product out rather than hoarding it. Even with these ingredients, successfully producing a general-purpose computer program, able to adapt through more than 40

---

51. "Stresses and Strains," *Spinoff*, 1998.
52. Federal Trade Commission news release, "MSC Software Settles FTC Charges by Divesting Nastran [sic] Software," Aug. 14, 2002, at http://www.ftc.gov/opa/2002/08/mscsoftware.shtm, accessed May 14, 2009.
53. John P. Gyekenyesi, "SCARE: A Post-Processor Program to MSC/NASTRAN," abstract from NASA Technical Reports Server at http://ntrs.nasa.gov; Karen F. Bartos and Michael A. Ernst, "Evaluation of MARC for the Analysis of Rotating Composite Blades," NASA TM-4423 (1993), p. 1; and Jim Loughlin, "FEM and Multiphysics Applications at NASA/GSFC," NASA Goddard Space Flight Center, in 2004 ANSYS Conference, May 24, 2004, Greenbelt, MD; view graph presentation on NASA Technical Report Server at http://ntrs.nasa.gov, accessed May 12, 2009.

years of changing hardware and software technology, was remarkable. Staying true to the guiding principles (general-purpose, modular, open-ended, etc.), even as difficult decisions had to be made and there was not time to develop every intended capability, was a crucial quality of the development team. In contrast, a team that gets sloppy under time pressure would not have produced a program such lasting value. NASTRAN may be one of the closest approaches ever achieved to 100-percent successful technology transition. Not every structural analyst uses NASTRAN, but certainly every modern structural analyst knows about it. Those who think they need it have access to copious information about it and multiple sources from which they can get it in various forms.

This state of affairs exists in part because of the remarkable nature of the product, and in part because of the priority that NASA places on the transition of its technology to other sectors. In preparation to address the other half of this paper—those many accomplishments that, though lesser than NASTRAN, also push the frontier forward in incremental steps—we now move to a discussion of those activities in which NASA engages for the specific purpose of accomplishing technology transition.

Dissemination and Distribution: NASA Establishes COSMIC

Transitioning technology to U.S. industry, universities, and other Government agencies is part of NASA's charter under the 1958 Space Act. Some such transfer happens "naturally" through conferences, journal publications, collaborative research, and other interactions among the technical community. However, NASA also has established specific, structured technology transfer activities. The NASA Technology Utilization (TU) program was established in 1963. The names of the program's components and activities have changed over time but have generally included the following:[54]

- Publications.
    - Tech briefs: notification of promising technologies.
    - Technology Utilization compilations.
    - Small-business announcements.
    - Technical Support Packages: more detailed

---

54. F. Douglas Johnson and Martin Kokus, "NASA Technology Utilization Program: A Summary of Cost Benefit Studies," NASA CR-201936 (1977), pp. i, 3. See also Johnson, Emily Miller, Nancy Gunderson, Panayes Gatseos, Charles F. Mourning, Thomas Basinger, and Kokus, "NASA Tech Brief Program: A Cost Benefit Evaluation," NASA CR-201938 (1977); and *Spinoff*, 1991, pp. 132–143.

information about a specific technology, provided on request.
- Industrial Applications Centers: university-based services to assist potential technology users in searching NASA scientific and technical information.
- Technology Utilization Officers (TUOs) at the NASA Centers to assist interested parties in identifying and understanding opportunities for technology transfer.
- An Applications Engineering process for developing and commercializing specific technologies, once interest has been established.
- The Computer Software Management and Information Center—a university-based center making NASA software and documentation available to industrial clients.

To expedite and enhance its technology transfer mandate, NASA established a Computer Software Management and Information Center at the University of Georgia at Athens. Within the first few years of the Technology Utilization program, it became apparent that many of the "technology products" being generated by NASA were computer programs. NASA therefore started COSMIC in 1966 to provide the services necessary to manage and distribute computer programs. These services included screening programs and documentation for quality and usability; announcing available programs to potential users; and storing, copying, and distributing the programs and documentation. In addition, as the collection grew, it was necessary to ensure that each new program added capability and was not duplicative with programs already being offered.[55]

After the first year of operation, COSMIC published a catalog of 113 programs that were fully operational and documented. Another 11 programs with incomplete documentation and 7 programs missing subroutines were also offered for customers who would find the data useful even in an incomplete state. Monthly supplements to the catalog added approximately 20 programs per month.[56] By 1971, COSMIC had distributed over 2,500 software packages and had approximately 900 computer programs

---

55. Joseph M. Carlson, "NASA Technology Transfer: The Computer Software Dissemination Program," in *NASTRAN: Users' Experiences,* Langley Research Center (1971), pp. 551–552.
56. Computer Software Management and Information Center (COSMIC), *A Directory of Computer Programs Available from COSMIC,* vol. 1, (Athens, GA: University of Georgia, 1967).

available. New additions were published in a quarterly *Computer Programs Abstracts Journal*. The collection expanded to include software developed by other Government agencies besides NASA. The Department of Defense (DOD) joined the effort in 1968, adding DOD-funded computer programs that were suitable for public release to the collection.[57] In 1981, there were 1,600 programs available.[58] Programs were also withdrawn, because of obsolescence or other reasons. During the early 1990s, the collection consisted of slightly more than 1,000 programs.[59] NASTRAN, when released publicly in 1970, was distributed through COSMIC, as were most of the other computer programs mentioned throughout this paper.

Customers included U.S. industry, universities, and other Government agencies. Customers received source code and documentation, and unlimited rights to copy the programs for their own use. Initially, the cost to the user was just the cost of the media on which the software and documentation were delivered. Basic program cost in 1967 was $75, furnished on cards (2,000 or less) or tape. Card decks exceeding 2,000 cards were priced on a case-by-case basis. Documentation could also be purchased separately, to help the user determine if the software itself was applicable to his or her needs. Documentation prices ranged from $1.50 for 25 pages or less, to $15 for 300 pages or more.[60] Purchase terms eventually changed to a lease/license format, and prices were increased somewhat to help defray the costs of developing and maintaining the programs. Nevertheless, the pricing was still much less than that of commercial software. A cost-benefit study, conducted in 1977 and covering the period from 1971–1976, noted that the operation of COSMIC during that period had only cost $1.7 million, against an estimated $43.5 million in benefit provided to users. During that period, there were 21,000 requests for computer program documentation, leading to 1,200 computer programs delivered.[61]

COSMIC operations continued through the 1990s. In 2001, custody of the COSMIC collection was transferred to a new organization, Open

---

57. Carlson, "NASA Technology Transfer," pp. 551–552.
58. COSMIC, "COSMIC: 1981 Catalog of Computer Programs," NASA CR-163916 (1981), abstract, on NASA Technical Reports Server at http://ntrs.nasa.gov, accessed Mar. 16, 2009.
59. COSMIC, "Monthly Progress Report," NASA CR-195809 (1994), pp. 1–3; *Innovation*, vol. 1, no. 6 (Nov./Dec. 1993), at http://ipp.nasa.gov/innovation/Innovation16/SoftwareWarehouse.html, accessed Feb. 11, 2009.
60. COSMIC, *A Directory of Computer Programs Available from COSMIC*, vol. 1 (Athens, GA: University of Georgia, 1967).
61. Johnson and Kokus, "Summary of Cost-Benefit Studies."

Channel Software (OCS). OCS and a related nonprofit organization, the Open Channel Foundation, were started in 1999 at the University of Chicago. Originally established to provide dissemination of university-developed software, this effort, like COSMIC, soon grew to include software from a broader range of academic and research institutions. The agreement to transfer the COSMIC collection to OCS was made through the Robert C. Byrd National Technology Transfer Center (NTTC), which itself was established in 1989 and had been working with NASA and other Government agencies to facilitate beneficial technology transfer and partnerships.[62]

Although COSMIC is no longer active, NASA continues to make new technical software available to universities, other research centers, and U.S. companies that can benefit from its use. User agreements are made directly with the Centers where the software is developed. User interfaces and documentation are typically not as polished as commercial software, but the level of technology is often ahead of anything commercially available, and excellent support is usually available from the research group that has developed the software. The monetary cost is minimal or, in many cases, zero.[63] Joint development agreements may be made if a potential user desires enhancements and is willing to participate in their development. Whether through COSMIC or by other means, most of the computer programs discussed in the following sections have been made available to U.S. industry and other interested users.

NASA Centers and Their Computational Structural Research

To gain a sense of the types of computational structures projects undertaken by NASA and the contributions of individual Centers to the Agency's efforts, it is necessary to examine briefly the computational structures analysis activities undertaken at each Center, reviewing representative projects, computer programs, and instances of technology transfer to industry—aircraft and otherwise. Projects included the development of

---

62. Open Channel Foundation, "About the Open Channel Foundation and Open Channel Software" at http://www.openchannelfoundation.org, accessed Feb. 10, 2009; "NTTC and Open Channel Software Launch NASA Software Applications Collection as Space Agency Celebrates Birthday," Oct. 10, 2001; "Open Channel Publishes the NASA COSMIC Collection," at http://www.openchannelfoundation.org/cosmic, Open Channel Foundation, accessed Feb. 12, 2009. A list of NASA-developed programs for finite element analysis is reported in Appendix D. As of this writing, the programs and related tasks are available through the Open Channel Foundation Web site. All except GEOFEST were originally released through COSMIC.

63. Interview of Chittur Venkatasubban by author, Wichita, KS, May 14, 2009.

new computer programs, the enhancement of existing programs, integration of programs to provide new capabilities, and, in some cases, just the development of methods to apply existing computer programs to new types of problems. The unique missions of the different Centers certainly influenced the research, but many valuable developments came from collaborative efforts between Centers and applying tools developed at one Center to the problems being worked at another.[64]

Ames Research Center

The Ames Research Center—with research responsibilities within aerodynamics, aeronautical and space vehicle studies, reentry and thermal protection systems, simulation, biomedical research, human factors, nanotechnology, and information technology—is one of the world's premier aerospace research establishments. It was the second NACA laboratory, established in 1939 as war loomed in Europe. The Center was built initially to provide for expansion of wind tunnel facilities beyond the space and power generation capacity available at Langley. Accordingly, in the computer age, Ames became a major center for computational fluid dynamics methods development.[65] Ames also developed a large and active structures effort, with approximately 50 to 100 researchers involved in the structural disciplines at any given time.[66] Areas of research include structural dynamics, hypersonic flight and reentry, rotorcraft, and multidisciplinary design/analysis/optimization. These last two are discussed briefly below.

---

64. Jennifer Ross-Nazzal, e-mail to author, Apr. 10, 2009; NASA, *9th NASTRAN Users' Colloquium*; Two Centers are not discussed in great detail because they undertake little work in computational structural analysis: the Kennedy Space Center and the Stennis Space Center. With responsibility for spacecraft launch operations, vehicle preparation, and integration, Kennedy maintains a minimal structural engineering staff, which is concerned primarily with supporting the Center's launch equipment. Thus, this Center is not active in structural analysis methods research and development, though it did host the *9th NASTRAN Users' Colloquium*. The primary mission of Stennis is to test a very specific type of hardware, namely the ground-testing of rocket engines. More recently, it has added research in remote sensing. But with the exception of a small number of studies related to the operation of the Center's own test equipment, the Stennis Space Center's involvement in research and development in computational methods is minimal. For more information, see John D. Anderson, Jr., "NASA and the Evolution of Computational Fluid Dynamics," a companion essay in this volume.

65. For more information, see Anderson, "NASA and the Evolution of Computational Fluid Dynamics."

66. Interview of John Gallman by author, Wichita, KS, Apr. 4, 2009.

In the early 1970s, a joint **NASA–U.S.** Army rotorcraft program led to a significant amount of rotorcraft flight research at Ames. "The flight research activity initially concentrated on control and handling issues. . . . Later on, rotor aerodynamics, acoustics, vibration, loads, advanced concepts, and human factors research would be included as important elements in the joint program activity."[67] As is typically the case, this effort impacted the direction of analytical work as well in rotor aeroelastics, aeroservoelastics, acoustics, rotor-body coupling, rotor air loads prediction, etc. For example, a "comprehensive analytical model" completed in 1980 combined structural, inertial, and aerodynamic models to calculate rotor performance, loads, noise, vibration, gust response, flight dynamics, handling qualities, and aeroelastic stability of rotorcraft.[68] Other efforts were less comprehensive and produced specialized methods for treating various aspects of the rotorcraft problem, such as blade aeroelasticity.[69] The General Rotorcraft Aeromechanical Stability Program (**GRASP**) combined finite elements with concepts used in spacecraft multibody dynamics problems, treating the helicopter as a structure with flexible, rotating substructures.[70]

Rotorcraft analysis has to be multidisciplinary, because of the many types of coupling that are active. Fixed wing aircraft have not always been treated with a multidisciplinary perspective, but the multidisciplinary analysis and optimization of aircraft is a growing field and one in which Ames has made many valuable contributions. The Advanced Concepts Branch, not directly associated with Structures & Loads but responsible for multidisciplinary vehicle design and optimization studies, has performed and/or sponsored much of this work.

A general-purpose optimization program, **CONMIN**, was developed jointly by Ames and by the U.S. Army Air Mobility Research &

---

67. Paul F. Borchers, James A. Franklin, and Jay W. Fletcher, *Flight Research at Ames: Fifty-Seven Years of Development and Validation of Aeronautical Technology, 1940–1997*, NASA SP-3300 (Washington, DC: NASA, 1998), p. 67.

68. Wayne Johnson, "A Comprehensive Analytical Model of Rotorcraft Aerodynamics and Dynamics, Part III: Program Manual," NASA TM-81184/AVRADCOM TR-80-A-7, Ames Research Center and U.S. Army Aviation Research and Development Command, June 1980, p. vii.

69. F.K. Straub, K.B. Sangha, and B. Panda, "Advance finite element modeling of rotor blade aeroelasticity," *American Helicopter Society Journal*, vol. 39, no. 2 (Apr. 1994), pp. 56–68.

70. A. Stewart Hopkins and Peter Likins, "Analysis of structures with rotating, flexible substructures," AIAA Paper 87-0951 (1987); D.H. Hodges, A.S. Hopkins, D.L. Kunz, and H.E. Hinnant, "Introduction to GRASP: General Rotorcraft Aeromechanical Stability Program—A Modern Approach to Rotorcraft Modeling," *42nd American Helicopter Society*, Washington, DC (1986).

Development Laboratory in 1973[71] and had been used extensively by NASA Centers and contractors through the 1990s. Garret Vanderplaats was the principal developer. Because it is a generic mathematical function minimization program, it can in principle drive any design/analysis process toward an optimum. CONMIN has been coupled with many different types of analysis programs, including NASTRAN.[72]

Aircraft Synthesis (ACSYNT) was an early example of a multidisciplinary aircraft sizing and conceptual design code. Like many early (and some current) total-vehicle sizing and synthesis tools, ACSYNT did not actually perform structural analysis but instead used empirically based equations to estimate the weight of airframe structure. ACSYNT was initially released in the 1970s and has been widely used in the aircraft industry and at universities. Collaboration between Ames and the Virginia Polytechnic Institute's CAD Laboratory, to develop a computer-aided design (CAD) interface for ACSYNT, eventually led to the commercialization of ACSYNT and the creation of Phoenix Integration, Inc., in 1995.[73] Phoenix Integration is currently a major supplier of analysis integration and multidisciplinary optimization software.

Tools such as ACSYNT are very practical, but it has also been a goal at Ames to couple the prediction of aerodynamic forces and loads to more rigorous structural design and analysis, which would give more insight into the effects of new materials or novel vehicle configurations. To this end, a code called ENSAERO was developed, combining finite element structural analysis capability with high-fidelity Euler (inviscid) and Navier-Stokes (viscous) aerodynamics solutions. "The code is capable of computing unsteady flows on flexible wings with vortical flows,"[74] and provisions were made to include control or thermal effects as well. ENSAERO was introduced in 1990 and developed and used throughout the 1990s.

---

71. Garret N. Vanderplaats, *CONMIN: A FORTRAN Program for Constrained Function Minimization, User's Manual*, (Moffett Field, CA: Ames Research Center and U.S. Army Air Mobility R&D Laboratory, 1973).

72. Ashish K. Sareen, Daniel P. Schrage, and T.S. Murthy, "Rotorcraft Airframe Structural Optimization for Combined Vibration and Fatigue Constraints," *47th American Helicopter Society (AHS) Forum, Phoenix, AZ* (1991).

73. "Aircraft Design Software," *Spinoff*, 1997, p. 107; Phoenix Integration, "About Phoenix," on Phoenix Integration Web site at *http://www.phoenix-int.com/about/company_profile.php*, accessed May 11, 2009.

74. G.P. Guruswamy, "ENSAERO: A multidisciplinary Program for Fluid/Structural Interaction Studies of Aerospace Vehicles," *Computing Systems in Engineering*, vol. 1, nos. 2–4 (1990), pp. 237–256.

In a cooperative project with Virginia Tech and McDonnell-Douglas Aerospace, ENSAERO was eventually coupled with NASTRAN to provide higher structural fidelity than the relatively limited structural capability intrinsic to ENSAERO.[75] Guru Guruswamy was the principal developer.

In the late 1990s, Juan Alonso, James Reuther, and Joaquim Martins, with other researchers at Ames, applied the adjoint method to the problem of combined aerostructural design optimization. The adjoint method, first applied to purely aerodynamic shape optimization in the late 1980s by Dr. Antony Jameson, is an approach to optimization that provides revolutionary gains in efficiency relative to traditional methods, especially when there are a large number of design variables. It is not an exaggeration to say that adjoint methods have revolutionized the art of aerodynamic optimization. Technical conferences often contain whole sessions on applications of adjoint methods, and several aircraft companies have made practical applications of the technique to the aerodynamic design of aircraft that are now in production.[76] Bringing this approach to aerostructural optimization is extremely significant.

Dryden Flight Research Center

NASA Dryden has a deserved reputation as a flight research and flight-testing center of excellence. Its personnel had been technically responsible for flight-testing every significant high-performance aircraft since the advent of the world's first supersonic research airplane, the Bell XS-1. When this facility first became part of the NACA, as the Muroc Flight Test Unit in the late 1940s, there was no overall engineering functional organization. There was a small team attached to each test aircraft, consisting of a project engineer, an engineer, and "computers"—highly skilled women mathematicians. There were also three supporting groups: Flight Operations (pilots, crew chiefs, and mechanics), Instrumentation, and Maintenance. By 1954, however, the High-Speed Flight Station (as it was then called) had been organized into four divisions: Research, Flight Operations, Instrumentation, and Administrative. The Research division included three branches: Stability & Control, Loads, and Performance.

---

75. Manoj K. Bhardwaj, "Aeroelastic Analysis of Modern Complex Wings Using ENSAERO and NASTRAN," *Progress Report,* Virginia Polytechnic Institute and State University (1995); and Rakesh K. Kapania, Bhardwaj, Eric Reichenbach, and Guruswamy, "Aeroelastic Analysis of Modern Complex Wings," AIAA-96-4011 (1996).

76. Author's experience.

Shortly thereafter, Instrumentation became Data Systems, to include Computing and Simulation (sometimes together, sometimes separately). There were changes to the organization, mostly gradual, after that, but these essential functions were always present from that time forward.[77] There are approximately 50 people in the structures, structural dynamics, and loads disciplines.[78]

Analysis efforts at Dryden include establishing safety of flight for the aircraft tested there, flight-test and ground-test data analysis, and the development and improvement of computational methods for prediction. Commercially available codes are used when they meet the need, and in-house development is undertaken when necessary. Methods development has been conducted in the fields of general finite element analysis, reentry problems, fatigue and structural life prediction, structural dynamics and flutter, and aeroservoelasticity.

Reentry heating has been an important problem at Dryden since the X15 program. Extensive thermal research was conducted during the NASA YF-12 flight project, which is discussed in a later section. One very significant application of thermal-structural predictive methods was the thermal modeling of the Space Shuttle orbiter, using the Lewis-developed Structural Performance and Redesign (SPAR) finite element code. Prior to first flight, the conditions of the boundary layer on various parts of the vehicle in actual reentry conditions were not known. SPAR was used to model the temperature distribution in the Shuttle structure, for three different cases of aerodynamic heating: laminar boundary layer, turbulent boundary layer, and separated flow. Analysis was based on the space transportation system—trajectory 1 (STS-1) flight profile—and results were compared with temperature time histories from the first mission. The analysis showed that the flight data were best matched under the assumption of extensive laminar flow on the lower surface, and partial laminar flow on the upper surface. This was one piece of evidence confirming the important realization that laminar boundary layers could exist, under conditions of practical interest for hypersonic flight.[79]

---

77. Richard P. Hallion and Michael H. Gorn, *On the Frontier: Experimental Flight at NASA Dryden* (Washington, DC: Smithsonian Books, 2002), pp. 344–354.
78. Martin Brenner, e-mail message to author, May 8, 2009.
79. W.L. Ko, R.D. Quinn, L. Gong, L.S. Schuster, and D. Gonzales, "Reentry Heat Transfer Analysis of the Space Shuttle Orbiter," in Howard B. Adelman, ed., *NASA Langley Research Center Computational Aspects of Heat Transfer in Structures*, NASA CP-2216 (1981), pp. 295–325.

Dryden has a unique thermal loads laboratory, large enough to house an SR-71 or similar-sized aircraft and heat the entire airframe to temperatures representative of high-speed flight conditions. This facility is used to calibrate flight instrumentation at expected temperatures and also to independently apply thermal and structural loads for the purpose of validating predictive methods or gaining a better understanding of the effects of each. It was built during the X15 program in the 1960s and is still in use today.

Aeroservoelastics—the interaction of air loads, flexible structures, and active control systems—has become increasingly important since the late 1970s. As active fly-by-wire control entered widespread use in high-performance aircraft, engineers at Dryden worked to integrate control system modeling with finite-element structural analysis and aerodynamic modeling. Structural Analysis Routines (STARS) and other programs were developed and improved from the 1980s through the present. Recent efforts have addressed the modeling of uncertainty and adaptive control.[80]

At Dryden, much of the technology transfer to industry comes not so much from the release of codes developed at Dryden, but from the interaction of the contractors who develop the aircraft with the technical groups at Dryden who participate in the analysis and testing. Dryden has been involved, for example, in aeroservoelastic analysis of the X-29; F15s and F18s in standard and modified configurations (including physical airframe modifications and/or modifications to the control laws); High Altitude Long Endurance (HALE) unpiloted vehicles, which have their own set of challenges, usually flying at lower speeds but also having longer and more flexible structures than fighter class aircraft; and many other aircraft types.

Glenn (Formerly Lewis) Research Center
Glenn is the primary Center for research on all aspects of aircraft and spacecraft propulsion, including engine-related structures. The structures area has typically consisted of approximately 50 researchers (not

---

80. K.K. Gupta, "STARS—A General-Purpose Finite Element Computer Program for Analysis of Engineering Structures," NASA RP-1129 (1984); Gupta, M.J. Brenner, and L.S. Voelker, "Development of an Integrated Aeroservoelastic Analysis Program and Correlation With Test Data," NASA TP-3120 (1991); and Brenner, e-mail message to author, May 8, 2009.

counting materials).[81] Structures research topics include: structures subjected to thermal loading, dynamic loading, and cyclic loading; spinning structures; coupled thermo-fluid-structural problems; structures with local plasticity and time-varying properties; probabilistic methods and reliability; analysis of practically every part of a turbine engine; Space Shuttle Main Engine (SSME) components; propeller and propfan flutter; failed blade containment analysis; and bird impact analysis. Some of the impact analysis research has been collaborative with Marshall Space Flight Center, which was interested in meteor and space debris impact effects on spacecraft.[82] Glenn has also collaborated extensively with Langley. In 1987, there was a joint Lewis-Langley Workshop on Computational Structural Mechanics (CSM) "to encourage a cooperative Langley-Lewis CSM program in which Lewis concentrates on engine structures applications, Langley concentrates on airframe and space structures applications, and all participants share technology of mutual interest."[83]

Glenn has been involved in NASTRAN improvements since NASTRAN was introduced in 1970 and hosted the sixth NASTRAN Users' Colloquium. Many of the projects at Glenn built supplemental capability for NASTRAN to handle the unique problems of propulsion system structural analysis: "The NASA Lewis Research Center has sponsored the development of a number of related analytical/computational capabilities for the finite element analysis program, NASTRAN. This development is based on a unified approach to representing and integrating the structural, aerodynamic, and aeroelastic aspects of the static and dynamic stability and response problems of turbomachines."[84]

The aircraft and spacecraft engine industries are naturally the primary customers of Glenn technology. However, no attempt is made here to document this technology transfer in detail. Other essays in this volume address advances in propulsion technology and high-temperature materials. Instead, attention is given here to those projects at Glenn that have advanced the general state of the art in computational structures

---

81. Interview of Charles Blankenship by author, Mar. 26, 2009.
82. Search of NASA Technical Reports Server at *http://ntrs.nasa.gov*, Apr. 20, 2009.
83. Nancy P. Sykes, ed., *NASA Workshop on Computational Structural Mechanics*, NASA CP-10012 (1989).
84. Open Channel Foundation, "AIRLOADS," on Open Channel Foundation Web site at *http://www.openchannelfoundation.org/projects/AIRLOADS*, accessed May 6, 2009.

methods and that have found other applications in addition to aerospace propulsion. These include SPAR, NESSUS, SCARE/CARE (and derivatives), ICAN, and MAC.

SPAR was a finite-element structural analysis system developed initially at NASA Lewis in the early 1970s and upgraded extensively through the 1980s. SPAR was less powerful than NASTRAN but relatively interactive and easy to use for tasks involving iterative design and analysis. Chrysler Corporation used SPAR for designing body panels, starting in the 1980s.[85] NASA Langley has made improvements to SPAR and has used it for many projects, including structural optimization, in conjunction with the Ames CONMIN program.[86] SPAR evolved into the EAL program, which was used for the structural portion of structural-optical analyses at Marshall.[87] Dryden Flight Research Center has used SPAR for Space Shuttle reentry thermal modeling.

Numerical Evaluation of Stochastic Structures under Stress (NESSUS) was the product of a Probabilistic Structural Analysis Methods (PSAM) project initiated in 1984 for probabilistic structural analysis of Shuttle and future spacecraft propulsion system components. The prime contractor was Southwest Research Institute (SwRI). NESSUS was designed for solving problems in which the loads, boundary conditions, and/or the material properties involved are best described by statistical distributions of values, rather than by deterministic (known, single) values. PSAM officially completed in 1995 with the delivery of NESSUS Version 6.2. SwRI was awarded another contract in 2002 for enhancements to NESSUS, leading to the release of Version 8.2 to NASA in December 2004 and commercially in 2005. Los Alamos National Laboratory has used NESSUS for weapon-reliability analysis under its Stockpile Stewardship program. Other applications included automotive collision analysis and prediction of the probability of spinal injuries during aircraft ejections, carrier landings, or emergency water landings. NESSUS is used in teaching and research at the University

---

85. "Auto Design," *Spinoff*, NASA, 1986, p. 72.
86. Gary L. Farley and Donald J. Baker, "Graphics and Composite Material Computer Program Enhancements for SPAR," NASA TM-80209 (1980); and J.L. Rogers, Jr., "Programming Structural Synthesis System," *NASA Tech Briefs*, vol. 10, no. 2 (May 1986).
87. Anees Ahmad and Lamar Hawkins, "Development of Software to Model AXAF-1 Image Quality," NASA CR-203978 (1996), p. 1.

of Texas at San Antonio.[88] In some applications, NESSUS is coupled with commercially available deterministic codes offering greater structural analysis capability, with NESSUS providing the statistically derived inputs.[89]

Ceramics Analysis and Reliability Evaluation of Structures (SCARE/CARES) was introduced as SCARE in 1985 and later renamed CARES. This program performed fast-fracture reliability and failure probability analysis of ceramic components. SCARE was built as a postprocessor to MSC/NASTRAN. Using MSC/NASTRAN output of the stress state in a component, SCARE performed the crack growth and structural reliability analysis of the component.[90] Upgrades and a very comprehensive program description and user's guide were introduced in 1990.[91] In 1993, an extension, CARES/LIFE, was developed to calculate the time dependence of the reliability of a component as it is subjected to testing or use. This was accomplished by including the effects of subcritical crack growth over time.[92] Another 1993 upgrade, CCARES (for CMC CARES), added the capability to analyze components made from ceramic matrix composite (CMC) materials, rather than just macroscopically isotropic materials.[93] CARES/PC, introduced in 1994 and made publicly available through COSMIC, ran on a personal computer but offered a more limited capability (it did not include fast-fracture calculations).[94]

R&D Magazine gave an R&D 100 Award jointly to NASA Lewis and to Philips Display Components for application of CARES/Life to the development of an improved television picture tube in 1995. "Cares/Life has been in high demand world-wide, although present technology

---

88. Cody R. Godines and Randall D. Manteufel, "Probabilistic Analysis and Density Parameter Estimation Within Nessus," NASA CR-2002-212008 (2002).
89. "Structural Analysis Made NESSUSary," *Spinoff*, NASA, 2005, p. 94.
90. John P. Gyekenyesi, "SCARE: A Post-Processor Program to MSC/NASTRAN," abstract from NASA Technical Reports Server at http://ntrs.nasa.gov.
91. N.N. Nemeth, et al., "Ceramics Analysis and Reliability," NASA TP-2916 (1990) pp. 1–3.
92. Nemeth, Lynn M. Powers, L.A. Janosik, and J.P. Gyekenyesi, "Lifetime reliability evaluation of structural ceramic parts with the CARES/LIFE computer program," in *34th AIAA/ASME/ASCE/AHS/ASC Structures, Structural Dynamics, and Materials Conference*, CA, Apr. 19–22, 1993; abstract from NASA Technical Reports Server at http://ntrs.nasa.gov.
93. Stephen F. Duffy and John P. Gyekenyesi, "CCARES: A computer algorithm for the reliability analysis of laminated CMC components," NASA TM-111096 (1993).
94. Nemeth, "CARES: Ceramics Analysis and Reliability Evaluation of Structures," NASA Lewis Research Center Report LEW-15168 (1994).

transfer efforts are entirely focused on U.S.-based organizations. Success stories can be cited in numerous industrial sectors, including aerospace, automotive, biomedical, electronic, glass, nuclear, and conventional power-generation industries."[95]

Integrated Composite Analyzer (ICAN) was developed in the early 1980s to perform design and analysis of multilayered fiber composites. ICAN considered hygrothermal (humidity-temperature) conditions as well as mechanical loads and provided results for stresses, stress concentrations, and locations of probable delamination.[96] ICAN was used extensively for design and analysis of composite space antennas and for analysis of engine components. Upgrades were developed, including new capabilities and a version that ran on a PC in the early 1990s.[97] ICAN was adapted (as ICAN/PART) to analyze building materials under a cost-sharing agreement with Master Builders, Inc., in 1995.[98]

Goodyear began working with Glenn in 1995 to apply Glenn's Micromechanics Analysis Code (MAC) to tire design. The relationship was formed, in part, as a result of Glenn's involvement with the Great Lakes Industrial Technology Center (GLITeC) and the Consortium for the Design and Analysis of Composite Materials. NASA worked with Goodyear to tailor the code to Goodyear's needs and provided onsite training. MAC was used to assess the effects of chord spacing, ply and belt configurations, and other tire design parameters. By 2002, Goodyear had several tires in production that had benefitted from the MAC design analysis capabilities. Dr. Steven Arnold was the Glenn point of contact in this effort.[99]

Goddard Space Flight Center

Goddard Space Flight Center was established in 1959, absorbing the U.S. Navy Vanguard satellite project and, with it, the mission of developing, launching, and tracking unpiloted satellites. Since that time, its roles and responsibilities have expanded to consider space science, Earth observation from space, and unpiloted satellite systems more broadly.

---

95. *Research & Technology,* NASA Lewis Research Center, Cleveland, OH, 1996, p. 100.
96. P.L.N. Murthy and C.C. Chamis, "ICAN: Integrated Composites Analyzer," *25th Structures, Structural Dynamics and Materials Conference, Palm Springs, CA* (1984).
97. NASA Technical Reports Server at *http://ntrs.nasa.gov,* accessed May 14, 2009.
98. *Research & Technology,* p. 76.
99. "A 'Tread' Ahead of the Competition," *Spinoff,* NASA, 2002, pp. 72–73.

Structural analysis problems studied at Goddard included definition of operating environments and loads applicable to vehicles, subsystems, and payloads; modeling and analysis of complete launch vehicle/payload systems (generic and for specific planned missions); thermally induced loads and deformation; and problems associated with lightweight, deployable structures such as antennas. Control-structural interactions and multibody dynamics are other related areas of interest.

Goddard's greatest contribution to computer structural analysis was, of course, the NASTRAN program. With public release of NASTRAN, management responsibility shifted to Langley. However, Goddard remained extremely active in the early application of NASTRAN to practical problems, in the evaluation of NASTRAN, and in the ongoing improvement and addition of new capabilities to NASTRAN: thermal analysis (part of a larger Structural-Thermal-Optical [STOP] program, which is discussed below), hydroelastic analysis, automated cyclic symmetry, and substructuring techniques, to name a few.[100]

Structural-Thermal-Optical analysis predicts the impact on the performance of a (typically satellite-based) sensor system due to the deformation of the sensors and their supporting structure(s) under thermal and mechanical loads. After NASTRAN was developed, a major effort began at GSFC to achieve better integration of the thermal and optical analysis components with NASTRAN as the structural analysis component. The first major product of this effort was the NASTRAN Thermal Analyzer. The program was based on NASTRAN and thereby inherited a great deal of modeling capability and flexibility. But, most importantly, the resulting inputs and outputs were fully compatible with NASTRAN: "Prior to the existence of the NASTRAN Thermal Analyzer, available general purpose thermal analysis computer programs were designed on the basis of the lumped-node thermal balance method. . . . They were not only limited in capacity but seriously handicapped by incompatibilities arising from the model representations [lumped-node versus finite-element]. The intermodal transfer of temperature

---

100. J.B. Mason, "The NASTRAN Hydroelastic Analyzer," NASA TM-X-65617 (1972); R.H. MacNeal, R.L. Harder, and Mason, "NASTRAN Cyclic Symmetry Capability," Goddard Space Flight Center, Sept. 1973, abstract on NASA Technical Reports Server at http://ntrs.nasa.gov, accessed Apr. 20, 2009; and W. R. Case, "Dynamic Substructure Analysis of the International Ultraviolet Explorer (IUE) Spacecraft," Goddard Space Flight Center, Sept. 1973, abstract on NASA Technical Reports Server at http://ntrs.nasa.gov, accessed Apr. 20, 2009.

data was found to necessitate extensive interpolation and extrapolation. This extra work proved not only a tedious and time-consuming process but also resulted in compromised solution accuracy. To minimize such an interface obstacle, the STOP project undertook the development of a general purpose finite-element heat transfer computer program."[101] The capability was developed by the MacNeal Schwendler Corporation under subcontract from Bell Aerospace. "It must be stressed, however, that a cooperative financial and technical effort between [Goddard and Langley] made possible the emergence of this capability."[102]

Another element of the STOP effort was the computation of "view factors" for radiation between elements: "In an in-house STOP project effort, GSFC has developed an IBM-360 program named 'VIEW' which computes the view factors and the required exchange coefficients between radiating boundary elements."[103] VIEW was based on an earlier view factor program, RAVFAC, but was modified principally for compatibility with NASTRAN and eventual incorporation as a subroutine in NASTRAN.[104] STOP is still an important part of the analysis of many of the satellite packages that Goddard manages, and work continues toward better performance with complex models, multidisciplinary design, and optimization capability, as well as analysis.

Jet Propulsion Laboratory

Jet Propulsion Laboratory (JPL) began as an informal group of students and staff from the California Institute of Technology (Caltech) who experimented with rockets before and during World War II; evolved afterward into the Nation's center for unpiloted exploration of the solar system and deep space, operating related tracking, and data acquisition systems; and was managed for NASA by Caltech.[105] Dr. Theodore von Kármán, then head of Caltech's Guggenheim Aeronautical Laboratory, shepherded this group to becoming a center of rocket

---

101. H. Lee and J.B. Mason, "NASTRAN Thermal Analyzer: A General Purpose Finite Element Heat Transfer Computer Program," in NASA Langley Research Center, *2nd NASTRAN Users' Colloquium* (1972), p. 444.
102. Ibid., p. 445.
103. Ibid., p. 449.
104. E.F. Puccinelli, "View Factor Computer Program (Program VIEW) User's Manual," Goddard Space Flight Center, July 1973, abstract on NASA Technical Reports Server at http://ntrs.nasa.gov, accessed Apr. 20, 2009.
105. JPL Annual Reports at http://www.jpl.nasa.gov/about/reports.cfm, accessed Apr. 25, 2009.

research for the Army. Upon NASA's formation in 1958, JPL came under NASA's responsibility.[106]

Consistent with its origins and Caltech's continuing role in its management, JPL's orientation has always emphasized advanced experimental and analytical research in various disciplines, including structures. JPL developed efficiency improvements for NASTRAN as early as 1971.[107] Other JPL research included basic finite element techniques, high-velocity impact effects, effect of spin on structural dynamics, geometrically nonlinear structures (i.e., structures that deflect sufficiently to significantly alter the structural properties), rocket engine structural dynamics, flexible manipulators, system identification, random processes, and optimization. The most notable of these are VISCEL, TEXLESP-S, and PID (AU-FREDI and MODE-ID).[108]

VISCEL (for Visco-Elastic and Hyperelastic Structures) and TEXLESP-S treat special classes of materials that general-purpose finite element codes typically cannot handle. VISCEL treats visco-elastic problems, in which materials exhibit viscosity (normally a fluid characteristic) as well as elasticity. VISCEL was introduced in 1971 and was adapted by industry over the next decade.[109] In 1982, the Shell Oil Company used VISCEL to validate a proprietary code that was in development for the design of plastic products.[110] In 1984, AiResearch was using VISCEL to analyze seals and similar components in aircraft auxiliary power units (APUs).[111]

JPL has been leading research in the structural dynamics of solid rockets almost since the laboratory was first established. TEXLESP-S was specifically developed for analysis of solid rocket fuels, which may be polymeric materials exhibiting such hyperelastic behavior. TEXLESP-S is a finite element code developed for large-strain (hyperelastic) problems, in which materials may be purely elastic but exhibit such large

---

106. Jet Propulsion Laboratory, "NASA Facts: Jet Propulsion Laboratory," NASA, *http://www.jpl.nasa.gov/news/fact_sheets/jpl.pdf*, accessed Aug. 13, 2009.
107. R. Levy and S. Wall, "Savings in NASTRAN Decomposition Time by Sequencing to Reduce Active Columns," in NASA Langley Research Center, *NASTRAN: Users' Experiences* (1971), pp. 627–631.
108. Search of NASA Technical Reports Server at *http://ntrs.nasa.gov*, Apr. 22, 2009.
109. F.A. Akyuz and E. Heer, VISCEL Computer Program *User Manual for Analysis of Linear Visco-elastic Structures*, vol. 1: *Users' Manual*, (Pasadena: NASA Jet Propulsion Laboratory, 1971).
110. NASA Scientific and Technical Information, "Computer Technology for Industry," *Spinoff Database* at *http://www.sti.nasa.gov/spinoff/database*, accessed April 25, 2009.
111. NASA Scientific and Technical Information, "Auxiliary Power Units," *Spinoff Database*.

strain deformations that the geometric configuration of the structure is significantly altered. (This is distinct from the small-strain, large-deflection situations that can occur, for example, with long flexible booms on spacecraft.)[112]

System Identification/Parameter Identification (PID, including AU-FREDI and MODE-ID) is the use of empirical data to build or tune a mathematical model of a system. PID is used in many disciplines, including automatic control, flight-testing, and structural analysis.[113] Ideally, excitation of the system is performed by systematically exciting specific modes. However, such controlled excitation is not always practical, and even under the best of circumstances, there is some uncertainty in the interpretation of the data. The MODE-ID program was developed in 1988 to estimate not only the modal parameters of a structure, but also the level of uncertainty with which those parameters have been estimated:

> Such a methodology is presented which allows the precision of the estimates of the model parameters to be computed. It also leads to a guiding principle in applications. Namely, when selecting a single model from a given class of models, one should take the most probable model in the class based on the experimental data. Practical applications of this principle are given which are based on the utilization of measured seismic motions in large civil structures. Examples include the application of a computer program MODE-ID to identify modal properties directly from seismic excitation and response time histories from a nine-story steel-frame building at JPL and from a freeway overpass bridge.[114]

---

112. Eric B. Becker and Trent Miller, "Final Report: Development of Non-Linear Finite Element Computer Code," NASA CR-179965 (1985).

113. James L. Beck, "Probabilistic System Identification in the Time Domain," in *USAF–NASA, USAF/NASA Workshop of Model Determination for Large Space Systems* (Pasadena: California Institute of Technology, 1988), abstract from NASA Technical Reports Server at http://ntrs.nasa.gov, accessed Apr. 24, 2009, notes: "Areas of application for system identification include the following: (1) Model Evaluation . . . (2) Model Improvement . . . (3) Empirical Modeling [using experimental data in the initial development of a model, when existing methods or information about the system are not sufficient]; and (4) Damage Detection and Assessment—continual or episodic updating of a structural model through vibration monitoring to detect and locate any structural damage."

114. Beck, "Probabilistic System Identification."

Another system identification program, Autonomous Frequency Domain Identification (AU-FREDI), was developed for the identification of structural dynamic parameters and the development of control laws for large and/or flexible space structures. It was furthermore intended to be used for online design and tuning of robust controllers, i.e., to develop control laws real time, although it could be modified for offline use as well. AU-FREDI was developed in 1989, validated in the Caltech/Jet Propulsion Laboratory's Large Spacecraft Control Laboratory and made publicly available.[115] This is just a small sample of the research that JPL has conducted and sponsored in system identification, control of flexible structures, integrated control/structural design, and related fields. While intended primarily for space structures, this research also has relevance for medicine, manufacturing technology, and the design and construction of large, ground-based structures.

Johnson Space Center

Johnson Space Center, a product of the "space age," is NASA's core center for human space flight, development of launch vehicles and systems, astronaut training, and human space flight operations. As a Center with significant hardware development and operational responsibilities, Johnson's activities in analysis methods have been "usually directed to specific problems relating to developing hardware that the Center is responsible for."[116]

Except for moderate downsizing in the 1980s and minor organizational changes such as separating Structures and Dynamics into two branches, the structures-related organization has been relatively stable over several decades. The Structural Engineering Division (ES) has approximately 120 employees divided into 5 branches: Structures, Dynamics, Thermal, Material, and Mechanisms. The Structures Branch (ES2) has responsibility for structural design, analysis (including computer methods), and testing.[117] Johnson has some very significant test facilities, including a tower that can hold a full Apollo or similar-sized

---

115. Y. Yam, "AU-FREDI—Autonomous Frequency Domain Identification," JPL, 1994, abstract from NASA Technical Reports Server at *http://ntrs.nasa.gov*, accessed Apr. 25, 2009.
116. According to Modlin, who was the Johnson representative on the Ad Hoc Group for Structural Analysis leading up to the development of NASTRAN, in an e-mail message to author, Apr. 23, 2009.
117. Ross-Nazzal, e-mail to author, Apr. 7, 2009.

vehicle and subject it to vibration testing.[118] Current directions at Johnson include sustaining activity for the Space Shuttle and the International Space Station (ISS), and new work related to the Orion spacecraft.[119]

With the emphasis on hardware and systems development, rather than on methods development, Johnson has favored the use of computer programs already available when they can meet the need. According to Modlin:

> Prior to NASTRAN we used the SAMIS program that was developed by JPL for stress and dynamics, our inputs regarding NASTRAN were directed to the NASTRAN office at NASA Langley after the program was delivered, but we did not do any development on our own. We and our contractor wanted to use NASTRAN on the Shuttle Orbiter, but required substructuring. This wasn't delivered in time [as a NASTRAN capability] so the contractor continued with ASKA. . . . Some programs developed in house relate to: Lunar landing, Apollo Crew Module water landing and flight loads. One more general program that has wide use is NASGRO (formerly FLAGRO), which was developed by Royce Forman. It is a fracture mechanics routine.[120]

Although this paper has not attempted to cover fracture mechanics, it is worth noting that NASGRO, originally developed for space applications, has been enhanced with "many features specifically implemented to suit the needs of the aircraft industry," because of increasing focus in the Federal Aviation Administration (FAA), NASA, and DOD on safety of aging aircraft.[121]

Other programs developed at Johnson or under Johnson sponsorship include TRASYS (Thermal Radiation Analysis System, 1973), FAMSOR (Frequencies and Modes of Shells Of Revolution, 1974), SNASOR (Static

---

118. Interview of Aleck C. Bond by Rebecca Wright (No. 3), Houston, TX, July 15, 1999, in *JSC Oral Histories* project at http://www.jsc.nasa.gov/history/oral_histories/oral_histories.htm, accessed Apr. 25, 2009.
119. Modlin, e-mail message to author, Apr. 23, 2009; and Ross-Nazzal, e-mail message to author, Apr. 7, 2009.
120. Modlin, e-mail message to author, Apr. 23, 2009.
121. Shivakumar Mettu, et al, "NASGRO 3.0—A Software for Analyzing Aging Aircraft," in NASA–FAA–DOD, *2nd Joint NASA/FAA/DoD Conference on Aging Aircraft*, pt. 2, (Washington, DC: NASA–FAA–DOD, 1999), pp. 792–801.

Nonlinear Analysis of Shells of Revolution, 1974), BUCKY (Plate buckling, 1992), and COMPAPP (Composite plate buckling, 1994).[122]

Langley Research Center

Langley was the first NACA laboratory, established in 1917. As such, it is the oldest and most distinguished of NASA aeronautics Centers, with a pedigree that dates to meetings held prior to the First World War to determine the future aeronautical laboratory structure of the Nation. Since the earliest days of American aviation, Langley has constantly anticipated, reacted, and adapted as necessary to meet the Nation's aeronautical research needs, reflecting its broad technical capabilities and expertise in areas such as aerodynamics, aircraft and spacecraft structures, flight dynamics, crew systems, space environmental physics, and life sciences.

Among the very earliest NACA technical reports were several concerning loads calculation and structural analysis, some of which are cited in the introduction to this paper. These papers, and others that followed throughout the era of the NACA, were widely used in the aircraft industry. By the time NASA was founded, Langley had become a major Center for all forms of aeronautics research, engineering, and analysis.

Through the 1980s and 1990s, Langley had approximately 150 technical professionals in the structural disciplines (not including Materials), covering both aircraft and spacecraft applications. This work was organized primarily in two divisions, Structural Mechanics (static problems) and Structural Dynamics, plus a separate Optimization Methods group of approximately 15 members.[123] Structural Mechanics included Composites, Computational Structural Mechanics, Thermal Structures, Structural Concepts, and AeroThermal Loads.[124] Structural Dynamics included Aeroelasticity, Unsteady Aerodynamics, Aeroservoelasticity, Landing and Impact Dynamics, Spacecraft Dynamics, and Interdisciplinary Research.[125] (Reorganizations sometimes changed the specific delineation of responsibilities.) Langley researchers pursued

---

122. Search of NASA Technical Reports Server at http://ntrs.nasa.gov, Apr. 20, 2009.
123. W.J. Stroud, (NASA Langley, retired), e-mail message to author, Mar. 29, 2009.
124. Kay S. Bales, "Structural Mechanics Division Research and Technology Plans for FY 1989 and Accomplishments for FY 1988," NASA TM-101592 (1989).
125. Eleanor C. Wynne, "Structural Dynamics Division Research and Technology Accomplishments for FY 1990 and Plans for FY 1991," NASA TM-102770 (1991), p. 1.

many separate computational structural analysis studies and efforts, but overall, the Center was particularly (and intimately) involved with NASTRAN, the Design Analysis Methods for Vibration (DAMVIBS) rotorcraft structural dynamics modeling program, and efforts at integration and optimization.

After NASTRAN was developed during the period from 1965 to 1970, management of it was transferred from Goddard to Langley. Accordingly, a major emphasis at Langley through the 1970s was the maintenance and continuing improvement of NASTRAN. The first four Users' Colloquia were held at Langley. While COSMIC handled the administrative aspects of NASTRAN distribution, the NSMO was responsible for technical management and coordinating NASTRAN development efforts across all Centers and many contractors. The program itself is discussed in greater detail elsewhere in this case.

The DAMVIBS research program, conducted from 1984 to 1991, reflected Langley's long-standing heritage of research on rotorcraft structural dynamics. DAMVIBS achieved concrete advances in the industry state of the art in helicopter structural dynamic modeling, analysis-to-test matching, and, perhaps most importantly, acceptance of and confidence in modeling as a useful tool in designing helicopter rotor-airframe systems for low vibration. Key NASA program personnel were William C. Walton, Jr., who spearheaded program concept and initial direction (he retired in 1984); Raymond G. Kvaternik, who furnished program direction after 1984; and Eugene C. Naumann, who supplied critical technical guidance. The industry participants were Bell Helicopter Textron, Boeing Helicopters, McDonnell-Douglas Helicopter Company, and Sikorsky Aircraft. The participants developed rotor-airframe finite element models, conducted ground vibration tests, made test/analysis comparisons, improved their models, and conducted further study into the "difficult components" that current state of the art rotorcraft analysis could not adequately model.[126]

Modeling "guides"—documented procedures—were identified from the start as a key element to the program:

> This program emphasized the planning of the modeling . . . the NASA Technical Monitor insisted on a well thought out

---

126. Raymond G. Kvaternik, "The NASA/Industry Design Analysis Methods for Vibration (DAMVIBS) Program—A Government Overview," in Kvaternik, ed., *A Government/Industry Summary of the Design Analysis Methods for Vibrations (DAMVIBS) Program*, NASA CP-10114 (1993), p. 9.

Case 8 | NASA and Computational Structural Analysis

plan of attack, accompanied by detailed preplanned instructions.... The plan was reviewed by other industry representatives prior to undertaking the actual modeling. Another unique feature was that at the end of the modeling, deviations from the planned guides due to cause were reported.[127]

All of the participants reported that finite element modeling could predict vibrations more accurately than previously realized but required more attention to detail in the modeling, with finer meshes and the inclusion of secondary components not normally modeled for static strength and stiffness analysis. The participants further reported on specific improvements to dynamic modeling practice resulting from the exercise and on the increased use and acceptance of such modeling in the design phase at each respective company.[128] As a result of DAMVIBS:

- Bell and Boeing incorporated **DAMVIBS** lessons into the modeling of their respective portions of the V-22.[129]
- Boeing made improvements the **NASTRAN** dynamic model of the **CH47D**, which was still in production, achieving greatly improved correlation to test data. Boeing

---

127. R. Gabel, P. Lang, and D. Reed, "The NASA/Industry Design Analysis Methods for Vibration (DAMVIBS) Program—Boeing Helicopters Airframe Finite Element Modeling," in Kvaternik, ed., *A Government/Industry Summary of the Design Analysis Methods for Vibrations (DAMVIBS) Program*, NASA CP-10114 (1993), p. 23.

128. For example, from Bell: "Structural optimization was found to be a useful tool and Bell is continuing with development of this methodology and integrating it into the design process to efficiently achieve minimum weight and vibration levels in future designs. The work that was accomplished under the NASA DAMVIBS program has had a major influence on the 'hardening' and growth of vibration technology in the helicopter industry...." From Sikorsky: "Prior to the DAMVIBS program, attempts to reduce [vibration] were usually limited to making modifications or adding vibration-control devices to an already designed and built airframe, in a trial-and-error fashion." The Sikorsky team noted that DAMVIBS-funded work had "brought for the first time the introduction of low-vibration design into the design cycle at Sikorsky." See James D. Cronkhite, "The NASA/Industry Design Analysis Methods for Vibration (DAMVIBS) Program—Bell Helicopter Textron Accomplishments," in Kvaternik, ed., *A Government/Industry Summary of the Design Analysis Methods for Vibrations (DAMVIBS) Program*, NASA CP-10114 (1993), pp. 11, 22; and William J. Twomey, "The NASA/Industry Design Analysis Methods for Vibration (DAMVIBS) Program—Sikorsky Aircraft—Advances Toward Interacting with the Airframe Design Process," in Kvaternik, ed., *A Government/Industry Summary of the Design Analysis Methods for Vibrations (DAMVIBS) Program*, NASA CP-10114 (1993), p. 47.

129. Gabel, Lang, and Reed, "DAMVIBS—Boeing," p. 33. 130. Ibid., p. 33–34.

credited Eugene Naumann of Langley with identifying many of the needed changes.[130]
- McDonnell-Douglas improved its dynamic models of existing and newly developed products, achieving improved correlation with test results.[131]
- Sikorsky developed an FEM model of the UH60A airframe "having a marked improvement in vibration-predicting ability."[132]
- Sikorsky also developed a new program (PAREDYM, programmed in NASTRAN DMAP language) that could automatically adjust an FEM model so that its modal characteristics would match test values.[133] PAREDYM then found use as a design tool: having the ability to modify a model of an existing design to better match test data, it also had the ability to modify a model of a new design not yet tested, to a set of desired modal characteristics. Designers could now specify a target (low) level of vibration response and let PAREDYM tune its model—essentially designing the airframe—to meet the goal. (The improvements would not be "free," however, as the program could add weight in the process.) After discovering this usage mode, the developers then added facilities for minimizing the weight impact to achieve a desired level of vibration improvement.[134]

DAMVIBS ended in 1991, though this did not mark an end to Langley's work on rotorcraft structural dynamics.[135] Rather, it reflected a shift in emphasis away from the traditional helicopter to other aeronautics and

---

130. Ibid., p. 33–34.
131. Mostafa Toossi, Richard Weisenburger, and Mostafa Hashemi-Kia, "The NASA/Industry Design Analysis Methods for Vibration (DAMVIBS) Program—McDonnell Douglas Helicopter Company Achievements," in Kvaternik, ed., *A Government/Industry Summary of the Design Analysis Methods for Vibrations (DAMVIBS) Program*, NASA CP-10114 (1993), p. 44.
132. Twomey, "DAMVIBS—Sikorsky," p. 47.
133. Ibid., p. 47.
134. Ibid., p. 52.
135. William T. Yeager, Jr., "A Historical Overview of Aeroelasticity Branch and Transonic Dynamics Tunnel Contributions to Rotorcraft Technology and Development," NASA TM-2001-211054 (2001), p. 38

astronautics research ventures as well.[136] As basic analysis capability had become relatively mature by around 1990, attention turned toward the integration of design, analysis, and optimization; to the integration of structural analysis with other disciplines; and to nondeterministic methods and the modeling of uncertainty.[137] Projects included further work on rotorcraft, aircraft aerostructural optimization, control-structural optimization for space structures, and nondeterministic or "fuzzy" structures, to name a few.[138] Many optimization projects at Langley used the CONMIN constrained function minimization program, developed at Ames, as the optimization driver, interfaced with various discipline-specific analysis codes developed at Langley or elsewhere.

In the 1970s, NASA Langley began what would prove to be some very significant multidisciplinary optimization (MDO) studies. Jaroslaw Sobieszczanski-Sobieski pioneered the Bi-Level Integrated System Synthesis (BLISS), a general approach that is applicable to design optimization in any set of disciplines and of any system, aircraft, or otherwise. His work at Langley, spanning from the 1970s to the present, is recognized throughout the aerospace industry and the MDO community. BLISS and related methods constitute one of the major classes of MDO techniques in widespread use today. Some of the early work on BLISS was concerned with improving the structural design process and addressing aerodynamic and structural problems concurrently. For example, in the late 1970s, Sobieszczanski-Sobieski developed methods for designing metal and/or composite wing structures of supersonic transports for minimum weight, including the effect of structural deformations on aeroelastic loads.[139]

This Langley work continued into the 1980s, when Langley researchers moved forward to apply the knowledge gained with BLISS to spacecraft, generating two other systems: the Integrated Design and Evaluation of Advanced Spacecraft (IDEAS) and Programming Structural Synthesis (PROSS). IDEAS did not perform optimization per se, but it did provide integration of design with analysis in multiple disciplines, includ-

---

136. Kvaternik, "The NASA/Industry Design Analysis Methods for Vibrations (DAMVIBS) Program: Accomplishments and Contributions," NASA TM-104192 (1991), p. 13.
137. W.J. Stroud, (NASA Langley, retired), e-mail message to author, Mar. 28, 2009.
138. Victor W. Sparrow and Ralph D. Buehrle, "Fuzzy Structures Analysis of Aircraft Panels in NASTRAN," AIAA Paper 2001-1320 (2001).
139. Jaroslaw Sobieszczanski-Sobieski, "An Integrated Computer Procedure for Sizing Composite Airframe Structures," NASA TP-1300 (1979).

ing structures and structural dynamics.[140] PROSS combined the Ames CONMIN optimizer with the SPAR structural analysis program (developed at NASA Lewis). PROSS was publicly released in 1983.[141] Several subsequent releases incorporated either new optimization strategies and/or improved finite element analysis.[142]

One of these was ST-SIZE, which started as a hypersonic vehicle structural-thermal design code. In 1996, Collier Research Corporation obtained an exclusive license from Langley for the ST-SIZE program. Under a new model for NASA technology transfer, Collier agreed to pay NASA royalties from sales of Collier's commercialized version of the code. This version, called HyperSizer (trademark of Collier Research Corporation), was intended to be applicable to a wide variety of uses, including office design and construction, marine systems, cargo containers, aircraft, and railcars. The program performed design, weight buildup, system-level performance assessments, structural analysis, and structural design optimization.[143] In 2003, *Spinoff* reported that this model had worked well and that Collier and NASA were still working together to enhance the program, specifically by incorporating further analysis codes from NASA Glenn Research Center: Micromechanics Analysis Code with Generalized Method Cells (MAC/GMC) and higher-order theory for functionally graded materials (HOTGFM). Both of these were developed collaboratively between Glenn, University of Virginia, Ohio Aerospace Institute, and Tel Aviv University.[144]

Marshall Space Flight Center

Consistent with its mission to develop spacecraft technologies and with its heritage as the site where Wernher von Braun and his team had

---

140. L. Bernard Garrett, "Interactive Modeling, Design and Analysis of Large Spacecraft," NASA Langley Research Center (1982), on NASA Technical Reports Server at *http://ntrs.nasa.gov*, accessed Apr. 4, 2009.
141. J.L. Rogers, Jr., J. Sobieszczanski-Sobieski, and R.B. Bhat, "Structural Optimization," *Tech Briefs*, vol. 7, no. 2, p. 184, May 1983; abstract on NASA Technical Reports Server at *http://ntrs.nasa.gov*, accessed Apr. 4, 2009.
142. Rogers, "NETS/PROSSS—Nets Coupled with the Programming System for Structural Synthesis," LAR-14818, Langley Research Center; abstract on NASA Technical Reports Server at *http://ntrs.nasa.gov*, accessed Apr. 4, 2009; and J.L. Rogers, Jr., "System for Structural Synthesis Combines Finite-Element Analysis and Optimization Programs," *NASA Tech Briefs*, vol. 8, no. 2 (Nov. 1984), p. 242.
143. "Structural Analysis and Design Software," *Spinoff*, NASA, 1997, p. 96.
144. "Efficient, Multi-Scale Design Takes Flight," *Spinoff*, NASA, 2003, pp. 68–69.

worked since 1950, Marshall Space Flight Center has always had a strong technical/analytical organization, engaged in science and engineering research as well as advanced design studies. Research areas have included basic finite element methods, shells, fluid-structure systems, and nonlinear structures, as well as quick-turnaround non-FEM methods for early design and feasibility studies.[145]

Applications have usually involved the structural and structural-dynamic problems of launch vehicles. As an example, computational techniques were used to help resolve "pogo" oscillations in both the first and second stages of the Saturn V launch vehicle. As the name implies, the pogo mode is a longitudinal tensile/compressive oscillation. Flight data from the unpiloted flight of the second Saturn V in 1968 showed severe vibrations from 125 to 135 seconds into the first-stage burn. The pogo mode is not always harmful, but in this case, there were concerns that it could upset the guidance system or damage the payload. The structural frequency was dependent on fuel load, and at a certain point in the flight, it would coincide with a natural frequency of the engine/fuel/oxygen system, causing resonance. Using the models to evaluate the effects of various design changes, the working group assigned to the task determined that accumulators in the liquid oxygen (LOX) lines would alter the engine frequency sufficiently to resolve the issue. Subsequently, engineers examining flight data from the Apollo 8, 9, and 13 missions noticed a similar occurrence in the second stage. This was studied and resolved using similar techniques.[146]

The first-stage pogo issue occurred at a point in the Apollo program when time was of the essence in identifying, analyzing, and resolving the problem. The computer models were most likely no more complex than they had to be to solve the problem at hand. Marshall Space Flight Center has continued to develop and use fairly simple codes for early conceptual studies. Simple, quick-turnaround tools developed at Marshall include Cylindrical Optimization of Rings, Skin and Stringers (CORSS, 1994) and the VLOADS launch loads and dynamics program (1997). VLOADS was developed as a Visual BASIC macro in Microsoft Excel. When released in COSMIC in 1997, it was also available in PC format.

---

145. Search of NASA Technical Reports Server at http://ntrs.nasa.gov, Apr. 20, 2009.
146. Andrew J. Dunar and Stephen P. Waring, *Power to Explore: A History of the Marshall Space Flight Center, 1960–1990*, NASA SP-4313 (Washington, DC: NASA, 1999), pp. 44–50.

It was distributed on a single 3.5-inch diskette.[147] This was a remarkable development from the days when the problem of launch vehicle dynamics occupied a sizable fraction of this Nation's computing power!

Like researchers at Langley, Marshall's personnel moved swiftly from single or limited application tools to finding ways to integrate them with other tools and processes and thereby achieve enhanced or previously unattainable capabilities. The Coupled Eulerian Lagrangian Finite Element (CELFE) code, developed collaboratively with NASA Lewis Research Center in 1978, included specialized nonlinear methods to calculate local effects of an impact. It was coupled to NASTRAN for calculation of the far-field response of the structure. Applications included space debris, micrometeor, and foreign object impact studies for aircraft engines.[148] Marshall developed an interface between the PATRAN finite element preprocessor (normally used with NASTRAN) and the NASA Langley STAGS shell analysis code in 1990.[149] Marshall sponsored Southwest Research Institute to develop an interface between Lewis-developed NESSUS probabilistic analysis and NASTRAN in 1996.[150] Both STAGS and NESSUS have been widely used outside NASA. This review of NASA Centers and their work on computational structural analysis has offered only a glimpse of the variety of structural problems that exist and the corresponding variety of methods developed and used at the various NASA Centers and then shared with industry.

Applying Computational Structural Analysis to Flight Research
We now turn to an area of activity that provides, for aviation, the ultimate proof of design techniques and predictive capabilities: flight-testing. While there are many fascinating projects that could be discussed, we will consider only five that had particular relevance to the subject at hand, either because they collected data that were specifically intended to provide validation of computational predictions of structural behavior, or because they demonstrated unique structural design approaches.

---

147. J.B. Graham and P.L. Luz, "Preliminary In-Flight Loads Analysis of In-Line Launch Vehicles Using the VLOADS 1.4 Program," NASA TM-1998-208472 (1998).
148. C. Chamis, "CELFE/NASTRAN code for the analysis of structures subjected to high velocity impact," NASA TM-79048 (1978); abstract on NASA Technical Report Server at *http://ntrs.nasa.gov*, accessed Apr. 26, 2009.
149. Neil Otte, "PATRAN-STAGS Translator (PATSTAGS)," NASA TM-100388 (1990).
150. Southwest Research Institute, "NESSUS/NASTRAN Interface," NASA CR-202778 (1996).

Two of these are the YF-12 Thermal Loads project and the Rotor Aerodynamic Limits survey, both of which collected data for validating and improving predictive methods. The remaining three are the Highly Maneuverable Aircraft Technology (HiMAT) digital fly-by-wire (DFBW) enhanced agility composite-structured canard demonstrator, the AD-1 oblique wing demonstrator, and the Grumman X-29 forward-swept wing (FSW) research aircraft. These three projects exercised, in progressively more challenging ways, the concept of aeroelastic tailoring: that is, predicting airframe flexibility and having enough confidence in those predictions to design an airplane that takes advantage of elastic deformation, rather than just trying to minimize it. In all of these, NASA-rooted computational structural prediction proved of great, and even occasionally, critical, significance.

The investigation of aircraft structural mechanics or, indeed, of almost any discipline, can be considered to include the following activities: investigation by basic theory, computational analysis or simulation, laboratory test, and flight test (or, more generally, any test of the final product in its actual operating environment). Many arguments have been had over which is the most valuable. This author is of the opinion—based on his experience in the practice of engineering, on a certain amount of historical research, and on the teaching and example of mentors and peers—that theory, computation, laboratory test, and flight test all constitute imperfect but complementary views of reality. Thus, until someone comes up with a way to know the exact state of stress and deflection in every part of a vehicle under actual operating conditions, we must form our understanding of reality as a composite image, using what information we can gain from each available source:

- Flight test, obviously, is the best representation we have of an aircraft in actual operational conditions. However, our ability to interrogate the system is most severely compromised in this activity. Many data parameters are not available unless special instrumentation is installed, if at all, and this is the most difficult environment in which to obtain stable, high-quality data.
- Laboratory test offers better visibility into the operation of specific parts of the system and better control of experimental parameters, at the price of some separation from true operational conditions.

- Computation offers even greater opportunity to interrogate the value of any data parameter at any time(s) and to simulate conditions that might be impossible, difficult, or dangerous to test. Computation also eliminates all physical complications of running the experiment and all physical sources of noise and uncertainty. But in stepping out of the physical world and into the analytical world, the researcher also becomes subject to the limited fidelity of his computational method: what effects are and are not included in the computation and how well the computation represents physical reality.
- Theory is sometimes the best source of insight and of understanding what parameters might be changed to obtain some desired effect, but it does not provide the detailed quantitative data necessary to implement the solution.

In this light, the following flight programs are discussed. Much more could be said about each of them. The present discussion is necessarily confined to their significance to the development or validation of loads and structural computation methods.

YF-12 Thermal Loads and Structural Dynamics

NASA operated two Lockheed YF-12As and one "YF-12C" (actually an early nonstandard SR-71A, although the Air Force at that time could not acknowledge that it was allowing NASA to operate an SR-71) between 1969 and 1979.[151] These aircraft were used for a variety of research projects. In some projects, the YF-12s were the test articles, exploring their performance, handling qualities, and propulsion system characteristics in various baseline or modified configurations and modes of operation. In other projects, the YF-12s were used as "flying wind tunnels" to carry test models and other experiments into the Mach 3+ flight environment. Testing directly related to structural analysis methods and/or loads prediction included a series of thermal-structural load tests from 1969 to 1972 and smaller projects concerning ventral fin loads and structural

---

151. NASA DFRC Fact Sheet: YF-12A, at http://www.nasa.gov/centers/dryden/news/FactSheets, accessed Aug. 13, 2009. For a detailed examination of this program, see a companion essay by William Flanagan.

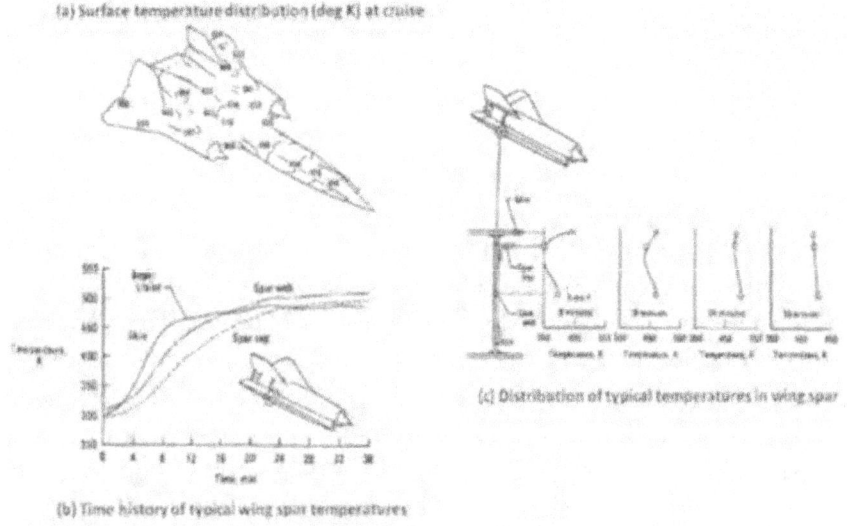

Temperature time histories from YF12 flight project. NASA.

dynamics.[152] The flight-testing was conducted at Dryden, which was also responsible for project management. Ames, Langley, and Lewis Research Centers were all involved in technical planning, analysis, and supporting research activities, coordinated through NASA Headquarters. The U.S. Air Force and Lockheed also provided support in various areas.[153] Gene Matranga of Dryden was the manager of the program before Berwin Kock later assumed that role.[154]

The thermal-structural loads project involved modeling and testing in Dryden's unique thermal load facility. The purpose was to correlate in-flight and ground-test measurements and analytical predictions of temperatures, mechanical loads, strains, and deflections. "In all the X-15 work, flight conditions were always transient. The vehicle went to high speed in a matter of two to three minutes. It slowed down in a matter of three to five minutes. . . . The YF-12, on the other hand, could stay at Mach 3 for 15 minutes. We could get steady-state temperature

---

152. Hallion and Gorn, *On the Frontier*, pp. 423–426.
153. James A. Albers, in James and Associates, eds., *YF-12 Experiments Symposium*, vol. 1, NASA CP-2054 (1978), p. 25.
154. Peter W. Merlin, *Mach 3+: NASA/USAF YF-12 Flight Research, 1969–1979*, NASA SP-2001-4525, No. 25 in the *Monographs in Aerospace History* series (Washington, DC: NASA, 2002), pp. 8, 95.

data, which would augment the X-15 data immeasurably."[155] The YF-12 testing showed that it could take up to 15 minutes for absolute temperatures in the internal structure to approach steady state, and, even then, the gradients—which have a strong effect on stresses because of differential expansion—did not approach steady state until close to 30 minutes into the cruise.[156]

NASTRAN and FLEXSTAB (a code developed by Boeing on contract to NASA Ames to predict aeroelastic effects on stability) were used to model the YF-12A's aeroelastic and aerothermoelastic characteristics. Alan Carter and Perry Polentz of NASA oversaw the modeling effort, which was contracted to Lockheed and accomplished by Al Curtis. This effort produced what was claimed to be the most extensive full-vehicle NASTRAN model developed up to that time. The computational models were used to predict loads and deflections, and also to identify appropriate locations for the strain gauges that would take measurements in ground- and flight-testing. The instrumentation included strain gauges, thermocouples, and a camera mounted on the fuselage to record airframe deflection in flight. Most of the flights, from Flight 11 in April 1970 through Flight 53 in February 1972, included data collection for this project, often mixed with other test objectives.[157] Subsequently, the aircraft ceased flying for more than a year to undergo ground tests in the high-temperature loads laboratory. The temperatures measured in flight were matched on the ground, using heated "blankets" placed over different parts of the airframe. Ground-testing with no aerodynamic load allowed the thermal effects to be isolated from the aerodynamic effects.[158]

There were also projects involving the measurement of aerodynamic loads on the ventral fin and the excitation of structural dynamic modes. The ventral fin project was conducted to provide improved understanding of the aerodynamics of low aspect ratio surfaces. FLEXSTAB was used in this effort but only for linear aerodynamic predictions. Ground tests had shown the fin to be stiff enough to be treated as a rigid surface. Measured load data were compared to the linear theory predictions and

---

155. Quoted in Merlin, *Mach 3+*, p. 90.
156. Jerald M. Jenkins and Albert E. Kuhl, "Recent Load Calibrations Experience with the YF 12 Airplane," in James and Associates, eds., *YF-12 Experiments Symposium*, vol. 1, NASA CP-2054 (1978), p. 49.
157. Merlin, *Mach 3+*, p. 92; Hallion and Gorn, *On the Frontier*, pp. 188–189 and 423–424.
158. Merlin, *Mach 3+*, p. 90.

to wind tunnel data.[159] For the structural dynamics tests, which occurred near the end of NASA's YF-12A program, "shaker vanes"—essentially oscillating canards—were installed to excite structural modes in flight. Six flights with shaker vanes between November 1978 and March 1979 "provided flight data on aeroelastic response, allowed comparison with calculated response data, and thereby validated analytical techniques."[160]

Experiences from the program were communicated to industry and other interested organizations in a YF-12 Experiments Symposium that was held at Dryden in 1978, near the end of the 10-year effort.[161] There were also briefings to Boeing, specifically intended to provide information that would be useful on the Supersonic Transport (SST) program, which was canceled in 1971.[162] There have been other civil supersonic projects since then—the High-Speed Civil Transport (HSCT)/High-Speed Research (HSR) efforts in the 1990s and some efforts related to supersonic business jets since 2000—but none have yet led to an operational civil supersonic aircraft.

Modern Rotor Aerodynamic Limits Survey

The Modern Rotor Aerodynamic Limits Survey was a 10-year program launched in 1984, which encompassed flight efforts in 1987 and 1993–1994. In 1987, a Sikorsky UH-60A Black Hawk was tested with conventional structural instrumentation installed on the rotor blades. Then:

> ... Sikorsky Aircraft was [subsequently] contracted to build a set of highly instrumented blades for the Black Hawk test aircraft: a pressure blade with 242 absolute pressure transducers and a strain-gauge blade with an extensive suite of strain gauges and accelerometers ... approximately 30 gigabytes of data were obtained in 1993–94 and installed in an electronic database that was immediately accessible to the domestic rotorcraft industry.[163]

---

159. Robert R. Meyer, Jr., and V. Michael DeAngelis, "Flight-Measured Aerodynamic Loads on a 0.92 Aspect Ratio Lifting Surface," in James and Associates, eds., *YF-12 Experiments Symposium*, vol. 1, NASA CP-2054 (1978), p. 77.
160. Merlin, *Mach 3+*, p. 42.
161. James and Associates, eds., *YF-12 Experiments Symposium*, vol. 1, NASA CP-2054 (1978).
162. Merlin, *Mach 3+*, pp. 93–94.
163. Borchers, et al., *Flight Research at Ames*, p. 71.

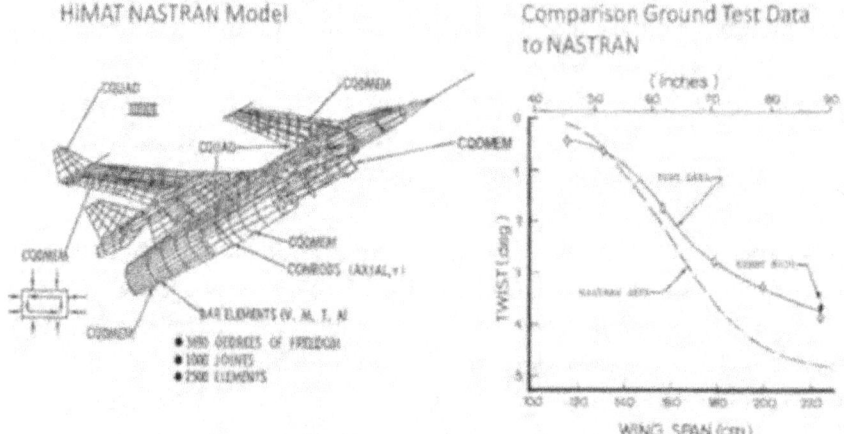

NASTRAN model and NASTRAN to static test comparison. NASA.

The two types of measurement systems are complementary. Strain gauges give an indication of the total load in a member, but little insight to the details of where and how the load is generated. The pressure taps show the distribution of the applied aerodynamic load, but only at given stations, so the total load estimate depends on how one computes the data through the unknown regions between the pressure transducers. The combination of both types of data is most useful to researchers trying to correlate computational loads predictions with the test data.

HiMAT

HiMAT was a small, unpiloted aircraft (23.5-feet long, 15.6-foot wingspan, weight just over 3,000 pounds) somewhat representative of a fighter type configuration, flown between 1979 and 1983, and developed to evaluate the following set of technologies and features:

- Close-coupled canard.
- Winglets.
- Digital fly-by-wire flight control.
- Composite structure.
- Aeroelastic tailoring.
- Supercritical airfoil.

It was intended that the benefits of these collected advances be shown together rather than separately and on an unpiloted platform, so that

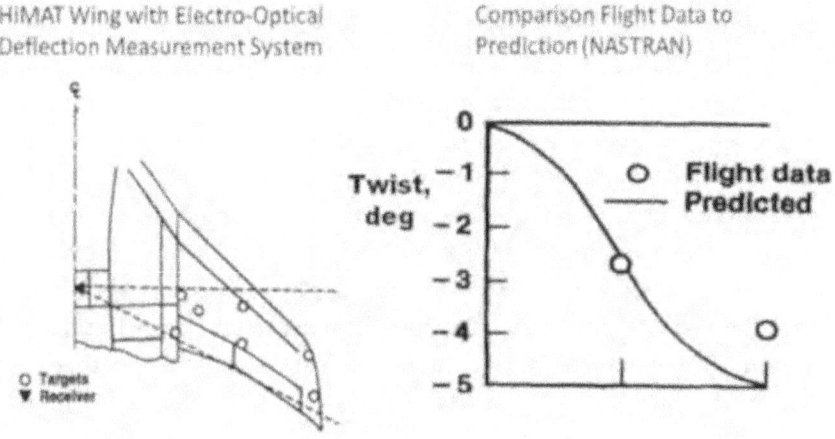

HiMAT Electro-Optical Flight Deflection Measurement System. NASA.

the vehicle could be tested more aggressively without danger to a pilot.[164]

"Aeroelastic tailoring" refers to the design of a structure to achieve aerodynamically favorable deformation under load, rather than the more traditional approach of simply minimizing deformation. The goal of aeroelastic tailoring on the HiMAT ". . . was to achieve an aero-dynamically favorable spanwise twist distribution for maneuvering flight conditions" in the canard and the outboard wing. "The NASTRAN program was used to compute structural deflections at each model grid point. Verification of these deflections was accomplished by performing a loads test prior to delivery of the vehicle to NASA." The ground-test loads were based on a sustained 8-g turn at Mach 0.9, which was one of the key performance design points of the aircraft. The NASTRAN model and a comparison between predicted and measured deflections are shown in the accompanying figure. Canard and wing twist were less than predicted. The difference was attributed to insufficient understanding of the matrix-dominated laminate material properties.[165]

The vehicle was also equipped with a system to measure deflections of the wing surface in flight. Light emitting diodes (LEDs)—referred to as targets—on the wing upper surface were detected by a photodiode

---

164. Hallion and Gorn, *On The Frontier*, pp. 276–278.
165. Richard C. Monaghan, "Description of the HiMAT Tailored Composite Structure and Laboratory Measured Vehicle Shape Under Load," NASA TM-81354 (1981), pp. 4–5, 7.

array mounted on the fuselage, at a location overlooking the wing. Three inboard targets were used to determine a reference plane, from which the deflection of the remaining targets could be measured. To measure wing twist, targets were positioned primarily in pairs along the front and rear wing spars.[166] The HiMAT wing had a relatively small number of targets—only two pairs besides the inboard reference set—so the in-flight measurements were not a detailed survey of the wing by any means. Rather, they provided measurement at a few key points, which could then be compared with the NASTRAN data and the ground loads test data. Target and receiver locations are illustrated here, together with a sample of the deflection data at the 8-g maneuver condition. In-flight deflection data showed similar twist to the ground-test data, indicating that the aerodynamic loads were well predicted.[167]

The HiMAT was an early step in the development of aeroelastic tailoring capability, providing a set of NASTRAN data, static load test data, and flight-test data, for surface deflection at a given loading condition. The project also proved out the electro-optical system for in-flight deflection measurements, which would later be used in the X-29 project.

AD-1 Oblique Wing Demonstrator

The AD-1 was a small and inexpensive demonstrator aircraft intended to investigate some of the issues of an oblique wing. It flew between 1979 and 1982. It had a maximum takeoff weight of 2,100 pounds and a maximum speed of 175 knots. It is an interesting case because (1) NASA had an unusually large role in its design and integration—it was essentially a NASA aircraft—and (2) because it provides a neat illustration of the prosecution of a particular objective through design, analysis, wind tunnel test, flight test, and planned follow-on development.[168]

The oblique wing was conceived by German aerodynamicists in the midst of the Second World War. But it was only afterward, through the

---

166. Walter J. Sefic and Karl F. Anderson, "NASA High Temperature Loads Calibration Laboratory," NASA TM-X-1868 (1969), pp. 3–4.
167. Glenn B. Gilyard, "The Oblique Wing Research Aircraft: A Test Bed for Unsteady Aerodynamic and Aeroelastic Research," in NASA Langley Research Center, Transonic Unsteady Aerodynamics and Aeroelasticity, pt. 2, Report N89-19247 (1987), p. 412; J.J. Burken, G.S. Alag, and G.B. Gilyard, "Aeroelastic Control of Oblique-Wing Aircraft," NASA TM-86808 (1986).
168. Robert E. Curry and Alex G. Sim, "In-Flight Total Forces, Moments, and Static Aeroelastic Characteristics of an Oblique-Wing Research Airplane," NASA TP-2224 (1984), p. 15.

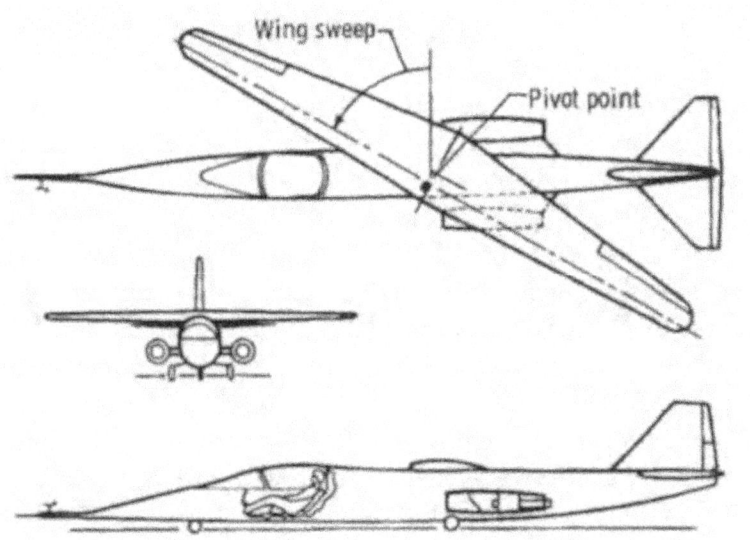

AD-1 three view. NASA.

brilliance and determination of NASA aerodynamicist Robert T. Jones that it advanced to actual flight. Indeed, Jones, father of the American swept wing, became one of the most persistent proponents of the oblique wing concept.[169] The principal advantage of the oblique wing is that it spreads both the lift and volume distributions of the wing over a greater length than that of a simple symmetrically swept wing. This has the effect of reducing both the wave drag because of lift and the wave drag because of volume, two important components of supersonic drag. With this theoretical advantage come practical challenges. The challenges fall into two broad categories: the effects of asymmetry on the flight characteristics (stability and handling qualities) of the vehicle, and the aeroelastic stability of the forward-swept wing. The research objectives of the AD-1 were primarily oriented toward flying qualities. The AD-1 was not intended to explore structural dynamics or divergence in depth, other

---

169. Jones continued to study it and advocate it until his death in 1999; the author had the pleasure of hearing him lecture on this topic in the aerodynamics class that Jones taught at Stanford University in the 1980s. For Jones, see statement of William Sears in Ames Research Center staff, "Collected Works of Robert T. Jones," NASA TM-X-3334 (1976), pp. vii-ix; Hallion, "Lippisch Gluhareff, and Jones: The Emergence of the Delta Planform and the Origins of the Sweptwing in the United States," *Aerospace Historian*, vol. 26, No. 1 (Mar. 1979), pp. 1–10; and Walter G. Vincenti, "Robert Thomas Jones," in *Biographical Memoirs*, vol. 86 (Washington, DC: National Academy of Sciences, 2005), pp. 3–21.

than establishing safety of flight. Mike Rutkowski analyzed the wing for flutter and divergence using NASTRAN and other methods.[170]

However, the project did make a significant accomplishment in the use of static aeroelastic tailoring. The fiberglass wing design by Ron Smith was tailored to bend just enough, with increasing g, to cancel out an aerodynamically induced rolling moment. Pure bending of the oblique wing increases the incidence (and therefore the lift) of the forward-swept tip and decreases the incidence (and lift) of the aft-swept tip. In a pullup maneuver, increasing lift coefficient (C_L), and load factor at a given flight condition, this would cause a rollaway from the forward-swept tip. At the same time, induced aerodynamic effects (the downwash/upwash distribution) increase the lift at the tip of an aft-swept wing. On an aircraft with only one aft-swept tip, this would cause a roll toward the forward-swept side. The design intent for the AD-1 was to have these two effects cancel each other as nearly as possible, so that the net change in rolling moment because of increasing g at a given flight condition would be zero. The design condition was $C_L = 0.3$ for 1-g flight at 170 knots, 12,500-foot altitude, and a weight of 1,850 pounds, with the wing at 60-degree sweep.[171]

An aeroelastically scaled one-sixth model was tested at full-scale Reynolds number in the Ames 12-Foot Pressure Wind Tunnel. A stiff aluminum wing was used for preliminary tests, then two fiberglass wings. The two fiberglass wings had zero sweep at the 25- and 30-percent chord lines, respectively, bracketing the full-scale AD-1 wing, which had zero sweep at 27-percent chord. The wings were tested at the design scaled dynamic pressure and at two lower values to obtain independent variation of wing load because of angle of attack and dynamic pressure at a constant angle of attack. Forces and moments were measured, and deflection was determined from photographs of the wing at test conditions.[172]

Subsequently, ". . . the actual wing deflection in bending and twist was verified before flight through static ground loading tests." Finally, in-flight measurements were made of total force and moment coefficients and of aeroelastic effects. Level-flight decelerations provided angle-of-attack sweeps at constant load, and windup turns provided angle-of-attack sweeps at constant "q" (dynamic pressure). Results

---

170. Steve Smith, interview with author, Apr. 6, 2009; and M.J. Rutkowski, "Aeroelastic stability analysis of the AD-1 Manned Oblique-Wing Aircraft," NASA TM-78439 (1977).
171. Curry and Sim, "In-Flight Total Forces," p. 3.
172. Ibid., pp. 3–4.

Case 8 | NASA and Computational Structural Analysis

were interpreted and compared with predictions. The simulator model, with aeroelastic effects included, realistically represented the dynamic responses of the flight vehicle.[173]

Provision had been made for mechanical linkage between the pitch and roll controls, to compensate for any pitch-roll coupling observed in flight. However, the intent of the aeroelastic wing was achieved closely enough that the mechanical interconnect was never used.[174] Roll trim was not needed at the design condition (60-degrees sweep) nor at zero sweep, where the aircraft was symmetric. At intermediate sweep angles, roll trim was required. The correction of this characteristic was not pursued because it was not a central objective of the project. Also, the airplane experienced fairly large changes in rolling moment with angle of attack beyond the linear range. Vortex lift, other local flow separations, and ultimately full stall of the aft-swept wing, occurred in rapid succession as angle of attack was increased from 8 to approximately 12 degrees. Therefore, it would be a severe oversimplification to say that the AD-1 had normal handling qualities.[175]

The AD-1 flew at speeds of 170 knots or less. On a large, high-speed aircraft, divergence of the forward-swept wing would also be a consideration. This would be addressed by a combination of inherent stiffness, aeroelastic tailoring to introduce a favorable bend-twist coupling, and, potentially, active load alleviation. The AD-1 project provided initial correlation of measured versus predicted wing bending and its effects on the vehicle's flight characteristics. NASA planned to take the next step with a supersonic oblique wing aircraft, using the same F-8 airframe that had been used for earlier supercritical wing tests. These studies delved deeper into the aeroelastic issues: "Preliminary studies have been performed to identify critical DOF [Degree of Freedom] for flutter model tests of oblique configurations. An 'oblique' mode has been identified with a 5 DOF model which still retains its characteristics with the three rotational DOF's. An interdisciplinary analysis code (STARS), which is capable of performing flutter and aeroservoelastic analyses, has been developed. The structures module has a large library of elements and in conjunction with numerical analysis routines, is capable of efficiently performing statics, vibration, buckling, and dynamic response

---

173. Ibid., pp. 3–4, 10.
174. Smith, interview with author.
175. Curry and Sim, "In-Flight Total Forces," pp. 7–8.

analysis of structures. . . . " The STARS code also included supersonic (potential gradient method) and subsonic (doublet lattice) unsteady aerodynamics calculations. " . . . Linear flutter models are developed and transformed to the body axis coordinate system and are subsequently augmented with the control law. Stability analysis is performed using hybrid techniques. The major research benefit of the OWRA [Oblique Wing Research Aircraft] program will be validation of design and analysis tools. As such, the structural model will be validated and updated based on ground vibration test (GVT) results. The unsteady aero codes will be correlated with experimentally measured unsteady pressures."[176]

While the OWRA program never reached flight, (NASA was ready to begin wing fabrication in 1987, expecting first flight in 1991), these comments illustrate the typical interaction of flight programs with analytical methods development and the progressive validation process that takes place. Such methods development is often driven by unconventional problems (such as the oblique wing example here) and eventually finds its way into routine practice in more conventional applications. For example, in the design of large passenger aircraft today, the loads process is typically iterated to include the effects of static aeroelastic deflections on the aerodynamic load distribution.[177]

X-29

The Grumman X-29 aircraft was an extraordinarily ambitious and productive flight-test program run between 1984 and 1992. It demonstrated a large (approximately 35 percent) unstable static margin in the pitch axis, a digital active flight control system utilizing three-surface pitch control (all-moving canards, wing flaps, and aft-mounted strake flaps), and a thin supercritical forward-swept wing, aeroelastically tailored to prevent structural divergence. The X-29 was funded by the Defense Advanced Research Projects Agency (DARPA) through the USAF Aeronautical Systems Division (ASD). Grumman was responsible for aircraft design and fabrication, including the primary structural analyses, although there was extensive involvement of NASA and the USAF in addressing the entire realm of unique technical issues on the project. NASA Ames Research Center/Dryden Flight Research Facility was the responsible test organization.[178]

---

176. Gilyard, "The Oblique Wing Research Aircraft."
177. Mendoza and Hazen, interview with author.
178. Sefic and Maxwell, "X-29A Overview," p. 2.

Careful treatment of aeroelastic stability was necessary for the thin FSW to be used on a supersonic, highly maneuverable aircraft. According to Grumman, "Automated design and analysis procedures played a major role in the development of the X-29 demonstrator aircraft." Grumman used one of its programs, called FASTOP, to optimize the X-29's structure to avoid aeroelastic divergence while minimizing the weight impact.[179]

In contrast to the AD-1, which allowed the forward-swept wing to bend along its axis, thereby increasing the lift at the forward tip, the X-29's forward-swept wings were designed to twist when bending, in a manner that relieved the load. This was accomplished by orienting the primary spanwise fibers in the composite skins at a forward "kick angle" relative to the nominal structural axis of the wing. The optimum angle was found in a 1977 Grumman feasibility study: "Both beam and coarse-grid, finite-element models were employed to study various materials and laminate configurations with regard to their effect on divergence and flutter characteristics and to identify the weight increments required to avoid divergence."[180] While a pure strength design was optimum at zero kick angle, an angle of approximately 10 degrees was found to be best for optimum combined strength and divergence requirements.

When the program reached the flight-test phase, hardware-in-the-loop simulation was integral to the flight program. During the functional and envelope expansion phases, every mission was flown on the simulator before it was flown in the airplane.[181] In flight, the X-29 No. 1 aircraft (of two that were built) carried extensive and somewhat unique instrumentation to measure the loads and deflections of the airframe, and particularly of the wing. This consisted of pressure taps on the left wing and canard, an optical deflection measurement system on the right wing, strain gages for static structural load measurement, and accelerometers for structural dynamic and buffet measurement.

The most unusual element of this suite was the optical system, which had been developed and used previously on the HiMAT demonstrator (see preceding description). Optical deflection data were sampled at a

---

179. Philip Mason, Edward Lerner, and Lawrence Sobel, "Applications of Integrated Design/Analysis Systems in Aerospace Structural Design," in NASA Langley Research Center, *Recent Advances in Multidisciplinary Analysis and Optimization*, pt. 1 (1989), p. 25.
180. Ibid., p. 26.
181. Sefic and Maxwell, "X-29A Overview," p. 3.

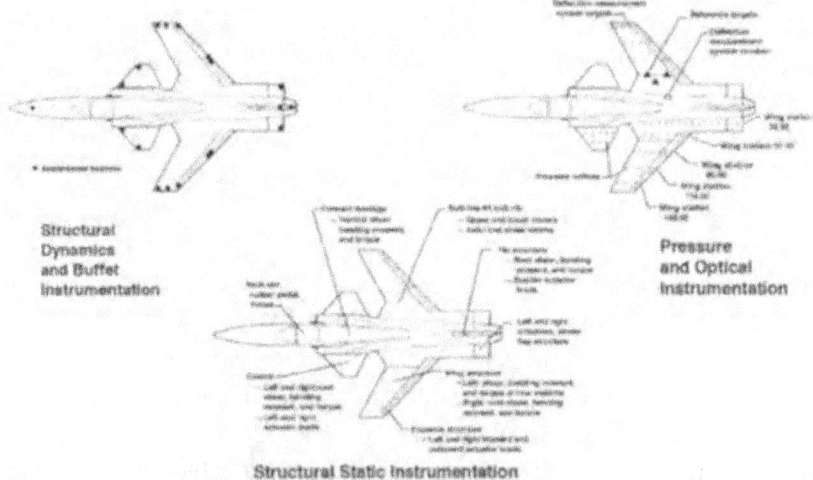

Loads measurement provisions on Grumman X-29A. NASA.

rate of 13 samples per channel per second. Data quality was reported to be very good, and initial results showed good match to predictions. In addition, pressure data from the 156 wing and 17 canard pressure taps was collected at a rate of 25 samples per channel per second. One hundred six strain gages provided static loads measurement as shown. Structural dynamic data from the 21 accelerometers was measured at 400 samples per channel per second. All data was transmitted to ground station and, during limited-envelope phase, to Grumman in Calverton, NY, for analysis.[182] "Careful analyses of the instrumentation requirements, flight test points, and maneuvers are conducted to ensure that data of sufficient quality and quantity are acquired to validate the design, fabrication, and test process."[183] The detailed analysis and measurements provided extensive opportunities to validate predictive methods.

The X-29 was used as a test case for **NASA's STARS** structural analysis computer program, which had been upgraded with aeroservoelastic analysis capability. In spite of the exhaustive analysis done ahead of time, there were, as is often the case, several "discoveries" made during flight test. Handling qualities at high alpha were considerably better than predicted, leading to an expanded high-alpha control and maneuverability investigation in the later phases of the project. The X-29

---

182. Ibid., pp. 3–4.
183. Ibid., p. 4.

No. 1 was initially limited to 21-degree angle of attack, but, during subsequent Phase II envelope expansion testing, its test pilots concluded it had "excellent control response to 45 deg. angle of attack and still had limited controllability at 67 deg. angle of attack."[184]

There were also at least two distinct types of aeroservoelastic phenomena encountered: buffet-induced modes and a coupling between the canard position feedback and the aircraft's longitudinal aerodynamic and structural modes were observed.[185] The modes mentioned involved frequencies between 11 and 27 hertz (Hz). Any aircraft with an automatic control system may experience interactions between the aircraft's structural and aerodynamic modes and the control system. Typically, the aeroelastic frequencies are much higher than the characteristic frequencies of the motion of the aircraft as a whole. However, the 35-percent negative static margin of the X-29A was much larger than any unstable margin designed into an aircraft before or since. As a consequence, its divergence timescale was much more rapid, making it particularly challenging to tune the flight control system to respond quickly enough to aircraft motions, without being excited by structural dynamic modes. Simply stated, the X-29A provided ample opportunity for aeroservoelastic phenomena to occur, and such were indeed observed, a contribution of the aircraft that went far beyond simply demonstrating the aerodynamic and maneuver qualities of an unstable forward-swept canard planform.[186]

In sum, each of these five advanced flight projects provides important lessons learned across many disciplines, particularly the validation of computer methods in structural design and/or analysis. The YF-12 project provided important correlation of analysis, ground-test data, and flight data for an aircraft under complex aerothermodynamic loading. The Rotor Aerodynamic Limits survey collected important data on helicopter rotors—a class of system often taken for granted yet one that

---

184. Quoted in Dryden Flight Research Center "Fact Sheets: X-29," http://www.nasa.gov/centers/dryden/news/FactSheets, accessed Mar. 19, 2009.
185. David F. Voracek and Robert Clarke, "Buffet Induced Structural/Flight Control System Interaction of the X 29A Aircraft," AIAA Paper 91-1053 (1991); and Michael W. Kehoe, Lisa J. Bjarke, and Edward J. Laurie, "An In-Flight Interaction of the X 29A Canard and Flight Control System," NASA TM-101718 (1990).
186. K.K. Gupta, M.J. Brenner, and L.S. Volker, "Integrated Aeroservoelastic Analysis Capability With X 29A Analytical Comparisons," AIAA Paper 87-0907 (1987).

represent an incredibly complex interaction of aerodynamic, aeroelastic, and inertial phenomena. The HiMAT, AD-1, and X-29 programs each advanced the state of the art in aeroelastic design as applied to nontraditional, indeed exotic, planforms featuring unstable design, composite structures, and advanced flight control concepts. Finally, the data required to validate structural analysis and design methods do not automatically come from the testing normally performed by aircraft developers and users. Special instrumentation and testing techniques are required. NASA has developed the facilities and the knowledge base needed for many kinds of special testing and is able assign the required priority to such testing. As these cases show, NASA therefore plays a key role in this process of gaining knowledge about the behavior of aircraft in flight, evaluating predictive capabilities, and flowing that experience back to the people who design the aircraft.

Computational Methods, Industrial Transfer, and the Way Ahead

Having surveyed the development of computational structural analysis within NASA, the contributions of various Centers, and key flight projects that tested and validated structural design and analysis methods in their ultimate application, we turn to the current state of affairs as of 2010 and future challenges.

Overall, even a cursory historical examination clearly indicates that the last four decades have witnessed revolutionary improvements in all of the following areas:

- Analysis capability.
- Complexity of structures that can be analyzed.
- Number of nodes.
- Types of elements.
- Complexity of processes simulated.
- Nonlinearity.
- Buckling.
- Other geometric nonlinearity.
- Material nonlinearity.
- Time-dependent properties.
- Yield or ultimate failure of some members.
- Statistical/nondeterministic processes.
- Thermal effects.
- Control system interactions.

- Usability.
- Execution time.
- Hardware improvements.
- Efficiency of algorithms.
- Adequate but not excessive model complexity.
- Robustness, diagnostics, and restart capability.
- Computing environment.
- Pre- and post-processing.

Before NASTRAN, capabilities generally available (i.e., not counting proprietary programs at the large aerospace companies) were limited to a few hundred nodes. In 1970, NASTRAN made it possible to analyze models with over 2,000 nodes. Currently, models with hundreds of thousands of nodes are routinely analyzed. The computing environment has changed just as dramatically, or more so: the computer used to be a shared resource among many users—sometimes an entire company, or it was located at a data center used by many companies—with punch cards for input and reams of paper for output. Now, there is a PC (or two) at every engineer's desk. NASTRAN can run on a PC, although some users prefer to run it on UNIX machines or other platforms.

Technology has thus come full circle: NASA now makes extensive use of commercial structural analysis codes that have their roots in NASA technology. Commercial versions of NASTRAN have essentially superseded NASA's COSMIC NASTRAN. That is appropriate, in this author's opinion, because it is not NASA's role to provide commercially competitive performance, user interfaces, etc. The existence and widespread use of these commercial codes indicates successful technology transition.

At the time of this writing, basic capability is relatively mature. Advances are still being made, but it is now possible to analyze the vast majority of macroscopic structural problems that are of practical interest in aeronautics and many other industries.

Improvements in the "usability" category are of greater interest to most engineers. Execution speed has improved orders of magnitude, but this has been partially offset by the corresponding orders-of-magnitude increases in model size. Engineers build models with hundreds of thousands of nodes, because they can.

Pre- and post-processing challenges remain. Building the model and interpreting the results typically take longer than actually running the analysis. It is by no means a trivial task to build a finite element model of a complex structure such as a complete airframe, or a major portion thereof. Some commercial software can generate finite element models automatically from CAD geometry. However, many practitioners in the aircraft industry prefer to have more involvement in the modeling process, because of the complexity of the analysis and the safety-critical nature of the task. The fundamental challenge is to make the modeling job easier, while providing the user with control when required and the ability to thoroughly check the resulting model.[187]

In 1982, Thomas Butler wrote, "I would compare the state of graphics pre- and post-processors today with the state that finite elements were in before NASTRAN came on the scene in 1964. Many good features exist. There is much to be desired in each available package."[188] Industry practitioners interviewed today have expressed similar sentiments. There is no single pre- or post-processing product that meets every need. Some users deliberately switch between different pre- and post-processing programs, utilizing the strengths of each for different phases of the modeling task (such as creating components, manipulating them, and visualizing and interrogating the finished model). A reasonable number of distinct pre- and post-processing systems maintain commercial competition, which many users consider to be important.[189]

As basic analysis capability has become well established, researchers step back and look at the bigger picture. Integration, optimization, and uncertainty modeling are common themes at many of the NASA Centers. This includes integration of design and analysis, of analysis and testing, and of structural analysis with analysis in other disciplines. NASA Glenn Research Center is heavily involved in nondeterministic analysis methods, life prediction, modeling of failure mechanisms, and modeling of composite materials, including high-temperature material systems for propulsion applications. Research at Langley spans many fields, including multidisciplinary analysis and optimization of aircraft and spacecraft, analysis and test correlation, uncertainty modeling and "fuzzy structures," and failure analysis.

In many projects, finite element analysis is being applied at the microscale to gain a better understanding of material behaviors. The

---

187. Thomas Christy and John Splichal (aircraft industry), interview by author, Wichita, KS, May 12, 2009; Jadic interview.
188. Butler, "Operating in the Age of NASTRAN," p. xx.
189. Mendoza, Hazen, Jadic, and Christy-Splichal interviews by author.

ability to perform such analysis is a noteworthy benefit coming from advances in structural analysis methods at the macroscopic level. Very real benefits to industry could result. The weight savings predicted from composite materials have been slow in coming, partly because of limitations on allowable stresses. In the civil aviation industry especially, such limitations are not necessarily based on the inherent characteristics of the material but on the limited knowledge of those characteristics. Analysis that gives insight into material behaviors near failure, documented and backed up by test results, may help to achieve the full potential of composite materials in airframe structures.

Applications of true optimization—such as rigorously finding the mathematical minimum of a "cost function"—are still relatively limited in the aircraft industry. The necessary computational tools exist. However, the combination of practical difficulties in automating complex analyses and a certain amount of cultural resistance has somewhat limited the application of true optimization in the aircraft industry up to the present time. There is untapped potential in this area. The path to reaching it is not necessarily in the development of better computer programs, but rather, in the development and demonstration of processes for the effective and practical use of capabilities that exist already. The DAMVIBS program (discussed previously in the section on the NASA Langley Research Center) might provide a model for how this kind of technology transfer can happen. In that program, industry teams essentially demonstrated to themselves that existing finite element programs could be useful in predicting and improving the vibration characteristics of helicopters—when coupled with some necessary improvements in modeling technique. All of the participants subsequently embraced the use of such methods in the design processes of their respective organizations. A comparable program could, perhaps, be envisioned in the field of structural and/or multidisciplinary optimization in aircraft design.[190]

Considering structural analysis as a stand-alone discipline, however, it can be stated without question that computational methods have been adopted throughout the aircraft industry. Specific processes vary between companies. Some companies perform more upfront optimization than others; some still test exhaustively, while others test

---

190. Jadic interview; and author's experience.

minimally. But the aircraft industry as a whole has embraced computational structural analysis and benefited greatly from it.

The benefits of computational structural analysis may not be adequately captured in one concise list, but they include the following:

- Improved productivity of analysis.
- Ability to analyze a more complete range of load cases.
- Ability to analyze a structure more thoroughly than was previously practical.
- Ability to correct and update analyses as designs and requirements mature.
- Improved quality and consistency of analysis.
- Improved performance of the end product. Designs can be improved through more cycles of design/analysis in the early stages of a project, and earlier identification of structural issues, than previously practical.
- Improved capabilities in related disciplines: thermal modeling and acoustic modeling, for example. Some aircraft companies utilize finite element models in the design stage of an aircraft to develop effective noise reduction strategies.
- Ability to analyze structures that could not be practically analyzed before. For example, composite and metallic airframes are different. Metal structures typically have more discrete load paths. Composite structures, such as honeycomb-core panels, have less distinct load paths and are less amenable to analysis by hand using classical methods. Therefore, finite element analysis enables airplanes to be built in ways that would not be possible (or, at least, not verifiable) otherwise.
- Reduced cost and increased utility of testing. Analysis does not replace all testing, but it can greatly enhance the amount of knowledge gained from a test. For example, modeling performed ahead of a test series can help identify the appropriate locations for strain gages, accelerometers, and other instrumentation, and aid in the interpretation of the resulting test data. The most difficult or costly types of testing can certainly be reduced. In a greatly simplified sense, the old paradigm is that

testing was the proof of the structure; now, testing validates the model, and the model proves the structure. Practically speaking, most aircraft companies practice something in between these two extremes.

NASA's contributions have included not only the development of the tools but also the development and dissemination of techniques to apply the tools to practical problems and the provision of opportunities—through unique test facilities and, ultimately, flight research projects—to prove, validate, and improve the tools.

In other industries also, there is now widespread use of computerized structural analysis for almost every conceivable kind of part that must operate under conditions of high mechanical and/or thermal stress. NASTRAN is used to analyze buildings, bridges, towers, ships, wind tunnels and other specialized test facilities, nuclear power plants, steam turbines, wind turbines, chemical processing plants, microelectronics, robotic systems, tools, sports equipment, cars, trucks, buses, trains, engines, transmissions, and tires. It is used for geophysical and seismic analysis, and for medical applications.

In conclusion, finite element analysis would have developed with or without NASA's involvement. However, by creating NASTRAN, NASA provided a centerpiece: a point of reference for all other development and an open-ended framework into which new capabilities could be inserted. This framework gradually collected the best or nearly best methods in every area. If NASTRAN had not been developed, different advances would have occurred only within proprietary codes used internally by different industrial companies or marketed by different software companies. There would have been little hope of consolidating all the important capabilities into one code or of making such capabilities available to the general user. NASTRAN brought high-powered finite element analysis within reach of many users much sooner than would have otherwise been the case. At the same time, the job of predicting every aspect of structural performance was by no means finished with the initial release of NASTRAN—nor is it finished yet. NASA has been and continues to be involved in the development of many new capabilities—developing programs and new ways to apply existing programs—and making the resulting tools and methods available to users in the aerospace industry and in many other sectors of the U.S. economy.

Appendix A:
NASTRAN Users' Colloquia, 1971–1993

Note: This appendix includes a list of the dates and locations of the NASTRAN Users' Colloquia and NASTRAN applications presented at the Colloquia by "nontraditional" users, i.e., industry other than aerospace, Government agencies other than NASA, and universities. Not every paper from these sources is listed, only those that represent applications. Many other papers were presented on modeling techniques, capability improvements, etc., which are not listed.

NASTRAN USERS' COLLOQUIA DATES AND LOCATIONS				
#	YEAR	DATE	LOCATION	CHAIRPERSON(S) / OTHER NOTES
1st	1971	Sept. 13–15	NASA Langley	J. Philip Raney (NASTSRAN SMO)
2nd	1972	Sept. 11–12	NASA Langley	J. Philip Raney
3rd	1973	Sept. 11–12	NASA Langley	<not available>
4th	1975	Sept. 9–11	NASA Langley	Deene J. Weidman
5th	1976	Oct. 5–6	NASA Ames	Deene J. Weidman
6th	1977	Oct. 4–6	NASA Lewis	Deene J. Weidman (Langley) and Christos Chamis (Lewis)
7th	1978	Oct. 4–6	NASA Marshall	Deene J. Weidman (Langley) Robert L. McComas (Marshall)
8th	1979	Oct. 30–31	NASA Goddard	Robert L. Brugh (COSMIC) Reginal
9th	1980	Oct. 22–23	NASA Kennedy	Robert L. Brugh (COSMIC) Henry Harris (KSC)
Note: From this point on, locations were no longer at NASA Centers, individual co/chairs are not identified in the proceedings, and the NASA Scientific & Technical Information (STI) Branch (or program) is listed in the proceedings as the responsible organization.				
10th	1982	May 13–14	New Orleans, LA	Co-chairs not identified.

11th	1983	May 2–6	San Francisco, CA	
12th	1984	May 7–11	Orlando, FL	
13th	1985	May 6–10	Boston, MA	
14th	1986	May 5–9	San Diego, CA	
15th	1987	May 4–8	Kansas City, MO	
16th	1988	Apr. 25–29	Arlington, VA	
17th	1989	Apr. 24–28	San Antonio, TX	
18th	1990	Apr. 23–27	Portland, OR	COSMIC, under the STI Branch.
19th	1991	Apr. 22–26	Williamsburg, VA	
20th	1992	Apr. 27–May 1	Colorado Springs, CO	
21st	1993	Apr. 26–30	Tampa, FL	

NONAEROSPACE INDUSTRY APPLICATIONS OF NASTRAN PRESENTED AT USERS' COLLOQUIA		
YEAR	COMPANY	DESCRIPTION
1972	Westenhoff and Novick	Analysis and design of on-grade railroad track support.
	General Motors	NASTRAN and in-house code for automotive structures.
	Westinghouse (Hanford)	Fuel handling machinery for reactors.
	Kleber-Colombes	Tires.
	Control Data Corp (CDC)	Structural analysis of 40-story building.
	Computer Sciences Corporation (CSC)	Structural dynamic and thermal analysis nuclear reactor vessel support system.
1975	B.F. Goodrich	Tires.
	Exxon	Petroleum processing machinery.
	Littleton Rsch & Eng, with CDC	Propeller-induced ship vibration.
	Westinghouse (Hanford)	Seismic analysis of nuclear reactor structures.
	Reactor Centrum Nederland & Hazameyer B.V.	Electromagnetic field problems.
	General Motors	Modeling and analysis of acoustic cavities.
1976	Sargent & Lundy (2 papers, 1 with CSC)	Deformations of thick cylinders (power plants); seismic analysis of nuclear power plant control panel.
	EBASCO Services, with Universal Analytics	Concrete cracking.
1977	Sperry Marine with Univ VA	Analysis of pressure vessels.
1978	Tennessee Eastman Co.	NASTRAN uses in petrochemical industry.
	EBASCO Services with Grumman (2)	Tokomak Fusion Test Reactor toroidal field coil and vacuum vessel structures.
	B.F. Goodrich	Rubber sonar dome window.
1979	B.F. Goodrich	Belt tensioning.
1980	Ontario Hydro	Seismic analysis.
	NKF Engineering	Problems involving enforced boundary motion.
	Tennessee Eastman	Analysis of heat-transfer fluid fill pipe failures.
1982	B.F. Goodrich	Bead area contact load at tire-wheel interface.

Case 8 | **NASA and Computational Structural Analysis**

1984	Tennessee Eastman	Support system for large compressor.
	Hughes Offshore	Bolted marine riser structure.
1985	John Deere	Use of COSMIC NASTRAN in design department.
1986	Texas Instruments	Nonlinear magnetic circuits.
1987	Texas Instruments	Forces on magnetized bodies.
	NKF Eng.	HVAC duct hanger systems.
1988	Tiernay Turbines	Stress and vibration analysis of gas turbine components.
	Texas Instruments	Magnetostatic nonlinear model of printhead.
1989	Deutsch Metal Components	General product line improvement (hydraulics, pneumatics, other power system components).
	Intergraph	NASTRAN in integrated conceptual design environ.
	Dynacs Eng.	Flexible multibody dynamics and control (NASTRAN with TREETOPS).
	Texas Instruments	Micromechanical deformable mirror.
1990	Analex Corp., with NASA Lewis	Low velocity impact analysis.
1991	Tennessee Eastman	Distillation tray structures.
1993	Butler Analyses	Seismic analysis.

OTHER GOVERNMENT AGENCY NASTRAN APPLICATIONS PRESENTED AT USERS' COLLOQUIA		
YEAR	GOVERNMENT AGENCY	DESCRIPTION
1971	Naval Air Dev Ctr	F-14A boron horizontal stabilizer static and dynamic.
	U.S. Army Air Mobility R&D Lab (USAAMRDL) with NASA Langley	NASTRAN in structural design optimization.
1975	Naval Weapons Center	Modeling and analysis of damaged wings.
	Naval Underwater Systems Center (NUSC)	Transient analysis of bodies with moving boundaries.
	(David Taylor) Naval Ship Rsch & Dev Ctr (DTNSRDC)	Dynamic analysis of submerged structures.
	Argonne Nat Lab	Fluid-coupled concentric cylinders (nuclear reactors).
1976	DTNSRDC	Underwater shock response.
	NUSC	Fluid-structure interactions.
	DTNSRDC	Submerged structures.
	USAAMRDL with Boeing Vertol	Thermal and structural analysis of helicopter transmission.
	U.S. Army, Watervliet	Crack problems.
1977	DTNSRDC	Finite element solutions for free surface flows.
	NUSC	Analysis of magnetic fields.
	U.S. Army, Watervliet	Large-deformation analysis of fiber-wrapped shells.
1978	Wright-Patterson AFB	Ceramic structures.
	DTNSRDC	Magnetostatic field problems.
1979	U.S. Army Armament Rsch & Dev Command (USAARDC) (2)	Stress concentrations in screw heads, elastic-plastic analysis.
	NUSC	Dynamically loaded periodic structures.
1980	NUSC, with A.O. Smith	Ring element dynamic stresses.
	USAARDC (2)	Simulated damage UH-1B tailboom, elastic-plastic analysis.

Case 8 | **NASA and Computational Structural Analysis**

Year	Organization	Topic
1982	DTNSRDC	Magnetic field problems.
	NUSC	Axisymmetric fluid structure interaction problems.
	USAARDC	Analysis of overloaded breech ring.
1983	DTNSRDC	Fluid-filled elastic piping systems.
	NUSC (2)	Wave propagation through plates (2).
	U.S. Army Benet Lab	Elastic-plastic analysis of annular plates.
1984	Dept. of the Navy	Acoustic scattering from submerged structures.
1985	WPAFB with Rockwell	NASTRAN in a computer aided design system.
	Naval Wpns Ctr	Missile inertia loads.
	WPAFB	Simulation of nuclear overpressures.
	DTNSRDC	Loss factors, frequency-dependent damping treatment.
	U.S. Army (Harry Diamond Lab) with Advanced Tech & Rsch	Transient analysis of fuze assembly.
	DTNSRDC	Magnetic heat pump.
1986	DTNSRDC (3)	Multidisciplinary design; acoustics (2).
	Naval Ocean Sys Ctr (2)	Stress concentrations; flutter of low aspect ratio wings.
	NUSC	Surface impedance analysis.
1987	DTNSRDC (2)	Computer animation of modal and transient vibrations; analysis of ship structures to underwater explosion shocks.
	DTNSRDC & NRL	Acoustic scattering.
	NUSC	Patrol boat subject to planning loads.
1988	David Taylor Rsch Ctr (DTRC—renamed)	Static preload effects in acoustic radiation and scattering.
1989	DTRC	Low frequency resonances of submerged structures.
1990	DTRC	Scattering from fluid-filled structures.
1991	Los Alamos Nat Lab	Computer animation of displacements.
	DTRC (2)	Transient fluid-structure interactions.
1992	Naval Surf. Warfare Ctr	Vibration and shock of laminated composite plates.
	DTRC	Acoustics of axisymmetric fluid regions.
1993	U.S.A.F. Wright Lab	Design optimization studies.

UNIVERSITY APPLICATIONS OF NASTRAN PRESENTED AT USERS' COLLOQUIA		
Year	University	Description
1971	Old Dominion Univ	Space Shuttle dynamics model.
1972	Old Dominion Univ	Vibrations of cross-stiffened ship's deck.
	Louisiana Tech Univ	NASTRAN as a teaching aid.
1975	Univ of MD	NASTRAN for simultaneous parabolic equations.
	Univ NB & Mayo Graduate School of Medicine, with IBM	Stress analysis of left ventricle of the heart.
	Univ MD, with Army, Frankford Arsenal	Nonlinear analysis of cartridge case neck separation malfunction.
1977	Univ VA	(with Sperry Marine, listed in "Other Industry" table)
1978	Univ MO Rolla	NASTRAN in education and research.
1982	Air Force Inst Tech	Elastic aircraft airloads.
1985	Univ of GA	Agricultural engineering—teaching and research.
	Clemson Univ	Plated bone fracture gap motion.
	Univ MO	Fillet weld stress.
1987	Univ of Naples, with NASA Langley	NASTRAN for prediction of aircraft interior noise.
1989	GWU, with DTRC	Electromagnetic fields and waves.

NASTRAN Reference Sources

At time of writing, these are available from the NASA Technical Reports Server at *http://ntrs.nasa.gov*:

NASTRAN: Users' Experiences, NASA TM-X-2378, 1971.
NASTRAN: Users' Experiences (2nd), NASA TM-X-2637, 1972.
NASTRAN: Users' Experiences (4th), NASA TM-X-3278, 1975.
NASTRAN: Users' Experiences (5th), NASA TM-X-3428, 1976.
Sixth NASTRAN Users' Colloquium, NASA CP-2018, 1977.
Seventh NASTRAN Users' Colloquium, NASA CP-2062, 1978.
Eight NASTRAN Users' Colloquium, NASA CP-2131, 1979.
Ninth NASTRAN Users' Colloquium, NASA CP-2151, 1980.
Tenth NASTRAN Users' Colloquium, NASA CP-2249, 1982.
Eleventh NASTRAN Users' Colloquium, NASA CP-2284, 1983.
Twelfth NASTRAN Users' Colloquium, NASA CP-2328, 1984.
Thirteenth NASTRAN Users' Colloquium, NASA CP-2373, 1985.
Fourteenth NASTRAN Users' Colloquium, NASA CP-2419, 1986.
Fifteenth NASTRAN Users' Colloquium, NASA CP-2481, 1987.
Sixteenth NASTRAN Users' Colloquium, NASA CP-2505, 1988.
Seventeenth NASTRAN Users' Colloquium, NASA CP-3029, 1989.
Eighteenth NASTRAN Users' Colloquium, NASA CP-3069, 1990.
Nineteenth NASTRAN Users' Colloquium, NASA CP-3111, 1991.
Twentieth NASTRAN Users' Colloquium, NASA CP-3145, 1992.
Twenty-First NASTRAN Users' Colloquium, NASA CP-3203, 1993.

Appendix B:
Miscellaneous NASA Structural Analysis Programs

Note: Miscellaneous computer programs, and in some cases test facilities or other related projects, that have contributed to the advancement of the state of the art in various ways are described here. In some cases, there simply was not room to include them in the main body of the paper; in others, there was not enough information found, or not enough time to do further research, to adequately describe the programs and document their significance. Readers are advised that these are merely examples; this is not an exhaustive list of all computer programs developed by NASA for structural analysis to the 2010 time period. Dates indicate introduction of capability. Many of the programs were subsequently enhanced. Some of the programs were eventually phased out.

1) FLEXSTAB (Ames, Dryden, and Langley Research Centers, 1970s)

FLEXSTAB was a method for calculating stability derivatives that included the effects of aeroelastic deformation. Originally developed in the early 1970s by Boeing under contract to NASA Ames, FLEXSTAB was also used and upgraded at Dryden. FLEXSTAB used panel-method aerodynamic calculations, which could be readily adjusted with empirical corrections. The structural effects were treated first as a steady deformation at the trim condition, then as "unsteady perturbations about the reference motion to determine dynamic stability by characteristic roots or by time histories following an initial perturbation or following penetration of a discrete gust flow field."[191] Comparisons between FLEXSTAB predictions and flight measurements were made at Dryden for the YF-12A, Shuttle, B1, and other aircraft. Initially developed for symmetric flight conditions only, FLEXSTAB was extended in 1981 to include nonsymmetric flight conditions.[192] In 1984, a procedure was developed to couple a NASTRAN structural model to the FLEXSTAB elastic-aircraft stability analysis.[193] NASA Langley and the Air Force

---

191. A.R. Dusto, et al., *A Method for Predicting the Stability Characteristics of an Elastic Airplane*, vol. 1: *FLEXSTAB Theoretical Description*, NASA CR-114712 (1974), p. xxi.
192. R.L. Sims, "User's Manual for FSLIP-3, FLEXSTAB Loads Integration Program," NASA TM-81364 (1981), p. 1.
193. Lawrence S. Schuster, "NASTRAN/FLEXSTAB Procedure for Static Aeroelastic Analysis," NASA TM-84897 (1984).

Flight Dynamics Laboratory also funded upgrades to FLEXSTAB, leading to the DYLOFLEX program, which added aeroservoelastic effects.[194]

2) ANSYMP Computer Program (Glenn Research Center, 1983)

ANSYMP was developed to capture the key elements of local plastic behavior without the overhead of a full nonlinear finite element analysis. "Nonlinear, finite-element computer programs are too costly to use in the early design stages for hot-section components of aircraft gas turbine engines. . . . This study was conducted to develop a computer program for performing a simplified nonlinear structural analysis using only an elastic solution as input data. The simplified method was based on the assumption that the inelastic regions in the structure are constrained against stress redistribution by the surrounding elastic material. Therefore the total strain history can be defined by an elastic analysis. . . . [ANSYMP] was created to predict the stress-strain history at the critical fatigue location of a thermomechanically cycled structure from elastic input data. . . . Effective [inelastic] stresses and plastic strains are approximated by an iterative and incremental solution procedure." ANSYMP was verified by comparison to a full nonlinear finite element code (MARC). Cyclic hysteresis loops and mean stresses from ANSYMP "were in generally good agreement with the MARC results. In a typical problem, ANSYMP used less than 1 percent of the central processor unit (CPU) time required by MARC to compute the inelastic solution."[195]

3) Structural Tailoring of Engine Blades (STAEBL, Glenn, 1985)

This computer program "was developed to perform engine fan blade numerical optimizations. These blade optimizations seek a minimum weight or cost design that satisfies realistic blade design constraints, by tuning one to twenty design variables. The STAEBL system has been generalized to include both fan and compressor blade numerical optimizations. The system analyses have been significantly improved through the inclusion of an efficient plate finite element analysis for blade stress and frequency determinations. Additionally, a finite element based approximate severe foreign object damage (FOD) analysis has been included. The new FOD analysis gives very accurate estimates of the full nonlinear

---

194. R.D. Miller, R.I. Kroll, and R.E. Clemmons, *Dynamic Loads Analysis System (DYLOFLEX) Summary*, vol. 1: *Engineering Formulation*, NASA CR-2846-1 (1979).
195. Albert Kaufman, "Simplified Method for Nonlinear Structural Analysis," NASA TP-2208 (1983), p. 2.

bird ingestion solution. Optimizations of fan and compressor blades have been performed using the system, showing significant cost and weight reductions, while comparing very favorably with refined design validation procedures."[196]

4) TRansfer ANalysis Code to Interface Thermal and Structural (3D TRANCITS, Glenn, 1985)

Transfer of data between different analysis codes has always been one of the challenges of multidisciplinary design, analysis, and optimization. Even if input and output format can be standardized, different types of analysis often require different types of information or different mesh densities, globally or locally. TRANCITS was developed to translate between heat transfer and structural analysis codes: "TRANCITS has the capability to couple finite difference and finite element heat transfer analysis codes to linear and nonlinear finite element structural analysis codes. TRANCITS currently supports the output of SINDA and MARC heat transfer codes directly. It will also format the thermal data output directly so that it is compatible with the input requirements of the NASTRAN and MARC structural analysis codes. . . . The transfer module can handle different elemental mesh densities for the heat transfer analysis and the structural analysis."[197] MARC is a commercial, general-purpose, nonlinear finite element code introduced by MARC Analysis and Research Corp. in the late 1970s. Because of its nonlinear analysis capabilities, MARC was used extensively at Glenn for engine component analyses and for other applications, such as the analysis of a space station strongback for launch loads in 1992.[198] Other commercial finite element codes used at Glenn included MSC/NASTRAN, which was used along with NASA's COSMIC version of NASTRAN.

5) COmposite Blade STRuctural ANalyzer (COBSTRAN, Glenn, 1989)

COBSTRAN was a preprocessor for NASTRAN, designed to generate finite element models of composite blades. While developed specifically

---

196. K.W. Brown, M.S. Hirschbein, and C.C. Chamis, "Finite Element Engine Blade Structural Optimization," AIAA Paper 85-0645 (1985).
197. R.L. Thompson and R.J. Maffeo, "A Computer Analysis Program for Interfacing Thermal and Structural Codes," NASA TM-87021 (1985).
198. Frank F. Monasa and Joseph M. Roche, "Collapse Analysis of a Waffle Plate Strongback for Space Station Freedom," NASA TM-105412 (1992).

for advanced turboprop blades under the Advanced Turboprop (ATP) project, it was subsequently applied to compressor blades and turbine blades. It could be used with both COSMIC NASTRAN and MSC/NASTRAN, and was subsequently extended to work as a preprocessor for the MARC nonlinear finite element code.[199]

6) BLAde SIMulation (BLASIM), 1992

BLASIM calculates dynamic characteristics of engine blades before and after an ice impact event. BLASIM could accept input geometry in the form of airfoil coordinates or as a NASTRAN-format finite element model. BLASIM could also utilize the ICAN program (discussed separately) to generate ply properties of composite blades.[200] "The ice impacts the leading edge of the blade causing severe local damage. The local structural response of the blade due to the ice impact is predicted via a transient response analysis by modeling only a local patch around the impact region. After ice impact, the global geometry of the blade is updated using deformations of the local patch and a free vibration analysis is performed. The effects of ice impact location, ice size and ice velocity on the blade mode shapes and natural frequencies are investigated."[201]

7) NPLOT (Goddard, 1982)

NPLOT was a product of research into the visualization of finite element models, which had been ongoing at Goddard since the introduction of NASTRAN. A fast, hidden line algorithm was developed in 1982 and became the basis for the NPLOT plotting program for NASTRAN, publicly released initially in 1985 and in improved versions into the 1990s.[202]

---

199. Karen F. Bartos and Michael A. Ernst, "Evaluation of MARC for the Analysis of Rotating Composite Blades," NASA TM-4423 (1993), p. 1.
200. E.S. Reddy and G.H. Abumeri, "Blade Assessment for Ice Impact (BLASIM), User's Manual, Version 1.0," NASA CR-19075 (1993), pp.1.1–1.2.
201. G.H. Abumeri, E.S. Reddy, P.L.N. Murthy, and C.C. Chamis, "Dynamic Analysis of a Pre-and-Post Ice Impacted Blade," NASA TM-105829 (1992).
202. G.K. Jones and K.J. McEntire, "NPLOT: An Interactive Plotting Program for NASTRAN Finite Element Models," in NASA, *13th NASTRAN Users' Colloquium* (1985), pp. 110–132;
K. McEntire, "NPLOT—NASTRAN Plot," program abstract, Jan. 1994, on NASA Technical Reports Server at *http://ntrs.nasa.gov*, accessed Apr. 20, 2009.

8) Integrated Modeling of Optical Systems (IMOS) (Goddard and JPL, 1990s)

A combined multidisciplinary code, IMOS was developed during the 1990s by Goddard and JPL: "Integrated Modeling of Optical Systems (IMOS) is a finite element-based code combining structural, thermal, and optical ray-tracing capabilities in a single environment for analysis of space-based optical systems."[203] IMOS represents a recent step in the continuing evolution of Structural-Thermal-Optical analysis capability, which has been an important activity at the Space Flight Centers since the early 1970s.

9) Dynamic Simulation of Controls & Structure (Goddard, 1970s–1990s)

Another important area of spacecraft structural modeling is in the interaction of control systems with flexible multibody structural systems. In a general sense, this is the spacecraft counterpart to aeroservoelasticity, although the driving mechanisms are very different. Dynamic Simulation of Controls & Structure (DISCOS) was developed in the late 1970s to perform this type of analysis. "The physical system undergoing analysis may be generally described as a cluster of contiguous flexible structures (bodies) that comprise a mechanical system, such as a spacecraft. The entire system (spacecraft) or portions thereof may be either spinning or nonspinning. Member bodies of the system may undergo large relative excursions, such as those of appendage deployment or rotor/stator motion. The general system of bodies is, by its inherent nature, a feedback system in which inertial forces (such as those due to centrifugal and Coriolis acceleration) and the restoring and damping forces are motion-dependent. . . . The DISCOS program can be used to obtain nonlinear and linearized time response of the system, interaction constant forces in the system, total system resonance properties, and frequency domain response and stability information for the system. DISCOS is probably the most powerful computational tool to date for the computer simulation of actively controlled coupled multi-flexible-body systems," according to the computer program abstract. The program was made available to approved licensees (for $1,000, in 1994) with the caveat that DISCOS " . . . is not easy to understand and effectively apply, but is not

---

203. Gregory Moore, "Integrated Modeling of Optical Systems (IMOS): An Assessment and Future Directions," Goddard Space Flight Center and Jet Propulsion Laboratory, in FEMCI Workshop 2001, abstract on NASA Technical Reports Server at http://ntrs.nasa.gov, accessed Apr. 20, 2009.

intended for simple problems. The **DISCOS** user is expected to have extensive working knowledge of rigid-body and flexible-body dynamics, finite-element techniques, numerical methods, and frequency-domain analysis." DISCOS was used extensively at least into the 1990s for spacecraft modeling.[204] In 1983, a program for bridging **DISCOS**, **NASTRAN**, and **SAMSAN**—a large order control system design program—was also publicly released.[205] A 1987 NASA-funded study by Honeywell (Space and Strategic Avionics Division) outlined some limitations of **DISCOS** and other contemporary multibody dynamics programs and made recommendations for future work in the field.[206] Also in the late 1980s, GSFC began collaborating with a research group at the University of Iowa that was developing similar multibody modeling capabilities for mechanical engineering applications. The National Science Foundation (NSF), U.S. Army Tank Automotive Command, and about 30 other Government and industry laboratories were involved in this project through the Industry/University Cooperative Research Center (I/UCRC) at the University of Iowa. Goals of the I/UCRC were to achieve mutual enhancement of capabilities in the modeling, simulation, and control of complex mechanical systems, including man/machine interactions applicable to manufacturing processes.[207]

10) Antenna Design and Analysis (JPL)

JPL operates both ground-based and space-borne antennas. These pose their own array (no pun intended) of structural challenges because of their large size, the need to maintain precision alignment, and, in the case of space-borne antennas, the need for extremely light weight. JPL was one of the early users of **NASTRAN**, using it to calculate gravity defor-

---

204. Harold P. Frisch, "DISCOS—Dynamic Interaction Simulation of Controls and Structures (IBM Version)," Computer Program abstract, Goddard Space Flight Center, Jan. 1994, on NASA Technical Reports Server at http://ntrs.nasa.gov, accessed Apr. 20, 2009.
205. Frisch, "NASDS-NASTRAN/DISCOS/SAMSAN DMAP Bridging Program," Computer Program abstract, Goddard Space Flight Center, Jan. 1994, on NASA Technical Reports Server at http://ntrs.nasa.gov, accessed Apr. 20, 2009.
206. D.W. Lips, "Approaches and Possible Improvements in the Area of Multibody Dynamics Modeling," NASA CR-179227 (1987).
207. Frisch, "Control system software, simulation, and robotic applications," in NASA Goddard Space Flight Center, *Technology 2000*, vol. 1 (Washington, DC: NASA, 1991), pp. 315–321.

mation effects on ground-based 210-foot-diameter antennas in 1971.[208] In 1976, JPL developed a simplified stiffness formulation (translational degrees of freedom only at the nodes) coupled with a structural member sizing capability for design optimization: "Computation times to execute several design/analysis cycles are comparable to the times required by general-purpose programs for a single analysis cycle."[209] In the late 1980s, JPL upgraded a set of 64-meter antennas to a new diameter of 70 meters. This project afforded "the rare opportunity to collect field data to compare with predictions of the finite-element analytical models." Static and dynamic tests were performed while the antenna structures were in a stripped-down configuration during the retrofit process. The data provided insight into the accuracy of the models that were used to optimize the original structural designs.[210]

11) TRASYS Radiative Heat Transfer (Johnson, 1980s–1990s)

TRASYS was developed by Martin Marietta for Johnson to calculate internode radiative heat transfer as well as heat transfer to a model from the surroundings. It was used extensively through the 1990s. Applications included propulsion analysis at Glenn Research Center and Structural-Thermal-Optical analysis (when integrated with NASTRAN for structural calculations and MACOS/IMOS and POPOS optical codes) at Marshall Space Center and JPL.[211] BUCKY was developed in-house. BUCKY was initially introduced as a basic plane stress and plate buckling program but was extensively developed during the 1990s to include plate bending, varying element thickness, varying edge and pressure loads, edge moments, plasticity, output formatting for visualization in I-DEAS (a CAD program developed by Structural Dynamics Research Corporation), three-dimensional axisymmetric capability, and improvements in execution time.[212]

---

208. M. Smoot Katow, "Static Analysis of the 64-m (210-ft) Antenna Reflector Structure," in NASA Langley Research Center, *NASTRAN: Users' Experiences* (Hampton, VA, NASA, 1971), pp. 123–129.
209. Roy Levy, "Computer-Aided Design of Antenna Structures and Components," JPL, in *2nd National Symposium on Computerized Structural Analysis and Design*, Washington, DC (1976); abstract in NASA Technical Report Server at http://ntrs.nasa.gov, accessed Apr. 24, 2009.
210. Levy, "Structural Optimization and Recent Large Ground Antenna Installations," in NASA, *Recent Advances in Multidisciplinary Analysis and Optimization*, pt. 3 (Washington, DC: NASA, 1989), pp. 1393–1416.
211. NASA Technical Reports Server at http://ntrs.nasa.gov, accessed Apr. 20, 2009.
212. James P. Smith, "BUCKY Instruction Manual Version 3.3," NASA TM-104793 (1994), p. 2.

12) Structural Analysis of General Shells (STAGS) (Marshall and Langley, 1960s–present)

Structural Analysis of General Shells (STAGS) evolved from early shell analysis codes developed by Lockheed Palo Alto Research Laboratory and sponsored by the NASA Marshall Space Flight Center between 1963 and 1968, with subsequent development funded primarily from Langley. B.O. "Bo" Almroth of Lockheed was the principal developer. The name STAGS seems to have first appeared around 1970.[213] Thus, STAGS initial development was nearly concurrent with that of NASTRAN. While NASTRAN development aimed to stem the proliferation of analysis codes, and of shell analysis codes in particular, NASTRAN did not initially provide the full capability needed to replace such codes. In particular, STAGS from the beginning included nonlinear capability that was found necessary in the accurate modeling of shells with cutouts. In the mid- to late 1970s, STAGS was released publicly, with user manuals. "Under contract with NASA, STAGS has been converted from being more or less a pure research tool into a code that is suitable for use by the public for practical engineering analysis. Suggestions from NASA-Langley have resulted in considerable enhancement of the code and are to some degree the cause of its increasing popularity. . . . User reaction consistently seems to indicate that the run time with STAGS is surprisingly low in comparison to comparable codes. A STAGS input deck is usually compact and time for its preparation is short."[214] STAGS continued to be enhanced through the 1980s (as STAGS-C1, actually a family of versions), offering unique capabilities for modeling total collapse of structures and problems that bifurcate into multiple possible solutions.[215] It was apparently popular and widely used. For example, in 1990, Engineering Dynamics, Inc., of Kenner, LA, used STAGS-C1 to model and verify a repair design for a damaged offshore oil platform.[216]

---

213. Norman F. Knight, Jr., and Charles C. Rankin, "STAGS Example Problems Manual," NASA CR-2006-214281 (2006), p. 5; B.O. Almroth and A.M. Holmes, "An experimental study of the strength and stability of thin monocoque shells with reinforced and unreinforced rectangular cutouts," NASA CR-115267 (1971), abstract on NASA Technical Reports Server at http://ntrs.nasa.gov, accessed May 10, 2009.
214. B.O. Almroth and F.A. Brogan, "The STAGS Computer Code," NASA CR-2950 (1978), pp. 1–2.
215. C.C. Rankin, P. Stehlin, and F.A. Brogan, "Enhancements to the STAGS Computer Code," NASA CR-4000 (1986), pp. 1, 3.
216. "Structural Analysis," Spinoff, NASA Technology Utilization Office, 1991, p. 57.

STAGS Version 5.0 was released in 2006, and STAGS is still used for failure analysis, analysis of damaged structures, and similar problems.[217]

13) Nonlinear Structures: PANES (1975) and AGGIE-I (1980) (Marshall)

Program for Analysis of Nonlinear Equilibrium and Stability (PANES) was developed for structural problems involving geometric and/or material nonlinear characteristics. AGGIE-I was a more comprehensive code capable of solving larger and more general problems, also involving geometric or material nonlinearities.[218]

14) Finite Element Modeling of Piping Systems (Stennis)

While Stennis is not active in structural methods research, there have been some activities applying finite element and structural health monitoring techniques to the complex fuel distribution systems at the facility. One such effort was presented at the 27th Joint Propulsion Conference in Sacramento, CA, in 1991: "A set of PC-based computational Dynamic Fluid Flow Simulation models is presented for modeling facility gas and cryogenic systems. . . . A set of COSMIC NASTRAN-based finite element models is also presented to evaluate the loads and stresses on test facility piping systems from fluid and gaseous effects, thermal chill down, and occasional wind loads. The models are based on Apple Macintosh software which makes it possible to change numerous parameters."[219] NASA was, in this case, its own spinoff technology customer.

---

217. Knight and Rankin, "STAGS Example Problems Manual"; Donald J. Baker, "Response of Damaged and Undamaged Tailored Extension-Shear-Coupled Composite Panels," *Journal of Aircraft*, vol. 43, no. 2 (2008), pp. 517–527.
218. Search of NASA Technical Reports Server at *http://ntrs.nasa.gov*, accessed Apr. 20, 2009.
219. L. Dequay, A. Lusk, and S. Nunez, "Integrated flow and structural modeling for rocket engine component test facility propellant systems," AIAA Paper 91-2402 (1991).

Appendix C:
Structural Analysis and Loads Prediction Facilities

Test facilities have an important role in verifying and improving analysis methods. A few test facilities that had a lot to do with the development and validation of structural analysis methods are described below. In addition to those described, other "landmark" test facilities include large-scale launch vehicle structural test facilities at Johnson and Marshall Space Centers, and the crash dynamics test facility at Langley Research Center.

Structural Dynamics Laboratory (Ames Research Center, 1965)

During the 1960s, Ames and Langley collaborated on some of the structural dynamics and buffet problems of spacecraft during ascent. (This collaboration occurred through some of the same meetings at NASA Headquarters that led to the development of NASTRAN.) To help assess the structural dynamic characteristics of boosters, and to build confidence in predictive methods, a large structural dynamics test facility was built at Ames (completed in 1965). This facility was large enough to hold a full-size Atlas or Titan II, had provisions for exciting the structural modes of the test article, and could be evacuated to test the structural damping characteristics in zero or reduced ambient air density.[220] The facility was also used for research on buffet during reentry and landing impacts.[221] Much of the structural dynamics research at Ames was discontinued or relocated during the early 1970s. The laboratory is long since deactivated, but the large, pentagonal tower still stands, housing a machine shop and a wind tunnel that can simulate Mars's atmosphere by evacuating the chamber and then filling to low pressure with CO_2.[222]

Thermal Loads Laboratory (Dryden Flight Research Center, 1960s)

A 1973 accounting of NASA research facilities listed only one major ground laboratory at Dryden: the High Temperature Loads Calibration Laboratory.[223] High supersonic and hypersonic flight research created a need (1) to test airframes on the ground under simultaneous thermal and structural loading conditions and (2) to calibrate loads instrumen-

---

220. Henry A. Cole, e-mail message to author, Apr. 19, 2009.
221. Glenn E. Bugos, e-mail to author, Apr. 3, 2009.
222. Smith interview by author (by telephone), Apr. 6, 2009.
223. Hallion and Gorn, *On The Frontier*, pp. 100, 358.

tation at elevated temperatures, so that the data obtained in flight could be reliably interpreted. These needs " . . . led to the construction of a laboratory for calibrating strain-gage installations to measure loads in an elevated temperature environment. The problems involved in measuring loads with strain gages . . . require the capability to heat and load aircraft under simulated flight conditions. . . . The laboratory has the capability of testing structural components and complete vehicles under the combined effects of loads and temperatures, and calibrating and evaluating flight loads instrumentation under [thermal] conditions expected in flight."[224]

The laboratory is housed in a hangarlike building with attached shop, offices, and control room. Capabilities included:

- Hangar-door opening 40 feet high by 136 feet wide.
- Unobstructed test area 150 by 120 by 40 feet allowed the testing of aircraft up to and including, for example, a YF-12 or SR-71.
- Ten megawatts of electrical heating power via quartz lamps and reflectors.
- Temperatures up to 3,000 °F.
- Hydraulic power of 4.5 gallons/minute at 3,000 pounds per square inch (psi) to apply loads.
- Fourteen channels closed-loop load or position control of up to 34 separate actuators.
- Sensors including strain gages, thermocouples, load cells, and position transducers.

Slots in the floor provided flexible locations for tiedown points, as well as routing for hydraulic and electrical power, instrumentation wiring, compressed air, or water (presumably for cooling). Closed-loop analog control of both mechanical load and heating was provided, to any desired preprogrammed time history.

The facility was used in the YF-12 thermal loads project (discussed elsewhere in this paper), in Space Shuttle structural verification at high

---

224. Sefic and Karl F. Anderson, "NASA High Temperature Loads Calibration Laboratory," NASA TM-X-1868 (1969), p. 1.

temperatures, and for a variety of other studies.[225] The loads laboratory made contributions to the validation of computational methods by providing the opportunity to compare computational predictions with test data obtained under known, controlled, thermal and structural loading conditions, applied together or independently as required. At time of this writing, the facility is still in use.[226]

Spin Rig (Glenn Research Center, 1983)

One particular facility of many, a spin test rig built at Lewis in 1983 is mentioned here because its stated purpose was not primarily the testing of engine parts to verify the parts but the testing of engine parts to verify analysis methods: "The Lewis Research Center spin rig was constructed to provide experimental evaluation of analysis methods developed under the NASA Engine Structural Dynamics Program. Rotors up to 51 cm (20 in.) in diameter can be spun to 16,000 rpm in vacuum by an air motor. Vibration forcing functions are provided by shakers that apply oscillatory axial forces or transverse moments to the shaft, by a natural whirling of the shaft, and by an air jet. Blade vibration is detected by strain gages and optical tip blade-motion sensors."[227]

---

225. "Fact Sheets: NASA Dryden's Contributions to Spaceflight," *http://www.nasa.gov/centers/dryden/news/FactSheets*, accessed Mar. 19, 2009.
226. Martin Brenner, e-mail message to author, May 8, 2009.
227. Gerald V. Brown, Robert E. Kielb, Erwin H. Meyn, Richard E. Morris, and Stephen J. Posta, "Lewis Research Center Spin Rig and Its Use in Vibration Analysis of Rotating Systems," NASA TP-2304 (1984), p. 1.

Appendix D:
NASA-Developed Finite Element Analysis

- NASTRAN (COSMIC versions). NASA case Nos. COS-10054, -10057, -10061, and -10064 through -10067.

- ICAN Integrated Composite Analyzer. Linear finite element analysis of multilayered fiber composites. NASA case No. LEW-15592. Lewis Research Center.

- AIRLOADS. Oscillatory airloads on turbosystem blades in nonuniform flow. NASA case no. LEW-14947. Lewis Research Center.

- ACTON AutoCAD to NASTRAN translator. NASA case no. GSC-13217. Goddard Space Flight Center.

- ACTOMP, AutoCAD to Mass Properties. Develops mass properties data from AutoCAD models. NASA case no. GSC-13228. Goddard Space Flight Center.

- GeoFEST, Geophysical Finite Element Simulation. Developed by Jet Propulsion Laboratory in 2002 and released directly through OCS.

- FECAP Finite Element Composite Analysis Program for microcomputers. NASA case no. LAR-14109. Langley Research Center.

- NASTPLT, plotting post-processor for NASTRAN. NASA case no. GSC-12833. Goddard Space Flight Center.

- STARS. Integrated multidisciplinary structural, fluids, aeroelastic, and aeroservoelastic analysis. NASA case no. FRC-09537. Dryden Flight Research Center.

Recommended Additional Readings
Reports, Papers, Articles, and Presentations:

Anees Ahmad and Lamar Hawkins, "Development of Software to Model AXAF-1 Image Quality," NASA CR-203978 (1996).

F.A. Akyuz and E. Heer, *VISCEL Computer Program User Manual for Analysis of Linear Viscoelastic Structures*, vol. 1: *Users' Manual* (Pasadena: NASA Jet Propulsion Laboratory, 1971).

B.O. Almroth and F.A. Brogan, "The STAGS Computer Code," NASA CR-2950 (1978).

B.O. Almroth and A.M. Holmes, "An experimental study of the strength and stability of thin monocoque shells with reinforced and unreinforced rectangular cutouts," NASA CR-115267 (1971).

Donald J. Baker, "Response of Damaged and Undamaged Tailored Extension-Shear-Coupled Composite Panels," *Journal of Aircraft*, vol. 43, no. 2 (2008), pp. 517–527.

Kay S. Bales, "Structural Mechanics Division Research and Technology Plans for FY 1989 and Accomplishments for FY 1988," NASA TM-101592 (1989).

Karen F. Bartos and Michael A. Ernst, "Evaluation of MARC for the Analysis of Rotating Composite Blades," NASA TM-4423 (1993).

James L. Beck, "Probabilistic System Identification in the Time Domain," in USAF–NASA, *USAF/NASA Workshop of Model Determination for Large Space Systems* (Pasadena: California Institute of Technology, 1988).

Eric B. Becker and Trent Miller, "Final Report: Development of Non-Linear Finite Element Computer Code," NASA CR-179965 (1985).

Stanley U. Benscoter and Richard H. MacNeal, "Equivalent Plate Theory for a Straight Multicell Wing," NACA TN-2786 (1952).

Manoj K. Bhardwaj, "Aeroelastic Analysis of Modern Complex Wings Using ENSAERO and NASTRAN," *Progress Report*, Virginia Polytechnic Institute and State University (1995).

K.W. Brown, M.S. Hirschbein, and C.C. Chamis, "Finite Element Engine Blade Structural Optimization," AIAA Paper 85-0645 (1985).

Gerald V. Brown, Robert E. Kielb, Erwin H. Meyn, Richard E. Morris, and Stephen J. Posta, "Lewis Research Center Spin Rig and Its Use in Vibration Analysis of Rotating Systems," NASA TP-2304 (1984).

Karen F. Bugos and Michael A. Ernst, "Evaluation of MARC for the Analysis of Rotating Composite Blades," NASA TM-4423 (1993).

J.J. Burken, G.S. Alag, and G.B. Gilyard, "Aeroelastic Control of Oblique-Wing Aircraft," NASA TM-86808 (1986).

Joseph M. Carlson, "NASA Technology Transfer: The Computer Software Dissemination Program," in *NASTRAN: Users' Experiences*, Langley Research Center (1971).

C.C. Chamis, "CELFE/NASTRAN code for the analysis of structures subjected to high velocity impact," NASA TM-79048 (1978).

Computer Software Management and Information Center (COSMIC), *A Directory of Computer Programs Available from COSMIC*, vol. 1 (Athens, GA: University of Georgia, 1967).

Computer Software Management and Information Center (COSMIC), "COSMIC: 1981 Catalog of Computer Programs," NASA CR-163916 (1981).

Computer Software Management and Information Center (COSMIC), "Monthly Progress Report," NASA CR-195809 (1994).

James D. Cronkhite, "The NASA/Industry Design Analysis Methods for Vibration (DAMVIBS) Program—Bell Helicopter Textron Accomplishments," in Raymond G. Kvaternik, ed., *A Government/*

Industry Summary of the Design Analysis Methods for Vibrations (DAMVIBS) Program, NASA CP-10114 (1993).

Robert E. Curry and Alex G. Sim, "In-Flight Total Forces, Moments, and Static Aeroelastic Characteristics of an Oblique-Wing Research Airplane," NASA TP-2224 (1984).

L. Dequay, A. Lusk, and S. Nunez, "Integrated flow and structural modeling for rocket engine component test facility propellant systems," AIAA Paper 91-2402 (1991).

Stephen F. Duffy and John P. Gyekenyesi, "CCARES: A computer algorithm for the reliability analysis of laminated CMC components," NASA TM-111096 (1993).

A.R. Dusto, et al., *A Method for Predicting the Stability Characteristics of an Elastic Airplane,* vol. 1: *FLEXSTAB Theoretical Description,* NASA CR-114712 (1974).

Gary L. Farley and Donald J. Baker, "Graphics and Composite Material Computer Program Enhancements for SPAR," NASA TM-80209 (1980).

R. Gabel, P. Lang, and D. Reed, "The NASA/Industry Design Analysis Methods for Vibration (DAMVIBS) Program—Boeing Helicopters Airframe Finite Element Modeling," in Raymond G. Kvaternik, ed., *A Government/Industry Summary of the Design Analysis Methods for Vibrations (DAMVIBS) Program,* NASA CP-10114 (1993).

L. Bernard Garrett, "Interactive Design and Analysis of Future Large Spacecraft Concepts," AIAA Paper 81-1177 (1981).

J.P. Giesing, T.P. Kalman, and W.P. Rodden, "Correction Factor Techniques for Improving Aerodynamic Prediction Methods," NASA CR-144967 (1976).

Glenn B. Gilyard, "The Oblique Wing Research Aircraft: A Test Bed for Unsteady Aerodynamic and Aeroelastic Research," in NASA Langley

Research Center, *Transonic Unsteady Aerodynamics and Aeroelasticity*, pt. 2, Report N89-19247 (1987).

Cody R. Godines and Randall D. Manteufel, "Probabilistic Analysis and Density Parameter Estimation Within Nessus," NASA CR-2002-212008 (2002).

J.B. Graham and P.L. Luz, "Preliminary In-Flight Loads Analysis of In-Line Launch Vehicles Using the VLOADS 1.4 Program," NASA TM-1998-208472 (1998).

K.K. Gupta, "STARS: A General-Purpose Finite Element Computer Program for Analysis of Engineering Structures," NASA RP-1129 (1984).

K.K. Gupta, M.J. Brenner, and L.S. Volker, "Integrated Aeroservoelastic Analysis Capability With X-29A Analytical Comparisons," AIAA Paper 87-0907 (1987).

K.K. Gupta, M.J. Brenner, and L.S. Voelker, "Development of an Integrated Aeroservoelastic Analysis Program and Correlation With Test Data," NASA TP-3120 (1991).

G.P. Guruswamy, "ENSAERO: A multidisciplinary Program for Fluid/Structural Interaction Studies of Aerospace Vehicles," *Computing Systems in Engineering*, vol. 1, nos. 2–4 (1990), pp. 237–256.

D.H. Hodges, A.S. Hopkins, D.L. Kunz, and H.E. Hinnant, "Introduction to GRASP: General Rotorcraft Aeromechanical Stability Program—A Modern Approach to Rotorcraft Modeling," *American Helicopter Society 42nd Annual Forum, Washington, DC, June 1986*.

A. Stewart Hopkins and Peter Likins, "Analysis of structures with rotating, flexible substructures," AIAA Paper 87-0951 (1987).

J.C. Hunsaker and E.B. Wilson, "Report on Behavior of Aeroplanes in Gusts," NACA TR-1 (1915).

Jerald M. Jenkins and Albert E. Kuhl, "Recent Load Calibrations Experience with the YF-12 Airplane," in James and Associates, eds., *YF-12 Experiments Symposium,* vol. 1, NASA CP-2054 (1978).

F. Douglas Johnson and Martin Kokus, "NASA Technology Utilization Program: A Summary of Cost Benefit Studies," NASA CR-201936 (1977).

F. Douglas Johnson, Emily Miller, Nancy Gunderson, Panayes Gatseos, Charles F. Mourning, Thomas Basinger, and Martin Kokus, "NASA Tech Brief Program: A Cost Benefit Evaluation," NASA CR-201938 (1977).

Wayne Johnson, "A Comprehensive Analytical Model of Rotorcraft Aerodynamics and Dynamics," pt. III: "Program Manual," NASA TM-81184/AVRADCOM TR-80-A-7 (1980).

G.K. Jones and K.J. McEntire, "NPLOT: An Interactive Plotting Program for NASTRAN Finite Element Models," in NASA, *13th NASTRAN Users' Colloquium* (1985).

Rakesh K. Kapania, Manoj K. Bhardwaj, Eric Reichenbach, and Guru P. Guruswamy, "Aeroelastic analysis of modern complex wings," AIAA-1996-4011 (1996).

Albert Kaufman, "Simplified Method for Nonlinear Structural Analysis," NASA TP-2208 (1983).

Michael W. Kehoe, Lisa J. Bjarke, and Edward J. Laurie, "An In-Flight Interaction of the X-29A Canard and Flight Control System," NASA TM-101718 (1990).

Norman F. Knight, Jr., and Charles C. Rankin, "STAGS Example Problems Manual," NASA CR-2006-214281 (2006).

W.L. Ko, R.D. Quinn, L. Gong, L.S. Schuster, and D. Gonzales, "Reentry Heat Transfer Analysis of the Space Shuttle Orbiter," in Howard B. Adelman, ed., *NASA Langley Research Center Computational Aspects of Heat Transfer in Structures,* NASA CP-2216 (1981), pp. 295–325.

Raymond G. Kvaternik, ed., *A Government/Industry Summary of the Design Analysis Methods for Vibrations (DAMVIBS) Program*, NASA CP-10114 (1993).

H. Lee and J.B. Mason, "NASTRAN Thermal Analyzer: A General Purpose Finite Element Heat Transfer Computer Program," in NASA Langley Research Center, *2nd NASTRAN Users' Colloquium* (1972).

Arthur W. Leissa, "The Historical Basis of the Rayleigh and Ritz Methods," in *7th International Symposium on Vibrations of Continuous Systems, Zakopane, Poland, July 2009*.

A.F. Leondis, "Large Advanced Space Systems (LASS) Computer Program," AIAA Paper 79-0904 (1979).

A.F. Leondis, "Large Advanced Space Systems Computer-Aided Design and Analysis Program," NASA CR-159191 (1980).

R. Levy and S. Wall, "Savings in NASTRAN Decomposition Time by Sequencing to Reduce Active Columns," in NASA Langley Research Center, *NASTRAN: Users' Experiences* (1971).

D.W. Lips, "Approaches and Possible Improvements in the Area of Multibody Dynamics Modeling," NASA CR-179227 (1987).

J.B. Mason, "The NASTRAN Hydroelastic Analyzer," NASA TM-X-65617 (1972).

Shivakumar Mettu, et al, "NASGRO 3.0—A Software for Analyzing Aging Aircraft," in NASA–FAA–DOD, *2nd Joint NASA/FAA/DoD Conference on Aging Aircraft*, pt. 2 (Washington, DC: NASA–FAA–DOD, 1999).

Robert R. Meyer, Jr., and V. Michael DeAngelis, "Flight-Measured Aerodynamic Loads on a 0.92 Aspect Ratio Lifting Surface," in James and Associates, eds., *YF-12 Experiments Symposium*, vol. 1, NASA CP-2054 (1978).

Roy G. Miller, "Torsion of Wing Trusses at Diving Speeds," NACA TR-104 (1921).

R.D. Miller, R.I. Kroll, and R.E. Clemmons, *Dynamic Loads Analysis System (DYLOFLEX) Summary*, vol. 1: *Engineering Formulation*, NASA CR-2846-1 (1979).

Richard C. Monaghan, "Description of the HiMAT Tailored Composite Structure and Laboratory Measured Vehicle Shape Under Load," NASA TM-81354 (1981).

Frank F. Monasa and Joseph M. Roche, "Collapse Analysis of a Waffle Plate Strongback for Space Station Freedom," NASA TM-105412 (1992).

P.L.N. Murthy and C.C. Chamis, "ICAN: Integrated Composites Analyzer," *25th Structures, Structural Dynamics and Materials Conference, Palm Springs, CA* (1984).

N.N. Nemeth, "CARES: Ceramics Analysis and Reliability Evaluation of Structures," NASA Lewis Research Center Report LEW-15168 (1994).

F.H. Norton and D.L. Bacon, "The Pressure Distribution Over the Horizontal Tail Surfaces of an Airplane II," NACA TR-119 (1921).

F.H. Norton and W.G. Brown, "Pressure distribution over the rudder and fin of an airplane in flight," NACA TR-149 (1923).

Neil Otte, "PATRAN-STAGS Translator (PATSTAGS)," NASA TM-100388 (1990).

Henry Pearson, William McGowan, and James Donegan, "Horizontal Tail Loads in Maneuvering Flight," NACA TR-1007 (1950).

J.P. Raney, D.J. Weidman, and H.M. Adelman, "NASTRAN: Status, Maintenance, and Future Development," in *First NASTRAN Users' Colloquium, NASA Langley Research Center* (1971).

C.C. Rankin, P. Stehlin, and F.A. Brogan, "Enhancements to the STAGS Computer Code," NASA CR-4000 (1986).

E.S. Reddy and G.H. Abumeri, "Blade Assessment for Ice Impact (BLASIM), User's Manual, Version 1.0," NASA CR-19075 (1993).

J.L. Rogers, Jr., "Programming Structural Synthesis System," *NASA Tech Briefs*, vol. 10, no. 2 (May 1986), p. 100.

M.J. Rutkowski, "Aeroelastic Stability Analysis of the AD-1 Manned Oblique-Wing Aircraft," NASA TM-78439 (1977).

Ashish K. Sareen, Daniel P. Schrage, and T.S. Murthy, "Rotorcraft Airframe Structural Optimization for Combined Vibration and Fatigue Constraints," *47th American Helicopter Society (AHS) Forum, Phoenix, AZ* (1991).

Lawrence S. Schuster, "NASTRAN/FLEXSTAB Procedure for Static Aeroelastic Analysis," NASA TM-84897 (1984).

Walter J. Sefic and Karl F. Anderson, "NASA High Temperature Loads Calibration Laboratory," NASA TM-X-1868 (1969).

R.L. Sims, "User's Manual for FSLIP-3, FLEXSTAB Loads Integration Program," NASA TM-81364 (1981).

James P. Smith, "BUCKY Instruction Manual Version 3.3," NASA TM-104793 (1994).

Jaroslaw Sobieszczanski-Sobieski, "An Integrated Computer Procedure for Sizing Composite Airframe Structures," NASA TP-1300 (1979).

Southwest Research Institute, "NESSUS/NASTRAN Interface," NASA CR-202778 (1996).

Victor W. Sparrow and Ralph D. Buehrle, "Fuzzy Structures Analysis of Aircraft Panels in NASTRAN," AIAA Paper 2001-1320 (2001).

F.K. Straub, K.B. Sangha, and B. Panda, "Advance finite element modeling of rotor blade aeroelasticity," *American Helicopter Society Journal*, vol. 39, no. 2 (Apr. 1994), pp. 56–68.

Nancy P. Sykes, ed., *NASA Workshop on Computational Structural Mechanics*, NASA CP-10012 (1989).

R.L. Thompson and R.J. Maffeo, "A Computer Analysis Program for Interfacing Thermal and Structural Codes," NASA TM-87021 (1985).

Mostafa Toossi, Richard Weisenburger, and Mostafa Hashemi-Kia, "The NASA/Industry Design Analysis Methods for Vibration (DAMVIBS) Program—McDonnell Douglas Helicopter Company Achievements," in Raymond G. Kvaternik, ed., *A Government/Industry Summary of the Design Analysis Methods for Vibrations (DAMVIBS) Program*, NASA CP-10114 (1993).

William J. Twomey, "The NASA/Industry Design Analysis Methods for Vibration (DAMVIBS) Program—Sikorsky Aircraft—Advances Toward Interacting with the Airframe Design Process," in Raymond G. Kvaternik, ed., *A Government/Industry Summary of the Design Analysis Methods for Vibrations (DAMVIBS) Program*, NASA CP-10114 (1993).

Garret N. Vanderplaats, *CONMIN: A FORTRAN Program for Constrained Function Minimization, User's Manual* (Moffett Field, CA: Ames Research Center and U.S. Army Air Mobility R&D Laboratory, 1973).

David F. Voracek and Robert Clarke, "Buffet Induced Structural/Flight Control System Interaction of the X-29A Aircraft," AIAA Paper 91-1053 (2009).

E.P. Warner, "Static Testing and Proposed Standard Specifications," NACA TN-6 (1920).

Deene J. Weidman, "NASTRAN Status and Plans," in *4th NASTRAN Users' Colloquium*, 1975.

Anita S. West, and William F. Hubka "Matrix Methods and Automation in Structural Engineering," NASA CR-71230 (1966).

A.W. Wilhite and J.J. Rehder, "AVID: A Design System for Technology Studies of Advanced Transportation Concepts," AIAA Paper 79-0872 (1979).

Eleanor C. Wynne, "Structural Dynamics Division Research and Technology Accomplishments for FY 1990 and Plans for FY 1991," NASA TM-102770 (1991).

William T. Yeager, Jr., "A Historical Overview of Aeroelasticity Branch and Transonic Dynamics Tunnel Contributions to Rotorcraft Technology and Development," NASA TM-2001-211054 (2001).

A.F. Zahm and L.H. Crook, "Airplane Stress Analysis," Report No. 82, Aerodynamical Laboratory, Bureau of Construction and Repair, U.S. Navy (1918).

Books and Monographs:

Paul F. Borchers, James A. Franklin, and Jay W. Fletcher, *Flight Research at Ames: Fifty-Seven Years of Development and Validation of Aeronautical Technology, 1940–1997*, NASA SP-3300 (Washington, DC: NASA, 1998).

John R. Brauer, *What Every Engineer Should Know about Finite Element Analysis* (New York: Marcel Dekker, 1993).

Glenn Bugos, *Atmosphere of Freedom: Sixty Years at the NASA Ames Research Center*, NASA SP-4314 (Washington, DC: NASA, 2000).

Paul E. Ceruzzi, *Beyond the Limits: Flight Enters the Computer Age* (Cambridge: MIT Press, 1989).

Paul E. Ceruzzi, *A History of Modern Computing* (Cambridge: MIT Press, 1999).

Edward R. Champion, Jr., and J. Michael Ensminger, *Finite Element Analysis with Personal Computers* (New York: Marcel Dekker, Inc., 1988).

Andrew J. Dunar and Stephen P. Waring, *Power to Explore: A History of the Marshall Space Flight Center, 1960–1990*, NASA SP-4313 (Washington, DC: NASA, 1999).

Richard P. Hallion and Michael H. Gorn, *On the Frontier: Experimental Flight at NASA Dryden* (Washington, DC: Smithsonian Books, 2002).

Caleb W. McCormick, *The NASTRAN Users' Manual*, NASA SP-222 (1970).

Peter W. Merlin, *Mach 3+: NASA/USAF YF-12 Flight Research, 1969–1979*, NASA SP-2001-4525, No. 25 in the *Monographs in Aerospace History* series (Washington, DC: NASA, 2002).

Victor Paschkis and Frederick Ryder, *Direct Analog Computers* (New York: John Wiley & Sons, Inc., 1968).

Alvin Renetzky and Barbara Flynn, *NASA Factbook*, (Orange, NJ: Academia Media, 1971).

Gene L. Waltman, *Black Magic and Gremlins: Analog Flight Simulations at NASA's Flight Research Center*, No. 20 in *Monographs in Aerospace History*, NASA SP-2000-4520 (Washington, DC: NASA, 2000).

The X-43A Hyper-X test vehicle drops away from its Boeing NB-52B mother ship in November 2004, beginning its flight to Mach 9.7. NASA.

CASE 9
High-Temperature Structures and Materials

T.A. Heppenheimer

Taking fullest advantage of the high-speed potential of rocket and air-breathing propulsion systems required higher-temperature structures. Researchers recognized that aerothermodynamics involved linking aerodynamic and thermodynamic understanding with the mechanics of thermal loading and deformation of structures. This drove use of new structural materials. NASA and other engineers would experiment with active and passive thermal protection systems, metals, and materials.

IN AEROSPACE ENGINEERING, high-temperature structures and materials solve two problems. They are used in flight above Mach 2 to overcome the elevated temperatures that occur naturally at such speeds. They also are extensively used at subsonic velocities, in building high-quality turbofan engines, and for the protection of structures exposed to heating.

Aluminum loses strength when exposed to temperatures above 210 degrees Fahrenheit (°F). This is why the Concorde airliner, which was built of this material, cruised at Mach 2.1 but did not go faster.[1] Materials requirements come to the forefront at higher speeds and escalate sharply as airplanes' speeds increase. The standard solutions have been to use titanium and nickel, and a review of history shows what this has meant.

Many people wrote about titanium during the 1950s, but to reduce it to practice was another matter. Alexander "Sasha" Kartveli, chief designer at Republic Aviation, proposed a titanium F-103 fighter, but his vision outreached his technology, and although started, it never flew. North American Aviation's contemporaneous Navaho missile program introduced chemical milling (etching out unwanted material) for aluminum as well as for titanium and steel, and was the first to use titanium skin in an aircraft. However, the version of Navaho that

---

1. Erik M. Conway, *High-Speed Dreams* (Baltimore: Johns Hopkins University Press, 2005), p. 57.

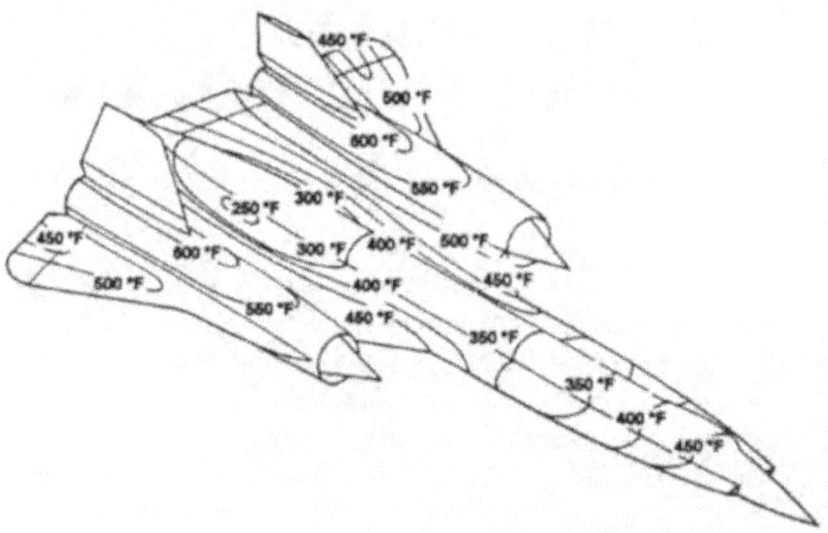

The Lockheed Blackbird experienced a wide range of upper surface temperatures, up to 600 °F. NASA.

was to use these processes never flew, as the program was canceled in 1957.[2]

The Lockheed A-12 Blackbird, progenitor of a family of exotic Mach 3.2 cruisers that included the SR-71, encountered temperatures as high as 1,050 °F, which required that 93 percent of its structural weight be titanium. The version selected was B-120 (Ti-13V-11Cr-3Al), which has the tensile strength of stainless steel but weighs only half as much. But titanium is not compatible with chlorine, cadmium, or fluorine, which led to difficulties. A line drawn on a sheet of titanium with a pen would eat a hole into it in a few hours. Boltheads tended to fall away from assemblies; this proved to result from tiny cadmium deposits made by tools. This brought removal of all cadmium-plated tools from toolboxes. Spot-welded panels produced during the summer tended to fail because the local water supply was heavily chlorinated to kill algae. The managers took to washing the parts in distilled water, and the problem went away.[3]

---

2. Richard A. DeMeis, "The Trisonic Titanium Republic," *Air Enthusiast*, No. 7, (July–Sept. 1978), pp. 198–213; Dale D. Myers, "The Navaho Cruise Missile: A Burst of Technology," *Acta Astronautica*, vol. 26 (Nov. 8–10, 1992), pp. 741–748.

3. David Robarge, Archangel: CIA's Supersonic A-12 Reconnaissance Aircraft (McLean, VA: CIA Center for the Study of Intelligence, 2007), pp. 11–12; see also Paul F. Crickmore, Lockheed SR-71 Blackbird (London: Osprey Publishing Ltd., 1986), pp. 90–92.

The SR-71 was a success. Its shop-floor practice with titanium at first was classified but now has entered the aerospace mainstream. Today's commercial airliners—notably the Boeing 787 and the Airbus A-380, together with their engines—use titanium as a matter of routine. That is because this metal saves weight.

Beyond Mach 4, titanium falters and designers must turn instead to alternatives. The X-15 was built to top Mach 6 and to reach 1,200 °F. In competing for the contract, Douglas Aircraft proposed a design that was to use magnesium, whose properties were so favorable that the aircraft would only reach 600 °F. But this concept missed the point, for managers wanted a vehicle that would cope successfully with temperatures of 1,200 °F. Hence it was built of Inconel X, a nickel alloy.[4]

High-speed flight represents one application of advanced metals. Another involves turbofans for subsonic flight. This application lacks the drama of Mach-breaking speeds but is far more common. Such engines use turbine blades, with the blade itself being fabricated from a single-crystal superalloy and insulated with ceramics. Small holes in the blade promote a circulation of cooler gas that is ducted downstream from high-pressure stages of the compressor. The arrangement can readily allow turbines to run at temperatures 750 °F above the melting point of the superalloy itself.[5]

The High-Speed Environment

During World War II the whole of aeronautics used aluminum. There was no hypersonics; the very word did not exist, for it took until 1946 for the investigator Hsue-shen Tsien to introduce it. Germany's V-2 was flying at Mach 5, but its nose cone was of mild steel, and no one expected that this simple design problem demanded a separate term for its flight regime.[6]

A decade later, aeronautics had expanded to include all flight speeds because of three new engines: the liquid-fuel rocket, the ramjet, and the variable-stator turbojet. The turbojet promised power beyond Mach 3, while the ramjet proved useful beyond Mach 4. The Mach 6 X-15 was under contract. Intermediate-range missiles were in development, with

---

4. "Evaluation Report on X-15 Research Aircraft Design Competition," Aug. 5, 1955, pp. 99–101, Record Group 255, Philadelphia Federal Records Center, National Archives and Records Service.
5. "Briefing: Rolls-Royce," *The Economist* (Jan. 10, 2009), pp. 60–62.
6. Willy Ley, *Rockets, Missiles, and Space Travel* (New York: Viking, 1957), p. 213.

ranges of 1,200 to 1,700 miles, and people regarded intercontinental missiles as preludes to satellite launchers.

A common set of descriptions presents the flight environments within which designers must work. Well beyond Mach 3, engineers accommodate aerodynamic heating through materials substitutions. The aircraft themselves continue to accelerate and cruise much as they do at lower speeds. Beyond Mach 4, however, cruise becomes infeasible because of heating. A world airspeed record for air-breathing flight (one that lasted for nearly the next half century) was set in 1958 with the Lockheed X-7, which was made of 4130 steel, at Mach 4.31 (2,881 mph). The airplane had flown successfully at Mach 3.95, but it failed structurally in flight at Mach 4.31, and no airplane has approached such performance in the past half century.[7]

No aircraft has ever cruised at Mach 5, and an important reason involves structures and materials. "If I cruise in the atmosphere for 2 hours," said Paul Czysz of McDonnell-Douglas, "I have a thousand times the heat load into the vehicle that the Shuttle gets on its quick transit of the atmosphere."[8] Aircraft indeed make brief visits to such speed regimes, but they don't stay there; the best approach is to pop out of the atmosphere and then return, the hallmark of a true transatmospheric vehicle.

At Mach 4, aerodynamic heating raises temperatures. At higher Mach, other effects are seen. A reentering intercontinental ballistic missile (ICBM) nose cone, at speeds above Mach 20, has enough kinetic energy to vaporize 5 times its weight in iron. Temperatures behind its bow shock reach 9,000 kelvins (K), hotter than the surface of the Sun. The research physicist Peter Rose has written that this velocity would be "large enough to dissociate all the oxygen molecules into atoms, dissociate about half of the nitrogen, and thermally ionize a considerable fraction of the air."[9]

---

7. Jay Miller, *The X-Planes, X-1 to X-45* (North Branch, MN: Specialty Press, 2001). See also Lee L. Peterson, "Evaluation Report on X-7A," Report AFMDC ADJ 57-8184, Oct. 3, 1957; and William A. Ritchie, "Evaluation Report on X-7A (System 601B)," Report AFMDC DAS 58-8129, Jan. 1959, copies in the Archives of the Air Force Historical Research Agency, Maxwell AFB, Montgomery, AL.
8. Author interview with Paul Czysz, Mar. 13, 1986.
9. *Time*, June 13, 1960, p. 70; P.H. Rose and W.I. Stark, "Stagnation Point Heat-Transfer Measurements in Dissociated Air." *Journal of the Aeronautical Sciences* (Feb. 1958), pp. 86–97; P.H. Rose, "Physical Gas Dynamics Research at the Avco Research Lab," AGARD Report 145, July 1957, p. 1.

Aircraft thus face a simple rule: they can cruise up to Mach 4 if built with suitable materials, but they cannot cruise at higher speeds. This rule applies not only to entry into Earth's atmosphere but also to entry into the atmosphere of Jupiter, which is far more demanding but which an entry probe of the Galileo spacecraft investigated in 1995, at Mach 50.[10]

Other speed limits become important in the field of wind tunnel simulation. The Government's first successful hypersonic wind tunnel was John Becker's 11-inch facility, which entered service in 1947. It approached Mach 7, with compressed air giving run times of 40 seconds.[11] A current facility, which is much larger and located at the National Aeronautics and Space Administration (NASA) Langley Research Center, is the Eight-Foot High-Temperature Tunnel—which also uses compressed air and operates near Mach 7.

The reason for such restrictions involves fundamental limitations of compressed air, which liquefies if it expands too much when seeking higher speeds. Higher speeds indeed are achievable but only by creating shock waves within an instrument for periods measured in milliseconds. Hence, the field of aerodynamics introduces an experimental speed limit of Mach 7, which describes its wind tunnels, and an operational speed limit of Mach 4, which sets a restriction within which cruising flight remains feasible. Compared with these velocities, the usual definition of hypersonics, describing flight beyond Mach 5, is seen to describe nothing in particular.

Metals, Ceramics, and Composites

Solid-state materials exist in one of these forms and may be reviewed separately. Metals and alloys, the latter being particularly common, exist usually as superalloys. These are defined as exhibiting excellent mechanical strength and creep resistance at high temperatures, good surface stability, and resistance to corrosion and oxidation. The base alloying element of a superalloy is usually nickel, cobalt, or nickel-iron. These three elements are compared in Table 1 with titanium.[12]

---

10. Richard E. Young, Martha A. Smith, and Charles K. Sobeck, "Galileo Probe: In Situ Observations of Jupiter's Atmosphere," *Science* (May 10, 1996), pp. 837–838.
11. James R. Hansen, *Engineer in Charge: A History of the Langley Aeronautical Laboratory, 1917–1958*, NASA SP-4305 (Washington: GPO, 1987), pp. 343–347.
12. Data from Matthew J. Donachie, Jr., *Superalloys Source Book* (Metals Park, OH: American Society for Metals, 1984), pp. 3–19.

TABLE 1: COMPARISON OF TITANIUM WITH SELECTED SUPER ALLOYS		
ELEMENT	NUMBER	MELTING POINT (K)
Titanium	22	1,941
Iron	26	1,810
Cobalt	27	1,768
Nickel	28	1,726

Superalloys generally are used at temperatures above 1,000 °F, or 810 K. They have been used in cast, rolled, extruded, forged, and powder-processed forms. Shapes produced have included sheet, bar, plate, tubing, airfoils, disks, and pressure vessels. These metals have been used in aircraft, industrial and marine gas turbines, nuclear reactors, aircraft skins, spacecraft structures, petrochemical production, and environmental-protection applications. Although developed for use at high temperatures, some are used at cryogenic temperatures. Applications continue to expand, but aerospace uses continue to predominate.

Superalloys consist of an austenitic face-centered-cubic matrix plus a number of secondary phases. The principal secondary phases are the carbides MC, M6C, M23C6, and the rare M7C3, which are found in all superalloy types, and the intermetallic compound Ni3(Al, Ti), known as gamma-prime, in nickel – and iron-nickel-base superalloys. The most important classes of iron-nickel-base and nickel-base superalloys are strengthened by precipitation of intermetallic compounds within a matrix. Cobalt-base superalloys are invariably strengthened by a combination of carbides and solid solution hardeners. No intermetallic compound possessing the same degree of utility as the gamma-prime precipitate—in nickel-base and iron-nickel-base superalloys—has been found to be operative in cobalt-base systems.

The superalloys derive their strength from solid solution hardeners and precipitating phases. In addition to those elements that promote solid solution hardening and promote the formation of carbides and intermetallics, elements including boron, zirconium, hafnium, and cerium are added to enhance mechanical or chemical properties.

TABLE 2: SELECTED ALLOYING ADDITIONS AND THEIR EFFECTS

ELEMENT	PERCENTAGES		EFFECT
	Iron-nickel- and nickel-base	Cobalt-base	
Chromium	5–25	19–30	Oxidation and hot corrosion resistance; solution hardening; carbides
Molybdenum, Tungsten	0–12	0–11	Solution hardening; carbides
Aluminum	0–6	0–4.5	Precipitation hardening; oxidation resistance
Titanium	0–6	0–4	Precipitation hardening; carbides
Cobalt	0–20	N/A	Affects amount of precipitate
Nickel	N/A	0–22	Stabilizes austenite; forms hardening precipitates
Niobium	0–5	0–4	Carbides; solution hardening; precipitation hardening (nickel-, iron-nickel-base)
Tantalum	0–12	0–9	Carbides; solution hardening; oxidation resistance

Table 2 presents a selection of alloying additions, together with their effects.[13] The superalloys generally react with oxygen, oxidation being the prime environmental effect on these alloys. General oxidation is not a major problem up to about 1,600 °F, but at higher temperatures, commercial nickel-and cobalt-base superalloys are attacked by oxygen. Below about 1,800 °F, oxidation resistance depends on chromium content, with Cr_2O_3 forming as a protective oxide; at higher temperatures, chromium and aluminum contribute in an interactive fashion to oxidation protection, with aluminum forming the protective Al_2O_3. Because the level of aluminum is often insufficient to provide

---

13. Donachie, *Introduction to Superalloys*, Ref. 13, p. 15.

long-term protection, protective coatings are often applied. Cobalt-base superalloys are readily welded using gas-metal-arc (GMA) or gas-tungsten-arc (GTA) techniques. Nickel- and iron-nickel-base superalloys are considerably less weldable, for they are susceptible to hot cracking, postweld heat treatment cracking, and strain-age cracking. However, they have been successfully welded using GMA, GTA, electron-beam, laser, and plasma arc methods. Superalloys are difficult to weld when they contain more than a few percentage points of titanium and aluminum, but superalloys with limited amounts of these alloying elements are readily welded.[14]

So much for alloys. A specific type of fiber, carbon, deserves discussion in its own right because of its versatility. It extends the temperature resistance of metals by having the unparalleled melting temperature of 6,700 °F. Indeed, it actually gains strength with temperature, being up to 50 percent stronger at 3,000 °F than at room temperature. It also has density of only 1.50 grams per cubic centimeter (g/cm^3). These properties allowed carbon fiber to serve in two path-breaking vehicles of recent decades. The Voyager aircraft, which flew around the world in 1986 on a single load of fuel, had some 90 percent of its structure made of carbon fibers in a lightweight matrix. The Space Shuttle also relies on carbon for thermal protection of the nose and wing leading edges.[15]

These areas needed particularly capable thermal protection, and carbon was the obvious candidate. It was lighter than aluminum and could be protected against oxidation with a coating. Graphite was initially the standard form, but it had failed to enter the aerospace mainstream. It was brittle and easily damaged, and it did not lend itself to use with thin-walled structures.

The development of a better carbon began in 1958 with Vought Missiles and Space Company (later LTV Aerospace) in the forefront. The work went forward with support from the Dyna-Soar and Apollo programs and brought the advent of an all-carbon composite consisting of graphite fibers in a carbon matrix. Existing composites had names such as carbon-phenolic and graphite-epoxy; this one was carbon-carbon.

It retained the desirable properties of graphite in bulk: lightweight, temperature resistance, and resistance to oxidation when coated. It had

---

14. Ibid., pp. 3–19.
15. Deborah D.L. Chung, *Carbon Fiber Composites* (Boston: Butterworth-Heinemann, 1994), pp. 5–10.

a very low coefficient of thermal expansion, which reduced thermal stress. It also had better damage tolerance than graphite.

Carbon-carbon was a composite. As with other composites, Vought engineers fabricated parts of this material by forming them as layups. Carbon cloth gave a point of departure, being produced by oxygen-free pyrolysis of a woven organic fiber such as rayon. Sheets of this fabric, impregnated with phenolic resin, were stacked in a mold to form the layup and then cured in an autoclave. This produced a shape made of laminated carbon cloth phenolic. Further pyrolysis converted the resin to its basic carbon, yielding an all-carbon piece that was highly porous because of the loss of volatiles. It therefore needed densification, which was achieved through multiple cycles of reimpregnation under pressure with an alcohol, followed by further pyrolysis. These cycles continued until the part had its specified density and strength.

The Shuttle's design specified carbon-carbon for the nose cap and leading edges, and developmental testing was conducted with care. Structural tests exercised their methods of attachment by simulating flight loads up to design limits, with design temperature gradients. Other tests, conducted within an arc-heated facility, determined the thermal responses and hot-gas leakage characteristics of interfaces between the carbon-carbon and the rest of the vehicle.

Additional tests used articles that represented substantial portions of the orbiter. An important test item, evaluated at NASA Johnson, reproduced a wing-leading edge and measured 5 by 8 feet. It had two leading-edge panels of carbon-carbon set side by side, a section of wing structure that included its main spars, and aluminum skin covered with thermal-protection tiles. It had insulated attachments, internal insulation, and internal seals between the carbon-carbon and the tiles. It withstood simulated air loads, launch acoustics, and mission temperature-pressure environments—not once but many times.[16]

There was no doubt that left to themselves, the panels of carbon-carbon that protected the leading edges would have continued to do so. Unfortunately, they were not left to themselves. During the ascent of the

---

16. Paul R. Becker, "Leading-Edge Structural Material System of the Space Shuttle," *American Ceramic Society Bulletin*, vol. 60, No. 11 (1981), pp. 1210–1214; L.J. Korb, C.A. Morant, R.M. Calland, and C.S. Thatcher, "The Shuttle Orbiter Thermal Protection System," *American Ceramic Society Bulletin*, vol. 60 (1981), pp. 1188–1193; Anon., "Technical Overview: Oxidation Resistant Carbon-Carbon for the Space Shuttle," Vought Missiles and Space Company, n.d. (c. 1970).

Shuttle Columbia, on January 16, 2003, a large piece of insulating foam detached itself from a strut that joined the external tank to the front of the orbiter. The vehicle at that moment was slightly more than 80 seconds into the flight, traveling at nearly Mach 2.5. This foam struck a carbon-carbon panel and delivered what proved to be a fatal wound. In words of the accident report:

> Columbia re-entered Earth's atmosphere with a preexisting breach in the leading edge of its left wing. This breach, caused by the foam strike on ascent, was of sufficient size to allow superheated air (probably exceeding 5,000 degrees Fahrenheit) to penetrate the cavity behind the RCC panel. The breach widened, destroying the insulation protecting the wing's leading edge support structure, and the superheated air eventually melted the thin aluminum wing spar. Once in the interior, the superheated air began to destroy the left wing. Finally, over Texas, the increasing aerodynamic forces the Orbiter experienced in the denser levels of the atmosphere overcame the catastrophically damaged left wing, causing the Orbiter to fall out of control.[17]

Three years of effort succeeded in securing the foam on future flights, and the Shuttle returned to flight in July 2006 with foam that stayed put. In contrast with the high tech of the Shuttle, carbon fibers also are finding use in such low-tech applications as automobiles. As with the Voyager round-the-world aircraft, what counts is carbon's light weight, which promotes fuel economy. The Graphite Car employs carbon fiber epoxy-matrix composites for body panels, structural members, bumpers, wheels, drive shafts, engine components, and suspension systems. A standard steel auto would weigh 4,000 pounds, but this car weighs only 2,750 pounds, for a saving in weight of nearly one-third.[18]

Superalloys thus represent the mainstream in aerospace materials, with composites such as carbon fiber extending their areas of use. There also are ceramics, but these are highly specialized. They cannot compete with the temperature resistance of carbon or with its light weight. They nevertheless come into play as insulators on turbine blades that protect the underlying superalloy. This topic will be discussed separately.

---

17. NASA, *Columbia Accident Investigation Report*, (Washington, DC: GPO, 2003), p. 12.
18. Chung, *Carbon Fiber Composites*, pp. 116–118.

Case 9 | High-Temperature Structures and Materials

Ablative and Radiative Structures

Atmosphere entry of satellites takes place above Mach 20, only slightly faster than the speed of reentry of an ICBM nose cone. The two phenomena nevertheless are quite different. A nose cone slams back at a sharp angle, decelerating rapidly and encountering heating that is brief but very severe. Entry of a satellite is far easier, taking place over a number of minutes.

To learn more about nose cone reentry, one begins by considering the shape of a nose cone. Such a vehicle initially has high kinetic energy because of its speed. Following entry, as it approaches the ground, its kinetic energy is very low. Where has it gone? It has turned into heat, which has been transferred both into the nose cone and into the air that has been disturbed by passage of the nose cone. It is obviously of interest to transfer as much heat as possible into the surrounding air. During reentry, the nose cone interacts with this air through its bow shock. For effective heat transfer into the air, the shock must be very strong. Hence the nose cone cannot be sharp like a church steeple, for that would substantially weaken the shock. Instead, it must be blunt, as H. Julian Allen of the National Advisory Committee for Aeronautics (NACA) first recognized in 1951.[19]

Now that we have this basic shape, we can consider methods for cooling. At the outset of the Atlas ICBM program, in 1953, the simplest method of cooling was the heat sink, with a thick copper shield absorbing the heat of reentry. An alternative approach, the hot structure, called for an outer covering of heat-resistant shingles that were to radiate away the heat. A layer of insulation, inside the shingles, was to protect the primary structure. The shingles, in turn, overlapped and could expand freely.

A third approach, transpiration cooling, sought to take advantage of the light weight and high heat capacity of boiling water. The nose cone was to be filled with this liquid; strong g-forces during deceleration in the atmosphere were to press the water against the hot inner skin. The skin was to be porous, with internal steam pressure forcing the fluid

---

19. See George W. Sutton, "The Initial Development of Ablation Heat Protection, An Historical Survey," *Journal of Spacecraft and Rockets*, (Jan.–Feb. 1982), pp. 3–11; and H. Julian Allen and A.J. Eggers, Jr., "A Study of the Motion and Aerodynamic Heating of Ballistic Missiles Entering the Earth's Atmosphere at High Supersonic Speeds," NACA TR-1381 (1953), which summarizes his early research and that of Alfred J. Eggers.

An Atlas ICBM with a low-drag ablatively cooled nose cone. USAF.

through the pores and into the boundary layer. Once injected, steam was to carry away heat. It would also thicken the boundary layer, reducing its temperature gradient and hence its rate of heat transfer. In effect, the nose cone was to stay cool by sweating.

Still, each of these approaches held difficulties. Transpiration cooling was poorly understood as a topic for design. The hot-structure concept raised questions of suitably refractory metals along with the prospect of losing the entire nose cone if a shingle came off. Heat sinks appeared to promise high weight. But they seemed the most feasible way to proceed, and early Atlas designs specified use of a heat-sink nose cone.[20]

Atlas was an Air Force program. A separate set of investigations was underway within the Army, which supported hot structures but raised problems with both heat sink and transpiration. This work anticipated the independent studies of General Electric's George Sutton, with both efforts introducing an important new method of cooling: ablation. Ablation amounted to having a nose cone lose mass by flaking off when hot. Such a heat shield could absorb energy through latent heat, when melting or evaporating, and through sensible heat, with its temperature rise. In addition, an outward flow of ablating volatiles thickened the boundary layer, which diminished the heat flow. Ablation promised all the advantages of transpiration cooling, within a system that could be considerably lighter and yet more capable, and that used no fluid.[21]

Though ablation proved to offer a key to nose cone reentry, experiments showed that little if any ablation was to be expected under the relatively mild conditions of satellite entry. But satellite entry involved high total heat input, while its prolonged duration imposed a new requirement for good materials properties as insulators. They also had to stay cool through radiation. It thus became possible to critique the usefulness of ICBM nose cone ablators for the new role of satellite entry.[22]

Heat of ablation, in British thermal units (BTU) per pound, had been a standard figure of merit. Water, for instance, absorbs nearly 1,000 BTU/lb when it vaporizes as steam at 212 °F. But for satellite entry, with little energy being carried away by ablation, head of ablation could be irrelevant. Phenolic glass, a fine ICBM material with a measured heat of 9,600 BTU/lb, was unusable for a satellite because it had an unacceptably high thermal conductivity. This meant that the prolonged thermal soak of a satellite entry could have enough time to fry a spacecraft.

---

20. C.E. Brown, W.J. O'Sullivan, and C.H. Zimmerman, "A Study of the Problems Related to High Speed, High Altitude Flight," NACA Langley, 1953; Jacob Neufeld, *The Development of Ballistic Missiles in the United States Air Force, 1945–1960* (Washington: Office of Air Force History, 1990).
21. Army Ballistic Missile Agency staff, *Re-Entry Studies*, vol. 1. (Redstone Arsenal, AL: ABMA, Nov. 25, 1958), pp. 2, 24–25, 31, 37–45, 61; Sutton, "Initial Development of Ablation Heat Protection," pp. 3–11.
22. E.R. Riddell and J.D. Teare, "The Differences Between Satellite and Ballistic Missile Re-Entry Problems," in Morton Alperin and Hollingsworth F. Gregory, eds., *Vistas in Astronautics*, vol. 2 (New York: Pergamon Press, 1959,) pp. 174–190.

Teflon, by contrast, had a measured heat only one-third as large. It nevertheless made a superb candidate because of its excellent properties as an insulator.[23]

Hence it became possible to treat the satellite problem as an extension of the ICBM problem. With appropriate caveats, the experience and research techniques of the ICBM program could carry over to this new realm. The Central Intelligence Agency was preparing to recover satellite spacecraft at the same time that the Air Force was preparing to fly full-size Atlas nose cones, with both being achieved in April 1959.

The Army flew a subscale nose cone to intermediate range in August 1958, which President Dwight Eisenhower displayed during a November news conference. The Air Force became the first to fly a nose cone to intercontinental range, in July 1958. Both flights carried a mouse, and both mice survived their reentry, but neither was recovered. Better success came the following April, when an Atlas launched the full-size RVX-l nose cone, and the Discoverer II reconnaissance spacecraft returned safely through the atmosphere—though it fell into Russian, not American, hands.[24]

Hot Structures: Dyna-Soar

Reentry of ICBM nose cones and of satellites takes place at nearly the same velocity. Reentry of spacecraft takes place at a standard velocity of Mach 25, but there are large differences in the technical means that have been studied for the thermal protection. During the 1960s, it was commonly expected that such craft would be built as hot structures. In fact, however, the thermal protection adopted for the Shuttle was the well-known "tiles," a type of reusable insulation.

The Dyna-Soar program, early in the '60s, was first to face this issue. Dyna-Soar used a radiatively cooled hot structure, with the primary or load-bearing structure being of René 41. Trusses formed the primary structure of the wings and fuselage, with many of their beams meeting at joints that were pinned rather than welded. Thermal gradients,

---

23. Leo Steg, "Materials for Re-Entry Heat Protection of Satellites," *American Rocket Society Journal* (Sept. 1960), pp. 815–822.

24. *Time* (Nov. 18, 1957), pp. 19–20; (Apr. 27, 1959), p. 16; Joel W. Powell, "Thor-Able and Atlas-Able," *Journal of the British Interplanetary Society* (May 1984), pp. 219–225; Kevin C. Ruffner, *Corona: America's First Satellite Program* (McLean, VA: Central Intelligence Agency, 1995); see also T.A. Heppenheimer, "Toward Transatmospheric Flight," (in this volume) for additional details.

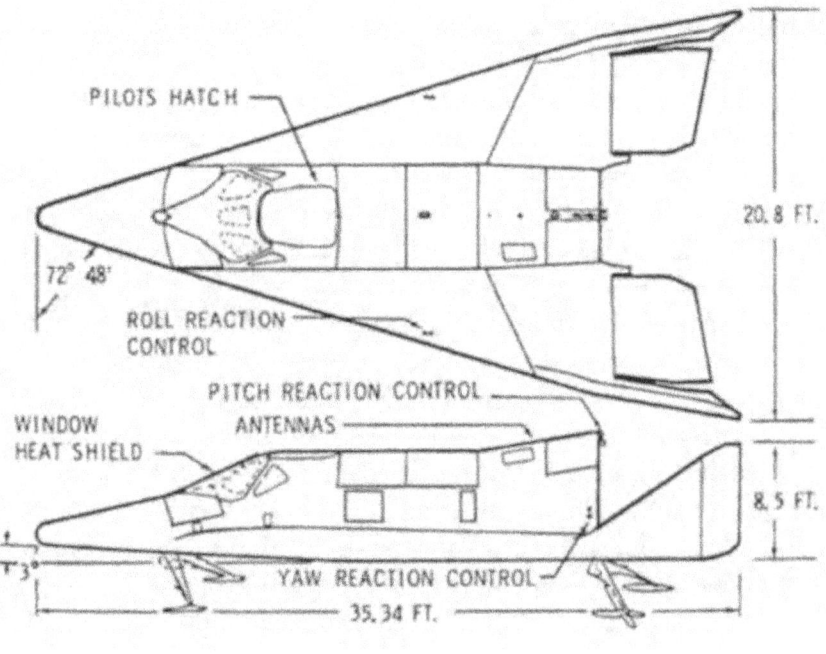

Schematic drawing of the Boeing X-20A Dyna-Soar. USAF.

imposing differential expansion on separate beams, caused these members to rotate at the pins. This accommodated the gradients without imposing thermal stresses. René 41 was selected as a commercially available superalloy that had the best available combination of oxidation resistance and high-temperature strength. Its yield strength, 130,000 pounds per square inch (psi) at room temperature, fell off only slightly at 1,200 °F and retained useful values at 1,800 °F. It could be processed as sheet, strip, wire, tubes, and forgings. Used as primary structure of Dyna-Soar, it supported a design specification that stated that the craft was to withstand at least four reentries under the most severe conditions permitted.

As an alloy, René 41 had a standard composition of 19 percent chromium, 11 percent cobalt, 10 percent molybdenum, 3 percent titanium, and 1.5 percent aluminum, along with 0.09 percent carbon and 0.006 percent boron, with the balance being nickel. It gained strength through age hardening, with the titanium and aluminum precipitating within the nickel as an intermetallic compound. Age-hardening weldments initially showed susceptibility to cracking, which occurred in parts that had been strained through welding or cold working. A new heat-treatment process permitted full aging without cracking, with the fabricated assemblies showing no significant tendency to develop cracks.[25]

As a structural material, the relatively mature state of René 41 reflected the fact that it had already seen use in jet engines. It nevertheless lacked the temperature resistance necessary for use in the metallic shingles or panels that were to form the outer skin of the vehicle, which were to reradiate the heat while withstanding temperatures as high as 3,000 °F. Here there was far less existing art, and investigators at Boeing had to find their way through a somewhat roundabout path. Four refractory or temperature-resistant metals initially stood out: tantalum, tungsten, molybdenum, and columbium. Tantalum was too heavy. Tungsten was not available commercially as sheet. Columbium also appeared to be ruled out, for it required an antioxidation coating, but vendors were unable to coat it without rendering it brittle. Molybdenum alloys also faced embrittlement because of recrystallization produced by a prolonged soak at high temperature in the course of coating formation. A promising alloy, Mo-0.5Ti, overcame this difficulty through addition of 0.07 percent zirconium. The alloy that resulted, Mo-0.5Ti-0.07Zr, was called TZM molybdenum. For a time it appeared as a highly promising candidate for all the other panels.[26]

Wing design also promoted its use, for the craft mounted a delta wing with leading-edge sweep of 73 degrees. Though built for hypersonic entry from orbit, it resembled the supersonic delta wings of contemporary aircraft such as the B-58 bomber. But this wing was designed using H. Julian Allen's blunt-body principle, with the leading edge being thickly rounded (that is, blunted) to reduce the rate of heating. The wing sweep then reduced equilibrium temperatures along the leading edge to levels compatible with the use of TZM.[27]

---

25. ASD staff, *Proceedings of 1962 X-20A (Dyna-Soar) Symposium*, vol. 3, *Structures and Materials* (Wright Patterson AFB: USAF Aeronautical Systems Division, 1963), DTIC AD-346912, pp. III-3-1-2 to -5, III-3-1-18 to -23, II-4-2-2 to -8. Re: René 41, see Howard J. Middendorf, "Materials and Processes for X-20A (Dyna-Soar)," Air Force Systems Command, June 1964, DTIC AD-449685, pp. 28, 41.
26. William Cowie, "Utilization of Refractory Metals on the X-20A (Dyna-Soar)," Air Force Systems Command, June 1964, DTIC: AD-609169; AD-449685, pp. 3–5, 20; AD-346912, pp. III-3-1-7, III-4-1-2 to -3.
27. DTIC AD-346912, p. III-3-1-8; Alvin Seiff and H. Julian Allen, "Some Aspects of the Design of Hypersonic Boost-Glide Aircraft," NACA RM-A55E26 (1955); and Clarence Geiger, "Strangled Infant: The X-20 Dyna-Soar," in Richard P. Hallion, ed., *Hypersonic Revolution*, vol. 1 (Bolling AFB: USAF, 1998).

Boeing's metallurgists nevertheless held an ongoing interest in columbium, because in uncoated form it showed superior ease of fabrication and lack of brittleness. A new Boeing-developed coating method eliminated embrittlement, putting columbium back in the running. A survey of its alloys showed that they all lacked the hot strength of TZM. Columbium nevertheless retained its attractiveness because it promised less weight. Based on coatability, oxidation resistance, and thermal emissivity, the preferred alloy was Cb-10Ti-5Zr, called D-36. It replaced TZM in many areas of the vehicle but proved to lack strength against creep at the highest temperatures. Moreover, coated TZM gave more of a margin against oxidation than coated D-36 did, again at the most extreme temperatures. D-36 indeed was chosen to cover most of the vehicle, including the flat underside of the wing. But TZM retained its advantage for such hot areas as the wing leading edges.[28]

The vehicle had some 140 running feet of leading edges and 140 square feet of associated area. This included leading edges of the vertical fins and elevons as well as of the wings. In general, D-36 served when temperatures during reentry did not exceed 2,700 °F, while TZM was used for temperatures between 2,700 and 3,000 °F. In accordance with the Stefan-Boltzmann law, all surfaces radiated heat at a rate proportional to the fourth power of the temperature. Hence for equal emissivities, a surface at 3,000 °F radiated 44 percent more heat than one at 2,700 °F.[29]

Panels of both TZM and D-36 demanded antioxidation coatings. These coatings were formed directly on the surfaces as metallic silicides (silicon compounds), using a two-step process that employed iodine as a chemical intermediary. Boeing introduced a fluidized-bed method for application of the coatings that cut the time for preparation while enhancing uniformity and reliability. In addition, a thin layer of silicon carbide, applied to the surface, gave the vehicle its distinctive black color. It enhanced the emissivity, lowering temperatures by as much as 200 °F.

It was necessary to show that complete panels could withstand aerodynamic flutter. A report of the Aerospace Vehicles Panel of the Air Force Scientific Advisory Board came out in April 1962 and singled out the problem of flutter, citing it as one that called for critical attention. The test program used two NASA wind tunnels: the 4-foot by 4-foot Unitary facility at Langley that covered a range of Mach 1.6 to 2.8

---

28. DTIC: AD-609169; AD-4496845, pp. 3–5, 20; AD-346912, pp. III-3-1-7, III-4-1-2 -3.
29. DTIC: AD-346912, p. III-3-1-8; AD-449685, p. iii.

and the 11-foot by 11-foot Unitary installation at Ames for Mach 1.2 to 1.4. Heaters warmed test samples to 840 °F as investigators started with steel panels and progressed to versions fabricated from René nickel alloy.

"Flutter testing in wind tunnels is inherently dangerous," a Boeing review declared. "To carry the test to the actual flutter point is to risk destruction of the test specimen. Under such circumstances, the safety of the wind tunnel itself is jeopardized." Panels under test were as large as 24 by 45 inches; flutter could have brought failure through fatigue, with parts of a specimen being blown through the tunnel at supersonic speed. Thus, the work started at dynamic pressures of 400 and 500 pounds per square foot (psf) and advanced over a year and a half to exceed the design requirement of close to 1,400 psf. Tests were concluded in 1962.[30]

Between the outer panels and the inner primary structure, a corrugated skin of René 41 served as the substructure. On the upper wing surface and upper fuselage, where the temperatures were no higher than 2,000 °F, the thermal-protection panels were also of René 41 rather than of a refractory. Measuring 12 by 45 inches, these panels were spot-welded directly to the corrugations of the substructure. For the wing undersurface and for other areas that were hotter than 2,000 °F, designers specified an insulated structure. Standoff clips, each with four legs, were riveted to the underlying corrugations and supported the refractory panels, which also were 12 by 45 inches in size.

The space between the panels and the substructure was to be filled with insulation. A survey of candidate materials showed that most of them exhibited a strong tendency to shrink at high temperatures. This was undesirable; it increased the rate of heat transfer and could create uninsulated gaps at seams and corners. Q-felt, a silica fiber from Johns Manville, also showed shrinkage. However, nearly all of it occurred at 2,000 °F and below; above 2,000 °F, further shrinkage was negligible. This meant that Q-felt could be "pre-shrunk" through exposure to temperatures above 2,000 °F for several hours. The insulation that resulted had density no greater than 6.2 pounds per cubic foot, one-tenth that of water. In addition, it withstood temperatures as high as 3,000 °F.[31]

---

30. DTIC AD-346912, pp. III-3-6-2 to -15 (quotes, pp. -8, -11); "Scientific Advisory Board Meme Report of the Aerospace Vehicles Panel on Dyna-Soar Panel Flutter," Apr. 20, 1962. See also Heppenheimer, "Toward Transatmospheric Flight."
31. DTIC: AD-449685, pp. 49–50; AD-609169, pp. 13, 29; AD-346912, pp. II-3-1-6 to -7, III-3-6-2, III-3-6-13 to -14.

TZM outer panels, insulated with Q-felt, proved suitable for wing leading edges. These were designed to withstand equilibrium temperatures of 2,825 °F and short-duration over-temperatures of 2,900 °F. But the nose cap faced temperatures of 3,680 °F along with a peak heat flux of 143 BTU/ft^2/sec. This cap had a radius of curvature of 7.5 inches, making it far less blunt than the contemporary Project Mercury heat shield that had a radius of 120 inches.[32] Its heating was correspondingly more severe. Reliable thermal protection of the nose was essential, so the program conducted two independent development efforts that used separate technical approaches. The firm of Chance Vought pursued the main line of activity, while Boeing also devised its own nose cap design.

The work at Vought began with a survey of materials that paralleled Boeing's review of refractory metals for the thermal-protection panels. Molybdenum and columbium had no strength to speak of at the pertinent temperatures, but tungsten retained useful strength even at 4,000 °F. But that metal could not be welded, while no coating could protect it against oxidation. Attention then turned to nonmetallic materials, including ceramics.

Ceramics of interest existed as oxides such as silica and magnesia, which meant that they could not undergo further oxidation. Magnesia proved to be unsuitable because it had low thermal emittance, while silica lacked strength. However, carbon in the form of graphite showed clear promise. It held considerable industrial experience; it was light in weight, while its strength actually increased with temperature. It oxidized readily but could be protected up to 3,000 °F by treating it with silicon, in vacuum and at high temperatures, to form a thin protective layer of silicon carbide. Near the stagnation point, the temperatures during reentry would exceed that level. This brought the concept of a nose cap with siliconized graphite as the primary material and with an insulated layer of a temperature-resistant ceramic covering its forward area. With graphite having good properties as a heat sink, it would rise in temperature uniformly and relatively slowly, while remaining below the 3,000 °F limit throughout the full time of the reentry.

Suitable grades of graphite proved to be available commercially from the firm of National Carbon. Candidate insulators included hafnia, thoria, magnesia, ceria, yttria, beryllia, and zirconia. Thoria was

---

32. DTIC AD-346912, pp. III-3-1-8, III-3-4-4; "How Mercury Capsule Design Evolved," *Aviation Week* (Sept. 21, 1959), pp. 52–59.

the most refractory but was very dense and showed poor resistance to thermal shock. Hafnia brought problems of availability and of reproducibility of properties. Zirconia stood out. Zirconium, its parent metal, had found use in nuclear reactors; the ceramic was available from the Zirconium Corporation of America. It had a melting point above 4,500 °F, was chemically stable and compatible with siliconized graphite, offered high emittance with low thermal conductivity, provided adequate resistance to thermal shock and thermal stress, and lent itself to fabrication.[33]

For developmental testing, Vought used two in-house facilities that simulated the flight environment, particularly during reentry. A ramjet, fueled with JP-4 and running with air from a wind tunnel, produced an exhaust with velocity up to 4,500 ft/sec and temperature up to 3,500 °F. It also generated acoustic levels above 170 decibels (dB), reproducing the roar of a Titan III booster and showing that samples under test could withstand the resulting stresses without cracking. A separate installation, built specifically for the Dyna-Soar program, used an array of propane burners to test full-size nose caps.

The final Vought design used a monolithic shell of siliconized graphite that was covered over its full surface by zirconia tiles held in place by thick zirconia pins. This arrangement relieved thermal stresses by permitting mechanical movement of the tiles. A heat shield stood behind the graphite, fabricated as a thick disk-shaped container made of coated TZM sheet metal and filled with Q-felt. The nose cap was attached to the vehicle with a forged ring and clamp that also were of coated TZM. The cap as a whole relied in radiative cooling. It was designed to be reusable; like the primary structure, it was to withstand four reentries under the most severe conditions permitted.[34]

The backup Boeing effort drew on that company's own test equipment. Study of samples used the Plasma Jet Subsonic Splash Facility, which created a jet with temperature as high as 8,000 °F that splashed over the face of a test specimen. Full-scale nose caps went into the Rocket Test Chamber, which burned gasoline to produce a nozzle exit velocity of 5,800 ft/sec and an acoustic level of 154 dB. Both installations were capable of long-duration testing, reproducing conditions during reentries that could last for 30 minutes.[35]

---

33. DTIC AD-346912, pp. III-4-5-3 to -6, III-4-5-16.
34. Ibid., pp. III-4-5-13 to -18, III-4-5-35.
35. Ibid., pp. III-3-4-4 to -6, III-3-4-20.

The Boeing concept used a monolithic zirconia nose cap that was reinforced against cracking with two screens of platinum-rhodium wire. The surface of the cap was grooved to relieve thermal stress. Like its counterpart from Vought, this design also installed a heat shield that used Q-felt insulation. However, there was no heat sink behind the zirconia cap. This cap alone provided thermal protection at the nose, through radiative cooling. Lacking pinned tiles and an inner shell, its design was simpler than that of Vought.[36]

Its fabrication bore comparison to the age-old work of potters, who shape wet clay on a rotating wheel and fire the resulting form in a kiln. Instead of using a potter's wheel, Boeing technicians worked with a steel die with an interior in the shape of a bowl. A paper honeycomb, reinforced with Elmer's Glue and laid in place, defined the pattern of stress-relieving grooves within the nose cap surface. The working material was not moist clay but a mix of zirconia powder with binders, internal lubricants, and wetting agents.

With the honeycomb in position against the inner face of the die, a specialist loaded the die by hand, filling the honeycomb with the damp mix and forming layers of mix that alternated with the wire screens. The finished layup, still in its die, went into a hydraulic press. A pressure of 27,000 psi compacted the form, reducing its porosity for greater strength and less susceptibility to cracks. The cap was dried at 200 °F, removed from its die, dried further, and then fired at 3,300 °F for 10 hours. The paper honeycomb burned out in the course of the firing. Following visual and x-ray inspection, the finished zirconia cap was ready for machining to shape in the attachment area, where the TZM ring-and-clamp arrangement waste anchor it to the fuselage.[37]

The nose cap, outer panels, and primary structure all were built to limit their temperatures through passive methods: radiation and insulation. Active cooling also played a role, reducing temperatures within the pilot's compartment and two equipment bays. These used a "water wall" that mounted absorbent material between sheet-metal panels to hold a mix of water and a gel. The gel retarded flow of this fluid, while the absorbent wicking kept it distributed uniformly to prevent hotspots.

---

36. Ibid., p. III-3-4-4; Geiger, "Strangled Infant," p. 357. See Heppenheimer, "Toward Transatmospheric Flight."
37. DTIC AD-346912, pp. III-4-6-5 to -7.

During reentry, heat reached the water walls as it penetrated into the vehicle. Some of the moisture evaporated as steam, transferring heat to a set of redundant water-glycol loops that were cooled by liquid hydrogen from an onboard supply. A catalytic bed combined the stream of warmed hydrogen with oxygen that again came from an onboard supply. This produced gas that drove the turbine of Dyna-Soar's auxiliary power unit, which provided both hydraulic and electric power to the craft.

A cooled hydraulic system also was necessary, to move the control surfaces as on a conventional airplane. The hydraulic fluid operating temperature was limited to 400 °F by using the fluid itself as an initial heat-transfer medium. It flowed through an intermediate water-glycol loop that removed its heat by being cooled with hydrogen. Major hydraulic components, including pumps, were mounted within an actively cooled compartment. Control-surface actuators, along with associated valves and plumbing, were insulated using inch-thick blankets of Q-felt. Through this combination of passive and active cooling methods, the Dyna-Soar program avoided a need to attempt to develop truly high-temperature arrangements, remaining instead within the state of the art.[38]

Specific vehicle parts and components brought their own thermal problems. Bearings, both ball and antifriction, needed strength to carry mechanical loads at high temperatures. For ball bearings, the cobalt-base superalloy Stellite 19 was known to be acceptable up to 1,200 °F. Investigation showed that it could perform under high load for short durations at 1,350 °F. Dyna-Soar nevertheless needed ball bearings qualified for 1,600 °F and obtained them as spheres of René 41 plated with gold. The vehicle also needed antifriction bearings as hinges for control surfaces, and here there was far less existing art. The best available bearings used stainless steel and were suitable only to 600 °F, whereas Dyna-Soar again faced a requirement of 1,600 °F. A survey of 35 candidate materials led to selection of titanium carbide with nickel as a binder.[39]

Antenna windows demanded transparency to radio waves at similarly high temperatures. A separate program of materials evaluation led to selection of alumina, with the best grade being available from the Coors Porcelain Company.[40]

---

38. Ibid., pp. 3-1-10 to -12; Geiger, "Strangled Infant," pp. 361–368.
39. DTIC: AD-346912, pp. III-4-6-9 to -11; AD-449685, pp. 63–65.
40. DTIC: AD-346912, pp. III-4-6-7 to -8; AD-449685, pp. 57–59.

Case 9 | High-Temperature Structures and Materials

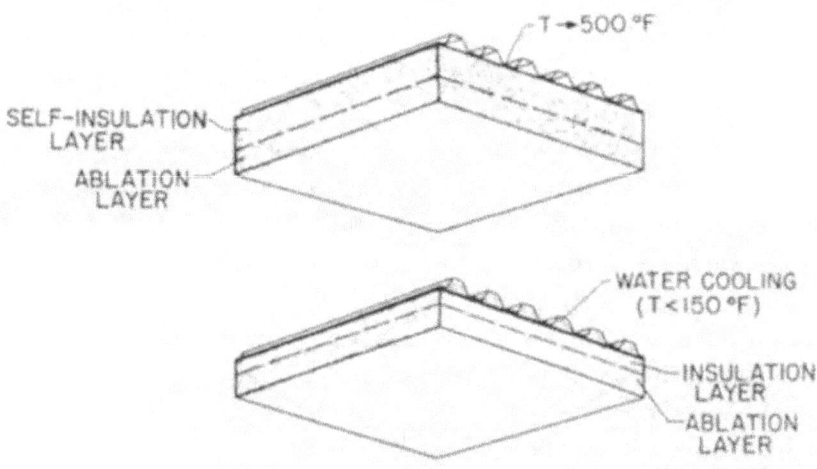

NASA concepts for passive and actively cooled ablative heat shields, 1960. NASA.

The pilot needed his own windows. The three main ones, facing forward, were the largest yet planned for a piloted spacecraft. They had double panes of fused silica, with infrared-reflecting surfaces on all surfaces except the outermost. This inhibited the inward flow of heat by radiation, reducing the load on the active cooling of the pilot's compartment. The window frames expanded when hot; to hold the panes in position, those frames were fitted with springs of René. The windows also needed thermal protection so they were covered with a shield of D-36. It was supposed to be jettisoned following reentry, around Mach 5, but this raised a question: what if it remained attached? The cockpit had two other windows, one on each side, which faced a less severe environment and were to be left unshielded throughout a flight. Over a quarter century earlier, Charles Lindbergh had flown the *Spirit of St. Louis* across the North Atlantic from New York to Paris using just side vision and a crude periscope. But that was a far cry from a plummeting lifting reentry vehicle. Now, test pilot Neil Armstrong flew Dyna-Soar–like approaches and landings in a modified Douglas F5D-1 fighter with side vision only and showed it was still possible.[41]

---

41. DTIC AD-346912, pp. III-3-1-10 to -11; Dennis Jenkins, *Space Shuttle* (Stillwater, MN: Voyageur Press, 2001).

The vehicle was to touch down at 220 knots. It lacked wheeled landing gear, for inflated rubber tires would have demanded their own cooled compartments. For the same reason, it was not possible to use a conventional oil-filled strut as a shock absorber. The craft therefore deployed tricycle landing skids. The two main skids, from Goodyear, were of Waspalloy nickel steel and mounted wire bristles of René 41. These gave a high coefficient of friction, enabling the vehicle to skid to a stop in a planned length of 5,000 feet while accommodating runway irregularities. In place of the usual oleo strut, a long rod of Inconel stretched at the moment of touchdown and took up the energy of impact, thereby serving as a shock absorber. The nose skid, from Bendix, was forged from René 41 and had an undercoat of tungsten carbide to resist wear. Fitted with its own energy-absorbing Inconel rod, the front skid had a reduced coefficient of friction, which helped to keep the craft pointing straight ahead during slide-out.[42]

Through such means, the Dyna-Soar program took long strides toward establishing hot structures as a technology suitable for operational use during reentry from orbit. The X-15 had introduced heat sink fabricated from Inconel X, a nickel steel. Dyna-Soar went considerably further, developing radiation-cooled insulated structures fabricated from René 41 and from refractory materials. A chart from Boeing made the point that in 1958, prior to Dyna-Soar, the state of the art for advanced aircraft structures involved titanium and stainless steel, with temperature limits of 600 °F. The X-15 with its Inconel X could withstand temperatures above 1,200 °F. Against this background, Dyna-Soar brought substantial advances in the temperature limits of aircraft structures.[43]

Hot Structures: ASSET

Dyna-Soar never flew, for Defense Secretary Robert S. McNamara canceled the program in December 1963. At that time, vehicles were well under construction but still were some 2½ years away from first flight.

---

42. DTIC AD-346912, pp. III-3-1-14 to -15; Geiger, "Strangled Infant," pp. 347–349, 360–361.
43. See Table 1 in Heppenheimer, "Toward Transatmospheric Flight," for specific advances by structural element. See also Terry L. Sunday and John R. London, "The X-20 Space Plane: Past Innovation, Future Vision," in John Becklake, ed., *History of Rocketry and Astronautics*, vol. 17 of the AAS History Series (San Diego: Univelt, 1995), pp. 253–284; Geiger, "Strangled Infant," pp. 344–346; R.L. Schleicher, "Structural Design of the X-15," *Journal of the Royal Aeronautical Society* (Oct. 1963), pp. 618–636.

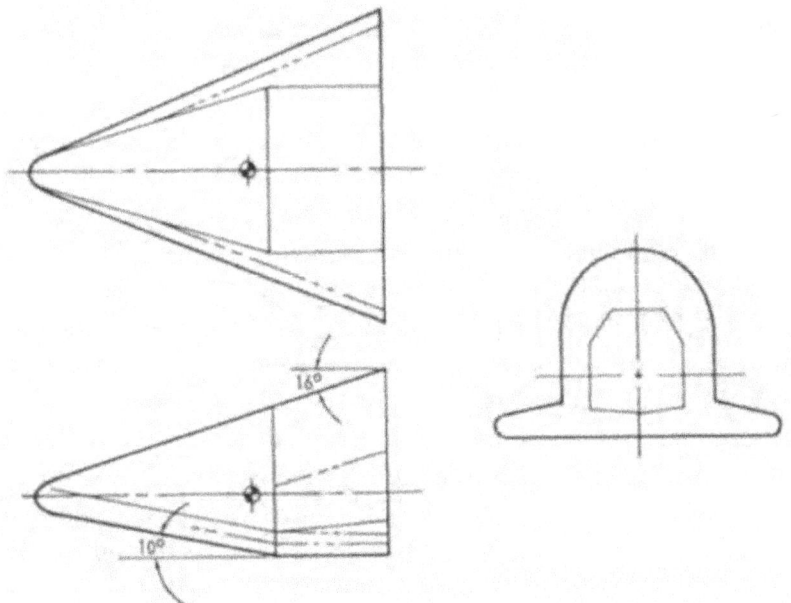

Geometric planform of the McDonnell ASSET reentry test vehicle. USAF.

Still its technology remained available for further development, and thus it fell to a related program, Aerothermodynamic/elastic Structural Systems Environmental Test (ASSET), to take up the hot structures cause and fly with them.[44]

As early as August 1959, the Flight Dynamics Laboratory at Wright-Patterson Air Force Base launched an in-house study of a small recoverable boost-glide vehicle that was to test hot structures during reentry. From the outset there was strong interest in problems of aerodynamic flutter. This was reflected in the ASSET concept name.

ASSET won approval as a program in January 1961. In April of that year, the firm of McDonnell Aircraft, which was already building Mercury spacecraft, won a contract to develop the ASSET flight vehicles. The Thor, which had been deployed operationally in England, was about to come home because it was no longer needed as a weapon. It became available for use as a launch vehicle.

ASSET took shape as a flat-bottomed wing-body craft that used a low-wing configuration joined to a truncated combined cone-cylinder

---

44. Hallion, *Hypersonic Revolution*, vol. 1, p. II-xvi (intro to Geiger case study on X-20).

body. It had a length of 59 inches and a span of 55 inches. Its bill of materials resembled that of Dyna-Soar, for it used TZM molybdenum to withstand 3,000 °F on the forward lower heat shield, graphite for similar temperatures on leading edges, and zirconia rods for the nose cap, which was rated at 4,000 °F. But ASSET avoided the use of René 41, with cobalt and columbium alloys being employed instead.[45]

ASSET was built in two varieties: the Aerothermodynamic Structural Vehicle (ASV) weighing 1,130 pounds and the Aerothermodynamic Elastic Vehicle (AEV) at 1,225 pounds. The AEVs were to study panel flutter along with the behavior of a trailing-edge flap, which represented an aerodynamic control surface in hypersonic flight. These vehicles did not demand the highest possible flight speeds and therefore flew with single-stage Thors as the booster. But the ASVs were built to study materials and structures in the reentry environment while taking data on temperatures, pressures, and heat fluxes. Such missions demanded higher speeds. These boost-glide craft therefore used the two-stage Thor-Delta launch vehicle, which resembled the Thor-Able that had conducted nose cone tests at intercontinental range in 1958.[46]

The program eventually conducted six flights:[47] several of these craft were to be recovered. Following standard practice, their launches were scheduled for the early morning, to give downrange recovery crews the maximum hours of daylight. That did not help ASV-1, the first flight in the program, which sank into the sea. Still, it flew successfully and returned good data. In addition, this flight set a milestone, for it was the first time in aviation history that a lifting reentry spacecraft had traversed the demanding hypersonic reentry corridor from orbit down to the lower atmosphere.[48]

ASV-2 followed, using the two-stage Thor-Delta, but it failed when the second stage did not ignite. The next one carried ASV-3, with this mission scoring a double achievement. It not only made a good flight downrange, but it was also successfully recovered. It carried a liquid-cooled double-wall test panel from Bell Aircraft along with a molybdenum heat-shield panel from Boeing, home of Dyna-Soar. ASV-3 also had a

---

45. "Advanced Technology Program: Technical Development Plan for Aerothermodynamic/Elastic Structural System Environmental Tests (ASSET)," Air Force Systems Command, Sept. 9, 1963.
46. Ibid., pp. 4, 11–13; Hallion, "ASSET," in *Hypersonic Revolution*, vol. 1, pp. 451, 464–465.
47. See Table 2 in Heppenheimer, "Toward Transatmospheric Flight," for a list of these flights and performance objectives.
48. Hallion, "ASSET," pp. 510–512 (quote, p. 512).

new nose cap. The standard ASSET type used zirconia dowels, 1.5 inches long by 0.5 inches in diameter, which were bonded together with a zirconia cement. The new cap, from International Harvester, had a tungsten base covered with thorium oxide and was reinforced with tungsten.

A company advertisement stated that it withstood reentry so well that it "could have been used again," and this was true for the craft as a whole. Historian Richard P. Hallion writes that "overall, it was in excellent condition. Water damage . . . caused some problems, but not so serious that McDonnell could not have refurbished and reflown the vehicle." The Boeing and Bell panels came through reentry without damage, and the importance of physical recovery was emphasized when columbium aft leading edges showed significant deterioration. They were redesigned, with the new versions going into subsequent AEV and ASV spacecraft.[49]

The next two flights were AEVs, each of which carried a flutter test panel and a test flap. AEV-1 returned only one high-Mach data point, at Mach 11.88, but this sufficed to indicate that its panel was probably too stiff to undergo flutter. Engineers made it thinner and flew a new one on AEV-2, where it returned good data until it failed at Mach 10. The flap experiment also showed value. It had an electric motor that deflected it into the airstream, with potentiometers measuring the force required to move it, and it enabled aerodynamicists to critique their theories. Thus one treatment gave pressures that were in good agreement with observations, whereas another did not.

ASV-4, the final flight, returned "the highest quality data of the ASSET program," according to the flight-test report. The peak speed of 19,400 ft/sec, Mach 18.4, was the highest in the series and was well above the design speed of 18,000 ft/sec. The long hypersonic glide covered 2,300 nautical miles and prolonged the data return, which presented pressures at 29 locations on the vehicle and temperatures at 39. An onboard system transferred mercury ballast to trim the angle of attack, increasing the lift-to-drag ratio (L/D) from its average of 1.2 to 1.4, and extending the trajectory. The only important problem came when the recovery parachute failed to deploy properly and ripped away, dooming ASV-4 to follow ASV-1 into the depths of the Atlantic.[50]

---

49. Ibid., pp. 512–516 (quote, p. 515); "ASSET ASV-3 Flight Test Report," McDonnell Aircraft Corp., Jan. 4, 1965, DTIC AD-357523; advertisement, *Aviation Week*, May 24, 1965, p. 62.
50. "Fourth ASSET Glider Gathers Data," *Aviation Week*, Nov. 2, 1964, pp. 25–26; "ASSET ASV-4 Flight Test Report," McDonnell Aircraft Corp., June 25, 1965, DTIC AD-366546; Hallion, "ASSET," pp. 516–519.

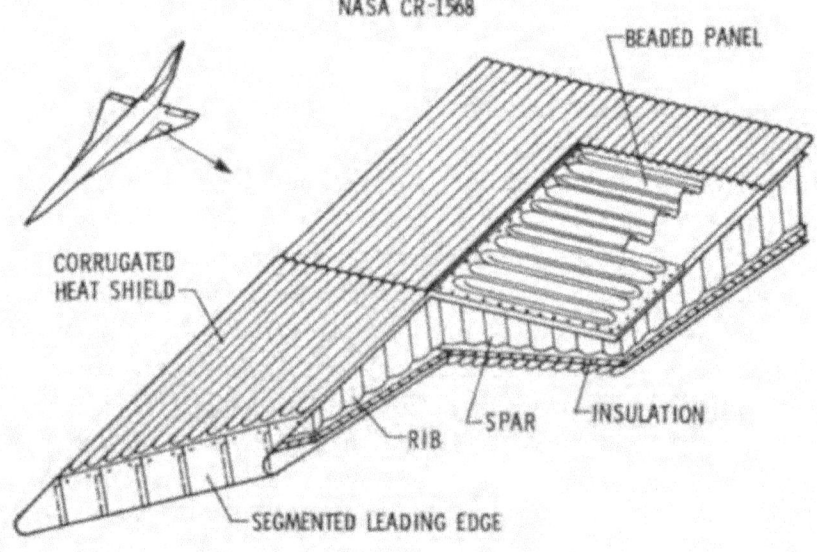

NASA concept for a hypersonic cruise wing structure formed of beaded, corrugated, and tubular structural panels, 1978. NASA.

On the whole, ASSET nevertheless scored a host of successes. It showed that insulated hot structures could be built and flown without producing unpleasant surprises, at speeds up to three-fourths of orbital velocity. It dealt with such practical issues of design as fabrication, fasteners, and coatings. In hypersonic aerodynamics, ASSET contributed to understanding of flutter and of the use of movable control surfaces. The program also developed and successfully used a reaction control system built for a lifting reentry vehicle. Only one flight vehicle was recovered in four attempts, but it complemented the returned data by permitting a close look at a hot structure that had survived its trial by fire.

Reusable Surface Insulation

Early in the 1960s, researchers at Lockheed introduced an entirely different approach to thermal protection, which in time became the standard. Ablatives were unrivalled for once-only use, but during that decade the hot structure continued to stand out as the preferred approach for reusable craft such as Dyna-Soar. As noted, it used an insulated primary or load-bearing structure with a skin of outer panels. These emitted heat by radiation, maintaining a temperature that was high but steady.

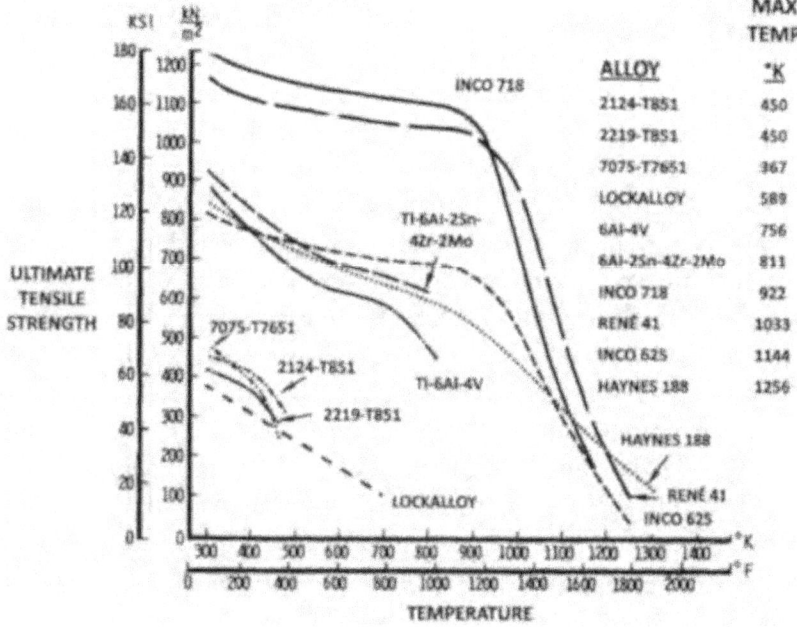

Strength versus temperature for various superalloys, including René 41, the primary structural material used on the X-20 Dyna-Soar. NASA.

Metal fittings supported these panels, and while the insulation could be high in quality, these fittings unavoidably leaked heat to the underlying structure. This raised difficulties in crafting this structure of aluminum and even of titanium, which had greater heat resistance. On Dyna-Soar, only René 41 would do.[51]

Ablatives avoided such heat leaks while being sufficiently capable as insulators to permit the use of aluminum. In principle, a third approach combined the best features of hot structures and ablatives. It called for the use of temperature-resistant tiles, made perhaps of ceramic, which could cover the vehicle skin. Like hot-structure panels, they would radiate heat, while remaining cool enough to avoid thermal damage. In addition, they were to be reusable. They also were to offer the excellent insulating properties of good ablators, preventing heat from reaching the underlying structure—which once more might be of aluminum. This concept became known as reusable surface insulation (RSI). In time, it gave rise to the thermal protection of the Shuttle.

51. F.M. Anthony, R.R. Fisher, and R.G. Helenbrook, "Selection of Space Shuttle Thermal Protection Systems," AIAA Paper 71-443 (1971).

RSI grew out of ongoing work with ceramics for thermal protection. Ceramics had excellent temperature resistance, light weight, and good insulating properties. But they were brittle and cracked rather than stretched in response to the flexing under load of an underlying metal primary structure. Ceramics also were sensitive to thermal shock, as when heated glass breaks when plunged into cold water. In flight, such thermal shock resulted from rapid temperature changes during reentry.[52]

Monolithic blocks of the ceramic zirconia had been specified for the nose cap of Dyna-Soar, but a different point of departure used mats of solid fiber in lieu of the solid blocks. The background to the Shuttle's tiles lay in work with such mats that took place early in the 1960s at Lockheed Missiles and Space Company. Key people included R.M. Beasley, Ronald Banas, Douglas Izu, and Wilson Schramm. A Lockheed patent disclosure of December 1960 gave the first presentation of a reusable insulation made of ceramic fibers for use as a heat shield. Initial research dealt with casting fibrous layers from a slurry and bonding the fibers together.

Related work involved filament-wound structures that used long continuous strands. Silica fibers showed promise and led to an early success: a conical radome of 32-inch diameter built for Apollo in 1962. Designed for reentry, it had a filament-wound external shell and a lightweight layer of internal insulation cast from short fibers of silica. The two sections were densified with a colloid of silica particles and sintered into a composite. This gave a nonablative structure of silica composite reinforced with fiber. It never flew, as design requirements changed during the development of Apollo. Even so, it introduced silica fiber into the realm of reentry design.

Another early research effort, Lockheat, fabricated test versions of fibrous mats that had controlled porosity and microstructure. These were impregnated with organic fillers such as Plexiglas (methyl methacrylate). These composites resembled ablative materials, though the filler did not char. Instead it evaporated or volatilized, producing an outward flow of cool gas that protected the heat shield at high heat-transfer rates. The Lockheat studies investigated a range of fibers that included silica, alumina, and boria. Researchers constructed multilayer composite structures of filament-wound and short-fiber materials that resembled the Apollo radome. Impregnated densities were 40 to 60 lb/

---

52. Wilson B. Schramm, Ronald P. Banas, and Y. Douglas Izu, "Space Shuttle Tile—The Early Lockheed Years," *Lockheed Horizons*, No. 13, 1983, pp. 2–15.

ft³, the higher density being close to that of water. Thicknesses of no more than an inch gave acceptably low back-face temperatures during simulations of reentry.

This work with silica-fiber ceramics was well underway during 1962. Three years later, a specific formulation of bonded silica fibers was ready for further development. Known as LI-1500, it was 89 percent porous and had a density of 15 lb/ft³, one-fourth that of water. Its external surface was impregnated with filler to a predetermined depth, again to provide additional protection during the most severe reentry heating. By the time this filler was depleted, the heat shield was to have entered a zone of more moderate heating, where the fibrous insulation alone could provide protection.

Initial versions of LI-1500, with impregnant, were intended for use with small space vehicles similar to Dyna-Soar that had high heating rates. Space Shuttle concepts were already attracting attention—the January 1964 issue of *Astronautics & Aeronautics*, the journal of the American Institute of Aeronautics and Astronautics, presents the thinking of the day—and in 1965 a Lockheed specialist, Maxwell Hunter, introduced an influential configuration called Star Clipper. His design employed LI-1500 for thermal protection.

Like other Shuttle concepts, Star Clipper was to fly repeatedly, but the need for an impregnant in LI-1500 compromised its reusability. But in contrast to earlier entry vehicle concepts, Star Clipper was large, offering exposed surfaces that were sufficiently blunt to benefit from H. Julian Allen's blunt-body principle. They had lower temperatures and heating rates, which made it possible to dispense with the impregnant. An unfilled version of LI-1500, which was inherently reusable, now could serve.

Here was the first concept of a flight vehicle with reusable insulation, bonded to the skin, which could reradiate heat in the fashion of a hot structure. However, the matted silica by itself was white and had low thermal emissivity, making it a poor radiator of heat. This brought excessive surface temperatures that called for thick layers of the silica insulation, adding weight. To reduce the temperatures and the thickness, the silica needed a coating that could turn it black for high emissivity. It then would radiate well and remain cooler.

The selected coating was a borosilicate glass, initially with an admixture of Cr_2O_3 and later with silicon carbide, which further raised the emissivity. The glass coating and the silica substrate were both silicon dioxide; this assured a match of their coefficients of thermal expansion, to prevent the coating from developing cracks under the temperature changes of reentry. The glass coating could soften at very high temperatures to heal minor nicks or scratches. It also offered true reusability, surviving repeated cycles to 2,500 °F. A flight test came in 1968 as NASA Langley investigators mounted a panel of LI-1500 to a Pacemaker reentry test vehicle along with several candidate ablators. This vehicle carried instruments and was recovered. Its trajectory reproduced the peak heating rates and temperatures of a reentering Star Clipper. The LI-1500 test panel reached 2,300 °F and did not crack, melt, or shrink. This proof-of-concept test gave further support to the concept of high-emittance reradiative tiles of coated silica for thermal protection.[53]

Lockheed conducted further studies at its Palo Alto Research Center. Investigators cut the weight of RSI by raising its porosity from the 89 percent of LI-1500 to 93 percent. The material that resulted, LI-900, weighed only 9 pounds per cubic foot, one-seventh the density of water.[54] There also was much fundamental work on materials. Silica exists in three crystalline forms: quartz, cristobalite, and tridymite. These not only have high coefficients of thermal expansion but also show sudden expansion or contraction with temperature because of solid-state phase changes. Cristobalite is particularly noteworthy; above 400 °F, it expands by more than 1 percent as it transforms from one phase to another. Silica fibers for RSI were to be glass, an amorphous rather than a crystalline state with a very low coefficient of thermal expansion and an absence of phase changes. The glassy form thus offered superb resistance to thermal stress and thermal shock, which would recur repeatedly during each return from orbit.[55]

The raw silica fiber came from Johns Manville, which produced it from high-purity sand. At elevated temperatures, it tended to undergo "devitrification," transforming from a glass into a crystalline state. Then, when cooling, it passed through phase-change temperatures and the fiber suddenly shrank, producing large internal tensile stresses. Some

---

53. Ibid.
54. Korb, et al., "The Shuttle Orbiter"; Wilson Schramm, "HRSI and LRSI—The Early Years," *American Ceramic Society Bulletin*, vol. 60 (1981), pp. 1194–1195.
55. Korb, et al., "The Shuttle Orbiter"; NASA, *NASA Space Shuttle Technology Conference*, vol. 2, "Structures and Materials," NASA TM-X-2273 (1971); Richard C. Thuss, Harry G. Thibault, and Arnold Hiltz, "The Utilization of Silica Based Thermal Insulation for the Space Shuttle Thermal Protection System," CASI 72A-10764, Oct. 1971.

fibers broke, giving rise to internal cracking within the RSI and degradation of its properties. These problems threatened to grow worse during subsequent cycles of reentry heating.

To prevent devitrification, Lockheed worked to remove impurities from the raw fiber. Company specialists raised the purity of the silica to 99.9 percent while reducing contaminating alkalis to as low as 6 parts per million. Lockheed proceeded to do these things not only in the laboratory but also in a pilot plant. This plant took the silica from raw material to finished tile, applying 140 process controls along the way. Established in 1970, the pilot plant was expanded in 1971 to attain a true manufacturing capability. Within this facility, Lockheed produced tiles of LI-1500 and LI-900 for use in extensive programs of test and evaluation. In turn, the increasing availability of these tiles encouraged their selection for Shuttle protection in lieu of a hot-structure approach.[56]

General Electric (GE) also became actively involved, studying types of RSI made from zirconia and from mullite, 3Al2O3+2SiO2, as well as from silica. The raw fibers were commercial grade, with the zirconia coming from Union Carbide and the mullite from Babcock and Wilcox. Devitrification was a problem, but whereas Lockheed had addressed it by purifying its fiber, GE took the raw silica from Johns Manville and tried to use it with little change. The basic fiber, the Q-felt of Dyna-Soar, also had served as insulation on the X-15. It contained 19 different elements as impurities. Some were present at a few parts per million, but others—aluminum, calcium, copper, lead, magnesium, potassium, sodium—ran from 100 to 1,000 parts per million. In total, up to 0.3 percent was impurity.

General Electric treated this fiber with a silicone resin that served as a binder, pyrolyzing the resin and causing it to break down at high temperatures. This transformed the fiber into a composite, sheathing each strand with a layer of amorphous silica that had a purity of 99.98 percent or more. This high purity resulted from that of the resin. The amorphous silica bound the fibers together while inhibiting their devitrification. General Electric's RSI had a density of 11.5 lb/ft^3, midway between that of LI-900 and LI-1500.[57]

Many Shuttle managers had supported hot structures, but by mid-1971 they were in trouble. In Washington, the Office of Management

---

56. Schramm, et al., "Space Shuttle Tile"; Schramm, "HRSI."
57. CASI 72A-10764; NASA TM-X-2273, pp. 39–93.

and Budget (OMB) now was making it clear that it expected to impose stringent limits on funding for the Shuttle, which brought a demand for new configurations that could cut the cost of development. Within weeks, the contractors did a major turnabout. They abandoned hot structures and embraced RSI. Managers were aware that it might take time to develop for operational use, but they were prepared to use ablatives for interim thermal protection and to switch to RSI once it was ready.[58]

What brought this dramatic change? The advent of RSI production at Lockheed was critical. This drew attention from Max Faget, a longtime NACA–NASA leader who had kept his hand in the field of Shuttle design, offering a succession of conceptual design configurations that had helped to guide the work of the contractors. His most important concept, designated MSC-040, came out in September 1971 and served as a point of reference. It used RSI and proposed to build the Shuttle of aluminum rather than René 41 or anything similar.[59]

Why aluminum? "My history has always been to take the most conservative approach," Faget explained subsequently. Everyone knew how to work with aluminum, for it was the most familiar of materials, but everything else carried large question marks. Titanium, for one, was literally a black art. Much of the pertinent shop-floor experience had been gained within the SR-71 program and was classified. Few machine shops had pertinent background, for only Lockheed had constructed an airplane—the SR-71—that used titanium hot structure. The situation was worse for columbium and the superalloys, for these metals had been used mostly in turbine blades. Lockheed had encountered serious difficulties as its machinists and metallurgists wrestled with titanium. With the Shuttle facing the OMB's cost constraints, no one cared to risk an overrun while machinists struggled with the problems of other new materials.[60]

NASA Langley had worked to build a columbium heat shield for the Shuttle and had gained a particularly clear view of its difficulties. It was heavier than RSI but offered no advantage in temperature resistance.

---

58. "Space Shuttle Program Definition: Phase B Extension Final Report," Grumman B35-43 RP-33 (Mar. 15, 1972); North American Rockwell 1971 Reports SV 71-50 and SV 71-59 (1971); "Interim Report to OMSF: Phase B System Study Extension," McDonnell-Douglas report (Sept. 1, 1971), p. 36.
59. Heppenheimer, *The Space Shuttle Decision*, NASA SP-4221 (Washington, DC: NASA, 1999); Jenkins, *Space Shuttle*, pp. 141–150.
60. Heppenheimer, *Turbulent Skies: The History of Commercial Aviation* (New York: John Wiley, 1995); author interview with Max Faget, Mar. 4, 1997.

In addition, coatings posed serious problems. Silicides showed promise of reusability and long life, but they were fragile and easily damaged. A localized loss of coating could result in rapid oxygen embrittlement at high temperatures. Unprotected columbium oxidized readily, and above the melting point of its oxide, 2,730 °F, it could burst into flame.[61] "The least little scratch in the coating, the shingle would be destroyed during reentry," Faget said. Charles Donlan, the Shuttle Program Manager at NASA Headquarters, placed this in a broader perspective in 1983:

> Phase B was the first really extensive effort to put together studies related to the completely reusable vehicle. As we went along, it became increasingly evident that there were some problems. And then as we looked at the development problems, they became pretty expensive. We learned also that the metallic heat shield, of which the wings were to be made, was by no means ready for use. The slightest scratch and you are in trouble.[62]

Other refractory metals offered alternatives to columbium, but even when proposing to use them, the complexity of a hot structure also militated against its selection. As a mechanical installation, it called for large numbers of clips, brackets, standoffs, frames, beams, and fasteners. Structural analysis loomed as a formidable task. Each of many panel geometries needed its own analysis, to show with confidence that the panels would not fail through creep, buckling, flutter, or stress under load. Yet this confidence might be fragile, for hot structures had limited ability to resist over-temperatures. They also faced the continuing issue of sealing panel edges against ingestion of hot gas during reentry.[63]

In this fashion, having taken a long look at hot structures, NASA did an about-face as it turned toward the RSI that Lockheed's Max Hunter had recommended as early as 1965. Then, in January 1972, President Richard Nixon gave his approval to the Space Shuttle program, thereby raising it to the level of a Presidential initiative. Within days, NASA's Dale Myers spoke

---

61. L.J. Korb and H.M. Clancy, "The Shuttle Thermal Protection System," CASI 81A-44344 (1981), pp. 232–249; Korb, et al., "The Shuttle Orbiter," p. 1189.
62. John Mauer interview of Charles Donlan, Oct. 19, 1983, *Oral History Series, Shuttle Interviews*, Box 1, JSC History Collection, University of Houston-Clear Lake; author interview of Faget, Mar. 4, 1997.
63. CASI 81A-44344, Ref. 62.

to a conference in Houston and stated that the Agency had made the basic decision to use RSI. Requests for proposal soon went out, inviting leading aerospace corporations to bid for the prime contract on the Shuttle orbiter, and North American won this $2.6-billion prize in July. However, the RSI wasn't Lockheed's. The proposal specified mullite RSI for the undersurface and forward fuselage, a design feature that had been held over from the company's studies of a fully reusable orbiter during the previous year.[64]

Still, was mullite RSI truly the one to choose? It came from General Electric and had lower emissivity than the silica RSI of Lockheed but could withstand higher temperatures. Yet the true basis for selection lay in the ability to withstand 100 reentries as simulated in ground test. NASA conducted these tests during the last 5 months of 1972, using facilities at its Ames, Johnson, and Kennedy Centers, with support from Battelle Memorial Institute.

The main series of tests ran from August to November and gave a clear advantage to Lockheed. That firm's LI-900 and LI-1500 went through 100 cycles to 2,300 °F and met specified requirements for maintenance of low back-face temperature and minimal thermal conductivity. The mullite showed excessive back-face temperatures and higher thermal conductivity, particularly at elevated temperatures. As test conditions increased in severity, the mullite also developed coating cracks and gave indications of substrate failure.

The tests then introduced acoustic loads, with each cycle of the simulation now subjecting the RSI to loud roars of rocket flight along with the heating of reentry. LI-1500 continued to show promise. By mid-November, it demonstrated the equivalent of 20 cycles to 160 decibels, the acoustic level of a large launch vehicle, and 2,300 °F. A month later, NASA conducted what Lockheed describes as a "sudden death shootout": a new series of thermal-acoustic tests, in which the contending materials went into a single large 24-tile array at NASA Johnson. After 20 cycles, only Lockheed's LI-900 and LI-1500 remained intact. In separate tests, LI-1500 withstood 100 cycles to 2,500 °F and survived a thermal overshoot to 3,000 °F, as well as an acoustic overshoot to 174 dB. Clearly, this was the material NASA wanted.[65]

---

64. "Sortie Module May Cut Experiment Cost," *Aviation Week* (Jan. 17, 1972), p. 17; North American Rockwell Report SV 72-19, 1972.
65. Schramm, et al., "Space Shuttle Tile," Ref. 53, pp. 11–14; William S. Hieronymus, "Two Reusable Materials Studied for Orbiter Thermal Protection," *Aviation Week* (Mar. 27, 1972), p. 48.

Case 9 | High-Temperature Structures and Materials

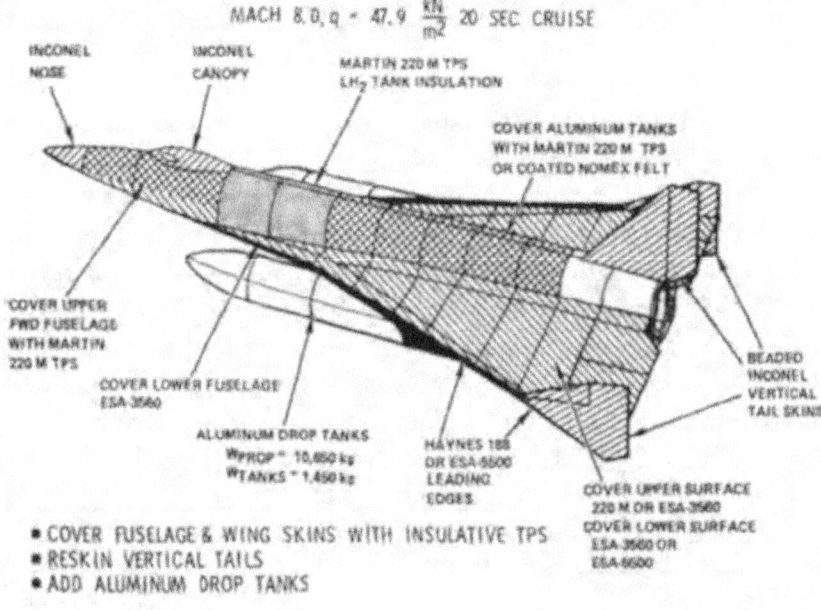

Thermal protection system for the proposed National Hypersonic Flight Research Facility, 1978. NASA.

As insulation, the tiles were astonishing. A researcher could heat one in a furnace until it was white hot, remove it, allow its surface to cool for a couple of minutes, and pick it up at its edges using his or her fingers, with its interior still at white heat. Lockheed won the thermal-protection subcontract in 1973, with NASA specifying LI-900 as the baseline RSI. The firm responded with preparations for a full-scale production facility in Sunnyvale, CA. With this, tiles entered the mainstream of thermal protection.

Cooling

Hypersonics has much to say about heating, so it is no surprise that it also has something to say about cooling. Active cooling merits only slight attention, as in the earlier discussion of Dyna-Soar. Indeed, two books on Shuttle technology run for hundreds of pages and give complete treatments of tiles for thermal protection—but give not a word about active cooling.[66]

---

66. Heppenheimer, *Development of the Space Shuttle, 1972–1981*, (Washington: Smithsonian Institution Press, 2002); Jenkins, *Space Shuttle*.

What the topic of cooling mostly comprises is the use of passive cooling, which allowed the Shuttle to be built of aluminum.

During the early 1970s, when there was plenty of talk of using a liquid-fueled booster from Marshall Space Flight Center, many designers considered building that booster largely of aluminum. This raised the question of how bare aluminum, without protection, could serve in a Shuttle booster. It was common understanding that aluminum airframes lost strength because of aerodynamic heating at speeds beyond Mach 2, with titanium being necessary at higher speeds. But this held true for aircraft in cruise, which faced their temperatures continually. Boeing's reusable booster was to reenter at Mach 7, matching the top speed of the X-15. Still, its thermal environment resembled a fire that does not burn the hand when one whisks it through quickly. Designers addressed the problem of heating on the vehicle's vulnerable underside by the simple expedient of using thicker metal construction to cope with anticipated thermal loads. Even these areas were limited in extent, with the contractors noting that "the material gauges (thicknesses) required for strength exceed the minimum heat sink gauges over the majority of the vehicle."[67]

McDonnell-Douglas went further. In mid-1971, it introduced its own orbiter, which lowered the staging velocity to 6,200 ft/sec. Its winged booster was 82 percent aluminum heat sink. Its selected configuration was optimized from a thermal standpoint, bringing the largest savings in the weight of thermal protection.[68] Then, in March 1972, NASA selected solid-propellant rockets for the boosters. The issue of their thermal protection now went away entirely, for these big solids used steel casings that were half an inch thick and that provided heat sink very effectively.[69]

Aluminum structure, protected by ablatives, also was in the forefront during the Precision Recovery Including Maneuvering Entry (PRIME) program. Martin Marietta, builder of the X-24A lifting body, also developed the PRIME flight vehicle, the SV-5D that later was referred to as the X-23. Although it was only 7 feet in length, it faithfully duplicated the shape of the X-24, even including a small bubble-like protrusion near the front that represented the cockpit canopy.

---

67. NASA SP-4221, pp. 335–340, 348–349; Grumman Report B35-43 RP-33; Michael L. Yaffee, "Program Changes Boost Grumman Shuttle," *Aviation Week* (July 12, 1963), pp. 36–39.
68. "External LH2 Tank Study Final Report," McDonnell-Douglas Report MDC E076-1, June 30, 1971.
69. NASA SP-4221, pp. 420–422; Heppenheimer, *Development*, p. 188.

Case 9 | High-Temperature Structures and Materials

PRIME complemented ASSET, with both programs conducting flight tests of boost-glide vehicles. However, while ASSET pushed the state of the art in materials and hot structures, PRIME used ablative thermal protection for a more straightforward design and emphasized flight performance. Accelerated to near-orbital velocities by Atlas launch vehicles, the PRIME missions called for boost-glide flight from Vandenberg Air Force Base (AFB) to locations in the western Pacific near Kwajalein Atoll. The SV-5D had higher L/D than Gemini or Apollo did, and, as with those NASA programs, it was to demonstrate precision reentry. The plans called for cross range, with the vehicle flying up to 710 nautical miles to the side of a ballistic trajectory and then arriving within 10 miles of its recovery point.

The piloted X-24A supersonic lifting body, used to assess the SV-5 shape's approach and landing characteristics, was built of aluminum. The SV-5D also used this material for both its skin and primary structure. It mounted both aerodynamic and reaction controls, the former consisting of right and left body-mounted flaps set well aft. Deflected symmetrically, they controlled pitch; deflected individually (asymmetrically), they produced yaw and roll. These flaps were beryllium plates that provided a useful thermal heat sink. The fins were of steel honeycomb, likewise with surfaces of beryllium sheet.

Most of the vehicle surface obtained thermal protection from ESA 3560 HF, a flexible ablative blanket of phenolic fiberglass honeycomb that used a silicone elastomer as the filler, with fibers of nylon and silica holding the ablative char in place during reentry. ESA 5500 HF, a high-density form of this ablator, gave added protection in hotter areas. The nose cap and the beryllium flaps used a different material: carbon-phenolic composite. At the nose, its thickness reached 3.5 inches.[70]

The PRIME program made three flights that took place between December 1966 and April 1967. All returned data successfully, with the third flight vehicle also being recovered. The first mission reached 25,300 ft/sec and flew 4,300 miles downrange, missing its target by only 900 feet. The vehicle executed pitch maneuvers but made no attempt at cross range. The next two flights indeed achieved cross range, respec-

---

70. Miller, *X-Planes*, pp. 256–261; "SV-5D PRIME Final Flight Test Summary," Martin Marietta Report ER-14465, Sept. 1967; William J. Normyle, "Manned Flight Tests to Seek Lifting-Body Technology," *Aviation Week* (May 16, 1966), pp. 64–75; John L. Vitelli and Richard P. Hallion, "Project PRIME: Hypersonic Reentry from Space," in Hallion, ed., *Hypersonic Revolution*, vol. 1, pp. 641, 648–649.

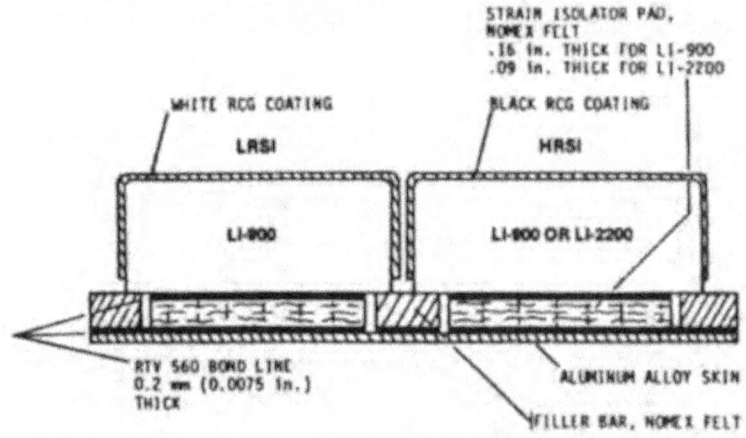

LRSI = Low Temperature Reusable Surface Insulation
HRSI = High Temperature Reusable Surface Insulation
RCG = Reaction Coated Glass
RTV = Room Temperature Vulcanizing Adhesive

Schematic of low- and high-temperature reusable surface insulation tiles, showing how they were bonded to the skin of the Space Shuttle. NASA.

tively of 500 and 800 miles, and the precision again was impressive. Flight 2 missed its aim point by less than 2 miles. Flight 3 missed by over 4 miles, but this still was within the allowed limit. Moreover, the terminal guidance radar had been inoperative, which probably contributed to the lack of absolute accuracy.[71]

A few years later, the Space Shuttle brought the question of whether its primary structure and skin should perhaps be built of titanium. Titanium offered a potential advantage because of its temperature resistance; hence, its thermal protection might be lighter. But the apparent weight saving was largely lost because of a need for extra insulation to protect the crew cabin, payload bay, and onboard systems. Aluminum could compensate for its lack of heat resistance because it had higher

---

71. Martin Marietta Report ER-14465, p. I-1; Vitelli and Hallion, "Project PRIME," in Hallion, ed., *Hypersonic Revolution*, vol. 1, pp. 694–702; B.K. Thomas, "USAF Nears Lifting Body Tests," *Aviation Week* (July 10, 1967), pp. 99–101.

thermal conductivity than titanium. It therefore could more readily spread its heat throughout the entire volume of the primary structure.

Designers expected to install RSI tiles by bonding them to the skin, and for this aluminum had a strong advantage. Both metals form thin layers of oxide when exposed to air, but that of aluminum is more strongly bound. Adhesive, applied to aluminum, therefore held tightly. The bond with titanium was considerably weaker and appeared likely to fail in operational use at around 500 °F. This was not much higher than the limit for aluminum, 350 °F, which showed that the temperature resistance of titanium did not lend itself to operational use.[72]

Turbine Blades

Turbine blades operate at speeds well below hypersonic, but this topic shares the same exotic metals that are used for flight structures at the highest speeds. It is necessary to consider how such blades use coatings to stay cool, an issue that represents another form of cooling. It also is necessary to consider directionally solidified and single-crystal castings for blades.

Britain's firm of Rolls-Royce has traditionally possessed a strong standing in this field, and *The Economist* has noted its activity:

> The best place to start is the surprisingly small, almost underwhelming, turbine blades that make up the heart of the giant engines slung beneath the wings of the world's biggest planes. These are not the huge fan blades you see when boarding, but are buried deep in the engines. Each turbine blade can fit in the hand like an oversized steak knife. At first glance it may not seem much more difficult to make. Yet they cost about $10,000 each. Rolls-Royce's executives like to point out that their big engines, of almost six tonnes, are worth their weight in silver—and that the average car is worth its weight in hamburger.[73]

Turbine blades are difficult to make because they have to survive high temperatures and huge stresses. The air inside big jet engines reaches about 2,900 °F in places, 750 degrees hotter than the melting point of

---

72. Scott Pace, "Engineering Design and Political Choice: The Space Shuttle 1969-72," master's thesis (Cambridge: Massachusetts Institute of Technology, May 1982), pp. 179–188.
73. "Briefing: Rolls-Royce," *The Economist* (Jan. 10, 2009), pp. 60–62.

the metal from which the turbine blades are made. Each blade is grown from a single crystal of alloy for strength and then coated with tough ceramics. A network of tiny air holes then creates a thin blanket of cool air that stops it from melting.

The study of turbine blades brings in the topic of thermal barrier coatings (TBC). By attaching an adherent layer of a material of low thermal conductivity to the surface of an internally cooled turbine blade, a temperature drop is induced across the thickness of the layer. This results in a drop in the temperature of the metal blade. Using this approach, temperature reductions of up to 340 °F at the metal surface have been estimated for 150-micron-thick yttria stabilized zirconia coatings. The rest of the temperature decrease is obtained by cooling the blade using air from the compressor that is ducted downstream to the turbine.

The cited temperature reductions reduce the oxidation rate of the bond coat applied to the blades and so delay failure by oxidation. They also retard the onset of thermal fatigue. One should note that such coatings are currently used only to extend the life of components. They are not used to increase the operating temperature of the engine.

Modern TBCs are required to not only limit heat transfer through the coating but to also protect engine components from oxidation and hot corrosion. No single coating composition appears able to satisfy these requirements. As a result, a "coating system" has evolved. Research in the last 20 years has led to a preferred coating system consisting of four separate layers to achieve long-term effectiveness in the high-temperature, oxidative, and corrosive environment in which the blades must function. At the bottom is the substrate, a nickel – or cobalt-based superalloy that is cooled from the inside using compressor air. Overlaying it is the bond coat, an oxidation-resistant layer with thickness of 75–150 microns that is typically of a NiCrAlY or NiCoCrAlY alloy. It essentially dictates the spallation failure of the blade. Though it resists oxidation, it does not avoid it; oxidation of this coating forms a third layer, the thermally grown oxide, with a thickness of 1 to 10 microns. It forms as Al_2O_3. The topmost layer, the ceramic topcoat, provides thermal insulation. It is typically of yttria-stabilized ZrO_2. Its thickness is characteristically about 300 microns when deposited by air plasma spray and 125 microns when deposited by electron beam physical vapor deposition (EB-PVD).[74]

---

74. Nitin P. Padture, Maurice Gelland, and Eric H. Jordan, "Thermal Barrier Coatings for Gas-Turbine Engine Applications," *Science*, vol. 296 (Apr. 12, 2002), pp. 280–284.

Yttria-stabilized zirconia has become the preferred TBC layer material for use in jet engines because of its low thermal conductivity and its relatively high thermal expansion coefficient, compared with many other ceramics. This reduces the thermal expansion mismatch with the metals of high thermal expansion coefficient to which it is applied. It also has good erosion resistance, which is important because of the entrainment of particles having high velocity in the engine gases. Robert Miller, a leading specialist, notes that NASA and the NACA, its predecessor, have played a leading role in TBC development since 1942. Flame-sprayed Rokide coatings, which extended the life of the X-15 main engine combustion chamber, represented an early success. Magnesia-stabilized zirconia later found use aboard the SR-71, allowing continuous use of the afterburner and sustained flight above Mach 3. By 1970, plasma-sprayed TBCs were in use in commercial combustors.[75]

These applications involved components that had no moving parts. For turbines, the mid-1970s brought the first "modern" thermal spray coating. It used yttria as a zirconia stabilizer and a bond coat that contained MCrAlY, and demonstrated that blade TBCs were feasible. C.W. Goward of Pratt & Whitney (P&W), writing of TBC experience with the firm's J75 engine, noted: "Although the engine was run at relatively low pressures, the gas turbine engine community was sufficiently impressed to prompt an explosive increase in development funds and programs to attempt to achieve practical utilization of the coatings on turbine airfoils."[76]

But tests in 1977 on the more advanced JT9D, also conducted at P&W, brought more mixed results. The early TBC remained intact on lower-temperature regions of the blade but spalled at high temperatures. This meant that further development was required. Stefan Stecura reported an optimum concentration of Y_2O_3 in ZrO_2 of 6–8 percent. This is still the state of the art. H.G. Scott reported that the optimum phase of zirconia was t'-ZrO_2. In 1987, Stecura showed that ytterbia-stabilized zirconia on a ytterbium-containing bond coat doubled the blade life and took it from 300 1-hour cycles to 600 cycles. Also at that

---

75. Robert A. Miller, "History of Thermal Barrier Coatings for Gas Turbine Engines," NASA TM-2009-215459 (2009).
76. G.W. Goward, "Seventeen Years of Thermal Barrier Coatings," in Department of Energy, *Proceedings of the Workshop on Coatings for Advanced Heat Engines, Castine, ME, July 27–30, 1987* (Washington, DC: U.S. Department of Energy, 1987).

time, P&W used a zirconia-yttria TBC to address a problem with endurance of vane platforms. A metallic platform, with no thermal barrier, showed burn-through and cracking from thermal-mechanical fatigue after 1,500 test cycles. Use of a TBC extended the service life to 18,000 hours or 2,778 test cycles and left platforms that were clean, uncracked, and unburned. P&W shared these results with NASA, which led to the TBC task in the Hot Section Technology (HOST) program. NASA collaborated with P&W and four other firms as it set out to predict TBC lifetimes. A preliminary NASA model showed good agreement between experiment and calculation. P&W identified major degradation modes and gave data that also showed good correlation between measured and modeled lives. Other important contributions came from Garrett Turbine Co. and General Electric. The late 1980s brought Prescribed Velocity Distribution (PVD) blades that showed failure when they were nearly out of the manufacturer's box. EV-PVD blades resolved this issue and first entered service in 1989 on South African Airways 747s. They flew from Johannesburg, a high-altitude airport with high mean temperatures where an airliner needed a heavy fuel load to reach London. EV-PVD TBCs remain the coating of choice for first-row blades, which see the hottest combustion gases. TBC research continues to this day, both at NASA and its contractors. Fundamental studies in aeronautics are important, with emphasis on erosion of turbine components. This work has been oriented toward rotorcraft and has brought the first EV-PVD coating for their blades. There also has been an emphasis on damping of vibration amplitudes. A new effort has dealt with environmental barrier coatings (EBCs), which Miller describes as "ceramic coatings, such as SiC, on top of ceramics."[77]

Important collaborations have included work on coatings for diesels, where thick TBCs permit higher operating temperatures that yield increased fuel economy and cleaner exhaust. This work has proceeded with Caterpillar Tractor Co. and the Army Research Laboratory.[78]

Studies of supersonic engines have involved cooperation with P&W and GE, an industrial interaction that Miller described as "a useful

---

77. Author interview with Robert A. Miller, June 10, 2009; see also Miller, NASA TM-2009-215459.
78. Dong-ming Zhu and Miller, "Investigation of Thermal High Cycle and Low Cycle Fatigue Mechanisms of Thick Thermal Barrier Coatings," *Materials Science and Engineering*, vol. A245 (1998), pp. 212–223.

reality check."[79] NASA has also pursued the Ultra Efficient Engine Technology program. Miller stated that it has not yet introduced engines for routine service but has led to experimental versions. This work has involved EBCs, as well as a search for low thermal conductivity. The latter can increase engine-operating temperatures and reduce cooling requirements, thereby achieving higher engine efficiency and lower emissions. At NASA Glenn, Miller and Dong-ming Zhu have built a test facility that uses a 3-kilowatt CO_2 laser with wavelength of 10.6 microns. They also have complemented conventional ZrO_2-Y_2O_3 coatings with other rare-earth oxides, including Nd_2O_3-Yb_2O_3 and Gd_2O_3-Yb_2O_3.[80]

Can this be reduced further? A promising approach involves development of new deposition techniques that give better control of TBC pore morphology. Air plasma spray deposition creates many intersplat pores between initially molten droplets, in what Miller described as "a messy stack of pancakes." By contrast, TBC layers produced by EB-PVD have a columnar microstructure with elongated intercolumnar pores that align perpendicular to the plane of the coating. Alternate deposition methods include sputtering, chemical vapor deposition (CVD), and sol-gel approaches. But these approaches involve low deposition rates that are unsuitable for economic production of coated blades. CVD and sol-gel techniques also require the use of dangerous and costly precursor materials. In addition, none of these approaches permit the precise control and manipulation of pore morphology. Thus, improved deposition methods that control this morphology do not now exist.

Blade Fabrication

The fabrication of turbine blades represents a related topic. No blade has indefinite life, for blades are highly stressed and must resist creep while operating under continuous high temperatures. Table 3 is taken from the journal *Metallurgia* and summarizes the stress to cause rupture in both wrought- and investment-case nickel-base superalloy.[81]

---

79. Miller interview.
80. Zhu and Miller, "Development of Advanced Low Conductivity Thermal Barrier Coatings," *International Journal of Applied Ceramic Technology*, vol. I (2004), pp. 86–94.
81. R. McCallum, "Casting Critical Components," *Superalloys Source Book*, pp. 286–291.

TABLE 3: STRESS TO CAUSE FAILURE OF VARIOUS ALLOYS		
TYPE OF ALLOY	STRESS TO CAUSE FAILURE AFTER:	
Wrought Alloys:	100 hours at 1400 °F, MPa	50 hours at 1750 °F, MPa
Nimonic 80	340	48
Nimonic 105	494	127
Nimonic 115	571	201
Investment-Cast Alloys:		
IN100	648	278
B1914	756	262
Mar-M246	756	309

An important development involved the introduction of directionally solidified (d.s.) castings. Their advent, into military engines in 1969 and commercial engines in 1974, brought significant increases in allowable metal temperatures and rotor speeds. D.s. blades and vanes were fabricated by pouring molten superalloy into a ceramic mold seated on a water-cooled copper chill plate. Grains nucleate on the chill surface and grow in a columnar manner parallel to a temperature gradient. These columnar grains fill the mold and solidify to form the casting.[82]

A further development involved single-crystal blades. More was required here than development of a solidification technique; it was necessary to consider as well the entire superalloy. It was to achieve a high melting temperature by containing no grain boundary-strengthening elements such as boron, carbon, hafnium, and zirconium. It would achieve high creep strength with a high gamma-prime temperature. A high temperature for solution heat treatment would also provide improved properties.

The specialized Alloy 454 had the best properties. It showed a complete absence of all grain boundary-strengthening elements and made significant use of tantalum, which suppressed a serious casting defect known as "freckling." Chromium and aluminum were included to protect against oxidation and hot corrosion. It had a composition of 12Ta+4W+10Cr+5Al+1.5Ti+5Co, balance Ni.

Single-crystal blades were fabricated using a variant of the cited d.s. arrangement. Instead of having the ceramic mold rest directly on the chill

---

82. J.E. Northwood, "Improving Turbine Blade Performance by Solidification Control," *Superalloys Source Book*, pp. 292–296.

Case 9 | High-Temperature Structures and Materials

A hypersonic scramjet configuration developed by Langley experts in the 1970s. The sharply swept double-delta layout set the stage for the National Aero-Space Plane program. NASA.

plate, it was separated from this plate by a helical single-crystal selector. A number of grains nucleated at the bottom of the selector, but most of them had their growth cut off by its walls, and only one grain emerged at the top. This grain was then allowed to grow and fill the entire mold cavity.

Creep-rupture tests showed that Alloy 454 had a temperature advantage of 75 to 125 °F over d.s. MAR-M200 + Hf, the strongest production-blade alloy. A 75 °F improvement in metal temperature capability corresponds to threefold improvement in life. Single-crystal Alloy 454 thus was chosen as the material for the first-stage turbine blades of the JT9D-7R4 series of engines that were to power the Boeing 767 and the Airbus A310 aircraft, with engine certification and initial production shipments occurring in July 1980.[83]

Materials Research and Development: The NASP Legacy
The National Aero-Space Plane (NASP) program had much to contribute to metallurgy, with titanium being a particular point. It has lately come to the fore in aircraft construction because of its high strength-to-density ratio, high corrosion resistance, and ability to withstand moderately

---

83. M. Gell, D.H. Duhl, and A.F. Giami, "The Development of Single Crystal Turbine Blades," *Superalloys Source Book*, pp. 297–306.

high temperatures without creeping. Careful redesign must be accomplished to include it, and it appears only in limited quantities in aircraft that are out of production. But newer aircraft have made increasing use of it, including the two largest manufacturers of medium- and long-range commercial jetliners, Boeing and Airbus, whose aircraft and their weight of titanium is shown in Table 4.

TABLE 4:	
BOEING AND AIRBUS AIRCRAFT MAKING SIGNIFICANT USE OF TITANIUM	
AIRCRAFT (INCLUDING THEIR ENGINES)	WEIGHT OF TI, IN METRIC TONS
Boeing 787	134
Boeing 777	59
Boeing 747	45
Boeing 737	18
Airbus A380	145
Airbus A350	74
Airbus A340	32
Airbus A330	18
Airbus A320	12

These numbers offer ample evidence of the increasing prevalence of titanium as a mainstream (and hence no longer "exotic") aviation material, mirroring its use in other aspects of the commercial sector. For example, in the 1970s, the Parker Pen Company used titanium in its T-1 line of ball pens and rollerballs, which it introduced in 1971. Production stopped in 1972 because of the high cost of the metal. But hammerheads fabricated of titanium entered service in 1999. Their light weight allows a longer handle, which increases the speed of the head and delivers more energy to the nail while decreasing arm fatigue. Titanium also substantially diminishes the shock transferred to the user because it generates much less recoil than a steel hammerhead.

In advancing titanium's use, techniques of powder metallurgy have been at the forefront. These methods give direct control of the microstructure of metals by forming them from powder, with the grains of powder sintering or welding together by being pressed in a mold at high temperature. A manufacturer can control the grain size independently of any heat-treating process. Powder metallurgy also overcomes restrictions on alloying by mixing in the desired additives as powdered ingredients.

Case 9 | High-Temperature Structures and Materials

Several techniques exist to produce the powders. Grinding a metal slab to sawdust is the simplest, though it yields relatively coarse grains. "Splat-cooling" gives better control. It extrudes molten metal onto the chilled rim of a rotating wheel that cools it instantly into a thin ribbon. This represents a quenching process that produces a fine-grained microstructure in the metal. The ribbon then is chemically treated with hydrogen, which makes it brittle so that it can be ground into a fine powder. Heating the powder then drives off the hydrogen.

The Plasma Rotating Electrode Process, developed by the firm of Nuclear Metals, has shown particular promise. The parent metal is shaped into a cylinder that rotates at up to 30,000 revolutions per minute (rpm) and serves as an electrode. An electric arc melts the spinning metal, which throws off droplets within an atmosphere of cool inert helium. The droplets plummet in temperature by thousands of degrees within milliseconds and their microstructures are so fine as to approach an amorphous state. Their molecules do not form crystals, even tiny ones, but arrange themselves in formless patterns. This process, called "rapid solidification," has brought particular gains in high-temperature strength.

Standard titanium alloys lose strength at temperatures above 700 to 900 °F. By using rapid solidification, McDonnell-Douglas raised this limit to 1,100 °F prior to 1986, when NASP got underway. Philip Parrish, the manager of powder metallurgy at the Defense Advanced Research Projects Agency (DARPA), notes that his agency spent some $30 million on rapid-solidification technology in the decade after 1975. In 1986, he described it as "an established technology. This technology now can stand along such traditional methods as ingot casting or drop forging."[84]

Eleven-hundred degrees nevertheless was not enough. But after 1990, the advent of new baseline configurations for the X-30 led to an appreciation that the pertinent areas of the vehicle would face temperatures no higher than 1,500 °F. At that temperature, advanced titanium alloys could serve in metal matrix composites (MMCs), with thin-gauge metals being reinforced with fibers.

A particular composition came from the firm of Titanium Metals and was designated Beta-21S. That company developed it specifically for the X-30 and patented it in 1989. It consisted of Ti along with 15Mo+2.8Cb+3Al+0.2Si. Resistance to oxidation proved to be its strong

---

84. Heppenheimer, "Making Planes from Powder," *High Technology* (Sept. 1986), pp. 54–55; author interview of Philip Parrish, Mar. 21, 1986.

suit, with this alloy showing resistance that was two orders of magnitude greater than that of conventional aircraft titanium. Tests showed that it could also be exposed repeatedly to leaks of gaseous hydrogen without being subject to embrittlement. Moreover, it lent itself readily to being rolled to foil-gauge thicknesses of 4 to 5 mils in the fabrication of MMCs.[85]

There also was interest in using carbon-carbon for primary structure. Here the property that counted was not its heat resistance but its light weight. In an important experiment, the firm of LTV fabricated half of an entire wing box of this material. An airplane's wing box is a major element of aircraft structure that joins the wings and provides a solid base for attachment of the fuselage fore and aft. Indeed, one could compare it with the keel of a ship. It extends to left and right of the aircraft centerline, with LTV's box constituting the portion to the left of this line. Built at full scale, it represented a hot-structure wing proposed by General Dynamics. It measured 5 by 8 feet, with a maximum thickness of 16 inches. Three spars ran along its length, five ribs were mounted transversely, and the complete assembly weighed 802 pounds.

The test plan called for it to be pulled upward at the tip to reproduce the bending loads of a wing in flight. Torsion or twisting was to be applied by pulling more strongly on the front or rear spar. The maximum load corresponded to having the X-30 execute a pullup maneuver at Mach 2.2 with the wing box at room temperature. With the ascent continuing and the vehicle undergoing aerodynamic heating, the next key event brought the maximum difference in the temperatures of the top and bottom of the wing box, with the former being at 994 °F and the latter being at 1,671 °F. At that moment, the load on the wing box corresponded to 34 percent of the Mach 2.2 maximum. Farther in the flight the wing box was to reach peak temperature, 1,925 °F, on the lower surface. These three points were to be reproduced through mechanical forces applied at the ends of the spars and through the use of graphite heaters.

But several important parts delaminated during their fabrication, which seriously compromised the ability of the wing box to bear its specified loads. Plans to impose the peak or Mach 2.2 load were abandoned, with the maximum planned load being reduced to the 34 percent

---

85. J. Sorensen, "Titanium Matrix Composites—NASP Materials and Structures Augmentation Program," AIAA Paper 90-5207 (1990); Stanley W. Kandebo, "Boeing 777 to Incorporate New Alloy Developed for NASP," *Aviation Week* (May 3, 1993), p. 36; "NASP Materials and Structures Program: Titanium Matrix Composites," McDonnell-Douglas, DTIC ADB-192559, (Dec. 31, 1991).

associated with the maximum temperature difference. For the same reason, the application of torsion was deleted from the test program. Amid these reductions in the scope of the structural tests, two exercises went forward during December 1991. The first took place at room temperature and successfully reached the mark of 34 percent without causing further damage to the wing box.

The second test, a week later, reproduced the condition of peak temperature difference while briefly applying the calculated load of 34 percent. The plan then called for further heating to the peak temperature of 1,925 °F. As the wing box approached this value, a difficulty arose because of the use of metal fasteners in its assembly. Some were made from coated columbium and were rated for 2,300 °F, but most were a nickel alloy that had a permissible temperature of 2,000 °F. However, an instrumented nickel-alloy fastener overheated and reached 2,147 °F. The wing box showed a maximum temperature of 1,917 °F at that moment, and the test was terminated because the strength of the fasteners now was in question. This test nevertheless counted as a success because it had come within 8 degrees of the specified temperature.[86]

Both tests thus were marked as having achieved their goals, but their merits were largely in the mind of the beholder. The entire project would have been far more impressive if it had avoided delamination, had successfully achieved the Mach 2.2 peak load, and had subjected the wing box to repeated cycles of bending, torsion, and heating. This effort stood as a bold leap toward a future in which carbon-carbon might take its place as a mainstream material, but it was clear that this future would not arrive during the NASP program. However, the all-carbon-composite airplane, as distinct from one of carbon-carbon, has now become a reality. Carbon alone has high temperature resistance, whereas carbon composite burns or melts readily. The airplane that showcases carbon composites is the White Knight 2, built by Burt Rutan's Scaled Composites firm as part of the Virgin Galactic venture that is to achieve commercial space flight. As of this writing, White Knight 2 is the world's largest all-carbon-composite aircraft in service; even its control wires are carbon composite. Its 140-foot-span wing is the longest single carbon composite aviation component ever fabricated.[87]

---

86. John Bradley, "Test and Evaluation Report for Carbon/Carbon Wing Box Component." General Dynamics, DTIC ADB-191627, (Feb. 21, 1992).
87. Leonard David, "Lift for Public Space Travel," *Aerospace America* (Feb. 2009), pp. 24–29.

Far below this rarefied world of transatmospheric tourism, carbon composites are becoming the standard material for commercial aviation. Aluminum has held this role up to Mach 2 since 1930, but after 80 years, Boeing is challenging this practice with its 787 airliner. By weight, the 787 is 50 percent composite, 20 percent aluminum, 15 percent titanium, 10 percent steel, and 5 percent other. The 787 is 80 percent composite by volume. Each of them contains 35 tons of composite reinforced with 23 tons of carbon fiber. Composites are used on fuselage, wings, tail, doors, and interior. Aluminum appears at wing and tail leading edges, with titanium used mainly on engines. The extensive application of composites promotes light weight and long range. The 787 can fly nonstop from New York City to Beijing. The makeup of the 787 contrasts notably with that of the Boeing 777. Itself considered revolutionary when it entered service in 1995, it nevertheless had a structure that was 50 percent aluminum and 12 percent composite. The course to all-composite construction is clear, and if the path is not yet trodden, nevertheless, the goal is clearly in sight. As in 1930, when all-metal structures first predominated in American commercial aviation, the year 2010 marks the same point for the evolution of commercial composite aircraft.

The NASP program also dealt with beryllium. This metal had only two-thirds the density of aluminum and possessed good strength, but its temperature range was restricted. The conventional metal had a limit of some 850 °F, while an alloy from Lockheed called Lockalloy, which contained 38 percent aluminum, was rated only for 600 °F. It had never become a mainstream material like titanium, but, for the X-30, it offered the advantage of high thermal conductivity. Work with titanium had greatly increased its temperatures of use, and there was hope of achieving similar results with beryllium.

Initial efforts used rapid-solidification techniques and sought temperature limits as high as 1,500 °F. These attempts bore no fruit, and from 1988 onward the temperature goal fell lower and lower. In May 1990, a program review shifted the emphasis away from high-temperature formulations toward the development of beryllium as a metal suitable for use at cryogenic temperatures. Standard forms of this metal became unacceptably brittle when only slightly colder than –100 °F, but cryoberyllium proved to be out of reach as well. By 1992, investigators were working with ductile alloys of beryllium and were sacrificing all prospect of use at temperatures beyond a few hundred degrees but were winning only modest improvements in low-temperature capability. Terence Ronald, the NASP materials director, wrote in 1995 of rapid-solidification versions with temperature limits as low as 500 °F, which was not what the X-30 needed to reach orbit.[88]

In sum, the NASP materials effort scored a major advance with Beta-21S, but the genuinely radical possibilities failed to emerge. These included carbon-carbon as primary structure along with alloys of beryllium that were rated for temperatures well above 1,000 °F. The latter, if available, might have led to a primary structure with the strength and temperature resistance of Beta-21S but with less than half the weight. Indeed, such weight savings would have ramified throughout the entire design, leading to a configuration that would have been smaller and lighter overall.

Generally, work with materials fell well short of its goals. In dealing with structures and materials, the contractors and the National Program Office established 19 program milestones that were to be accomplished by September 1993. A General Accounting Office program review, issued in December 1992, noted that only six of them would indeed be completed.[89] This slow progress encouraged conservatism in drawing up the bill of materials, but this conservatism carried a penalty.

When the scramjets faltered in their calculated performance and the X-30 gained weight while falling short of orbit, designers lacked recourse to new and very light materials, such as beryllium and carbon-carbon, that might have saved the situation. With this, NASP spiraled to its end. The future belonged to other less ambitious but more attainable programs, such as the X-43 and X-51. They, too, would press the frontier of aerothermodynamic structural design, as they pioneered the hypersonic frontier.

---

88. Rockwell International, "High Conductivity Composites: Executive Summary, Copper Materials, Beryllium Materials, Coatings, Ceramic Materials and Joining," DTIC ADB-191898 (Mar. 1993); Terence Ronald, "Status and Applications of Materials Developed for NASP," AIAA Paper 95-6131 (1995).

89. Frank C. Conahan, "National Aero-Space Plane: Restructuring Further Research and Development Efforts," General Accounting Office Report NSIAD-93-71 (Dec. 3, 1992), table, p. 20.

Recommended Additional Readings
Reports, Papers, Articles, and Presentations:

H. Julian Allen and A.J. Eggers, Jr., "A Study of the Motion and Aerodynamic Heating of Ballistic Missiles Entering the Earth's Atmosphere at High Supersonic Speeds," NACA TR-1381 (1953).

Anon., "How Mercury Capsule Design Evolved," *Aviation Week* (Sept. 21, 1959).

Anon., "Fourth ASSET Glider Gathers Data," *Aviation Week* (Nov. 2, 1964).

Anon., "Sortie Module May Cut Experiment Cost," *Aviation Week* (Jan. 17, 1972).

F.M. Anthony, R.R. Fisher, and R.G. Helenbrook, "Selection of Space Shuttle Thermal Protection Systems," AIAA Paper 71-443 (1971).

Paul R. Becker, "Leading-Edge Structural Material System of the Space Shuttle," *American Ceramic Society Bulletin*, vol. 60, No. 11 (1981), pp. 1210–1214.

C.E. Brown, W.J. O'Sullivan, and C.H. Zimmerman, "A Study of the Problems Related to High Speed, High Altitude Flight," NACA Langley Research Center, 1953.

Frank C. Conahan, "National Aero-Space Plane: Restructuring Further Research and Development Efforts," General Accounting Office Report NSIAD-93-71 (Dec. 3, 1992).

William Cowie, "Utilization of Refractory Metals on the X-20A (Dyna-Soar)," Air Force Systems Command, June 1964, DTIC AD-609169.

Leonard David, "Lift for Public Space Travel," *Aerospace America* (Feb. 2009), pp. 24–29.

Richard A. DeMeis, "The Trisonic Titanium Republic," *Air Enthusiast*, No. 7 (July–Sept. 1978), pp. 198–213.

Grumman Corporation, "Space Shuttle Program Definition: Phase B Extension Final Report," B35-43 RP-33 (Mar. 15, 1972).

T.A. Heppenheimer, "Making Planes from Powder," *High Technology* (Sept. 1986), pp. 54–55.

William S. Hieronymus, "Two Reusable Materials Studied for Orbiter Thermal Protection," *Aviation Week* (Mar. 27, 1972), p. 48.

Stanley W. Kandebo, "Boeing 777 to Incorporate New Alloy Developed for NASP," *Aviation Week* (May 3, 1993), p. 36.

L.J. Korb, C.A. Morant, R.M. Calland, and C.S. Thatcher, "The Shuttle Orbiter Thermal Protection System," *American Ceramic Society Bulletin*, vol. 60 (1981), pp. 1188–1193.

Martin Marietta, "SV-5D PRIME Final Flight Test Summary," Report ER-14465 (Sept. 1967).

McDonnell Aircraft Corp., "ASSET ASV-3 Flight Test Report," Jan. 4, 1965, DTIC AD-357523.

McDonnell Aircraft Corp., "ASSET ASV-4 Flight Test Report," June 25, 1965, DTIC AD-366546.

McDonnell-Douglas, "External LH2 Tank Study Final Report," MDC E076-1, June 30, 1971.

McDonnell-Douglas, "Interim Report to OMSF: Phase B System Study Extension," Sept. 1, 1971.

McDonnell-Douglas, "NASP Materials and Structures Program: Titanium Matrix Composites," Dec. 31, 1991.

Howard J. Middendorf, "Materials and Processes for X-20A (Dyna-Soar)," Air Force Systems Command, June 1964, DTIC AD-449685.

Robert A. Miller, "History of Thermal Barrier Coatings for Gas Turbine Engines," NASA TM-2009-215459 (2009).

Dale D. Myers, "The Navaho Cruise Missile: A Burst of Technology," *Acta Astronautica*, vol. 26 (Nov. 8–10, 1992), pp. 741–748.

William J. Normyle, "Manned Flight Tests to Seek Lifting-Body Technology," *Aviation Week* (May 16, 1966), pp. 64–75.

Scott Pace, "Engineering Design and Political Choice: The Space Shuttle 1969–72," master's thesis (Cambridge: Massachusetts Institute of Technology, May 1982).

Nitin P. Padture, Maurice Gelland, and Eric H. Jordan, "Thermal Barrier Coatings for Gas-Turbine Engine Applications," *Science,* vol. 296 (Apr. 12, 2002), pp. 280–284.

Lee L. Peterson, "Evaluation Report on X-7A," Report AFMDC ADJ 57-8184, Oct. 3, 1957.

Joel W. Powell, "Thor-Able and Atlas-Able," *Journal of the British Interplanetary Society* (May 1984), pp. 219–225.

William A. Ritchie, "Evaluation Report on X-7A (System 601B)," Report AFMDC DAS 58-8129, Jan. 1959.

Rockwell International, "High Conductivity Composites: Executive Summary, Copper Materials, Beryllium Materials, Coatings, Ceramic Materials and Joining" (Mar. 1993), DTIC ADB-191898.

Terence Ronald, "Status and Applications of Materials Developed for NASP," AIAA Paper 95-6131 (1995).

P.H. Rose, "Physical Gas Dynamics Research at the Avco Research Lab," NATO Advisory Group for Aeronautical Research and Development Report 145, July 1957.

P.H. Rose and W.I. Stark, "Stagnation Point Heat-Transfer Measurements in Dissociated Air," *Journal of the Aeronautical Sciences* (Feb. 1958), pp. 86–97.

R.L. Schleicher, "Structural Design of the X-15," *Journal of the Royal Aeronautical Society* (Oct. 1963), pp. 618–636.

Wilson B. Schramm, Ronald P. Banas, and Y. Douglas Izu, "Space Shuttle Tile—The Early Lockheed Years," *Lockheed Horizons*, No. 13 (1983), pp. 2–15.

Wilson Schramm, "HRSI and LRSI—The Early Years," *American Ceramic Society Bulletin*, vol. 60 (1981), pp. 1194–1195.

Alvin Seiff and H. Julian Allen, "Some Aspects of the Design of Hypersonic Boost-Glide Aircraft," NACA RM-A55E26 (1955).

J. Sorensen, "Titanium Matrix Composites—NASP Materials and Structures Augmentation Program," AIAA Paper 90-5207 (1990).

Leo Steg, "Materials for Re-Entry Heat Protection of Satellites," *American Rocket Society Journal* (Sept. 1960), pp. 815–822.

George W. Sutton, "The Initial Development of Ablation Heat Protection, An Historical Survey," *Journal of Spacecraft and Rockets* (Jan.–Feb. 1982), pp. 3–11.

B.K. Thomas, "USAF Nears Lifting Body Tests," *Aviation Week* (July 10, 1967), pp. 99–101.

Richard C. Thuss, Harry G. Thibault, and Arnold Hiltz, "The Utilization of Silica Based Thermal Insulation for the Space Shuttle Thermal Protection System," CASI 72A-10764, Oct. 1971.

Vought Missiles and Space Company, "Technical Overview: Oxidation Resistant Carbon-Carbon for the Space Shuttle," n.d. (c. 1970).

Michael L. Yaffee, "Program Changes Boost Grumman Shuttle," *Aviation Week* (July 12, 1963), pp. 36–39.

Richard E. Young, Martha A. Smith, and Charles K. Sobeck, "Galileo Probe: In Situ Observations of Jupiter's Atmosphere," *Science* (May 10, 1996), pp. 837–838.

U.S. Air Force Aeronautical Systems Division staff, *Proceedings of 1962 X-20A (Dyna-Soar) Symposium*, vol. 3, *Structures and Materials* (Wright Patterson AFB: HQ ASD, 1963).

U.S. Air Force Systems Command staff, "Advanced Technology Program: Technical Development Plan for Aerothermodynamic/Elastic Structural System Environmental Tests (ASSET)," HQ AFSC, Sept. 9, 1963.

U.S. Army Ballistic Missile Agency staff, *Re-Entry Studies*, vol. 1 (Redstone Arsenal, AL: HQ ABMA, Nov. 25, 1958).

U.S. Department of Energy, *Proceedings of the Workshop on Coatings for Advanced Heat Engines, Castine, ME, July 27–30, 1987* (Washington, DC: HQ DoE, 1987).

U.S. National Aeronautics and Space Administration staff, *NASA Space Shuttle Technology Conference*, vol. 2, *Structures and Materials*, NASA TM-X-2273 (1971).

Dong-ming Zhu and Robert A. Miller, "Investigation of Thermal High Cycle and Low Cycle Fatigue Mechanisms of Thick Thermal Barrier Coatings," *Materials Science and Engineering*, vol. A245 (1998), pp. 212–223.

Dong-ming Zhu and Robert A. Miller, "Development of Advanced Low Conductivity Thermal Barrier Coatings," *International Journal of Applied Ceramic Technology*, vol. 1 (2004), pp. 86–94.

Books and Monographs:

Morton Alperin and Hollingsworth F. Gregory, eds., *Vistas in Astronautics*, vol. 2 (New York: Pergamon Press, 1959).

John Becklake, ed., *History of Rocketry and Astronautics*, vol. 17 of the *AAS History Series* (San Diego: Univelt, 1995).

Deborah D.L. Chung, *Carbon Fiber Composites* (Boston: Butterworth-Heinemann, 1994).

Erik M. Conway, *High-Speed Dreams* (Baltimore: Johns Hopkins University Press, 2005).

Paul F. Crickmore, *Lockheed SR-71 Blackbird* (London: Osprey Publishing Ltd., 1986).

Matthew J. Donachie, Jr., *Superalloys Source Book* (Metals Park, OH: American Society for Metals, 1984).

Richard P. Hallion, ed., *The Hypersonic Revolution: Case Studies in the History of Hypersonic Technology*, vols. 1 and 2 (Bolling AFB: USAF, 1998).

James R. Hansen, *Engineer in Charge: A History of the Langley Aeronautical Laboratory, 1917–1958*, NASA SP-4305 (Washington: GPO, 1987).

T.A. Heppenheimer, *Turbulent Skies: The History of Commercial Aviation* (New York: John Wiley, 1995).

T.A. Heppenheimer, *The Space Shuttle Decision*, NASA SP-4221 (Washington, DC: NASA, 1999).

T.A. Heppenheimer, *Development of the Space Shuttle, 1972–1981* (Washington: Smithsonian Institution Press, 2002).

Dennis Jenkins, *Space Shuttle* (Stillwater, MN: Voyageur Press, 2001).

Willy Ley, *Rockets, Missiles, and Space Travel* (New York: Viking, 1957).

Jay Miller, *The X-Planes, X-1 to X-45* (North Branch, MN: Specialty Press, 2001).

NASA, *Columbia Accident Investigation Report* (Washington: GPO, 2003).

Jacob Neufeld, *The Development of Ballistic Missiles in the United States Air Force, 1945–1960* (Washington: Office of Air Force History, 1990).

David Robarge, *Archangel: CIA's Supersonic A-12 Reconnaissance Aircraft* (McLean, VA: CIA Center for the Study of Intelligence, 2007).

Kevin C. Ruffner, *Corona: America's First Satellite Program* (McLean, VA: Central Intelligence Agency, 1995).

THIS PAGE INTENTIONALLY BLANK

The AFTI/F-16, shown here on a 1991 test flight, flew 15 years and over 700 research flights at NASA Dryden. NASA.

CASE 10
Fly-By-Wire: Making the Electric Jet

Albert C. Piccirillo

Fly-by-wire (FBW) technology pioneered by NASA has enabled the design of highly unconventional airframe configurations. Continuing NASA FBW research has validated integrated digital propulsion and flight control systems. FBW has been applied to civil aircraft, improving their safety and efficiency. Lessons learned from NASA FBW research have been transferred to maritime design, and Agency experts have supported the U.S. Navy in developing digital electronic ship control.

THE EVOLUTION OF ADVANCED AIRCRAFT equipped with computerized flight and propulsion control systems goes back a long way and involved many players, both in the United States and internationally. During the Second World War, use of electronic sensors and subsystems began to become pervasive, adding new mission capabilities as well as increasing complexity. Autopilots were coupled to flight control systems, and electric trim was introduced. Hydraulically or electrically boosted flight control surfaces, along with artificial feel and stability augmentation systems, soon followed as did electronic engine control systems. Most significantly, the first uses of airborne computers emerged, a trend that soon resulted in their use in aircraft and missiles for their flight and mission control systems. Very early on, there was a realization that traditional mechanical linkages between the cockpit flight controls and the flight control surfaces could be replaced by a computer-controlled fly-by-wire (FBW) approach in which electric signals were transmitted from the pilot's controls to the control surface actuators by wire. This approach was understood to have the potential to reduce mechanical complexity, lower weight, and increase safety and reliability. In addition, the processing power of the computer could be harnessed to enable unstable aircraft designs to be controlled. Properly tailored, these unstable designs could enable new aircraft concepts to be implemented that could fully exploit the advantages of active flight control. Such aircraft would be more maneuverable and lighter, have better range, and also allow for the fully integrated control of aircraft, propulsion, navigation, and mission systems to optimize overall capability.

Two significant events unfolded during the 1960s that fostered the move to electronic flight control systems. The space race was a major influence, with most space systems relying on computerized fly-by-wire control systems for safe and effective operation. The Vietnam war provided a strong impetus for the development of more survivable aircraft systems as well for new aircraft with advanced performance features. The National Aeronautics and Space Administration (NASA) and the Air Force aggressively responded to these challenges and opportunities, resulting in the rapid transition of digital computer technology from the space program into aircraft fly-by-wire applications as exemplified by the Digital Fly-By-Wire F-8 and the AFTI/F-16 research programs. Very quickly after, a variety of flight research programs were implemented. These programs provided the basis for development and fielding of numerous military and civil aircraft equipped with advanced digital fly-by-wire flight control systems. On the civil aviation side, safety has been improved by preventing aircraft flight envelope limitations from being exceeded. Operating efficiency has been greatly enhanced and major weight savings achieved from fly-by-wire and related electronic flight control system components. Integrated flight and propulsion control systems precisely adjust throttles and fuel tank selections. Rudder trim drag because of unbalanced engine thrust is reduced. Fuel is automatically transferred between tanks throughout the aircraft to optimize center of gravity during cruise flight, minimizing elevator trim drag. In the case of advanced military aircraft, electronically controlled active flight, propulsion, and mission systems have been fully integrated, providing revolutionary improvements in capabilities. Significantly, new highly unstable aircraft configurations are providing unprecedented levels of mission performance along with very low radar signatures, capabilities that have been enabled by exploiting digital fly-by-wire flight and propulsion control systems pioneered in NASA.

Aircraft Flight Control: Beginnings to the 1950s

Early aviation pioneers gradually came to realize that an effective flight control system was necessary to control the forces and moments acting on an aircraft. The creation of such a system of flight control was one of the great accomplishments of the Wright brothers, who used a combination of elevator, rudder, and wing warping to achieve effective three-axis flight control in their 1903 Flyer. As aircraft became larger and faster, wing warping was replaced by movable ailerons to control motion around the roll axis. This basic three-axis flight control system is still used today. It enables the pilot to maneuver the aircraft about its pitch, yaw, and roll axes and, in conjunction with engine power adjustments, to control velocity and acceleration as well. For many generations after the dawn of manned, controllable powered flight, a system of direct mechanical linkages between the pilot's cockpit controls and the aircraft's control surfaces was used to both assist in stabilizing the aircraft as well as to change its flight path or maneuver. The pilot's ability to "feel" the forces being transmitted to his flight controls, especially during rapid maneuvering, was critically important, because many early aircraft were statically unstable in pitch with the pilot having to exert a constant stabilizing influence with his elevator control. Well into the First World War, many aircraft on both sides of the conflict had poor stability and control characteristics, issues that would continue to challenge aircraft designers well into the jet age. Wartime experience showed that adequate stability and positive aircraft handling qualities, coupled with high performance (as exemplified by parameters such as low wing loading, high power-to-weight ratio, and good speed, turn, and climb rates) played a major role in success in combat between fighters.

By the Second World War, electrically operated trim tabs located on aerodynamic control surfaces and other applications of electrical control and actuation were emerging.[1] However, as aircraft performance increased and airframes grew larger and heavier, it became increasingly harder for pilots to maneuver their aircraft because of high aero-

---

1. An early example was the innovative Focke-Wulf FW 190 fighter of World War II fame. Its landing gear and flaps were electrically controlled and actuated by small reliable high-torque electric motors. A variable incidence electrically controlled and actuated movable horizontal stabilizer maintained trim about the longitudinal (pitch) axis and was regularly used to relieve the heavy control forces encountered during steep dives (the FW 190 exceeded Mach 0.8 during testing). Another highly advanced FW 190 feature was the BMW 801 Kommandogerät (command equipment). This consisted of a mechanical-hydraulic analog computer that automatically adjusted engine fuel flow, variable pitch propeller setting, supercharger setting, fuel mixture, and ignition timing in response to pilot commands via the single throttle lever. This pioneering step in computerized integrated propulsion control systems greatly simplified engine control. See Albert C. Piccirillo, "Electric Aircraft Pioneer—The Focke-Wulf Fw 190," Society of Automotive Engineers (SAE) Paper 965631, Oct. 1996.

dynamic forces on the control surfaces. World War II piston engine fighters were extremely difficult to maneuver in pitch and roll and often became uncontrollable as compressibility effects were encountered as speeds approached about Mach 0.8. The introduction of jet propulsion toward the later stages of the Second World War further exacerbated this controllability problem. Hydraulically actuated fight control surfaces were introduced to assist the pilot in moving the control surfaces at higher speeds. These "boosted" control surface actuators were connected to the pilot's flight controls through a system consisting of cables, pulleys, and cranks, with hydraulic lines now also being routed through the airframe to power the control surface actuators.[2]

With a boosted flight control system, the pilot's movements of the cockpit flight controls opens or closes servo valves in the hydraulic system, increasing or decreasing the hydraulic pressure powering the actuators that move the aircraft control surfaces. Initially, hydraulic boost augmented the force transmitted to the control surfaces by the pilot; such an approach is referred to as partial boost. However, fully boosted flight controls quickly became standard on larger aircraft as well as on those aircraft requiring high maneuverability at high indicated airspeeds and Mach numbers. The first operational U.S. jet fighter, the Lockheed P-80 Shooting Star, used electric pitch and roll trim and had hydraulically boosted ailerons to provide roll effectiveness at higher airspeeds. The first jet-powered U.S. bomber to enter production, the four-engine North American B-45, flew for the first time in March 1947. It had hydraulically boosted flight control surfaces and an electrically actuated trim tab on the elevator that was used to maintain longitudinal trim. Despite their undeniable benefits, boosted flight control systems could also produce unanticipated hazards. The chief test pilot for the Langley Aeronautical Laboratory of what was then the National Advisory Committee for Aeronautics (NACA), Herbert "Herb" Hoover, was killed in the crash of a B-45 on August 14, 1952, when the aircraft disintegrated during a test mission near Barrowsville, VA.[3]

---

2. Control cables were replaced with rigid pushrods in German World War II flight control systems, for example in the FW 190 and the Fiesler Fi-103 pulsejet-powered cruise missile (the V-1). Pushrods minimized effects produced by control cables stretching under load.
3. "NASA Dryden Flight Research Center Pilot Biographies." http://www.nasa.gov/centers/dryden/news/Biographies/Pilots/index.html, accessed July 21, 2009. Herb Hoover was the first civilian pilot to fly faster than the speed of sound. He exceeded Mach 1 in the Bell XS-1 on Mar. 10, 1948.

As a NASA report noted: "The aerodynamic power of the trim-tab-elevator combination [on the B-45] was so great that, in the event of an inadvertent maximum tab deflection, the pilot's strength was insufficient to overcome the resulting large elevator hinge moments if the hydraulic boost system failed or was turned off. Total in-flight destruction of at least one B-45, the aircraft operated by NACA, was probably caused by this combination of circumstances that resulted in a normal load factor far greater than the design value."[4]

The Air Force/NACA Bell XS-1 rocket-powered research aircraft was equipped with an electrically trimmed adjustable horizontal stabilizer. It enabled the pilot to maintain pitch control as the conventional elevator lost effectiveness as the speed of sound was approached.[5] Using this capability, U.S. Air Force (USAF) Capt. Chuck Yeager exceeded Mach 1 in level flight in the XS-1 in October 1947, followed soon after by a North American XP-86 Sabre (although in a dive). Hydraulically boosted controls and fully movable horizontal tails were rapidly implemented on operational high-performance jet aircraft, an early example being the North American F-100.[6] To compensate for the loss of natural feedback to the pilot with fully boosted flight controls, various devices such as springs and bob weights were integrated into the flight control system. These "artificial feel" devices provided force feedback to the pilot's controls that was proportional to changes in airspeed and acceleration. Industry efforts to develop boosted fight control surfaces directly benefited from NACA flight-test efforts of the immediate postwar period.

Fly-By-Wire: The Beginnings

The Second World War witnessed the first applications of computer-controlled fly-by-wire flight control systems. With fly-by-wire, primary control surface movements were directed via electrical signals transmitted by wires rather than by the use of mechanical linkages. The German Army's A-4 rocket (the famous V-2 that postwar was the basis for both U.S. and Soviet efforts to move into space) used an electronic

---

4. Lawrence K. Loftin, "Quest for Performance: The Evolution of Modern Aircraft," Part II, ch. 12, Science and Technology Branch, NASA SP-468 (1985).
5. Richard P. Hallion, "The Air Force and the Supersonic Breakthrough," published in *Technology and the Air Force: A Retrospective Assessment*, Air Force History and Museums Program, Washington, DC, 1997.
6. Hallion, "On the Frontier: Flight Research at Dryden, 1946–1981," NASA SP-4303 (1984).

analog computer that modeled the differential equations governing the missile's flight control laws. The computer-generated electronic signals were transmitted by wire to direct movement of the actuators that drove graphite vanes located in the rocket motor exhaust. The thrust of the rocket engine was thus vectored as required to stabilize the V-2 missile at lower airspeeds until the aerodynamic control surfaces on the fins became effective.[7] Postwar, a similar analog computer-controlled fly-by-wire thrust vectoring approach was used in the U.S. Army Redstone missile, perhaps not surprisingly, because Redstone was predominantly designed by a team of German engineers headed by Wernher von Braun of V-2 fame. The Redstone would be used to launch the Mercury space capsule that carried Alan Shepard (the first American into space) in 1961.

The German Mistle (Mistletoe) composite aircraft of late World War II was probably the first example of the use of fly-by-wire for flight control in a manned aircraft application. Mistle consisted of a fighter (usually a Focke-Wulf FW 190) mounted on a support structure on a Junkers Ju 88 bomber.[8] The Ju 88 was equipped with a 3,500-pound warhead and was intended to be flown to the vicinity of its target by the FW 190 pilot, at which time he would separate from the bomber and evade enemy defenses while the Ju 88 flew into its target. Potentiometers at the base of the FW 190 pilot's control stick generated electrical commands that were transmitted via wire through the support structure to the bomber. These electrical commands activated electric motors that moved the system of pushrods leading to the Ju 88 control surfaces.[9]

Another electronic flight control system innovation related to the fly-by-wire concept had its origins in electronic feedback flight control research that began in Germany in the late 1930s and was published by Ernst Heinkel and Eduard Fischel in 1940. Their research was used in the 1944 development of a directional stability augmentation system for the Luftwaffe's heavily armed and armored Henschel Hs 129 ground

---

7. Michael J. Neufeld, "The Rocket and the Reich; Peenemünde and the Coming of the Ballistic Missile Era," The Smithsonian Institution, published by The Free Press, a division of Simon and Schuster, New York, 1995. James E. Tomayko, "Computers Take Flight: A History of NASA's Pioneering Digital Fly-by Wire Project," NASA SP-2000-4224 (2000).
8. Much like the Space Shuttle is carried on top of the NASA Boeing 747s when it is ferried from Edwards AFB to Cape Kennedy, FL.
9. Tomayko, "Blind Faith: The United States Air Force and the Development of Fly-By-Wire Technology," Technology and the Air Force: A Retrospective Assessment, Air Force History and Museums Program, Washington, DC, 1997.

attack aircraft to compensate for an inherent Dutch roll[10] instability that affected strafing accuracy with its large-caliber, low-rate-of-fire antitank cannon.[11] This consisted of modifying the rudder portion of the flight control system for dual mode operation. The rudder was split into two sections, with the lower portion directly linked to the pilot's flight controls. The upper section was electromechanically linked to a gyroscopic yaw rate sensor that automatically provided rudder corrections as yawing motions were detected.[12] This was the first practical aircraft yaw damper. Northrop incorporated electronic stability augmentation devices into its YB-49 flying wing bomber that first flew in late 1947 in an attempt to compensate for serious directional stability problems. After the war, the NACA Ames Aeronautical Laboratory conducted extensive flight research into artificial stability. An NACA-operated Grumman F6F-3 Hellcat was modified to incorporate roll and yaw rate servos that provided stability augmentation, with flight tests beginning in 1948. In the following years, a number of other aircraft were modified by the NACA at Ames for variable stability research, including several variants of the North American F-86.[13] By the 1950s, most high-performance swept wing jet-powered aircraft were designed with electronic stability augmentation devices.

Early Aircraft Fly-By-Wire Applications

By the 1950s, fully boosted flight controls were common, and the potential benefits of fly-by-wire were becoming increasingly apparent. Beginning during the Second World War and continuing postwar, fly-by-wire and power-by-wire flight control systems had been fielded in various target drones and early guided missiles.[14] However, most aircraft designers were reluctant to completely abandon mechanical linkages to

---

10. Dutch roll is a term commonly used to describe an out of phase combination of yawing and rolling. Dutch roll instability can be improved by the use of a yaw damper.
11. The Hs 129 was armed with a variety of cannon that could include a 75-millimeter antitank gun.
12. Norman C. Weingarten, "History of In-Flight Simulation & Flying Qualities Research at CALSPAN," *AIAA Journal of Aircraft*, vol. 42, no. 2, March/April 2005.
13. Paul F. Borchers, James A. Franklin, and Jay W. Fletcher, "Flight Research at Ames, 1940–1997: Fifty-Seven Years of Development and Validation of Aeronautical Technology," NASA SP-3300 (1998).
14. Rowland F. Pocock, "German Guided Missiles of the Second World War," Arco Publishing Company, Inc., New York, 1967. These included Matador, Snark, BOMARC, Rascal, plus many others developed in a number of countries. See Gavin D. Jenny, James W. Morris, and Vernon R. Schmitt, "Fly-by-Wire, A Historical and Design Perspective," The Society of Automotive Engineers, 1998.

flight control surfaces in piloted aircraft, an attitude that would undergo an evolutionary change over the next two decades as a result of a broad range of NACA–NASA, Air Force, and foreign research efforts.

Beginning in 1952, the NACA Langley Aeronautical Laboratory began an effort oriented to exploring various aspects of fly-by-wire, including the use of a side stick controller.[15] By 1954, flight-testing began with what was perhaps the first jet-powered fly-by-wire research aircraft, a modified former U.S. Navy Grumman F9F-2 Panther carrier-based jet fighter used as an NACA research aircraft. The primary objective of the NACA effort was to evaluate various automatic flight control systems, including those based on rate and normal acceleration feedback. Secondary objectives were to evaluate use of fly-by-wire with a side stick controller for pilot inputs. The existing F9F-2 hydraulic flight control system, with its mechanical linkages, was retained with the NACA designing an auxiliary flight control system based on a fly-by-wire analog concept. A small, 4-inch-tall side stick controller was mounted at the end of the right ejection seat armrest. The controller was pivoted at the bottom and was used for both lateral (roll) and longitudinal (pitch) control. Only 4 pounds of force were required for full stick deflection. The control friction normally present in a hydromechanical system was completely eliminated by the electrically powered system. Additionally, the aircraft's fuel system was modified to enable fuel to be pumped aft to destabilize the aircraft by moving the center of gravity rearward. Another modification was the addition of a steel container mounted on the lower aft fuselage. This carried 250 pounds of lead shot to further destabilize the aircraft. In an emergency, the shot could be rapidly jettisoned to restabilize the aircraft. Fourteen pilots flew the modified F9F-2, including NACA test pilots William Alford[16] and Donald L. Mallick.[17] Using only the side stick controller, the pilots conducted

---

15. An electrical side stick controller was developed in Germany during the Second World War to guide air-launched Henshel Hs 293 and Ruhrstahl Fritz X missiles against ships maneuvering at sea. Command signals were transmitted from the controller to the missiles via radio link. A similar guidance approach was used on the surface-to-air Wasserfall (Waterfall) command-guided antiaircraft missile. Examples of these weapons are on display at the National Air and Space Museum's Udvar-Hazy Center at Dulles Airport near Washington, DC, and at the National Museum of the U.S. Air Force in Dayton, OH.
16. Bill Alford was killed on Oct. 12, 1959, in the crash of a British Blackburn Buccaneer during a visiting test pilot evaluation flight from the Aircraft and Armament Experimental Establishment at Boscombe Down in the United Kingdom.
17. Donald L. Mallick, with Peter W. Merlin, "The Smell of Kerosene: A Test Pilot's Odyssey," NASA SP-4108 (2003).

takeoffs, stall approaches, acrobatics, and rapid precision maneuvers that included air-to-air target tracking, ground strafing runs, and precision approaches and landings. The test pilots quickly became used to flying with the side stick and found it comfortable and natural to use.[18]

In mid-1956, after interviewing aircraft flight control experts from the Air Force Wright Air Development Center's Flight Control Laboratory, *Aviation Week* magazine concluded:

> The time may not be far away when the complex mechanical linkage between the pilot's control stick and the airplane's control surface (or booster valve system) is replaced with an electrical servo system. It has long been recognized that this "fly-by-wire" approach offered attractive possibilities for reducing weight and complexity. However, airplane designers and pilots have been reluctant to entrust such a vital function to electronics whose reliability record leaves much to be desired.[19]

Even as the *Aviation Week* article was published, several noteworthy aircraft were under development that would incorporate various fly-by-wire approaches in their flight control systems. In 1956, the British Avro Vulcan B.2 bomber flew with a partial fly-by-wire system that operated in conjunction with hydraulically boosted, mechanically activated flight controls. The supersonic North American A-5 Vigilante Navy carrier-based attack bomber flew in 1958 with a pseudo-fly-by-wire flight control system. The Vigilante served the fleet for many years, but its highly complex design proved very difficult to maintain and operate in an aircraft carrier environment. By the mid-1960s, the General Dynamics F-111 was flying with triple-redundant, large-authority stability and command augmentation systems and fly-by-wire-controlled wing-mounted spoilers.[20]

On the basic research side, the delta winged British Short S.C.1, first flown in 1957, was a very small, single-seat Vertical Take-Off and Landing

---

18. S.A. Sjoberg, "Some Experience With Side Controllers," *Research Airplane Committee Report on the Progress of the X-15 Project*, pp. 167–174; Conference held at NACA Langley Aeronautical Laboratory, Langley Field, VA, Oct. 25–26, 1956.
19. "Automatic Flight Control System Sought," *Aviation Week*, Aug. 6, 1956, pp. 275–284.
20. Duane McRuer and Dunstan Graham, "A Flight Control Century: Triumphs of the Systems Approach," Paper 617, *Journal of Guidance, Control and Dynamics*, vol. 27, no. 2, pp. 161–173, AIAA, 2003.

(VTOL) aircraft. It incorporated a triply redundant fly-by-wire flight control system with a mechanical backup capability. The outputs from the three independent fly-by-wire channels were compared, and a failure in a single channel was overridden by the other two. A single channel failure was relayed to the pilot as a warning, enabling him to switch to the direct (mechanical) control system. The S.C.1 had three flight control modes, as described below, with the first two only being selectable prior to takeoff.[21]

- Full fly-by-wire mode with aerodynamic surfaces and nozzles controlled electrically via three independent servo motors with triplex fail-safe operation in conjunction with three analog autostabilizer control systems.
- A hybrid mode in which the reaction nozzles were servo/autostabilizer (fly-by-wire) controlled and the aerodynamic surfaces were linked directly to the pilot's manual controls.
- A direct mode in which all controls were mechanically linked to the pilot control stick.

The S.C.1 weighed about 8,000 pounds and was powered by four vertically mounted Rolls-Royce RB.108 lift engines, providing a total vertical thrust of 8,600 pounds. One RB.108 engine mounted horizontally in the rear fuselage provided thrust for forward flight. The lift engines were mounted vertically in side-by-side pairs in a central engine bay and could be swiveled to produce vectored thrust (up to 23 degrees forward for acceleration or –12 degrees for deceleration). Variable thrust nose, tail, and wingtip jet nozzles (powered by bleed air from the four lift engines) provided pitch, roll, and yaw control in hover and at low speeds during which the conventional aerodynamic controls were ineffective. The S.C.1 made its first flight (a conventional takeoff and landing) on April 2, 1957. It demonstrated tethered vertical flight on May 26, 1958, and free vertical flight on October 25, 1958. The first transition from vertical flight to conventional flight was made April 6, 1960.[22]

---

21. J.K.B. Illingworth and H.W. Chinn, "Variable Stability and Control Tests on the S.C.1 Aircraft in Jet-Borne Flight, with Particular Emphasis on Reference to Desirable VTOL Flying Qualities," Royal Aircraft Establishment, Bedford, U.K., Her Majesty's Stationery Office, London, 1969.

22. D. Lean and H.W. Chinn, "Review of General Operating Experience with a Jet-Lift VTOL Research Aircraft (Short S.C.1)," Aeronautical Research Council Current Paper C.P. No. 832, Her Majesty's Stationery Office, London, 1965.

During 10 years of flight-testing, the two S.C.1 aircraft made hundreds of flights and were flown by British, French, and NASA test pilots. A Royal Aircraft Establishment (RAE) report summarizing flight-test experience with the S.C.1 noted: "Of the visiting pilots, those from NASA [Langley's John P. "Jack" Reeder and Fred Drinkwater from Ames] flew the aircraft 6 or 7 times each. They were pilots of very wide experience, including flight in other VTOL aircraft and variable stability helicopters, which was of obvious assistance to them in assessing the S.C.1."[23] On October 2, 1963, while hovering at an altitude of 30 feet, a gyro input malfunction in the flight control system produced uncontrollable pitch and roll oscillations that caused the second S.C.1 test aircraft (XG 905) to roll inverted and crash, killing Shorts test pilot J.R. Green. The aircraft was then rebuilt for additional flight-testing. The first S.C.1 (XG 900) was used for VTOL research until 1971 and is now part of the Science Museum aircraft collection at South Kensington, London. The second S.C.1 (XG 905) is in the Flight Experience exhibit at the Ulster Folk and Transport Museum in Northern Ireland, near where the aircraft was originally built by Short Brothers.

The Canadian Avro CF-105 Arrow supersonic interceptor flew for the first time in 1958. Revolutionary in many ways, it featured a dual channel, three-axis fly-by-wire flight control system designed without any mechanical backup flight control capability. In the CF-105, the pilot's control inputs were detected by pressure-sensitive transducers mounted in the pilot's control column. Electrical signals were sent from the transducers to an electronic control servo that operated the valves in the hydraulic system to move the various flight control surfaces. The CF-105 also incorporated artificial feel and stability augmentation systems.[24] In a highly controversial decision, the Canadian government canceled the Arrow program in 1959 after five aircraft had been built and flown. Although only about 50 flight test hours had been accumulated, the Arrow had reached Mach 2.0 at an altitude of 50,000 feet. During its development, NACA Langley Aeronautical Laboratory assisted the CF-105 design team in a number of areas, including aerodynamics, performance, stability, and control. After the program was terminated,

---

23. Illingworth and Chinn, "Variable Stability and Control Tests on the S.C.1 Aircraft in Jet-Borne Flight, with Particular Emphasis on Reference to Desirable VTOL Flying Qualities."

24. J.C. Floyd, "The Canadian Approach to All-Weather Interceptor Development, Fourteenth British Commonwealth Lecture," *The Journal of the Royal Aeronautical Society*, vol. 62, No. 576, Dec. 1958.

many Avro Canada engineers accepted jobs with NASA and British or American aircraft companies.[25] Although it never entered production and details of its pioneering flight control system design were reportedly little known at the time, the CF-105 presaged later fly-by-wire applications.

NACA test data derived from the F9F-2 fly-by-wire experiment were used in development of the side stick controllers in the North American X-15 rocket research plane, with its adaptive flight control system.[26] First flown in 1959, the X-15 eventually achieved a speed of Mach 6.7 and reached a peak altitude of 354,200 feet. One of the two side stick controllers in the X-15 cockpit (on the left console) operated the reaction thruster control system, critical to maintaining proper attitude control at high Mach numbers and extreme altitudes during descent back into the higher-density lower atmosphere. The other controller (on the right cockpit console) operated the conventional aerodynamic flight control surfaces. A CALSPAN NT-33 variable stability test aircraft equipped with a side stick controller and an NACA-operated North American F-107A (ex-USAF serial No. 55-5120), modified by NACA engineers with a side stick flight control system, were flown by X-15 test pilots during 1958–1959 to gain side stick control experience prior to flying the X-15.[27]

Interestingly, the British VC10 jet transport, which first flew in 1962, has a quad channel flight control system that transmits electrical signals directly from the pilot's flight controls or the aircraft's autopilot via electrical wiring to self-contained electrohydraulic Powered Flight Control Units (PFCUs) in the wings and tail of the aircraft, adjacent to the flight control surfaces. Each VC10 PFCU consists of an individual small self-contained hydraulic system with an electrical pump and small reservoir. The PFCUs move the control surfaces based on electrical signals provided to the servo valves that are electrically connected to the cockpit flying controls.[28] There are no mechanical linkages or hydraulic lines between the pilot and the PFCUs. The PFCUs drive the primary flight

---

25. Thirty-one former CF-105 engineers were hired by NASA with several going on to hold important positions within the NASA Mercury, Gemini, and Apollo programs. Chris Gainor, "Arrows to the Moon: Avro's Engineers and the Space Race," (Burlington: Apogee Books, 2001).
26. "Experience with the X-15 Adaptive Flight Control System," NASA TN-D-6208, NASA Flight Research Center, Edwards, CA (March 1971).
27. Hallion, "On the Frontier Flight Research at Dryden, 1946–1981."
28. Allan F. Damp, "Evaluation Tests on Boulton-Paul VC-10 Aileron Integrated Flight Control Actuator," Commercial Division, Boeing Company, Renton, WA, Mar. 10, 1970.

control surfaces that consist of split rudders, ailerons, and elevators on separate electrical circuits. Thus, the VC10 has many of the attributes of fly-by-wire and power-by-wire flight control systems. It also features a backup capability that allows it to be flown using the hydraulically boosted variable incidence tail plane and differential spoilers that are operated via conventional mechanical linkages and separate hydraulic systems.[29] The VC10K air refueling tanker was still in Royal Air Force (RAF) service as of 2009, and the latest Airbus airliner, the A380, uses the PFCU concept in its fly-by-wire flight control system.

The Anglo-French Concorde supersonic transport first flew in 1969 and was capable of transatlantic sustained supercruise speeds of Mach 2.0 at cruising altitudes well above 50,000 feet. In support of the Concorde development effort, a two-seat Avro 707C delta winged flight research aircraft was modified as a fly-by-wire technology testbed with a side stick controller. It flew 200 hours on fly-by-wire flight trails at the U.K. at Farnborough until September 1966.[30] Concorde had a dual channel analog fly-by-wire flight control system with a backup mechanical capability. The mechanical system served in a follower role unless problems developed with the fly-by-wire control elements of the system, in which case it was automatically connected. Pilot movements of the cockpit controls operated signal transducers that generated commands to the flight control system. These commands were processed by an analog electrical controller that included the aircraft autopilot. Mechanically operated servo valves were replaced by electrically controlled ones. Much as with the CF-105, artificial feel forces were electrically provided to the Concorde pilots based on information generated by the electronic controller.[31]

Space Race and the War in Vietnam: Emphasis on FBW Accelerates

During the 1960s, two major events would unfold in the United States that had very strong influence on the development and eventual

---

29. Molly Neal, "VC10: Vickers-Armstrongs' Long-range Airliner," *Flight International*, May 10, 1962.
30. Ray Sturtivant, "British Research and Development Aircraft: Seventy Years at the Leading Edge," Haynes/Foulis, 1990; Kyrill Von Gersdorff, "Transfer of German Aeronautical Knowledge After 1945," in Hirschel, Ernst Heinrich, Horst Prem, and Gero Madelung, "Aeronautical Research in Germany (From Lilienthal to Today)," *Springer-Verlag*, Berlin Heidelberg, 2004.
31. B.S. Wolfe, "The Concorde Automatic Flight Control System: A Description of the Automatic Flight Control System for the Anglo-French Supersonic Transport and its Development to Date," *Aircraft Engineering and Aerospace Technology*, vol. 39, issue 5, 1967.

introduction into operational service of advanced computer-controlled fly-by-wire flight control systems. Early in his administration, President John F. Kennedy had initiated the NASA Apollo program with the goal of placing a man on the Moon and safely bringing him back to Earth by the end of the decade. The space program, and Apollo in particular, would lead to major strides in the application of the digital computer to manage and control sensors, systems, and advanced fly-by-wire vehicles (eventually including piloted aircraft). During the same period, America became increasingly involved in the expanding conflict in South Vietnam, an involvement that rapidly escalated as the war expanded into a conventional conflict with dimensions far beyond what was originally foreseen. As combat operations intensified in Southeast Asia, large-scale U.S. strike missions began to be flown against North Vietnam. In response, the Soviet Union equipped North Vietnamese forces with improved air defense weapons, including advanced fighters, air-to-air and surface-to-air missiles, and massive quantities of conventional antiaircraft weapons, ranging in caliber from 12.7 to 100 millimeters (mm). U.S. aircraft losses rose dramatically, and American warplane designs came under increasing scrutiny as the war escalated.[32] Analyses of combat data revealed that many aircraft losses resulted from battle damage to hydromechanical flight control system components. Traditionally, primary and secondary hydraulic system lines had been routed in parallel through the aircraft structure to the flight control system actuators. In the Vietnam combat, experience revealed that loss of hydraulic fluid because of battle damage often led to catastrophic fires or total loss of aircraft control. Aircraft modification programs were developed to reroute and separate primary and secondary hydraulic lines to reduce the possibility of a total loss of fluid given a hit. Other changes to existing aircraft flight control systems improved survivability, such as a modification to the F-4 that froze the horizontal tail in the neutral position to prevent the aircraft from going out of control when hydraulic fluid was lost.[33] However, there was an increasing body of opinion that felt a

---

32. During the war in Southeast Asia, 8,961 U.S. aircraft were lost. Of these, Air Force losses totaled 2,251, the Navy 859, the Marine Corps 463, and the Army 5,388 (mostly helicopters). The McDonnell-Douglas F-4 Phantom was the predominant fighter/attack aircraft used by the USAF, the USMC, and the Navy during the Vietnam war. A total of 765 F-4s were lost.

33. Robert E. Ball, "The Fundamentals of Aircraft Combat Survivability Analysis and Design," American Institute of Aeronautics and Astronautics (AIAA), New York, 1985.

new approach to flight control system design was necessary and technically feasible.

The Air Force JB-47E Fly-By-Wire Project

The USAF Flight Dynamics Laboratory at Wright Patterson Air Force Base (AFB), OH, sponsored a number of technology efforts and flight-test programs intended to increase the survivability of aircraft flight control system components such as fly-by-wire hydraulic actuators. Beginning in 1966, a Boeing B-47E bomber was progressively modified (being redesignated JB-47E) to incorporate analog computer-controlled fly-by-wire actuators for both pitch and roll control, with pilot inputs being provided via a side stick controller. The program spanned three phases. For Phase I testing, the JB-47E only included fly-by-wire in its pitch axis. This axis was chosen because the flight control system in the standard B-47E was known to have a slow response in pitch because of the long control cables to the elevator stretching under load. Control signals to the pitch axis FBW actuator were generated by a transducer attached to the pilot's control column. The pilot had a simple switch in the cockpit that allowed him to switch between the standard hydromechanical flight control system (which was retained as a backup) and the computer-controlled FBW system. Modified thus, the JB-47E flew for the first time, in December 1967. Test pilots reported that the modified B-47 had better handling qualities then were attainable with the standard B-47E elevator control system, especially in high-speed, low-level flight.[34]

Phase II of the JB-47E program added fly-by-wire roll control and a side stick controller that used potentiometers to measure pilot input. By the end of the flight-test program, over 40 pilots had flown the FBW JB-47E. The Air Force chief test pilot during Phase II, Col. Frank Geisler, reported: "In ease of control there is no comparison between the standard system and the fly-by-wire. The fly-by-wire is superior in every aspect concerning ease of control. . . . It is positive, it is rapid—it responds well—and best of all the feel is good."[35] Before the JB-47E Phase III flight-test program ended in early 1969, a highly reliable four-channel redundant

---

34. Gavin D. Jenny, "JB-47E Fly-By-Wire Flight Test Program (Phase I)," Air Force Flight Dynamics Laboratory TR-69-40, Wright-Patterson AFB, OH, Sept. 1969.

35. Gavin D. Jenny, James W. Morris, and Vernon R. Schmitt, "Fly-by-Wire, A Historical and Design Perspective," The Society of Automotive Engineers (SAE), 1998.

electrohydraulic actuator had been installed in the pitch axis and successfully evaluated.[36] By this time, the Air Force had already initiated Project 680J, the Survivable Flight Control System (SFCS), which resulted in the prototype McDonnell-Douglas YF-4E Phantom aircraft being modified into a testbed to evaluate the potential benefits of fly-by-wire in a high-performance, fighter-type aircraft.[37] The SFCS YF-4E was intended to validate the concept that dispersed, redundant fly-by-wire flight control elements would be less vulnerable to battle damage, as well as to improve the performance of the flight control system and increase overall mission effectiveness.

Project 680J: Survivable Flight Control System YF-4E

In mid-1969, modifications began to convert the prototype McDonnell-Douglas YF-4E (USAF serial No. 62-12200) for the SFCS program. A quadruple-redundant analog computer-based three-axis fly-by-wire flight control system with integrated hydraulic servo-actuator packages was incorporated and side stick controllers were added to both the front and back cockpits. Roll control was pure fly-by-wire with no mechanical backup. For initial testing, the Phantom's mechanical flight control system was retained in the pitch and yaw axes as a safety backup. On April 29, 1972, McDonnell-Douglas test pilot Charles P. "Pete" Garrison flew the SFCS YF-4E for the first time from the McDonnell-Douglas factory at Lambert Field in St. Louis, MO. The mechanical flight control system was used for takeoff with the pilot switching to the fly-by-wire system during climb-out. The aircraft was then flown to Edwards AFB for a variety of additional tests, including low-altitude supersonic flights. After the first 27 flights, which included 23 hours in the full three-axis fly-by-wire configuration, the mechanical flight control system was disabled. First flight in the pure fly-by-wire configuration occurred January 22, 1973. The SFCS YF-4E flew as a pure fly-by-wire aircraft for the remainder of its flight-test program, ultimately completing over 100 flights.[38]

Whereas the earlier phases of the flight-test effort were primarily flown by McDonnell-Douglas test pilots, the next aspect of the SFCS

---

36. Tomayko, "Blind Faith: The United States Air Force and the Development of Fly-By-Wire Technology," *Technology and the Air Force: A Retrospective Assessment*, Air Force History and Museums Program, Washington, DC, 1997.
37. Michael L. Yaffee, "Survivable Controls Gain Emphasis," *Aviation Week*, Feb. 2, 1970.
38. Jenny, "JB-47E Fly-By-Wire Flight Test Program (Phase I)."

program was focused on an Air Force evaluation of the operational suitability of fly-by-wire and an assessment of the readiness of the technology for transition into new aircraft designs. During this phase, 15 flights were accomplished by two Air Force test pilots (Lt. Col. C.W. Powell and Maj. R.C. Ettinger), who concluded that fly-by-wire was indeed ready and suitable for use in new designs. They also noted that flying qualities were generally excellent, especially during takeoffs and landings, and that the pitch transient normally encountered in the F-4 during rapid deceleration from supersonic to subsonic flight was nearly eliminated. Another aspect of the flight-test effort involved so-called technology transition and demonstration flights in the SFCS aircraft. At this time, the Air Force had embarked on the Lightweight Fighter (LWF) program. One of the two companies developing flight demonstrator aircraft (General Dynamics)had elected to use fly-by-wire in its new LWF design (the YF-16). A block of 11 flights in the SFCS YF-4E was allocated to three pilots assigned to the LWF test force at Edwards AFB (Lt. Col. Jim Ryder, Maj. Walt Hersman, and Maj. Mike Clarke). Based on their experiences flying the SFCS YF-4E, the LWF test force pilots were able to provide valuable inputs into the design, development, and flight test of the YF-16, directly contributing to the dramatic success of that program. An additional 10 flights were allocated to another 10 pilots, who included NASA test pilot Gary E. Krier and USAF Maj. Robert Barlow.[39] Earlier, Krier had piloted the first flight of a digital fly-by-wire (DFBW) flight control system in the NASA DFBW F-8C on May 25, 1972. That event marked the first time that a piloted aircraft had been flown purely using a fly-by-wire flight control system without any mechanical backup provisions. Barlow, as a colonel, would command the Air Force Flight Dynamics Laboratory during execution of several important fly-by-wire flight research efforts. The Air Force YF-16 and the NASA DFBW F-8 programs are discussed in following sections.

The Precision Aircraft Control Technology YF-4E Program

In order to evaluate the use of computer-controlled fly-by-wire systems to actively control a relaxed stability aircraft, the SFCS YF-4E was further modified under the Air Force Precision Aircraft Control Technology (PACT) program. Movable canard surfaces were mounted

---

39. Ibid.

ahead of the wing and above the YF-4E's inlets. A dual channel electronic fly-by-wire system with two hydraulic systems directed the canard actuators. The canards, along with the capability to manage internal fuel to move the center of gravity of the aircraft aft, effectively reduced the static stability margin to as low as –7.5 percent, that is, fully unstable, in the pitch axis. Relaxing static stability by moving the center of gravity aft reduces trim drag and decreases the downward force that the horizontal tail needs to produce to trim the aircraft. However, as center of gravity moves aft, an aircraft becomes less and less stable in the pitch axis, leading to a need to provide artificial stability augmentation. A negative static margin means that the aircraft is unstable, it cannot maintain stable flight and will be uncontrollable without artificial stability augmentation. During its test program, the PACT aircraft was flown 34 times, primarily by McDonnell-Douglas test pilots. The Pact aircraft demonstrated significant performance gains that included an increase in the 4 g maneuvering ceiling of over 4,000 feet (to 50,000 feet) along with an increase in turning radius.[40] The approach used by the PACT YF-4E to create a relaxed stability research aircraft was soon mirrored by several foreign flight research programs that are discussed in a separate section. In January 1979, the PACT YF-4E aircraft was delivered by Army helicopter from the McDonnell-Douglas factory in St. Louis to the Air Force Museum at Wright-Patterson AFB, OH, for permanent display.[41] By this time, the Air Force had initiated the Control Configured Vehicle (CCV) F-16 project. This was followed by the Advanced Fighter Technology Integration (AFTI) F-16 program. Those programs would carry forward the fly-by-wire explorations initiated by the PACT YF-4E.

European FBW Research Efforts

By the late 1960s, several European research aircraft using partial fly-by-wire flight control systems were in development. In Germany, the supersonic VJ-101 experimental Vertical Take-Off and Landing fighter technology demonstrator, with its swiveling wingtip mounted afterburning turbojet engines, and the Dornier Do-31 VTOL jet transport used analog computer-controlled partial fly-by-wire flight control systems. American test pilots were intimately involved with both programs. George W. Bright flew the VJ-101 on its first flight in 1963, and NASA test

---

40. Ibid.
41. Tomayko, "Computers Take Flight: A History of NASA's Pioneering Digital Fly-by Wire Project."

pilot Drury W. Wood, Jr., headed the cooperative U.S.–German Do-31 flight-test program that included representatives from NASA Langley and NASA Ames. Wood flew the Do-31 on its first flight in February 1967. He received the Society of Experimental Test Pilots' Ivan C. Kinchloe Award in 1968 for his role on the Do-31 program.[42] By that time, NASA test pilot Robert Innis was chief test pilot on the Do-31 program. The German VAK-191B VTOL fighter technology flight demonstrator flew in 1971. Its triply redundant analog flight control system assisted the pilot in operating its flight control surfaces, engines, and reaction control nozzles, but the aircraft retained a mechanical backup capability. Later in its flight-test program, the VAK-191B was used to support development of the partial fly-by-wire flight control system that was used in the multinational Tornado multirole combat aircraft that first flew in August 1974.[43]

In the U.K., a Hawker Hunter T.12 two-seat jet trainer was converted into a fly-by-wire testbed by the Royal Aircraft Establishment. It incorporated a three-axis, quadruplex analog Integrated Flight Control System (IFCS) and a "sidearm" controller. The mechanical backup flight control system was retained.[44] First flown in April 1972, the Hunter was eventually lost in a takeoff accident.

In the USSR, a Sukhoi Su-7U two-seat jet fighter trainer was modified with forward destabilizing canards as the Projekt 100LDU fly-by-wire testbed. It first flew in 1968 in support of the Sukhoi T-4 supersonic bomber development effort. Fitted with a quadruple redundant fly-by-wire flight control system with a mechanical backup capability, the four-engine Soviet Sukhoi T-4 prototype first flew in August 1972. Reportedly, the fly-by-wire flight control system provided much better handling qualities than the T-4's mechanical backup system. Four T-4 prototypes were built, but only the first aircraft ever flew. Designed for Mach 3.0, the T-4 never reached Mach 2.0 before the program was canceled after only 10 test flights and about 10 hours of flying time.[45] In 1973–1974, the Projekt

---

42. In a Dec. 30, 2006, e-mail, Drury Wood wrote: "I was the project test pilot on this airplane. Flew all of the test rigs and full size over 600 flights. Made the record flights, Awarded Bundesverdienstkreuz am Bande and Society of Experimental Test Pilots' highest award, Kinchloe."
43. Ulrich Butter, "Control, Navigation, Avionics, Cockpit," in Ernest Heinrich Hirschel, Horst Prem, and Gero Madelung, Aeronautical Research in Germany (From Lilienthal to Today), *Springer-Verlag*, Berlin Heidelberg, 2004.
44. "RAE Electric Hunter," *Flight International*, June 28, 1973, pp. 1010–1011.
45. Yefim Gordon, "An Industry of Prototypes: Sukhoi T-4, Russia's Mach 3 Bomber," *Wings of Fame*, vol. 9, Aerospace Publishing Limited, London, 1997.

100LDU testbed was used to support development of the fly-by-wire system flight control system for the Sukhoi T-10 supersonic fighter prototype program. The T-10 was the first pure Soviet fly-by-wire aircraft with no mechanical backup; it first flew on May 27, 1977. On July 7, 1978, the T-10-2 (second prototype) entered a rapidly divergent pitch oscillation at supersonic speed. Yevgeny Solovyev, distinguished test pilot and hero of the Soviet Union, had no chance to eject before the aircraft disintegrated.[46] In addition to a design problem in the flight control system, the T-10's aerodynamic configuration was found to be incapable of providing required longitudinal, lateral, and directional stability under all flight conditions. After major redesign, the T-10 evolved into the highly capable Sukhoi Su-27 family of supersonic fighters and attack aircraft.[47]

Understanding of FBW Benefits

By the early 1970s, the full range of benefits that could be possible by the use of fly-by-wire flight control had become ever more apparent to aircraft designers and pilots. Relevant technologies were rapidly maturing, and various forms of fly-by-wire flight control had successfully been implemented in missiles, aircraft, and spacecraft. Fly-by-wire had many advantages over more conventional flight control systems, in addition to those made possible from the elimination of mechanical linkages. A computer-controlled fly-by-wire flight control system could generate integrated pitch, yaw, and roll control instructions at very high rates to maintain the directed flight path. It would automatically provide artificial stability by constantly compensating for any flight path deviations. When the pilot moved his cockpit controls, commands would be automatically be generated to modify the artificial stability enough to enable the desired maneuvers to be accomplished. It could also prevent the pilot from commanding maneuvers that would exceed established aircraft limits in either acceleration or angle of attack. Additionally, the

---

46. In an interview published in Krylya Rodiny, No. 8, Aug. 1989, under the title "An Aircraft of the 21st Century," the chief designer at the Sukhoi OKB (Experimental Design Bureau), Mikhail Petrovich Simonov, commented: "It is also time to talk about those for whom risk, courage, and a willingness to devote their lives to learning the unknown. . . . I am talking about the test pilots of our design bureau. . . . Back then we did not know how the frequency responses of control match the human capabilities. Zhenya [Solovyev] ended up in a resonant mode and was killed and the aircraft was destroyed."

47. Gordon and Peter Davidson, "Sukhoi Su-27 Flanker," Specialty Press, 2006.

flight control system could automatically extend high-lift devices, such as flaps, to improve maneuverability.

Conceptual design studies indicated that active fly-by-wire flight control systems could enable new aircraft to be developed that featured smaller aerodynamic control surfaces. This was possible by reducing the inherent static stability traditionally designed into conventional aircraft. The ability to relax stability while maintaining good handling qualities could also lead to improved agility. Agility is a measure of an aircraft's ability to rapidly change its position. In the 1960s, a concept known as energy maneuverability was developed within the Air Force in an attempt to quantify agility. This concept states that the energy state of a maneuvering aircraft can be expressed as the sum of its kinetic energy and its potential energy. An aircraft that possesses higher overall energy inherently has higher agility than another aircraft with lower energy. The ability to retain a high-energy state while maneuvering requires high excess thrust and low drag at high-lift maneuvering conditions.[48] Aircraft designers began synthesizing unique conceptual fighter designs using energy maneuver theory along with exploiting an aerodynamic phenomenon known as vortex lift.[49] This approach, coupled with computer-controlled fly-by-wire flight control systems, was felt to be a key to unique new fighter aircraft with very high agility levels.

Neutrally stable or even unstable aircraft appeared to be within the realm of practical reality and were the subject of ever increasing interest and widespread study in NASA and the Air Force, as well as in foreign governments and the aerospace industry. Often referred to at the time as Control Configured Vehicles, such aircraft could be optimized for specific missions with fly-by-wire flight control system characteristics

---

48. David C. Aronstein and Albert C. Piccirillo, "The Lightweight Fighter Program: A Successful Approach to Fighter Technology Transition," AIAA, 1996. M.J. Wendl, G.G. Grose, J.L. Porter, and V.R. Pruitt, "Flight/Propulsion Control Integration Aspects of Energy Management," Society of Automotive Engineers, Paper No. 740480, 1974.

49. U.S. industry interest in "vortex lift" increased during the early 1960s as a result of NASA Langley aerodynamic studies of foreign delta wing aircraft such as the Concorde supersonic transport and the combination canard/delta wing Swedish AJ-37 Viggen fighter. Langley's vortex lift research program was led by Edward Polhamus, with researcher Linwood McKinney studying the favorable effects of vortexes on lift produced by strong leading-edge vortex flow off slender lifting surfaces. See Chambers, Joseph R., *Partners in Freedom: Contributions of the Langley Research Center to U.S. Military Aircraft of the 1990s*, Monographs in Aerospace History No. 19, NASA, Washington, DC, 2000.

designed to improve aerodynamic performance, maneuverability, and agility while reducing airframe weight. Other CCV possibilities included the ability to control structural loads while maneuvering (maneuver load control) and the potential for implementation of unconventional control modes. Maneuver load control could allow new designs to be optimized, for example, by using automated control surface deflections to actively modify the spanwise lift distribution to alleviate wing bending loads on larger aircraft. Unconventional or decoupled control modes would be possible by using various combinations of direct-force flight controls to change the aircraft flight path without changing its attitude or, alternatively, to point the aircraft without changing the aircraft flight path. These unconventional flight control modes were felt at the time to provide an improved ability to point and fire weapons during air combat.[50]

In summary, the full range of benefits possible through the application of active fly-by-wire flight control in properly tailored aircraft design applications was understood to include:

- Enhanced performance and improved mission effectiveness made possible by the incorporation of relaxed static stability and automatically activated high-lift devices into mission-optimized aircraft designs to reduce drag, optimize lift, and improve agility and handling qualities throughout the flight and maneuvering envelope.
- New approaches to aircraft control, such as the use of automatically controlled thrust modulation and thrust vectoring fully integrated with the movement of the aircraft's aerodynamic flight control surfaces and activation of its high-lift devices.
- Increased safety provided by automatic angle-of-attack and angle-of-sideslip suppression as well as automatic limiting of normal acceleration and roll rates. These measures protect from stall and/or loss of control, prevent inadvertent overstressing of the airframe, and give the pilot maximum freedom to focus on effectively maneuvering the aircraft.
- Improved survivability made possible by the elimination

---

50. Aronstein and Piccirillo, "The Lightweight Fighter Program: A Successful Approach to Fighter Technology Transition."

of highly vulnerable hydraulic lines and incorporation of fault tolerant flight control system designs and components.
- Greatly improved flight control system reliability and lower maintenance costs resulting from less mechanical complexity and automated built-in system test and diagnostic capabilities.
- Automatic flight control system reconfiguration to allow safe flight, recovery, and landing following battle damage or system failures.

Digital Fly-By-Wire: The Space Legacy

Both the Mercury and Gemini capsules controlled their reaction control thrusters via electrical commands carried by wire. They also used highly reliable computers specially developed for the U.S. manned space flight program. During reentry from space on his historic 1962 Mercury mission, the first American in space, Alan Shepard, took manual control of the spacecraft attitude, one axis at a time, from the automatic attitude control system. Using the Mercury direct side controller, he "hand-flew" the capsule to the retrofire attitude of 34 degrees pitch-down. Shepard reported that he found that the spacecraft response was about the same as that of the Mercury simulator at the NASA Langley Research Center.[51] The success of fly-by-wire in the early manned space missions gave NASA confidence to use a similar fly-by-wire approach in the Lunar Landing Research Vehicle (LLRV), built in the early 1960s to practice lunar landing techniques on Earth in preparation for the Apollo missions to the Moon. Two LLRVs were built by Bell Aircraft and first flown at Dryden in 1964. These were followed by three Lunar Landing Training Vehicles (LLTVs) that were used to train the Apollo astronauts. The LLTVs used a triply redundant fly-by-wire flight control system based on the use of three analog computers. Pure fly-by-wire in their design (there was insufficient weight allowance for a mechanical backup capability), they proved invaluable in preparing the astronauts for actual landings on the surface of the Moon, flying until November 1972.[52] A total of 591 flights were accomplished, during which one LLRV and two LLTVs crashed in

---

51. Robert B. Voas, "Manned Control of Mercury Spacecraft," *Astronautics*, vol. 7, no. 3, Mar. 1962, p. 18.
52. "LLRV Fact Sheet," FS-2002-09-026-DFRC, NASA Dryden Flight Research Center.

spectacular accidents but fortunately did so without loss of life.[53] During this same period, digital computers were demonstrating great improvements in processing power and programmability. Both the Apollo Lunar Module and the Command and Service Module used full-authority digital fly-by-wire controls. Fully integrated into the fly-by-wire flight control systems used in the Apollo spacecraft, the Apollo digital computer provided the astronauts with the ability to precisely maneuver their vehicles during all aspects of the lunar landing missions. The success of the Apollo digital computer in these space vehicles led to the idea of using this computer in a piloted flight research aircraft.

By the end of 1969, many experts within NASA and especially at the NASA Flight Research Center at Edwards Air Force Base were convinced that digital-computer-based fly-by-wire flight control systems would ultimately open the way to dramatic improvements in aircraft design, flight safety, and mission effectiveness. A team headed by Melvin E. Burke—along with Dwain A. Deets, Calvin R. Jarvis, and Kenneth J. Szalai—proposed a flight-test program that would demonstrate exactly that. The digital fly-by-wire proposal was evaluated by the Office of Advanced Research and Technology (OART) at NASA Headquarters. A strong supporter of the proposal was Neil Armstrong, who was by then the Deputy Associate Administrator for Aeronautics. Armstrong had been the first person to step on the Moon's surface, in July 1969 during the Apollo 11 mission, and he was very interested in fostering transfer of technology from the Apollo program into aeronautics applications. During discussion of the digital fly-by-wire proposal with Melvin Burke and Cal Jarvis, Armstrong strongly supported the concept and reportedly commented: "I just went to the Moon with one." He urged that they contact the Massachusetts Institute of Technology (MIT) Draper Laboratory to evaluate the possibility of using modified Apollo hardware and software.[54] The Flight Research Center was authorized to modify a fighter type aircraft with a digital fly-by-wire system. The modification would be based on the Apollo computer and inertial sensing unit.

The NASA Digital Fly-By-Wire F-8 Program

A former Navy F-8C Crusader fighter was chosen for modification, with the goal being to both validate the benefits of a digital fly-by-wire aircraft

---

53. LLRV No. 2 is on display at the Dryden Flight Research Center.
54. Tomayko, "Computers Take Flight: A History of NASA's Pioneering Digital Fly-by Wire Project."

flight control system and provide additional confidence on its use. Mel Burke had worked with the Navy to arrange for the transfer of four LTV F-8C Crusader supersonic fighters to the Flight Research Center. One would be modified for the F-8 Super Cruise Wing project, one was converted into the F-8 DFBW Iron Bird ground simulator, another was modified as the DFBW F-8, and one was retained in its basic service configuration and used for pilot familiarization training and general proficiency flying. When Burke left for a job at NASA Headquarters, Cal Jarvis, a highly experienced engineer who worked on fly-by-wire systems on the X-15 and LLRV programs, took over as program manager. In March 1971, modifications began to create the F-8 DFBW Iron Bird simulator. The Iron Bird effort was planned to ensure that development of the ground simulator always kept ahead of conversion efforts on the DFBW flight-test aircraft. This, the very first F-8C built for the Navy in 1958 (bureau No. 1445546), carried the NASA tail No. 802 along with a "DIGITAL FLY-BY-WIRE" logo painted in blue on its fuselage sides.

F-8 DFBW: Phase I

In implementing the DFBW F-8 program, the Flight Research Center chose to remove all the mechanical linkages and cables to the flight control surfaces, thus ensuring that the aircraft would be a pure digital fly-by-wire system from the start. The flight control surfaces would be hydraulically activated, based on electronic signals transmitted via circuits that were controlled by the digital flight control system (DFCS). The F-8C's gun bays were used to house auxiliary avionics, the Apollo Display and Keyboard (DSKY) unit,[55] and the backup analog flight control system. The Apollo digital guidance computer, its related cooling system, and the inertial platform that also came from the Apollo program were installed in what had been the F-8C avionics equipment bay. The reference information for the digital flight control system was provided by the Apollo Inertial Management System (IMS). In the conversion of the F-8 to the fly-by-wire configuration, the original F-8 hydraulic actuator slider values were replaced with specially developed secondary actuators. Each secondary actuator had primary and backup modes. In the primary mode, the digital computer sent analog position signals for a single actuation cylinder. The cylinder was controlled by a dual self-

---

55. The DSKY had been developed for the Apollo program and enabled input and output to the digital computer system. It was used during Phase I of the DFBW F-8 program.

monitoring servo valve. One valve controlled the servo; the other was used as a model for comparison. If the position values differed by a predetermined amount, the backup was engaged. In the backup mode, three servo cylinders were operated in a three-channel, force-summed arrangement.[56]

The triply redundant backup analog-computer-based flight control system—known as the Backup Control System (BCS)—used an independent power supply and was based on the use of three Sperry analog computers.[57] In the event of loss of electrical power, 24-volt batteries could keep the BCS running for about 1 hour. Flight control was designed to revert to the BCS if any inputs from the primary digital control system to the flight control surface actuators did not match up; if the primary (digital) computer self-detected internal failures, in the event of electrical power loss to the primary system; and if inputs to secondary actuators were lost. The pilot had the ability to disengage the primary flight control system and revert to the BCS using a paddle switch mounted on the control column. The pilot could also vary the gains[58] to the digital flight control system using rotary switches in the cockpit, a valuable feature in a research aircraft intended to explore the development of a revolutionary new flight control system.

The control column, rudder pedals, and electrical trim switches from the F-8C were retained. Linear Differential Variable Transformers (LDVTs) installed in the base of the control stick were used to detect pilot control inputs. They generated electrical signals to the flight control system to direct aircraft pitch and yaw changes. Pilot inputs to the rudder pedals were detected by LDVTs in the tail of the aircraft. There were two LDVTs in each aircraft control axis, one for the primary (digital) flight control system and one for the BCS. The IMS supplied the flight control system with attitude, velocity, acceleration, and position change references that were compared to the pilot's control inputs; the flight control computer would then calculate required control surface position changes to maneuver the aircraft as required.

By the end of 1971, software for the Phase I effort was well along, and the aircraft conversion was nearly complete. Extensive testing of the aircraft's flight control systems was accomplished using the Iron Bird, and

---

56. Tomayko, "Computers Take Flight: A History of NASA's Pioneering Digital Fly-by Wire Project."
57. These Sperry-developed analog computers were also used in the Air Force's YF-4E fly-by-wire project, which was in progress at the same time as NASA's DFBW F-8 effort. Ibid.
58. Gain is a measure of the sensitivity of the aircraft to command inputs to the flight control system.

Case 10 | Fly-By-Wire: Making the Electric Jet

For the DFBW F-8 program, the Flight Research Center removed all mechanical linkages and cables to the flight control surfaces. NASA.

planned test mission profiles were evaluated. On May 25, 1972, NASA test pilot Gary Krier made the first flight ever of an aircraft under digital computer control, when he took off from Edwards Air Force Base. Envelope expansion flights and tests of the analog BCS followed with supersonic flight being achieved by mid-June. Problems were encountered with the stability augmentation system especially, in formation flight because of the degree of attention required by the pilot to control the aircraft in the roll axis. As airspeeds approached 400 knots, control about all axes became too sensitive. Despite modifications, roll axis control remained a problem with lag encountered between control stick movement and aircraft response. In September 1972, Tom McMurtry flew the aircraft, finding that the roll response was highly sensitive and could lead to lateral pilot-induced oscillations (PIOs). By May 1973, 23 flights had been completed in the Phase I DFBW program. Another seven flights were accomplished in June and July, during which different gain combinations were evaluated at various airspeeds.

In August 1973, the DFBW F-8 was modified to install an YF-16 side stick controller.[59] It was connected to the analog BCS only. The center stick installation was retained. Initially, test flights by Gary Krier and Tom McMurtry were restricted to takeoff and landing using the center control stick, with transition to the BCS and side stick control being made at altitude. Aircraft response and handling qualities were rated as highly positive. A wide range of maneuvers, including takeoffs and landings, were accomplished by the time the side stick evaluation was completed in October 1973. The two test pilots concluded that the YF-16 side stick control scheme was feasible and easy for pilots to adapt to. This inspired high confidence in the concept and resulted in the incorporation of the side stick controller into the YF-16 flight control design. Subsequently, four other NASA test pilots flew the aircraft using the side stick controller in the final six flights of the DFBW F-8 Phase I effort, which concluded in November 1973. Among these pilots was General Dynamics chief test pilot Phil Oestricher, who would later fly the YF-16 on its first flight in January 1974. The others were NASA test pilots William H. Dana (a former X-15 pilot), Einar K. Enevoldson, and astronaut Kenneth Mattingly. During Phase I flight-testing, the Apollo digital computer maintained its reputation for high reliability and the three-channel analog backup fly-by-wire system never had to be used.

DFBW F-8: Phase II

On November 16, 1973, the DFBW team received a NASA group achievement award for its highly impressive accomplishments during the Phase I effort. By that time, planning was well underway for the Phase II effort, with the first version of the software specification having already been issued in April 1973. Whereas Phase I had verified the feasibility of flight control using a digital computer, Phase II was intended to develop a more practical approach to the implementation of digital flight control, one that could be used to justify the incorporation of digital technology into production designs for both military and commercial use. In the Phase

---

59. As noted earlier, the NACA had evaluated a side stick controller as early as 1952. Side stick controllers had been successfully used in the NASA F9F-2, F-107A, and X-15, as well as the Mercury, Gemini, and Apollo space vehicles, and they were planned to be used in the upcoming Space Shuttle. The Air Force had flight-tested a side stick controller in a B-47E and a C-141 in the late 1960s and was planning on its use in the fly-by-wire project 680J YF-4E Survivable Flight Control System (SFCS) test aircraft.

II design, the single channel Apollo computer-based flight control system was replaced with a triply redundant flight control system approach using three International Business Machines (IBM) AP-101 digital computers. The challenge was how to program this multicomputer system to act as a single computer in processing flight control laws and directing aircraft maneuvers while functioning independently for purposes of fault tolerance.[60] The 32-bit IBM AP-101 computer had been selected for use in the Space Shuttle. It consumed 370 watts of power, weighed about 50 pounds, and had 32,000 words of memory.[61] The DFBW program decided to also use the AP-101 computer in its Phase II effort, and a purchase contact with IBM was signed in August 1973. However, the reliability of the AP-101 computer, as measured by mean time between failures, left much to be desired. The computer would turn out to require major redesign, and it never came close to meeting its reliability projections. As Ken Szalai recently commented: "the IBM AP-101 computer was one of the last of the 'beasts.' It was big and ran hot. The circuit boards tended to fail as temperatures increased. This was found to be due to thermal expansion causing the layers within the circuit boards to separate breaking their electrical connections." Szalai recounted that he notified the Space Shuttle team as soon as the issue was discovered. They were surprised, as they had never seen a similar problem with the AP-101. The reason soon became apparent. The AP-101s installed in the F-8 Iron Bird were being tested in a non–air-conditioned hangar; Space Shuttle flight control system testing had been in a 50 degree Fahrenheit (50 °F) cooled laboratory environment. When the Space Shuttle was tested on the flight line in typical outside temperatures encountered at Dryden, similar reliability problems were encountered. IBM subsequently changed the thermal coating process used in the manufacture of the AP-101 circuit boards, a measure that partly resolved the AP-101's reliability problems.[62]

Software for Phase II was also larger and more complex than that used in Phase I because of the need for new pilot interface devices. Flight control modes still included the direct (DIR) mode, the stability augmentation (SAS) mode, and the control-augmentation (CAS) mode. A pitch maneuver-load-control feature was added to the CAS mode, and a digital autopilot was fitted that incorporated Mach hold, altitude-hold,

---

60. Kenneth J. Szalai, telephone conversation with author, Mar. 11, 2009.
61. Tomayko, "Computers Take Flight: A History of NASA's Pioneering Digital Fly-by Wire Project."
62. Szalai, telephone conversation.

and heading-hold selections. The software gradually matured to the point where pilots could begin verification evaluations in the Iron Bird simulator in early 1976. By July, no anomalies were reported in the latest software release, with the direct and stability-augmentation modes considered flight-ready. The autopilot and control-augmentation mode still required more development, but they were not necessary for first flight.

The backup analog flight control system was also redesigned for Phase II, and the secondary actuators were upgraded. Sperry supplied an updated version of the Phase I Backup Control System using the same technology that had been used in the Air Force's YF-4E project. Signals from the analog computers were now force-summed when they reached the actuators, resulting in a quicker response. The redesigned secondary actuators provided 20 percent more force, and they were also more reliable. The hydraulic actuators used in Phase I had two sources of hydraulic pressure for the actuators; in those chosen for Phase II, there were three hydraulic sources that corresponded with the three channels in each of the primary and secondary flight control systems. The secondary electronic actuators had three channels, with one dedicated to each computer in the primary system. The actuators were shared by the analog computer bypass system in the event of failure of the primary digital system.

The final Phase II design review occurred in late May 1975, with both the Iron Bird and aircraft 802 undergoing modification well into 1976. By early April, Gary Krier was able to fly the Iron Bird simulator with flight hardware and software. Handling qualities were generally rated as very good, but actuator anomalies and transients were noted, as were some problems with the latest software releases. After these issues were resolved, a flight qualification review was completed on August 20. High-speed taxi tests beginning 3 days later, then, on August 27, 1976, Gary Krier took off on the first flight of the Phase II program. On the second Phase II flight, one of the AP-101 computers failed with the aircraft at supersonic speed. An uneventful landing was accomplished with the flight control system remaining in the primary flight control mode. This was in accordance with the established flight-test procedure in the event of a failure of one of the primary computers. Flight-testing was halted, and all AP-101s were sent back to IBM for refurbishment. After 4 months, the AP-101s were back at Dryden, but another AP-101 computer failure occurred on the very next flight. Again, the primary digital flight control system handled the failure well, and flights were soon being accomplished without incident, providing ever increasing confidence in the system.

In the spring of 1977, the DFBW F-8 was modified to support the Space Shuttle program. It flew eight times with the Shuttle Backup Flight System's software test package running in parallel with the F-8 flight control software. Data from this package were downlinked as the F-8 pilots flew a series of simulated Shuttle landing profiles. Later in 1977, the unpowered Space Shuttle Enterprise was being used to evaluate the flight characteristics of the Space Shuttle during approach and landing in preparation for full-up shuttle missions. During the Shuttle Approach and Landing Test (ALT) program, the Enterprise was carried aloft atop the NASA 747 Shuttle carrier aircraft. After release, the Shuttle's handling qualities and the responsiveness of its digital fly-by-wire system were evaluated. On the fifth and last of the shuttle ALT flights in October 1977, a pilot-induced oscillation developed just as the Enterprise was landing. The DFBW F-8C was then used in a project oriented to duplicating the PIO problem encountered on the Shuttle during a series of flights in 1978 that were initially flown by Krier and McMurtry. They were joined by Einar K. Enevoldson and John A. Manke, who had extensive experience flying NASA lifting body vehicles. The lifting body vehicles used side stick controllers and had approach characteristics that were similar to those of the Space Shuttle.

Flying simulated Shuttle landing profiles with the DFBW F-8, the pilots gathered extremely valuable data that supported the Shuttle program in establishing sampling rates and control law execution limits. The DFBW F-8 flight control software had been modified to enable the pilot to vary transport delay times to evaluate their effect on control response. Transport delay is the elapsed time between pilot movement of his cockpit control and the actual movement of the flight control surfaces. It is a function of several factors, including the time needed to do analog-to-digital conversion, the time required to execute the appropriate flight control law, length of the electrical wires to the actuators, and the lag in response of the hydraulic system. If transport delay is too long, the pilot may direct additional control surface movement while his initial commands are in the process of being executed by the flight control system. This can result in overcontrol. Subsequent attempts to correct the overshoot can lead to a series of alternating overshoots or oscillations that are commonly referred to as a PIO. The range of transport delay times within which the Shuttle would be unlikely to encounter a PIO was determined using the DFBW F-8, enabling Dryden to develop a PIO suppression filter for the Shuttle. The PIO suppression filter was successfully evaluated in the F-8, installed in the Shuttle prior to its first mission into space, and proved to effectively eliminate the PIO issue.[63]

During Phase II, 169 flights were accomplished with several other test pilots joining the program, including Stephen D. Ishmael, Rogers Smith, and Edward Schneider. In addition to its previously noted accomplishments, the DFBW F-8 successfully evaluated adaptive control law approaches that would later become standard in many FBW aircraft. It was used in the Optimum Trajectory Research Experiment (OPTRE). This involved testing data uplink and downlink between the F-8 and a computer in the then-new Remotely Piloted Vehicle Facility. This experiment demonstrated that an aircraft equipped with a digital flight control system could be flown using control laws that were operating in ground-based digital computers. The F-8 conducted the first in-flight evaluations of an automatic angle-of-attack limiter and maneuvering flaps. These features are now commonly used on nearly all military and commercial aircraft with fly-by-wire flight controls. The DFBW F-8 also successfully tested an approach that used a backup software system known as the Resident Backup System (REBUS) to survive potential software faults that could cause all three primary flight control system computers to fail. The REBUS concept was later used in other experimental aircraft, as well as in production fly-by-wire flight control systems. The final flight-test effort of the DFBW program involved the development of a methodology called analytical redundancy management. In this concept, dynamic and kinematic relationships between dissimilar sensors and measurements were used to detect and isolate sensor failures.[64]

DFBW F-8: An Appreciation

The NASA DFBW F-8 had conclusively proven that a highly redundant digital flight control system could be successfully implemented and all aspects of its design validated.[65] During the course of the program, the DFBW F-8 demonstrated the ability to be upgraded to take advantage of emerging state-of-the-art technologies or to meet evolv-

---

63. Szalai, Calvin R. Jarvis, Gary E. Krier, Vincent A. Megna, Larry D. Brock, and Robert N. O'Donnell, "Digital Fly-by-Wire Flight Control Validation Experience," NASA TM-72860 (Dec. 1978). R.E. Bailey, M.F. Schaeffer, R.E. Smith, and J.F. Stewart, "Flight Test Experience With Pilot-Induced-Oscillation Suppression Filters," NASA TM-86028, NASA Dryden Research Center (Jan. 1984).
64. Digital Fly-By-Wire, "The All-Electric Airplane," NASA Dryden TF-2001-02 DFRC.
65. Szalai, et al., "Digital Fly-By-Wire Flight Control Validation Experience."

ing operational requirements. It proved that digital fly-by-wire flight control systems could be adapted to the new design and employment concepts that were evolving in both the military and in industry at the time. Perhaps the best testimony to the unique accomplishments of the F-8 DFBW aircraft and its NASA flight-test team is encapsulated in the following observations of former NASA Dryden director Ken Szalai:

> DFBW systems are 'old hat' today, but in 1972, only Apollo astronauts had put their life and missions into the hands of software engineers. We considered the F-8 DFBW a very high risk in 1972. That fact was driven home to us in the control room when we asked the EAFB [Edwards Air Force Base] tower to close the airfield, as was preplanned with the USAF, for first flight. It was the first time this 30-year-old FCS [Flight Control System] engineer had heard that particular radio call. . . . The project was both a pioneering effort for the technology and a key enabler for extraordinary leaps in aircraft performance, survivability, and superiority. The basic architecture has been used in numerous production systems, and many of the F-8 fault detection and fault handling/recovery technology elements have become 'standard equipment.' . . . In the total flight program, no software error/fault ever occurred in the operational software, synchronization was never lost in hundreds of millions of sync cycles, it was never required to transfer to the analog FBW backup system, there were zero nuisance channel failures in all the years of flying, and many NASA and visiting guest pilots easily flew the aircraft, including Phil Oestricher before the first YF-16 flight.[66]

In retrospect, the NASA DFBW F-8C is of exceptional interest in the history of aeronautics. It was the first aircraft to fly with a digital fly-by-wire flight control system, and it was also the first aircraft to fly without any mechanical backup flight controls. Flown by Ed Schneider, the DFBW F-8 made its last flight December 16, 1985, completing 211 flights. The aircraft is now on display at the NASA Dryden Flight Research Center. Its sustained record of success over a 13-year period provided a

---

66. Szalai, e-mail to the author, Mar. 11, 2009.

high degree of high confidence in the use of digital computers in fly-by-wire flight control systems. The DFBW F-8C also paved the way for a number of other significant NASA, Air Force, and foreign research programs that would further explore and expand the application of digital computers to modern flight control systems, providing greatly improved aircraft performance and enhanced flight safety.

The Lightweight Fighter Program and the YF-16

In addition to the NASA F-8 DFBW program, several other highly noteworthy efforts involving the use of computer-controlled fly-by-wire flight control technology occurred during the 1970s. The Air Force had initiated the Lightweight Fighter program in early 1972. Its purpose was "to determine the feasibility of developing a small, light-weight, low-cost fighter, to establish what such an aircraft can do, and to evaluate its possible operational feasibility."[67] The LWF effort was focused on demonstrating technologies that provided a direct contribution to performance, were of moderate risk (but sufficiently advanced to require prototyping to reduce risk), and helped hold both procurement and operating costs down. Two companies, General Dynamics (GD) and Northrop, were selected, and each was given a contract to build two flight-test prototypes. These would be known as the YF-16 and the YF-17. In its YF-16 design, GD chose to use an analog-computer-based quadruplex fly-by-wire flight control system with no mechanical backup. The aircraft had been designed with a negative longitudinal static stability margin of between 7 percent and 10 percent in subsonic flight—this indicated that its center of gravity was aft of the aerodynamic center by a distance of 7 to 10 percent of the mean aerodynamic chord of the wing. A high-speed, computer-controlled fly-by-wire flight control system was essential to provide the artificial stability that made the YF-16 flyable. The aircraft also incorporated electronically activated and electrically actuated leading edge maneuvering laps that were automatically configured by the flight control system to optimize lift-to-drag ratio based on angle of attack, Mach number, and aircraft pitch rate. A side stick controller was used in place of a conventional control column.[68]

---

67. "General Dynamics and Northrop to Build Lightweight Fighter Prototypes," *Interavia*, July 1972, p. 693.
68. C. Droste and J. Walker, "The General Dynamics Case Study on the F-16 Fly-By-Wire Flight Control System," *AIAA Professional Study Series*, AIAA, New York, June 1998.

Following an exceptionally rapid development effort, the first of the two highly maneuverable YF-16 technology demonstrator aircraft (USAF serial No. 72-1567) had officially first flown in February 1974, piloted by General Dynamics test pilot Phil Oestricher. However, an unintentional first flight had actually occurred several weeks earlier, an event that is discussed in a following section as it relates to developmental issues with the YF-16 fly-by-wire flight control system. During its development, NASA had provided major assistance to GD and the Air Force on the YF-16 in many technical areas. Fly-by-wire technology and the side stick controller concept originally developed by NASA were incorporated in the YF-16 design. The NASA Dryden DFBW F-8 was used as a flight testbed to validate the YF-16 side stick controller design. NASA Langley also helped solve numerous developmental challenges involving aerodynamics and control laws for the fly-by-wire flight control system. The aerodynamic configuration had been in development by GD since 1968. Initially, a sharp-edged strake fuselage forebody had been eliminated from consideration because it led to flow separation; however, rounded forward fuselage cross sections caused significant directional instability at high angles of attack. NASA aerodynamicists conducted wind tunnel tests at NASA Langley that showed the vortexes generated by sharp forebody strakes produced a more stable flow pattern with increased lift and improved directional stability. This and NASA research into leading – and trailing-edge flaps were used by GD in the development of the final YF-16 configuration, which was intensively tested in the Langley Full-Scale Wind Tunnel at high angle-of-attack conditions.[69]

During NASA wind tunnel tests, deficiencies in stability and control, deep stall, and spin recovery were identified even though GD had predicted the configuration to be controllable at angles of attack up to 36 degrees. NASA wind tunnel testing revealed serious loss of directional stability at angles of attack higher than 25 degrees. As a result, an automatic angle of attack limiter was incorporated into the YF-16 flight control system along with other changes designed to address deep stall and spin issues. Ensuring adequate controllability at higher angles of attack also required further research on the ability of the YF-16's fly-by-wire flight control system to automatically limit certain other flight parameters during energetic air combat maneuvering. The YF-16's all-moving

---

69. Chambers, *Partners in Freedom*.

horizontal tails provided pitch control and also were designed to operate differentially to assist the wing flaperons in rolling the aircraft. The ability of the horizontal tails and longitudinal control system to limit the aircraft's angle of attack during maneuvers with high roll rates at low airspeeds was critically important. Rapid rolling maneuvers at low airspeeds and high angles of attack were found to create large nose-up trim changes because of inertial effects at the same time that the aerodynamic effectiveness of the horizontal tails was reduced.[70]

An important aspect of NASA's support to the YF-16 flight control system development involved piloted simulator studies in the NASA Langley Differential Maneuvering Simulator (DMS). The DMS provided a realistic means of simulating two aircraft or spacecraft operating with (or against) each other (for example, spacecraft conducting docking maneuvers or fighters engaged in aerial combat against each other). The DMS consisted of two identical fixed-base cockpits and projection systems, each housed inside a 40-foot-diameter spherical projection screen. Each projection system consisted of a sky-Earth projector to provide a horizon reference and a system for target-image generation and projection. The projectors and image generators were gimbaled to allow visual simulation with completely unrestricted freedom of motion. The cockpits contained typical fighter cockpit instruments, a programmable buffet mechanism, and programmable control forces, plus a g-suit that activated automatically during maneuvering.[71] Extensive evaluations of the YF-16 flight control system were conducted in the DMS using pilots from NASA, GD, and the Air Force, including those who would later fly the aircraft. These studies verified the effectiveness of the YF-16 fly-by-wire flight control system and helped to identify critical flight control system components, timing schedules, and feedback gains necessary to stabilize the aircraft during high angle-of-attack maneuvering. As a result, gains in the flight control system were modified, and new con-

---

70. Later, during high angle-of-attack flight-testing of an early production F-16, the aircraft entered a stabilized deep-stall condition following a series of rolls in a vertical climbing maneuver. The test pilot was unable to recover with normal aerodynamic controls and used the anti-spin parachute installed for high angle-of-attack testing. NASA worked with the Air Force and the contractor to develop a fix that involved a "pitch rocker" technique to force the aircraft out of the deep stall. The approach was incorporated into the production F-16 flight control system as a pilot selectable emergency recovery mode. In addition, the horizontal tail area of the production F-16 was increased by about 25 percent. Ibid.
71. B.R. Ashworth and William M. Kahlbaum, Jr., "Description and Performance of the Langley Differential Maneuvering Simulator," NASA TN-D-7304, NASA Langley Research Center (June 1973).

trol elements—such as a yaw rate limiter, a rudder command fadeout, and a roll rate limiter—were developed and evaluated.[72]

Despite the use of the DMS and the somewhat similar GD Fort Worth domed simulator to develop and refine the YF-16 flight control system, nearly all flight control functions, including roll stick force gradient, were initially too sensitive. This contributed to the unintentional YF-16 first flight by Phil Oestricher at Edwards AFB on January 20, 1974. The intent of the scheduled test mission on that day was to evaluate the aircraft's pretakeoff handling characteristics. Oestricher rotated the YF-16 to a nose-up attitude of about 10 degrees when he reached 130 knots, with the airplane still accelerating slightly. He made small lateral stick inputs to get a feel for the roll response but initially got no response, presumably because the main gear were still on the ground. At that point, he slightly increased angle of attack, and the YF-16 lifted off the ground. The left wing then dropped rather rapidly. After a right roll command was applied, it went into a high-frequency pilot-induced oscillation. Before the roll oscillation could be stopped, the aft fin of the inert AIM-9 missile on the left wingtip lightly touched the runway, the right horizontal tail struck the ground, and the aircraft bounced on its landing gear several times, resulting in the YF-16 heading toward the edge of the runway. Oestricher decided to take off, believing it impossible to stay on the runway. He touched down 6 minutes later and reported: "The roll control was too sensitive, too much roll rate as a function of stick force. Every time I tried to correct the oscillation, I got full-rate roll." The roll control sensitivity problem was corrected with adjustments to the control gain logic. Stick force gradients and control gains continued to be refined during the flight-test program, with the YF-16 subsequently demonstrating generally excellent control characteristics. Oestricher later said that the YF-16 control problem would have been discovered before the first flight if better visual displays had been available for flight simulators in the early 1970s.[73] Lessons from the YF-16 and DFBW F-8 simulation experiences helped NASA, the Air Force, and industry refine the way that preflight simulation was structured to support new fly-by-wire flight control systems development. Another flight control issue that arose during

---

72. Chambers, *Partners in Freedom*.
73. Joe Stout, "What a Wonderful Airplane: YF-16 First Flight," *Code One Magazine*, General Dynamics, July 1992.

the YF-16 flight-test program involved an instability caused by interaction of the active fly-by-wire flight control system with the aeroelastic properties of the airframe. Flutter analysis had not accounted for the effects of active flight control. Closed loop control systems testing on the ground had used simulated aircraft dynamics based on a rigid airframe modeling assumption. In flight, the roll sensors detected aeroelastic vibrations in the wings, and the active flight control system attempted to apply corrective roll commands. However, at times these actually amplified the airframe vibrations. This problem was corrected by reducing the gain in the roll control loop and adding a filter in the feedback patch that suppressed the high-frequency signals from structural vibrations. The fact that this problem was also rapidly corrected added confidence in the ability of the fly-by-wire flight control system to be reconfigured. Another change made as a result of flight test was to fit a modified side stick controller that provided the pilot with some small degree of motion (although the control inputs to the flight control system were still determined by the amount of force being exerted on the side stick, not by its position).[74]

Three days after its first official flight on February 2, 1974, the YF-16 demonstrated supersonic windup turns at Mach 1.2. By March 11, it had flown 20 times and achieved Mach 2.0 in an outstanding demonstration of the high systems reliability and excellent performance that could be achieved with a fly-by-wire flight control system. By the time the 12-month flight-test program ended January 31, 1975, the two YF-16s had flown a total of 439 flight hours in 347 flights, with the YF-16 Joint Test Force averaging over 30 sorties per month. Open communications between NASA, the Air Force, and GD had been critical to the success of the YF-16 development program. In highlighting this success, Harry J. Hillaker, GD Vice President and Deputy Program Director for the F-16, noted the vital importance of the "free exchange of experience from the U.S. Air Force Laboratories and McDonnell-Douglas 680J projects on the F-4 and from NASA's F-8 fly-by-wire research program."[75] The YF-16 would serve as the basis for the extremely successful family of F-16 mul-

---

74. Aronstein and Piccirillo, "The Lightweight Fighter Program: A Successful Approach to Fighter Technology Transition."
75. H.J. Hillaker, "The F-16: A Technology Demonstrator, a Prototype, and a Flight Demonstrator," *Proceedings of AIAA Aircraft Prototype and Technology Demonstrator Symposium*, AIAA, New York, 1983, pp. 113–120.

tinational fighters; over 4,400 were delivered from assembly lines in five countries by 2009, and production is expected to continue to 2015. While initial versions of the production F-16 (the A and B models) used analog computers, later versions (starting with the F-16C) incorporated digital computers in their flight control systems.[76] Fly-by-wire and relaxed static stability gave the F-16 a major advantage in air combat capability over conventional fighters when it was introduced, and this technology still makes it a viable competitor today, 35 years after its first flight.

The F-16's main international competition for sales at the time was another statically unstable full fly-by-wire fighter, the French Dassault Mirage 2000, which first flew in 1978. Despite the F-16 being selected for European coproduction, over 600 Mirage 2000s would also eventually be built and operated by a number of foreign air forces. The other technology demonstrator developed under the LWF program was the Northrop YF-17. It featured a conventional mechanical/hydraulic flight control system and was statically stable. When the Navy decided to build the McDonnell-Douglas F/A-18, the YF-17 was scaled up to meet fleet requirements. Positive longitudinal static stability was retained, and a primary fly-by-wire flight control system was incorporated into the F/A-18's design. The flight control system also had an electric backup that enabled the pilot to transmit control inputs directly to the control surfaces, bypassing the flight control computer but using electrical rather than mechanical transmission of signals. A second backup provided a mechanical linkage to the horizontal tails only. These backup systems were possible because the F/A-18, like the YF-17, was statically stable about all axes.[77]

F-16 CCV

By the mid-1970s, the Air Force Flight Dynamics Laboratory had initiated a Control Configured Vehicle flight research program to investigate the use of nonconventional (often called "decoupled") movements of aircraft flight control surfaces to enable maneuvers in one plane without movement in another. An example of such a maneuver would be a wings-level turn without having to bank the aircraft. The very first General Dynamics YF-16 technology demonstrator aircraft (USAF serial No. 72-1567) was selected for modification. Rebuilt in December 1975 and fitted with twin

---

76. Aronstein and Piccirillo, "The Lightweight Fighter Program: A Successful Approach to Fighter Technology Transition."
77. Ibid.

vertical canards underneath the air intake, it became known as the F-16/CCV. Its flight controls were modified to enable the wing trailing edge flaperons to move in combination with the all-moving stabilator. In addition, the fuel system in the YF-16 was modified to enable the aircraft center of gravity to be adjusted in flight by transferring fuel between tanks, thus allowing the stability of the aircraft to be varied. The YF-16/CCV flew for the first time on March 16, 1976, piloted by GD test pilot David J. Thigpin. On June 24, 1976, while being flown by David Thigpin, it was seriously damaged in a crash landing. Engine power had been lost on final approach, and the landing gear collapsed in the subsequent hard landing. Repairs to the aircraft would take over 6 months. The F-16 CCV returned to flight in the spring of 1977. It would complete 87 flights and 125 flying hours before the research program ended, with the last flight F-16 CCV flight on June 30, 1977.[78]

The F-16 CCV anticipated CCV flight-test approaches that were soon undertaken by a number of foreign countries. Flight research projects in Germany, the U.K., and Japan converted existing military aircraft into CCV testbeds. Fitted with computer-controlled fly-by-wire flight control systems, these projects provided experience and insights that enabled these countries to incorporate fly-by-wire flight control into their next generations of advanced civil and military aircraft. These foreign CCV projects are discussed in a separate section of this report. Experience gained with the F-16 CCV served as the basis for the subsequent Flight Dynamics Laboratory AFTI/F-16 program, which would yield valuable insights into many issues associated with developing advanced DFBW flight control systems and result in significantly improved capabilities being incorporated into U.S. military aircraft.

YC-14

The Air Force Boeing YC-14 Short Take-Off and Landing (STOL) jet transport technology demonstrator flew for the first time on August 9, 1976, from Boeing Field in Seattle, WA, during the period between Phase I and Phase II of the NASA DFBW F-8 program. Two prototypes were built with the second aircraft flying in October 1976. The YC-14 is noteworthy in that it was the first aircraft to fly with a fault-tolerant multichannel redundant digital fly-by-wire flight control system.

---

78. Stout, "What a Wonderful Airplane: YF-16 First Flight." Joseph F. Baugher, "General Dynamics YF-16/CCV," *American Military Aircraft*, Mar. 31, 2000.

A mechanical backup flight control capability was retained. The full authority triply redundant digital fly-by-wire flight control system, designed by the British Marconi Company, performed computational commands for pitch, roll, and yaw that were used to control the elevator, aileron, and rudder actuation systems. The reconfigurable computer architecture divided the basic control path into three subfunctional elements with these elements replicated to provide fault tolerance. The internal element redundancy management function was intended to detect and isolate faulty elements and perform the necessary reconfiguration. The input signal selection methodology was intended to guarantee that all three computers used the same numbers and thus produced identical output values. During normal operation, the overall system output value was selected as the midvalue of the three individual values. The system would continue to operate in the event of a failure of one computer by taking the average of the output of the two remaining computers. If they disagreed, both were disabled and the aircraft reverted to the backup manual control system.[79]

The YC-14 was also noteworthy in that it used optical data links to exchange data between the triply redundant computers. The optical communications medium was chosen to eliminate electromagnetic interference effects, electrical grounding loop problems, and the potential propagation of electrical malfunctions between channels. Optical coupling was used to maintain interchannel integrity. Each sensor's output was coupled to the other channels so that each computer had data from each of the other sensors. Identical algorithms in each computer were used. They consolidated the data, enabling equalization and fault detection/isolation of the inputs. The computers were synchronized to avoid sampling time differences and to assure that all computers were receiving identical data inputs.[80]

An important observation involving redundant computer-controlled fly-by-wire flight control systems was derived from the YC-14 flight-test experience. As noted above, the system was designed to ensure that all computers used the same sensor input values and should therefore produce identical outputs. However, a significant fault in the digital

---

79. L. Martin and D. Gangsaas, "Testing the YC-14 Flight Control System Software," *AIAA Journal of Guidance and Control*, July–Aug. 1978.
80. H.A. Rediess and E.C. Buckley, "Technology Review of Flight Crucial Flight Control Systems (Application of Optical Technology)," NASA CR-172332 (Supplement 1) (Sept. 1984).

flight control software was encountered during flight-testing that had not been detected during ground laboratory testing. The software fault resulted in incorrect tracking of control law computations in each of the three flight control channels, with each channel performing signal selections on a different set of values. This resulted in different input data for the three channels. Although the discrepancies between each channel's inputs were small, the cumulative effect led to large tracking errors between flight control channels when airborne.[81]

Following cancellation of the Air Force YC-14A program in 1979, the two prototypes were placed in storage at Air Force's Aerospace Maintenance and Regeneration Group (AMARC) at Davis-Monthan AFB, AZ, in April 1980. The first prototype is now displayed at the Pima Air and Space Museum in Tucson, AZ.

Advanced Fighter Technology Integration F-16 Program

The USAF Flight Dynamics Laboratory began the Advanced Fighter Technology Integration program in the late 1970s. Overall objectives of this joint Air Force and NASA research program were to develop and demonstrate technologies and assess alternative approaches for use in future aircraft design. In December 1978, the F-16 was selected for modification as the AFTI/F-16. General Dynamics began conversion of the sixth preproduction F-16A (USAF serial No. 75-0750) at its Forth Worth, TX, factory in March 1980. The aircraft had originally been built in 1978 for the F-16 full-scale development effort. GD built on earlier experience with its F-16 CCV program. The twin canted movable canard ventral fins from the F-16 CCV were installed under the inlet of the AFTI/F-16. In addition, a dorsal fairing was fitted to the top of the fuselage to accommodate extra avionics equipment. A triply redundant, asynchronous, multimode, digital flight control system with an analog backup was installed in the aircraft. The DFCS was integrated with improved avionics and had different control modes optimized for air-to-air combat and air-to-ground attack. The Stores Management System (SMS) was responsible for signaling requests for mode change to the DFCS. Other modifications included provision for a six-degree-of-freedom Automated Maneuvering Attack System (AMAS), a 256-word-capacity Voice-Controlled Interactive Device (VCID) to control the avionics

---

81. Martin, et al., "Testing the YC-14 Flight Control System Software."

Case 10 | Fly-By-Wire: Making the Electric Jet

During Phase I, five test pilots from NASA, the Air Force, and the Navy flew the AFTI/F-16 at NASA Dryden in California. NASA.

suite, and a helmet-mounted target designation sight that could automatically slave the forward-looking infrared (FLIR) device and the radar to the pilot's head movements.[82] First flight of the modified aircraft in the AFTI/F-16 configuration occurred on July 10, 1982, from Carlswell AFB, TX, with GD test pilot Alex V. Wolfe at the controls. Following contractor testing, the aircraft was flown to Edwards AFB for AFTI/F-16 test effort. This was organized into two phases; Phase I was a 2-year effort focused on evaluating the DFCS, with a follow-on Phase II oriented to assessing the AMAS and other technologies.

AFTI Phase I Testing

Phase I flight-testing was conducted by the AFTI/F-16 Joint Test Force from the NASA Dryden Flight Research Facility at Edwards AFB, CA, from July 10, 1982, through July 30, 1983. During this phase, five test pilots from NASA, the Air Force, and the U.S. Navy flew the aircraft. Initial flights checked out the aircraft's stability and control systems. Handling qualities were assessed in air-to-air and air-to-ground scenarios, as well

---

82. Stephen D. Ishmael and Donald R. McMonagle, "AFTI/F-16 Flight Test Results and Lessons," NASA TM-84920 (Oct. 1983).

673

as in-formation flight and during approach and landing. The Voice Command System allowed the pilot to change switch positions, display formats, and modes simply by saying the correct word. Initial tests were of the system's ability to recognize words, with later testing conducted under increasing levels of noise, vibrations, and g-forces. Five pilots flew a total of 87 test sorties with the Voice Command System, with a general success rate approaching 90 percent. A prototype helmet-mounted sight was also evaluated. On July 30, 1983, the AFTI/F-16 aircraft was flown back to the General Dynamics facility at Fort Worth, TX, for modification for Phase II. During the Phase I test effort, 118 flight-test sorties were flown, totaling about 177 flight hours. In addition to evaluating the DFCS, the potential operational utility of task-tailored flight modes (that included decoupling of aircraft attitude and flight path) was also assessed. During these unconventional maneuvers, the AFTI/F-16 demonstrated that it could alter its nose position without changing flight path and change its flight path without changing aircraft attitude. The aircraft also performed coordinated horizontal turns without banking or sideslip.[83] NASA test pilot Bill Dana recounted: "In Phase I we evaluated non-classic flight control modes. By deflecting the elevators and flaps in various relationships, it was possible to translate the aircraft vertically without changing pitch attitude or to pitch-point the airplane without changing your altitude. You could also translate laterally without using bank and yaw-point without translating the aircraft, by using rudder and canard inputs programmed together in the flight control computer."[84]

Phase II Testing

From mid-1983 through mid-1984, components for the Automated Maneuvering Attack System and related avionics systems were installed into the AFTI/F-16 at GD in Fort Worth in preparation for the Phase II effort. Precision electrical-optical tracking pods were installed in the wing root area on both sides of the aircraft. First flight of the AFTI/F-16 in the AMAS configuration was on July 31, 1984, with Phase II flight-testing at Edwards beginning shortly after the aircraft returned to Dryden on August 6, 1984. Beginning in September 1984 and continuing through April 1987, improved sensors, integrated fire and flight control, and enhancements

---

83. Ibid.
84. Gary Creech, "AFTI/F-16 Retires After 22 Years," *The Dryden Express*, Dryden Flight Research Center, vol. 43, issue 2, Feb. 23, 2001.

in pilot-vehicle interface were evaluated. During Phase II testing, the system demonstrated automatic gun tracking of airborne targets and accurate delivery of unguided bombs during 5-g curvilinear toss bomb maneuvers from altitudes as low as 200 feet. An all-attitude automatic ground collision avoidance capability was demonstrated,[85] as was the Voice Command System (for interfacing with the avionics system), a helmet-mounted sight (used for high off bore sight target cueing), and a digital terrain system with color moving map.[86] The sortie rate during Phase II was very high. From the start of the AMAS tests in August 1984 to the completion of Phase II in early 1987, 226 flights were accomplished, with 160 sorties being flown during 1986. To manage this high sortie rate, the ground maintenance crews worked a two-shift operation.

Follow-On AFTI/F-16 Testing

Following Phase II in 1987, the forward fuselage-mounted ventral fins were removed and the AFTI/F-16 was flown in support of other test efforts and new aircraft programs, such as evaluating strike technologies proposed for use in the next generation ground attack aircraft, which eventually evolved into the Joint Strike Fighter (JSF) program.

Strike Technology Testbed

In the summer of 1991, a flight-test effort oriented to close air support and battlefield air interdiction began. The focus was to demonstrate technologies to locate and destroy ground targets day or night, good weather or bad, while maneuvering at low altitudes. The AFTI/F-16 was modified with two forward-looking infrared sensors mounted in turrets on the upper fuselage ahead of the canopy. The pilot was equipped with a helmet-mounted sight that was integrated with the infrared sensors. As he moved his head, they followed his line of sight and transmitted their images to eyepieces mounted in his helmet. The nose-mounted canards used in earlier AFTI/F-16 testing were removed. Testing emphasized giving pilots the capability to fly their aircraft and attack targets in darkness or bad weather. To assist in this task, a digital terrain map was stored

---

85. James Blaylock, Donald Swihart, and William Urshel, "Integration of Advanced Safety Enhancements for F-16 Terrain Following," AIAA-1987-2906.
86. Charles A. Baird and Franklin B. Snyder, with introduction by Lt. Mark Bierele, AFTI/F-16 Program Office, "Terrain-Aided Altitude Computations on the AFTI/F-16," Harris Corporation, Melbourne, FL, Aug. 1990.

in the aircraft computer. Advanced terrain following was also evaluated. This used the AFTI/F-16's radar to scan terrain ahead of the aircraft and automatically fly over or around obstacles. The pilot could select minimum altitudes for his mission. The system would automatically calculate that the aircraft was about to descend below this altitude and initiate a 5 g pullup maneuver. The advanced terrain following system was connected to the Automated Maneuvering Attack System, enabling the pilot to delivery weapons from altitudes as low as 500 feet in a 5 g turn. An automatic Pilot Activated Recovery System was integrated with the flight control system. If the pilot became disoriented at night or in bad weather, he could activate a switch on his side controller. This caused the flight control computer to automatically recover the aircraft putting it into a wings-level climb. Many of these technologies have subsequently transitioned into upgrades to existing fighter/attack aircraft.[87]

The final incarnation of this unique aircraft would be as the AFTI/F-16 power-by-wire flight technology demonstrator.

Power-By-Wire Testbed

During 1997, NASA Dryden had evaluated a single electrohydrostatic actuator installation on the NASA F-18 Systems Research Aircraft (SRA), with the primary goal being the flight demonstration of power-by-wire technology on a single primary flight control surface. The electrohydrostatic actuator, provided by the Air Force, replaced the F-18's standard left aileron actuator and was evaluated throughout the aircraft's flight envelope out to speeds of Mach 1.6. Numerous mission profiles were accomplished that included a full series of aerobatic maneuvers. The electrohydrostatic actuator accumulated 23.5 hours of flight time on the F-18 SRA between January and July 1997. It performed as well as the standard F-18 actuator and was shown to have more load capability than required by the aileron actuator specification for the aircraft.[88]

At about the same time, a Joint Strike Fighter/Integrated Subsystems Technology program had been formed to reduce the risk of selected

---

87. Finley Barfield, Duke Browning, and Judith Probert, "All Terrain Ground Collision Avoidance and Maneuver Terrain Following for Automated Low Level Night Attack," *IEEE AES Systems Magazine*, Mar. 1993.

88. Robert Navarro, "Performance of an Electro-Hydrostatic Actuator on the F-18 Systems Research Aircraft," NASA TM-97-206224, NASA Dryden Flight Research Center (1997).

technology candidates, in particular the power-by-wire approach that was intended to replace cumbersome hydraulic actuation systems with all-electrical systems for flight surface actuation. A key to this effort was the AFTI F-16, which was modified to replace all of the standard hydraulic actuators on the primary flight control surfaces with electrohydrostatic actuators (EHAs) to operate the flaperons, horizontal tails, and rudder. Each electrohydrostatic actuator uses an internal electric motor to drive an integral hydraulic pump, thus it relies on local hydraulics for force transmission (similar to the approach used with the Powered Flight Control Units on the Vickers VC10 aircraft discussed earlier).[89]

In a conventional F-16, the digital fly-by-wire flight control system sends out electrical command signals to each of the flight control actuators. These electrical signals drive the control valves (located with the actuators) that schedule the fluid from the high-pressure hydraulic pump to position the flight control surfaces. Dual engine-driven 3,000 pounds per square inch (psi) hydraulic systems power each primary control surface actuator to drive the control surfaces to the desired position. The standard F-16 hydraulic actuators operate continuously at 3,000 psi, and power is dumped into the actuators, whether it is needed or not.[90] In straight and level flight (where most aircraft operate most of their time, including even high-performance fighters), the actual electrical power requirement of the actuation system is low (only about 500 watts per actuator), and excess energy is dissipated as heat and is transferred into the fuel system.[91]

With the electrohydrostatic power design tested in the AFTI/F-16, the standard fly-by-wire flight control system was relatively unchanged. However, the existing F-16 hydraulic power system was removed and replaced by a new power-by-wire system, consisting of an engine-driven Hamilton Sundstrand dual 270-volt direct current (DC) electrical power generation system (to provide redundancy) and Parker Aerospace electrohydrostatic actuators on the flaperons, rudder, and horizontal stabilizer. The new electrical system powers five dual power electronics units, one for each flight control surface actuator. Each power electronics unit regulates the DC electrical power that drives dual motor/pumps that are self-contained in each electrohydrostatic actuator. The dual

---

89. James W. Ramsey, "Power-by-Wire," *Avionics Magazine*, May 1, 2001.
90. Droste, et al., "The General Dynamics Case Study on the F-16 Fly-By-Wire Flight Control System."
91. Ramsey, "Power-by-Wire."

motor/pumps convert electrical power into hydraulic power, allowing the piston on the actuators to move the control surfaces. The electrohydrostatic actuators operate at pressures ranging from 300 to 3,000 psi, providing power only on demand and generating much less heat. An electrical distribution and electrical actuation system simplifies secondary power and thermal management systems, because the need to provide secondary and emergency backup sources of hydraulic power for the flight control surfaces is eliminated. The electrohydrostatic system also provides more thermal margin, which can be applied to cooling other high-demand systems (such as avionics and electronic warfare), or, alternatively, the thermal management system weight and volume can be reduced making new aircraft designs smaller, lighter, and more affordable. Highly integrated electrical subsystems, including power-by-wire, reportedly could reduce takeoff weight by 6 percent, vulnerable area by 15 percent, procurement cost by 5 percent, and total life-cycle cost by 2 to 3 percent, compared with current fighters based on Air Force and industry studies. The power-by-wire approach is now being used in the Lockheed Martin F-35 Lightning II, with the company estimating a reduction in aircraft weight of as much as 700 pounds because of weight reductions in the hydraulic system, the secondary power system, and the thermal management system, made possible because the electrical power-by-wire system produces less heat than the traditional hydraulic system that it replaces.[92]

The modified power-by-wire AFTI/F-16 was the first piloted aircraft of any type to fly with a totally electric control surface actuation system with no hydraulic or mechanical backup flight control capability of any kind. It was designed to have the same flight control system responses as an unmodified F-16. After the first power-by-wire AFTI/F-16 flight on October 24, 2000, at Fort Worth, Lockheed Martin test pilot Steve Barter stated aircraft handling qualities with the power-by-wire modifications were indistinguishable from that of the unmodified AFTI/F-16. The aircraft was subsequently flown about 10 times, with flight control effectiveness of the power-by-wire system demonstrated during supersonic flight. Test pilots executed various flying quality maneuvers, including high-g turns, control pulses (in pitch, roll, and yaw), doublet inputs, and sideslips. The tests also included simulated low-altitude attack

---

92. Ibid.

missions and an evaluation of the electrostatic actuator and generator subsystems and their thermal behavior under mission loads.[93]

NASA Dryden hosted the AFTI/F-16 program for 16 years, from 1982 to 1998. During that time, personnel from Dryden composed 50 percent of the AFTI joint test team. Dryden pilots who flew the AFTI/F-16 included Bill Dana, Dana Purifoy, Jim Smolka, Rogers Smith, and Steve Ishmael. Dryden responsibilities, in addition to its host role, included flight safety, operations, and maintenance. Mark Skoog, who served as the USAF AFTI/F-16 project manager for many years and later became a NASA test pilot, commented: "AFTI had the highest F-16 sortie success rate on base, due to Dryden maintenance personnel having tremendous expertise in tailoring their operations to the uniqueness of the vehicle. That includes all the other F-16s based at Edwards during those years, none of which were nearly as heavily modified as the AFTI."[94] A good summary of the AFTI/F-16's accomplishments was provided by NASA test pilot Dana Purifoy: "Flying AFTI was a tremendous opportunity. The aircraft pineered many important technologies including glass cockpit human factors, automated ground collision avoidance, integrated night vision capability and on-board data link operations. All of these technologies are currently being implemented to improve the next generation of both civil and military aircraft."[95] The AFTI F-16's last flight at Dryden was on November 4, 1997. Over a period of 15 years, it made over 750 flights and was flown by 23 pilots from the U.S. Air Force, NASA, the U.S. Marine Corps, and the Swedish Air Force. The AFTI F-16 then served as an Air Force technology testbed. Experience and lessons learned were used to help develop the production DFBW flight control system used in the F-16. The F-16, the F-22, and the F-35, in particular, directly benefited from AFTI/F-16 research and technology maturation efforts. After 22 years as a research aircraft for NASA and the Air Force, the AFTI F-16 was flown to Wright-Patterson AFB, OH, on January 9, 2001, for display at the Air Force Museum.[96]

Technology Transfer and Lessons Learned

Flight-test results from the AFTI/F-16 program were exceptionally well-documented by NASA and widely published in technical papers,

---

93. Ramsey, "Power-by-Wire."
94. Creech, "AFTI/F-16 Retires After 22 Years."
95. Ibid.
96. "AFTI F-16 Fact Sheet," National Museum of the U.S. Air Force, 2001.

memorandums, and presentations.[97] These provide invaluable insights into the problems, issues, and achievements encountered in this relatively early attempt to integrate an asynchronous digital flight control system into a high-performance military jet fighter. As the definition implies, in an asynchronous flight control system design, the redundant channels run autonomously. Each computer samples sensors and evaluates flight control laws independently. Each separately sends command signals to an averaging or selection device that is used to drive the flight control actuators. In this DFCS implementation, the unsynchronized individual computers can sample the sensors at slightly different times. Thus, they can obtain readings that may differ quite appreciably from one another, especially if the aircraft is maneuvering aggressively. Flight control law gains can further amplify these input differences, causing even larger differences between the results that are submitted to the output selection algorithm.[98]

During ground qualification of the AFTI/F-16, it was found that these differences sometimes resulted in a channel being declared failed when no real failure had occurred.[99] An even more serious shortcoming of asynchronous flight control systems can occur when the control laws contain decision points. Sensor noise and sampling variations may cause independent channels within the DFCS to take different paths at the decision points and to produce widely divergent outputs.[100] This occurred on AFTI/F-16 flight No. 44. Two channels in the DFCS declared each other failed; the analog backup was not selected because simultaneous failure of two DFCS channels had not been anticipated. The pilot could not reset the system, and the aircraft was flown home on the single remaining DFCS channel. In this case, all protective redun-

---

97. Stephen D. Ishmael, Dale A. Mackall, and Victoria A. Regenie, "Design Implications From AFTI/F16 Flight Test," NASA TM-86026, NASA Ames Research Center, Dryden Flight Research Facility, Edwards, CA (1984); Ishmael, et al., "AFTI/F-16 Flight Test Results and Lessons"; Mackall, "Development and Flight Test Experiences with a Flight-Crucial Digital Control System," NASA TP-2857, NASA Ames Research Center, Dryden Flight Research Facility, Edwards, CA (Nov. 1988); Mackall, "AFTI/F-16 Digital Flight Control System Experience," in Gary P. Beasley, ed., *NASA Aircraft Controls Research 1983*, pp. 469–487, NASA CP-2296 (1984), *Proceedings of Workshop Held at NASA Langley Research Center*, Oct. 25–27, 1983.
98. John Rushby, "Formal Methods and the Digital Systems Validation for Airborne Systems," NASA CR-4551, NASA Langley Research Center (Dec. 1, 1993).
99. Mackall, "Development and Flight Test Experiences with a Flight-Crucial Digital Control System."
100. Rushby, "Formal Methods and the Digital Systems Validation for Airborne Systems."

dancy had been lost, yet an actual hardware failure had not occurred. Several other difficulties and failure indications were observed during the flight-test program that were traced to asynchronous operation, allowing different channels to take different paths at certain selection points. The software was subsequently modified to introduce voting at some of these software selection points.[101]

NASA Observations

NASA observations on some of the more serious issues encountered in early testing of the AFTI/F-16 asynchronous digital flight control system are worthy of note. For example, an unknown failure in the Stores Management System on flight No. 15 caused it to request DFCS mode changes at a rate of 50 times per second. The DFCS could not keep up and responded at a rate of 5 mode changes per second. The pilot reported that the aircraft felt like it was in severe turbulence. The flight was aborted, and the aircraft landed safely. Subsequent analysis showed that if the aircraft had been maneuvering at the time, the DFCS would have failed. A subsequent software modification improved the DFCS's immunity to this failure mode.[102]

A highly significant flight control law anomaly was encountered on AFTI/F-16 flight No. 36. Following a planned maximum rudder "step and hold" input by the pilot, a 3-second departure from controlled flight occurred. Sideslip angle exceeded 20 degrees, normal acceleration fluctuated from –4 g to +7 g, angle of attack varied between –10 and +20 degrees, and the aircraft rolled 360 degrees. Severe structural loads were encountered with the vertical tailfin exceeding its design load. During the out-of-control situation, all control surfaces were operating at rate limits, and failure indications were received from the hydraulics and canard actuators. The failures were transient and reset after the pilot regained control. The problem was traced to a fault in the programmed flight control laws. It was determined that the aerodynamic model used to develop the control laws did not accurately model the nonlinear nature of yaw stability variations as a function of higher sideslip angles. The same inaccurate control laws were also used in the real-time AFTI/F-16 ground flight simulator. An additional complication was caused when the side fuselage-mounted air-data probes were blanked by the canard

---

101. Mackall, "Development and Flight Test Experiences with a Flight-Crucial Digital Control System."
102. Ibid.

at the high angles of attack and sideslip encountered. This resulted in incorrect air data values being passed to the DFCS. Operating asynchronously, the different flight control system channels took different paths through the flight control laws. Analysis showed these faults could have caused complete failure of the DFCS and reversion to analog backup.[103] Subsequently, the canards were removed from the command path to prevent the AFTI/F-16 from obtaining higher yaw angles.

AFTI/F-16 flight-testing revealed numerous other flight control problems of a similar nature. These prompted NASA engineer Dale Mackall to report: "The asynchronous design of the [AFTI/F-16] DFCS introduced a random, unpredictable characteristic into the system. The system became untestable in that testing for each of the possible time relationships between the computers was impossible. This random time relationship was a major contributor to the flight test anomalies. Adversely affecting testability and having only postulated benefits, asynchronous operation of the DFCS demonstrated the need to avoid random, unpredictable, and uncompensated design characteristics." Mackall also provided additional observations that would prove to be highly valuable in developing, validating, and certifying future software-intensive digital fly-by-wire flight control system designs. Urging more formal approaches and rigorous control over the flight control system software design and development process, Mackall reported:

> The criticality and number of anomalies discovered in flight and ground tests owing to design oversights are more significant than those anomalies caused by actual hardware failures or software errors. . . . As the operational requirements of avionics systems increase, complexity increases. . . . If the complexity is required, a method to make system designs more understandable, more visible, is needed . . . qualification of such a complex system as this, to some given level of reliability, is difficult . . . the number of test conditions becomes so large that conventional testing methods would require a decade for completion. The fault-tolerant design can also affect overall system reliability by being made too complex and by adding characteristics which are random in nature, creating an untestable design.[104]

---

103. Ibid.
104. Ibid.

Aircraft Certification Contributions

Certification of new aircraft with digital fly-by-wire flight control systems, especially for civilian airline service, requires software designs that provide highly reliable, predictable, and repeatable performance. For this reason, NASA experts concluded that a comprehensive understanding of all possible software system behaviors is essential, especially in the case of highly complex systems. This knowledge base must be formally documented and accurately communicated for both design and system certification purposes. This was highlighted in a 1993 research paper sponsored by NASA and the Federal Aviation Administration (FAA) that noted:

> This formal documentation process would prove to be a tremendously difficult and challenging task. It was only feasible if the underlying software was rationally designed using principles of abstraction, layering, information-hiding, and any other technique that can advance the intellectual manageability of the task. This calls strongly for an architecture that promotes separation of concerns (whose lack seems to be the main weakness of asynchronous designs), and for a method of description that exposes the rationale for design decisions and that allows, in principle, the behavior of the system to be calculated (i.e., predicted or, in the limit, proved) . . . formal methods can make their strongest contribution to quality assurance for ultra-dependable systems: they address (as nothing else does) [NASA engineer Dale] Mackall's plea for 'a method to make system designs more understandable, more visible.'[105]

Formal software development methodologies for critical aeronautical and space systems developments have been implemented within NASA and are contained in certification guidebooks and other documents for use by those involved in mission critical computer and software systems.[106] Designed to help transition Formal Methods from experimental use into

---

105. Rushby, "Formal Methods and the Certification of Critical Systems," SRI CSL-93-07, SRI International (Nov. 1993).
106. "Formal Methods Specification and Verification Guidebook for Software and Computer Systems, Volume I: Planning and Technology Insertion," NASA TP-98-208193 (1998); "Formal Methods Specification and Analysis Guidebook for the Verification of Software and Computer Systems, Volume II: A Practitioner's Companion," NASA GB-001-97 (1997).

practical application for critical software requirements and systems design within NASA, they discuss technical issues involved in applying Formal Methods techniques to aerospace and avionics software systems. Dryden's flight-test experience and the observations obtained from flight-testing of such systems were exceptionally well-documented and would prove to be highly relevant to NASA, the FAA, and military service programs oriented to developing Formal Methods and structured approaches in the design, development, verification, validation, testing, and certification of aircraft with advanced digital flight control systems.[107] The NASA DFBW F-8 and AFTI/F-16 experiences (among many others) were also used as background by Government and industry experts tasked with preparing the FAA Digital Systems Validation Handbook. Today, the FAA uses Formal Methods in the specification and verification of software and hardware requirements, designs, and implementations; in the identification of the benefits, weaknesses, and difficulties in applying these Formal Methods to digital systems used in critical applications; and in support of aircraft software systems certification.

YA-7D DIGITAC

Digital Flight Control for Tactical Aircraft (DIGITAC) was a joint program between the Air Force Flight Dynamics Laboratory (AFFDL) at Wright-Patterson AFB, OH, and the USAF Test Pilot School (TPS) at Edwards AFB. Its purpose was to develop and demonstrate digital flight control technology for potential use in future tactical fighter and attack aircraft, including the feasibility of using digital flight control computer technology to optimize an airplane's tracking and handling qualities for a full range of weapons delivery tasks. The second prototype LTV YA-7D (USAF serial No. 67-14583) was selected for modification as the DIGITAC testbed by replacing the analog computer of the YA-7D Automated Flight Control System (AFCS) with the DIGITAC digital multimode flight control system that was developed by the AFFDL. The mechanical flight control system in the YA-7D was unchanged and was retained as a backup capability.

The YA-7D's flight control system was eventually upgraded to DIGITAC II configuration. DIGITAC II used military standard data buses and transferred critical flight control data between individual computers and between computers and remote terminals. The data buses used

---

107. Szalai, et al., "Digital Fly-By-Wire Flight Control Validation Experience."

were dual channel wire and dual channel fiber optic and were selectable in the cockpit by the pilot to allow him to either fly-by-wire or fly-by-light. Alternately, for flight-test purposes, the pilot was able to implement one wire channel and one fiber optic channel. During early testing, the channel with the multifiber cables (consisting of 210 individual fibers) encountered numerous fiber breakage problems during normal ground maintenance. The multifiber cable design was replaced by single-fiber cables with tough protective shields, a move that improved data transmission qualities and nearly eliminated breakage issues. The DIGITAC fly-by-light system flew 290 flights during a 3-year period, performing flawlessly with virtually no maintenance. It was so reliable that it was used to fly the aircraft on all routine test missions. The system performance and reliability was considered outstanding, with the technical approach assessed as ready for consideration for use in production aircraft.[108]

The DIGITAC YA-7D provided the TPS with a variable stability testbed aircraft for use in projects involving assessments of advanced aircraft flying qualities. Results obtained from these projects contributed to the flying qualities database in many areas, including degraded-mode flight control cross-coupling, control law design, pro versus adverse yaw studies, and roll-subsistence versus roll-time-delay studies. Under a TPS project known as Have Coupling, the YA-7D DIGITAC aircraft was used to investigate degradation to aircraft handling qualities that would occur in flight when a single pitch control surface (such as one side of the horizontal stabilizer) was damaged or impaired. An asymmetric flight control situation would result when a pure pitch motion was commanded by the pilot, with roll and yaw cross-coupling motions being produced. For the Have Coupling tests, various levels of cross-coupling were programmed into the DIGITAC aircraft. The resulting data provided a valuable contribution to the degraded flight control mode handling qualities body of knowledge. This included the interesting finding that with exactly the same amounts of cross-coupling present, pilot ratings of aircraft handling qualities in flight-testing were significantly different compared with those rating obtained in the ground-based simulator.[109]

---

108. Rediess, et al., "Technology Review of Flight Crucial Flight Control Systems (Application of Optical Technology)."
109. E. Robert Lemble, "DIGITAC—A Unique Digital Flight Control Testbed Aircraft," AIAA-1990-1288, *SFTE, DGLR, and SETP Biannual Flight Test Conference, 5th*, Ontario, CA, May 22–24, 1990.

The TPS operated the YA-7D DIGITAC aircraft for over 15 years, beginning in 1976. It made significant contributions to advances in flight control technology during investigations involving improved directional control, the effect of depressed roll axis on air-to-air tracking, and airborne verification of computer-simulated flying qualities. The DIGITAC aircraft was used to conduct the first Air Force flight tests of a digital fight control system, and it was also used to flight-test the first fiber-optical fly-by-light DFCS. Other flight-test firsts included the integration of a dynamic gun sight and the flight control system and demonstrations of task-tailored multimode flight control laws.[110] The DIGITAC YA-7D is now on display at the Air Force Flight Test Center Museum at Edwards AFB.

Highly Maneuverable Aircraft Technology

The Highly Maneuverable Aircraft Technology (HiMAT) program provides an interesting perspective on the use of unmanned research aircraft equipped with digital fly-by-wire flight control systems, one that is perhaps most relevant to today's rapidly expanding fleet of unpiloted aircraft whose use has proliferated throughout the military services over the past decade. HiMAT research at Dryden was conducted jointly by NASA and the Air Force Flight Dynamics Laboratory at NASA Dryden between 1979 and 1983. The project began in 1973, and, in August 1975, Rockwell International was awarded a contract to construct two HiMAT vehicles based on the use of advanced technologies applicable to future highly maneuverable fighter aircraft. Designed to provide a level of maneuverability that would enable a sustained 8 g turn at 0.9 Mach at an altitude of 25,000 feet, the HiMAT vehicles were approximately half the size of an F-16. Wingspan was about 16 feet, and length was 23.5 feet. A GE J85 turbojet that produced 5,000 pounds of static thrust at sea level powered the vehicle that could attain about Mach 1.4. Launched from the NASA B-52 carrier aircraft, the HiMAT weighed about 4,000 pounds, including 660 pounds of fuel. About 30 percent of the airframe consisted of experimental composite materials, mainly fiberglass and graphite epoxy. Rear-mounted swept wings, a digital flight control system, and controllable forward canards enabled exceptional maneuverability with a turn radius about half of a conventional piloted fighter. For example, at Mach 0.9 at 25,000 feet, the HiMAT could

---

110. Dennis R. Furman, "USAF Test Pilot School Use of DIGITAC in Systems Testing," Society of Automotive Engineers Document No. 851827, Oct. 1985.

Case 10 | Fly-By-Wire: Making the Electric Jet

Research on the HiMAT remotely piloted test vehicle was conducted by NASA and the Air Force Flight Dynamics Laboratory between 1979 and 1983. NASA.

sustain an 8-g turn, while F-16 capability under the same conditions is about 4.5 g.[111]

Ground-based, digital fly-by-wire control systems, developed at Dryden on programs such as the DFBW F-8, were vital to success of the HiMAT remotely piloted research vehicle approach. NASA Ames Research Center and Dryden worked closely with Rockwell International in design and development of the two HiMAT vehicles and their ground control system, rapidly bringing the test vehicles to flight status. Many tests that would have been required for a more conventional piloted research aircraft were eliminated, an approach largely made possible by extensive use of computational aerodynamic design tools developed at Ames. This resulted in drastic reductions in wind tunnel testing but caused the need to devote several HiMAT flights to obtain stability and control data needed for refinements to the digital flight control system.[112]

The HiMAT flight-test maneuver autopilot was based on a design developed by Teledyne Ryan Aeronautical, then a well-known man-

---

111. Robert W. Kempel and Michael R. Earls, "Flight Control Systems Development and Flight Test Experience with the HiMAT Research Vehicles," NASA TP-2822 (June 1988).
112. "HiMAT: Highly Maneuverable Aircraft Technology," NASA Dryden FS-2002-06-025-DFRC (2002).

ufacturer of target drones and remotely piloted aircraft. Teledyne also developed the backup flight control system.[113] Refining the vehicle control laws was an extremely challenging task. Dryden engineers and test pilots evaluated the contractor-developed flight control laws in a ground simulation facility and then tested them in flight, making adjustments until the flight control system performed properly. The HiMAT flight-test maneuver autopilot provided precise, repeatable control, enabling large quantities of reliable test data to be quickly gathered. It proved to be a broadly applicable technique for use in future flight research programs.[114]

Launched from the NASA B-52 at 45,000 feet at Mach 0.68, the HiMAT vehicle was remotely controlled by a NASA research pilot in a ground station at Dryden, using control techniques similar to those in conventional aircraft. The flight control system used a ground-based computer interlinked with the HiMAT vehicle through an uplink and downlink telemetry system. The pilot used proportional stick and rudder inputs to command the computer in the primary flight control system. A television camera mounted in the cockpit provided visual cues to the pilot. A two-seat Lockheed TF-104G aircraft was used to chase each HiMAT mission. The F-104G was equipped with remote control capability, and it could take control of the HiMAT vehicle if problems developed at the ground control site. A set of retractable skids was deployed for landing, which was accomplished on the dry lakebed adjacent to Dryden. Stopping distance was about 4,500 feet. During one of the HiMAT flight tests, a problem was encountered that resulted in a landing with the skids retracted. A timing change had been made in the ground-based HiMAT control system and in the onboard software that used the uplinked landing gear deployment command to extend the skids. Additionally, an onboard failure of one uplink receiver contributed to cause the anomaly. The timing change had been thoroughly tested with the onboard flight software. However, subsequent testing determined that the flight software operated differently when an uplink failure was present.[115]

---

113. Kempel, "Flight Experience with a Backup Flight Control System for the HiMAT Research Vehicle," AIAA Paper 82-1541 (Aug. 1982).
114. E.L. Duke, F.P. Jones, and R.B Roncoli, "Development of a Flight Test Maneuver Autopilot for a Highly Maneuverable Aircraft," NASA TP-2218 (1986).
115. Mackall, "Development and Flight Test Experiences with a Flight-Crucial Digital Control System," p. 112.

HiMAT research also brought about advances in digital flight control systems used to monitor and automatically reconfigure aircraft flight control surfaces to compensate for in-flight failures. HiMAT provided valuable information on a number of other advanced design features. These included integrated computerized flight control systems, aeroelastic tailoring, close-coupled canards and winglets, new composite airframe materials, and a digital integrated propulsion control system. Most importantly, the complex interactions of this set of then-new technologies to enhance overall vehicle performance were closely evaluated. The first HiMAT flight occurred July 27, 1979. The research program ended in January 1983, with the two vehicles completing a total of 26 flights, during which 11 hours of flying time were recorded.[116] The two HiMAT research vehicles are today on exhibit at the NASA Ames Research Center and the Smithsonian Institution National Air and Space Museum.

International CCV Flight Research Efforts

As we have seen earlier, as far back as the Second World War and continuing through the 1950s and 1960s, the Europeans in particular were very active in exploiting the benefits to be gained from the use of fly-by-wire flight control systems in aircraft and missile systems. Experimental fly-by-wire research aircraft programs in Europe and Japan rapidly followed, sometimes nearly paralleled, and even occasionally led NASA and Air Force fly-by-wire research programs, often with the assistance of U.S. flight control system companies. As with U.S. programs, foreign efforts focused on the application of digital fly-by-wire flight control systems in conjunction with modifications to existing service aircraft to create unstable CCV testbeds. Foreign CCV research efforts conclusively validated the benefits attainable from integration of digital computers into fly-by-wire flight control systems and provided experience and confidence in their use in new aircraft designs that have increasingly become multinational.

German CCV F-104G

Capitalizing on their earlier experience with analog fly-by-wire flight control research, by early 1975 the Germans had begun a flight research

---

116. Dwain A. Deets, V. Michael DeAngelis, and David P. Lux, "HiMAT Flight Program: Test Results and Program Assessment Overview," NASA TM-86725 (June 1986).

program to investigate the flying qualities of a highly unstable high-performance aircraft equipped with digital flight controls. For this purpose, they modified a Luftwaffe Lockheed F-104G to incorporate a quadruplex digital flight control system. Known as the CCV F-104G, it featured a canard (consisting of another F-104G horizontal tail) mounted at a fixed negative incidence angle of 4 degrees, on the upper fuselage behind the cockpit and a large jettisonable weight carried under the aft fuselage. These features, in conjunction with internal fuel transfer, were capable of moving the aircraft's center of gravity rearward to create a negative stability margin of up to 20 percent. The CCV F-104G flew for the first time in 1977 from the German flight research center at Manching, with flight-testing of the aircraft in the canard configuration beginning in 1980. The CCV F-104G test program ended in 1984 after 176 flights.[117]

U.K. Jaguar ACT

In the U.K., the Royal Aircraft Establishment began an effort oriented to producing a CCV testbed in 1977. For this purpose, an Anglo-French Jaguar strike fighter was modified by British Aerospace (BAe) to prove the feasibility of active control technology. Known as the Jaguar Active Control Technology (ACT), the aircraft's mechanical flight control system was entirely removed and replaced with a quad-redundant digital fly-by-wire control system that used electrical channels to relay instructions to the flight control surfaces. The initial flight of the Jaguar ACT with the digital FBW system was in October 1981. As with the CCV F-104G, ballast was added to the aft fuselage to move the center of gravity aft and destabilize the aircraft. In 1984, the Jaguar ACT was fitted with rounded oversized leading-edge strakes to move the center of lift of the aircraft forward, further contributing to pitch instability. It first flew in this configuration in March 1984. Marconi developed the Jaguar ACT flight control system. It included an optically coupled data transmission link that was essentially similar to the one that they had developed for the U.S. Air Force YC-14 program (an interesting example of the rapid proliferation of advanced aerspace technology between nations).[118]

---

117. Butter, "Control, Navigation, Avionics, Cockpit," *Aeronautical Research in Germany (From Lilienthal to Today)*, Springer-Verlag, Berlin Heidelberg, 2004.

118. Rediess, et al., "Technology Review of Flight Crucial Flight Control Systems (Application of Optical Technology)."

Flight-testing began in 1981, with the test program ending in 1984 after 96 flights.[119]

French Mirage NG

Although not intended purely as a research aircraft, the French Dassault Mirage 3NG (Nouvelle Generation) was a greatly modified Mirage IIIE single-engine jet fighter that was used to demonstrate the improved air combat performance advantages made possible using relaxed static stability and fly-by-wire. One prototype was built by Dassault; modifications included destabilizing canards, extended wing root leading edges, an analog-computer-controlled fly-by-wire flight control system (based on that used in the production Mirage 2000 fighter), and the improved Atar 9K-50 engine. The Mirage 3NG first flew in December 1982, demonstrating significant performance improvements over the standard operational Mirage IIIE. These were claimed to include a 20–25-percent reduction in takeoff distance, a 40-percent improvement in time to reach combat altitude, a nearly 10,000-foot increase in supersonic ceiling, and similarly impressive gains in acceleration, instantaneous turn rate, and combat air patrol time.

Japanese CCV T-2

In Japan, the CCV approach that was taken involved modification of a Mitsubishi T-2 jet training aircraft. Horizontal canards were fitted to reduce static stability, and an all-movable vertical surface was added to the forward fuselage to enable direct side force control investigations. The existing wing-mounted flaps were modified to enable direct lift control and maneuver load control studies. A triply redundant digital fly-by-wire flight control system was installed with quadruplex pilot force sensors used to sense stick and rudder pedal forces and aircraft motion sensors. Aircraft motion sensors (such as pitch, roll, and yaw rate gyros, and vertical and lateral acceleration sensors) were also quadruplex. The original mechanical flight control system was retained as a backup mode. Three identical digital computers processed sensor signals, and the resultant command signals were used to control the horizontal stabilizer, leading and trailing edge flaps, rudder, and vertical canard. Electrohydraulic actuators converted electrical

---

119. C.J. Yeo, "The Fly-by-Wire Jaguar," *Society of Experimental Test Pilots (SETP) 27th Symposium Proceedings*, Beverly Hills, CA, Sept. 28–Oct. 1, 1983, pp. 193–214.

signals into mechanical inputs for the control surface actuators. The CCV T-2 first flew in August 1983. After 24 flights by Mitsubishi, the aircraft was delivered to the Japanese Technical Research Development Institute (TRDI) at Gifu Air Base in March 1984 for government flight-testing, which was completed in March 1986.[120]

These research programs (along with the Soviet Projekt 100LDU testbed discussed earlier) provided invaluable hands-on experience with state-of-the-art flight control technologies. Data from the Jaguar ACT and the CCV F-104G supported the Experimental Aircraft Program (EAP) and contributed to the technology base for the Anglo-German-Italian-Spanish Eurofighter multirole fighter, now known as the Typhoon. Many other advanced aircraft development programs, including the French Rafale, the Mitsubishi F-2 fighter, the Russian Su-27 family of fighters and attack aircraft, and the entire family of Airbus airliners, were the beneficiaries of these research efforts. In addition, the importance of the infusion of technology made possible by open dissemination of NASA technical publications should not be underestimated.

X-29

The Grumman X-29 research aircraft played a very interesting role in the evolution of modern fly-by-wire flight control systems. Exotic in appearance, with its forward-swept wings and large movable canard control surfaces, the X-29 was highly unstable about the longitudinal axis with a static stability margin of –35 percent. This level of instability probably indicates that the X-29 represents the most unstable piloted aircraft that has ever been successfully flown. Not only did it fly, but it also demonstrated good controllability at very high angles of attack. This degree of success was only possible through the use of a very advanced fly-by-wire flight control system that employed a combination of both digital computers in the primary system and analog computers in the backup system. The program began in 1977, with the Defense Advanced Research Projects Agency (DARPA) and the Air Force Flight Dynamics Laboratory jointly soliciting industry proposals for a research aircraft designed to investigate the forward-swept wing concept in a high-performance aircraft application. In December 1981, Grumman Aircraft

---

120. Katsuhei Shibata and Hideaki Ohmiya, "The T-2 Control Configured Vehicle Development, Integration and Flight Test," AIAA-1988-3882, published in *Technical Papers*, pt. 1 (A89-18051 05-06) AIAA/IEEE Digital Avionics Systems Conference, San Jose, CA, Oct. 17–20, 1988, pp. 177–184.

The X-29 represents the most unstable piloted aircraft that has ever successfully flown. NASA.

was selected to build two aircraft, which were designated X-29. The most unique and visually obvious aspect of the design was the forward-swept wings that incorporated a thin supercritical airfoil, but there were many other areas of the X-29 design that embodied advanced technology. The aircraft used advanced composite materials and unusual construction approaches. Its control system made use of variable camber wing surfaces, aft fuselage-mounted strake flaps, fully movable canards mounted on the sides of the engine inlets, and a computerized fly-by-wire flight control system to maintain control of the otherwise highly unstable aircraft.[121]

In constructing the X-29s, Grumman used the forward fuselage and nose landing gear from two Northrop F-5A fighters. Control surface actuators and the main landing gear came from the F-16. The unique aspect of the X-29 airframe, its forward-swept wing, was developed by Grumman. Because of the major differences between the X-29s and the F-5As, the modified aircraft were assigned new USAF serial numbers, becoming 82-0003 and 82-0049. The main difference between the two

---

121. Terrill W. Putnam, "X-29 Flight-Research Program," NASA TM-86025, Ames Research Center-Dryden Flight Research Facility, Edwards, CA (Jan. 1984).

X-29s was the emergency spin parachute system mounted at the base of the rudder on the second aircraft. The X-29 flight research program was conducted by the NASA Dryden Flight Research Center in two phases and included participation by the Air Force Flight Test Center and the Grumman Corporation.[122] The joint NASA–Air Force portion of the X-29 test program extended from 1984 to 1991; the Air Force conducted a follow-on investigation of vortex flow control (VFC) that lasted into 1992.

The X-29's thin supercritical forward-swept wing presented significant design challenges. The typical stall pattern of an aft-swept wing, from wingtip to root, is reversed for a forward-swept wing, which stalls from the root to the tip. The aerodynamic lift forces on the outer portions of a forward-swept wing produce a twisting moment that tends to force the leading edge further upward. This increases the angle of attack at the wingtips, causing even further twisting that, if uncontrolled, can lead to structural failure of the wing, a phenomenon known as aeroelastic divergence. To deal with this problem, Grumman made use of state-of-the-art composite materials in designing the wing external skin, which it laminated in a way that produced an inherent coupling between wing bending and torsion loads, a concept known as aeroelastic tailoring. At increasing angles of attack (higher lift), the structural characteristics incorporated into composite laminates were designed to ensure that the wing twisted to counter the upward twist produced by aerodynamic loads. The key to the design was balancing the aerodynamic aspects of the wing's configuration with the structural characteristics of the composite laminate to control potential aeroelastic divergence. The wing substructure and the basic airframe itself were made of aluminum and titanium.[123]

The X-29 featured an unusual combination of flight control surfaces. These consisted of forward-mounted canards that contributed positive lift and provided primary control about the pitch axis. Wing flaperons (combination flaps and ailerons) could change the camber of the wing and also functioned as ailerons for roll control. The actuators used to control wing camber were mounted externally in streamlined fairings at the trailing edge of the wing because of the thinness of the supercritical airfoil. The strake flaps on each side of aft fuselage augmented the canards, proving additional pitch control. The control surfaces were electronically linked to a triple-redundant digital fly-by-wire

---

122. Ibid.
123. Ibid.

flight control system (with analog backup) that provided artificial stability necessary for controlling the inherently unstable forward-swept wing, close-coupled canard design used on the X-29. Each of the three digital flight control computers had an analog backup. If one of the digital computers failed, the remaining two took over. If two of the digital computers failed, the flight control system switched to the analog mode. If one of the analog computers failed, the two remaining analog computers took over.[124]

Grumman chief test pilot Charles A. "Chuck" Sewell flew the first X-29 at Edwards AFB on December 14, 1984.[125] During the Phase I research effort, X-29 aircraft No. 1 was used exclusively, flying 242 times. Its wingtips remained unstalled up to the 21-degree angle of attack allowed in Phase I testing. This limitation was due to the fact that an anti-spin parachute was not installed on the aircraft. The aeroelastic tailored wing prevented structural divergence of the wing, and the digital flight control system functioned safely and reliably. Flight control laws and control surface effectiveness combined to provide good pilot handling qualities during maneuvering flight. The aircraft's supercritical airfoil contributed to enhanced cruise and maneuver performance in the transonic regime.[126] However, overall drag reduction was not as great as had been predicted for the configuration. On December 13, 1985, with NASA test pilot Steve Ishmael at the controls, the X-29 became the first aircraft with a forward-swept wing to fly beyond the speed of sound, reaching Mach 1.03 in level flight.[127] Other test pilots who flew the X-29 during Phase I of the joint test program were NASA test pilot Rogers Smith, Lt. Col. Theodore "Ted" Wierzbanowski and Maj. Harry Walker from the Air Force, and Navy Cdr. Ray Craig.

---

124. Robert Clarke, John J. Burken, John T. Bosworth, and Jeffery E. Bauer, "X-29 Flight Control Lessons Learned," NASA TM-4598 (June 1984).
125. Sewell was later killed in the crash of a World War II-era Grumman TBM Avenger torpedo bomber during a takeoff from Danielston, CT, on Aug. 4, 1988. With Grumman since 1969, he had flown with the USMC in Korea and Vietnam and had over 10,000 flying hours on 140 different aircraft types.
126. Developed by NASA and originally tested on the NASA F-8 Super Critical Wing (SCW) aircraft at Dryden in the 1970s, a supercritical airfoil has a flattened upper surface compared to a conventional airfoil shape. This shape delays the onset of shock waves on the upper wing surface and reduces their strength, theoretically resulting in a decrease in overall drag.
127. "X-29 Fact Sheet," NASA Dryden Flight Research Center, Mar. 1, 2008.

The second X-29 aircraft, modified to incorporate an anti-spin parachute and its deployment mechanism, was used during Phase II testing to investigate the aircraft's high-angle-of-attack characteristics and the potential usefulness of the forward-swept wing and canard configuration on military fighter plane designs. First flown on May 23, 1989, it would eventually fly 120 research flights and demonstrate control and maneuvering qualities that were better in many cases than the predictions derived from computational methods and simulation models. NASA, Air Force, and Grumman project pilots reported that the X-29 had excellent control response up to angles of attack of 45 degrees, with limited controllability still available at up to 67 degrees angle of attack. Phase II flight-testing defined an allowable X-29 flight envelope that extended to Mach 1.48, an altitude of just over 50,000 feet, an angle of attack of up to 50 degrees at 1 g and 35 degrees at airspeeds up to 300 knots. Much of the X-29's high-angle-of-attack capability was attributed to the quality of the flight control laws that were cooperatively developed by NASA and the Air Force. These had initially been developed using results obtained from extensive wind tunnel testing and predictions derived from radio-controlled flight tests of a 22-percent scale drop model at NASA Langley Research Center.[128] Flight control system engineers at NASA Dryden and the Air Force Flight Test Center at Edwards used these as the basis for detailed flight control system design. This design used a combination of pitch rate and angle of attack in developing the longitudinal control laws. Selectable gain was included in the flight control system design, and this was used by the X-29 test pilots during flying qualities assessments and evaluations of the effects of control law gain changes at higher angles of attack. Prior to the start of flight-testing, wing rock was estimated to restrict the available angle of attack to less than about 35 degrees. However in flight-testing, wing rock amplitude was found to be less than half of what had been predicted, allowing the roll rate to aileron gain to be lowered to one-fourth of the value that had been derived from preflight data using the subscale free flight model. The available flight envelope was extended to 67 degrees angle of attack at 1 g. Maneuvering flight about all axes was cleared up to an angle of attack of 45 degrees in 1 g flight. The reduced wing rock that had been observed in flight was apparently due to higher

---

128. Chambers, *Partners in Freedom*.

roll damping and increased aileron control power for large aileron deflections.[129]

Preflight predictions of the X-29's pitch capabilities matched flight-test results up through 40 degrees angle of attack. Differences in nose-up pitching moment above an angle of attack of 40 degrees were found to require more canard deflection than predicted. Large yaw asymmetries led to several instances in which the aircraft tended to stabilize at very high nose-up pitch angles during maneuvers at angles of attack above 50 degrees, with the aircraft at an aft center of gravity. Modifications to provide additional nose-down pitch authority were not possible because of physical limits on canard deflection. Maximum pitch rates were limited by the high level of static instability inherent in the X-29 design and control surface rate limits. New actuators with at least a 50-percent higher actuation rate would have been required to achieve pitch rates comparable to those of an operational fighter like the F/A-18. The full wingspan flaperons were found to provide good roll control that was not affected by the fact that the X-29 did not use wing leading-edge maneuvering flaps. Pilot-selectable variable gain capability was used during examination of airplane stability and maneuverability. Basic fighter maneuvers were flown, and roll and yaw gains were increased to improve roll performance. A gain that provided maximum rudder authority resulted in the best pilot comments. Roll coordination was better than anticipated, with rudder effectiveness also higher than preflight predictions at angles of attack between 20 and 40 degrees. Yaw asymmetries developed above 40 degrees angle of attack. Diminished aileron and rudder power was not sufficient to overpower these asymmetries. Increasing gain further produced rudder saturation, resulting in uncoordinated turns, a result that was disliked by test pilots, even though this actually resulted in better roll performance.[130]

Flight-test data from Phase II, the high-angle-of-attack and military utility phase of the X-29 program, satisfied the program's primary objective. The technologies demonstrated in the program had potential to improve future fighter aircraft mission performance, and the forward-swept wings/movable canard configuration provided excellent control response at up to 45 degrees angle of attack. Very importantly, the X-29A program provided a significant pool of knowledge that

---

129. Jeffrey E. Bauer, Robert Clarke, and John J. Burken, "Flight Test of the X-29A at High Angle of Attack: Flight Dynamics and Controls," NASA TP-3537 (Feb. 1995).
130. Ibid.

was very useful background for fine-tuning ground-based predictive techniques for high-angle-of-attack aircraft.[131] One significant potential safety issue with the flight control system, the danger of sensor selection thresholds being set too wide, was discovered during the X-29 test program. The flight control system used three sources of air data in its computations. These air data sources were the nose probe and two probes mounted one on each side of the forward fuselage. The selection algorithm in the flight control system used the data from the nose probe as the primary source, provided it was within some threshold of the data from both side probes. However, the data selection threshold was intentionally large to accommodate known data errors in certain flight modes because of the position of the side sensors in the airflow. Long after the start of flight-testing, ground simulation revealed that the nose probe could, in some circumstances, furnish erroneous information at very low flight speeds, causing the X-29 to go out of control. Although this fault was successfully identified through ground simulation, 162 flights had already been flown before it was detected and corrected.[132]

In 1992, the Air Force began a follow-on program with X-29 No. 2 that investigated the use of vortex flow control as a means of providing increased aircraft control at very high angles of attack, at which normal rudder control is ineffective. Wind tunnel tests had showed that injection of air into the vortexes coming off the nose of the aircraft would change the direction of vortex flow. The forces created on the nose of the aircraft could be used to control directional (yaw) stability. The second X-29 aircraft was modified to incorporate two high-pressure nitrogen tanks, related control valves with two small nozzle jets on the forward upper portion of the nose. The nozzles injected air into the vortexes, which flowed off the nose of the aircraft at high angles of attack. From May to August 1992, 60 test flights were flown. Data from these flights were used to determine that VFC was more effective than expected in generating yaw forces, but it was less successful in providing control when sideslip was present, and it did little to decrease roll oscillations.[133] The two X-29 aircraft flew a total of 436 flights, 254 by the first, and 182 by the second. The former is exhibited at the National Museum of the

---

131. Ibid.; Chambers, *Partners in Freedom*.
132. Mackall, "Development and Flight Test Experiences with a Flight-Crucial Digital Control System."
133. R. Guyton and F. Luria, "Flight Testing of Pneumatic Forebody Vortex Control on the X-29 Technology Demonstrator," Society of Automotive Engineers (SAE) No. 922008, Oct. 1992.

The X-31 aircraft, showing thrust vectoring paddles. NASA.

United States Air Force, while the latter remained at Edwards and is on exhibit at the Dryden Flight Research Center.[134]

X-31 Enhanced Fighter Maneuverability Demonstrator

The X-31 was the first international experimental aircraft development program in which the U.S. participated. Two X-31 Enhanced Fighter Maneuverability (EFM) demonstrator aircraft were designed and constructed by Rockwell International Corporation's North American Aircraft Division and Deutsche Aerospace. Assigned U.S. Navy bureau Nos. 164584 and 164585, the aircraft would be used to obtain data that could be applied to the design of highly maneuverable next-generation fighters. During the conceptual phase of the program, the personnel examined the application of EFM technologies and defined the requirements for the demonstrator aircraft. Next, the preliminary design of the demonstrator and the manufacturing approach were defined. Technical experts from the U.S. Navy, German Federal Ministry of Defense, and

---

134. Hallion and Michael H. Gorn, *On the Frontier: Experimental Flight at NASA Dryden*, (Washington: Smithsonian Books, 2003), Appendix W, X-29 flight test chronology.

NASA evaluated all aspects of the design. Detail design and fabrication followed, with the two aircraft being assembled at the Rockwell International (now Boeing) facility at Palmdale, CA. Both aircraft were required to fly a limited flight-test program at Rockwell. The first aircraft flew its first flight on October 11, 1990, piloted by Rockwell chief test pilot Ken Dyson. The second aircraft made its first flight on January 19, 1991, with Deutsche Aerospace chief test pilot Dietrich Seeck at the controls.[135]

The X-31 had a digital fly-by-wire flight control system that included four digital flight control computers with no analog or mechanical backup. Three synchronous main computers drove the flight control surfaces. The fourth computer served as a tiebreaker in case the three main computers produced conflicting commands. Three thrust vectoring paddles were mounted on the X-31's aft fuselage adjacent to the engine nozzle. Directed by the DFBW flight control system, the paddles were moved in and out of the exhaust flow with the resultant thrust vectoring augmenting the aerodynamic control surfaces in pitch and yaw control to improve maneuverability. Made of an advanced carbon-fiber-reinforced composite material, the paddles could sustain temperatures of up to 1,500 degrees Celsius. The X-31 also had movable forward canards for pitch control. As a result of controllability issues identified during the X-31 flight-test program, fixed strakes between the trailing edge of the wing and the engine exhaust were incorporated. They provided additional nose-down pitch control at very high angles of attack. Another fix that was found necessary was the addition of small fixed-nose strakes to help control sideslip.[136]

During flight-test operations at the Rockwell Aerospace facility, the two X-31s flew 108 test missions, validating the use of thrust vectoring to compensate for loss of aerodynamic control at high angles of attack and expanding the poststall envelope up to 40 degrees angle of attack. The poststall envelope refers to the region in which the aircraft demonstrated an ability to maintain controlled flight beyond the normal X-31 stall angle of attack of 30 degrees. X-31 flight operations moved to NASA Dryden in February 1992, with the first flight under International Test Organization (ITO) management occurring in April 1992. The ITO initially included about 110 people from NASA, the U.S. Navy, the U.S. Air Force, Rockwell Aerospace, the Federal Republic of Germany, and

---

135. "X-31 Enhanced Fighter Maneuverability Demonstrator Fact Sheet," NASA Dryden Flight Research Center, 2001.
136. Chambers, *Partners in Freedom*.

Daimler-Benz. The ITO staff was eventually reduced to approximately 60 people. Overall management of the X-31 program came under by the Defense Advanced Research Projects Agency, with NASA responsible for flight-test operations, aircraft maintenance, and research engineering after the project moved to Dryden. The ITO director and NASA's X-31 project manager at Dryden was Gary Trippensee. Pilots included NASA pilot Rogers Smith, U.S. Navy Cdr. Al Groves, German pilots Karl Lang and Dietrich Seeck, Rockwell International pilot Fred Knox, and Air Force Flight Test Center pilot Lt. Col. Jim Wisneski. By July 1992 the X-31 flight envelope was being expanded in preparation for military utility evaluations that would fly the aircraft against nonthrust vectored fighters to evaluate effectiveness in simulated air combat. Thrust vectoring effectiveness at supersonic speed was evaluated out to Mach 1.28 at an altitude of 35,000 feet.

In early flight-testing, the X-31 flight control system went into a reversionary mode four times in the first nine flights because of disagreement between the two air data sources.[137] The X-31 was very sensitive to sideslip. This caused difficulties for the flight control system at higher angles of attack. Below 30 degrees, the nose boom updated the inertial navigation unit with air data. Above angles of attack of 30 degrees, the inertial navigation unit began calculating erroneous sideslip angles as a result of changes in the relative wind vector. To resolve this problem, a so-called Kiel probe replaced the standard NASA Pitot tube to calculate airflow. The Kiel probe was bent 10 degrees downward from the standard pitot configuration. In addition, the sideslip vane was rotated downward 20 degrees relative to the nose boom to compensate for a yawing oscillation that occurred at an angle of attack of 62 degrees. These changes resulted in accurate air data being provided to the inertial navigation unit throughout the X-31 flight envelope with false sideslip readings at high angles of attack eliminated.[138]

Throughout the process of envelope expansion, many modifications to the flight control laws were required because actual aerodynamics of the aircraft were somewhat different from wind tunnel predictions. When the pilots started flying at angles of attack above 50 degrees, they

---

137. Rushby, "Formal Methods and the Certification of Critical Systems," SRI-CSL-93-07, SRI International, Nov. 1993. Rushby, "Formal Methods and the Digital Systems Validation for Airborne Systems," NASA CR-4551 (Dec. 1, 1993).
138. "X-31 Enhanced Fighter Maneuverability Demonstrator Fact Sheet."

encountered erratic lateral lurching movements. In an attempt to counter this phenomenon, narrow, 1/4-inch-wide strips of grit were attached to the sides of the nose boom and the radome. These effectively changed the vortex flow across the forward fuselage of the aircraft, reducing the randomness of the lurches and enabling expansion of the flight envelope to the design angle of attack limit of 70 degrees at 1 g. However, pilots encountered unintentional departures from controlled flight as the aircraft approached poststall angles of attack of 60 degrees during Split-S maneuvers.[139] The asymmetric yawing moment encountered during this maneuver was beyond the capability of the thrust vectoring system to maintain adequate control.[140] Testing in the Langley full-scale wind tunnel resulted in nose strakes and a modified slightly blunter nose tip design that were fitted to the two aircraft, allowing resumption of the flight-test program. The nose strakes were 6/10 of an inch wide and 20 inches long and forced more symmetric transition of forebody vortexes. The blunted nose tip reduced yaw asymmetries.[141]

Poststall pitch control effectiveness, especially with the X-31 center of gravity at the aft allowable design location, was initially marginal.[142] In these high-angle-of-attack conditions, test pilots rated aircraft response as unsatisfactory. NASA Langley conducted wind tunnel tests of various approaches intended to provide increased nose-down pitch control at high angles of attack. Sixteen different modifications were rapidly tested in the full-scale wind tunnel, with Langley recommending that a pair of strakes 6 inches wide and 65 inches long be mounted along the sides of the aft fuselage to assist in nose-down recovery. These were incorporated on the X-31, with subsequent flight-testing confirming greatly improved nose-down pitch control.[143] Positive control at 70 degrees angle of attack with a controlled roll around the aircraft velocity vector was demonstrated November 6, 1992. On April 29, 1993, a minimum radius 180-degree post-

---

139. A Split-S maneuver consists of a 180-degree roll to inverted flight followed by a pull-through to erect level flight with the aircraft ending up positioned 180 degrees from its initial heading at a lower altitude.
140. Patrick C. Stoliker and Bosworth, "Evaluation of High-Angle-of-Attack Handling Qualities for the X-31A Using Standard Evaluation Maneuvers," NASA TM-104322 (1996).
141. Chambers, *Partners in Freedom*.
142. Al Groves, Fred Knox, Rogers Smith, and Jim Wisneski, "X-31 Flight Test Update," *37th Symposium Proceedings, Society of Experimental Test Pilots (SETP)*, Lancaster, CA, 1993, pp. 100–116.
143. Chambers, *Partners in Freedom*.

stall "Herbst Maneuver" was accomplished for the first time.[144] During the final phase of evaluation, the X-31s engaged in simulated air combat scenarios against F/A-18s. During these scenarios, the X-31s were able to outmaneuver the F/A-18s purely through use of poststall maneuvers and without use of thrust vectoring. X-31 test pilots did not support trading off basic fighter characteristics to acquire poststall maneuvering capabilities but concluded that improved pitch pointing and velocity vector maneuvering possible with thrust vector control did provide additional options during close-in combat. Thrust vectoring, combined with fully controllable poststall maneuvering, enabled X-31 pilots to position their aircraft in ways that adversary pilots could not counter, but it had to be used selectively and rapidly to be effective.[145]

In 1994, software was installed in the X-31 to simulate the feasibility of stabilizing a tailless aircraft at both subsonic and supersonic speed using thrust vectoring. The aircraft was modified to enable the pilot to destabilize the aircraft with the rudder to lower stability levels to those that would have been encountered if the aircraft had a reduced-size vertical tail. For this purpose, the rudder control surface was used to cancel the stabilizing effects of the vertical tail, and yaw thrust vector commands were applied by the flight control system to restabilize and control the aircraft. The X-31 was flown in the quasi-tailless mode supersonically at 38,000 feet at Mach 1.2, and maneuvers involving roll and yaw doublets, 30-degree bank-to-bank rolls, and windup turns to 2 g were flown. During subsonic testing, simulated precision carrier landing approaches and ground attack profiles were successfully evaluated. The quasi-tailless flight-test experiment demonstrated the feasibility of tailless and reduced-tail highly maneuverable fighter/attack aircraft designs. Such designs could have reduced drag and lower weight as well as reduced radar and visual detectability. It determined that thrust vectoring is a viable flight control effector that can replace the functions provided by a vertical tail and rudder control surface. Potential disad-

---

144. A nonconventional poststall turning maneuver named in honor of Dr.-Ing. Wolfgang Herbst, who, as director of advanced design and technology at the Military Aircraft Division of Messerschmitt-Bölkow-Blohm in Germany, had emphasized designs capable of poststall maneuvering. The author of numerous technical articles and papers dealing with preliminary aircraft design, aerodynamics, and maneuverability, Herbst died in the crash of a replica FW 190 fighter that he was flying in 1993.
145. Dave Canter, "X-31 Post-Stall Envelope Expansion and Tactical Utility Testing," *Fourth High Alpha Conference*, NASA Dryden Flight Research Center, July 12–14, 1994 (NASA CP-10143), vol. 2.

vantages include the added weight, complexity, and reliability issues associated with a thrust vectoring system. Additionally, flight conditions that require lower engine thrust settings (such as approach and landing) may necessitate provision of additional aerodynamic high-drag devices to enable high-thrust settings to be maintained, ensuring adequate thrust vectoring control. Early integration of such considerations into the overall design process, along with an increased level of interaction between propulsion and flight control systems, is required in order to derive the maximum benefit from reduced or tailless aircraft that rely on thrust vectoring for stability and control.[146]

The No. 1 X-31 aircraft was lost on its 292nd flight on January 19, 1995. German test pilot Karl Lang had just finished a series of test maneuvers and was in the process of recovering back to a landing at Edwards. At an altitude of 20,000 feet, he observed discrepancies in the air data displays along with a master caution light. The aircraft then began a series of diverging pitch oscillations and became uncontrollable. Lang ejected safely at an altitude of 18,000 feet, and the aircraft crashed in an unpopulated desert area just north of Edwards. The crash was determined to have resulted from an unanticipated single-point failure in the nose-mounted Kiel probe that provided critical airspeed and altitude data to the aircraft flight control system computers. These data were critical to safe flight, yet the Kiel probe did not include provision for electrical de-icing, presumably because the aircraft would only be flown in clear desert weather conditions. However, during descent to recovery back to Edwards, ice accumulated in the unheated X-31 pitot tube, resulting in the flight control system automatically configuring the aircraft control surfaces for what it assumed were lower airspeed conditions. Unanticipated movements of the flight control surfaces caused the aircraft to begin oscillating about all axes followed by an uncontrolled pitch-up to an angle of attack of over 90 degrees.[147] The subsequent X-31 accident investigation board recommended that training be conducted on the system safety analysis process, that procedures be implemented to ensure that all test team members receive configuration change notices, and that improvements be made in the remaining X-31 to prevent similar single-point failures.[148]

---

146. Bosworth and P.C. Stoliker, "The X-31A Quasi-Tailless Flight Test Results," NASA TP-3624 (1996).
147. Don Haley, "Ice Cause of X-31 Crash," NASA R-5-203, Dryden Flight Research Center (Nov. 7, 1995).
148. Jay Levine, "X-31's loss," *Dryden Flight Research Center X-Press*, Jan. 2004.

Case 10 | Fly-By-Wire: Making the Electric Jet

A panel that included former Dryden Research Center director Ken Szalai met at Dryden in early 2004 to review the X-31 accident. The panel noted that the primary contributing factor was the installation of the unheated Kiel probe in place of the original heated Pitot tube. The lack of electrical de-icing capability on the Kiel probe had not been considered a safety risk because X-31 mission rules prohibited flight in precipitation or clouds. However, there was no stipulation specifically restricting flight during potential icing conditions, despite simulations that showed icing of the Pitot static system could lead to loss of control.[149] Information had been distributed among the X-31's test pilots and flight-test engineers explaining the Pitot tube change, but a formal process was not in place to ensure that everyone fully understood the implications of the change. Test pilot Lang had noticed anomalies on his cockpit instrumentation and, assuming the presence of icing, told the control room that he was switching on Pitot heat. Shortly afterward, he advised that he was leaving the Pitot heat on for descent and approach to landing. The ground controller then told Lang that the pitot heat switch in the cockpit was not functional. Discrepancies between the X-31's airspeed and altitude readouts were being observed in the control room, but that information was not shared with the entire control room staff. There was a redundant source of air data and a pilot-selectable alternative control mode that could have saved the aircraft if better communications had existed. Dryden X-31 project manager Gary Trippensee noted that complacency is the enemy of success in flight research; prior to the accident, 523 successful X-31 research missions had been flown.[150]

In 2000, the remaining X-31 was brought back from long-term storage at NASA Dryden, where it had been since 1995, and reconfigured for another round of flight-testing for the Vectoring, Extremely Short Takeoff and Landing Control and Tailless Operation Research (VECTOR) program. This program would explore the use of thrust vectoring for extremely short takeoff and landing (ESTOL), with a focus on the aircraft carrier environment. An international Cooperative Test Organization was created for the VECTOR program. U.S. participants/partners were the Navy, Boeing, General Electric, and NASA.[151]

---

149. The Kiel probe was also found to be more susceptible to ice accumulation, a fact discovered after the accident through wind tunnel testing.
150. Ibid.
151. "X-31 VECTOR Program Phase I Begins," NASA Dryden PR-98-09 (Mar. 9, 1998)

The Swedish government was represented by Volvo and SAAB, with the German Ministry of Defense and DASA (Daimler-Benz consortium) from Germany. The X-31 aircraft was modified to incorporate a Swedish RM-1 engine, the same powerplant used in the Saab JAS-39 Gripen fighter.[152] On February 24, 2001, flown by U.S. Navy Cdr. Vivian Ragusa, the upgraded X-31 took to the air for the first time from Naval Air Station (NAS) Patuxent River.[153] German test pilot Rüdiger "Rudy" Knöpfel, U.S. Marine Corps Maj. Cody Allee, and Navy Lt. J.R. Hansen would fly most of the subsequent ESTOL test program.[154] The VECTOR X-31 went on to accomplish over 2 years' of flight-testing, culminating in the final ESTOL flight by Maj. Allee on April 29, 2003.

From April 22 to 29, 2003, the VECTOR X-31 flew 11 test flights, during which fully automated, high-angle-of-attack approaches to landing were conducted. The automated flight control system utilized inputs from a special Global Positioning System (GPS)-based navigation system to maneuver the aircraft to a precise spot above the runway. Known as the Integrity Beacon Landing System (IBLS), it was supplemented by two virtual satellites, or "pseudolites," on both sides of the runway. Precise spatial position and flight attitude data were inputs for the automatic approach control and landing system used in the VECTOR X-31. An ESTOL approach began with the pilot flying into the area covered by the pseudolites; after entering an engagement box, the automatic approach and landing system was activated. The aircraft then assumed a high-angle-of-attack approach attitude and followed a curvilinear path to the touchdown point. Just before touchdown, with the thrust vectoring paddles less than 2 feet above the runway, the X-31A automatically reduced its attitude back down to the normal 12-degree angle of attack for landing. An autothrottle system from an F/A-18 and a special autopilot developed by the VECTOR team were coupled with the flight control system to provide the integrated flight and propulsion control capability used to automatically derotate the aircraft from its steep final approach attitude to touchdown attitude at 2 feet above the runway.

---

152. The RM-1 was license manufactured by Volvo in Sweden and was a derivative of the General Electric GE F404 engine used in the F/A-18.
153. James Darcy, "Successful First Flight for VECTOR X-31 at Pax," NAS Patuxent River Release 01-040 (Feb. 28, 2001).
154. "EADS/Boeing X-31," *Flugrevue Magazine*, Dec. 28, 2002.

Case 10 | Fly-By-Wire: Making the Electric Jet

On the final flight of the VECTOR program, the angle of attack during landing approach was 24 degrees (twice the angle of attack on a normal landing approach). Approach airspeed was 121 knots, or about 30 percent lower than the normal 175 knots, and the resultant landing distance was only 1,700 feet, compared to the normal landing distance of nearly 8,000 feet. Maj. Allee commented on the experience of riding along on a VECTOR X-31 automatic approach and landing: "There are no g forces and you sit leaning somewhat backwards in the ejection seat while the nose is pointing sharply upwards. . . . At angle of attacks greater than 15 degrees the pilot cannot see the runway except on the screen on the right-hand side of the instrument panel. . . . Whereas on a normal landing the landscape flashes by, now everything takes place as if in slow motion."[155]

Another technical accomplishment demonstrated during the VECTOR X-31 program was the successful test of an advanced Flush Air Data System (FADS). Based on data collected by a dozen sensors located around the nose of the aircraft, the FADS provided accurate air data, including airspeed, altitude, angle of attack, and yaw angle, to the flight control system at angles of attack up to 70 degrees all the way out to supersonic speed.[156]

The two X-31 aircraft completed a total of 580 flights, a record for an X-plane program. Of these, 559 were research missions and 21 were flown in Europe in support of the 1995 Paris Air Show. Fourteen pilots from NASA, the U.S. Navy, the U.S. Marine Corps, the U.S. Air Force, the German Air Force, Rockwell International, and Deutsche Aerospace flew the aircraft during the original program at Palmdale and Dryden, with two U.S. pilots (one Navy and one Marine Corps) and a German pilot flying the VECTOR X-31 test program at Patuxent River. The surviving X-31, U.S. Navy Bureau No. 164585, flew 288 times, making its last flight on April 29, 2003. This aircraft is now on display at the Deutsches Museum annex at Oberschleißheim, near Munich, and it will eventually be returned to the United States. The other X-31, bureau No. 164584, had flown 292 times before it was lost on January 19, 1995.

Digital Electronic Engine Control

NASA pioneered in the development and validation of advanced computer-controlled electronic systems to optimize engine performance

---

155. Karl Schwartz, "Successful ESTOL Landings with the X-31A," *Flugreview*, July 2003.
156. J.R. Wilson, "X-31 Finds a Shorter Path to Success," *Aerospace America*, Aug. 2003.

The YF-12C used twin Pratt & Whitney J58 afterburning engines. NASA.

across the full flight envelope while also improving reliability. One such system was the Digital Electronic Engine Control (DEEC), whose genesis can be traced back to NASA Dryden work on the integrated flight and engine control system developed and evaluated in a joint NASA–Air Force program that used two Mach 3+ Lockheed YF-12C aircraft. The YF-12C was a cousin of the SR-71 strategic reconnaissance aircraft, and both aircraft used twin Pratt & Whitney J58 afterburning engines. As the SR-71 neared Mach 3, a significant portion of the engine thrust was produced from the supersonic shock wave that was captured within each engine inlet and exited through the engine nozzle. A serious issue with the operational SR-71 fleet was so-called engine inlet unstarts. These occurred when the airflow into the inlet was not properly matched to that of the engine. This caused the standing shock wave normally located in the inlet to be expelled out the front of the SR-71's inlet, causing insufficient pressure and airflow for normal engine operations. The result was a sudden loss of thrust on the affected engine. The resulting imbalance in thrust between the two SR-71 engines caused violent yawing, along with pitching and rolling motions. Studies showed that strong vortexes produced by each of the forward fuselage chines passed directly into the inlets during the yawing motion produced by an unstart. NASA efforts supported development of a computerized automatic inlet sensing and cone control system and helped to optimize the ratio of air passing through the engine to that leaving the inlet through the forward bypass doors. Dryden successfully integrated the engine inlet control, auto-throttle, air data, and navigation functions to improve overall performance, with aircraft range being increased 7 percent. Handling qualities were also improved, and the frequency of engine inlet unstarts was greatly reduced. Pratt & Whitney and the Air Force incorporated the improvements demonstrated by Dryden into the entire SR-71 fleet in 1983.[157] The Dryden YF-12C made its last NASA flight on October 31, 1979. On November 7, 1979, it was ferried to the Air Force Museum at Wright-Patterson AFB, OH, where it is now on display.[158]

The broad objective of the DEEC program, conducted by NASA Dryden between 1981 and 1983, was to demonstrate and evaluate the system on a turbofan engine in a high-performance fighter across its full flight envelope. The program was a joint effort between Dryden, Pratt & Whitney, the Air Force, and NASA Lewis Research Center (now the NASA Glenn Research Center). The DEEC had been commercially developed by Pratt & Whitney based on its experience with the J58 engine during the NASA YF-12 flight research program. It integrated a variety of engine functions to improve performance and extend engine life. The DEEC system was tested on an F100 engine mounted in the left engine bay of a NASA Dryden McDonnell-Douglas F-15 fighter. Engine-mounted and fuel-cooled, the DEEC was a single-channel digital controller. Engine inputs to the DEEC included compressor face static pressure and temperature, fan and core rotation speed, burner pressure, turbine inlet temperature, turbine discharge pressure, throttle position, afterburner fuel flow, and fan and compressor speeds. Using these inputs, the DEEC computer set the variable vanes, positioned the compressor air bleed, controlled gas-generator and augmentor fuel flows, adjusted the augmentor segment-sequence valve, and controlled the exhaust nozzle position. Thirty test missions that accumulated 35.5 flight hours were flown during the 2-year test program, which covered the operational

---

157. "YF-12 Flight Research Program," NASA Dryden Technology Fact Sheet TF-2004-17, NASA Dryden Flight Research Center (2004).
158. "Lockheed YF-12 NASA Fact Sheet," FS-2002-09-047 DFRC, NASA Dryden Flight Research Center (2002).

envelope of the F-15 at speeds up to Mach 2.36 and altitudes up to 60,000 feet. The DEEC evaluation included nearly 1,300 throttle and afterburner transients, more than 150 air starts, maximum accelerations and climbs, and the full spectrum of flight maneuvers. An engine nozzle instability that caused stalls and blowouts was encountered when operating in afterburner at high altitudes. This instability had not been predicted in previous computer simulations or during ground-testing in NASA high-altitude test facilities. The instability problem was eventually resolved, and stall-free engine operation was demonstrated across the entire F-15 flight envelope. Faster throttle response, improved engine airstart capability, and an increase of more than 10,000 feet in the altitude that could be attained in afterburner without pilot restrictions on throttle use were achieved.[159]

DEEC-equipped engines were then installed on several operational USAF F-15s for service testing, during which they showed major improvements in reliability and maintainability. Mean time between failures was doubled, and unscheduled engine removals were reduced by a factor of nine. As a result, DEEC-equipped F100 engines were installed in all USAF F-15 and F-16 aircraft. The DEEC was a major event in the history of jet engine propulsion control and represented a significant transition from hydromechanical to digital-computer-based engine control. Performance improvements made possible by the DEEC included faster throttle responses, improved air-start capability, and an altitude increase of over 10,000 feet in afterburner without pilot restrictions on throttle use. Following the successful NASA test program, the DEEC went into standard use on F100 engines in the Boeing F-15 and the Lockheed F-16. Pratt & Whitney also incorporated digital engine control technology in turbofan engines used on some Boeing commercial jetliners. The lineage of similar digital engine control units used on other engines can be traced to the results of NASA's DEEC test and evaluation program.[160]

Highly Integrated Digital Electronic Control

The Highly Integrated Digital Electronic Control (HIDEC) evolved from the earlier DEEC research effort. Major elements of the HIDEC were

---

159. L.K. Myers, D.A. Mackall, and F.W. Burcham, Jr., "Flight Test Results of a Digital Electronic Engine Control System in an F-15 Airplane," AIAA Paper 82-1080 (June 1982).
160. "Digital Electronic Engine Control," NASA Dryden Technology Fact Sheet TF-2004-03, Dryden Flight Research Center (2004).

a Digital Electronic Flight Control System (DEFCS), engine-mounted DEECs, an onboard general-purpose computer, and an integrated architecture that provided connectivity between components. The HIDEC F-15A (USAF serial No. 71-0287) was modified to incorporate DEEC-equipped F100 engine model derivative (EMD) engines. A dual channel Digital Electronic Flight Control System augmented the standard hydromechanical flight control system in the F-15A and replaced its analog control augmentation system. The DEFCS was linked to the aircraft data buses to tie together all other electronic systems, including the aircraft's variable geometry engine inlet control system.[161] Over a span of about 15 years, the HIDEC F-15 would be used to develop several modes of integrated propulsion and flight control systems. These integrated modes were Adaptive Engine Control System, Performance Seeking Control, Self-Repairing Flight Control System, and the Propulsion-Only Flight Control System. They are discussed separately in the following sections.[162]

Adaptive Engine Control System

The Adaptive Engine Control System (ADECS) improved engine performance by exploiting the excess stall margin originally designed into the engines using capabilities made possible with the integrated computerized flight and engine control systems.[163] ADECS used airframe and engine data to allow the engine to operate at higher performance levels at times when inlet distortion was low and the full engine stall margin is not needed. Initial engineering work on ADECS began in 1983, with research flights beginning in 1986. Test results showed thrust improvements of between 8 and 10 percent depending on altitude. Fuel flow reductions of between 7 and 17 percent at maximum afterburning thrust at an altitude of 30,000 feet were recorded. Rate of climb increased 14 percent at 40,000 feet. Time required to climb from 10,000 feet to 40,000 feet dropped 13 percent. Acceleration improved between 5 and 24 percent at intermediate and maximum power settings, depending on altitude. No unintentional engine stalls were encountered in the test program.

---

161. Frank W. Burcham, Lawrence P. Myers, and Ronald J. Ray, "Predicted Performance Benefits of an Adaptive Digital Engine Control System on an F-15 Airplane," NASA TM-85916 (Jan. 1985).
162. "The HIDEC Program," NASA Dryden Research Center Technology Fact Sheet, 1999.
163. Engine stall margin is a measure of the amount that engine operating pressure is reduced to provide a margin of safety to help prevent engine compressor stall due to excessive pressure.

ADECS technology has been incorporated into the Pratt & Whitney F119 engine used on the Air Force F-22 Raptor.[164]

Performance Seeking Control

The Performance Seeking Control (PSC) effort followed the Adaptive Electronic Control System project. Previous engine control modes utilized on the HIDEC aircraft used stored schedules of optimum engine pressure ratios based an average engine on a normal standard day. Using digital flight control, inlet control, and engine control systems, PSC used highly advanced computational techniques and control laws to identify the actual condition of the engine components and optimize the overall propulsion system for best efficiency based on actual engine and flight conditions that the aircraft was encountering, ensuring the highest engine and maneuvering performance in all flight environments. PSC testing with the HIDEC aircraft began in 1990. Results of flight-testing with PSC included increased fuel efficiency, improved engine thrust during accelerations and climbs, and increased engine service life achieved by reductions in turbine inlet temperature. Flight-testing demonstrated turbine inlet temperature reductions of more than 160 °F. Such large operating temperature reductions can significantly extend the life of jet engines. Additionally, improvements in thrust of between 9 percent and 15 percent were observed in various flight conditions, including acceleration and climb.[165] PSC also included the development of methodologies within the digital engine control system designed to detect engine wear and impending failure of certain engine components. Such information, coupled with normal preventative maintenance, could assist in implementing future fail-safe propulsion systems.[166] The flight demonstration and evaluation of the PSC system at NASA Dryden directly contributed to the rapid transition of the technology into operational use. For example, PSC technology has been applied to the F100 engine

---

164. Jennifer L. Baer-Riedhart and Robert J. Landy, "Highly Integrated Digital Electronic Control-Digital Flight Control, Aircraft Model Identification and Adaptive Engine Control," NASA TM-86793 (Mar. 1987); "The HIDEC Program," NASA Dryden Flight Research Center Technology Fact Sheet, 1999.
165. Glen B. Gilyard and John S. Orme, "Subsonic Flight Test Evaluation of a Performance Seeking Algoritm on an F-15 Airplane," NASA TM-4400 (Aug. 1992).
166. Baer-Riedhart, et al., "Highly Integrated Digital Electronic Control-Digital Flight Control, Aircraft Model Identification and Adaptive Engine Control"; "The HIDEC Program," NASA Dryden Flight Research Center Technology Fact Sheet, 1999.

used in the F-15 Eagle, the F119 engine in the F-22 Raptor, and the F135 engine for the F-35 Lightning II.

Self-Repairing Flight Control System

The Self-Repairing Flight Control System (SRFCS) consists of software integrated into an aircraft's digital flight control system that is used to detect failures or damage to the aircraft control surfaces. In the event of control surface damage, the remaining control surfaces are automatically reconfigured to maintain control, enabling pilots to complete their mission and land safely. The program, sponsored by the U.S. Air Force, demonstrated the ability of a flight control system to identify the failure of a control surface and reconfigure commands to other control devices, such as ailerons, rudders, elevators, and flaps, to continue the aircraft's mission or allow it to be landed safely. As an example, if the horizontal elevator were damaged or failed in flight, the SRFCS would diagnose the failure and determine how the remaining flight control surfaces could be repositioned to compensate for the damaged or inoperable control surface. A visual warning to the pilot was used to explain the type of failure that occurred. It also provided revised aircraft flight limits, such as reduced airspeed, angle of attack, and maneuvering loads. The SRFCS also had the capability of identifying failures in electrical, hydraulic, and mechanical systems. Built-in test and sensor data provided a diagnostic capability and identified failed components or system faults for subsequent ground maintenance repair. System malfunctions on an aircraft with a SRFCS can be identified and isolated at the time they occur and then repaired as soon as the aircraft is on the ground, eliminating lengthy postflight maintenance troubleshooting.[167]

The SRFCS was flown 25 times on the HIDEC F-15 at NASA Dryden between December 1989 and March 1990, with somewhat mixed results. The maintenance diagnostics aspect of the system was a general success, but there were frequent failures with the SRFCS. Simulated control system failures were induced, with the SRFCS correctly identifying every failure that it detected. However, it only sensed induced control system failures 61 percent of the time. The overall conclusion was that the SRFCS concept was promising, but it needed more development if it was to be successfully implemented into production aircraft.

---

167. "F-15 Flight Research Facility," NASA Dryden Flight Research Center Fact Sheet FS-022, 1999.

NASA test pilot Jim Smolka flew the first SRFCS flight, on December 12, 1989, with test engineer Gerard Schkolnik in the rear cockpit; other SRFCS test pilots were Bill Dana and Tom McMurtry.[168]

Propulsion Controlled Aircraft System

Initiated in 1989, the Propulsion Controlled Aircraft (PCA) system was developed and flight-tested at NASA Dryden, with the goal being to help pilots land safely in the event that flight control components were disabled. PCA automatically provides computer-controlled variations in engine thrust that give pilots adequate pitch, yaw, and roll authority to fly their aircraft. The PCA system was tested and initially demonstrated on the HIDEC F-15. In simulator studies, NASA demonstrated the PCA concept on more than a dozen other commercial and military aircraft. The PCA system integrates the aircraft flight control and engine control computers to manage engine thrust and ensure adequate aircraft control. When the PCA system is activated, moving the control column aft causes the engine thrust to be automatically increased, and the aircraft begins to climb. Forward movement of the control column results in reduced thrust, and descent begins. Right or left movements of the control column produce differential engine thrust, resulting in the aircraft yawing in the direction of the desired turn. Flight-testing with the HIDEC F-15 was carried out at landing approach speeds of 150 knots with the flaps down and between 170 and 190 knots with the flaps retracted. At the conclusion of testing, the HIDEC F-15 accomplished a successful landing using the PCA system on April 21, 1993, after a flight in which the pilot used only engine power to turn, climb, and descend for approach to the runway.[169]

The NASA Dryden F-15A HIDEC testbed had originally been obtained from the Air Force in January 1976. During its career with NASA, it was involved in more than 25 advanced research projects involving aerodynamics, performance, propulsion control, systems integration, instrumentation development, human factors, and flight-test techniques before its last flight at Dryden, in October 1993.[170]

---

168. Tomayko, "The Story of Self-Repairing Flight Control Systems," *Dryden Historical Study*, No. 1, Christian Gelzer, ed., 2003.
169. "F-15 Flight Research Facility."
170. "F-15 HIDEC (Highly Integrated Digital Electronic Control)," NASA Dryden Flight Research Center Fact Sheet, 1999.

A similar propulsion controlled aircraft approach was later evaluated and publicly demonstrated using a modified three-engine McDonnell-Douglas MD-11 jet airliner in a cooperative program between NASA, McDonnell-Douglas, Pratt & Whitney, and Honeywell. Pratt & Whitney modified the engine control software, and Honeywell designed the software for the MD-11 flight control computer. Standard autopilot controls already present in the aircraft were used along with the Honeywell PCA software in the reprogrammed MD-11 flight control computers. NASA Ames performed computer simulations in support of the PCA program. On August 29, 1995, NASA Ames test pilot Gordon Fullerton successfully landed the PCA-modified MD-11 at Edwards AFB with an engine out after activating the aircraft's auto-land system.[171] Simulator testing of a PCA system continued using a NASA Ames–FAA motion-based Boeing 747 simulator, with pilots making about 50 landings in the simulator. Additional simulation research by Ames resulted in further tests of the PCA system on B-747, B-757, MD-11, and C-17 aircraft. NASA Dryden test pilots flew simulated tests of the system in August 1998 in the NASA Ames Advanced Concepts Simulator. Ten pilots were involved in these tests, with 20 out of 20 attempted landings successfully accomplished. PCA technology can be used on current or future aircraft equipped with digital flight control systems.[172]

NF-15B Advanced Control Technology: Air Force S/MTD

NASA Dryden used an NF-15B research aircraft on various research projects from 1993 through early 2009. Originally designated the TF-15, it was the first two-seat F-15 Eagle built by McDonnell-Douglas, the sixth F-15 off the assembly line, and the oldest F-15 flying up to its retirement. First flown in July 1973, the aircraft was initially used for F-15 developmental testing and evaluation as part of the F-15 combined test force at Edwards AFB in the 1970s. In the 1980s, the aircraft was extensively modified for the Air Force's Short Takeoff and Landing Maneuver Technology Demonstrator (S/MTD) program. Modifications included the integration of a digital fly-by-wire control system, canards mounted on the engine inlets ahead of the wings,[173] and two-dimensional thrust-vectoring, thrust-reversing nozzles. The vectoring nozzles redirected

---

171. Tomayko, "The Story of Self-Repairing Flight Control Systems."
172. "Propulsion Controlled Aircraft," NASA Dryden Fact Sheet FS-041, 1999.
173. The canards used on the NF-15B were based on the F/A-18 horizontal stabilators.

The F-15 ACTIVE during yaw vectoring tests in 1996. NASA.

engine exhaust either up or down, giving greater pitch control and additional aerodynamic braking capability. Designated NF-15B to reflect its status as a highly modified research aircraft, the aircraft was used in the S/MTD program from 1988 until 1993. During Air Force S/MTD testing, a 25-percent reduction in takeoff roll was demonstrated with thrust-reversing, enabling the aircraft to stop in just 1,650 feet. Takeoffs using thrust-vectoring produced nose rotation speeds as low as 40 knots, resulting in greatly reduced takeoff distances. Additionally, thrust-reversing produced extremely rapid in-flight decelerations, a feature valuable during close-in combat.[174]

NASA Advanced Control Technology for Integrated Vehicles

In 1994, after the conclusion of Air Force S/MTD testing, the aircraft was transferred to NASA Dryden for the NASA Advanced Control Technology for Integrated Vehicles (ACTIVE) research project. ACTIVE was oriented to determining if axisymmetric vectored thrust could contribute to drag reduction and increased fuel economy and range compared with conventional aerodynamic controls. The project was a collaborative effort between NASA, the Air Force Research Laboratory, Pratt & Whitney,

---

174. Chambers, *Partners in Freedom*.

and Boeing (formerly McDonnell-Douglas). An advanced digital flight fly-by-wire control system was integrated into the NF-15B, which was given NASA tail No. 837. Higher-thrust versions of the Pratt & Whitney F100 engine with newly developed axisymmetric thrust-vectoring engine exhaust nozzles were installed. The nozzles could deflect engine exhaust up to 20 degrees off centerline. This allowed variable thrust control in both pitch and yaw, or combinations of the two axes. An integrated propulsion and flight control system controlled both aerodynamic flight control surfaces and the engines. New cockpit controls and electronics from an F-15E aircraft were also installed in the NF-15B. The first supersonic flight using yaw vectoring occurred in early 1996. Pitch and yaw thrust vectoring were demonstrated at speeds up to Mach 2.0, and yaw vectoring was used at angles of attack up to 30 degrees. An adaptive performance software program was developed and successfully tested in the NF-15B flight control computer. It automatically determined the optimal setting or trim for the thrust-vectoring nozzles and the aerodynamic control surfaces to minimize aircraft drag. An improvement of Mach 0.1 in level flight was achieved at Mach 1.3 at 30,000 feet with no increase in engine thrust. The ACTIVE NF-15B continued investigations of integrated flight and propulsion control with thrust-vectoring during 1997 and 1998, including an experiment that combined thrust vectoring with aerodynamic controls during simulated ground attack missions. Following completion of the ACTIVE project, the NF-15B was used as a testbed for several other NASA Dryden research experiments, which included the efforts described below.[175]

High Stability Engine Control

NASA Lewis (now Glenn) Research Center evaluated an automated computerized engine control system that sensed and responded to high levels of engine inlet airflow turbulence to prevent sudden in-flight engine compressor stalls and potential engine failures. Known as High Stability Engine Control (HISTEC), the system used a high-speed digital processor to evaluate airflow data from engine sensors. The technology involved in the HISTEC approach was intended to control distortion at the engine face. The HISTEC system included two major functional subelements: a Distortion Estimation System (DES) and a Stability Management Control

---

175. "NASA NF-15B Research Aircraft Fact Sheet," Dryden Flight Research Center, Mar. 11, 2009.

(SMC). The DES is an aircraft-mounted, high-speed computer processor. It uses state-of-the-art algorithms to estimate the amount and type of distortion present at the engine face based on measurements from pressure sensors in the engine inlet near the fan. Maneuver information from the digital flight control system and predictive angle-of-attack and angle-of-yaw algorithms are used to provide estimates of the type and extent of airflow distortion likely to be encountered by the engine. From these inputs, the DES calculates the effects of the engine face distortion on the overall propulsion system and determines appropriate fan and compressor pressure ratio commands. These are then passed to the SMC as inputs. The SMC performs an engine stability assessment using embedded stall margin control laws. It then issues actuator commands to the engine to best accommodate the estimated distortion.[176]

A dozen flights were flown on the ACTIVE F-15 aircraft at Dryden from July 15 to August 26, 1997, to validate the HISTEC concept, during which the system successfully directed the engine control computer to automatically command engine trim changes to adjust for changes in inlet turbulence level. The result was improved engine stability when inlet airflow was turbulent and increased engine performance when the airflow was stable.[177]

Intelligent Flight Control System

Beginning in 1999, the NF-15B supported the Intelligent Flight Control System (IFCS) neural network project. This was oriented to developing a flight control system that could identify aircraft characteristics through the use of neural network technology in order to optimize performance and compensate for in-flight failures by automatically reconfiguring the flight control system. IFCS is an extension of the digital fly-by-wire flight control system and is intended to maintain positive aircraft control under certain failure conditions that would normally lead to loss of control. IFCS would automatically vary engine thrust and reconfigure flight control surfaces to compensate for in-flight failures. This is accomplished through the use of upgrades to the digital flight control system software that incorporate self-learning neural network technology. A

---

176. John C. DeLaat, George W. Gallops, Laura J. Kerr, Robert P. Kielb, John S. Orme, Robert D. Southwick, and Mark G. Walsh, "High Stability Engine Control (HISTEC) Flight Test Results," NASA TM-1998-208481 (July 1998).
177. Ibid.

neural network that could train itself to analyze flight properties of an aircraft was developed, integrated into the NASA NF-15B, and evaluated in flight testing. The neural network "learns" aircraft flight characteristics in real time, using inputs from the aircraft sensors and from error corrections provided by the primary flight control computer. It uses this information to create different aircraft flight characteristic models. The neural network learns to recognize when the aircraft is in a stable flight condition. If one of the flight control surfaces becomes damaged or nonresponsive, the IFCS detects this fault and changes the flight characteristic model for the aircraft. The neural network then drives the error between the reference model and the actual aircraft state to zero. Dryden test pilot Jim Smolka flew the first IFCS test mission on March 19, 1999, with test engineer Gerard Schkolnik in the rear cockpit.[178]

The NF-15B IFCS test program provided the opportunity for a limited flight evaluation of a direct adaptive neural network-based flight control system.[179] This effort was led by the Dryden Flight Research Center, with collaboration from the Ames Research Center, Boeing, the Institute for Scientific Research at West Virginia University, and the Georgia Institute of Technology.[180] John Bosworth was the NASA Dryden IFCS chief engineer. Flight-testing of the direct adaptive neural network-based flight control system began in 2003 and evaluated the outputs of the neural network. The neural network had been pretrained using flight characteristics obtained for the F-15 S/MTD aircraft from wind tunnel testing. During this phase of testing, the neural network did not actually provide any flight control inputs in-flight. The outputs of the neural network were run directly to instrumentation for data collection purposes only.

In 2005, a fully integrated direct adaptive neural-network-based flight control system demonstrated that it could continuously provide error corrections and measure the effects of these corrections in order to learn new flight models or adjust existing ones. To measure the aircraft state, the neural network took a large number of inputs from the roll, pitch, and yaw axes and the aircraft's control surfaces. If differences were detected between the measured aircraft state and the flight model, the neural network adjusted the outputs from the primary flight computer

---

178. Tomayko, "The Story of Self-Repairing Flight Control Systems."
179. Peggy S. Williams-Hayes, "Flight Test Implementation of a Second Generation Intelligent Flight Control System," NASA TM-2005-213669 (Nov. 2005).
180. "Intelligent Flight Control System," NASA Dryden Fact Sheet FS-076, Feb. 13, 2006.

to bring the differences to zero before they were sent to the actuator control electronics that moved the control surfaces.[181] IFCS software evaluations with the NF-15B included aircraft handling qualities maneuvers, envelope boundary maneuvers, control surface excitations for real-time parameter identification that included pitch, roll, and yaw doublets, and neural network performance assessments.[182] During NF-15B flight-testing, a simulated failure was introduced into the right horizontal stabilizer that simulated a frozen pitch control surface. Handling qualities were evaluated with and without neural network adaptation. The performance of the adaptation system was assessed in terms of its ability to decouple roll and pitch response and reestablish good onboard model tracking. Flight-testing with the simulated stabilator failure and the adaptive neural network flight control system adaptation showed general improvement in pitch response. However, a tendency for pilot-induced roll oscillations was encountered.[183]

Concurrent with NF-15B IFCS flight-testing, NASA Ames conducted a similar neutral network flight research program using a remotely controlled Experimental Air Vehicle (EAV) equipped with an Intelligent Flight Controller (IFC). Aerodynamically, the EAV was a one-quarter-scale model of the widely used Cessna 182 Skylane general aviation aircraft. The EAV was equipped with two electrical power supplies, one for the digital flight control system that incorporated the neural-network IFC capability and one for the avionics installation that included three video cameras to assist the pilots with situation awareness. Several pilots flew the EAV during the test program. Differences in individual pilot control techniques were found to have a noticeable effect on the performance of the Intelligent Flight Controller. Interestingly, IFCS flight-testing with the NF-15B aircraft uncovered many of the same issues related to the controller that the EAV program found. IFCS was determined to provide increased stability margins in the presence of large destabilizing failures. The adaptive system provided better closed-loop behavior with

---

181. M.G. Perhinschi, J. Burken, M.R. Napolitano, G. Campa, and M.L. Fravolini, "Performance Comparison of Different Neural Augmentation for the NASA Gen - 2 IFCS F-15 Control Laws," *Proceedings of the 2004 American Control Conference*, Boston, MA, 2004.
182. Creech, "NASA Dryden Neural Network Passes Milestone," NASA Dryden Press Release, Sept. 3, 2003.
183. Bosworth and Williams-Hayes, "Flight Test Results from the NF-15B Intelligent Flight Control System (IFCS) Project with Adaptation to a Simulated Stabilator Failure," AIAA Paper 2007-2818, 2007.

improved matching of the onboard reference model. However, the convergent properties of the controller were found to require improvement because continued maneuvering caused continued adaptation change. During ground simulator evaluation of the IFCS, a trained light-plane pilot was able to successfully land a heavily damaged large jet airliner despite the fact that he had no experience with such an aircraft. Test data from the IFCS program provided a basis for analysis and understanding of neural network-based adaptive flight control system technology as an option for implementation into future aircraft.[184]

After a 35-year career, during which it had flown with McDonnell-Douglas, the Air Force, and NASA, the NF-15B was retired following its final flight, on January 30, 2009. During its 14 years at NASA Dryden, the aircraft had flown 251 times. The NF-15B will be on permanent display with a group of other retired NASA research aircraft at Dryden.[185]

The Continuing Legacy of FBW Research in Aircraft Development

Fly-by-wire technology developed by NASA and the Air Force served as the basis for flight control systems in several generations of military and civilian aircraft. Many of these aircraft featured highly unconventional airframe configurations that would have been unflyable without computer-controlled fly-by-wire systems. An interesting example was the then highly classified Lockheed Have Blue experimental stealth technology flight demonstrator. This very unusual aircraft first flew in 1977 and was used to validate the concept of using a highly faceted airframe to provide a very low radar signature. Unstable about multiple axes, Have Blue was totally dependent on its computer-controlled fly-by-wire flight control system that was based on that used in the F-16. Its success led to the rapid development and early deployment of the stealthy Lockheed F-117 attack aircraft that first flew in 1981 and was operational in 1983.[186] More advanced digital fly-by-wire flight control systems enabled an entirely new family of unstable, aerodynamically refined "stealth" combat aircraft to be designed and deployed. These

---

184. Cory Ippolito, John Kaneshige, and Yoo-Hsiu Yeh, "Neural Adaptive Flight Control Testing on an Unmanned Experimental Aerial Vehicle," AIAA Paper 2007-2827, 2003; Tomayko, "The Story of Self-Repairing Flight Control Systems."
185. "NASA NF-15B Research Aircraft Fact Sheet."
186. Aronstein and Piccirillo, "Have Blue and the F-117A, Evolution of the "Stealth Fighter," American Institute of Aeronautics and Astronautics, Reston, VA, 1997.

include the Northrop B-2 Spirit flying wing bomber and Lockheed's F-22 Raptor and F-35 Lightning II fighters with their highly integrated digital propulsion and flight control systems.

Knowledge of the benefits and confidence in the use of digital fly-by-wire technology are today widespread across the international aerospace industry. Nearly all new military aircraft—including fighters, bombers, and cargo aircraft, as well as commercial airliners, both U.S. and foreign—have reaped immense benefits from the legacy of NASA's pioneering digital fly-by-wire flight and propulsion control efforts. On the airlift side, the Air Force's Boeing C-17 was designed with a quad-redundant digital fly-by-wire flight control system.[187] In Europe, Airbus Industrie was an early convert to digital fly-by-wire and the increasing use of electronic subsystems. All of its airliners, starting with the A320 in 1987, were designed with fully digital fly-by-wire flight control architectures along with side stick controllers.[188] Reliance on complex and heavy hydraulic systems is being reduced as companies increase the emphasis on electrically powered flight controls. With this approach, both electrical and self-contained electrohydraulic actuators are controlled by the digital flight control system's computers. The benefits are lower weight, reduced maintenance cost, the ability to provide redundant electrical power circuits, and improved integration between the flight control system and the aircraft's avionics and electrical subsystems. Electric flight control technology reportedly resulted in a weight reduction of 3,300 pounds in the A380 compared with a conventional hydromechanical flight control system.[189] Boeing introduced fly-by-wire with its 777, which was certified for commercial airline service in 1995. It has been in routine airline service with its reliable digital fly-by-wire flight control system ever since. In addition to a digital fly-by-wire flight control system, the next Boeing airliner, the 787, incorporates some electrically powered and operated flight control elements (the spoilers and horizontal stabilizers). These are designed to remain functional in the event of either total hydraulic systems failure or flight control computer failure, allowing the pilots to maintain control in pitch, roll, and yaw and safely land the aircraft.

---

187. Brian W. Kowal, Carl J. Scherz, and Richard Quinliven, "C-17 Flight Control System Overview," *IEEE Aerospace and Electronic Systems Magazine*, July 1992.

188. C. Favre, "Fly-by-Wire for Commercial Aircraft: the Airbus Experience," *International Journal of Control*, vol. 59, no. 1, 1994, pp. 139–157.

189. "All-Electric Aircraft Research Speeds Up," *Aerospace America*, Jan. 2009, pp. 4–8.

Today, the tremendous benefits made possible by the use of digital fly-by-wire in vehicle control systems have migrated into a variety of applications beyond the traditional definition of aerospace systems. As a significant example, digital fly-by-wire ship control systems are now operational in the latest U.S. Navy warships, such as the Seawolf and Virginia class submarines. NASA experts, along with those from the FAA and military and civil aviation agencies, supported the Navy in developing its fly-by-wire ship control system certification program.[190] Thus, the vision of early advocates of digital fly-by-wire technology within NASA has been fully validated. Safe and efficient, digital fly-by-wire technology is today universally accepted with its benefits available to the military services, airline travelers, and the general public on a daily basis.

---

190. William Palmer, "Submarine Fly-by-Wire Ship Control System Certification Program Leverages Aviation Industry Best Practice," *Wavelengths Online, The Official Publication of the Naval Surface Warfare Center, Carderock Division*, Oct. 1, 2004.

Recommended Additional Readings
Reports, Papers, Articles, and Presentations:

"All-Electric Aircraft Research Speeds Up," *Aerospace America,* Jan. 2009.

B.R. Ashworth and William M. Kahlbaum, Jr., "Description and Performance of the Langley Differential Maneuvering Simulator," NASA TN-D-7304 (June 1973).

"Automatic Flight Control System Sought," *Aviation Week,* Aug. 6, 1956.

Jennifer L. Baer-Riedhart and Robert J. Landy, "Highly Integrated Digital Electronic Control-Digital Flight Control, Aircraft Model Identification and Adaptive Engine Control," NASA TM-86793 (Mar. 1987).

R.E. Bailey, M.F. Schaeffer, R.E. Smith, and J.F. Stewart, "Flight Test Experience With Pilot-Induced-Oscillation Suppression Filters," NASA TM-86028 (Jan. 1984).

Charles A. Baird and Franklin B. Snyder, with introduction by Lt. Mark Bierele, "Terrain-Aided Altitude Computations on the AFTI/F-16," Harris Corporation, Melbourne, FL, Aug. 1990.

Finley Barfield, Duke Browning, and Judith Probert, "All Terrain Ground Collision Avoidance and Maneuver Terrain Following for Automated Low Level Night Attack," *IEEE AES Systems Magazine,* March 1993.

Joseph F. Baugher, "General Dynamics YF-16/CCV," *American Military Aircraft,* Mar. 31, 2000.

Jeffrey E. Bauer, Robert Clarke, and John J. Burken, "Flight Test of the X-29A at High Angle of Attack: Flight Dynamics and Controls," NASA TP-3537 (Feb. 1995).

James Blaylock, Donald Swihart, and William Urshel, "Integration of Advanced Safety Enhancements for F-16 Terrain Following," AIAA-1987-2906 (1997).

John T. Bosworth and P.C. Stoliker, "The X-31A Quasi-Tailless Flight Test Results," NASA TP-3624 (1996).

John T. Bosworth and Peggy S. Williams-Hayes, "Flight Test Results from the NF-15B Intelligent Flight Control System (IFCS) Project with Adaptation to a Simulated Stabilator Failure," AIAA Paper 2007-2818 (2007).

Frank W. Burcham, Lawrence P. Myers, and Ronald J. Ray, "Predicted Performance Benefits of an Adaptive Digital Engine Control System on an F-15 Airplane," NASA TM-85916 (Jan. 1985).

Ulrich Butter, "Control, Navigation, Avionics, Cockpit," *Aeronautical Research in Germany (From Lilienthal to Today),* Springer-Verlag, Berlin Heidelberg, 2004.

Dave Canter, "X-31 Post-Stall Envelope Expansion and Tactical Utility Testing," *Fourth High Alpha Conference,* NASA Dryden Flight Research Center, July 12–14, 1994, NASA CP-10143, vol. 2 (1994).

Robert Clarke, John J. Burken, John T. Bosworth, and Jeffery E. Bauer, "X-29 Flight Control Lessons Learned," NASA TM-4598 (June 1984).

Gary Creech, "AFTI/F-16 Retires After 22 Years," *The Dryden Express,* vol. 43, issue 2, Feb. 23, 2001.

Gary Creech, "NASA Dryden Neural Network Passes Milestone," NASA Dryden Press Release, Sept. 3, 2003.

Allan F. Damp, "Evaluation Tests on Boulton-Paul VC-10 Aileron Integrated Flight Control Actuator," Commercial Division, Boeing Company, Renton, WA, Mar. 10, 1970.

James Darcy, "Successful First Flight for VECTOR X-31 at Pax," NAS Patuxent River Release 01-040, Feb. 28, 2001.

John C. DeLaat, George W. Gallops, Laura J. Kerr, Robert P. Kielb, John S. Orme, Robert D. Southwick, and Mark G. Walsh, "High Stability Engine Control (HISTEC) Flight Test Results," NASA TM-1998-208481 (July 1998).

Dwain A. Deets, V. Michael DeAngelis, and David P. Lux, "HiMAT Flight Program: Test Results and Program Assessment Overview," NASA TM-86725 (June 1986).

Dwain A. Deets and Kenneth J. Szalai, "Design and Flight Experience with a Digital Fly-By-Wire Control System in an F-8 Airplane," NATO Advisory Group for Aeronautical Research and Development, Conference Paper AGARD-CP-137 (1974).

E.L. Duke, F.P. Jones, and R.B. Roncoli, "Development of a Flight Test Maneuver Autopilot for a Highly Maneuverable Aircraft," NASA TP-2218 (1986).

"EADS/Boeing X-31," *Flugrevue Magazine,* Dec. 28, 2002.

"Experience with the X-15 Adaptive Flight Control System," NASA TN-D-6208, NASA Flight Research Center, Edwards, CA (Mar. 1971).

C. Favre, "Fly-by-Wire for Commercial Aircraft: the Airbus Experience," *International Journal of Control,* vol. 59, no. 1, 1994.

J.C. Floyd, "The Canadian Approach to All-Weather Interceptor Development, Fourteenth British Commonwealth Lecture," *The Journal of the Royal Aeronautical Society,* vol. 62, no. 576, Dec. 1958.

Dennis R. Furman, "USAF Test Pilot School Use of DIGITAC in Systems Testing," Society of Automotive Engineers (SAE) Document No. 851827, Oct. 1985.

"General Dynamics and Northrop to Build Lightweight Fighter Prototypes," *Interavia,* July 1972.

Glen B. Gilyard and John S. Orme, "Subsonic Flight Test Evaluation of a Performance Seeking Algoritm on an F-15 Airplane," NASA TM-4400 (Aug. 1992).

Yefim Gordon, "An Industry of Prototypes: Sukhoi T-4, Russia's Mach 3 Bomber," *Wings of Fame,* vol. 9, Aerospace Publishing Limited, London (1997).

Al Groves, Fred Knox, Rogers Smith, and Jim Wisneski, "X-31 Flight Test Update," *37th Symposium Proceedings,* Society of Experimental Test Pilots (SETP), Lancaster, CA, 1993.

R. Guyton and F. Luria, "Flight Testing of Pneumatic Forebody Vortex Control on the X-29 Technology Demonstrator," Society of Automotive Engineers (SAE) No. 922008, Oct. 1992.

Don Haley, "Ice Cause of X-31 Crash," NASA Press Release 5-203, Dryden Flight Research Center, Nov. 7, 1995.

Richard P. Hallion, "The Air Force and the Supersonic Breakthrough," *Technology and the Air Force: A Retrospective Assessment,* Air Force History and Museums Program, Washington (1997).

H.J. Hillaker, "The F-16: A Technology Demonstrator, a Prototype, and a Flight Demonstrator," *Proceedings of AIAA Aircraft Prototype and Technology Demonstrator Symposium,* AIAA, New York, 1983.

J.K.B. Illingworth and H.W. Chinn, "Variable Stability and Control Tests on the S.C.1 Aircraft in Jet-Borne Flight, with Particular Emphasis on Reference to Desirable VTOL Flying Qualities," Royal Aircraft Establishment, Bedford, U.K., Her Majesty's Stationery Office, London, 1969.

Cory Ippolito, John Kaneshige, and Yoo-Hsiu Yeh, "Neural Adaptive Flight Control Testing on an Unmanned Experimental Aerial Vehicle," AIAA Paper 2007-2827, 2007.

Stephen D. Ishmael, Dale A. Mackall, and Victoria A. Regenie, "Design Implications From AFTI/F-16 Flight Test," NASA TM-86026, 1984.

Stephen D. Ishmael and Donald R. McMonagle, "AFTI/F-16 Flight Test Results and Lessons," NASA TM-84920 (Oct. 1983).

Gavin D. Jenny, "JB-47E Fly-By-Wire Flight Test Program (Phase I)," AFFDL TR-69-40, Air Force Flight Dynamics Laboratory, Wright-Patterson AFB, OH, Sept. 1969.

Robert W. Kempel, "Flight Experience with a Backup Flight Control System for the HiMAT Research Vehicle," AIAA Paper 82-1541 (Aug. 1982).

Robert W. Kempel and Michael R. Earls, "Flight Control Systems Development and Flight Test Experience with the HiMAT Research Vehicles," NASA TP-2822 (June 1988).

Brian W. Kowal, Carl J. Scherz, and Richard Quinliven, "C-17 Flight Control System Overview," *IEEE Aerospace and Electronic Systems Magazine*, July 1992.

D. Lean and H.W. Chinn, "Review of General Operating Experience with a Jet-Lift VTOL Research Aircraft (Short S.C.1)," Aeronautical Research Council Current Paper No. 832, Her Majesty's Stationary Office, London (1965).

Robert E. Lemble, "DIGITAC—A Unique Digital Flight Control Testbed Aircraft," AIAA-1990-1288, *SFTE, DGLR, and SETP Biannual Flight Test Conference, 5th,* Ontario, CA, May 22–24, 1990.

Jay Levine, "X-31's loss," *Dryden Flight Research Center X-Press,* Jan. 2004.

Dale A. Mackall, "AFTI/F-16 Digital Flight Control System Experience," in Gary P. Beasley, ed., *NASA Aircraft Controls Research 1983,* pp. 469–487, NASA CP-2296, 1984; Proceedings of Workshop Held at NASA Langley Research Center, Oct. 25–27, 1983.

Dale A. Mackall, "Development and Flight Test Experiences with a Flight-Crucial Digital Control System," NASA TP-2857 (Nov. 1988).

D.L. Martin and D. Gangsaas, "Testing the YC-14 Flight Control System Software," *AIAA Journal of Guidance and Control,* July–Aug. 1978.

Duane McRuer and Dunstan Graham, "A Flight Control Century: Triumphs of the Systems Approach," *Journal of Guidance, Control and Dynamics,* vol. 27, no. 2, pp. 161–173, AIAA, 2003.

L.K. Myers, D.A. Mackall, and F.W. Burcham, Jr., "Flight Test Results of a Digital Electronic Engine Control System in an F-15 Airplane," AIAA Paper 82-1080 (June 1982).

Robert Navarro, "Performance of an Electro-Hydrostatic Actuator on the F-18 Systems Research Aircraft," NASA TM-97-206224 (1997).

Molly Neal, "VC10: Vickers-Armstrongs' Long-range Airliner," *Flight International,* May 10, 1962.

William Palmer, "Submarine Fly-by-Wire Ship Control System Certification Program Leverages Aviation Industry Best Practice," *Wavelengths Online,* Naval Surface Warfare Center, Carderock Division, Oct. 1, 2004.

M.G. Perhinschi, J. Burken, M.R. Napolitano, G. Campa, and M.L. Fravolini, "Performance Comparison of Different Neural Augmentation for the NASA Gen – 2 IFCS F-15 Control Laws," *Proceedings of the 2004 American Control Conference,* Boston, MA, 2004.

Terrill W. Putnam, "X-29 Flight-Research Program," NASA TM-86025 (Jan. 1984).

"RAE Electric Hunter," *Flight International,* June 28, 1973.

James W. Ramsey, "Power-by-Wire," *Avionics Magazine,* May 1, 2001.

H.A. Rediess and E.C. Buckley, "Technology Review of Flight Crucial Flight Control Systems (Application of Optical Technology)," NASA CR-172332 (Supplement 1), Sept. 1984.

John Rushby, "Formal Methods and the Certification of Critical Systems," SRI-CSL-93-07, SRI International, Nov. 1993.

John Rushby, "Formal Methods and the Digital Systems Validation for Airborne Systems," NASA CR-4551 (Dec. 1, 1993).

Karl Schwartz, "Successful ESTOL Landings with the X-31A," *Flugreview*, July 2003.

Katsuhei Shibata and Hideaki Ohmiya, "The T-2 Control Configured Vehicle Development, Integration and Flight Test," AIAA-1988-3882, published in *Technical Papers, Part 1 (A89-18051 05-06) AIAA/IEEE Digital Avionics Systems Conference*, San Jose, CA, Oct. 17–20, 1988.

S.A. Sjoberg, "Some Experience With Side Controllers," *Research Airplane Committee Report on the Progress of the X-15 Project*, Conference held at NACA Langley Aeronautical Laboratory, Langley Field, VA, Oct. 25–26, 1956.

Patrick C. Stoliker and John T. Bosworth, "Evaluation of High-Angle-of-Attack Handling Qualities for the X-31A Using Standard Evaluation Maneuvers," NASA TM-104322 (1996).

Joe Stout, "What a Wonderful Airplane: YF-16 First Flight," *Code One Magazine*, July 1992.

Kenneth J. Szalai, Calvin R. Jarvis, Gary E. Krier, Vincent A. Megna, Larry D. Brock, and Robert N. O'Donnell, "Digital Fly-by-Wire Flight Control Validation Experience," NASA TM-72860 (Dec. 1978).

L.W. Taylor, Jr., and E.J. Adkins, "Adaptive Flight Control Systems—Pro and Con," NASA TM-X-56008 (1964).

L.W. Taylor, Jr., and G.B. Merrick, "X-15 Airplane Stability Augmentation System," NASA TN-D-1157 (1962).

L.W. Taylor, Jr., and J.W. Smith, "An Analysis of the Limit-Cycle and Structural-Resonance Characteristics of the X-15 Stability Augmentation System," NASA TN-D-4287 (1967).

James E. Tomayko, "Blind Faith: The United States Air Force and the Development of Fly-By-Wire Technology," *Technology and the Air

Force: A Retrospective Assessment, Air Force History and Museums Program, Washington, DC, 1997.

Robert A. Tremant, "Operational Experience and Characteristics of the X-15 Flight Control System," NASA TN-D-1402 (1962).

Robert B. Voas, "Manned Control of Mercury Spacecraft," *Astronautics,* vol. 7, no. 3, Mar. 1962.

Kyrill Von Gersdorff, "Transfer of German Aeronautical Knowledge After 1945," *Aeronautical Research in Germany (From Lilienthal to Today),* Springer-Verlag, Berlin Heidelberg, 2004.

Norman C. Weingarten, "History of In-Flight Simulation & Flying Qualities Research at CALSPAN," *AIAA Journal of Aircraft,* vol. 42, no. 2, Mar./Apr. 2005.

M.J. Wendle, G.G. Grose, J.L. Porter, and V.R. Pruitt, *Flight/Propulsion Control Integration Aspects of Energy Management,* SAE Paper Number 740480, 1974.

Peggy S. Williams-Hayes, "Flight Test Implementation of a Second Generation Intelligent Flight Control System," NASA TM-2005-213669, Nov. 2005.

J.R. Wilson, "X-31 Finds a Shorter Path to Success," *Aerospace America,* Aug. 2003.

B.S. Wolfe, "The Concorde Automatic Flight Control System: A Description of the Automatic Flight Control System for the Anglo-French Supersonic Transport and its Development to Date," *Aircraft Engineering and Aerospace Technology,* vol. 39, issue 5, 1967.

"X-31 VECTOR Program Phase I Begins," NASA Dryden Press Release 98-09, Mar. 9, 1998.

Michael L. Yaffee, "Survivable Controls Gain Emphasis," *Aviation Week,* Feb. 2, 1970.

C.J. Yeo, "The Fly-by-Wire Jaguar," *Society of Experimental Test Pilots (SETP) 27th Symposium Proceedings,* Beverly Hills, CA, Sept. 28–Oct. 1, 1983.

Books and Monographs:

David C. Aronstein, Michael J. Hirschberg and Albert C. Piccirillo, *Advanced Tactical Fighter to F-22 Raptor* (Reston: AIAA, 1998).

David C. Aronstein and Albert C. Piccirillo, *Have Blue and the F-117A, Evolution of the Stealth Fighter* (Reston: AIAA, 1997).

David C. Aronstein and Albert C. Piccirillo, *The Lightweight Fighter Program: A Successful Approach to Fighter Technology Transition* (Reston: AIAA, 1996).

Robert E. Ball, *The Fundamentals of Aircraft Combat Survivability Analysis and Design* (New York: AIAA, 1985).

Paul F. Borchers, James A. Franklin, and Jay W. Fletcher, *Flight Research at Ames, 1940–1997: Fifty Seven Years of Development and Validation of Aeronautical Technology,* NASA SP-3300 (Washington: NASA, 1998).

Joseph R. Chambers, *Partners in Freedom: Contributions of the Langley Research Center to U.S. Military Aircraft of the 1990s,* SP-2000-4519 (Washington: NASA, 2000).

C. Droste and J. Walker, *The General Dynamics Case Study on the F-16 Fly-By-Wire Flight Control System* (New York: AIAA, undated).

Richard P. Hallion, *On the Frontier: Flight Research at Dryden, 1946–1981,* SP-4303 (Washington: NASA, 1984).

E.H. Herschel, H. Prem, and G. Madelung, *Aeronautical Research in Germany (From Lilienthal to Today)* (Berlin Heidelberg: Springer-Verlag, 2004).

Dennis R. Jenkins, *X-15: Extending the Frontiers of Flight,* SP-2007-562 (Washington: NASA, 2007).

Gavin D. Jenny, James W. Morris, and Vernon R. Schmitt, *Fly-by-Wire, A Historical and Design Perspective* (Warrendale: SAE, 1998).

Lawrence K. Loftin, *Quest for Performance: The Evolution of Modern Aircraft, Part II: The Jet Age, Chapter 12: Jet Bomber and Attack Aircraft*, NASA SP-468 (Washington: NASA, 1985).

Donald L. Mallick, with Peter W. Merlin, *The Smell of Kerosene: A Test Pilot's Odyssey,* NASA SP-4108 (Washington: NASA, 2003).

NASA Office of Safety and Mission Assurance, *Formal Methods Specification and Verification Guidebook for Software and Computer Systems, Volume I: Planning and Technology Insertion,"* P-98-208193 (Washington: NASA, 1998).

NASA Office of Safety and Mission Assurance, *Formal Methods Specification and Analysis Guidebook for the Verification of Software and Computer Systems, Volume II: A Practitioner's Companion,* NASA GB-001-97 (Washington: NASA, 1997).

Michael J. Neufeld, *The Rocket and the Reich; Peenemünde and the Coming of the Ballistic Missile Era* (Washington: Smithsonian Institution, 1995).

Albert C. Piccirillo, *German V/STOL Fighter Program,* (Reston: AIAA, 1997).

Rowland F. Pocock, *German Guided Missiles of the Second World War,* (New York: Arco publishing company, 1967)

Ray Sturtivant, *British Research and Development Aircraft: Seventy Years at the Leading Edge* (UK: Haynes/Foulis, 1990).

James E. Tomayko, *Computers Take Flight: A History of NASA's Pioneering Digital Fly-by Wire Project,* NASA SP-2000-4224 (Washington: NASA, 2000).

James E. Tomayko, *The Story of Self-Repairing Flight Control Systems,* Dryden Historical Study No.1, Christian Gelzer, ed. (Washington: NASA, 2003).

An NACA cowling on a Northrop A-17A at the Agency's Langley Laboratory in 1939. The NACA cowling was its signal contribution to integrating propulsive and aerodynamic design. NASA.

CASE 11

Advancing Propulsive Technology

James Banke

Ensuring proper aircraft propulsion has been a powerful stimulus. In the interwar years, the NACA researched propellers, fuels, engine cooling, supercharging, and nacelle and cowling design. In the postwar years, the Agency refined gas turbine propulsion technology. NASA now leads research in advancing environmentally friendly and fuel-conserving propulsion, thanks to the Agency's strengths in aerodynamic and thermodynamic analysis, composite structures, and other areas.

EACH DAY, OUR SKIES FILL with general aviation aircraft, business jets, and commercial airliners. Every 24 hours, some 2 million passengers worldwide are moved from one airport to the next, almost all of them propelled by relatively quiet, fuel-efficient, and safe jet engines.[1]

And no matter if the driving force moving these vehicles through the air comes from piston-driven propellers, turboprops, turbojets, turbofans—even rocket engines or scramjets—the National Aeronautics and Space Administration (NASA) during the past 50 years has played a significant role in advancing that propulsion technology the public counts on every day.

Many of the advances seen in today's aircraft powerplants can trace their origins to NASA programs that began during the 1960s, when the Agency responded to public demand that the Government apply major resources to tackling the problems of noise pollution near major airports. Highlights of some of the more noteworthy research programs to reduce noise and other pollution, prolong engine life, and increase fuel efficiency will be described in this case study.

But efforts to improve engine efficiency and curb unwanted noise actually predate NASA's origins in 1958, when its predecessor, the National Advisory Committee for Aeronautics (NACA), served as the Nation's preeminent laboratory for aviation research. It was during the 1920s that

---

1. William H. More, ed., *National Transportation Statistics* (Washington, DC: U.S. Department of Transportation, 2009), p. 72.

Air pollution is evident as this Boeing B-47B takes off in 1954 with the help of its General Electric J47 jet engines and Rocket Assisted Take Off solid rocket motors. U.S. Air Force.

the NACA invented a cowling to surround the front of an airplane and its radial engine, smoothing the aerodynamic flow around the aircraft while also helping to keep the engine cool. In 1929, the NACA won its first Collier Trophy for the breakthrough in engine and aerodynamic technology.[2]

During World War II, the NACA produced new ways to fix problems discovered in higher-powered piston engines being mass-produced for wartime bombers. NACA research into centrifugal superchargers was particularly useful, especially on the R-1820 Cyclone engines intended for use on the Boeing B-17 Flying Fortress, and later with the Wright R-3350 Duplex Cyclone engines that powered the B-29.

Basic research on aircraft engine noise was conducted by NACA engineers, who reported their findings in a paper presented in 1956 to the 51st Meeting of the Acoustical Society of America in Cambridge, MA. It would seem that measurements backed up the prediction that the noise level of the spinning propeller depended on several variables,

---

2. Roger E. Bilstein, *Orders of Magnitude: A History of the NACA and NASA, 1915–1990*, NASA SP-4406 (Washington, DC: NASA, 1989), p. 9.

including the propeller diameter, how fast it is turning, and how far away the recording device is from the engine.[3]

As the jet engine made its way from Europe to the United States and designs for the basic turboprop, turbojet, and turbofan were refined, the NACA during the early 1950s began one of the earliest noise-reduction programs, installing multitube nozzles of increasing complexity at the back of the engines to, in effect, act as mufflers. These engines were tested in a wind tunnel at Langley Research Center in Hampton, VA. But the effort was not effective enough to prevent a growing public sentiment that commercial jet airliners should be seen and not heard.

In fact, a 1952 Presidential commission chaired by the legendary pilot James H. Doolittle predicted that aircraft noise would soon turn into a problem for airport managers and planners. The NACA's response was to form a Special Subcommittee on Aircraft Noise and pursue a three-part program to understand better what makes a jet noisy, how to quiet it, and what, if any, impact the noise might have on the aircraft's structure.[4]

As the NACA on September 30, 1958, turned overnight into the National Aeronautics and Space Administration on October 1, the new space agency soon found itself with more work to do than just beating the Soviet Union to the Moon.

Noise Pollution Forces Engine Improvements

Fast-forward a few years, to a time when Americans embraced the promise that technology would solve the world's problems, raced the Soviet Union to the Moon, and looked forward to owning personal family hovercraft, just like they saw on the TV show *The Jetsons*. And during that same decade of the 1960s, the American public became more and more comfortable flying aboard commercial airliners equipped with the modern marvel of turbojet engines. Boeing 707s and McDonnell-Douglas DC-8s, each with four engines bolted to their wings, were not only a common sight in the skies over major cities, but their presence could also easily be heard by anyone living next to or near where the planes took off and landed. Boeing 727s and 737s soon followed. At the same

---

3. Edward M. Kerwin, Jr., "Procedures for Estimating the Near Field Noise of Rotating Aircraft Propellers," presented at the *Fifty-First Meeting of the Acoustical Society of America, Cambridge, MA, June 17–23, 1956.*
4. J.H. Doolittle, *The Airport and Its Neighbors, The Report of the President's Airport Commission* (Washington, DC: U.S. Government Printing Office, 1952), p. 45.

A jet engine is prepared for a test in 1967 as part of an early noise research program at Lewis Research Center. NASA.

time that commercial aviation exploded, people moved away from the metropolis to embrace the suburban lifestyle. Neighborhoods began to spring up immediately adjacent to airports that originally were built far from the city, and the new neighbors didn't like the sound of what they hearing.[5]

---

5. Alain Depitre, "Aircraft Noise Certification History/Development," presented at the *ICAO Noise Certification Workshop*, Montreal, 2004, p. 3.

By 1966, the problem of aircraft noise pollution had grown to the point of attracting the attention of President Lyndon Johnson, who then directed the U.S. Office of Science and Technology to set a new national policy that said:

> The FAA and/or NASA, using qualified contractors as necessary, (should) establish and fund . . . an urgent program for conducting the physical, psycho-acoustical, sociological, and other research results needed to provide the basis for quantitative noise evaluation techniques which can be used . . . for hardware and operational specifications.[6]

As a result, NASA began dedicating resources to aggressively address aircraft noise and sought to contract much of the work to industry, with the goals of advancing technology and conducting research to provide lawmakers with the information they needed to make informed regulatory decisions.[7]

During 1968, the Federal Aviation Administration (FAA) was given authority to implement aircraft noise standards for the airline industry. Within a year, the new standards were adopted and called for all new designs of subsonic jet aircraft to meet certain criteria. Aircraft that met these standards were called Stage 2 aircraft, while the older planes that did not meet the standards were called Stage 1 aircraft. Stage 1 aircraft over 75,000 pounds were banned from flying to or from U.S. airports as of January 1, 1985. The cycle repeated itself with the establishment of Stage 3 aircraft in 1977, with Stage 2 aircraft needing to be phased out by the end of 1999. (Some of the Stage 2 aircraft engines were modified to meet Stage 3 aircraft standards.) In 2005, the FAA adopted an even stricter noise standard, which is Stage 4. All new aircraft designs submitted to the FAA on or after July 5, 2005, must meet Stage 4 requirements. As of this writing, there is no timetable for the mandatory phaseout of Stage 3 aircraft.[8]

---

6. *Alleviation of Jet Aircraft Noise Near Airports* (Washington, DC: U.S. Office of Science and Technology, 1966), p. 8.
7. Newell D. Sanders, *Aircraft Engine Noise Reduction*, NASA SP-311 (Washington, DC: NASA, 1972), p. 2.
8. David M. Bearden, *Noise Abatement and Control: An Overview of Federal Standards and Regulations* (Washington, DC: Congressional Research Service, 2006), p. 3.

With every new set of regulations, the airline industry required upgrades to its jet engines, if not wholesale new designs. So having already helped establish reliable working versions of each of the major types of jet engines—i.e., turboprop, turbojet, and turbofan—NASA and its industry partners began what has turned out to be a continuing 50-year-long challenge to constantly improve the design of jet engines to prolong their life, make them more fuel efficient, and reduce their environmental impact in terms of air and noise pollution. With this new direction, NASA set in motion three initial programs.[9]

NASA's first major new program was the Acoustically Treated Nacelle program, managed by the Langley Research Center. Engines flying on Douglas DC-8 and Boeing 707 aircraft were outfitted with experimental mufflers, which reduced noise during approach and landing but had negligible effect on noise pollution during takeoff, according to program results reported during a 1969 conference at Langley.[10]

The second was the Quiet Engine program, which was managed by the Lewis Research Center in Cleveland (Lewis became the Glenn Research Center on March 1, 1999). Attention here focused on the interior design of turbojet and turbofan engines to make them quieter by as much as 20 decibels. General Electric (GE) was the key industry partner in this program, which showed that noise reduction was possible by several methods, including changing the rotational speed of the fan, increasing the fan bypass ratio, and adjusting the spacing of rotating and stationary parts.[11]

The third was the Steep Approach program, which was jointly managed by Langley and the Ames Research Center/Dryden Flight Research Facility, both in California. This program did not result in new engine technology but instead focused on minimizing noise on the ground by developing techniques for pilots to use in flying steeper and faster approaches to airports.[12]

---

9. *U.S. Government Support of the U.S. Commercial Aircraft Industry*, Prepared for the Commission of the European Communities (Washington, DC: Arnold and Porter, 1991), pp. 37–43.
10. Sanders, *Aircraft Engine Noise Reduction*, NASA SP-311 (Washington, DC: NASA, 1972), p. 2.
11. M.J. Benzakein, S.B. Kazin, and F. Montegani, "NASA/GE Quiet Engine 'A,'" AIAA Paper 72-657 (1972).
12. Vicki L. Golich and Thomas E. Pinelli, *Knowledge Diffusion in the U.S. Aerospace Industry* (London: Alex Publishing, 1998), p. 61.

Case 11 Advancing Propulsive Technology

Quiet Clean Short Haul Experimental Engine
A second wave of engine-improvement programs was initiated in 1969 and continued throughout the 1970s, as the noise around airports continued to be a social and political issue and the FAA tightened its environmental regulations. Moreover, with the oil crisis and energy shortage later in the decade adding to the forces requiring change, the airline industry once again turned to NASA for help in identifying new technology.

At the same time, the airline industry was studying the feasibility of introducing a new generation of commuter airliners to fly between cities along the Northeast corridor of the United States. To make these routes attractive to potential passengers, new airports would have to be built as close to the center of cities such as Boston, New York, and Philadelphia. For aircraft to fly into such airports, which would have shorter runways and strict noise requirements, the airliners would have to be capable of making steep climbs after takeoff, quick turns without losing control, and steep descents on approach to landing, accommodating short runways and meeting the standards for Stage 2 noise levels.[13]

In terms of advancing propulsion technology, NASA's answer to all of these requirements was the Quiet Clean Short Haul Experimental Engine. Contracts were awarded to GE to design, build, and test two types of high-bypass fanjet engines: an over-the-wing engine and an under-the-wing engine. Self-descriptive as to their place on the airplane, both turbofans were based on the same engine core used in the military F-101 fighter jet. Improvements to the design included noise-reduction features evolved from the Quiet Engine program; a drive-reduction gear to make the fan spin slower than the central shaft; a low-pressure turbine; advanced composite construction for the inlet, fan frame, and fan exhaust duct; and a new digital control system that allowed flight computers to monitor and control the jet engine's operation with more precision and quicker response than a pilot could.[14]

In addition to those "standard" features on each engine, the under-the-wing engine tried out a variable pitch composite low-pressure fan with a 12 to 1 ratio—both features were thought to be valuable in reducing noise, although the variable pitch proved challenging for the GE

---

13. Robert V. Garvin, "Starting Something Big: The Commercial Emergence of GE Aircraft Engines," AIAA Paper 72-657 (1999), pp. 162–165.
14. A.P. Adamson, "Quiet Clean Short-Haul Experimental Engine (QCSEE) Design Rationale," SAE Paper 750605 (1975).

The giant General Electric GE90 jet engine that powers the Boeing 777 benefited from the Energy Efficient Engine project. General Electric.

team leading the research. Two pitch change mechanisms were tested, one by GE and the other by Hamilton Standard. Both worked well in controlled test conditions but would need a lot of work before they could go into production.[15]

The over-the-wing engine incorporated a higher fan pressure and a 10 to 1 bypass ratio, a fixed pitch fan, a variable area D-shaped fan exhaust nozzle, and low tip speeds on the fans. Both engines directed their exhaust along the surface of the wing, which required modifications to handle the hot gas and increase lift performance.[16]

The under-the-wing engine was test-fired for 153 hours before it was delivered to NASA in August of 1978, while the over-the-wing engine received 58 hours of testing and was received by NASA during July of 1977. Results of the tests proved that the technology was sound and, when configured to generate 40,000 pounds of thrust, showed a reduction in

15. Garvin, "Starting Something Big," pp. 162–165.
16. C.C. Ciepluch, "A Review of the QCSEE Program," NASA TM-X-71818 (1975).

noise of 8 to 12 decibels, or about 60- to 75-percent quieter than the quietest engines flying on commercial airliners at that time. The new technologies also resulted in sharp reductions in emissions of carbon monoxide and unburned hydrocarbons.[17]

Unfortunately, the new generation of Short Take-Off and Landing (STOL) commuter airliners and small airports near city centers never materialized, so the new engine technology research managed and paid for by NASA but conducted mostly by its industry partners never found a direct commercial application. But there were many valuable lessons learned about the large-diameter turbofans and their nacelles, information that was put to good use by GE years later in the design and fabrication of the GE90 engine that powers the Boeing 777 aircraft.[18]

Aircraft Energy Efficiency Program

Approved in 1975 and begun in 1976, the Aircraft Energy Efficiency (ACEE) program was managed by NASA and funded through 1983, as yet another round of research and development activities were put in work to improve the state of the art of aircraft structural and propulsion design. And once again, the program was aimed at pushing the technological envelope to see what might be possible. Then, based on that information, new Government regulations could be enacted, and the airline industry could decide if the improvements would offer a good return on its investment. The answer, as it turned out, was an enthusiastic yes, as the overall results of the program led directly to the introduction of the Boeing 757 and 767.[19]

Driving this particular program was the rapid increase in fuel costs since 1973 and the accompanying energy crisis, which was brought on by the Organization of Arab Petroleum Exporting Countries' decision to embargo all shipments of oil to the United States. This action began in October 1973 and continued to March 1974. As a result of this and other economic influences, the airlines saw their fuel prices as a percentage of direct operating costs rise from 25 percent to as high as 50 percent within a few weeks. With the U.S. still vulnerable to a future oil embargo, along with general concerns about an energy shortage, the

---

17. Ciepluch and W.S. Willis, "QCSEE—The Key to Future Short-Haul Air Transport," *ICAO Bulletin* 34 (1979).
18. "A Giant Step in Jetliner Propulsion," *Spinoff 1996* (Washington, DC: NASA, 1996), pp. 56–57.
19. *U.S. Government Support of the U.S. Commercial Aircraft Industry* (1991).

Federal Government reacted by ordering NASA to lead an effort to help find ways for airlines to become more profitable. Six projects were initiated under the ACEE program, three of which had to do with the aircraft structure and three of which involved advancing engine technology. The aircraft projects included Composite Structures, Energy Efficient Transport, and Laminar Flow Control. The propulsion technology projects included Engine Component Improvement, Energy Efficient Engine, and Advanced Turboprop—all three of which are detailed next.[20]

Engine Component Improvement Project

The Engine Component Improvement project was tasked with enhancing performance and lowering fuel consumption of several existing commercial aircraft jet engines, in particular Pratt & Whitney's JT8D and JT9D engines and GE's CF6. The specific goals included:

- Improving the current versions of the engines without requiring a brand-new design or engine replacement.
- Reducing the amount of fuel a typical jet engine would use on any given flight by 5 to 6 percent.
- Significantly slowing the pace at which the engine's components would naturally degrade and cause a loss of performance over time.

To do this, researchers tried and tested several ideas, including reducing the clearance between rotating parts, lowering the amount of cooling air that is passed through the engine, and making refinements to the aerodynamic design of certain engine parts to raise their efficiency. All together, engineers identified 16 concepts to incorporate into the engines.[21]

Ultimately, as a result of the Engine Component Improvement efforts, engine parts were incorporated that could resist erosion and warping, better seals were introduced, an improved compressor design was used, and ceramic coatings were added to the gas turbine blades to increase their performance. Tests of the improvements were so promising that many were put into production before the program ended, benefiting the

---

20. Peter G. Batterton, "Energy Efficient Engine Program Contributions to Aircraft Fuel Conservation," NASA TM-83741 (1984).
21. Louis J. Williams, *Small Transport Aircraft Technology* (Honolulu: University Press of the Pacific, 2001), pp. 37–39.

Case 11 | Advancing Propulsive Technology

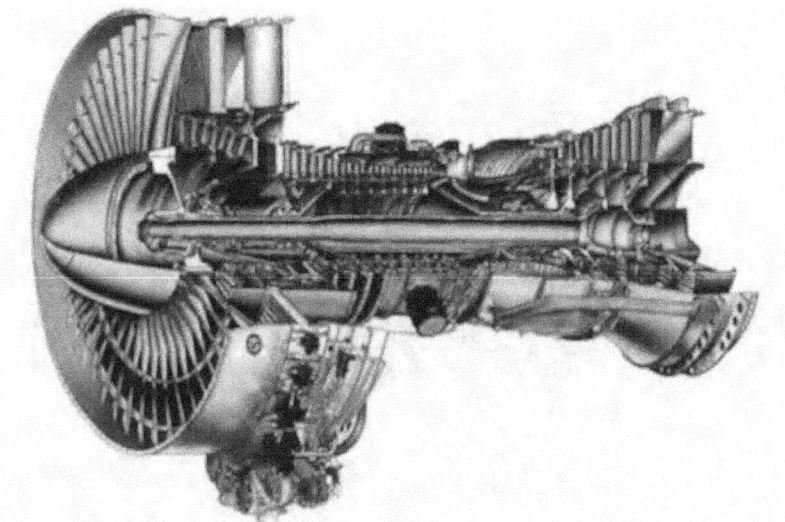

The classic Pratt & Whitney JT9D engine interior and its major components: the fan, compressor, combustion chamber, turbine, and nozzle. Pratt & Whitney.

workhorse airliners at the time, namely the McDonnell-Douglas DC-9 and DC-10, as well as the Boeing 727, 737, and 747.[22]

Energy Efficient Engine Project

Taking everything learned to date by NASA and the industry about making turbo machinery more fuel efficient, the Energy Efficient Engine (E Cubed) project sought to further reduce the airlines' fuel usage and its effect on direct operating costs, while also meeting future FAA regulations and Environmental Protection Agency exhaust emission standards for turbofan engines. Research contracts were awarded to GE and Pratt & Whitney, which initially focused on the CF6-50C and JT9D-7A engines, respectively. The program ran from 1975 to 1983 and cost NASA about $200 million.[23]

Similar to the goals for the Engine Component Improvement project, the E Cubed goals included a 12-percent reduction in specific fuel consumption (SFC), which is a measure of the ratio between the mass of fuel used to the output power of the jet engine—much like a miles per gallon measurement for automobiles. Other goals of the E Cubed effort included a 5-percent reduction in direct operating costs and a

---

22. *U.S. Government Support of the U.S. Commercial Aircraft Industry.*
23. Lawrence. E. Macioce, John W. Schaefer, and Neal T. Saunders, "The Energy Efficient Engine Project," NASA TM-81566 (1980).

A high-pressure 14 to 1 ratio compressor rotor for a prototype Energy Efficient Engine on display in 1984 at Lewis Research Center. NASA.

50-percent reduction in the rate at which the SFC worsens over time as the engine ages. In addition to making these immediate improvements, it was hoped that a new generation of fuel-conservative turbofan engines could be developed from this work.[24]

Highlighting that program was development of a new type of compressor core and an advanced combustor made up of a doughnut-shaped ring with two zones—or domes—of combustion. During times when low power is needed or the engine is idling, only one of the two zones is lit up. For higher thrust levels, including full power, both domes are ignited. By creating a dual combustion option, the amount of fuel being burned can be more carefully controlled, reducing emissions of smoke, carbon monoxide, and hydrocarbons by 50 percent, and nitrogen oxides by 35 percent.[25]

As part of the development of the new compressor in particular, and the E Cubed and Engine Component Improvement programs in general, the Lewis Research Center developed first-generation computer programs for use in creating the new engine. The software helped engineers

---

24. Saunders, "Advanced Component Technologies for Energy-Efficient Turbofan Engines," NASA TM-81507 (1980).
25. Guy Norris and Mark Wagner, *Boeing 777: The Technological Marvel* (Osceola, WI: MBI Publishing, 2001).

with conceptualizing the aerodynamic design and visualizing the flow of gases through the engine. The computer programs were credited with making it possible to design more fuel-efficient compressors with less tip and end-wall pressure losses, higher operating pressure ratios, and the ability to use fewer blades. The compressors also helped to reduce performance deterioration, surface erosion, and damage from bird strikes.[26]

History has judged the E Cubed program as being highly successful, in that the technology developed from the effort was so promising—and proved to meet the objectives for reducing emissions and increasing fuel efficiency—that both major U.S. jet engine manufacturers, GE and Pratt & Whitney, moved quickly to incorporate the technology into their products. The ultimate legacy of the E Cubed program is found today in the GE90 engine, which powers the Boeing 777. The E Cubed technology is directly responsible for the engine's economical fuel burn, reduced emissions, and low maintenance cost.[27]

Advanced Turboprop Project—Yesterday and Today

The third engine-related effort to design a more fuel-efficient powerplant during this era did not focus on another idea for a turbojet configuration. Instead, engineers chose to study the feasibility of reintroducing a jet-powered propeller to commercial airliners. An initial run of the numbers suggested that such an advanced turboprop promised the largest reduction in fuel cost, perhaps by as much as 20 to 30 percent over turbofan engines powering aircraft with a similar performance. This compared with the goal of a 5-percent increase in fuel efficiency for the Engine Component Improvement program and a 10- to 15-percent increase in fuel efficiency for the E Cubed program.[28]

But the implementation of an advanced turboprop was one of NASA's more challenging projects, both in terms of its engineering and in securing public acceptance. For years, the flying public had been conditioned to see the fanjet engine as the epitome of aeronautical advancement. Now they had to be "retrained" to accept the notion that a turbopropeller engine could be every bit as advanced, indeed, even more advanced, than the conventional fanjet engine. The idea was to have a jet engine firing

---

26. *NASA Glenn Research Center at Lewis Field: Achieving the Extraordinary* (Cleveland: NASA, 1999), p. 24.
27. "A Giant Step in Jetliner Propulsion," *Spinoff 1996* (Washington, DC: NASA, 1996).
28. Williams, *Small Transport Aircraft Technology*, p. 38.

as usual with air being compressed and ignited with fuel and the exhaust expelled after first passing through a turbine. But instead of the turbine spinning a shaft that turned a fan at the front of the engine, the turbines would be spinning a shaft, which fed into a gearbox that turned another shaft that spun a series of unusually shaped propeller blades exterior to the engine casing.[29]

Begun in 1976, the project soon grew into one of the larger NASA aeronautics endeavors in the history of the Agency to that point, eventually involving 4 NASA Field Centers, 15 university grants, and more than 40 industrial contracts.[30]

Early on in the program, it was recognized that the major areas of concern were going to be the efficiency of the propeller at cruise speeds, noise both on the ground and within the passenger cabin, the effect of the engine on the aerodynamics of the aircraft, and maintenance costs. Meeting those challenges were helped once again by the computer-aided, three-dimensional design programs created by the Lewis Research Center. An original look for an aircraft propeller was devised that changed the blade's sweep, twist, and thickness, giving the propellers the look of a series of scimitar-shaped swords sticking out of the jet engine. After much development and testing, the NASA-led team eventually found a solution to the design challenge and came up with a propeller shape and engine configuration that was promising in terms of meeting the fuel-efficiency goals and reduced noise by as much as 65 decibels.[31]

In fact, by 1987, the new design was awarded a patent, and the NASA–industry group was awarded the coveted Collier Trophy for creating a new fuel-efficient turboprop propulsion system. Unfortunately, two unexpected variables came into play that stymied efforts to put the design into production.[32]

The first had to do with the public's resistance to the idea of flying in an airliner powered by propellers—even though the blades were still

---

29. Roy D. Hager and Deborah Vrabel, *Advanced Turboprop Project*, NASA SP-495 (Washington, DC: NASA, 1988), p. 5.
30. Mark D. Bowles and Virginia P. Dawson, "The Advanced Turboprop Project: Radical Innovation in a Conservative Environment," in *From Engineering Science to Big Science, The NACA and NASA Collier Trophy Research Project Winners*, NASA SP-4219 (Washington, DC: NASA, 1998), p. 323.
31. Glenn A. Mitchell, "Experimental Aerodynamic Performance of Advanced 40 Degree-Swept, 10-Blade Propeller Model at Mach 0.6 to 0.85," NASA TM-88969 (1988).
32. Bowles and Dawson, "The Advanced Turboprop Project," p. 323.

Case 11 | Advancing Propulsive Technology

A General Electric design for an Unducted Fan engine is tested during the early 1980s. General Electric.

being turned by a jet engine. It didn't matter that a standard turbofan jet also derived most of its thrust from a series of blades—which did, in fact, look more like a fan than a series of propellers. Surveys showed passengers had safety concerns about an exposed blade letting go and sending shrapnel into the cabin, right where they were sitting. Many passengers also believed an airliner equipped with an advanced turboprop was not as modern or reliable as pure turbojet engine. Jets were in; propellers were old fashioned. The second thing that happened was that world fuel prices dropped to the lower levels that preceded the oil embargo and the very rationale for developing the new turboprop in the first place. While fuel-efficient jet engines were still needed, the "extra mile" in fuel efficiency the advanced turboprop provided was no longer required. As a result, NASA and its partners shelved the technology and waited to use the archived files another day.[33]

---

33. Ibid.

The story of the Advanced Turboprop project had one more twist to it. While NASA and its team of contractor engineers were working on their new turboprop design, engineers at GE were quietly working on their own design, initially without NASA's knowledge. NASA's engine was distinguished by the fact that it had one row of blades, while GE's version featured two rows of counter-rotating blades. GE's design, which became known as the Unducted Fan (UDF), was unveiled in 1983 and demonstrated at the 1985 Paris Air Show. A summary of the UDF's technical features is described in a GE-produced report about the program:

> The engine system consists of a modified F404 gas generator engine and counterrotating propulsor system, mechanically decoupled, and aerodynamically integrated through a mixing frame structure. Utilization of the existing F404 engine minimized engine hardware, cost, and timing requirements and provided an engine within the desired thrust class. The power turbine provides direct conversion of the gas generator horsepower into propulsive thrust without the requirement for a gearbox and associated hardware. Counterrotation utilizes the full propulsive efficiency by recovering the exit swirl between blade stages and converting it into thrust.[34]

Although shelved during the late 1980s, the Alternate Turboprop and UDF technology and concept is being explored again as part of programs such as the Ultra-High Bypass Turbofan and Pratt & Whitney's Geared Turbofan. Neither engine is routinely flying yet on commercial airliners. But both concepts promise further reductions in noise, increases in fuel efficiency, and lower operating costs for the airline—goals the aerospace community is constantly working to improve upon.

Several concepts are under study for an Ultra-High Bypass Turbofan, including a modernized version of the Advanced Turboprop that takes advantage of lessons learned from GE's UDF effort. NASA has teamed with GE to start testing an open-rotor engine. For the NASA tests at Glenn Research Center, GE will run two rows of counter-rotating fan blades, with 12 blades in the front row and 10 blades in the back row. The composite fan blades are one-fifth subscale in size. Tests in

---

34. "Full Scale Technology Demonstration of a Modern Counterrotating Unducted Fan Engine Concept: Design Test," NASA CR-180867 (1987).

a low-speed wind tunnel will simulate low-altitude aircraft speeds for acoustic evaluation, while tests in a high-speed wind tunnel will simulate high-altitude cruise conditions in order to evaluate blade efficiency and performance.[35]

"The tests mark a new journey for GE and NASA in the world of open rotor technology. These tests will help to tell us how confident we are in meeting the technical challenges of an open-rotor architecture. It's a journey driven by a need to sharply reduce fuel consumption in future aircraft," David Joyce, president of GE Aviation, said in a statement.[36]

In an Ultra-High Bypass Turbofan, the amount of air going through the engine casing but not through the core compressor and combustion chamber is at least 10 times greater than the air going through the core. Such engines promise to be quieter, but there can be tradeoffs. For example, an Ultra-High Bypass Engine might have to operate at a reduced thrust or have its fan spin slower. While the engine would meet all the goals, it would fly slower, thus making passengers endure longer trips.

In the case of Pratt & Whitney's Geared Turbofan engine, the idea is to have an Ultra-High Bypass Ratio engine, yet spin the fan slower (to reduce noise and improve engine efficiency) than the core compressor blades and turbines, all of which traditionally spin at the same speed, as they are connected to the same central shaft. Pratt & Whitney designed a gearbox into the engine to allow for the central shaft to turn at one speed yet turn a second shaft connected to the fan at another speed.[37]

Alan H. Epstein, a Pratt & Whitney vice president, testifying before the House Subcommittee on Transportation and Infrastructure in 2007, explained the potential benefits the company's Geared Turbofan might bring to the aviation industry:

> The Geared Turbofan engine promises a new level of very low noise while offering the airlines superior economics and environmental performance. For aircraft of 70 to 150 passenger size, the Geared Turbofan engine reduces the fuel burned,

---

35. Deb Case and Rick Kennedy, "GE and NASA To Begin Wind-Tunnel Testing This Summer of Open Rotor Jet Engine Systems," GE Aviation News Release (Evendale, OH: General Electric, 2009).
36. Ibid.
37. Jeff Schweitzer, "An Overview of Recent Collaboration Research with NASA in Ultra High Bypass Technology," presented at the *NASA Fundamental Aeronautics 2007 Annual Meeting*, New Orleans, Oct. 30–Nov. 1, 2007.

and thus the CO2 produced, by more than 12% compared to today's aircraft, while reducing cumulative noise levels about 20dB below the current Stage 4 regulations. This noise level, which is about half the level of today's engines, is the equivalent difference between standing near a garbage disposal running and listening to the sound of my voice right now.[38]

Pratt & Whitney's PW1000G engine incorporating a geared turbofan is selected to be used on the Bombardier CSeries and Mitsubishi Regional Jet airliners beginning in 2013. The engine was first flight-tested in 2008, using an Airbus A340-600 airliner out of Toulouse, France.[39]

Digital Electronic Engine Controls

As one set of NASA and contractor engineers worked on improving the design of the various types of jet engines, another set of researchers representing another science discipline were increasingly interested in marrying the computer's capabilities to the operation of a jet engine, much in the same way that fly-by-wire systems already were in use with aircraft flight controls.

Beginning with that first Wright Flyer in 1903, flying an airplane meant moving levers and other mechanical contrivances that were directly connected by wires and cables to control the operation of the rudder, elevator, wing surfaces, instruments, and engine. When Chuck Yeager broke the sound barrier in 1947 in the X-1, if he wanted to go up, he pulled back on the yoke and cables directly connecting the stick to the elevator, which made that aerosurface move to effect a change in the aircraft's attitude. The rockets propelling the X-1 were activated with a switch throw that closed an electrical circuit whose wiring led directly from the cockpit to the engines. As planes grew bigger, so did their control surfaces. Aircraft such as the B-52 bomber had aerosurfaces as big as the entire wings of smaller airplanes—too bulky and heavy for a single pilot to move using a simple cable/pulley system. A hydraulic system was required and "inserted" between the pilot's input on the yoke and the control surface needing to be moved. Meanwhile, engine

---

38. Alan H. Epstein, *Statement Before the Subcommittee on Aviation Committee on Transportation and Infrastructure, U.S. House of Representatives Hearing on Aviation and Environment: Noise,* Washington, DC, Oct. 24, 2007.
39. "Pratt & Whitney Pure Power PW1000G Engines," Pratt & Whitney S16154.9.08 (2008).

operation remained more or less "old fashioned," with all parameters such as fuel flow and engine temperatures reported to the cockpit on dials the pilot could read, react to, and then make changes by adjusting the throttle or other engine controls.

With the introduction of digital computers and the miniaturization of their circuits—a necessity inspired, in part, by the reduced mass requirements of space flight—engineers began to consider how the quick-thinking electronic marvels might ease the workload for pilots flying increasingly more complex aircraft designs. In fact, as the 1960s transitioned to the 1970s, engineers were already considering aircraft designs that could do remarkable maneuvers in the sky but were inherently unstable, requiring constant, subtle adjustments to the flight controls to keep the vehicle in the air. The solution—already demonstrated for spacecraft applications during Project Apollo—was to insert the power of the computer between the cockpit controls and the flight control surfaces—a concept known as fly-by-wire. A pilot using this system and wanting to turn left would move the control stick to the left, apply a little back pressure, and depress the left rudder pedal. Instead of a wire/cable system directly moving the related aerosurfaces, the movement of the controls would be sensed by a computer, which would send electronic impulses to the appropriate actuators, which in turn would deflect the ailerons, elevator, and rudder.[40]

Managed by NASA's Dryden Flight Research Facility, the fly-by-wire system was first tested without a backup mechanical system in 1972, when a modified F-8C fighter took off from Edwards Air Force Base in California. Testing on this aircraft, whose aerodynamics were known and considered stable, proved that fly-by-wire could work and be reliable. In the years to follow, the system was used to allow pilots to safely fly unstable aircraft, including the B-2 bomber, the forward-swept winged X-29, the Space Shuttle orbiter, and commercial airliners such as the Airbus A320 and Boeing 777.[41]

As experienced was gained with the digital flight control system and computers shrunk in size and grew in power, it didn't take long for propulsion experts to start thinking about how computers could monitor

---
40. C.R. Jarvis, "An Overview of NASA's Digital Fly-By-Wire Technology Development Program," NASA 75N18246 (1975).
41. James E. Tomayko, *Computers Take Flight: A History of NASA's Pioneering Digital Fly-By-Wire Project*, NASA SP-4224 (Washington, DC: NASA, 2000), p. vii.

engine performance and, by making many adjustments in every variable that affects the efficiency of a jet engine, improve the powerplant's overall capabilities.

The first step toward enabling computer control of engine operations was taken by Dryden engineers in managing the Integrated Propulsion Control System (IPCS) program during the mid-1970s. A joint effort with the U.S. Air Force, the IPCS was installed on an F-111E long-range tactical fighter-bomber aircraft. The jet was powered by twin TF30 afterburning turbofan engines with variable-geometry external compression inlets. The IPCS effort installed a digital computer to control the variable inlet and realized significant performance improvements in stallfree operations, faster throttle response, increased thrust, and improved range flying at supersonic speeds. During this same period, results from the IPCS tests were applied to NASA's YF-12C Blackbird, a civilian research version of the famous SR-71 Blackbird spy plane. A digital control system installed on the YF-12C successfully tested, monitored, and adjusted the engine inlet control, autothrottle, air data, and navigation functions for the Pratt & Whitney-built engines. The results gave the aircraft a 7-percent increase in range, improved handling characteristics, and lowered the frequency of inlet unstarts, which happen when an engine shock wave moves forward of the inlet and disrupts the flow of air into the engine, causing it to shutdown. Seeing how well this computer-controlled engine worked, Pratt & Whitney and the U.S. Air Force in 1983 chose to incorporate the system into their SR-71 fleet.[42]

The promising future for more efficient jet engines from developing digitally controlled integrated systems prompted Pratt & Whitney, the Air Force, and NASA (involving both Dryden and Lewis) to pursue a more robust system, which became the Digital Electronic Engine Control (DEEC) program.

Pratt & Whitney actually started what would become the DEEC program, using its own research and development funds to pay for configuration studies beginning during 1973. Then, in 1978, Lewis engineers tested a breadboard version of a computer-controlled system on an engine in an altitude chamber. By 1979, the Air Force had approached NASA and asked if Dryden could demonstrate and evaluate a DEEC system using an F100 engine installed in a NASA F-15, with flight tests beginning in

---

42. James F. Stewart, Frank W. Burcham, Jr., and Donald H. Gatlin, "Flight-Determined Benefits of Integrated Flight-Propulsion Control Systems," NASA TM-4393 (1992), pp. 2–4.

Case 11 | Advancing Propulsive Technology

The Digital Electronic Engine Control system was tested on a Pratt & Whitney F100 turbojet, similar to the one shown here undergoing a hot fire on a test stand. Pratt & Whitney.

1981. At every step in the test program, researchers took advantage of lessons learned not only from the IPCS exercise but also from a U.S. Navy-funded effort called the Full Authority Digital Engine Control program, which ran concurrently to the IPCS program during the mid-1970s.[43]

A NASA Dryden fact sheet about the control system does a good job of explaining in a concise manner the hardware involved, what it monitored, and the resulting actions it was capable of performing:

> The DEEC system tested on the NASA F-15 was an engine mounted, fuel-cooled, single-channel digital controller that received inputs from the airframe and engine to control a wide range of engine functions, such as inlet guide vanes, compressor stators, bleeds, main burner fuel flow, afterburner fuel flow and exhaust nozzle vanes.

---

43. T.W. Putnam, "Digital Electronic Engine Control History," NASA 86N25344 (1984), p. 2.

Engine input measurements that led to these computer-controlled functions included static pressure at the compressor face, fan and core RPM, compressor face temperature, burner pressure, turbine inlet temperature, turbine discharge pressure, throttle position, afterburner fuel flow, fan and compressor speeds and an ultra violet detector in the afterburner to check for flame presence.

Functions carried out after input data were processed by the DEEC computer included setting the variable vanes, positioning compressor start bleeds, controlling gas-generator and augmentation of fuel flows, adjusting the augmenter segment-sequence valve, and controlling the exhaust nozzle position. These actions, and others, gave the engine—and the pilot—rapid and stable throttle response, protection from fan and compressor stalls, improved thrust, better performance at high altitudes, and they kept the engine operating within its limits over the full flight envelope.[44]

When incorporated into the F100 engine, the DEEC provided improvements such as faster throttle responses, more reliable capability to restart an engine in flight, an increase of more than 10,000 feet in altitude when firing the afterburners, and the capability of providing stallfree operations. And with the engine running more efficiently thanks to the DEEC, overall engine and aircraft reliability and maintainability were improved as well.[45]

So successful and promising was this program that even before testing was complete the Air Force approved widespread production of the F100 control units for its F-15 and F-16 fighter fleet. Almost at the same time, Pratt & Whitney added the digital control technology in its PW2037 turbofan engines for the then-new Boeing 757 airliner.[46]

With the DEEC program fully opening the door to computer control of key engine functions, and with the continuing understanding of fly-by-wire systems for aircraft control—along with steady improvements in making computers faster, more capable, and smaller—the next logi-

---

44. "The DEEC," NASA TF-2004-03-DFRC (2004).
45. "Digital Electronic Engine Control (DEEC) Flight Evaluation in an F-15 Airplane," NASA CP-2298 (1984).
46. Christian Gelzer, "60 Years of Cutting-Edge Flight Research Marked at NASA Dryden," Dryden News Release 06-37 (2006).

cal step was to combine together computer control of engines and flight controls. This was done initially with the Adaptive Engine Control System (ADECS) program accomplished between 1985 and 1989, followed by the Performance Seeking Control (PSC) program that performed 72 flight tests between 1990 and 1993. The PSC system was designed to handle multiple variables in performance, compared with the single-variable control allowed in ADECS. The PSC effort was designed to optimize the engine and flight controls in four modes: minimum fuel flow at constant thrust, minimum turbine temperature at constant thrust, maximum thrust, and minimum thrust.[47]

The next evolution in the combining of computer-controlled flight and engine controls— a legacy of the original DEEC program—was inspired in large part by the 1989 crash in Sioux City, IA, of a DC-10 that had lost all three of its hydraulic systems when there was an uncontained failure of the aircraft's No. 2 engine. With three pilots in the cockpit, no working flight controls, and only the thrust levels available for the two remaining working engines, the crew was able to steer the jet to the airport by using variable thrust. During the landing, the airliner broke apart, killing 111 of the 296 people on board.[48]

Soon thereafter, Dryden managers established a program to thoroughly investigate the idea of a Propulsion Controlled Aircraft (PCA) using variable thrust between engines to maintain safe flight control. Once again, the NASA F-15 was pressed into service to demonstrate the concept. Beginning in 1991 with a general ability to steer, refinements in the procedures were made and tested, allowing for more precise maneuvering. Finally, on April 21, 1993, the flight tests of PCA concluded with a successful landing using only engine power to climb, descend, and maneuver. Research continued using an MD-11 airliner, which successfully demonstrated the technology in 1995.[49]

Numerical Propulsion System Simulation

NASA and its contractor colleagues soon found another use for computers to help improve engine performance. In fact, looking back at the history

---

47. John S. Orne, "Performance Seeking Control Program Overview," NASA 95N33011 (1995), p. 32.
48. "Aircraft Accident Report: United Airlines Flight 232, McDonnell Douglas DC-10-10, Sioux Gateway Airport, Sioux City, Iowa, July 19, 1989," NTSB AAR-90-06 (1989).
49. Tom Tucker, *Touchdown: The Development of Propulsion Controlled Aircraft at NASA Dryden* (Washington, DC: NASA, 1999).

11

A computer system known as Propulsion Controlled Aircraft is tested aboard an MD-11 airliner in 1995 at the Dryden Flight Research Center. NASA.

of NASA's involvement with improving propulsion technology, a trilogy of major categories of advances can be suggested based on the development of the computer and its evolution in the role that electronic thinkers have played in our culture.

Part one of this story includes all the improvements NASA and its industry partners have made with jet engines before the computer came along. Having arrived at a basic operational design for a turbojet engine—and its relations, the turboprop and turbofan—engineers sought to improve fuel efficiency, reduce noise, decrease wear, and otherwise reduce the cost of maintaining the engines. They did this through such efforts as the Quiet Clean Short Haul Experimental Engine and Aircraft Energy Efficiency program, detailed earlier in this case study. By tinkering with the individual components and testing the engines on the ground and in the air for thousands of hours, incremental advances were made.[50]

Part two of the story introduces the capabilities made available to engineers as computers became powerful enough and small enough to be incorporated into the engine design. Instead of requiring the pilot to manually make occasional adjustments to the engine operation in

---

50. "Propulsion/ACEE," NASA FACTS-93/8-81 (1981).

flight depending on what the instruments read, a small digital computer built into the engine senses thousands of measurements per minute and caused an equal number of adjustments to be made to keep the powerplant performing at peak efficiency. With the Digital Electronic Engine Control, engines designed years before behaved as though they were fresh off the drawing boards, thanks to their increased capabilities.[51]

Having taken engine designs about as far as it was thought possible, the need for even more fuel-efficient, quieter, and capable engines continued. Unfortunately, the cost of developing a new engine from scratch, building it, and testing it in flight can cost millions of dollars and take years to accomplish. What the aerospace industry needed was a way to take advantage of the powerful computers available at the dawn of the 21st century to make the engine development process less expensive and timelier. The result was part three of NASA's overarching story of engine development: the Numerical Propulsion System Simulation (NPSS) program.[52]

Working with the aerospace industry and academia, NASA's Glenn Research Center led the collaborative effort to create the NPSS program, which was funded and operated as part of the High Performance Computing and Communications program. The idea was to use modern simulation techniques and create a virtual engine and test stand within a virtual wind tunnel, where new designs could be tried out, adjustments made, and the refinements exercised again without costly and time-consuming tests in the "real" world. As stated in a 1999 industry review of the program, the NPSS was built around inclusion of three main elements: "Engineering models that enable multi-disciplinary analysis of large subsystems and systems at various levels of detail, a simulation environment that maximizes designer productivity and a cost-effective, high-performance computing platform."[53]

In explaining to the industry the potential value of the program during a 2006 American Society of Mechanical Engineers conference in

---

51. Jennifer L. Baer-Riedhart and Robert J. Landy, "Highly Integrated Digital Electronic Control—Digital Flight Control, Aircraft Model Identification and Adaptive Engine Control," NASA TM-86793 (1987).
52. John K. Lytle, "The Numerical Propulsion System Simulation: A Multidisciplinary Design System for Aerospace Vehicles," NASA TM-1999-209194 (1999), p. 1.
53. John Lytle, Greg Follen, Cynthia Naiman, Austin Evans, Joseph Veres, Karl Owen, and Isaac Lopez, "Numerical Propulsion System Simulation (NPSS) 1999 Industry Review," NASA TM-2000-209795 (2000), p. 7.

Spain, a NASA briefer from Glenn suggested that if a standard turbojet development program for the military—such as the F100—took 10 years, $1.5 billion, construction of 14 ground-test engines, 9 flight-test engines, and more than 11,000 hours of engine tests, the NPSS program could realize a:

- 50-percent reduction in tooling cost.
- 33-percent reduction in the average development engine cost.
- 30-percent reduction in the cost of fabricating, assembling, and testing rig hardware.
- 36-percent reduction in the number of development engines.
- 60-percent reduction in total hardware cost.[54]

A key—and groundbreaking—feature of NPSS was its ability to integrate simulated tests of different engine components and features, and run them as a whole, fully modeling all aspects of a turbojet's operation. The program did this through the use of the Common Object Request Broker Architecture (CORBA), which essentially provided a shared language among the objects and disciplines (mechanical, thermodynamics, structures, gas flow, etc.) being tested so the resulting data could be analyzed in an "apples to apples" manner. Through the creation of an NPSS developer's kit, researchers had tools to customize the software for individual needs, share secure data, and distribute the simulations for use on multiple computer operating systems. The kit also provided for the use of CORBA to "zoom" in on the data to see specific information with higher fidelity.[55]

Begun in 1997, the NPSS team consisted of propulsion experts and software engineers from GE, Pratt & Whitney, Boeing, Honeywell, Rolls-Royce, Williams International, Teledyne Ryan Aeronautical, Arnold Engineering Development Center, Wright-Patterson AFB, and NASA's

---

54. Ann K. Sehra, "The Numerical Propulsion System Simulation: A Vision for Virtual Engine Testing," presented at the *American Society of Mechanical Engineers TURBO EXPO, Barcelona, Spain, May 8–11, 2006.*

55. Cynthia G. Naiman and Gregory J. Follen, "Numerical Propulsion System Simulation—A Common Tool for Aerospace Propulsion Being Developed," Research and Technology Report 2000 (Cleveland: NASA, 2001).

Glenn Research Center. By the end of the 2000 fiscal year, the NPSS team had released Version 1.0.0 on schedule. According to a summary of the program produced that year:

> (The new software) can be used as an aero-thermodynamic zero-dimensional cycle simulation tool. The capabilities include text-based input syntax, a sophisticated solver, steady-state and transient operation, report generation, a built-in object-oriented programming language for user-definable components and functions, support for distributed running of external codes via CORBA, test data reduction, interactive debug capability and customer deck generation.[56]

Additional capabilities were added in 2001, including the ability to support development of space transportation technologies. At the same time, the initial NPSS software quickly found applications in aviation safety, ground-based power, and alternative energy devices, such as fuel cells. Moreover, project officials at the time suggested that with the further development of the software, other applications could be found for the program in the areas of nuclear power, water treatment, biomedicine, chemical processing, and marine propulsion. NPSS proved to be so capable and promising of future applications that NASA designated the program a cowinner of the NASA Software of the Year Award for 2001.[57]

Work to improve the capabilities and expand the applications of the software continued, and, in 2008, NASA transferred NPSS to a consortium of industry partners, and, through a Space Act Agreement, it is currently offered commercially by Wolverine Ventures, Inc., of Jupiter, FL. Now at Version 1.6.5, NPSS's features include the ability to model all types of complex systems, plug-and-play interfaces for fluid properties, built-in plotting package, interface to higher fidelity legacy codes, multiple model views, command language interpreter with language sensitive text editor, comprehensive component solver, and variable setup controls. It also can operate on Linux, Windows, and UNIX platforms.[58]

---

56. Ibid.
57. Laurel J. Strauber and Cynthia G. Naiman, "Numerical Propulsion System Simulation (NPSS): An Award Winning Propulsion System Simulation Tool," Research and Technology Report 2001 (Cleveland: NASA, 2002).
58. "NPSS User Guide, Software Release: NPSS 1.6.5," NASA NPSS-User (2008), pp. 1-1 to 1-2.

Originally begun as a virtual tool for designing new turbojet engines, NPSS has since found uses in testing rocket engines, fuel cells, analog controls, combined cycle engines, thermal management systems, airframe vehicles preliminary design, and commercial and military engines.[59]

Ultra Efficient Engine Technology Program

With the NPSS tool firmly in place and some four decades of experience incrementally improving the design, operation, and maintenance of the jet engine, it was time to go for broke and assemble an ultra-bright team of engineers to come up with nothing short of the best jet engine possible.

Building on the success of technology development programs such as the Quiet Clean Short Haul Experimental Engine and Energy Efficient Engine project—all of which led directly to the improvements and production of turbojet engines now propelling today's commercial airliners—NASA approached the start of the 21st century with plans to take jet engine design to accomplish even more impressive feats. In 1999, the Aeronautics Directorate of NASA began the Ultra Efficient Engine Technology (UEET) program—a 5-year, $300-million effort—with two primary goals. The first was to find ways that would enable further improvements in engine efficiency to reduce fuel burn and, as a result, carbon dioxide emissions by yet another 15 percent. The second was to continue developing new materials and configuration schemes in the engine's combustor to reduce emissions of nitrogen oxides (NOx) during takeoff and landings by 70 percent relative to the standards detailed in 1996 by the International Civil Aviation Organization.[60]

NASA's Glenn Research Center led the program, with participation from three other NASA Centers: Ames, Langley, and the Goddard Space Flight Center in Greenbelt, MD. Also involved were GE, Pratt & Whitney, Honeywell, Allison/Rolls-Royce, Williams International, Boeing, and Lockheed Martin.[61]

---

59. Edward J. Hall, Joseph Rasche, Todd A. Simons, and Daniel Hoyniak, "NPSS Multidisciplinary Integration and Analysis," NASA CR-2006-213890 (2006).
60. Joe Shaw, "Ultra-Efficient Engine Technology Project Continued to Contribute to Breakthrough Technologies," Research and Technology Report 2002 (Cleveland: NASA, 2003).
61. Lori A. Manthey, "NASA Glenn Research Center UEET (Ultra-Efficient Engine Technology) Program: Agenda and Abstracts," NASA RTOP-714-01-4A (2001).

The program was comprised of seven major projects, each of which addressed particular technology needs and exploitation opportunities.[62] The Propulsion Systems Integration and Assessment project examined overall component technology issues relevant to the UEET program to help furnish overall program guidance and identify technology shortfalls.[63] The Emissions Reduction project sought to significantly reduce NOx and other emissions, using new combustor concepts and technologies such as lean burning combustors with advanced controls and high-temperature ceramic matrix composite materials.[64] The Highly Loaded Turbomachinery project sought to design lighter-weight, reduced-stage cores, low-pressure spools and propulsors for more efficient and environmentally friendly engines, and advanced fan concepts for quieter, lighter, and more efficient fans.[65] The Materials and Structures for High Performance project sought to develop and demonstrate high-temperature material concepts such as ceramic matrix composite combustor liners and turbine vanes, advanced disk alloys, turbine airfoil material systems, high-temperature polymer matrix composites, and innovative lightweight materials and structures for static engine structures.[66] The Propulsion-Airframe Integration project studied propulsion systems and engine locations that could furnish improved engine and environmental benefits without compromising the aerodynamic performance of the airplane; lowering aircraft drag itself constituted a highly desirable means of reducing fuel burn, and, hence, CO_2 emissions will develop advanced technologies to yield lower drag propulsion system integration with the airframe for a wide range of vehicle classes. Decreasing drag improves air vehicle performance and efficiency, which

---

62. Manthey, "Ultra-Efficient Engine Technology (UEET) Program," Research and Technology Report 2001 (NASA, 2002).
63. Ronald C. Plybon, Allan VanDeWall, Rajiv Sampath, Mahadevan Balasubramaniam, Ramakrishna Mallina, and Rohinton Irani, "High Fidelity System Simulation of Multiple Components in Support of the UEET Program," NASA CR-2006-214230 (2006).
64. Kathleen M. Tacina and Changlie Wey, "NASA Glenn High Pressure Low NOx Emissions Research," NASA TM-2008-214974 (2008).
65. Michael T. Tong and Scott M. Jones, "An Updated Assessment of NASA Ultra-Efficient Engine Technologies," presented at *17th International Symposium on Airbreathing Engines, Munich, Germany, Sept. 4–9, 2006.*
66. James A. DiCarlo, Hee Mann Yun, Gregory N. Morscher, and Ramakrishna T. Bhatt, "High-Performance SiC/SiC Ceramic Composite Systems Developed for 1315 C (2400 F) Engine Components," Research and Technology Report 2003 (Cleveland: NASA, 2004).

reduces fuel burn to accomplish a particular mission, thereby reducing the CO2 emissions.[67] The Intelligent Propulsion Controls Project sought to capitalize upon breakthroughs in electronic control technology to improve propulsion system life and enhance flight safety via integrating information, propulsion, and integrated flight propulsion control technologies.[68] Finally, the Integrated Component Technology Demonstrations project sought to evaluate the benefits of off-the-shelf propulsion systems integration on NASA, Department of Defense, and aeropropulsion industry partnership efforts, including both the UEET and the military's Integrated High Performance Turbine Engine Technology (IHPTET) programs.[69]

By 2003, the 7 project areas had come up with 10 specific technology areas that UEET would investigate and incorporate into an engine that would meet the program's goals for reducing pollution and increasing fuel burn efficiency. The technology goals included:

1. Advanced low-NOx combustor design that would feature a lean burning concept.
2. A highly loaded compressor that would lower system weight, improve overall performance, and result in lower fuel burn and carbon dioxide emissions.
3. A highly loaded, high-pressure turbine that could allow a reduction in the number of high-pressure stages, parts count, and cooling requirements, all of which could improve fuel burn and lower carbon dioxide emissions.
4. A highly loaded, low-pressure turbine and aggressive transition duct that would use flow control techniques that would reduce the number of low-pressure stages within the engine.
5. Use of a ceramic matrix composite turbine vane that would allow high-pressure vanes to operate at a higher

---

67. Cecile M. Burg, Geoffrey A. Hill, Sherilyn A. Brown, and Karl A. Geiselhart, "Propulsion Airframe Aeroacoustics Technology Evaluation and Selection Using a Multi-Attribute Decision Making Process and Non-Deterministic Design," AIAA Paper 2004-4436 (2004).
68. Sanjay Garg, "NASA Glenn Research in Controls and Diagnostics for Intelligent Aerospace Propulsion Systems," presented at the *Integrated Condition Management 2006 Conference*, Anaheim, CA, Nov. 14–16, 2006.
69. Mary Jo Long-Davis, "Integrated Components Technology Demonstrations Overview," NTRS Document ID 200.502.14062 (2001).

inlet temperature, which would reduce the amount of engine cooling necessary and result in lower carbon dioxide emissions.
6. The same ceramic matrix composite material would be used to line the combustor walls so it could operate at a higher temperature and reduce NOx emissions.
7. Coat the turbine airfoils with a ceramic thermal barrier material to allow the turbines to operate at a higher temperature and thus reduce carbon dioxide emissions.
8. Use advanced materials in the construction of the turbine airfoil and disk. Specifically, use a lightweight single crystal superalloy to allow the turbine blades and vanes to operate at a higher temperature and reduce carbon dioxide emissions, as well as a dual microstructure nickel-base superalloy to manufacture turbine disks tailored to meet the demands of the higher-temperature environment.
9. Determine advanced materials and structural concepts for an improved, lighter-weight impact damage tolerance and noise-reducing fan containment case.
10. Develop active tip clearance control technology for use in the fan, compressor, and turbine to improve each component's efficiency and reduce carbon dioxide emissions.[70]

In 2003, the UEET program was integrated into NASA's Vehicle Systems program to enable the enginework to be coordinated with research into improving other areas of overall aircraft technology. But in the wake of policy changes associated with the 2004 decision to redirect NASA's space program to retire the Space Shuttle and return humans to the Moon, the Agency was forced to redirect some of its funding to Exploration, forcing the Aeronautics Directorate to give up the $21.6 million budgeted for UEET in fiscal year 2005, effectively canceling the biggest and most complicate jet engine research program ever attempted. At the same time, NASA was directed to realign its jet engine research to concentrate on further reducing noise.[71]

---

70. Michael T. Tong and Scott M. Jones, "An Updated Assessment of NASA Ultra-Efficient Engine Technologies," ISABE-2005-1163 (2005), p. 3.
71. John W. Douglass, "NASA Aeronautics Research Funding: The Wrong Direction," *Space News*, Mar. 28, 2005, opinion page.

Nevertheless, results from tests of UEET hardware showed promise that a large, subsonic aircraft equipped with some of the technologies detailed above would have a "very high probability" of achieving the program goals laid out for reducing emissions of carbon dioxide and other pollutants. The data remain for application to future aircraft and engine schemes.[72]

Damage-Tolerant Fan Casing

While most eyes were on the big picture of making major engine advancements through the years, some very specific problems were addressed with programs that are just as interesting to consider as the larger research endeavors. The casings that surround the jet engine's turbomachinery are a case in point.

With the 1989 crash of United Airlines Flight 232 at Sioux City, IA, aviation safety officials became more interested in finding new materials capable of containing the resulting shrapnel created when a jet engine's blade or other component breaks free. In the case of the DC-10 involved in this particular crash, the fan disk of the No. 2 engine—the one located in the tail—separated from the engine and caused the powerplant to explode, creating a rain of shrapnel that could not be contained within the engine casing. The sharp metal fragments pierced the body of the aircraft and cut lines in all three of the aircraft's hydraulic systems. As previously mentioned in this case study, the pilots on the DC-10 were able to steer their aircraft to a nearly controlled landing. The incident inspired NASA pilots to refine the idea of using only jet thrust to maneuver an airplane and undertake the Propulsion Controlled Aircraft program, which took full advantage of the earlier Digital Electronic Engine Control research. The Iowa accident also sent structures and materials experts off on a hunt to find a way to prevent accidents like this in the future.

The United Flight 232 example notwithstanding, the challenge for structures engineers is to design an engine casing that will contain a failed fan blade within the engine so that it has no chance to pierce the passenger compartment wall and threaten the safety of passengers or cause a catastrophic tear in the aircraft wall. Moreover, not only does the casing have to be strong enough to withstand any blade or shrapnel impacts, it must not lose its structural integrity during an emergency

---

72. Tong and Jones, "An Updated Assessment of NASA Ultra-Efficient Engine Technologies," p. 1.

engine shutdown in flight. A damaged engine can take some 15 seconds to shut down, during which time cracks from the initial blade impacts can propagate in the fan case. Should the fan case totally fail, the resulting breakup of the already compromised turbomachinery could be catastrophic to the aircraft and all aboard.[73]

As engineers considered the use of composite materials, two methods for containing blade damage within the engine casing were now available: the new softwall and the traditional hardwall. In the softwall concept, the casing was made of a sandwich-type aluminum structure overwound with dry aramid fibers. (Aramid fibers were introduced commercially by DuPont during the early 1960s and were known by the trade name Nomex.) The design allows broken blades and other shrapnel to pass through the "soft" aluminum and be stopped and contained within the aramid fiber wrap. In the hardwall approach, the casing is made of aluminum only and is built as a rigid wall to reflect blade bits and other collateral damage back into the casing interior. Of course that vastly increases the risk that the shrapnel will be ingested through the engine and cause even greater damage, perhaps catastrophic. While that risk exists with the softwall design, it is not as substantial. Another benefit of the hardwall is that it maintains its structural soundness, or ductility, during a breakup of an engine. A softwall also features some amount of ductility, but the energy-absorbing properties of the aramid fibers is the major draw.[74]

In 1994, NASA engineers at the Lewis Research Center began looking into better understanding engine fan case structures and conducted impact tests as part of the Enabling Propulsion Materials program. Various metallic materials and new ideas for lightweight fan containment structures were studied. By 1998, the research expanded to include investigations into use of polymer composites for engine fan casings. As additional composite materials were made available, NASA researchers sought to understand their properties and the appropriateness of those materials in terms of containment capability, damage tolerance, commercial viability, and understanding any potential risk not yet identified for their use on jet engines.[75]

---

73. C.L. Stotler and A.P. Coppa, "Containment of Composite Fan Blades," NASA CR-159544 (1979).
74. Bob Griffiths, "Composite Fan Blade Containment Case: Innovative Use of Carbon-Fiber Braid Yields a Ductile Structure that Resists Blade Impact," *High Performance Composites* (May 1, 2005).
75. Ibid.

In 2001, NASA awarded a Small Business Innovation Research (SBIR) grant to A&P Technology, Inc., of Cincinnati to develop a damage-tolerant fan casing for a jet engine. Long before composites came along, the company's expertise was in braiding materials together, such as clotheslines and candlewicks. A&P—working together with the FAA, Ohio State University, and the University of Akron—was able to rapidly develop a prototype composite fan case that could be compared to the metal fan case. Computer simulations were key to the effort and serendipitously provided an opportunity to grow the industry's understanding and ability to use those very same simulation capabilities. First, well understood metallic casings undergoing a blade-out scenario were modeled, and the computer tested the resulting codes to reproduce the already-known results. Then came the trick of introducing code that would represent A&P's composite casing and its reaction to a blade-out situation. The process was repeated for a composite material wrapped with a braided fiber material, and results were very promising.[76]

The composite casing proposed by A&P used a triaxial carbon braid, which has a toughness superior to aluminum but is lighter, which helps ease fuel consumption. In tests of debris impact, the braided laminate performed better than the metal casing, because in some cases, the composite structure absorbed the energy of the impact as the debris bounced off the wall, and in other cases where the shrapnel penetrated the material, the damage to the wall was isolated to the impact point and did not spread. In a metal casing that was pierced, the resulting hole would instigate several cracks that would continue to propagate along the casing wall, appearing much like the spiderweb of cracks that appear on an automobile windshield when it is hit with a small stone on the freeway.

NASA continues to study the use of composite casings to better understand the potential effects of aging and/or degradation following the constant temperature, vibration, and pressure cycles a jet engine experiences during each flight. There also is interest in studying the effects of higher operating temperatures on the casing structure for possible use on future supersonic jets. (The effect of composite fan blades on casing containment also has been studied.)[77]

---

76. "Damage-Tolerant Fan Casings for Jet Engines," *Spinoff 2006* (Washington, DC: NASA, 2006), p. 14.
77. C.L. Stotler and A.P. Coppa, "Containment of Composite Fan Blades," NASA CR-159544 (1979).

Case 11 | Advancing Propulsive Technology

A General Electric GEnx engine with a composite damage-tolerant fan casing is checked out before eventual installation on the new Boeing 787. General Electric.

While composites have found many uses in commercial and military aviation, the first use of an all-composite engine casing, provided by A&P, is set to be used on GE's GEnx turbojet designed for the Boeing 787. The braided casing weighs 350 pounds less per engine, and, when other engine installation hardware to handle the lighter powerplants is considered, the 787 should weigh 800 pounds less than a similarly equipped airliner using aluminum casings. The weight reduction also should provide a savings in fuel cost, increased payload, and/or a greater range for the aircraft.[78]

Conclusion and a Look Ahead

For more than 50 years now, NASA has methodically and, for the most part, quietly advanced the state of art of propulsion technology. With the basic design of the jet engine unchanged since it was invented during World War II, modern jet engines incorporate every lesson learned during NASA's past five decades of research. As a result, jet engines are

---

78. "Damage-Tolerant Fan Casings for Jet Engines."

now quieter, safer, more fuel-efficient, less expensive to operate, and less polluting, while being easier to maintain. And thanks to advancements in computers and simulations, new engines can be tested for thousands of hours at a time without ever bending one piece of aluminum or braiding a square yard of composite material.

So what's in store for propulsion technology during the next few decades? More improvements with every possible variable of engine operations are still possible, with future advances more closely linked to new aircraft designs, such as the blended wing and body in which the engines may be more fully integrated into the structure of the aircraft.

In a feature story written in April 2009 for NASA's Aeronautics Research Mission Directorate Web site, this author interviewed several key Agency officials who are considering what the future holds for engine development and making plans for what the Agency's approach will be for managing the effort. Here is that look ahead.

NASA Researchers Work to Reduce Noise in Future Aircraft Design

It's a noisy world out there, especially around the Nation's busiest airports, so NASA is pioneering new technologies and aircraft designs that could help quiet things down a bit. Every source of aircraft noise, from takeoff to touchdown, is being studied for ways to reduce the racket, which is expected to get worse as officials predict that air traffic will double in the next decade or so.

"It's always too noisy. You have to always work on making it quieter," said Edmane Envia, an aerospace engineer at NASA's Glenn Research Center in Cleveland. "You always have to stay a step ahead to fulfill the needs and demands of the next generation of air travel."[79]

Noise reduction research is part of a broader effort by NASA's Aeronautics Research Mission Directorate in Washington to lay a technological foundation for a new generation of airplanes that are not as noisy, fly farther on less fuel, and may operate out of airports with much shorter runways than exist today. There are no clear solutions yet to these tough challenges, neither is there a shortage of ideas from NASA researchers who are confident positive results eventually will come.[80]

---

79. Interview of Envia by Jim Banke, Cape Canaveral, Feb. 4, 2009.
80. Jeffrey J. Berton, Envia, and Casey L. Burley, "An Analytical Assessment of NASA's N1 Subsonic Fixed Wing Project Noise Goal," NASA LF99-8609 (2009).

"Our goal is to have the technologies researched and ready, but ultimately it's the aircraft industry, driven by the market, that makes the decision when to introduce a particular generation of aircraft," Envia said.

NASA organized its research to look three generations into the future, with conceptual aircraft designs that could be introduced 10, 20, or 30 years from now. The generations are called N+1, N+2, and N+3. Each generation represents a design intended to be flown a decade or so later than the one before it and is to feature increasingly sophisticated methods for delivering quieter aircraft and jet engines.[81]

"Think of the Boeing 787 Dreamliner as N and the N+1 as the next generation aircraft after that," Envia said.

The N+1 is an aircraft with familiar parts, including a conventional tube-shaped body, wings, and a tail. Its jet engines still are attached to the wings, as with an N aircraft, but those engines might be on top of the wings, not underneath. Conceptual N+2 designs throw out convention and basically begin with a blank computer screen, with design engineers blending the line between the body, wing, and engines into a more seamless, hybrid look. What an N+3 aircraft might look like is anyone's guess right now. But with its debut still 30 years away, NASA is sponsoring research that will produce a host of ideas for consideration. The Federal Aviation Administration's current guidelines for overall aircraft noise footprints constitute the design baseline for all of NASA's N aircraft concepts. That footprint summarizes in a single number, expressed as a decibel, the noise heard on the ground as an airplane lands, takes off, and then cuts back on power for noise abatement. The noise footprint extends ahead and behind the aircraft and to a certain distance on either side. NASA's design goal is to make each new aircraft generation quieter than today's airplanes by a set number of decibels. The N+1 goal is 32 decibels quieter than a fully noise compliant Boeing 737, while the N+2 goal is 42 decibels quieter than a Boeing 777. So far, the decibel goal for the N+1 aircraft has been elusive.[82]

"What makes our job very hard is that we are asked to reduce noise but in ways that do not adversely impact how high, far or fast an airplane is capable of flying," Envia said.

---

81. Envia, "Progress Toward SFW N+1 Noise Goal," presented at the *NASA Fundamental Aeronautics Program 2nd Annual Meeting*, Atlanta, Oct. 7, 2008.
82. Beth Dickey, "NASA Awards Future Vehicle Aircraft Research Contracts," NASA Contract Release C08-60 (2008).

NASA researchers have studied changes in the operation, shape, or materials from which key noise contributors are made. The known suspects include the airframe, wing flaps, and slats, along with components of the jet engine, such as the fan, turbine, and exhaust nozzle. While some reductions in noise can be realized with some design changes in these components, the overall impact still falls short of the N+1 goal by about 6 decibels. Envia said that additional work with design and operation of the jet engine's core may make up the difference, but that a lot more work needs to be done in the years to come. Meanwhile, reaching the N+2 goals may or may not prove easier to achieve.[83]

"We're starting from a different aircraft configuration, from a clean sheet, that gives you the promise of achieving even more aggressive goals," said Russell Thomas, an aerospace engineer at Langley Research Center. "But it also means that a lot of your prior experience is not directly applicable, so the problem gets a lot harder from that point of view. You may have to investigate new areas that have not been researched heavily in the past."[84]

Efforts to reduce noise in the N+2 aircraft have focused on the airframe, which blends the wing and fuselage together, greatly reducing the number of parts that extend into the airflow to cause noise. Also, according to Thomas, the early thinking on the N+2 aircraft is that the jet engines will be on top of the vehicle, using the airplane body to shield most of the noise from reaching the ground.

"We're on course to do much more thorough research to get higher quality numbers, better experiments, and better prediction methods so we can really understand the acoustics of this new aircraft configuration," Thomas said.

As for the N+3 aircraft, it remains too early to say how NASA researchers will use technology not yet invented to reduce noise levels to their lowest ever.

"Clearly significant progress has been made over the years and airplanes are much quieter than they were 20 years ago," Envia said, noting that further reductions in noise will require whole new approaches to aircraft design. "It is a complicated problem and so it is a worthy challenge to rise up to."

---

83. Don Weir, ed., "Engine Validation of Noise and Emission Reduction Technology Phase 1," NASA CR-2008-215225 (2008).
84. Interview of Russell Thomas by Banke, Cape Canaveral, Feb. 4, 2009.

Recommended Additional Readings
Technical Reports:

"ACEE Composite Structures Technology: Review of Selected NASA Research on Composite Materials and Structures," NASA CP-2321 (1984).

"Advanced Aerodynamics, Selected NASA Research," NASA CP-2208 (1981).

A.P. Adamson, "Quiet Clean Short-Haul Experimental Engine (QCSEE) Design Rationale," SAE Paper 750605 (1975).

"Aircraft Accident Report: United Airlines Flight 232, McDonnell Douglas DC-10-10, Sioux Gateway Airport, Sioux City, Iowa, July 19, 1989," NTSB AAR-90-06 (1989).

R.J. Antl and J.E. Mcaulay, "Improved Components for Engine Fuel Savings," NASA TM-81577 (1980).

Jennifer L. Baer-Riedhart and Robert J. Landy, "Highly Integrated Digital Electronic Control—Digital Flight Control, Aircraft Model Identification and Adaptive Engine Control," NASA TM-86793 (1987).

Peter G. Batterton, "Energy Efficient Engine Program Contributions to Aircraft Fuel Conservation," NASA TM-83741 (1984).

R.S. Beitler and G.W. Bennett, "Energy Efficient Engine: Control System Component Performance Report," NASA CR-174651 (1984).

M.J. Benzakein, S.B. Kazin, and F. Montegani, "NASA/GE Quiet Engine 'A,'" AIAA Paper 72-657 (1972).

Jeffrey J. Berton, Edmane Envia, and Casey L. Burley, "An Analytical Assessment of NASA's N1 Subsonic Fixed Wing Project Noise Goal," NASA LF99-8609 (2009).

Frank W. Burcham, Jr., L.P. Myers, and K.R. Walsh, "Flight Evaluation Results for a Digital Electronic Engine Control in an F-15 Airplane," AIAA Paper 83-2703 (1983).

Frank W. Burcham, Jr., and Ronald J. Ray, "The Value of Early Flight Evaluation of Propulsion Concepts Using the NASA F-15 Research Airplane," NASA TM-100408 (1987).

Cecile M. Burg, Geoffrey A. Hill, Sherilyn A. Brown, and Karl A. Geiselhart, "Propulsion Airframe Aeroacoustics Technology Evaluation and Selection Using a Multi-Attribute Decision Making Process and Non-Deterministic Design," AIAA Paper 2004-4436 (2004).

T.K. Cho and F.W. Burcham, Jr., "Preliminary Flight Evaluation of F100 Engine Model Derivative Airstart Capability in an F-15 Airplane," NASA TM-86031 (1984).

C.C. Ciepluch, "A Review of the QCSEE Program," NASA TM-X-71818 (1975).

C.C. Ciepluch, "Overview of the QCSEE Program," NTRS 197.800.16123 (1976).

C.C. Ciepluch and W.S. Willis, "QCSEE—The Key to Future Short-Haul Air Transport," ICAO Bulletin No. 34 (1979).

D.B. Crawford and F.W. Burcham, Jr., "Effect of Control Logic Modifications on Airstart Performance of F100 Engine Model Derivative Engines in an F-15 Airplane," NASA TM-85900 (1984).

Donald Y. Davis and Marshall Stearns, "Energy Efficient Engine: Flight Propulsion System Final Design and Analysis," NASA CR-168219 (1985).

"The DEEC," NASA TF-2004-03-DFRC (2004).

"Digital Electronic Engine Control (DEEC) Flight Evaluation in an F-15 Airplane," NASA CP-2298 (1984).

W.A. Fasching, "The CF6 Engine Performance Improvement," NASA CR-165612 (1982).

"Full Scale Technology Demonstration of a Modern Counterrotating Unducted Fan Engine Concept: Design Test," NASA CR-180867 (1987) and CR-180868 (1987).

W.O. Gaffin, "The JT8D and JT9D Engine Component Improvement: Performance Improvement Program," NASA CR-167965 (1982).

Robert V. Garvin, "Starting Something Big: The Commercial Emergence of GE Aircraft Engines," AIAA Paper 72-657 (1999).

D.E. Gray and William B. Gardner, "Energy Efficient Engine Program Technology Benefit/Cost Study. Volume 1: Executive Summary," NASA CR-174766 (1983).

Edward J. Hall, Joseph Rasche, Todd A. Simons, and Daniel Hoyniak, "NPSS Multidisciplinary Integration and Analysis," NASA CR-2006-213890 (2006).

Scott M. Jones, "An Introduction to Thermodynamic Performance Analysis of Aircraft Gas Turbine Engine Cycles Using the Numerical Propulsion System Simulation Code," NASA TM-2007-214690 (2007).

Theo G. Keith, Jr., "Aeroelastic Stability and Response of Rotating Structures," NASA CR-191803 (1993).

C.R. Jarvis, "An Overview of NASA's Digital Fly-By-Wire Technology Development Program," NASA 75N18246 (1975).

Anatole P. Kurkov, "Optical Measurement of Unducted Fan Blade Deflections," NASA TM-100966 (1988).

S.J. Licata and F.W. Burcham, Jr., "Airstart Performance of a Digital Electronic Engine Control System in an F-15 Airplane," NASA TM-84908 (1983).

Mary Jo Long-Davis, "Integrated Components Technology Demonstrations Overview," NTRS Document ID 200.502.14062 (2001).

Isaac Lopez, Gregory J. Follen, Richard Guiterrez, Ian Foster, Brian Ginsburg, Olle Larsson, Stuart Martin, Steven Tuecke, and David Woodford, "NPSS on NASA's Information Power Grid: Using CORBA and Globus to Coordinate Multidisciplinary Aeroscience Applications," NASA TM-2000-209956 (2000).

John K. Lytle, "The Numerical Propulsion System Simulation: A Multidisciplinary Design System for Aerospace Vehicles," NASA TM-1999-209194 (1999).

John K. Lytle, "The Numerical Propulsion System Simulation: Concept to Product," NTRS 199.802.19326 (1997).

John Lytle, Greg Follen, Cynthia Naiman, Austin Evans, Joseph Veres, Karl Owen, and Isaac Lopez, "Numerical Propulsion System Simulation (NPSS) 1999 Industry Review," NASA TM-2000-209795 (2000).

John Lytle, Greg Follen, Cynthia Naiman, Joseph Veres, Karl Owen, and Isacc Lopez, "2000 Numerical Propulsion System Simulation Review," NASA CP-2001-210673 (2001).

Lawrence E. Macioce, John W. Schaefer, and Neal T. Saunders, "The Energy Efficient Engine Project," NASA TM-81566 (1980).

Lori Manthey, "NASA Glenn Research Center UEET (Ultra-Efficient Engine Technology) Program: Agenda and Abstracts," NASA RTOP-714-01-4A (2001).

J.E. Mcaulay, "Engine Component Improvement Program—Performance Improvement," AIAA Paper 80-0223 (1980).

Edward T. Meleason and K.O. Johnson, "Unducted-Fan Engine," LEW-14429 (1987).

Glenn A. Mitchell, "Experimental Aerodynamic Performance of Advanced 40 Degree-Swept, 10-Blade Propeller Model at Mach 0.6 to 0.85," NASA TM-88969 (1988).

Lester D. Nichols and Christos C. Chamis, "Numerical Propulsion System Simulation: An Interdisciplinary Approach," NASA TM-105181 (1991).

"NPSS User Guide, Software Release: NPSS 1.6.5," *NASA NPSS-User* (2008).

Dean Olson, "Pratt and Whitney Space Propulsion NPSS Usage," E-14760 (2004).

John S. Orne, "Performance Seeking Control Program Overview," NASA 95N33011 (1995).

Khary I. Parker, James L. Felder, Thomas M. Lavella, Colleen A. Withrow, Albert Y. Lehmann, and V.A. William, "Integrated Control Modeling for Propulsion Systems Using NPSS," NASA TM-2004-212945 (2004).

Ronald C. Plybon, Allan VanDeWall, Rajiv Sampath, Mahadevan Balasubramaniam, Ramakrishna Mallina, and Rohinton Irani, "High Fidelity System Simulation of Multiple Components in Support of the UEET Program," NASA CR-2006-214230 (2006).

"Pratt & Whitney Pure Power PW1000G Engines," Pratt & Whitney S16154.9.08 (2008).

"Propulsion/ACEE," NASA FACTS-93/8-81 (1981).

T.W. Putnam, "Digital Electronic Engine Control History," NASA 86N25344 (1984).

"Quiet Clean Short-Haul Experimental Engine (QSCEE) Preliminary Analyses and Design Report," vol. 1, NASA CR-134838 (1974).

A.A. Saunders, Jr., and A.C. Hoffman, "Application of Digital Controls on the Quiet Clean Short Haul Experimental Engines," AIAA Paper 79-1203 (1979).

Neal T. Saunders, "Advanced Component Technologies for Energy-Efficient Turbofan Engines," NASA TM-81507 (1980).

John R. Szuch, Dale J. Arpasi, and Anthony J. Strazisar, "Enhancing aeropropulsion Research with High-Speed Interactive Computing," NASA TM-104374 (1991).

H.D. Sowers and W.E. Coward, "Quiet, Clean, Short-Haul, Experimental Engine (QCSEE) Under-The-Wing (UTW) Engine Acoustic Design," NASA CR-135267 (1978).

H.D. Sowers and W.E. Coward, "Quiet, Clean, Short-Haul Experimental Engine (QCSEE) Over-The-Wing (OTW) Engine Acoustic Design," NASA CR-135268 (1978).

James F. Stewart, Frank W. Burcham, Jr., and Donald H. Gatlin, "Flight-Determined Benefits of Integrated Flight-Propulsion Control Systems," NASA TM-4393 (1992).

C.L. Stotler and A.P. Coppa, "Containment of Composite Fan Blades," NASA CR-159544 (1979).

Kathleen M. Tacina and Changlie Wey, "NASA Glenn High Pressure Low NOx Emissions Research," NASA TM-2008-214974 (2008).

Michael T. Tong and Scott M. Jones, "An Updated Assessment of NASA Ultra-Efficient Engine Technologies," ISABE-2005-1163 (2005).

Mark G. Turner, John A. Reed, Robert Ryder, and Joseph P. Veres, "Multi-Fidelity Simulation of a Turbofan Engine With Results Zoomed Into Mini-Maps for a Zero-D Cycle Simulation," NASA TM-2004-213076 (2004).

Don Weir, ed., "Engine Validation of Noise and Emission Reduction Technology Phase 1," NASA CR-2008-215225 (2008).

Conference Presentations:

Alain Depitre, "Aircraft Noise Certification History/Development," presented at *ICAO Noise Certification Workshop, Montreal, 2004*.

Edmane Envia, "Progress Toward SFW N+1 Noise Goal," presented at the *NASA Fundamental Aeronautics Program 2nd Annual Meeting, Atlanta, Oct. 7, 2008*.

Sanjay Garg, "NASA Glenn Research in Controls and Diagnostics for Intelligent Aerospace Propulsion Systems," presented at the *Integrated Condition Management 2006 Conference, Anaheim, CA, Nov. 14–16, 2006*.

Edward M. Kerwin, Jr., "Procedures for Estimating the Near Field Noise of Rotating Aircraft Propellers," presented at the *Fifty-First Meeting of the Acoustical Society of America, Cambridge, MA, June 17–23, 1956*.

Jeff Schweitzer, "An Overview of Recent Collaboration Research with NASA in Ultra High Bypass Technology," presented at the *NASA Fundamental Aeronautics 2007 Annual Meeting, New Orleans, Oct. 30–Nov. 1, 2007*.

Ann K. Sehra, "The Numerical Propulsion System Simulation: A Vision for Virtual Engine Testing," presented at the *American Society of Mechanical Engineers TURBO EXPO, Barcelona, Spain, May 8–11, 2006*.

Michael T. Tong and Scott M. Jones, "An Updated Assessment of NASA Ultra-Efficient Engine Technologies," presented at *17th International Symposium on Airbreathing Engines, Munich, Germany, Sept. 4–9, 2006*.

Books and Monographs:

Alleviation of Jet Aircraft Noise Near Airports (Washington, DC: U.S. Office of Science and Technology, 1966).

Arnold & Porter, *U.S. Government Support of the U.S. Commercial Aircraft Industry, Prepared for the Commission of the European Communities* (Washington, DC: Arnold & Porter, 1991).

Ronald H. Aungier, *Turbine Aerodynamics: Axial-Flow and Radial-Flow Turbine Design and Analysis* (New York: ASME Press, 2006).

David M. Bearden, *Noise Abatement and Control: An Overview of Federal Standards and Regulations* (Washington, DC: Congressional Research Service, 2006).

Roger E. Bilstein, *Orders of Magnitude: A History of the NACA and NASA, 1915–1990*, NASA SP-4406 (Washington, DC: NASA, 1989).

Charles D. Bright, *The Jet Makers: The Aerospace Industry from 1945 to 1972* (Lawrence, KS: Regents Press of Kansas, 1978).

Mark D. Bowles and Virginia P. Dawson, "The Advanced Turboprop Project: Radical Innovation in a Conservative Environment," in *From Engineering Science to Big Science, The NACA and NASA Collier Trophy Research Project Winners*, NASA SP-4219 (Washington, DC: NASA, 1998).

Walter J. Boyne and Donald S. Lopez, eds., *The Jet Age: Forty Years of Jet Aviation* (Washington, DC: Smithsonian Institution Press, 1979).

Margaret Conner, *Hans von Ohain: Elegance in Flight* (Reston, VA: AIAA, 2001).

Edward W. Constant, *The Origins of the Turbojet Revolution* (Baltimore: The Johns Hopkins University Press, 1980).

N.A. Cumpsty, *Compressor Aerodynamics* (Malabar, FL: Krieger Publishing Company, 2004 ed.).

"Damage-Tolerant Fan Casings for Jet Engines," *Spinoff 2006* (Washington, DC: NASA, 2006).

Virginia P. Dawson, *Engines and Innovation: Lewis Laboratory and American Propulsion Technology*, NASA SP-4306 (Washington, DC: NASA, 1991).

James A. DiCarlo, Hee Mann Yun, Gregory N. Morscher, and Ramakrishna T. Bhatt, "High-Performance SiC/SiC Ceramic Composite Systems Developed for 1315 C (2400 F) Engine Components," *Research and Technology Report 2003* (Cleveland: NASA, 2004).

Michael Donne, *Leader of the Skies, Rolls-Royce: The First Seventy-Five Years* (London: Frederick Muller Limited, 1981).

J.H. Doolittle, *The Airport and Its Neighbors, The Report of the President's Airport Commission* (Washington, DC: U.S. Government Printing Office, 1952).

Jeffrey L. Ethell, *Fuel Economy in Aviation*, NASA SP-462 (Washington, DC: NASA, 1983).

General Electric, *Seven Decades of Progress: A Heritage of Aircraft Turbine Technology* (Fallbrook, CA: Aero Publishers, Inc., 1979).

"A Giant Step in Jetliner Propulsion," *Spinoff 1996* (Washington, DC: NASA, 1996).

Vicki L. Golich and Thomas E. Pinelli, *Knowledge Diffusion in the U.S. Aerospace Industry* (London: Alex Publishing, 1998).

John Golley, *Genesis of the Jet: Frank Whittle and the Invention of the Jet Engine* (Ramsbury, Marlborough: The Crowood Press, 1997).

Bob Griffiths, "Composite Fan Blade Containment Case: Innovative Use of Carbon-Fiber Braid Yields a Ductile Structure that Resists Blade Impact," *High Performance Composites* (May 1, 2005).

Bill Gunston, *The Development of Jet and Turbine Aero Engines* (Minneapolis: Motorbooks International, 2006).

Roy D. Hager and Deborah Vrabel, *Advanced Turboprop Project*, NASA SP-495 (Washington, DC: NASA, 1988).

T.A. Heppenheimer, *Turbulent Skies: The History of Commercial Aviation* (New York: John Wiley & Sons, Inc., 1986).

Klaus Hunecke, *Jet Engines: Fundamentals of Theory, Design and Operation* (Minneapolis: Motorbooks International, 2003).

Lori A. Manthey, "Ultra-Efficient Engine Technology (UEET) Program," *Research and Technology Report 2001* (NASA, 2002).

Jack D. Mattingly, William H. Heiser, and David T. Pratt, *Aircraft Engine Design* (Reston, VA: AIAA, 2002).

Andrew Nahum, *Frank Whittle: Invention of the Jet* (Duxford, UK: Icon Books, Ltd., in association with the Science Museum, 2004).

Cynthia G. Naiman and Gregory J. Follen, "Numerical Propulsion System Simulation—A Common Tool for Aerospace Propulsion Being Developed," *Research and Technology Report 2000* (Cleveland: NASA, 2001).

NASA Glenn Research Center at Lewis Field: Achieving the Extraordinary (Cleveland: NASA, 1999).

Guy Norris and Mark Wagner, *Boeing 777: The Technological Marvel* (Osceola, WI: MBI Publishing, 2001).

Brian H. Rowe with Martin Ducheny, *The Power to Fly: An Engineer's Life* (Reston, VA: AIAA, 2005).

Newell D. Sanders, *Aircraft Engine Noise Reduction*, NASA SP-311 (Washington, DC: NSA, 1972).

Herb Saravanamuttoo, Gordon Rogers, and Henry Cohen, *Gas Turbine Theory*, 5th ed. (Upper Saddle River, NJ: Prentice Hall, 2001).

Robert Schlaifer and S.D. Heron, *Development of Aircraft Engines and Fuels* (Boston: Harvard University Graduate School of Business Administration, 1950).

Joe Shaw, "Ultra-Efficient Engine Technology Project Continued to Contribute to Breakthrough Technologies," *Research and Technology Report 2002* (Cleveland: NASA, 2003).

James St. Peter, *The History of Aircraft Gas Turbine Engine Development in the United States . . . A Tradition of Excellence* (Atlanta: International Gas Turbine Institute of The American Society of Mechanical Engineers, 1999).

Laurel J. Strauber and Cynthia G. Naiman, "Numerical Propulsion System Simulation (NPSS): An Award Winning Propulsion System Simulation Tool," *Research and Technology Report 2001* (Cleveland: NASA, 2002).

James E. Tomayko, *Computers Take Flight: A History of NASA's Pioneering Digital Fly-By-Wire Project*, NASA SP-4224 (Washington, DC: NASA, 2000).

Tom Tucker, *Touchdown: The Development of Propulsion Controlled Aircraft at NASA Dryden* (Washington, DC: NASA, 1999).

Frank Whittle, *Jet: The Story of a Pioneer* (New York: Philosophical Library, 1954).

Louis J. Williams, *Small Transport Aircraft Technology* (Honolulu: University Press of the Pacific, 2001).

The F-16XL-2 was one of two test aircraft that NASA Dryden used in the 1990s to test the application of laminar flow at supersonic speeds. NASA.

CASE 12

Leaner and Greener: Fuel Efficiency Takes Flight

Caitlin Harrington

Decades of NASA research have led to breakthroughs in understanding the physical processes of pollution and determining how to secure unprecedented levels of propulsion and aerodynamic efficiency to reduce emissions. Goaded by recurring fuel supply crises, NASA has responded with a series of research plans that have dramatically improved the efficiency of gas turbine propulsion systems, the lift-to-drag ratio of new aircraft designs, and myriad other challenges.

ALTHOUGH NASA'S AERONAUTICS BUDGET has fallen dramatically in recent years,[1] the Agency has nevertheless managed to spearhead some of America's biggest breakthroughs in fuel-efficient and environmentally friendly aircraft technology. The National Aeronautics and Space Administration (NASA) has engaged in major programs to increase aircraft fuel efficiency that have laid the groundwork for engines, airframes, and new energy sources—such as alternative fuel and fuel cells—that are still in use today. NASA's research on aircraft emissions in the 1970s also was groundbreaking, leading to a widely accepted view at the national—and later, global—level that pollution can damage the ozone layer and spawning a series of efforts inside and outside NASA to reduce aircraft emissions.[2]

This case study will explore NASA's efforts to improve the fuel efficiency of aircraft and also reduce emissions, with a heavy emphasis on the 1970s, when the energy crisis and environmental concerns created a national demand for "lean and green" airplanes.[3] The launch of

---

1. The Administration's fiscal year 2010 budget request for NASA is $507 million, compared with the 1998 aeronautics budget of $1.5 billion. *Opening Remarks of Senator Barbara Mikulski, Chairwoman of the Senate Appropriations Subcommittee on Commerce, Justice, Science, and Related Agencies. Hearing of the Fiscal Year 2010 Budget Request for NASA.* U.S. Congress. May 21, 2009.
2. Joseph Chambers, *Innovation in Flight* (SP-2005-4539), p. 46.
3. The author thanks Robert Arrighi, archivist, Wyle Information Systems, History Program at Glenn, for providing instrumental documents, and Air Force Capt. Jarrett S. Lee, for providing insights and support.

Sputnik in 1957 and the resulting space race with the Soviet Union spurred the National Advisory Committee for Aeronautics (NACA)—subsequently restructured within the new National Aeronautics and Space Administration—to shift its research heavily toward rocketry—at the expense of aeronautics—until the mid-1960s.[4] But as commercial air travel grew in the 1960s, NASA began to embark on a series of ambitious programs that connected aeronautics, energy, and the environment. This case study will discuss some of NASA's most important programs in this area.

Key propulsion initiatives to be discussed include the Energy Efficient Engine program—perhaps NASA's greatest contribution to fuel-efficient flight—as well as later efforts to increase propulsion efficiency, including the Advanced Subsonic Technology (AST) initiative and the Ultra Efficient Engine Technology (UEET) program. Another propulsion effort that paved the way for the development of fuel-efficient engine technology was the Advanced Turboprop, which led to current NASA and industry attempts to develop fuel-efficient "open rotor" concepts.

In addition to propulsion research, this case study will also explore several NASA programs aimed at improving aircraft structures to promote fuel efficiency, including initiatives to develop supercritical wings and winglets and efforts to employ laminar flow concepts. NASA has also sought to develop alternative fuels to improve performance, maximize efficiency, and minimize emissions; this case study will touch on liquid hydrogen research conducted by NASA's predecessor—the NACA—as well as subsequent attempts to develop synthetic fuels to replace hydrocarbon-based jet fuel.

NASA's Involvement in Energy Efficiency and Emissions Reduction

The goal of improving aircraft fuel efficiency is one shared by aerospace engineers everywhere: with increased efficiency come the exciting possibilities of reduced fuel costs and increased performance in terms of speed, range, or payload. American engineers recognized the potential early on and were quick to create a center of gravity for their efforts to improve the fuel efficiency of aircraft engines. The NACA established the Aircraft Engine Research Laboratory—later known as NASA Lewis and then NASA Glenn—in 1941 in Cleveland, OH, as the Nation's nerve

---

4. Virginia P. Dawson, Engines and Innovation: Lewis Laboratory and American Propulsion Technology, NASA SP-4306 (1991), ch. 9.

center for propulsion research.[5] The lab first worked on fast fixes for piston engines in production for use in World War II, but it later moved on to pursue some of America's most forward-leaning advances in jet and rocket propulsion.[6] Improving fuel efficiency was naturally at the center of the laboratory's propulsion research, and many of NASA's most important fuel-saving engine concepts and technology originated there.[7] While NASA Glenn spearheaded the majority of aircraft fuel efficiency research, NASA Langley also played a critical role in the development of new fuel-saving aircraft structures.[8]

NASA's efforts to develop aircraft technology that both increased fuel efficiency and reduced emissions reached their nadir in the 1970s. From the time of Sputnik to the late 1960s, space dominated NASA's focus, particularly the drive to land on the Moon. But in the late 1960s, and particularly after introduction of the wide-body Boeing 747, the Agency turned increasing attention toward air transport, consistent with air transport itself dramatically increasing as a means of global mobility. Government and airline interest in improving jet fuel efficiency was high. However, NASA Lewis struggled to reenter the air-breathing propulsion game because the laboratory had lost much of its aeronautics expertise during the Sputnik crisis and now faced competition for Government support.[9] Aircraft engine companies had developed their own research facilities, and the U.S. Air Force (USAF) had completed its propulsion wind tunnel facility at Arnold Engineering Development Center in Tullahoma, TN, in 1961.[10] NASA scientists and engineers needed a new aeronautics niche. Luckily for them, they found it with the arrival of the oil embargo of 1973 and the coinciding emergence of a national awareness of environmental concerns. NASA's "clean and green" research agenda had been born.

The Organization of the Petroleum Exporting Countries (OPEC) oil embargo led Americans to realize that the Nation's economy and military

---

5. Neal T. Saunders and Arthur J. Glassman, "Lewis Aeropropulsion Technology: Remembering the Past and Challenging the Future," NASA Lewis Research Center, 1991.
6. Ibid.
7. Ibid.
8. Phone interview of Dennis Huff by Caitlin Harrington, Mar. 31, 2009. See also Valerie J. Lyons and Arlene S. Levine, "An Overview of NASA's Contributions to Energy Technology," NASA AIAA-2008-5641 (Jan. 2009).
9. Ibid.
10. Ibid. See also Arnold Engineering Development Center, *Propulsion and Wind Tunnel Facility*, http://www.nimr.org/systems/images/pwt.html, accessed July 29, 2009.

were far too dependent on foreign sources of energy. In 1973, 64 percent of U.S. oil imports came from OPEC countries.[11] The airline industry was particularly hard hit; jet fuel prices jumped from 12 cents to over $1 per gallon, and annual fuel expenditures increased to $1 billion—triple the earnings of airlines.[12] During the oil crisis, fuel accounted for half the airlines' operating costs,[13] and those operating costs were rising faster than the rate of inflation and faster than efficiencies in the airlines' own operations could reduce them.[14] The airline lobby descended on Capitol Hill, warning that its struggles to maintain profitability in the face of rising fuel costs were a bellwether for the Nation's entire economy. Lawmakers turned to NASA to for help.

In 1975, the U.S. Senate asked NASA to create the Aircraft Energy Efficiency (ACEE) program, with the twin goals of lowering the fuel burn of existing U.S. commercial aircraft and building new fuel-efficient aircraft to match foreign competition.[15] The 10-year, $670 million ACEE yielded two of NASA's greatest contributions to aircraft fuel-efficiency research. The most significant was the Energy Efficient Engine (E Cubed) program, which spawned technology still used in gas turbine engines today. The second key element of ACEE was the Advanced Turboprop (ATP), a bold plan to build an energy-efficient open-rotor engine. The open-rotor concept never made it into the mainstream, but aircraft propulsion research today still draws from ATP concepts, as this case study will later explain. Other technology developed under ACEE led to improved aerodynamic structures and laminar flow, as well as the design of supercritical wings, winglets, and composites.

Around the same time as ACEE, NASA began to sharpen its focus on the reduction of aircraft emissions. Space exploration had opened the Nation's eyes to the fragility of the planet and the potential impact that

---

11. Statistics available from Energy Information Administration, *U.S. Imports by Country of Origin*, http://tonto.eia.doe.gov/dnav/pet/pet_move_impcus_a2_nus_epc0_im0_mbblpd_a.htm, accessed July 29, 2009.
12. Mark D. Bowles and Virginia Dawson, "The Advanced Turboprop Project: Radical Innovation in a Conservative Environment," *From Engineering Science to Big Science*, Pamela Mack, ed., (Washington, DC: NASA, 1998), p. 7.
13. Ibid.
14. General Electric, "Keeping a Bold Promise," GE Aviation, Evendale, OH, Sept. 12, 1983.
15. "The Energy Conservative Engines Program at Lewis," *Lewis News*, Aug. 3, 1979, History Office archives, NASA Glenn Research Center.

Case 12 | Leaner and Greener: Fuel Efficiency Takes Flight

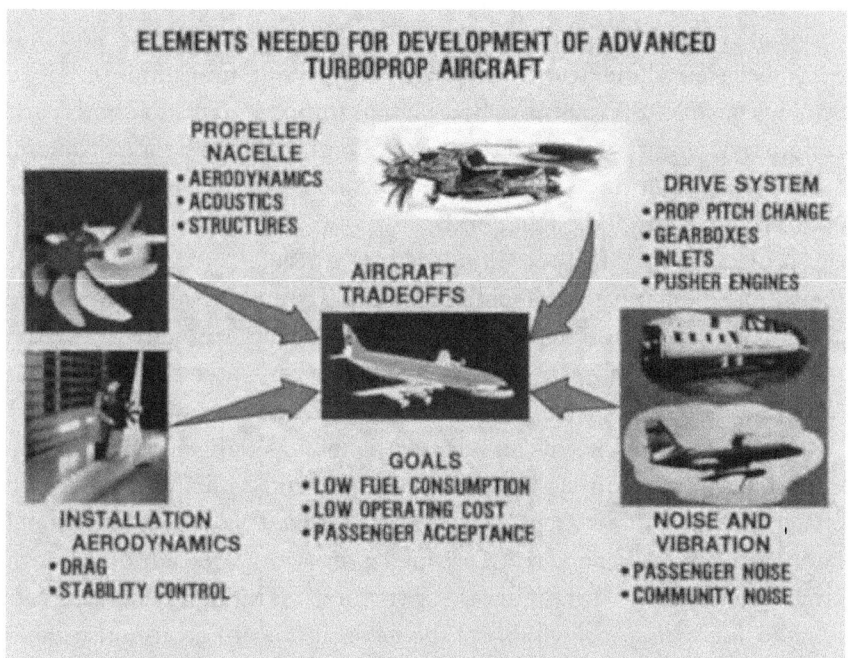

NASA slide circa 1985 showing Advanced Turboprop (ATP) requirements. NASA.

humans could have on the environment.[16] The U.S. Congress pushed NASA to become increasingly involved in projects to study the impact of stratospheric flight on the ozone layer following the cancellation of the Supersonic Transport (SST) in 1971. The Agency provided high-altitude research aircraft, balloons, and sounding rockets for the Climactic Impact Assessment Program (CIAP), which was launched by the Department of Transportation (DOT) to examine whether the environmental concerns that helped kill the SST were valid.[17]

DOT and NASA's CIAP research led to the discovery that aircraft emissions could, in fact, damage the ozone layer. CIAP results showed that nitrogen oxides would indeed cause ozone depletion if hundreds of Concorde and Tu-144 aircraft—the Concorde's Russian cousin—were to fly as planned. Following the release of CIAP, Congress then called on NASA to conduct further research into the impacts of stratospheric flight on the ozone layer, prompting NASA and DOT to move forward with a

---

16. Lyons and Levine, "An Overview of NASA's Contributions to Energy Technology."
17. Erick M. Conway, *High Speed Dreams* (Baltimore: Johns Hopkins University Press, 2005), p. 166.

series of studies that by the 1980s were pointing to the conclusion that SSTs were less dangerous to the ozone layer than first thought.[18] The findings gave NASA reason to believe that improvements in combustor technology might be enough to effectively mitigate the ozone problem.

After conducting its breakthrough ozone research, NASA has fairly consistently included clean combustor goals in many of its aeronautics projects in an effort to reduce aircraft emissions (examples include the Ultra Efficient Engine Technology program and Advanced Subsonic Technology program). Today, NASA has broadened its aeronautics research to focus not only on NOx (the collective term for water vapor, nitrogen oxide, and nitrogen dioxide), but also carbon dioxide and other pollutants.[19]

NASA's research in this area is seen as increasingly important as the view that aircraft emissions harm air quality and contribute to climate change becomes more widely accepted. The United Nations International Panel on Climate Change (IPCC) issued a report in 2007 stating that aircraft emissions account for about 2 percent of all human-generated carbon dioxide emissions, which are the most significant greenhouse gas.[20] The report also found that aviation accounts for about 3 percent of the potential warming effect of global emissions that could impact Earth's climate.[21] The report forecasts that by 2050, the aviation industry (including aircraft emissions) will produce about 3 percent of global carbon dioxide and 5 percent of the potential warming effect generated by human activity.[22]

In addition to NASA's growing interest in climate change, the Agency's research on improving the fuel efficiency of aircraft has also continued at a relatively steady pace over the years, although it has seemed to fluctuate to some extent in relation to oil prices. The oil shocks of the 1970s spurred a flurry of activity, from the E Cubed to the ATP and alternative fuels research. But interest in ambitious aircraft fuel-efficiency programs seemed to wane during the 1990s, when oil prices were low. Now

---

18. Chambers, *Innovation in Flight*.
19. Phone interview with Dan Bulzan, NASA's Associate Principle Investigator for the Subsonic Fixed Wing and Supersonic Aeronautics Project, by Harrington, Mar. 12, 2009.
20. International Panel on Climate Change, "Fourth Assessment Report," 2007.
21. Ibid.
22. IPCC's forecasts are based on projections regarding the future economy and the development of new energy-saving aircraft technology, so they are inherently uncertain. See U.S. Government Accountability Office, "Aviation and Climate Change: Aircraft Emissions Expected to Grow, but Technological and Operational Improvements and Government Policies Can Help Control Emissions," GAO-09-544 (Jan. 2009).

that oil prices are high again, however, fuel-efficiency programs seem to be back in vogue. (Several alternative fuels research efforts now underway at NASA will be discussed later in this case study.)

One example of the correlation between oil prices and the level of NASA's interest in fuel-efficiency programs is the ATP, NASA's ambitious plan to return to open-rotor engines. The concept never made it into mainstream use, partly because of widespread concerns that open-rotor engines are too noisy for commercial airline passengers,[23] but also partly because fuel prices began to fall and there was no longer a demand for expensive but highly energy-efficient engines. "We were developing the ATP in the late '70s and early '80s during the fuel crisis. And while fuel prices went up, they didn't continue to escalate like we originally thought they might, so the utility just went down; it just wasn't cost effective," said John Baughman, Manager of Military Advanced System design at General Electric (GE).[24] With oil prices once again on the rise today, however, there are several new initiatives underway that take off where E Cubed and the ATP left off.

Lean and Clean Propulsion Systems

NASA's efforts to improve engine design stand out as the Agency's greatest breakthroughs in "lean and green" engine development because of their continuing relevance today. Engineers are constantly seeking to increase efficiency to make their engines more attractive to commercial airlines: with increased efficiency comes reduced fuel costs and increased performance in terms of speed, range, or payload.[25] Emissions have also remained a concern for commercial aviation. The International Civil Aviation Organization (ICAO) has released increasingly strict standards for NOx emissions since 1981.[26] The Environmental Protection

---

23. Phone interview of David Ed Crow by Harrington, Pratt & Whitney Customer Training Center, Hartford, CT, Mar. 27, 2009. For noise concerns related to Advanced Turboprop, see also Bowles and Dawson, "The Advanced Turboprop Project," *From Engineering Science to Big Science*, Pamela Mack, ed., p. 27.
24. Phone interview of John Baughman, General Electric's Manager of Military Advanced Systems Design, by Harrington, Apr. 24, 2009.
25. For example, see U.S. Patent No. 4,550,561, "Method for Improving the Fuel Efficiency of a Gas Turbine Engine," Nov. 5, 1985, awarded to NASA Lewis researcher George A. Coffinberry and assigned to the U.S. Government.
26. International Civil Aviation Organization, Environment Section: *Civil Aircraft Emissions*, http://www.icao.int/icao/en/env/aee.htm, accessed July 29, 2009.

Agency has adopted emissions standards to match those of ICAO and also has issued emissions standards for aircraft and aircraft engines under the Clean Air Act.[27]

NASA's most important contribution to fuel-efficient aircraft technology to date has arguably been E Cubed, a program focused on improving propulsion systems mainly to increase fuel efficiency. The end goal was not to produce a production-ready fuel-efficient engine, but rather to develop technologies that could—and did—result in propulsion efficiency breakthroughs at major U.S. engine companies. These breakthroughs included advances in thermal and propulsive efficiency, as well as improvements in the design of component engine parts. Today, General Electric and Pratt & Whitney (P&W) continue to produce engines and evaluate propulsion system designs based on research conducted under the E Cubed program.

The U.S. Government's high expectations for E Cubed were reflected in the program's budget, which stood at about $250 million, in 1979 dollars.[28] The money was divided between P&W and GE, which each used the funding to sweep its most cutting-edge technology into a demonstrator engine that would showcase the latest technology for conserving fuel, reducing emissions, and mitigating noise. Lawmakers funded E Cubed with the expectation that it would lead to a dramatic 12-percent reduction in specific fuel consumption (SFC), a term to describe the mass of fuel needed to provide a certain amount of thrust for a given period.[29] Other E Cubed goals included a 5-percent reduction in direct operating costs, a 50-percent reduction in the rate of performance deterioration, and further reductions in noise and emissions levels compared to other turbofan engines at the time.[30]

The investment paid off in spades. What began as a proposal on Capitol Hill in 1975 to improve aircraft engine efficiency ended in 1983[31] with GE and P&W testing engine demonstrators that improved SFC between 14 and 15 percent, exceeding the 12-percent goal. The dem-

---

27. U.S. Government Accountability Office, "Aviation's Effects on the Global Atmosphere Are Potentially Significant and Expected to Grow," GAO/RCED-00-57 (Feb. 2000) available at http://www.gao.gov/archive/2000/rc00057.pdf, accessed July 29, 2009.
28. "The Energy Conservative Engines Program at Lewis," Lewis News, Aug. 3, 1979.
29. Ibid.
30. Ibid.
31. General Electric, "Keeping a Bold Promise," GE Aviation, Evendale, OH, Sept. 12, 1983.

onstrators were also able to achieve a reduction in emissions. A NASA report from 1984 hailed E Cubed for helping to "keep American engine technology at the forefront of the world market."[32] Engineers involved in E Cubed at both GE and P&W said the technology advances were game changing for the aircraft propulsion industry.

"The E Cubed program is probably the single biggest impact that NASA has ever had on aircraft propulsion," GE's John Baughman said. "The improvements in fuel efficiency and noise and emissions that have evolved from the E Cubed program are going to be with us for years to come."[33] Ed Crow, former Senior Vice President of Engineering at P&W, agreed that E Cubed marked the pinnacle of NASA's involvement in improving aircraft fuel efficiency. "This was a huge program," he said. "It was NASA and the Government's attempt to make a huge step forward."[34]

E Cubed spurred propulsion research that led to improved fuel efficiency in three fundamental ways:

First, E Cubed allowed both GE and P&W to improve the thermal efficiency of their engine designs. Company engineers were able to significantly increase the engine-pressure ratio, which means the pressure inside the combustor becomes much higher than atmospheric pressure. They were able to achieve the higher pressure ratio by improving the efficiency of the engine's compressor, which condenses air and forces it into the combustor.

In fact, one of the most significant outcomes of the E Cubed program was GE's development of a new "E Cubed compressor" that dramatically increased the pressure ratio while significantly reducing the number of compression stages. If there are too many stages, the engine can become big, heavy, and long; what is gained in fuel efficiency may be lost in the weight and cost of the engine. GE's answer to that problem was to develop a compressor that had only 10 stages and produced a pressure ratio of about 23 to 1, compared to the company's previous compressors, which had 14 stages and produced a pressure ratio of 14 to 1.[35] That compressor is still in use today in GE's latest engines, including the GE-90.[36]

---

32. Jeffrey Ethell, *"Fuel Economy in Aviation,"* NASA SP-462 (1983), p. 42.
33. Harrington, Baughman interview.
34. Harrington, Crow interview.
35. Harrington, Baughman interview.
36. Ibid.

P&W's E Cubed demonstrator had a bigger, 14-stage compressor, but the company was able to increase the pressure ratio by modifying the compressor blades to allow for increased loading per stage. P&W's engines prior to E Cubed had pressure ratios around 20 to 1; P&W's E Cubed demonstrator took pressure ratios to about 33 to 1, according to Crow.[37]

The second major improvement enabled by E Cubed research was a substantial increase in propulsive efficiency. Air moves most efficiently through an engine when its velocity doesn't change much. The way to ensure that the velocity remains relatively constant is to maximize the engine's bypass ratio: in other words, a relatively large mass of air must bypass the engine core—where air is mixed with fuel—and go straight out the back of the engine at a relatively low exhaust speed. Both GE and P&W employed more efficient turbines and improved aerodynamics on the fan blades to increase the bypass ratio to about 7 to 1 (compared with about 4 to 1 on P&W's older engines).[38]

Finally, E Cubed enabled major improvements in engine component parts. This was critical, because other efficiencies can't be maximized unless the engine parts are lightweight, durable, and aerodynamic. Increasing the pressure ratio, for example, leads to very high temperatures that can stress the engine. Both P&W and GE developed materials and cooling systems to ensure that engine components did not become too hot.

In addition to efforts to improve fuel efficiency, E Cubed gave both GE and P&W opportunities to build combustors that would reduce emissions. E Cubed emissions goals were based on the Environmental Protection Agency's 1981 guidelines and called for reductions in carbon monoxide, hydrocarbons, NOx, and smoke. Both companies developed their emissions-curbing combustor technology under NASA's Experimental Clean Combustor program, which ran from 1972 to 1976. Their main efforts were focused on controlling where and in what proportions air and fuel were mixed inside the combustor. Managing the fuel/air mix inside the combustor is critical to maximize combustion efficiency (and reduce carbon dioxide emissions as a natural byproduct) and to ensure that temperatures do not get so high that NOx is generated. GE tackled the mixing issue by developing a dual annular

---

37. Harrington, Crow interview.
38. Ethell, "Fuel Economy in Aviation," p. 30.

combustor, while P&W went with a two-stage combustor that had two in-line combustor zones to control emissions.[39]

Ultimately, E Cubed provided the financial backing required for both GE & P&W to pursue propulsion technology that has fed into their biggest engine lines. GE's E Cubed compressor technology is used to power three types of GE engines, including the GE90-115B, which powers the Boeing 777-300ER and holds the world record for thrust.[40] Other GE engines incorporating the E Cubed compressor include the GP-7200, which recently went into service on the Airbus A380, and the GE-NX, which is about to enter service on the Boeing 787.[41] P&W also got some mileage out of the technologies developed under E Cubed. The company's E Cubed demonstrator engine served as the inspiration for the PW2037, which fed into other engine designs that today power the Boeing 757 commercial airliner (the engine is designated PW2000) and the U.S. military's C-17 cargo aircraft (the engine is designated F117).[42]

High-Speed Research

When NASA decided to start a High-Speed Research (HSR) program in 1990, it quickly decided to draw in the E Cubed combustor research to address previous concerns about emissions. The goal of HSR was to develop a second generation of High-Speed Civil Transport (HSCT) aircraft with better performance than the Supersonic Transport project of the 1970s in several areas, including emissions. The project sought to lay the research foundation for industry to pursue a supersonic civil transport aircraft that could fly 300 passengers at more than 1,500 miles per hour, or Mach 2, crossing the Atlantic or Pacific Ocean in half the time of subsonic jets. The program had an aggressive NOx goal because there were still concerns, held over from the days of the SST in the 1970s, that a super-fast, high-flying jet could damage the ozone layer.[43]

NASA's Atmospheric Effects of Stratospheric Aircraft project was used to guide the development of environmental standards for the new HSCT exhaust emissions. The study yielded optimistic findings:

---

39. Ibid., p. 31–42.
40. Harrington, Baughman interview.
41. Ibid.
42. Harrington, Crow interview.
43. NASA, *NASA's High Speed Research Program*, http://oea.larc.nasa.gov/PAIS/HSR-Overview2.html, accessed July 29, 2009.

there would be negligible environmental impact from a fleet of 500 HSCT aircraft using advanced technology engine components.[44] The HSR set a NOx emission index goal of 5 grams per kilogram of fuel burned, or 90 percent better than conventional technology at the time.[45]

NASA sought to meet the NOx goal primarily through major advancements in combustion technologies. The HSR effort was canceled in 1999 because of budget constraints, but HSR laid the groundwork for future development of clean combustion technologies under the AST and UEET programs discussed below.

Advanced Subsonic Technology Program and UEET

NASA started a project in the mid-1990s known as the Advanced Subsonic Technology program. Like HSR before it, the AST focused heavily on reducing emissions through new combustor technology. The overall objective of the AST was to spur technology innovation to ensure U.S. leadership in developing civil transport aircraft. That meant lowering NOx emissions, which not only raised concern in local airport communities but also by this time had become a global concern because of potential damage to the ozone layer. The AST sought to spur the development of new low-emissions combustors that could achieve at least a 50-percent reduction in NOx from 1996 International Civil Aviation Organization standards. The AST program also sought to develop techniques that would better measure how NOx impacts the environment.[46]

GE, P&W, Allison Engines, and AlliedSignal engines all participated in the project.[47] Once again, the challenge for these companies was to control combustion in such a way that it would minimize emissions. This required carefully managing the way fuel and air mix inside the combustor to avoid extremely hot temperatures at which NOx would be created, or at least reducing the length of time that the gases are at their hottest point.

---

44. Ibid.
45. Committee on High Speed Research, Aeronautics and Space Engineering Board, Commission on Engineering and Technical Systems, National Research Council, *U.S. Supersonic Aircraft: Assessing NASA's High Speed Research Program* (Washington, DC: National Academies Press, 1997), p. 53.
46. "Advanced Subsonic Technology Program Plan," Office of Aero-Space Technology, NASA (May 1999).
47. John Rohde, "Overview of the NASA AST and UEET Emissions Reductions Projects," Mar. 1, 2002, http://www.techtransfer.berkeley.edu/aviation02downloads/JohnRohde.pdf, accessed July 29, 2009.

Ultimately the AST emissions reduction project achieved its goal of reducing NOx emissions by more than 50 percent over the ICAO standard, a feat that was accomplished not with actual engine demonstrators but with a "piloted airblast fuel preparation chamber."[48]

Despite their relative success, however, NASA's efforts to improve engine efficiency and reduce emissions began to face budget cuts in 2000. Funding for NASA's Atmospheric Effects of Aviation project, which was the only Government program to assess the effects of aircraft emissions at cruise altitudes on climate change, was canceled in 2000.[49] Investments in the AST and the HSR also came to an end. However, NASA did manage to salvage parts of the AST aimed at reducing emissions by rolling those projects into the new Ultra Efficient Engine Technology program in 2000.[50]

UEET was a 6-year, nearly $300 million program managed by NASA Glenn that began in October 1999 and included participation from NASA Centers Ames, Goddard, and Langley; engine companies GE Aircraft Engines, Pratt & Whitney, Honeywell, Allison/Rolls Royce, and Williams International; and airplane manufacturers Boeing and Lockheed Martin.[51]

UEET sought to develop new engine technologies that would dramatically increase turbine performance and efficiency. It sought to reduce NOx emissions by 70 percent within 10 years and 80 percent within 25 years, using the 1996 International Civil Aviation Organization guidelines as a baseline.[52] The UEET project also sought to reduce carbon dioxide emissions by 20 percent and 50 percent in the same timeframes, using 1997 subsonic aircraft technology as a baseline.[53] The dual goals posed a major challenge because current aircraft engine technologies typically require a tradeoff between NOx and carbon emissions; when engines are designed to minimize carbon dioxide emissions, they tend to generate more NOx.

In the case of the UEET project, improving fuel efficiency was expected to lead to a reduction in carbon dioxide emissions by at least

---

48. Ibid.
49. U.S. Government Accountability Office, "Aviation's Effects on the Global Atmosphere Are Potentially Significant and Expected to Grow."
50. "Advanced Subsonic Technology Program Plan."
51. NASA, "Ultra Efficient Engine Technology Program," June 25, 2001, http://www.grc.nasa.gov/WWW/RT/RT2000/2000/2100shaw.html, accessed July 29, 2009.
52. Joe Shaw and Catherine Pettie, "NASA Ultra Efficient Engine Technology Project Overview," NASA Glenn Research Center NASA CP-2004-212963, vol. 1.
53. Ibid.

8 percent: the less fuel burned, the less carbon dioxide released.[54] The UEET program was expected to maximize fuel efficiency, requiring engine operations at pressure ratios as high as 55 to 1 and turbine inlet temperatures of 3,100 degrees Fahrenheit (°F).[55] However, highly efficient engines tend to run at very hot temperatures, which lead to the generation of more NOx. Therefore, in order to reduce NOx, the UEET program also sought to develop new fuel/air mixing processes and separate engine component technologies that would reduce NOx emissions 70 percent from 1996 ICAO standards for takeoff and landing conditions and also minimize NOx impact during cruise to avoid harming Earth's ozone layer.

Under UEET, NASA worked on ceramic matrix composite (CMC) combustor liners and other engine parts that can withstand the high temperatures required to maximize energy efficiency and reduce carbon emissions while also lowering NOx emissions. These engine parts, particularly combustor liners, would need to endure the high temperatures at which engines operate most efficiently without the benefit of cooling air. Cooling air, which is normally used to cool the hottest parts of an engine, is unacceptable in an engine designed to minimize NOx, because it would create stoichiometric fuel-air mixtures—meaning the number of fuel and air molecules would be optimized so the gases would be at their hottest point—thereby producing high levels of NOx in regions close to the combustor liner.[56]

NASA's sponsorship of the AST and the UEET also fed into the development of two game-changing combustor concepts that can lead to a significant reduction in NOx emissions. These are the Lean Pre-mixed, Pre-vaporized (LPP) and Rich, Quick Mix, Lean (RQL) combustor concepts. P&W and GE have since adopted these concepts to develop combustors for their own engine product lines. Both concepts focus on improving the way fuel and air mix inside the engine to ensure that core temperatures do not get so high that they produce NOx emissions.

GE has drawn from the LPP combustor concept to develop its Twin Annular Pre-mixing Swirler (TAPS) combustor. Under the LPP concept,

---

54. NASA, Industry Roundtable: Aerospace Industry Enterprise, NASA Headquarters, *http://www.aeronautics.nasa.gov/events/showcase/environ.htm?goto=index.htm*, accessed July 29, 2009.
55. Ibid.
56. Committee on High Speed Research, Aeronautics and Space Engineering Board, Commission on Engineering and Technical Systems, National Research Council, *U.S. Supersonic Aircraft: Assessing NASA's High Speed Research Program* (1997), p. 55.

air from the high-pressure compressor comes into the combustor through two swirlers adjacent to the fuel nozzles. The swirlers premix the fuel and combustion air upstream from the combustion zone, creating a lean (more air than fuel) homogenous mixture that can combust inside the engine without reaching the hottest temperatures, at which NOx is created.[57]

NASA support also helped lay the groundwork for P&W's Technology for Advanced Low Nitrogen Oxide (TALON) low-emissions combustor, which reduces NOx emissions through the RQL process. The front end of the combustor burns very rich (more fuel than air), a process that suppresses the formation of NOx. The combustor then transitions in milliseconds to burning lean. The air must mix very rapidly with the combustion products from the rich first stage to prevent NOx formation as the rich gases are diluted.[58] The goal is to spend almost no time at extremely hot temperatures, at which air and fuel particles are evenly matched, because this produces NOx.[59]

Today, NASA continues to study the difficult problem of increasing fuel efficiency and reducing NOx, carbon dioxide, and other emissions. At NASA Glenn, researchers are using an Advanced Subsonic Combustion Rig (ASCR), which simulates gas turbine combustion, to engage in ongoing emissions testing. P&W, GE, Rolls Royce, and United Technologies Corporation are continuing contracts with NASA to work on low-emissions combustor concepts.

"The [ICAO] regulations for NOx keep getting more stringent," said Dan Bulzan, NASA's associate principle investigator for the subsonic fixed wing and supersonic aeronautics project. "You can't just sit there with your old combustor and expect to meet the NOx emissions regulations. The Europeans are quite aggressive and active in this area as well. There is a competition on who can produce the lowest emissions combustor."[60]

Advanced Turboprop Project

Another significant program to emerge from NASA's ACEE program was the Advanced Turboprop project, which lasted from 1976 to 1987.

---

57. U.S. Supersonic Aircraft, p. 54.
58. Ibid.
59. Harrington, Crow interview.
60. Harrington, Bulzan interview.

Like E Cubed, the ATP was largely focused on improving fuel efficiency. The project sought to move away from the turbofan and improve on the open-rotor (propeller) technology of 1950s. Open rotors have high bypass ratios and therefore hold great potential to dramatically increase fuel efficiency. NASA believed an advanced turboprop could lead to a reduction in fuel consumption of 20 to 30 percent over existing turbofan engines with comparable performance and cabin comfort (acceptable noise and vibration) at a Mach 0.8 and an altitude of 30,000 feet.[61]

There were two major obstacles to returning to an open-rotor system, however. The most fundamental problem was that propellers typically lose efficiency as they turn more quickly at higher flight speeds. The challenge of the ATP was to find a way to ensure that propellers could operate efficiently at the same flight speeds as turbojet engines. This would require a design that allowed the fan to operate at slow speeds to maximize efficiency while the turbine operates fast to achieve adequate thrust. Another major obstacle facing NASA's ATP was the fact that turboprop engines tend to be very noisy, making them less than ideal for commercial airline use. NASA's ATP sought to overcome the noise problem and increase fuel efficiency by adopting the concept of swept propeller blades.

The ATP generated considerable interest from the aeronautics research community, growing from a NASA contract with the Nation's last major propeller manufacturer, Hamilton Standard, to a project that involved 40 industrial contracts, 15 university grants, and work at 4 NASA research Centers—Lewis, Langley, Dryden, and Ames. NASA engineers, along with a large industry team, won the Collier Trophy for developing a new fuel-efficient turboprop in 1987.[62]

NASA initially contracted with Allison, P&W, and Hamilton Standard to develop a propeller for the ATP that rotated in one direction. This was called a "single rotation tractor system" and included a gearbox, which enabled the propeller and turbines to operate at different speeds. The NASA/industry team first conducted preliminary ground-testing. It combined the Hamilton Standard SR-7A prop fan with the Allison turbo shaft engine and a gearbox and performed 50 hours of success-

---

61. Bowles and Dawson, "The Advanced Turboprop Project," p. 10.
62. *NACA/NASA Research and Development Projects Receiving the Robert J Collier Trophy*, http://history.nasa.gov/Timeline/collier.html, accessed July 29, 2009.

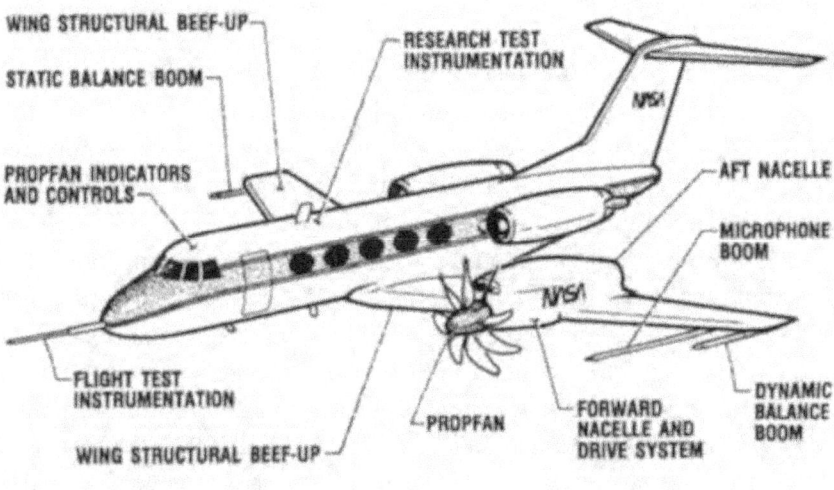

Schematic drawing of the NASA propfan testbed, showing modifications and features proposed for the basic Grumman Gulfstream airframe. NASA.

ful stationary tests in May and June 1986.[63] Next, the engine parts were shipped to Savannah, GA, and reassembled on a modified Gulfstream II with a single-blade turboprop on its left wing. Flight-testing took place in 1987, validating NASA's predictions of a 20 to 30 percent fuel savings.[64]

Meanwhile, P&W's main rival, GE, was quietly developing its own approach to the ATP known as the "unducted fan." GE released the design to NASA in 1983, and NASA Headquarters instructed NASA Lewis to cooperate with GE on development and testing.[65] Citing concerns about weight and durability, P&W decided not to use a gearbox to allow the propellers and the turbines to turn at different speeds.[66] Instead, the company developed a counter-rotating pusher system. They mounted two counter-rotating propellers on the rear of the plane, which pushed it into flight. They also put counter-rotating blade rows in the turbine. The counter-rotating turbine blades were turning relatively

---

63. Bowles and Dawson, "The Advanced Turboprop Project."
64. Edwin J. Graber, "Overview of NASA PTA Propfan Flight Test Program," NASA Glenn Research Center, Nov. 1, 1987, CASI Doc. 19880006423; and B.H. Little, D.T. Poland, H.W. Bartel, C.C. Withers, and P.C. Brown, "Propfan Test Assessment (PTA) Final Project Report," NASA CR-185138 (1989).
65. Bowles and Dawson, "The Advanced Turboprop Project."
66. Harrington, Baughman interview.

slowly to accommodate the fan, but because they were turning in opposite directions, their relative speed was high and therefore highly efficient.[67]

GE performed ground tests of the unducted fan in 1985 that showed a 20 percent fuel-conservation rate.[68] Then, in 1986, a year before the NASA/industry team flight test, GE mounted the unducted fan—the propellers and the fan mounted behind an F404 engine—on a Boeing 727 airplane and conducted a successful flight test.[69]

Mark Bowles and Virginia Dawson have noted in their analysis of the ATP that the competition between the two ATP concepts and industry's willingness to invest in the open-rotor technology fostered public acceptance of the turboprop concept.[70] But despite the growing momentum and the technical success of the ATP project, the open rotor was never adopted for widespread use on commercial aircraft. P&W's Crow said that the main reason was that it was just too noisy.[71] "This was clearly more fuel-efficient technology, but it was not customer friendly at all," said Crow. Another problem was that the rising fuel prices that had spurred NASA to work on energy-efficient technology were now going back down. There was no longer a favorable ratio of cost to develop turboprop technology versus savings in fuel burn.[72] "In one sense of the word it was a failure," said Crow. "Neither GE nor Pratt nor Boeing nor anyone else wanted us to commercialize those things."

Nevertheless, the ATP yielded important technological breakthroughs that fed into later engine technology developments at both GE and P&W. Crow said the ATP set the stage for the development of P&W's latest engine, the geared turbofan.[73] That engine is not an open-rotor system, but it does use a gearbox to allow the fan to turn more slowly than the turbines. The fan moves a large amount of air past the engine core without changing the velocity of the air very much. This enables a high bypass ratio, thereby increasing fuel efficiency; the bypass ratio is 8 to 1 in the 14,000–17,000-pound thrust class and 12 to 1 in the 17,000–23,000-pound thrust class.[74]

---

67. Ibid.
68. James J. Haggerty, "Propfan Update," *Aerospace*, fall/winter 1986, pp. 10–11.
69. Graber, "Overview of NASA PTA"; Harrington, Baughman interview.
70. Bowles and Dawson, "The Advanced Turboprop Project."
71. Harrington, Crow interview.
72. Bowles and Dawson, "The Advanced Turboprop Project."
73. Harrington, Crow interview.
74. United Technologies, Inc., *Leading Change: The Geared Turbofan Engine*, Pratt & Whitney Customer Training Center, Hartford, CT, 2009.

GE renewed its ATP research to compete with P&W's geared turbofan, announcing in 2008 that it would consider both open rotor and encased engine concepts for its new engine core development program, known as E Core. The company announced an agreement with NASA in the fall of 2008 to conduct a joint study on the feasibility of an open-rotor engine design. In 2009, GE plans to revisit its original open-rotor fan designs to serve as a baseline. GE and NASA will then conduct wind tunnel tests using the same rig that was used for the ATP.[75] Snecma, GE's 50/50 partner in CFM International—an engine manufacturing partnership—will participate in fan blade design testing. GE says the new E Core design—whether it adopts an open rotor or not—aims to increase fuel efficiency 16 percent above the baseline (a conventional turbofan configuration) in narrow-body and regional aircraft.[76]

Another major breakthrough resulting from the ATP was the development of computational fluid dynamics (CFD), which allowed engineers to predict the efficiency of new propulsion systems more accurately. "What computational fluid dynamics allowed us to do was to design a new air foil based on what the flow field needed rather than proscribing a fixed air foil before you even get started with a design process," said Dennis Huff, NASA's Deputy Chief of the Aeropropulsion Division. "It was the difference between two- and three-dimensional analysis; you could take into account how the fan interacted with nacelle and certain aerodynamic losses that would occur. You could model numerically, whereas the correlations before were more empirically based."[77] Initially, companies were reluctant to embrace NASA's new approach because they distrusted computational codes and wanted to rely on existing design methods, according to Huff. However, NASA continued to verify and validate the design methods until the companies began to accept them as standard practice. "I would say by the time we came out of the Advanced Turboprop project, we had a lot of these aerodynamic CFD tools in place that were proven on the turboprop, and we saw the companies developing codes for the turbo engine," Huff said.[78]

---

75. Harrington, Baughman interview.
76. General Electric, "GE Launches Engine Core for New Commercial Airplanes," July 13, 2008, available at *http://www.businesswire.com/portal/site/home/permalink/?ndmViewId=news_view&newsId=20080713005020*, accessed July 29, 2009.
77. Harrington, Huff interview.
78. Harrington, Huff interview.

Aircraft Materials and Structures

While refinements in engine design have been the cornerstone of NASA's efforts to improve fuel efficiency, the Agency has also sought to improve airframe structures and materials. The ACEE included not only propulsion improvement programs but also efforts to develop lightweight composite airframe materials and new aerodynamic structures that would increase fuel efficiency. Composite materials, which consist of a strong fiber such as glass and a resin that binds the fibers together, hold the potential to dramatically reduce the weight—and therefore the fuel efficiency—of aircraft.

Initially, Boeing began to investigate composite materials, using fiberglass for major parts such as the radome on the 707 and 747 commercial airliners.[79] Starting around 1962, composite sandwich parts comprised of fiberglass-epoxy materials were applied to aircraft such as the Boeing 727 in a highly labor-intensive process.[80] The next advance in composites was the use of graphite composite secondary aircraft structures, such as wing control surfaces, wing trailing and leading edges, vertical fin and stabilizer control surfaces, and landing gear doors.[81]

NASA research on composite materials began to gain momentum in 1972, when NASA and the Air Force undertook a study known as Long Range Planning Study for Composites (RECAST) to examine the state of existing composites research. The RECAST study found two major obstacles to the use of composites: high costs and lack of confidence in the materials.[82]

However, by 1976, interest in composite materials had picked up steam because they are lighter than aluminum and therefore have the potential to increase aircraft fuel efficiency. Research on composites was formally wrapped into ACEE in the form of the Composite Primary Aircraft Structures program. NASA hoped that research on composites would yield a fuel savings for large aircraft of 15 percent by the 1990s.

NASA's efforts under ACEE ultimately led the aircraft manufacturing industry to normalize the use of composites in its manufacturing

---

79. Chambers, "Concept to Reality: Contributions of the NASA Langley Research Center to Civil Aircraft of the 1990s," NASA SP 2003-4529 (2003), http://oea.larc.nasa.gov/PAIS/Concept2Reality/composites.html, accessed July 29, 2009.
80. Ibid.
81. Ibid.
82. Ethell, "Fuel Economy in Aviation," p. 59.

processes, driving down costs and making composites far more common in aircraft structures. "Ever since the ACEE program has existed, manufacturers have been encouraged by the leap forward they have been able to make in composites," Jeffrey Ethell, the late aviation author and analyst, wrote in his 1983 account NASA's fuel-efficiency programs. "They have moved from what were expensive, exotic materials to routine manufacture by workers inexperienced in composite structures."[83] Today, composite materials have widely replaced metallic materials on parts of an aircraft's tail, wings, fuselage, engine cowlings, and landing gear doors.[84]

NASA research under ACEE also led to the development of improved aerodynamic structures and active controls. This aspect of ACEE was known as the Energy Efficient Transport (EET) program. Aerodynamic structures can improve the way that the aircraft's geometry affects the airflow over its entire surface. Active controls are flight control systems that can use computers and sensors to move aircraft surfaces to limit unwanted motion or aerodynamic loads on the aircraft structure and to increase stability. Active controls lighten the weight of the aircraft, because they replace heavy hydraulic lines, rods, and hinges. They also allow for reductions in the size and weight of the wing and tail. Both aerodynamic structures and active controls can increase fuel efficiency because they reduce weight and drag.[85]

One highly significant aerodynamic structure that was explored under ACEE was the supercritical wing. During the 1960s and 1970s, Richard Whitcomb, an aeronautical engineer at NASA Langley Research Center, led the development of the new airfoil shape, which has a flattened top surface to reduce drag and tends to be rounder on the bottom, with a downward curve at the trailing edge to increase lift. ACEE research at NASA Dryden led to the finding that the supercritical wing could lead to increased cruising speed and flight range, as well as an

---

83. Ibid., p. 71.
84. "NASA Technology," Dryden Flight Research Center, Oct. 21, 2008, available at *http://www.nasa.gov/centers/dryden/news/X-Press/stories/2008/10_08_technology.html*, accessed July 29, 2009.
85. Ethell, "Fuel Economy in Aviation," p. 77; see also N.A. Radovich, D. Dreim, D.A. O'Keefe, L. Linner, S.K. Pathak, J.S. Reaser, D. Richardson, J. Sweers, and F. Conner, "Study for the Optimization of a Transport Aircraft Wing for Maximum Fuel Efficiency," vol. 1: "Methodology, Criteria, Aeroelastic Model Definition, and Results," NASA CR-172551 (1985).

increase in fuel efficiency of about 15 percent over conventional-wing aircraft. Supercritical wings are now in widespread use on modern subsonic commercial transport aircraft.[86]

Whitcomb also conducted research on winglets, which are vertical extensions of wingtips that can improve an aircraft's fuel efficiency and range. He predicted that adding winglets to transport-size aircraft would lead to improved cruising efficiencies between 6 and 9 percent. In 1979 and 1980, flight tests involving a U.S. Air Force KC-135 aerial refueling tanker demonstrated an increased mileage rate of 6.5 percent.[87] The first big commercial aircraft to feature winglets was the MD-11, built by McDonnell-Douglas, which is now a part of Boeing. Today, winglets can be are commonly found on many U.S.- and foreign-made commercial airliners.[88]

Laminar flow is another important fuel-saving aircraft concept spearheaded by NASA. Aircraft designed to maximize laminar flow offer the potential for as much as a 30-percent decrease in fuel usage, a benefit that can be traded for increases in range and endurance. The idea behind laminar flow is to minimize turbulence in the boundary layer—a layer of air that skims over the aircraft's surface. The amount of turbulence in the boundary layer increases along with the speed of the aircraft's surface and the distance air travels along that surface. The more turbulence, the more frictional drag the aircraft will experience. In a subsonic transport aircraft, about half the fuel required to maintain level flight in cruise results from the necessity to overcome frictional drag in the boundary layer.[89]

There are two types of methods used to achieve laminar flow: active and passive. Active Laminar Flow Control (LFC) seeks to reduce turbulence in the boundary layer by removing a small amount of fluid (air) from the boundary layer. Active LFC test sections on an aircraft wing contain tiny holes or slots that siphon off the most turbulent air by using an internal suction system. Passive laminar flow does not involve a suc-

---

86. Dryden Flight Research Center, "NASA Technology." Whitcomb's work is thoroughly discussed by Jeremy Kinney in a companion case study.
87. Dryden Flight Research Center, "NASA Dryden Technology Facts: Winglets," *http://www.nasa.gov/centers/dryden/about/Organizations/Technology/Facts/TF-2004-15-DFRC.html*, accessed July 29, 2009.
88. Ibid.
89. Albert L. Braslow, *A History of Suction-Type Laminar Flow Control with Emphasis on Flight Research*, Monographs in Aerospace History, No. 13 (Washington, DC: NASA History Division, Office of Policy and Plans, NASA Headquarters, 1999), p. 1.

Case 12 | Leaner and Greener: Fuel Efficiency Takes Flight

An F-16XL flow visualization test. This F-16 Scamp model was tested in the NASA Langley Research Center Basic Aerodynamics Research Tunnel. This was a basic flow visualization test using a laser light sheet to illuminate the smoke. NASA.

tion system to remove turbulent air; instead, it relies on careful contouring of the wing's surface to reduce turbulence.[90]

In 1990, NASA and Boeing sponsored flight tests of a Boeing 757 that used a hybrid of both active and passive LFC. The holes or slots used in active LFC can get clogged with bugs. As a result, NASA and Boeing used

---

90. Dryden Flight Research Center, "Fact Sheet: F-16XL Laminar Flow Research Aircraft," updated Mar. 1, 2008, http://www.nasa.gov/centers/dryden/news/FactSheets/FS-023-DFRC.html, accessed July 29, 2009.

a hybrid LFC system on the 757 that limited the air extraction system to the leading edge of the wing, followed by a run of the natural laminar flow.[91] Based on the flight tests, engineers calculated that the application of hybrid LFC on a 300-passenger, long-range subsonic transport could provide a 15-percent reduction in fuel burned, compared with a conventional equivalent.[92]

NASA laminar flow research continued to evolve, with NASA Dryden conducting flight tests on two F-16 test aircraft known as the F-16XL-1 and F-16XL-2 in the early and mid-1990s. The purpose was to test the application of active and passive laminar flow at supersonic speeds. Technical data from the tests are available to inform the development of future high-speed aircraft, including commercial transports.[93]

Today, laminar flow research continues, although active LFC, required for large transport aircraft, has not yet made its way into widespread use on commercial aircraft. However, NASA is continuing work in this area. NASA's subsonic fixed wing project, the largest of its four aeronautics programs, is working on projects to reduce noise, emissions, and fuel burn on commercial-transport-size aircraft by employing several technology concepts, including laminar flow control. The Agency is hoping to develop technology to reduce fuel burn for both a next generation of narrow-body aircraft (N+1) and a next generation of hybrid wing/body aircraft (N+2).[94] NASA is expected to conduct wind tunnel tests of two hybrid wing body (also known as blended wing body) aircraft known as N2A and N2B in 2011. Those aircraft, which will incorporate hybrid LFC, are expected to reduce fuel burn by as much as 40 percent.[95]

Together with this research on emissions and fuel burn has come a heightened awareness on reducing aircraft noise. One example of a very

---

91. Mary L. Sandy and H. Keith Henry, "Flight Tests Prove Concept for Jet Liner Fuel Economy," NASA Headquarters Press Release, Aug. 23, 1990, http://www.nasa.gov/home/hqnews/1990/90-115.txt, accessed July 29, 2009.
92. Braslow, *A History of Suction-Type Laminar Flow Control with Emphasis on Flight Research*, p. 32.
93. Dryden Flight Research Center, "Fact Sheet: F-16XL Laminar Flow Research Aircraft."
94. Graham Warwick, "NASA Steps Up Research into Fuel Efficient Aircraft," *Flight International*, June 14, 2007, available at http://www.flightglobal.com/articles/2007/06/14/214635/nasa-steps-up-research-into-efficient-aircraft.html, accessed July 29, 2009.
95. Warwick, "NASA Pushes Blended Wing/Body," *Aviation Week*, Jan. 13, 2009, available at http://www.aviationweek.com/aw/generic/story.jsp?id=news/Body011309.xml&headline=NASA%20Pushes%20Blended%20Wing/Body%20&channel=space, accessed July 29, 2009.

beneficial technical "fix" to the noise problem is the chevron exhaust nozzle, so called because it has a serrated edge resembling a circular saw blade, or a series of interlinked chevrons. The exhaust nozzle chevron has become a feature of recent aircraft design, though how to best configure chevron shapes to achieve maximum noise-reduction benefit without losing important propulsive efficiencies is not yet a refined science. The takeoff noise reduction benefits, when "traded off" against potential losses in cruise efficiency, clearly required continued study, in much the same fashion that, in the piston-engine era, earlier NACA engineers grappled with assessing the benefits of the controllable-pitch propeller and the best way to configure early radial engine cowlings. As that resulted in the emergence of the NACA cowling as a staple and indeed, design standard, for future aircraft design, so too, presumably, will NASA's work lead to better understanding of the benefits and design tradeoffs that must be made for chevron design.[96]

Hydrogen Research Leads to Rockets and Fuel Cells

While most of NASA's aircraft fuel-efficiency research grew out of the reality jolt of the 1970s oil crisis and environmental concerns, there is at least one notable exception. Researchers first began to investigate liquid hydrogen as an alternative to hydrocarbon-based fuel in the mid-1950s because they suspected major performance efficiencies could be gained.[97] The Lewis Flight Propulsion Laboratory issued a seminal report in April 1955—although it was not declassified until September 1962—suggesting that liquid hydrogen might have a positive impact on the performance of high-altitude military aircraft (subsonic and supersonic bombers, fighters, and reconnaissance aircraft flying at 75,000 to 85,000 feet).[98] While the report raised the aviation community's awareness of the potential for hydrogen as a fuel source, it did not lead to widespread use in aircraft because of technical problems with using

---

96. Travis L. Turner, Randolph H. Cabell, Roberto J. Cano, and Richard J. Silcox, "Testing of SMA-Enabled Active Chevron Prototypes Under Representative Flow Conditions," NASA Langley Research Center, Report LAR-17332, Paper 6928-36, CASI Doc. 20080014174 (2008); Vance Dippold, III, "CFD Analyses and Jet-Noise Predictions of Chevron Nozzles with Vortex Stabilization," NASA TM-2008-215150 (2008).
97. Abe Silverstein and Eldon Hall, "Research Memorandum: Liquid Hydrogen as Jet Fuel for High Altitude Aircraft," Lewis Flight Propulsion Laboratory, Cleveland, OH, N63-12541, (Apr. 15, 1955).
98. Ibid.

The Martin B-57B light bomber that Lewis modified to test hydrogen propulsion, shown on the snowy ramp of Lewis Laboratory (now Glenn Research Center) in 1958. NASA.

hydrogen inside an aircraft. Again, this early interest in hydrogen did not reflect environmental or conservation concerns, but rather, a desire to achieve much higher flight vehicle performance.

In the report, two NACA researchers—Abe Silverstein and Eldon Hall—argued the case for using liquid hydrogen, noting it has a special advantage as an aviation fuel: a high heating value. This means that it takes less hydrogen fuel than hydrocarbon fuel to achieve the same thrust. That advantage could prove particularly important at high altitudes, the researchers noted, where maximizing fuel efficiency is critical to make up for other penalties associated with flying high.

Indeed, one of the downsides of a high-altitude flight regime—in which atmospheric pressure is low—is that it generally requires heavy, high-pressure-ratio engines to ensure combustion is sustained and thrust levels are adequate. The NACA report speculated that it might be possible to use lighter engines—albeit less-efficient ones, with lower pressure ratios—if liquid hydrogen were used for fuel. Liquid hydrogen requires less combustion volume than hydrocarbons do, making shorter and lighter engines feasible. And, with its high heating value, liquid hydrogen fuel generates more thrust per pound than hydrocarbons do, even if it's being used in a light engine running at a lower pressure ratio. The report posited that if every pound of weight saved by using a lighter

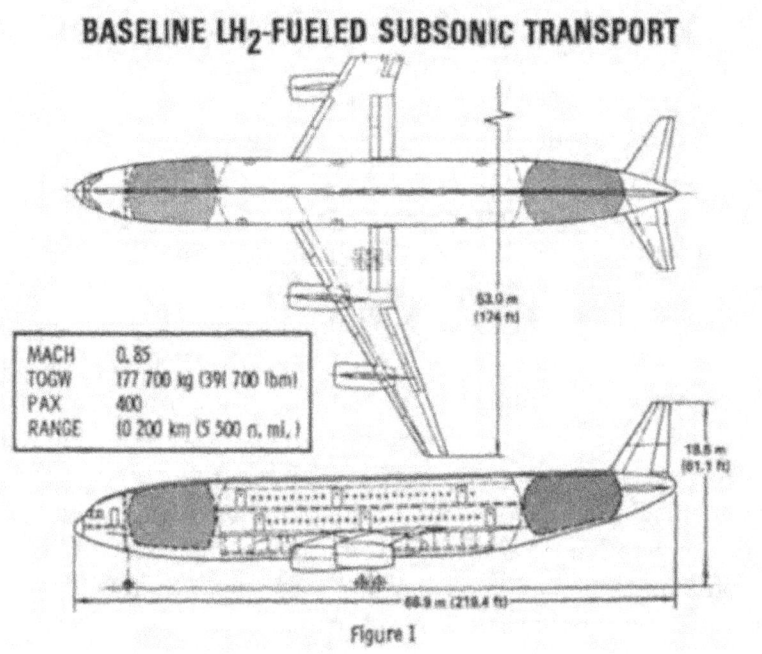

A 1978 NASA conceptual design study for a liquid-hydrogen-powered jetliner. Note the large volume required by liquid hydrogen, compared with conventional hydrocarbon-fueled airliners. NASA.

engine could be replaced by a pound of liquid hydrogen fuel, an aircraft could be over twice as effective in extending its range at high altitudes.

After Silverstein and Hall issued their report, the NACA conducted experiments showing that hydrogen had a high combustion efficiency in a turbojet combustor even in low-pressure conditions. In 1956, NACA researchers at Lewis made "three completely successful flights" using liquid hydrogen in one engine of a modified Martin B-57B jet bomber, thereby effectively demonstrating that liquid hydrogen could be handled and jettisoned safely and was feasible for use in aircraft.[99] Meanwhile, from 1956 to 1958, the U.S. Air Force began work on a secret project, known as Suntan, to develop a high-altitude, hydrogen-fueled aircraft with performance superior to the secret U-2 spy plane of the Counter Intelligence Agency (CIA).

The use of liquid hydrogen in aircraft would have marked a major breakthrough in terms of high altitude flight because engine weight is

---

99. Donald R. Mulholland, Loren W. Acker, Harold H. Christenson, and William V. Gough, "Flight Investigation of a Liquid-Hydrogen Fuel System," NACA RM-E57F19a (1957).

"the single most important variable determining the height at which an airplane can fly."[100] Liquid hydrogen offered the potential to fly at high altitudes at an extended range. Despite its potential, however, neither the NACA nor the Air Force was able to convince enough stakeholders in Government and industry that liquid hydrogen was a viable candidate for aviation fuel.

Hydrogen's excellent combustion qualities raised questions about whether it could be safely transported or carried inside aircraft. To be sure, NACA flight tests demonstrated that safe handling of hydrogen fuel on the ground and in the air was possible. Also, the Air Force conducted tests in which liquid hydrogen tanks under pressure were ruptured, and it found that in many cases the hydrogen quickly escaped without ignition. Yet concerns about safety persisted, and, in a tight budget climate, hydrogen-fuelled aircraft lost out to other priorities. After receiving a full briefing on Suntan, Gen. Curtis E. Lemay, the former head of Strategic Air Command who had moved up to Vice Chief of Staff in July 1957, raised concerns about safety. "What," he said, "put my pilots up there with a bomb?"[101]

Others questioned whether using liquid hydrogen would truly yield big gains in aircraft range at high altitudes. Hydrogen has a high volume—10 times that of hydrocarbons—which means that the aircraft fuselage has to be bigger and weigh more to accommodate the fuel. Silverstein and Hall argued that there would be more room for hydrogen fuel tanks in high-altitude aircraft, which would need larger wings and therefore a bigger fuselage to provide lift in the thin air of the upper atmosphere (a bigger fuselage would mean more room for hydrogen fuel tanks). But while it might have been possible to extend the range of aircraft because of the increased efficiency of liquid hydrogen, others questioned whether hydrogen-fuelled aircraft would still be fairly limited by the tremendous amount of fuel storage capacity that hydrogen requires. Kelly Johnson, the Lockheed Martin engineer who designed the U-2 and the hydrogen-fueled CL-400 for the Air Force's Suntan project, said he could see a range growth of only 3 percent from adding more hydrogen fuel storage capacity to his CL-400 design. "We have crammed the max-

---

100. Silverstein and Hall, "Research Memorandum: Liquid Hydrogen as Jet Fuel for High Altitude Aircraft."
101. John L. Sloop, "Liquid Hydrogen as a Propulsion Fuel," NASA SP-4404 (Washington, DC: NASA, 1978).

imum amount of hydrogen in the fuselage that it can hold. You do not carry hydrogen in the flat surfaces of the wing," he said.[102]

While liquid hydrogen is highly energetic, it has far less energy density than hydrocarbon fuels. Thus, to get an equivalent amount of energy from hydrogen requires a much greater volume. Accordingly, a hydrogen airplane would have extremely large fuel tanks, which, having to be supercold as well, pose significant technical challenges to aircraft designers. Researchers have not yet found a way to overcome the challenges associated with hydrogen's large volume, which forces aircraft design compromises and requires complex ground transportation, storage distribution, and vent capture system. Moreover, hydrogen is not a viable source of energy in itself; producing it requires the use of other sources of energy—such as electric power produced by nuclear fusion as well as a large source of clean water. However, in one respect, hydrogen could "pay back" this "debt," for it could be used to enrich the production process of synthetic fuel, achieving similar production efficiencies while reducing the amount of water and coal traditionally required for enrichment.[103]

Despite these technical challenges, NASA's research on the use of hydrogen to power aircraft did lead to some important findings: namely, that hydrogen is a potentially promising turbojet fuel in a high-altitude, low-speed flight regime. These conditions favor a fuel that can operate efficiently in low-pressure conditions. High altitudes also favor a large-volume aircraft, helping to offset the disadvantage of hydrogen's low density. Given these attractive characteristics, the prospect of using hydrogen as an aircraft propellant has continued to resurface in the past decade, especially when the cost of hydrocarbon-based fuel rises. For example, NASA's Zero CO2 research project sought to eliminate carbon dioxide and lower NOx emissions by converting propulsion systems to hydrogen fuel.[104] One new propulsion technology that NASA engineers considered as part of Zero CO2 was the use of fuel cells, which are discussed below. A NASA Glenn Web page updated as recently as 2008 says that the Combustion Branch of NASA's Propulsion Division is still study-

---

102. Ibid.
103. D. Daggett, O. Hadaller, R. Hendricks, and R. Walther, "Alternative Fuels and Their Potential Impact on Aviation," NASA TM-2006-214365 (2006); see also C.G. Jay, "Hydrogen Enrichment of Synthetic Fuel," NASA Tech Briefs, vol. 3, no. 1 (June 1, 1978), NTTC Doc. 19780000039.
104. Lyons and Levine, "An Overview of NASA's Contributions to Energy Technology."

ing hydrogen combustion to demonstrate that hydrogen can be used as an aviation fuel to minimize emissions.[105]

The NACA's early research on hydrogen-fuelled aircraft also created an awareness of hydrogen as a potential fuel source that did not exist before Silverstein and Hall embarked on their study. This awareness helped lead to important breakthroughs in rocketry and fuel cell research. In particular, research on the use of hydrogen in air-breathing aircraft laid the groundwork for the successful development of hydrogen-fueled rockets in the mid-1950s. In fact, Silverstein and Hall's research helped to inform NASA's decision in 1959 to use liquid hydrogen as a propellant in the upper stage of the Saturn launch vehicle. That decision was one of the keys to the success of the Apollo Moon landing missions of the 1960s and 1970s.[106]

The NACA's early efforts to draw attention to hydrogen as a power source also led to the development of fuel cells for the Apollo and Gemini capsules. Apollo employed the world's first fuel cells, which used hydrogen and oxygen to generate onboard power for Apollo command and service modules. Fuel cells are essentially plastic membranes treated with a special catalyst; hydrogen seeps into the membrane and meets up with the oxygen inside to generate electricity and water. The fuel cells used on the Apollo proved so successful that they were once again employed on the Space Shuttle orbiter.

Fuel Cells for Aircraft

Observing the success of fuel cells in space flight, NASA researchers in the late 1990s began to explore the potential for fuel cells to power aircraft. The cells were attractive to NASA's aeronautics directorate because they have near-zero emissions, are quiet and efficient, and can work in any environment where the temperature is lower than the cell's operating temperature. Valerie Lyons, NASA's Division Chief of Power and In-Space Propulsion, said she began pushing for NASA aeronautics to pursue fuel cell research about 10 years ago. "I would venture to say they hadn't really looked at it before that," she said. "When I looked at the fuel cell area, I said, 'This is pretty interesting, can we use this?'"[107]

---

105. NASA Glenn Research Center Propulsion Systems Division Combustion Branch, updated July 24, 2008, available at http://www.grc.nasa.gov/WWW/combustion, accessed July 29, 2009.
106. Sloop, "Liquid Hydrogen as a Propulsion Fuel."
107. Harrington, Lyons phone interview, Apr. 27, 2009.

One of NASA's main fuel cell initiatives was the Next Generation Clean Aircraft Power (NEXCAP) program, which sought to use the cells as auxiliary power units (APUs) for aircraft.[108] The APUs could be used in onboard electrical systems to power a grounded aircraft, providing an alternative to wasting fuel and producing emissions by drawing on power from an idling engine. NASA researchers hoped that the APU research would eventually lead to the design and test flight in 2030 of an electric airplane, which would rely on fuel cells for propulsion.[109] While NASA's electric airplane never came to fruition, Boeing flew the first piloted airplane powered by hydrogen fuel cells in 2008.[110] NASA researchers maintain a strong interest in the potentiality of fuel cells to meet the future energy needs of the Nation.

The 1970s and the Rise of Synthetic Fuels

NASA's interest in alternative fuels did not end with liquid hydrogen; synthetic fuel research, joined with research on new, more aerodynamically efficient aircraft configurations, took off in the 1970s and 1980s, as rising oil prices and a growing concern about mankind's (and aviation's) impact on the environment pushed researchers to seek alternatives to oil-based fuel.[111]

In 1979, NASA Langley released an aircraft fuel study that compared liquid hydrogen, liquid methane, and synthetic aviation kerosene derived from coal or oil shale.[112] The study took into account factors including cost, capital requirements, and energy resources required to make the fuel. These factors were considered in light of the practicality of using the fuel in terms of the fuel production processes, transportation, storage, and its suitability for use on aircraft. Environmental emissions and safety aspects of the fuel also were considered. The study concluded that all three fuels met the criteria, but that synthetic aviation kerosene was the most attractive because it was the least expensive.[113]

---

108. Anita Liang, Chief, Aeropropulsion Projects Office, "Emerging Fuel Cell Developments at NASA for Aircraft Applications," Glenn Research Center, 2003, available at http://www.netl.doe.gov/publications/proceedings/03/seca/Anita%20Liang.pdf, accessed July 29, 2009.
109. Ibid.
110. Boeing, "Boeing Successfully Flies First Fuel-Cell Powered Plane," Apr. 3, 2008, http://www.boeing.com/news/releases/2008/q2/080403a_nr.html, accessed July 29, 2009.
111. See, for example, A.L. Nagel, W.J. Alford, and J.F. Dugan, "Future Long-Range Transports: Prospects for Improved Fuel Efficiency," NASA TM-X-72659 (1975).
112. Robert D. Wicofski, "Comparison of Alternate Fuel for Aircraft," NASA TM-80155 (Sept. 1979).
113. Ibid.

Despite the promising findings of NASA's study, however, synthetic fuel never made it into mainstream production. The fuel's capital costs are still relatively high when compared with oil-based jet fuel, because new synthetic fuel production plants have to be built to produce the fuel.[114] Private industry has been hesitant to get into this business, fearing it would not make a return on its investment. If oil prices were to drop—as they did in the mid-1980s— companies that invested in synthetic aircraft fuel production would find it difficult to compete with cheap oil-based jet fuel.

Regardless of industry's hesitation, Government efforts to develop and test alternative fuels are springing to life again as a result of a return to high oil prices and a growing concern about the impact of emissions on air quality and climate change. The U.S. Air Force has engaged in a systematic process to certify all of its aircraft to fly on a 50/50 blend of oil-based jet fuel and synthetic fuel. Air Force officials hoped that testing and flying their own aircraft on synthetic fuels would encourage commercial airlines to do the same, believing that if the service and airline industry could create a buyer's market for synthetic fuel, then the energy industry might be more amenable to investing the money required to build synthetic fuel plants for mass production.[115]

NASA has also begun testing the performance and emissions of two synthetic fuels derived from coal and natural gas. While the Air Force's interest in alternative fuels is largely related to concerns about oil price volatility and the national security risks of relying on foreign oil suppliers, NASA has embarked on alternative fuels research largely to study the potential for reducing emissions. NASA's research effort, which is being conducted at NASA Dryden, seeks to closely measure particulate levels. "Even though there are no current regulations for particulates, we see particulates as being very important," said Bulzan, who is leading the alternative fuels effort. "They are very important to local air quality when the aircraft is taking off and landing at the airport, and they can also generate cloud formation that can affect global warming."[116]

---

114. A plant that produces Fischer-Tropsch synthetic fuel, which is being used by the U.S. Air Force to power jet engines, requires five times the capital costs as an oil refinery. See Gerrard Cowan and Harrington, "Adding Fuel to the Fire: Energy and Defence," *Jane's Defence Weekly*, Nov. 21, 2008.
115. Harrington, "USAF Promotes Fuel Alternative," *Jane's Defence Weekly*, Mar. 16, 2007.
116. NASA has had a long-standing interest in such research. See, for example, R. Bradford, W.T. Atkins, R.M. Bass, R. Dascher, J. Dunkin, N. Luce, W. Seward, and D. Warren, "Coal Conversion and Synthetic-Fuel Production," *NASA Tech Briefs*, vol. 5, no. 1 (Aug. 1, 1980), p. 56.

Both the USAF and NASA are using synthetic fuel derived from a process developed by the Germans in World War II known as Fischer-Tropsch. In this process, a mixture of carbon monoxide and hydrogen is used to create liquid hydrocarbons for fuel. NASA Dryden's latest alternative fuels testing, which took place in early 2009, involved fueling a grounded DC-8 with both 100-percent synthetic fuel and a 50/50 blend. The test results are being compared with baseline tests of hydrocarbon fuel emissions tests performed in the DC-8 in 2004. Air Force researchers were on hand to help measure the emissions.[117]

NASA and the Air Force are also working with Boeing to explore the possibility of using biofuel, which may prove to be cleaner than fuel derived from the Fischer-Tropsch process. The main obstacle to biofuel use at this time is the fact that it is difficult to procure in large quantities. For example, algae are attractive feedstock for biofuel, but the problem lies in being able to grow enough. NASA has begun to take on the feedstock problem by setting up a Greenlab Research Facility at NASA Glenn, where NASA researchers are seeking to optimize the growing conditions for algae and halophytes, which are plants tolerant of salt water.[118]

In conclusion, the oil crisis and growing environmental awareness of the 1970s presented a critical opportunity for NASA to reclaim its mantle at the forefront of aeronautics research. NASA-led programs in fuel-efficient engines, aircraft structures, and composites—as well as the Agency's contribution to computational fluid dynamics—planted the seeds that gave private industry the confidence and technological know-how to pursue bold aircraft fuel-efficiency initiatives on its own. Without NASA's E Cubed program, U.S. engine companies may not have had the financial resources to develop their fuel-saving, emissions-reducing TAPS and TALON combustors. E Cubed also spawned the open-rotor engine concept, which is still informing engine fuel-efficiency efforts today. The turbulent 1970s also created the opening for NASA Langley's Richard Whitcomb to proceed full throttle with efforts to develop supercritical wings and winglets that have revolutionized fuel-efficient airframe design. And NASA's research on alternative fuels during the 1970s, if stillborn, nevertheless set the stage for the Agency to play a significant role in the Government's revitalized alternative fuels research that came with the dawning of the 21st century.

---

117. Harrington, Bulzan interview.
118. Ibid.

Addressing the Nation's scientific leadership in 2009, President Barack Obama compared the energy challenge facing America to the shock of Sputnik in 1957, declaring it the Nation's new "great project."[119] Reflecting the increasing emphasis and rising priorities of Federal environmental research, NASA had received funding to support global climate studies, while NASA's aeronautics research received additional funding to "improve aircraft performance while reducing noise, emissions, and fuel consumption."[120] Clearly, NASA's experience in energy and aeronautics positioned the Agency well to continue playing a major role in these areas.

As the Agency enters the second decade of the 21st century, much remains to be done to increase aircraft fuel efficiency, but much, likewise, has already been accomplished. To NASA's aeronautics researchers, inheritors of a legacy of accomplishment in flight, the energy and environmental challenges of the new century constitute an exciting stimulus, one as profoundly intriguing as any of the other challenges—supersonic flight and landing on the Moon among them—that the NACA and NASA have faced before. Those challenges, too, had appeared daunting. But just as creative NACA–NASA research overcame them, those in the Agency charged with responsibility for pursuing the energy and environmental challenges of the new century were confident that they, and the Agency, would once again see their efforts crowned with success.

---

119. President Barack Obama, *Remarks by the President at the National Academy of Sciences Annual Meeting*, (Washington, DC: Office of the Press Secretary, Apr. 27, 2009). Available at *http://www.whitehouse.gov/the_press_office/Remarks-by-the-President-at-the-National-Academy-of-Sciences-Annual-Meeting*, accessed July 29, 2009.

120. The White House, "NASA Budget Highlights: The National Aeronautics and Space Administration 2010 Budget," updated Apr. 28, 2009, available at *http://www.nasa.gov/news/budget/index.html*, accessed July 29, 2009.

Recommended Additional Readings
Reports, Papers, Articles, and Presentations:

Theodore G. Ayers and James B. Hallissy, "Historical Background and Design Evolution of the Transonic Aircraft Technology Supercritical Wing," NASA TM-81356 (1981).

Darrell H. Baldwin and Bradford S. Linscott, "The Federal Wind Program at NASA Lewis Research Center," NASA TM-83480 (1983).

R. Bradford, W.T. Atkins, R.M. Bass, R. Dascher, J. Dunkin, N. Luce, W. Seward, and D. Warren, "Coal Conversion and Synthetic-Fuel Production," *NASA Tech Briefs*, vol. 5, no. 1 (Aug. 1, 1980), p. 56.

B.P. Collins, A.L. Haines, and C.J. Wales, "A Concept for a Fuel Efficient Flight Planning Aid for General Aviation," NASA CR-3533 (1982).

Gerrard Cowan and Caitlin Harrington, "Adding Fuel to the Fire: Energy and Defence," *Jane's Defence Weekly*, Nov. 21, 2008.

D. Daggett, O. Hadaller, R Hendricks, and R. Walther, "Alternative Fuels and Their Potential Impact on Aviation," NASA TM-2006-214365 (2006).

John H. Del Frate and Gary B. Cosentino, "Recent Flight Test Experience with Uninhabited Aerial Vehicles at the NASA Dryden Flight Research Center," NASA TM-1998-206546 (1998).

Vance Dippold, III, "CFD Analyses and Jet-Noise Predictions of Chevron Nozzles with Vortex Stabilization," NASA TM-2008-215150 (2008).

Edwin J. Graber, "Overview of NASA PTA Propfan Flight Test Program," NASA Lewis [Glenn] Research Center, Nov. 1, 1987, CASI Doc 198.800.06423.

Caitlin Harrington, "USAF Promotes Fuel Alternative," *Jane's Defence Weekly*, Mar. 16, 2007.

C.G. Jay, "Hydrogen Enrichment of Synthetic Fuel," *NASA Tech Briefs*, vol. 3, no. 1 (June 1, 1978), NTTC Doc. 197.800.00039.

B.H. Little, D.T. Poland, H.W. Bartel, C.C. Withers, and P.C. Brown, "Propfan Test Assessment (PTA) Final Project Report," NASA CR-185138 (1989).

Valerie J. Lyons and Arlene S. Levine, "An Overview of NASA's Contributions to Energy Technology," NASA AIAA-2008-5641 (Jan. 2009).

Donald R. Mulholland, Loren W. Acker, Harold H. Christenson, and William V. Gough, "Flight Investigation of a Liquid-Hydrogen Fuel System," RM E57F19a (1957).

A.L. Nagel, W.J. Alford, and J.F. Dugan, "Future Long-Range Transports: Prospects for Improved Fuel Efficiency," NASA TM-X-72659 (1975).

Palmer C. Putnam, "Wind Power: Yesterday, Today, and Tomorrow," NASA CP-2230 (1982).

N.A. Radovich, D. Dreim, D.A. O'Keefe, L. Linner, S.K. Pathak, J.S. Reaser, D. Richardson, J. Sweers, and F. Conner, "Study for the Optimization of a Transport Aircraft Wing for Maximum Fuel Efficiency," vol. 1: "Methodology, Criteria, Aeroelastic Model Definition, and Results," NASA CR-172551 (1985).

Travis L. Turner, Randolph H. Cabell, Roberto J. Cano, and Richard J. Silcox, "Testing of SMA-Enabled Active Chevron Prototypes Under Representative Flow Conditions," NASA Langley Research Center, Report LAR-17332, Paper 6928-36, CASI Doc. 200.800.14174.

U.S. Federal Energy Administration, "Project Independence Blueprint. Task Force. Report. Solar Energy," U.S. Department of Commerce, National Technical Information Service (1974).

U.S. Government Accountability Office, "Aviation's Effects on the Global Atmosphere Are Potentially Significant and Expected to Grow," GAO/RCED-00-57 (Feb. 2000).

U.S. National Research Council [Committee on High Speed Research, Aeronautics and Space Engineering Board of the Commission on

Engineering and Technical Systems], *U.S, Supersonic Aircraft: Assessing NASA's High Speed Research Program* (Washington: NRC, 1997).

Robert D. Wicofski, "Comparison of Alternate Fuel for Aircraft," NASA TM-80155 (1979).

Books and Monographs:

Joseph R. Chambers, *Partners in Freedom: Contributions of the Langley Research Center to U.S. Military Aircraft of the 1990s,* SP-2000-4519 (Washington: NASA, 2000).

Erick M. Conway, *High Speed Dreams* (Baltimore: Johns Hopkins University Press, 2005).

Virginia P. Dawson, *Engines and Innovation: Lewis Laboratory and American Propulsion Technology*, SP-4306 (Washington, DC: NASA, 1991).

Jeffrey Ethell, *Fuel Economy in Aviation*, NASA SP-462 (1983).

Alexander Laufer, et al., *Shared Voyage: Learning and Unlearning from Remarkable Projects*, NASA History Series, NASA SP-2005-4111, (Washington, DC: NASA, 2005).

Anita Liang, *Emerging Fuel Cell Developments at NASA for Aircraft Applications* (Cleveland: NASA Glenn Research Center, 2003).

Pamela Mack, ed., *From Engineering Science to Big Science,* NASA SP-4219 (Washington, DC: NASA, 1998).

Palmer C. Putnam, *Power from the Wind* (New York: Van Nostrand Reinhold Company, 1948).

Robert W. Righter, *Wind Energy in America: A History* (Norman, OK: University of Oklahoma Press, 1996).

John L. Sloop, *Liquid Hydrogen as a Propulsion Fuel*, NASA SP-4404 (Washington. DC: NASA, 1978).

A full-pitch, fiberglass Vestas wind turbine in Rome in 2009. The Danish company remains the largest in the wind turbine industry. R.P. Hallion.

CASE 13

Good Stewards: NASA's Role in Alternative Energy

Bruce I. Larrimer

Consistent with its responsibilities to exploit aeronautics technology for the benefit of the American people, NASA has pioneered the development and application of alternative energy sources. Its work is arguably most evident in wind energy and solar power for high-altitude remotely piloted vehicles. Here, NASA's work in aerodynamics, solar power, lightweight structural design, and electronic flight controls has proven crucial to the evolution of novel aerospace craft.

THIS CASE STUDY REVIEWS two separate National Aeronautics and Space Administration (NASA) programs that each involved research and development (R&D) in the use of alternative energy. The first part of the case study covers NASA's participation in the Federal Wind Energy Program from 1974 through 1988. NASA's work in the wind energy area included design and fabrication of large horizontal-axis wind turbine (HAWT) generators, and the conduct of supporting research and technology projects. The second part of the case study reviews NASA's development and testing of high-altitude, long-endurance solar-powered unmanned aerial vehicles (UAVs). This program, which ran from 1994 through 2003, was part of the Agency's Environmental Research and Aircraft Sensor Technology Program.

Wind Energy Program and Large Horizontal-Axis Wind Turbines (1974–1988)

The energy crisis of the 1970s brought about renewed interest in the development of alternative energy sources, including harnessing wind power for the generation of electricity. This renewed interest led to the establishment of the Federal Wind Energy Program in 1974 as part of the Nation's solar energy program. The initial program overview, technical analysis, and objectives were formalized by the Project Independence Interagency Solar Task Force that was formed in April 1974 and chaired by the National Science Foundation (NSF). Approximately 100 individuals—representing various Government agencies, universities, research laboratories, private industries, and consulting firms—participated in the task force project. Thirteen of the participants were from NASA, including six from NASA Lewis (now NASA John H. Glenn Research Center). The task force's final findings were outlined in the November 1974 "Project Independence Blueprint" report. The task force identified the six following "most promising" technologies for converting solar energy to a variety of useful energy forms: (1) solar heating and cooling of buildings, (2) solar thermal energy conversion, (3) wind energy conversion, (4) bioconversion to fuels, (5) ocean thermal energy conversion, and (6) photovoltaic electric power systems. The task force noted that the objective of the wind energy conversion part of the program was to improve the efficiency of wind turbine systems in a variety of applications and to reduce their costs. In regard to site selection, the task force concluded that the first attainment of economic viability in the United States would occur in areas such as the Great Plains, Alaska, the Great Lakes, the Atlantic and Pacific coasts, New England, and Hawaii. It concluded that the key to large-scale application of wind energy conversion systems was the reduction of costs through advanced technology, new materials, mass production, and the use of field fabrication techniques. Finally, the task force noted that a closely monitored program of proof-of-concept experiments was expected to reduce cost and constraint uncertainties.[1]

As a prelude to the formation of the wind energy program, NASA Lewis made significant contributions to a wind energy workshop that reviewed both the current status of wind energy and assessed the potential of wind power. This workshop was held as part of the Research Applied to National Needs (RANN) project that led to the National Science Foundation's role in the initial planning of a 5-year sustained wind energy program. In January 1975, the wind energy program was transferred to the newly formed Energy Research and Development Administration (ERDA), which was incorporated into the newly formed U.S. Department of Energy (DOE) in 1977.

Pursuant to the initial agreement between NASA and the NSF, which had no research centers of its own, NASA's Lewis Research Center at

---

1. Federal Energy Administration, under direction of National Science Foundation, "Project Independence Blueprint," Solar Energy PB 248 507 (reproduced by National Technical Information Service, U.S. Department of Commerce, Springfield, VA, Nov. 1974), pp. I-3, I-20-22.

Lewis Field in Cleveland, OH, was given overall project management for the portion of the Wind Energy Program that involved the development and fabrication of large experimental horizontal-axis wind turbines. NASA Lewis's responsibilities also included the conduct of supporting research and technology for the wind turbine conversion systems. This sponsorship continued under the Department of Energy once DOE took over the Federal Wind Energy Program. Louis Divone, who initially selected NASA Lewis to participate in the program, was the wind energy program manager for the NSF and later for ERDA and DOE. The program goal was the development of the technology for safe, reliable, and environmentally acceptable large wind turbine systems that could generate significant amounts of electricity at costs competitive with conventional electricity-generating systems.

NASA Lewis engineers were very interested in getting involved in the Wind Energy Program and realized early on that they could make significant contributions because of the Research Center's long experience and expertise in propeller and power systems, aerodynamics, materials, and structures testing. The selection of NASA Lewis also represented an interesting historical context. Over 85 years earlier, in 1887–1888, in Cleveland, OH, an engineer by the name of Charles F. Brush constructed a 60-foot, 80,000-pound wind-electric dynamo that is generally credited as being the first automatically operating wind turbine for electricity generation. Brush's wind turbine, which supplied power for his home for up to 10 years, could produce a maximum 12,000 watts of direct current that charged 12 batteries that in turn ran 350 incandescent lights, 2 arc lights, and a number of electric motors. His dynamo made 50 revolutions to 1 revolution of the wind wheel, which consisted of 144 wooden blades and was 56 feet in diameter, accounting for 1,800 square feet of total blade surface swept area. The wind dynamo had an automatic regulator that prevented the power from running above 90 volts at any speed. Brush later dismantled his wind dynamo, apparently without attempting to develop a unit that could feed into a central power network.[2]

The use of wind power to generate electricity achieved a degree of success in rural and remote areas of the United States in the 1920s and 1930s. These generators, however, were small, stand-alone wind-

---

2. See Robert W. Righter, *Wind Energy in America: A History* (Norman, OK: University of Oklahoma Press, 1996), pp. 44–49 and 106.

electric systems such as those designed and marketed by Marcellus and Joseph Jacobs, who built three-bladed systems, and the Windcharger Corporation, which built two-bladed generators. Most of these efforts were abandoned in the 1940s and 1950s because of the expansion of electrical utility networks, especially in response to passage of the Rural Electrification Act of 1937 and the availability of low-cost fossil fuels.

The first American effort to build a large wind turbine to feed into a power network was undertaken by Palmer Cosslett Putnam. This effort was funded by the S. Morgan Smith Company, which constructed and installed a 1.25-megawatt wind turbine at Grandpa's Knob, VT. Prior to fabrication of his turbine, Putnam considered a number of questions that were still being debated years later, including whether to build a vertical- or horizontal-axis wind turbine; if horizontal, how many blades should there be; whether the generator should be aloft or on the ground; whether the drive should be mechanical or hydraulic; whether the tower should rotate or be stationary; and what size generator should be used. He noted that examples of all of these configurations existed in writings on wind power. Putnam, with the concurrence of both Beauchamp and Burwell Smith, decided on using the horizontal-axis, two-bladed stainless steel configuration, with a mechanically driven synchronous generator mounted aloft. He then concluded that the optimum size of a wind turbine generator (WTG) was close to 2 megawatts and noted that studies indicated that increased efficiency appeared to be flat between 2 and 3 megawatts.[3] The Smith-Putnam wind turbine, which supplied power to the Central Vermont Public Service Corporation's power network, started operations on October 19, 1941, and operated intermittently for a total electric generation period of approximately 16 months. A bearing failure caused a blade separation accident, and the project was terminated in March 1945. While the turbine was not rebuilt, the system's operation demonstrated that wind could be harnessed on a large scale to produce electricity. The power company, as well as others, envisioned that wind turbines would operate in conjunction with hydroelectric power systems.[4]

In the late 1950s, a German engineer by the name of Ulrich Hütter also built a smaller, 100-kilowatt wind turbine generator (the Hütter-

---

3. Palmer C. Putnam, "Wind Power: Yesterday, Today, and Tomorrow," NASA CP-2230 (Proceedings of a workshop sponsored by U.S. Department of Energy, Wind Energy Technology Division, and NASA Lewis Research Center, Cleveland, OH, July 28–30, 1981).
4. Righter, *Wind Energy in America*, pp. 126–137.

Allgaier wind turbine) that was tied into a power utility grid. Hütter's machine used a 112-foot-diameter, two-bladed downwind rotor with full span pitch control. The blades were mounted on a teetered hub. In preparation to commence work on its own wind turbines, NASA purchased the plans from Hütter and considered or incorporated a number of design criteria and features of both the Smith-Putnam and Hütter-Allgaier wind turbine generators.[5] NASA also participated in a joint NASA–Danish financing of the restoration of the wind turbine, which was completed in 1977. NASA Lewis later did aerodynamics testing and modeling of the Gedser wind turbine information using the Mod-0 testbed turbine.

NASA Lewis's involvement in wind energy leading up to its selection to oversee the wind turbine development portion of the Federal Wind Energy Program included designing, at the request of Puerto Rico, a wind turbine to generate electricity for the Island of Culebra. This project grew out of an unrelated NASA Lewis 1972 project to take wind measurements in Puerto Rico. Later on, under the Wind Energy Program, NASA returned to Puerto Rico to build one of the Agency's first-generation (Mod-0A) wind turbine machines. NASA Lewis's involvement in the Wind Energy Program also was enhanced by its research of past wind energy projects and its projection of the future feasibility of using wind power to generate electricity for U.S. power utility networks. NASA's overview and findings were presented as a paper at a symposium held in Washington, DC, that brought together past developers of wind turbines, including Palmer Putnam, Beauchamp Smith, Marcellus Jacobs, and Ulrich Hütter, as well as a new group of interested wind energy advocates.

1973 RANN Symposium Sponsored by the National Science Foundation
In reviewing the current status and potential of wind energy, Ronald Thomas and Joseph M. Savino, both from NASA's Lewis Research Center, in November 1973 presented a paper at the Research Applied to National Needs Symposium in Washington, DC, sponsored by the National Science Foundation. The paper reviewed past experience with wind generators, problems to be overcome, the feasibility of wind power to help meet energy needs, and the planned Wind Energy Program. Thomas and Savino pointed out that the Dutch had used windmills for years to provide power for pumping water and grinding grain; that the Russians built

---

5. "MOD-0A 200 Kilowatt Wind Turbine Generator Design and Analysis Report," executive summary, DOE/NASA/0163-1; NASA CR-165127; AESD-TME-3051 (Aug. 1980), p. 2.

a 100-kilowatt generator at Balaclava in 1931 that feed into a power network; that the Danes used wind as a major source of power for many years, including the building of the 200-kilowatt Gedser mill system that operated from 1957 through 1968; that the British built several large wind generators in the early 1950s; that the Smith-Putnam wind turbine built in Vermont in 1941 supplied power into a hydroelectric power grid; and that Germans did fine work in the 1950s and 1960s building and testing machines of 10 and 100 kilowatts. The two NASA engineers noted, however, that in 1973, no large wind turbines were in operation.

Thomas and Savino concluded that preliminary estimates indicated that wind could supply a significant amount of the Nation's electricity needs and that utilizing energy from the wind was technically feasible, as evidenced by the past development of wind generators. They added, however, that a sustained development effort was needed to obtain economical systems. They noted that the effects of wind variability could be reduced by storage systems or connecting wind generators to fossil fuel or hydroelectric systems, or dispersing the generated electricity throughout a large grid system. Thomas and Savino[6] recommended a number of steps that the NASA and National Science Foundation program should take, including: (1) designing, building, and testing modern machines for actual applications in order to provide baseline information for assessing the potential of wind energy as an electric power source, (2) operating wind generators in selected applications for determining actual power costs, and (3) identifying subsystems and components that might be further reduced in costs.[7]

NASA–Industry Wind Energy Program Large Horizontal-Axis Wind Turbines

The primary objective of the Federal Wind Energy Program and the specific objectives of NASA's portion of the program were outlined in a followup technical paper presented in 1975 by Thomas, Savino, and Richard L. Puthoff. The paper noted that the overall objective of the

---

6. In 2007, Purdue University recognized Joseph Savino's work on wind energy and other accomplishments during his 41 years with NASA by awarding him its Outstanding Mechanical Engineer Award.

7. Ronald L. Thomas and Joseph M. Savino, "Status of Wind Energy Conversion," NASA TM-X-71523 (technical paper presented at *RANN Symposium Sponsored by the National Science Foundation, Washington, DC, Nov. 18–20, 1973*), pp. 1–5.

program was "to develop the technology for practical cost-competitive wind-generator conversion systems that can be used for supplying significant amounts of energy to help meet the nation's energy needs."[8] The specific objectives of NASA Lewis's portion of the program were to: (1) identify cost-effective configurations and sizes of wind-conversion systems; (2) develop the technology needed to produce cost-effective, reliable systems; (3) design wind turbine generators that are compatible with user applications, especially with electric utility networks; (4) build up industry capability in the design and fabrication of wind turbine generators; and (5) transfer the technology from the program to industry for commercial application. To satisfy these objectives, NASA Lewis divided the development function into the three following areas: (1) design, fabrication, and testing of a 100-kilowatt experimental wind turbine generator; (2) optimizing the wind turbines for selected user operation; and (3) supporting research and technology for the systems.

The planned workload was divided further by assignment of different tasks to different NASA Research Centers and industry participants. NASA Lewis would provide project management and support in aerodynamics, instrumentation, structural dynamics, data reduction, machine design, facilities, and test operations. Other NASA Research Centers would provide consulting services within their areas of expertise. For example, Langley worked on aeroelasticity matters, Ames consulted on rotor dynamics, and Marshall provided meteorology support. Initial industry participants included Westinghouse, Lockheed Corporation, General Electric, Boeing, and Kaman Aerospace.

In order to undertake its project management role, NASA Lewis established the Center's Wind Power Office, which consisted initially of three operational units—one covering the development of an experimental 100-kilowatt wind turbine, one handling the industry-built utility-operated wind turbines, and one providing supporting research and technology. The engineers in these offices basically worked together in a less formal structure, crossing over between various operational areas. Also, the internal organization apparently underwent several changes during the program's existence. For example, in 1976, the program was

---

8. R. Thomas, R. Puthoff, and J. Savino, "Plans and Status of the NASA-Lewis Research Center Wind Energy Project," (technical paper to be presented at the *Joint Power Conference cosponsored by the Institute of Electronic and Electrical Engineers and American Society of Mechanical Engineers*, Portland, OR, Sept. 28–Oct. 1, 1975), p. 1.

NASA Mod-0 testbed wind turbine, Plum Brook Station, Sandusky, OH. NASA.

directed by the Wind Power Office as part of the Solar Energy Branch. The first two office managers were Ronald Thomas and William Robbins. By 1982, the organization consisted of a Wind Energy Project Office, which was once again under the supervision of Thomas and was part of the Wind and Stationary Power Division. The office consisted of a project development and support section under the supervision of James P. Couch (who managed the Mod-2 project), a research and technology section headed by Patrick M. Finnegan, and a wind turbine analysis section under the direction of David A. Spera. By 1984, the program organization had changed again with the Wind Energy Project Office, which was under the supervision of Darrell H. Baldwin, becoming part of the Energy Technology Division. The office consisted of a technology section under Richard L. Puthoff and an analysis section headed by David A. Spera. The last NASA Lewis wind energy program manager was Arthur Birchenough.

NASA's Experimental (Mod-0) 100-Kilowatt Wind Turbine Generator (1975–1987)

Between 1974 and 1988, NASA Lewis led the U.S. program for large wind turbine development, which included the design and installation of 13 power-utility-size turbines. The 13 wind turbines included an initial testbed turbine designated the Mod-0 and 3 generations of followup wind turbines designated Mod-0A/Mod-1, Mod-2, and Mod-5. As noted in the Project Independence task force report, the initial 100-kilowatt wind turbine project and related supporting research was to be performed in-house by NASA Lewis, while the remaining 100-kilowatt systems, megawatt systems, and large-scale multiunit systems subprograms were to be performed by contractors under NASA Lewis direction. Each successive generation of technology increased reliability and efficiency while reducing the cost of electricity. These advances were made by gaining a better understanding of the system-design drivers, improving the analytical design tools, verifying design methods with operating field data, and incorporating new technology and innovative designs. However, before these systems could be fabricated and installed, NASA Lewis needed to design and construct an experimental testbed wind turbine generator.

NASA's first experimental wind turbine (the Mod-0) was constructed at Plum Brook Station in Sandusky, OH, and first achieved rated speed and power in December 1975. The initial design of the Mod-0 drew upon some of the previous information from the Smith-Putnam and Hütter-Allgaier turbines. The primary objectives of the Mod-0 wind turbine generator were to provide engineering data for future use as a base for the entire Federal Wind Energy Program and to serve as a testbed for the various components and subsystems, including the testing of different design concepts for blades, hubs, pitch-change mechanisms, system controls, and generators. Also, a very important function of the Mod-0 was to validate a number of computer models, codes, tools, and control algorithms.

The Mod-0 was an experimental 100-kilowatt wind turbine generator that at a wind speed of 18 miles per hour (mph) was expected to generate 180,000 kilowatthours of electricity per year in the form of 440-volt, 3-phase, 60-cycle alternating current output. The initial testbed system, which included two metal blades that were each 62-feet long from hub to blade tip located downwind of the tower, was mounted on a 100-foot, four-legged steel lattice (pinned truss design) tower. The drive train and rotor were in a nacelle with a centerline 100 feet above ground.

The blades, which were built by Lockheed and were based on NASA's and Lockheed's experience with airplane wing designs, were capable of pitch change (up and down movement) and full feather (angle of the blade change so that wind resistance is minimized). The hub was of the rigid type, meaning that the blades were bolted to the main shaft. A yaw (deviation from a straight path) control aligned the wind turbine with the wind direction, and pitch control was used for startup, shutdown, and power control functions. When the wind turbine was in a shutdown mode, the blades were feathered and free to slowly rotate. The system was linked to a public utility power network through an automatic synchronizer that converted direct current to alternating current.[9]

A number of lessons were learned from the Mod-0 testbed. One of the first problems involved the detection of larger than expected blade bending incidents that would have eventually caused early fatigue failure of the blades. The blade bending occurred for both the flatwise (out-of-plane) and edgewise (in-plane) blade positions. Followup study of this problem determined that high blade loads that resulted in the bending of the blades were caused by impulses applied to the blade each time it passed through the wake of the tower. Basically, the pinned truss design of the tower was blocking the airflow to a much greater degree than anticipated. The cause of this problem, which related to the flatwise load factors, was confirmed by site wind measurements and wind tunnel tower model tests. The initial measure taken to reduce the blocking effect was to remove the stairway from the tower. Eventually, however, NASA developed the soft tube style tower that later became the standard construction method for most wind turbine towers. Followup study of the edgewise blade loads determined that the problem was caused by excessive nacelle yawing (side-to-side) motion. This problem was addressed by replacing a single yaw drive, which aligns the rotor with the wind direction, with a dual yaw drive, and by adding three brakes to the yaw system to provide additional stiffness.[10]

Both of the above measures reduced the bending problems below the predicted level. Detection of these problems on the testbed Mod-0

---

9. Richard L. Puthoff, "Fabrication and Assembly of the ERDA/NASA 100-Kilowatt Experimental Wind Turbine," NASA TM-X-3390 (1976), pp. 1-9; and Thomas, Puthoff, and Savino, "Plans and Status of the NASA-Lewis Research Center Wind Energy Project," pp. 3–4.
10. Ronald L. Thomas and Richard M. Donovon, "Large Wind Turbine Generators." NASA TM-73767 (1978), pp. 3–5.

resulted in reevaluation of the analytical tools and the subsequent redesign of the wind turbine that proved extremely important in the design of NASA's subsequent horizontal-axis large wind turbines. In regard to other operational testing of the Mod-0 system, NASA engineers determined that the wind turbine controls for speed, power, and yaw worked satisfactorily and that synchronization to the power utility network was successfully demonstrated. Also, the startup, utility operation, shutdown, and standby subsystems worked in a satisfactory manner. Finally, the Mod-0 was used to check out remote operation that was planned for future power utility systems. In summary, the Mod-0 project satisfied its primary objective of providing the entire Federal Wind Energy Program with early operations and performance data and through continued experience with testing new concepts and components. While NASA Lewis was ready to move forward with fabrication of its next level Mod-0A and Mod-1 wind turbines, the Mod-0 testbed continued to provide valuable testing of new configurations, components, and concepts for over 11 more years.

First Generation DOE–NASA Wind Turbine Systems (Mod-0A and Mod-1) (1977–1982)

The Mod-0 testbed wind turbine system was upgraded from 100 kilowatts to a 200-kilowatt system that became the Mod-0A. Installation of the first Mod-0A system was completed in November 1977, with one additional machine installed each year through 1980 at four locations: Clayton, NM; Culebra, PR; Block Island, RI; and Oahu, HI. This first generation of wind turbines completed its planned experimental operations in 1982 and was removed from service.

The basic components and systems of the Mod-0A consisted of the rotor- and pitch-change mechanism, drive train, nacelle equipment, yaw drive mechanism and brake, tower and foundation, electrical system and components, and control systems. The rotor consisted of the blades, hub, pitch-change mechanism, and hydraulic system. The drive train included the low-speed shaft, speed increaser, high-speed shaft, belt drive, fluid coupling, and rotor blades. The electrical system and components were the generator, switchgear, transformer, utility connection, and slip rings. The control systems were the blade pitch, yaw, generator control, and safety system.[11]

---

11. "MOD-0A 200 Kilowatt Wind Turbine Generator Design and Analysis Report," executive summary, DOE/NASA/0163-1; NASA CR-165127; AESD-TME-3051 (Aug. 1980), p. 1.

Similar to the Mod-0 testbed, the Mod-0A horizontal-axis machines had a 125-foot-diameter downwind rotor mounted on a 100-foot rigid pinned truss tower. However, this more powerful first generation of turbines had a rated power of 200 kilowatts at a wind speed of 18 miles per hour and made 40 revolutions per minute. The turbine had two aluminum blades that were each 59.9 feet long. The Westinghouse Electric Corporation was selected, by competitive bidding, as the contractor for building the Mod-0A, and Lockheed was selected to design and build the blades. NASA and Westinghouse personnel were involved in the installation, site tests, and checkout of the wind turbine systems.

The primary goal of the Mod-0A wind turbine was to gain experience and obtain early operation performance data with horizontal-axis wind turbines in power utility environments, including resolving issues relating to power generation quality, and safety, and procedures for system startup, synchronization, and shutdown. This goal included demonstrating automatic operation of the turbine and assessing machine compatibility with utility power systems, as well as determining reliability and maintenance requirements. To accomplish this primary goal, small power utility companies or remote location sites were selected in order to study problems that might result from a significant percentage of power input into a power grid. NASA engineers also wanted to determine the reaction of the public and power utility companies to the operation of the turbines. The Mod-0A systems were online collectively for over 38,000 hours, generating over 3,600 megawatthours of electricity into power utility networks. NASA determined that while some early reliability and rotor-blade life problems needed to be corrected, overall the Mod-0A wind turbine systems accomplished the engineering and research objectives of this phase of the program and made significant contributions to second- and third-generation machines that were to follow the Mod-0A and Mod-1 projects. Interface of the Mod-0A with the power utilities demonstrated satisfactory operating results during their initial tests from November 1977 to March 1978. The wind turbine was successfully synchronized to the utility network in an unattended mode. Also, dynamic blade loads during the initial operating period were in good agreement with the calculation using the MOSTAB computer code. Finally, successful testing on the Mod-0 provided the database that led the way for private development of a wide range of small wind turbines that were placed in use during the late 1980s.[12]

Closely related to the Mod-0A turbine was the Mod-1 project, for which planning started in 1976, with installation of the machine taking place in May 1979. In addition to noise level and television interference testing (see below), the primary objective of the Mod-1 program was to demonstrate the feasibility of remote utility wind turbine control. Three technical assessments were planned to evaluate machine performance, interface with the power utility, and examine the effects on the environment. This system was a one-of-a-kind prototype that was much larger than the Mod-0A, with a rated power of 2,000 kilowatts (later reduced to 1,350) and a blade swept diameter of 200 feet. The Mod-1 was the largest wind turbine constructed up to that time. Considerable testing was done on the Mod-1 because the last experience with megawatt-size wind turbines was nearly 40 years earlier with the Smith-Putnam 1.25-megawatt machine, a very different design. Full-span blade pitch was used to control the rotor speed at a constant 35 revolutions per minute (later reduced to 23 rpm). The machine was mounted on a steel tubular truss tower that was 12 feet square at the top and 48 feet square at the bottom. General Electric was the prime contractor for designing, fabricating, and installing the Mod-1. The two steel blades were manufactured by the Boeing Engineering and Construction Company. There was also a set of composite rotor blades manufactured by the Kaman Aerospace Corporation that was fully compatible for testing on the Mod-1 system. The wind turbine, which was in Boone, NC, was tested with the Blue Ridge Electrical Membership Corporation from July 1979 to January 1981. The machine, operating in fully automatic synchronized mode, fed into the power network within utility standards.[13]

One of the testing objectives of this first-generation prototype was to determine noise levels and any potential electromagnetic interference with microwave relay, radio, and television associated with

---

12. Louis V. Divone, "Evolution of the Modern Wind Turbine," ch. 3, in David A. Spera, ed., *Wind Turbine Technology: Fundamental Concepts of Wind Turbine Engineering* (New York: ASME Press, 1994) pp. 116–117.

13. D.A. Spera, L.A. Viterna, T.R. Richards, and H.E. Neustadter, "Preliminary Analysis of Performance and Loads Data from the 2-Megawatt MOD-1 Wind Turbine Generator," NASA TM-81408 (1979), pp. 1–3; L.A. Viterna, "The NASA-LeRC Wind Turbine Sound Prediction Code," presented at *Second DOE/NASA Wind Turbine Dynamics Workshop, Cleveland, Feb. 24–26, 1981*.

mountainous terrain. These potential problems were among those identified by an initial study undertaken by NASA Lewis, General Electric, and the Solar Energy Research Institute. An analytical model developed at NASA Lewis of acoustic emissions from the rotor recommended that the rotor speed be reduced from 35 to 23 revolutions per minute, and the 2,000-kilowatt generator was replaced with a 1,350-kilowatt, 1,200-rpm generator. This change to the power train made a significant reduction in measured rotor noise. During the noise testing, however, the Mod-1, like the Mod-0A, experienced a failure in the low-speed shaft of the drive train and, because NASA engineers determined that both machines had accomplished their purposes, they were removed from the utility sites. Lessons learned from the engineering studies and testing of the first-generation wind turbine systems indicated the need for technological improvements to make the machines more acceptable for large utility applications. These lessons proved valuable in the design, construction, and operation of the next generation of DOE–NASA wind turbines. Other contributions from the Mod-1 program included low-cost wind turbine design concepts and metal and composite blade design and fabrication. Also, computer codes were verified for dynamic and loads analysis.

Although the Mod-1 was a one-of-kind prototype, there was a conceptual design that was designated as the Mod-1A. The conceptual design incorporated improvements identified during the Mod-1 project but, because of schedule and budget constraints, were not able to be used in fabrication of the Mod-1 machine. One of the improvements involved ideas to lessen the weight of the wind turbine. Also, one of the proposed configurations made use of a teetered hub and upwind blades with partial span control. Although the Mod-1A was not built, many of the ideas were incorporated into the second- and third-generation DOE–NASA wind turbines.

Second-Generation DOE–NASA Wind Turbine Systems (Mod-2)

While the primary objectives of the Mod-0, Mod-0A, and Mod-1 programs were research and development, the primary goal of the second-generation Mod-2 project was for direct and efficient commercial application. The Mod-2 program was designed to determine the potential cost-effectiveness of megawatt-sized remote site operation wind turbines when located in areas of moderate (14 mph) winds. Significant changes from the Mod-0 and Mod-1 included use of a soft-shell-type

Case 13 | Good Stewards: NASA's Role in Alternative Energy

DOE–NASA Mod-2 megawatt wind turbine cluster, Goldendale, WA. NASA.

tower, an epicyclic gearbox, a quill shaft to attenuate torque and power oscillations, and a rotor designed primarily to commercial steel fabrication standards. Other significant changes were the switch from fixed to a teetered (pivot connection) hub rotor, which reduced rotor fatigue, weight, and cost; use of tip control rather than full span control; and orienting the rotor upwind rather than downwind, which reduced rotor fatigue and resulted in a 2.5-percent increase in power produced by the system. Each of these changes resulted in a favorable decrease in the cost of electricity. One of the more important changes, as noted in a Boeing conference presentation, was the switch from the stiff truss type tower to a soft shell tower that weighed less, was much cheaper to fabricate, and enabled the use of heavy but economical and reliable rotor designs.[14]

Four primary Mod-2 wind turbine units were designed, built, and operated under the second-generation phase of the DOE–NASA program. The first three machines were built as a cluster at Goldendale, WA, where the Department of Energy selected the Bonneville Power

---

14. Richard R. Douglas, "Large Wind Turbine Design Characteristics and R&D Requirements," NASA CP-2106; DOE Publication CONF-7904111 (workshop held at Cleveland, OH, Apr. 24–26, 1979), pp. 61–65.

DOE–NASA Mod-2 wind turbine nacelle and rotor. NASA.

Administration as the participating utility. The operation of several wind turbines at one site afforded NASA the opportunity to study the effects of single and multiple wind turbines operating together while feeding into a power network. The Goldendale project demonstrated the successful operation of a cluster of large NASA Mod-2 horizontal-axis wind turbines operating in an unattended mode within a power grid. For construction of these machines, DOE–NASA awarded a competitively bid contract in 1977 to Boeing. The first of the three wind turbines started operation in November 1980, and the two additional machines went into service between March and May 1981. As of January 1985, the three-turbine cluster had generated over 5,100 megawatthours of electricity while synchronized to the power grid for over 4,100 hours. The Mod-2 machines had a rated power of 2.5 megawatts, a rotor-blade diameter of 300 feet, and a hub height (distance of the center of blade rotation to the ground) of 300 feet. Boeing evaluated a number of design options and tradeoffs, including upwind or downwind rotors, two- or three-bladed rotors, teetered or rigid hubs, soft or rigid towers, and a number of different drive train and power generation configurations. A fourth 2.5-megawatt Mod-2 wind turbine was purchased by the Department of the Interior, Bureau of Reclamation, for installation near Medicine Bow, WY, and a fifth turbine unit was purchased by Pacific Gas and Electric for operation in Solano County, CA.[15]

System Verification Units

In addition to the DOE–NASA units, NASA Lewis participated with the Bureau of Reclamation in the experimentation with two other turbines near Medicine Bow, WY. Both of these machines were designated as system verification units (SVU) because of their purpose of verifying the concept of integrating wind turbine generators with hydroelectric power networks. This was viewed as an important step in the Bureau of Reclamation's long-range program of supplementing hydroelectric power generation with wind turbine power generation. One of the two turbines was a new design developed by the Hamilton Standard Division of United Technologies Corp., a 4-megawatt WTS-4 system, in the Medicine Bow area. A Swedish company, Karlskronavarvet (KKRV), was selected as a major subcontractor responsible for the design and fabrication of the nacelle hardware. The WTS-4 had a two-blade fiberglass downwind rotor that was 256.4 feet in diameter. For over 20 years, this 4-megawatt machine remained the largest power rated wind turbine generator ever built. In a reverse role, an additional 3-megawatt version of the same machine was built for the Swedish government, with KKRV as the prime contractor and Hamilton Standard as the subcontractor.[16]

The other SVU turbine was a Mod-2 design. While NASA engineers determined that the initial Mod-2 wind turbine generator performance was acceptable, they noted areas where improvement was needed. The problems encountered were primarily hardware-oriented and were attributed to fabrication or design deficiencies. Identification of these problems led to a number of modifications, including changes in the hydraulic, electric, and control systems; rework of the rotor hub flange; addition of a forced-lubrication system; and design of a new low-speed shaft.

---

15. Darrell H. Baldwin and Jerry Kennard, "Development of Large, Horizontal-Axis Wind Turbines," NASA TM-86950 (1985), pp. 3–5 and Table 1.
16. Ibid., p. 4; and telephone interview of Larry Viterna by author, Apr. 27, 2009.

DOE–NASA Mod-5B 3.2-megawatt wind turbine, Oahu, HI. NASA.

Third-Generation Advanced Multimegawatt Wind Turbines—The Mod-5 Program (1980–1988)

The third-generation (Mod-5) program, which started in 1980, was intended to incorporate the experiences from the earlier DOE–NASA wind turbines, especially the Mod-2 experiences, into a final proof-of-concept system for commercial use by an electric utility company. Two construction contracts were awarded to build the Mod-5 turbines—one unit to General Electric, which was designated the Mod-5A, and one unit to Boeing, which was designated the Mod-5B. As intermediate steps between the Mod-2 and Mod-5, two conceptual studies were undertaken for fabrication of both an advanced large wind turbine designated the Mod-3 and a medium turbine designated the Mod-4. Likewise, both a large-scale Mod-5 and medium-scale Mod-6 were planned as the final Wind Energy Program turbines. The Mod-3 and Mod-4 studies, however, were not carried through to construction of the turbines, and the Mod-6 program was canceled because of budget constraints and changing priorities resulting from a decline in oil prices following the end of the oil crisis of the 1970s. Also, General Electric chose not to proceed beyond the design phase with its Mod-5A. As a result, only the Boeing Mod-5B was constructed and placed into power utility service.[17]

Although its design was never built, General Electric did complete the detailed design work and all of the significant development tests and documented the entire Mod-5A program. The planned Mod-5A system contained many interesting features that NASA Lewis chose to preserve for future reference. The Mod-5A wind turbine was expected to generate electricity at a cost competitive with conventional forms of power generation once the turbines were in volume production. The program was divided into three phases: conceptual design, which was completed in March 1981; preliminary design, which was completed in May 1982; and final design, which was started in June 1982. The Mod-5A was planned to have a 7.3-megawatt generator, a 400-foot-diameter two-bladed teetered rotor, and hydraulically actuated ailerons over the outboard 40 percent of the blade span to regulate the power and control shutdown. The blades were to be made of epoxy-bonded wood laminates. The yaw drive was to include a hydraulically actuated disk brake system, and the tower was to be a soft-designed welded steel plate cylindrical shell with a conical base. The Mod-5A was designed to operate in wind speeds of between 12 and 60 mph at hub height. The system was designed for automatic unattended operation and for a design life of 30 years.[18]

The Mod-5B, which was the only Mod-5 unit built, was physically the world's largest wind turbine generator. The Mod-5B represented very advanced technology, including an upwind teetered rotor, compact planetary gearbox, pitchable tip blade control, soft-shell-type tower, and a variable-speed electrical induction generator/control system. Variable speed control enabled the turbine speed to vary with the wind speed, resulting in an increase energy capture and a decrease in fatigue loads on the drive train. The system underwent a number of design changes before the final fabricated version was built. For example, the turbine originally was planned to have a blade swept diameter of 304 feet. This was increased to 420 feet and finally reduced to 320 feet because of the use of blade steel tips and control improvements. Also, the turbine generator was planned initially to be rated at 4.4 megawatts. This

---

17. Divone, "Evolution of Modern Wind Turbines," ch. 3, in Spera, *Wind Turbine Technology*, p. 130.
18. General Electric Company, Advanced Energy Programs Department, "MOD-5A Wind Turbine Generator Program Design Report," vol. 1, executive summary, NASA CR-174734 (1984), pp. 1–6.

was increased to 7.2 megawatts and then decreased to the final version 3.2 megawatts because of development of better tip control and load management. The rotor weighed 319,000 pounds and was mounted on a 200-foot tower. Extensive testing of the Mod-5B system was conducted, including 580 hours of operational testing and 660 hours of performance and structural testing. Performance testing alone generated over 72 reports reviewing test results and problems resolved.[19]

The Mod-5B was the first large-scale wind turbine to operate successfully at variable rotational speeds, which varied from 13 to 17.3 revolutions per minute depending on the wind speed. In addition, the Mod-5B was the first large wind turbine with an apparent possibility of lasting 30 years. The turbine, with a total system weight of 1.3 million pounds, was installed at Kahuku on the north shore of Oahu, HI, in 1987 and was operated first by Hawaiian Electric Incorporated and later by the Makani Uwila Power Corporation. The turbine started rated power rotation July 1, 1987. In January 1988, the Mod-5B was sold to the power utility, which continued to operate the unit as part of its power generation network until the small power utility ceased operations in 1996. In 1991, the Mod-5B produced a single wind turbine record of 1,256 megawatthours of electricity. The Mod-5B was operated in conjunction with 15 Westinghouse 600-kilowatt wind turbines. While the Westinghouse turbines were not part of the NASA program, the design of the turbines combined successful technology from NASA's Mod-0A and Mod-2 programs.[20]

The Mod-5B, which represented a significant decrease over the Mod-2 turbines in the cost of production of electricity, was designed for the sole purpose of providing electrical power for a major utility network. To achieve this goal, a number of changes were made over the Mod-2 systems, including changes in concepts, size, and design refinements. These changes were reflected in more than 20 engineering studies, which addressed issues such as variable pitch versus fixed pitch, optimum machine size, steel shell versus truss tower, blade aerodynamics, material selection, rotor control, tower height, cluster optimization, and

---

19. Boeing Aerospace Company, "MOD-5B Wind Turbine System Final Report," vol. 1, executive summary, NASA CR-180896 (1988), p. 2–2, and vol. 3, "Acceptance Testing," NASA CR-180898 (1988), pp. v–vi, 1–1, 8–1, and B–1.
20. Divone, "Evolution of Modern Wind Turbines," in Spera, *Wind Turbine Technology*, pp. 130–131.

gearbox configuration. For example, the studies indicated that loads problem was the decisive factor with regard to the use of a partial span variable pitch system rather than a fixed pitch rotor system, dynamic simulation led to selection of the variable speed generator, analysis of operational data enabled a significant reduction in the weight and size of the gearbox, and the development of weight and cost trend data for use in size optimization studies resulted in the formulation of machine sizing programs.[21]

A number of design elements resulted in significant contributions to the success of the Mod-5B wind turbine. Aerodynamic improvement over the Mod-2, including improvements in vortex generators, trailing edge tabs, and better shape control, resulted in an 18-percent energy capture increase. Improved variable speed design resulted in an increase of greater than 7 percent (up to as high as 11 percent) over an equivalent synchronous generator system. Both cycloconverter efficiency and control optimization of rotor speed versus wind speed proved to be better than anticipated. Use of the variable speed generator system to control power output directly, as opposed to the pitch power control on the Mod-2, substantially reduced blade activity, especially at below rated power levels. The variable speed design also resulted in a substantial reduction in structural loads. Adequate structural integrity was demonstrated for all stress measurement locations. Lessons learned during the earlier operation of the Mod-2 systems resulted in improved yaw and pitch systems. Extensive laboratory simulation of control hardware and software likewise reduced control problems compared with Mod-2 systems.[22] In summary, the Mod-5B machine represented a reliable proof-of-concept large horizontal-axis wind turbine conversion system capable of long-life production of electricity into a power grid system, thus fulfilling the DOE–NASA program objectives.

The Mod-5B was the last DOE–NASA wind turbine generator built under the Federal Wind Energy Program. In his paper on the Mod-5B wind turbine system, Boeing engineer R.R. Douglass noted the following size versus cost problem relating to the purchase of large wind turbines faced by power utility companies:

---

21. R.R. Douglas, Boeing Engineering and Construction Company, "Conceptual Design of the 7 Megawatt Mod-5B Wind Turbine Generator Final Report," N83-19272 (1982), pp. 1–7; and Boeing Aerospace Company, "Mod-5B Wind Turbine System Final Report," vol. 3, pp. v, vi, 1-1, 8-1, and B-1.
22. "Mod-5B Wind Turbine Generator Final Report," vol. 1, pp. 7-1 and 7-2.

... large scale commercialization of large wind turbines suffers from the chicken and egg syndrome. That is, costs of units are so high when produced one or two at a time on prototype tooling that the utilities can scarcely afford to buy them. On the other hand, industry cannot possibly afford to invest the huge capital required for an automated high rate production capability without an established order base. To break this log jam will require a great deal of cooperation between government, industry, and the utilities.[23]

Boeing noted, however, in its final Mod-5B report that: "In summary the Mod-5B demonstrated the potential to generate at least 11 percent more revenue at a given site than the original design goal. It also demonstrated that multi-megawatt class wind turbines can be developed with high dependability which ultimately should show up in reduced operation and maintenance costs."[24]

NASA's Wind Turbine Supporting Research and Technology Contributions

A very significant NASA Lewis contribution to wind turbine development involved the Center's Supporting Research and Technology (SR&T) program. The primary objectives of this component of NASA's overall wind energy program were to gather and report new experimental data on various aspects of wind turbine operation and to provide more accurate analytical methods for predicting wind turbine operation and performance. The research and technology activity covered the four following areas: (1) aerodynamics, (2) structural dynamics and aeroelasticity, (3) composite materials, and (4) multiple wind turbine system interaction. In the area of aerodynamics, NASA testing indicated that rounded blade tips improved rotor performance as compared with square rotor tips, resulting in an increase in peak rotor efficiency by approximately 10 percent. Also in the aerodynamics area, significant improvements were made in the design and fabrication of the rotor blades. Early NASA rotor blades used standard airfoil shapes from the aircraft industry, but wind turbine rotors operated over a significantly wider range of angles of attack (angles between the centerline of the blade and incoming airstream). The rotor

---

23. Douglass, "Conceptual Design of the 7 Megawatt Mod-5 Wind Turbine Generator," p. 6.
24. Boeing Aerospace Company, "Mod-5B Wind Turbine System Final Report," vol. 1, executive summary, NASA CR-180896 (Mar. 1988), p. 8–1.

blades also needed to be designed to last up to 20 or 30 years, which represented a challenging problem because of the extremely high number of cyclic loads involved in operating wind turbines. To help solve these problems, NASA awarded development grants to the Ohio State University to design and wind tunnel test various blade models, and to the University of Wichita to wind tunnel test a rotor airfoil with ailerons.[25]

In the structural dynamics area, NASA was presented with problems related to wind loading conditions, including wind shear (variation of wind velocity with altitude), nonuniform wind gusts over the swept rotor area, and directional changes in the wind velocity vector field. NASA overcame this problem by developing a variable speed generator system that permitted the rotor speed to vary with the wind condition, thus producing constant power.

Development work on the blade component of the wind turbine systems, including selecting the material for fabrication of the blades, represents another example of supporting technology. As noted above, NASA Lewis brought considerable structural design expertise in this area to the wind energy program as a result of previous work on helicopter rotor blades. Early in the program, NASA tested blades made of steel, aluminum, and wood. For the 2-megawatt Mod-1 phase of the program, however, NASA Lewis decided to contract with the Kaman Aerospace Corporation for the design, manufacture, and ground-testing of two 100-foot fiberglass composite blades. NASA provided the general design parameters, as well as the static and fatigue load information, required for Kaman to complete the structural design of the blades. As noted in Kaman's report on the project, the use of fiberglass, which later became the preferred material for most wind turbine blades, had a number of advantages, including nearly unlimited design flexibility in adopting optimum planform tapers, wall thickness taper, twist, and natural frequency control; resistance to corrosion and other environmental effects; low notch sensitivity with slow failure propagation rate; low television interference; and low cost potential because of adaptability to highly automated production methods.[26]

---

25. Darrell H. Baldwin and Bradford S. Linscott, "The Federal Wind Program at NASA Lewis Research Center," NASA TM-83480 (1983), pp. 5–9.
26. W.R. Batesole, Kaman Aerospace Corporation, "Fiberglass Composite Blades for the 2 MW Mod-1 Wind Turbine Generator," NASA CR-2230, Cleveland, OH, July 28–30, 1981, pp. 215–226.

The above efforts resulted in a significant number of technical reports, analytical tests and studies, and computer models based upon contributions of a number of NASA, university, and industry engineers and technicians. Many of the findings grew out of tests conducted on the Mod-0 testbed wind turbine at Plum Brook Station. One is work done by Larry A. Viterna, a senior NASA Lewis engineer working on the wind energy project, in aerodynamics. In studying wind turbine performance at high angles of attack, he developed a method (often referred to as the Viterna method or model) that is widely used throughout the wind turbine industry and is integrated into design codes that are available from the Department of Energy. The codes have been approved for worldwide certification of wind turbines. Tests with the Mod-0 and Gedser wind turbines formed the basis for his work on this analytical model, which, while not widely accepted at the time, later gained wide acceptance. Twenty-five years later, in 2006, NASA recognized Larry Viterna and Bob Corrigan, who assisted Viterna on data testing, with the Agency's Space Act Award from the Inventions and Contributions Board.[27]

Benefits of NASA's "Good Stewardship" Regarding the Agency's Participation in the Federal Wind Energy Program

NASA Lewis's involvement in the Federal Wind Energy Program from 1974 through 1988 brought a high degree of engineering experience and expertise to the project that had a lasting impact on the development of use of wind energy both in the United States and internationally. During this program, NASA developed the world's first multimegawatt horizontal-axis wind turbines, the dominant wind turbine design in use throughout the world today.

NASA Lewis was able to make a quick start and contribution to the program because of the Research Center's longstanding experience and expertise in aerodynamics, power systems, materials, and structures. The first task that NASA Lewis accomplished was to bring forward and document past efforts in wind turbine development, including work undertaken by Palmer Putnam (Smith-Putnam wind turbine), Ulrich Hütter (Hütter-Allgaier wind turbine), and the Danish

---

27. Viterna and David C. Janetzke, "Theoretical and Experimental Power From Large Horizontal-Axis Wind Turbines," NASA TM-82944 (Sept. 1982); and "Wind Energy Reaps Rewards," NASA release dated May 31, 2006, http://www.nasa.gov/vision/earth/technologies/wind_turbines_prt.htm, accessed Aug. 20, 2009.

Gedser mill. This information and database served both to get NASA Lewis involved in the Wind Energy Program and to form an initial data and experience foundation to build upon. Throughout the program, NASA Lewis continued to develop new concepts and testing and modeling techniques that gained wide use within the wind energy field. It documented the research and development efforts and made this information available for industry and others working on wind turbine development.

Lasting accomplishments from NASA's program involvement included development of the soft shell tubular tower, variable speed asynchronous generators, structural dynamics, engineering modeling, design methods, and composite materials technology. NASA Lewis's experience with aircraft propellers and helicopter rotors had quickly enabled the Research Center to develop and experiment with different blade designs, control systems, and materials. A significant blade development program advanced the use of steel, aluminum, wood epoxy composites, and later fiberglass composite blades that generally became the standard blade material. Finally, as presented in detail above, NASA was involved in the development, building, and testing of 13 large horizontal-axis wind turbines, with both the Mod-2 and Mod-5B turbines demonstrating the feasibility of operating large wind turbines in a power network environment. With the end of the energy crisis of the 1970s and the resulting end of most U.S. Government funding, the electric power market was unable to support the investment in the new large wind turbine technology. Development interest moved toward the construction and operation of smaller wind turbine generators for niche markets that could be supported where energy costs remained high.

NASA Lewis's involvement in the wind energy program started winding down in the early 1980s, and, by 1988, the program was basically turned over to the Department of Energy. With the decline in energy prices, U.S. turbine builders generally left the business, leaving Denmark and other European nations to develop the commercial wind turbine market.

While NASA Lewis had developed a 4-megawatt wind turbine in 1982, Denmark had developed systems with power levels only 10 percent of that at that time. However, with steady public policy and product development, Denmark had captured much of the $15 billion world market by 2004.

TABLE 1			
COMPARATIVE WIND TURBINE TECHNOLOGICAL DEVELOPMENT, 1981–2007			
TURBINE TYPE	Nibe A	NASA WTS-4	Vestas
YEAR	1981	1982	2007
COUNTRY OF ORIGIN	Denmark	United States	Denmark
POWER (IN KW)	630	4,000	1,800
TIP HEIGHT (FEET)	230	425	355
POWER REGULATION	Partial pitch	Full pitch	Full pitch
BLADE NUMBER	3	2	3
BLADE MATERIAL	Steel/fiberglass	Fiberglass	Fiberglass
TOWER STRUCTURE	Concrete	Steel tubular	Steel tubular
Source: Larry A. Viterna, NASA.			

Most of the technology developed by NASA, however, continued to represent a significant contribution to wind power generation applicable both to large and small wind turbine systems. In recent years, interest has been renewed in building larger-size wind turbines, and General Electric, which was involved in the DOE–NASA wind energy program, has now become the largest U.S. manufacture of wind power generators and, in 2007, was among world's top three manufacturers of wind turbine systems. The Danish company Vestas remained the largest company in the wind turbine field. GE products currently include 1.5-, 2.5-, and, for offshore use, 3.6-megawatt systems. New companies, such as Clipper Wind Power, with its manufacturing plant in Cedar Rapids, IA, and Nordic Windpower also have entered the large turbine fabrication business in the United States. Clipper, which is a U.S.–U.K. company, installed its first system at Medicine Bow, WY, which was the location of a DOE–NASA Mod-2 unit. In the first quarter of 2007, the company installed eight commercial 2.5-megawatt Clipper Liberty machines. Nordic Windpower, which represents a merger of Swedish, U.S., and U.K. teams, markets its 1-megawatt unit that encompasses a two-bladed teetered rotor that evolved from the WTS-4 wind turbine under the NASA Lewis program.

In summary, NASA developed and made available to industry significant technology and turbine hardware designs through its "good stewardship" of wind energy development from 1974 through 1988. NASA thus played a leading role in the international development and utilization of wind power to help address the Nation's energy needs today. In doing so, NASA Lewis fulfilled its primary wind program goal of developing and transferring to industry the technology for safe, reliable, and environmentally acceptable large wind turbine systems capable of generating significant amounts electricity at cost competitive prices. In 2008, the United States achieved the No. 1 world ranking for total installed capacity of wind turbine systems for the generation of electricity.

Solar Propulsion for High-Altitude Long-Endurance Unmanned Aerial Vehicles

Another area of NASA involvement in the development and use of alternative energy was work on solar propulsion for High-Altitude Long-Endurance (HALE) unmanned aerial vehicles (remotely piloted vehicles). Work in this area evolved out of the Agency's Environmental Research Aircraft and Sensor Technology (ERAST) program that started in 1994. This program, which was a joint NASA/industry effort through a Joint Sponsored Research Agreement (JSRA), was under the direction of NASA's Dryden Flight Research Center. The primary objectives of the ERAST program were to develop and transfer advanced technology to an emerging American unmanned aerial vehicle industry, and to conduct flight demonstrations of the new technologies in controlled environments to validate the capability of UAVs to undertake operational science missions. A related and important aspect of this mission was the development, miniaturization, and integration of special purpose sensors and imaging equipment for the solar-powered aircraft. These goals were in line with both the revolutionary vehicles development aspect of NASA's Office of Aerospace Technology aeronautics blueprint and with NASA's Earth Science Enterprise efforts to expand scientific knowledge of the Earth system using NASA's unique capabilities from the standpoint of space, aircraft, and onsite platforms.[28]

Specific program objectives were to develop UAV capabilities for flying at extremely high altitudes and for long periods of time; demonstrate payload capabilities and sensors for atmospheric research; address and resolve UAV certification and operational issues; demonstrate the UAV's usefulness to scientific, Government, and civil customers; and foster the emergence of a robust UAV industry in the United States.[29]

---

28. "ERAST: Environmental Research Aircraft and Sensor Technology," NASA FS-2002-08-020 DFRC, pp. 1–3.
29. Mishap Investigation Board, "Investigation of the Helios Prototype Aircraft Mishap," vol. 1 (Mishap Report), NASA (Jan. 2004), p. 16.

The ERAST program envisioned missions that included remote sensing for Earth science studies, hyperspectral imaging for agriculture monitoring, tracking of severe storms, and serving as telecommunications relay platforms. Related missions called for the development and testing of lightweight microminiaturized sensors, lightweight materials, avionics, aerodynamics, and other forms of propulsion suitable for extreme altitudes and flight duration.[30]

The ERAST program involved the development and testing of four generations of solar-powered UAVs, including the Pathfinder, the Pathfinder Plus, the Centurion, and the Helios Prototype. Because of budget limitations, the Helios Prototype was reconfigured in what could be considered a fifth-generation test vehicle for long-endurance flying (see below). Earlier UAVs, such as the Perseus, Theseus, and Proteus, relied on gasoline-powered engines. The first solar-powered UAV was the RAPTOR/Pathfinder, also known as the High-Altitude Solar (HALSOL) aircraft, which that was originally developed by the U.S. Ballistic Missile Defense Organization (BMDO—now the Missile Defense Agency) as part of a classified Government project and subsequently turned over to NASA for the ERAST program. In addition to BMDO's interest in having NASA take over solar vehicle development, a workshop held in Truckee, CA, in 1989 played an important role in the origin of the ERAST program.

The Truckee Workshop and Conference Report

In July 1989, NASA Ames sponsored a workshop on requirements for the development and use of very high-altitude aircraft for atmospheric research. The primary objectives of the workshop were to assess the scientific justification for development of new aircraft that would support stratospheric research beyond the altitudes attainable by NASA's ER-2 aircraft and to determine the aircraft characteristics (ceiling, altitude, payload capabilities, range, flight duration, and operational capabilities) required to perform the stratospheric research missions. Approximately 35 stratospheric scientists and aircraft design and operations experts attended the conference, either as participants or as observers. Nineteen of these attendees were from NASA (1 for NASA Langley, 16 from NASA Ames, and 2 representing both NASA Dryden and Ames); 4 were from universities and institutes, including Harvard University and Pennsylvania

---

30. "Pathfinder Solar-Powered Aircraft" NASA, Dryden Flight Research Center Fact Sheet, FS-034.

State University; and 6 represented aviation companies, including Boeing Aerospace, Aurora Flight Sciences and Lockheed. Crofton Farmer, representing the Jet Propulsion Laboratory, served as workshop chair, and Philip Russell, from NASA Ames, was the workshop organizer and report editor. The attendees represented a broad range of expertise, including 9 aircraft and development experts, 3 aircraft operations representatives, 2 aeronautical science experts, 2 Earth science specialists, 1 instrument management expert (Steven Wegener from NASA Ames, who later directed the science and payload projects for the solar UAV program), 1 general management observer, and 17 stratospheric scientists.[31]

The workshop considered pressing scientific questions that required advanced aircraft capabilities in order to accomplished a number of proposed science related missions, including: (1) answering important polar vortex questions, including determining what causes ozone loss above the dehydration region in Antarctica and to what extent the losses are transmitted to the middle latitudes; (2) determining high-altitude photochemistry in tropical and middle latitudes; (3) determining the impact and degree of airborne transport of certain chemicals; and (4) studying volcanic, stratospheric cloud/aerosol, greenhouse, and radiation balance. The workshop concluded that carrying out the above missions would require flights at a cruise altitude of 100,000 feet, the ability to make a round trip of between 5,000 and 6,000 nautical miles, the capability to fly into the polar night and over water more than 200 nautical miles from land, and carry a payload equal to or greater than the ER-2. The workshop report noted that experience with satellites pointed out the need for increased emphasis on correlative measurements for current and future remote sensing systems. Previously, balloons had provided most of this information, but balloons presented a number of problems, including a low frequency of successful launches, the small number of available launch sites worldwide, the inability to follow selected paths, and the difficulty in recovering payloads. The workshop concluded with the following finding:

> We recommend development of an aircraft with the capacity to carry integrated payloads similar to the ER-2 to significantly higher altitude preferably with greater range. It is important

---

31. "Global Stratospheric Change: Requirements for a Very-High Altitude Aircraft for Atmospheric Research," (workshop held in Truckee, CA, July 15–16, 1989), NASA CP-10041 (1989), p. xi.

that the aircraft be able to operate over the ocean and in the polar night. This may dictate development of an autonomous or remotely piloted plane. There is a complementary need to explore strategies that would allow payloads of reduced weight to reach even higher altitude, enhancing the current capability of balloons.[32]

High-altitude, long-duration vehicle development, along with development of reduced weight instrumentation, both became goals of the ERAST program.

Joint Sponsored Research Agreement: "The ERAST Alliance"

The ERAST program was organized pursuant to a unique arrangement known as a Joint Sponsored Research Agreement (JSRA).[33] This type of agreement was authorized by the National Aeronautics and Space Act of 1958, and the specific ERAST agreement was authorized under the NASA Administrator's delegations of March 31, 1992, and February 25, 1994. The purpose of the agreement was to: (1) develop and demonstrate UAV flight capability at altitudes up to 100,000 feet and up to 4 days' duration; (2) further develop payload integration capabilities responsive to the data collection and measurement requirements of the atmospheric community; (3) research activity toward further resolution of UAV certification and civil operational issues; (4) further demonstrate UAV viability to scientific, Government, and civil users, leading to increased applications for UAVs; and (5) effect technology transfers to the parties to develop a robust United States UAV industry capable of asserting the lead as the premier provider of UAVs for government and civil users worldwide. The agreement, which became effective in September 1994, provided for the terms and conditions of the arrangement, various participation categories,

---

32. "Global Stratospheric Change," workshop held in Truckee, CA, p. 2.
33. The ERAST program was the first large NASA program to use a Joint Sponsored Research Agreement. The JSRA type arrangement was first researched as a new form of contracting option in the 1980s at NASA Ames. Ames had previously used this type of joint arrangement on three smaller projects. Jack Glazer, a NASA Ames attorney, started research in 1988 on the possibility of using the "Other Transaction Authority" clause in the Space Act of 1958 for application to collaborative projects. Glazer recommended creating AmTech as a nonprofit company to document JSRA legal and policy issues and to establish prototype projects. (See "The Promise of ERAST," NASA Dryden, Oct. 2004.)

preliminary budgets for the first 5 years, and operational and reporting requirements.[34]

The agreement established an Alliance Council, which was to meet at least twice a year, to coordinate with NASA Dryden's ERAST Project Office on planning the research and development and flight-testing to be performed. The joint agreement provided for ERAST management through a NASA program manager and a NASA ERAST project manager. Required reports included an annual report, R&D/technical reports, monthly progress reports, intellectual property reports, commercialization reports, and management and financial reports. While actual program expenditures could and did vary, NASA's projected financial commitments from 1994 through 2000 were $2.8 million for 1994, $5.75 million for 1995, $6.05 million for 1996, $6.35 million for 1997, $6.70 million for 1998, $7.25 million for 1999, and $7.25 million for 2000. Finally, the terms and conditions of the agreement provided for extensions through December 31, 2000.[35]

The above arrangement, however, actually remained in effect after 2000. In 2002, NASA entered into a followup joint agreement with AeroVironment, Inc., of Simi Valley, CA, including its SkyTower subsidiary. The new agreement was intended to streamline existing efforts to merge solar-powered UAV development into a single solar-electric platform program with the goal of developing multiple aircraft. This collaborative effort included continued development of the Helios Prototype.[36]

Industry partners that participated in the JSRA program included four primary companies—AeroVironment, Inc. (builder of the four solar prototype UAVs), Aurora Flight Sciences (manufacture of the Perseus B), General Atomics Aeronautical Systems (builder of the Altus 2), and Scaled Composites (developer of the Proteus). American Technology Alliances (AmTech) served as facilitator for the alliance, and Karen Risa Robbins, a founder of AmTech, played a primary role in development and acceptance of the ERAST JSRA. Of the above companies, AeroVironment was the primary one involved in the solar-powered part of the ERAST pro-

---

34. "Joint Sponsored Research Agreement: Environmental Research Aircraft & Sensor Technology: 'ERAST Alliance' for High Altitude, Long Endurance Unmanned Aerial Vehicles," (Office of Aeronautics, NASA, Aug. 1994), pp. 1 and 39.
35. Ibid. p. 44.
36. "NASA Dryden, AeroVironment Sign Solar/Electric Aircraft Pact," NASA Dryden News Release 02-48 (Aug. 19, 2002).

gram. There were up to 28 participants in the alliance, including small businesses, universities, and nonprofit organizations. NASA also worked closely with the Federal Aviation Administration to address a program goal of resolving issues related to operation of UAVs in the National Airspace System, including development of "see and avoid" sensors and "over-the-horizon" communications equipment. Under the joint agreement, NASA was able to provide program management and oversight, flight-test facilities, operational support, and project funding. The funding aspect of the joint NASA-industry effort was facilitated because the program was permitted to use Federal Acquisition Regulations as guidelines rather than as rules. Furthermore, NASA safety regulations were not required to be specifically followed.[37] As ERAST project manager, NASA Dryden was responsible for the setting of priorities, determination of technical approaches toward meeting project objectives, project funding and oversight, coordination of facilities for UAV operations, development and coordination of payloads for test flights, and foresight to ensure that actions taken by ERAST alliance partners satisfied NASA's future needs for UAVs. Each company in the alliance made contributions to the project through combinations of money and services. The ERAST program, however, required only nominal funding by the companies, and, in order to further commercial development of HALE UAVs, NASA offered the companies ownership of all hardware developed by the program.[38]

Jenny Baer-Riedhart, NASA Dryden ERAST Program Manager for the first 4 years, described the NASA-industry working relationship under the joint agreement as follows:

> NASA and the companies agreed on business plans at the annual alliance meeting. Each year at this meeting, I laid out the requirements for the program, based on input from all of the parties. Together, we evaluated our working business plan against these requirements. We set programmatic milestones, as well as milestones for each of the companies.[39]

---

37. Mishap Investigation Board, "Investigation of Helios Prototype Aircraft Mishap," p. 16.
38. Alexander Laufer, Todd Post, Edward J. Hoffman, and Ronald A. Heifotz, *Shared Voyage: Learning and Unlearning from Remarkable Projects* (Washington, DC: NASA History Division, 2005), ch. 4, "Flying High on Spirit: The Pathfinder Solar-Powered Airplane," p. 125.
39. Ibid., p. 130.

Baer-Riedhart added that NASA and the companies put funding into a shared bank account from which AmTech, acting as the go-between for the companies and NASA, distributed the funds to the parties. She noted that, at first, the companies wanted to get their own money, build their own aircraft, and have a flyoff, but that NASA's vision from the start of the alliance was for the companies to get together to build one aircraft. Jeffrey Bauer, who was the Chief Engineer at Dryden and later served as the last ERAST Program Manager, credits the success of the program to the structure and partnerships that formed the alliance, noting: "One of the major attributes of the program is the alliance of government and industry. ERAST is not a contract. We work collectively to develop what's best for the group and community."[40] John Del Frate, Dryden solar-power aircraft manager, commenting on the alliance, stated: "The technology early on was immature. We knew there would be problems, but the foundation of the program was built on the premise that we were allowed to take risks, and that made it very successful."[41] In addition to Baer-Riedhart, Del Frate, and Bauer, other NASA Dryden senior ERAST program/project managers included James Stewart, John Sharkey, and John Hicks.

Adding an industry perspective to the working relationship between the ERAST alliance partners, Ray Morgan, then vice president of AeroVironment, a company that had over 13 years of experience developing UAVs, noted: "Like most new relationships, the alliance went through an initial courtship phase, followed by a few spats, before it settled into an ongoing relationship that worked, more or less, for the good of all." Morgan added that NASA brought considerable expertise to the program, including vast experience in developing and testing unique air vehicles at high altitudes.[42]

One area of NASA expertise that Morgan specifically noted was the advice that NASA provided AeroVironment regarding how best to implement redundant systems for critical components, especially where the systems must automatically determine which sensors are working properly and which ones are not. AeroVironment had used "single thread" systems across major components for the first Pathfinder prototype, meaning that failure of one component would likely cause

---

40. Jay Levine, ed., "The X-Press," vol. 44, issue 2, Dryden Flight Center (May 8, 2002), p. 3.
41. Levine, ed., "The X-Press," vol. 45, issue 4, Dryden Flight Center (June 2003).
42. Laufer, et al., *Shared Voyage*, section 2, ch. 4, p. 129.

failure of the UAV.⁴³ The utilization of redundant systems also extended to other components of vehicle operation. For example, the control systems for the solar-powered UAVs were remotely piloted through a dual radio frequency data link with the vehicle's automatic control system, which likewise achieved redundancy through the use of two identical flight computers, uplink receivers, and downlink transmitters. In addition, there was a triple set of airspeed sensors and dual Global Positioning System (GPS) receivers.⁴⁴ Even the fuel cells were originally to be completely redundant, but this plan was abandoned because of budget limitations and fuel cell development problems. This need for redundant systems in UAVs was reinforced by NASA Dryden's experience with testing UAVs, including a number of program mishaps. The NASA team realized that the chance of mission success was greatly improved through the use of redundant systems for the UAVs.

Pathfinder—First-Generation ERAST Solar Program Test Vehicle (1994–1997)

The first-generation ERAST solar-test program HALE vehicle was the Pathfinder, which was designed and built by AeroVironment. AeroVironment's earlier solar aircraft projects included the building of the piloted Gossamer Albatross and a scaled-down version known as the Gossamer Penguin. This experience assisted the company in building the Pathfinder UAV. In addition to Ray Morgan, who was vice president of AeroVironment, the company team included a number of experienced engineers and technicians, including William Parks, who was the company's chief engineer for the Centurion and Helios Prototype UAVs, and Robert Curtin and Kirk Flittie, who both served as project and later as program managers. The program brought honors for Bob Curtin and Ray Morgan, who both received the *Aviation Week* Laurel Award in 1996.⁴⁵

Pathfinder, which initially was battery-powered, was a remotely piloted flying wing that demonstrated a number of technologies, including lightweight composite structures, low wing loading flying-wing configuration, redundant and fault tolerant flight control systems,

---

43. Ibid., p.131.
44. Derek L. Lisoski and Mark B. Tischler, "Solar Powered Stratosphere Research Aircraft—Flight Test and Systems Identification," presented at the *RTO SCI Symposium held in Madrid, Spain, May 5–7, 1998*, and published in RTO MP-11, p. 27-4.
45. Laufer, et al., *Shared Voyage*, section 4, ch. 4, p. 224, and information from John Del Frate.

Pathfinder aircraft in flight. NASA.

high-efficiency electric motors, thermal control systems for high-altitude flight, and a high specific power solar array. The remotely piloted Pathfinder had 6 electric motors that each weighed 13 pounds and consisted of a fixed-pitch 79-inch propeller and a solid-state motor with internal power electronics, nacelle, and cooling fins. Differential power to two wingtip motors on either side was used for lateral control. Wing dihedral (upsweep) provided roll stability, and 26 elevator control surfaces were attached to the wing's trailing edge for pitch control. Pathfinder's solar array generated approximately 8,000 watts near solar noon. The solar UAV could obtain an airspeed of between 15 and 25 mph and a cruising speed of between 17 to 20 mph. The vehicle had a length of 12 feet, a wingspan of 98.4 feet, a wing chord (front to rear distance) of 8 feet, a gross weight of approximately 560 pounds, a payload capacity of up to 100 pounds, a wing aspect ratio (the ratio between the wingspan and the wing chord) of 12 to 1, and a power-off glide ratio of 18 to 1. Pathfinder had a maximum bank rate of 5 degrees and a maximum turn rate of 3 degrees per second at sea level and 1.7 degrees at 60,000 feet.[46]

---

46. Lisoski and Tischler, "Solar Powered Stratospheric Research Aircraft—Flight Test and Systems Identification," pp. 27-1 through 27-5; and "Pathfinder Solar-Powered Aircraft," NASA Dryden Flight Research Center, FS-034.

To gain some introductory understanding and experience with the challenges and nuances of solar cell operation, prior to the official start of the ERAST program and the transfer of Pathfinder from BMDO, the project team had arranged for some solar cell flight tests on local sorties over the Edwards dry lake. Pathfinder itself was not equipped with solar arrays on early test flights because the ERAST alliance did not want to risk any damage to the expensive solar cell arrays until Pathfinder's flying capabilities could be tested. Pathfinder's gross weight of 560 pounds produced a very low wing-loading load distribution of less than 0.64 pounds per square foot that significantly increased sensitivity to winds during takeoff and landing. This necessitated special training for the ground controllers, especially during takeoff and landing of the UAV. Pathfinder's first foray to high altitude took it to 50,500 feet and proved immensely productive. "We learned tons from that flight," John Del Frate recalled afterward, noting, "There were a lot of naysayers that were quieted after that flight."[47] Unfortunately, afterward the vehicle dramatically demonstrated its sensitivity to wind, being seriously damaged in its hangar when ground crews opened both hangar doors, thus creating a draft that blew Pathfinder into a jet that was in the same hangar.

After a number of developmental flights and further modifications at Dryden, Pathfinder was transported to the Island of Kauai, which offered a more favorable wind environment and a greater operational area with less competing air traffic. From testing at Edwards Air Force Base, the NASA-industry alliance learned that weather factors—including wind, turbulence, cloud cover, humidity, temperature, and pressure—were critical in attempting to fly the wing-loaded Pathfinder at high altitudes. In addition, the team noted that the UAVs would probably not be flying in the same conditions as found in standard atmosphere reference tables because testing indicated a surprising variance in actual temperature in comparison with the tables. The team also noted that the higher a solar UAV flies, the greater the downwind drift distance if activation of a flight termination system (FTS) is required.[48] These factors required careful study of historical weather patterns to determine the optimum

---

47. Review comments of Del Frate to author, Aug. 4, 2009.
48. As Del Frate recalls: "Flight termination can vary from aircraft to aircraft. We chose to not cripple the aircraft if we had a problem, but rather we would cut power to the motors and use a small chute off-center to provide asymmetrical drag to put PF into a descending spiral in order to not destroy the aircraft. As a result, the wind aloft could carry us quite a ways downwind."

site to attempt to set a world record UAV altitude flight. Accordingly, NASA selected the Navy's Pacific Missile Range Facility in Hawaii as the location to test the high-altitude capabilities of the solar UAVs.[49]

In Hawaii, Pathfinder was flown for seven additional flights, one of which in 1997 set a world record of over 71,500 feet for a high-altitude flight by a propeller-driven aircraft. This broke a 1995 Pathfinder test altitude flight record of 50,500 feet, which had earned NASA recognition as 1 of the 10 most memorable record flights of 1995. Pathfinder test flights also demonstrated solar-powered HALE vehicles' potential as platforms for environmental monitoring and technical demonstration missions by gaining additional information relating to the Island of Kauai's terrestrial and coastal ecosystems. These science missions, which employed specially build lightweight sensor systems (see below), included detection of forest nutrient status, forest regrowth after damage from Hurricane Iniki in 1992, sediment concentrations in coastal waters, and assessment of coral reef conditions. Experience from Pathfinder test flights, in combination with other UAV testing, also yielded a number of lessons regarding hardware reliability, including the following recommended procedures: (1) testing the airframe structure as much as possible before flight, particularly the composite airframe joint bondings; (2) testing the vehicle's systems in an altitude chamber because of the extreme cold and low-pressure conditions encountered by high-altitude science aircraft; (3) recognizing that UAVs, like aircraft, have a tendency to gain weight; (4) maintaining strict configuration control; and (5) ensuring that a redundant system is functional before switching from the primary system.[50]

Pathfinder Plus—Second-Generation ERAST Program Test Vehicle (1997–1998)

NASA and AeroVironment upgraded the Pathfinder UAV to an improved configuration known as the Pathfinder Plus, which had a longer wingspan, two additional electric motors for a total of eight motors, and improved solar cells. The two additional motors were more efficient than the six carried forward from Pathfinder. The Pathfinder Plus UAV, which was intended to serve as a "transitional" aircraft between Pathfinder and the next-generation Centurion UAV, had a wingspan of 121 feet that

---

49. Del Frate and Gary B. Cosentino, "Recent Flight Test Experience with Uninhabited Aerial Vehicles at NASA Dryden Flight Research Center," NASA TM-1998-206546 (Apr. 1998), p. 5.
50. Ibid.

Pathfinder Plus flight in Hawaii. NASA.

included use of four of the five wing sections from the original Pathfinder. The new center wing section contained more efficient solar cells that converted approximately 19 percent of the solar energy into electrical power for the vehicle's motors, avionics, and communications system. This compared with an efficiency rating of only 14 percent for the solar cells on the remaining four wing sections from the original Pathfinder. The addition of the fifth wing section enabled Pathfinder Plus to generate 12,500 watts, as compared with the original Pathfinder's 7,500 watts. Pathfinder Plus had a gross weight of 700 pounds, up to a 150-pound payload capacity, an aspect ratio of 15 to 1, a wing chord of 8 feet, and a power-off glide ratio of 21 to 1. Pathfinder Plus enabled higher-altitude flights and was used to qualify the next-generation Centurion wing panel structural design, airfoil, and SunPower Corporation's solar array. Several flight tests were conducted in Hawaii. On its last flight, on August 6, 1998, Pathfinder Plus set a world altitude record of 80,201 feet for solar-power- and propeller-driven aircraft. These flight tests demonstrated the power, aerodynamic, and systems technologies needed for the Centurion and confirmed the practical utility of using high-altitude remotely controlled solar powered aircraft for commercial purposes.[51]

---

51. "NASA Dryden Fact Sheet—Pathfinder Solar-Powered Aircraft," Dryden Flight Research Center, FS-034, pp. 2–3.

ERAST Pathfinder Sensor Technology Development

As noted in the last two letters of the ERAST acronym, the development of sensor technology for the program constituted a major program goal. Science activities of the Pathfinder missions were coordinated by NASA's Ames Research Center. Ames developed and tested a number of scientific instruments, including two imaging sensors—a high spectral resolution Digital Array Scanned Interferometer (DASI) and an Airborne Real-Time Imaging System (ARTIS). Steven Wegener of NASA Ames served as project manager for the science and sensor program. Dougal Maclise was payload project manager, Steven Dunagan was the team leader of the DASI project, and Stan Ault was team leader for the ARTIS project. DASI, which weighs less than 25 pounds and was mounted under Pathfinder's wing, is a remote sensing instrument that looks at reflected spectral intensities from the Earth. The ARTIS payload was built around a color infrared six-megapixel digital camera. Both sensors were designed to be small, lightweight, and interactive in accordance with ERAST program goals of miniaturizing flight payloads. Both sensor systems also were designed to complement high-altitude studies of atmospheric ozone, land-cover changes, and natural hazard studies conducted by NASA's Earth Resources Survey ER-2 aircraft. The Pathfinder's imaging systems featured a remote interactive operation and near-real-time transmission of images to ground stations and the Internet. This capability improved the speed, quality, and efficiency of data collection, analysis, and interpretation. The NASA team noted that the rapid availability of information from these systems could aid in fast decision making during natural disasters.[52]

The science and sensor aspects of the ERAST program promoted new solar UAV payloads and missions, including disaster management with the Global Disaster Information Network, over-the-horizon and real-time technologies, support of Earth science enterprises, high-resolution mapping, and promotion of the Commercial Remote Sensing program partnership.[53] The advantages of UAVs over satellites and piloted aircraft include: (1) long-range capability, including the ability to fly to remote locations

---

52. See, for example, "Pathfinder Solar-Powered Aircraft Begins Science Missions," NASA Release: 97-45, http://www.nasa.gov/centers/dryden/news/NewsReleases/1997/97-45.html, accessed Aug. 20, 2009.

53. Stephen S. Wegener and James Brass, "Environmental Research Aircraft and Sensor Technology—New Technologies for Earth Science," Earth Science Enterprise, *Atmospheric Chemistry and Dynamics*, p. 177.

and cover large areas; (2) long-endurance capability, including the ability to fly longer and revisit areas on a frequent basis; (3) high-altitude capability, including the ability to fly above weather or danger; (4) slow-speed flight, including the capability to stay near one location; and (5) elimination of pilot exposure, thus enabling long duration or dangerous flights.[54]

Solar Cells and Fuel Cells for Solar-Powered ERAST Vehicles

NASA had first acquired solar cells from Spectralab but chose cells from SunPower Corporation of Sunnyvale, CA, for the ERAST UAVs. These photovoltaic cells converted sunlight directly into electricity and were lighter and more efficient than other commercially available solar cells at that time. Indeed, after NASA flew Helios, SunPower was selected to furnish high-efficiency solar concentrator cells for a NASA Dryden ground solar cell test installation, spring-boarding, as John Del Frate recalled subsequently, "from the technology developed on the PF+ and Helios solar cells."[55] The Dryden solar cell configuration consisted of two fixed-angle solar arrays and one sun-tracking array that together generated up to 5 kilowatts of direct current. Field-testing at the Dryden site helped SunPower lower production costs of its solar cells and identify uses and performance of its cells that enabled the company to develop large-scale commercial applications, resulting in the mass-produced SunPower A-300 series solar cells.[56] SunPower's solar cells were selected for use on the Pathfinders, Centurion, and Helios Prototype UAVs because of their high-efficiency power recovery (more than 50-percent higher than other commercially available cells) and because of their light weight. The solar cells designed for the last generation of ERAST UAVs could convert about 19 percent of the solar energy received into 35 kilowatts of electrical current at high noon on a summer day. The solar cells on the ERAST vehicles were bifacial, meaning that they could absorb sunlight on both sides of the cells, thus enabling the UAVs to catch sunrays reflected upward when flying above cloud covers, and were specially developed for use on the aircraft.

---

54. S.M. Schoenung and S.S. Wegener, "Meteorological and Remote Sensing Applications of High Altitude Unmaned Aerial Vehicles," presented at the *Fourth International Airborne Remote Sensing Conference and Exposition/21st Canadian Symposium on Remote Sensing*, Ottawa, Ontario, Canada, June 21–24, 1999, sections 1.0 ("Introduction") and 2.0 ("Applications").
55. Del Frate, e-mail to author, Aug. 4, 2009.
56. See "Originating Technology/NASA Contribution," pp. 66–67, *Environment and Resources Management*.

While solar cell technology satisfied the propulsion problem during daylight hours, a critical problem relating to long-endurance backup systems remained to be solved for flying during periods of darkness. Without solving this problem, solar UAV flight would be limited to approximately 14 hours in the summer (much less, of course, in the dark of winter), plus whatever additional time could be provided by the limited (up to 5 hours for the Pathfinder) backup batteries. Although significant improvements had been made, batteries failed to satisfy both the weight limitation and long duration power generation requirements for the solar-powered UAVs.

As an alternative to batteries, the ERAST alliance tested a number of different fuel cells and fuel cell power systems. An initial problem to overcome was how to develop lightweight fuel cells because only 440 pounds of Helios's takeoff weight of 1,600 pounds were originally planned to be allocated to a backup fuel cell power system. Helios required approximately 120 kilowatthours of energy to power the craft for up to 12 hours of flight during darkness, and, fortunately, the state of fuel cell technology had advanced far enough to permit attaining this; earlier efforts back to the early 1980s had been frustrated because fuel cell technology was not sufficiently developed at that time. The NASA-industry team later determined, as part of the ERAST program, that a hydrogen-oxygen regenerative fuel cell system (RFCS or regen system) was the hoped for solution to the problem, and substantial resources were committed to the project.

RFCSs are closed systems whereby some of the electrical power produced by the UAV's solar array during daylight hours is sent to an electrolyzer that takes onboard water and disassociates the water into hydrogen gas and oxygen gas, both of which are stored in tanks aboard the vehicle. During periods of darkness, the stored gases are recombined in the fuel cell, which results in the production of electrical power and water. The power is used to maintain systems and altitude. The water is then stored for reuse the following day. This cycle theoretically would repeat on a 24-hour basis for an indefinite time period. NASA and AeroVironment also considered, but did not use, a reversible regen system that instead of having an electrolyzer and a fuel cell used only a reversible fuel cell to do the work of both components.[57]

---

57. Michael A. Dornheim, "The Quest for Perpetual Flight: Special Fuel Cells Key to Months-long Flight," *Aviation Week & Space Technology* (Feb. 28, 2000), pp. 58–61; telephone interview of Del Frate by author (Apr. 27, 2009), along with followup e-mail.

As originally planned, Helios was to carry two separate regen fuel cell systems contained in two of four landing gear pods. This not only disbursed the weight over the flying wing, but also was in keeping with the plan for redundant systems. If one of the two fuel cells failed, Helios could still stay aloft for several days, albeit at a lower altitude. Contracts to make the fuel cell and electrolyzer were given to two companies—Giner of Waltham, MA, and Lynntech, Inc., of College Station, TX. Each of the two systems was planned to weigh 200 pounds, including 27 pounds for the fuel cell, 18 pounds for the electrolyzer, 40 pounds for oxygen and hydrogen tanks, and 45 pounds for water. The remaining 70 pounds consisted of plumbing, controls, and ancillary equipment.[58]

While the NASA–AeroVironment team made a substantial investment in the RFCS and successfully demonstrated a nearly closed system in ground tests, it decided that the system was not yet ready to satisfy the planned flight schedule. Because of these technical difficulties and time and budget deadlines, NASA and AeroVironment agreed in 2001 to switch to a consumable hydrogen-air primary fuel cell system for the Helios Prototype's long-endurance ERAST mission. The fuel cells were already in development for the automotive industry. The hydrogen-air fuel cell system required Helios to carry its own supply of hydrogen. In periods of darkness, power for the UAV would be produced by combining gaseous hydrogen and air from the atmosphere in a fuel cell. Because of the low air density at high altitudes, a compressor needed to be added to the system. This system, however, would operate only until the hydrogen fuel was consumed, but the team thought that the system could still provide multiple days of operation and that an advanced version might be able to stay aloft for up to 14 days. The installation plan was likewise changed. The fuel cell was now placed in one pod with the hydrogen tanks attached to the lower surface of the wing near each wingtip. This modification, of course, dramatically changed Helios's structural loadings, transforming it from a span-loaded flying wing to a point-loaded vehicle.[59]

Centurion—Third-Generation ERAST Program Test Vehicle (1996–1998)

The Centurion, which was built in 1998 by AeroVironment, represented third-generation advancement on the technology developed in the Pathfinder and Pathfinder Plus UAVs. The Centurion, however, was still

---

58. Dornheim, "The Quest for Perpetual Flight," *Aviation Week & Space Technology*, pp. 58–61.
59. Telephone interview of Del Frate by author (Apr. 2, 2009), and followup e-mail.

Case 13 | Good Stewards: NASA's Role in Alternative Energy

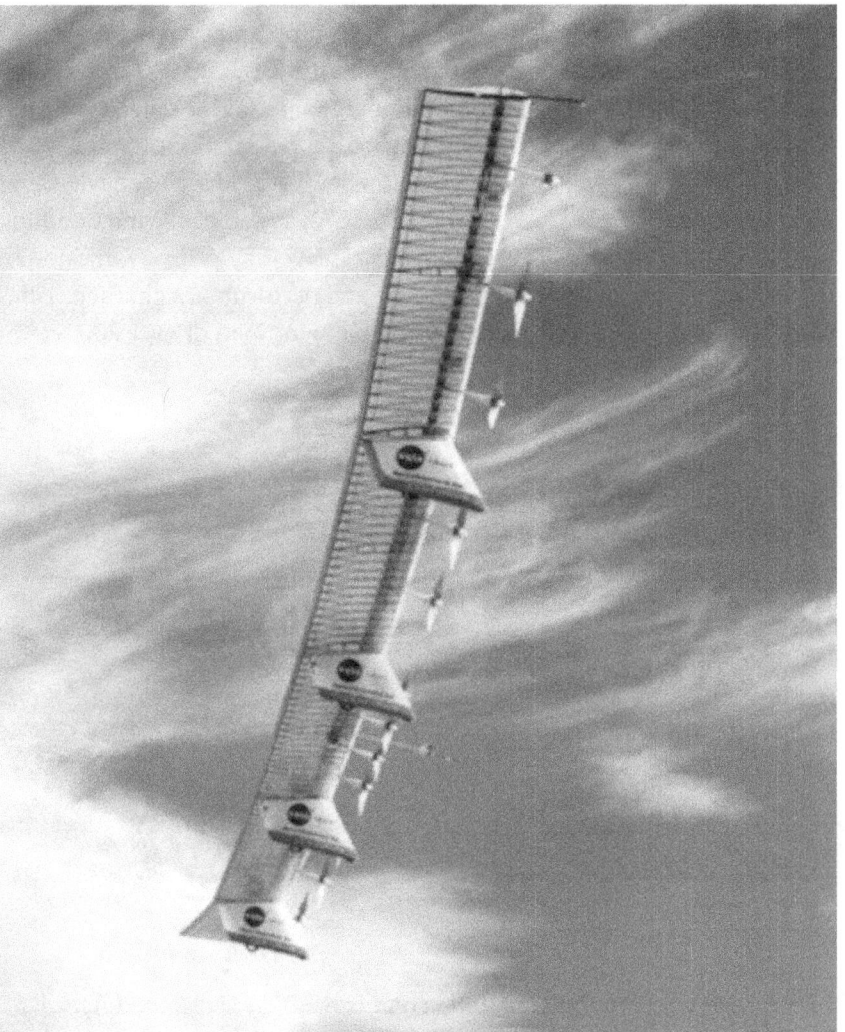

Centurion in flight. NASA.

considered to be a prototype demonstrator for future solar-powered vehicles that could stay airborne for long periods of time. Prior to construction of the full-scale Centurion UAV, a quarter-scale model was constructed to verify design predictions. The Centurion had a five-section wingspan of 206 feet, which was more than twice the wingspan of Pathfinder; incorporated a redesigned airfoil; and had a length of 12 feet and an aspect ratio of 26 to 1. The aircraft, which had 14 direct-current electric motors, was to be powered by bifacial solar cells that covered 80 percent of the upper wing surface and had a maximum output of 31 kilowatts. The Centurion had a cruising speed of between 17 and 21 mph and could carry a payload of approximately 100 pounds to an altitude of 100,000 feet, or 600 pounds to 80,000 feet. The primary building material consisted of carbon fiber and graphite epoxy composite structure, Kevlar, Styrofoam leading edge, and plastic film covering. Centurion flew three low-altitude developmental test flights on battery power at NASA Dryden, verifying handling qualities, performance, and structural integrity. The primary mission of the Centurion was to verify the handling and performance characteristics of an ultra-lightweight all-wing UAV with a wingspan of over 200 feet.[60]

Helios Prototype checkout flight in Hawaii. NASA.

Helios Prototype—Fourth-Generation and Last ERAST Unmanned Aerial Vehicle

The Helios Prototype, which resulted from modifications of the Centurion and renaming of the aircraft to Helios, was a proof-of-concept flying wing that because of budget limitations had two different configurations: a high-altitude configuration designated the HP-01 (1998–2002) and a long-endurance configuration designated the HP-03 (2003).[61] The two primary objectives for the Helios Prototypes were to demonstrate sustained flight at an altitude of 100,000 feet and the ability to fly nonstop

---

60. "Centurion," NASA Dryden Flight Research Center Fact Sheet FS-056.
61. NASA never tested the Helios configuration (both HP-01 and HP-03) at Edwards, only testing Centurion. (Information from Del Frate, Aug. 4, 2009.)

Case 13 | Good Stewards: NASA's Role in Alternative Energy

Helios Prototype taking off for record altitude flight. NASA.

for at least 24 hours, including at least 14 hours above 50,000 feet. Initial low-altitude test flights were conducted under battery power at Edwards Air Force Base in 1999. Afterward, Helios was equipped with high-efficiency photovoltaic solar cells and underwent high-altitude flight-testing in the summer of 2001 at the U.S. Navy Pacific Missile Range Facility in Hawaii. On August 13, 2001, following further upgrading of systems, the high-altitude configuration (HP-01) reached an altitude of 96,863 feet, setting a world record for sustained horizontal flight by a winged aircraft and effectively satisfying the first Helios objective of high-altitude performance. The previous record was held by the Air Force's SR-71A. The plan for the long-endurance configuration was to use solar cells to power the electric motors and subsystems during daylight hours and a modified hydrogen-air fuel cell system during the night. The vehicle also was equipped with backup lithium batteries and was battery-powered for its first six low-altitude test flights at Dryden.

Helios used wing dihedral (wing upsweep), engine power, elevator control surfaces, and a stability augmentation and control system to provide aerodynamic stability and control. Helios added a sixth wing panel to the Centurion, giving the remotely powered aircraft a wingspan of 247 feet, which was longer than the wingspans of either the U.S. Air Force C-5 transport or the Boeing 747 commercial airliner. The aircraft had a length of 12 feet, a wing chord of 8 feet, an empty weight of 1,322 pounds, an aspect ratio of 31 to 1, and wing loading of 0.835 pounds per square foot. The craft, which had 14 (only 10 on long-endurance HPO3 version) brushless direct-current electric motors each rated at 2 horsepower (1.5 kW) that each drove a lightweight, 79-inch-diameter propeller, was powered by 62,120 bifacial solar cells covering the upper wing surface. Bifacial solar cells enabled Helios to convert solar energy into electricity when illuminated from either above or below, which enabled the vehicle to absorb reflected energy when flying above cloud cover. The Helios had a cruising speed of between 19 and 27 mph and could carry a payload of up to 726 pounds, including ballast, instruments, experiments, and a supplemental energy system. Helios was designed to operate at up to 100,000 feet, which is above 99 percent of the Earth's atmosphere, with a typical endurance mission at 50,000 to 70,000 feet. Like the Centurion, the Helios Prototype was constructed mostly of composite material, including carbon fiber, graphite epoxy, Kevlar, Styrofoam, and a thin transparent plastic film. The main tubular wing spar, which was made of carbon fiber and also wrapped with Nomex and Kevlar, was thicker on the top and bottom to absorb the constant bending motions that occurred during flight. Under-wing pods were attached to carry the landing gear, battery backup system, fuel cells, flight control computers, and data instruments. The Helios flight control surfaces consisted of 72 trailing-edge elevators that provided pitch control. The fixed landing gear consisted of two wheels on each pod.[62]

At an October 13, 1999, flight demonstration at the Dryden Flight Center, Ray Morgan—vice president of AeroVironment, which developed the Helios—noted that the ultimate intention is for Helios-type UAVs to "stay in the stratosphere for months at a time and act as an 11-mile-high tower." Morgan added that the Helios had a "very unique, slow, and stable flight characteristic . . . that means we can fly in fairly tight circles and be, essentially, a geo-stationary platform in the sky—and that has lots of potential application."[63] Some of commercial tasks envisioned for future solar-powered HALE aircraft to follow the Helios Prototype were storm tracking studies, atmospheric sampling, spectral imaging for agricultural and natural resource purposes, pipeline mon-

---

62. "Helios Prototype," Dryden Flight Research Center Fact Sheet FS-068.
63. John D. Hunley and Yvonne Kellogg, compilers, *Proceedings* of addresses, sessions and workshops of the NASA ERAST Exclusive Preview sponsored by NASA Dryden Flight Research Center, Edwards, CA, Oct. 13, 1999, ERAST: Scientific Applications and Technology Commercialization (Sept. 2000), p. 4.

itoring, and telecommunications platforms. Morgan further noted that reliability of the Helios was obtained in two ways: simplicity, because the UAV is simply a flying wing, and redundancy. Each motor pylon turned 2 horsepower into 10 pounds of thrust. There was only one moving part in each pylon. There were no brushes, gearboxes, or mechanism for variation in the pitch required to operate the UAV.[64] Furthermore, while the Helios had 72 elevators, the UAV could use differential thrust to turn the aircraft without use of the elevators.

The Helios HP03 long-endurance configuration, which included a hydrogen-air fuel cell pod, made its first high-altitude flight June 7, 2003. The test objectives of this flight were: (1) to demonstrate the readiness of the aircraft systems, flight support equipment and instrumentation, and flight procedures; (2) to validate the handling and aeroelastic stability of the aircraft; (3) to demonstrate the operation of the fuel cell system and gaseous hydrogen storage tanks; and (4) to provide training for personnel to staff future multiday flights. The flight was planned to be a 30-hour endurance flight, but because of leakage in the coolant and compressed air line, the flight had to be aborted after 15 hours. Flight data, however, verified that the vehicle was aeroelastically stable under the flight conditions expected for the long-endurance flight demonstration.

On its second test flight, on June 26, 2003, following further modifications based on the first flight, the Helios prototype crashed on the Navy's Pacific Missile Range Facility in Hawaii. This flight was intended to evaluate the vehicle's hydrogen fuel cell system designed to power the UAV during periods of darkness, which would be necessary for long-duration flights. Approximately 30 minutes into its flight and at an altitude of only 2,800 feet, the aircraft encountered turbulence and morphed into a high dihedral (wings upswept) configuration that, over time, took on an increasingly alarming U-like appearance. This caused Helios to become unstable in pitch, and the big UAV began to nose up and down. During one descending swoop, the big UAV exceeded its maximum permissible airspeed, causing the wing's leading edge secondary structure on the outer panels to fail and the solar cells and skin on the upper surface to rip off. Shedding bits and pieces of structure, the Helios progressively disintegrated as it plunged downward, being destroyed when it impacted the ocean.[65]

---

64. Ibid., p. 6.
65. Mishap Investigation Board, "Investigation of the Helios Prototype Aircraft Mishap," passim.

The Mishap Investigation Board (MIB) concluded that: "In summary, there was no evidence that suggests that there was prior structural failure(s) that would have contributed to the vehicle developing the higher-than-normal dihedral deflections [wing upsweep] that caused the unstable pitch oscillation and subsequent failure of the vehicle."[66] The report, in noting the lessons learned, identified both the proximate and root causes of the loss of Helios. The proximate cause of the crash was determined to be the "high dynamic pressure reached by the aircraft during the last cycle of the unstable pitch oscillation leading to failure of the vehicle's secondary structure." Two root causes, however, that contributed to or created the proximate cause of the accident were: (1) the lack of adequate analysis methods, which led to an inaccurate risk assessment of the effects of the configuration change from the HPO1 (high altitude) to the HPO3 (long endurance) version of the Helios and (2) configuration changes to the aircraft, which altered the Helios from a spanloader to a highly point-loaded mass distribution on the same structure, significantly reducing design robustness and margins of safety.[67]

The report noted that the Helios UAV represented "a nonlinear stability and control problem involving complex interactions among the flexible structure, unsteady aerodynamics, flight control system, propulsion system, the environmental conditions, and vehicle flight dynamics." As a result of the investigation, the MIB made a number of key recommendations for future UAV development. These included:

1. Develop more advanced, multidisciplinary "time-domain" analysis methods appropriate to highly flexible, "morphing" vehicles;
2. Develop ground-test procedures and techniques appropriate to UAV class vehicles to validate new analysis methods and predictions;
3. Improve the technical insight for highly complex projects, using the expertise available from all NASA Research Centers;
4. Develop multidisciplinary models, which can describe the nonlinear dynamic behavior of aircraft modifications or perform incremental flight-testing; and

---

66. Ibid., p. 45.
67. Ibid., pp. 10, 54, 87–89, and Table 10.1.

5. Provide adequate resources to future programs for more incremental flight-testing when large configuration changes significantly deviate from the initial design concept.

As already noted, because of budget constraints, the NASA-industry team basically had to use one prototype vehicle for two tests: high altitude and long endurance. Adding to this factor, as noted in the MIB report, the switch from a regenerative fuel cell to a hydrogen-air fuel cell, which was necessitated by technical problems with the RFCS, combined with project time and budget deadlines, caused a critical change in the load factor on the wing structure. The hydrogen-air fuel cell system was more point loaded because of having to use a third pod at the centerline of the vehicle to hold the 520-pound primary hydrogen-air fuel cell. The originally planned for regenerative fuel cell system would have required only two pods, each one-third of the distance from the vehicle's centerline to the wingtip.[68]

Even with the loss of the Helios Prototype, the MIB noted both the success and the challenge of the solar UAV program, as reflected in the following statement from the Mishap report:

> During the course of this investigation the MIB discovered that the AV [AeroVironment]/NASA technical team had created most of the world's knowledge in the area of High Altitude-Long Endurance (HALE) aircraft design, development, and test. This has placed the United States in a position of world leadership in this class of vehicle, which has significant strategic implications for the nation. The capability afforded by such vehicles is real and unique, and can enable the use of the stratosphere for many government and commercial applications. The MIB also found that this class of vehicle is orders of magnitude more complex than it appears but that the AV/NASA technical team had identified and solved the toughest technical problems. Although more knowledge can and should be pursued as recommended in this report, an adequate knowledge base now exists to design, develop, and deploy operational HALE systems.[69]

---

68. Ibid., pp. 10, 89, and Table 10-1. Both HP-01 and HP-03 had a total of five under-wing pods; the HP-03 configuration replaced the original center pod with the fuel cell pod.
69. Ibid., p. 10.

ERAST Program Overview and Accomplishments

The ERAST program, which started in 1994 and ended in 2003, following the loss of the Helios Prototype, accomplished almost all of its primary objectives, including development of a solar-powered UAV capable of flying at very high altitudes; demonstrating payload capabilities and sensors for atmospheric research; resolving solar-powered UAV operational issues; and demonstrating UAV's usefulness to scientific, Government, and civil customers. The ERAST program also demonstrated that the unique joint NASA-industry ERAST alliance under a Joint Sponsored Research Agreement worked well and that good cooperation under this agreement led to efficient use of resources and expedited reaching the project milestones required to satisfy the program's budget and time constraints. In regard to the alliance partner manufacturing the solar cells for the four generations of solar-powered UAVs, SunPower Corporation was able to make significant increases in efficiency and lower the cost of its commercial solar cells, which led to its successful mass-produced series A-300 solar cells. The one primary objective that remained unfulfilled at the close of the ERAST NASA-industry program was obtaining the long-endurance capability to fly for multiple days or longer. As with the earlier solar UAV program in the 1980s, a backup power system that would enable solar UAVs to fly in periods of darkness remained the critical problem to be solved. Significant progress in this area, however, was made and work continued at NASA Glenn under the Low Emissions Alternative Power (LEAP) program on the goal of perfecting a lightweight regenerative fuel cell system. NASA Glenn made its first closed-loop (system completely sealed) regenerative fuel cell demonstration in September 2003 and demonstrated five contiguous back-to-back charge-discharge cycles at full power under semiautonomous control in July 2005.[70]

---

70. D.J. Bents, "Solar Airplanes and Regenerative Fuel Cells," presentation to the *43rd annual I.R.I.S. show in Mayfield Heights, OH, Oct. 9, 2007.*

Recommended Additional Readings
Reports, Papers, Articles, and Presentations:

T.S. Anderson, et al., "MOD-0A 200-Kilowatt Wind Turbine Generator Design and Analysis Report," executive summary, NASA CR-165127 (1980).

Darrell H. Baldwin and Jerry Kennard, "Development of Large, Horizontal-Axis Wind Turbines," NASA TM-86950 (1985).

Darrell H. Baldwin and Bradford S. Linscott, "The Federal Wind Program at NASA Lewis Research Center," NASA TM-83480 (1983).

A.G. Birchenough, et al., "Operating Experience with the 200 KW MOD-0A Wind Turbine Generators," NASA Lewis Wind Energy Project Office, NASA 83N-15916 (1983).

Boeing Aerospace Company, "MOD-5B Wind Turbine System Final Report," vol. 1 (executive summary), NASA CR-180896 (1988), and vol. 3 ("Acceptance Testing") NASA CR-180898 (1988).

John H. Del Frate and Gary B. Cosentino, "Recent Flight Test Experience with Uninhabited Aerial Vehicles at the NASA Dryden Flight Research Center," NASA TM-1998-206546 (1998).

Michael A. Dornheim, "AeroVironment Pushes Limits of Solar Flight," *Aviation Week & Space Technology* (May 4, 1998).

Michael A. Dornheim, "Giant Centurion Starts Flight Test," *Aviation Week & Space Technology* (Nov. 16, 1998).

Richard R. Douglas, "The Boeing MOD-2 at 2.5 MW," workshop held at Lewis Research Center, Cleveland, OH, Apr. 24–26, 1979, NASA CP-2106 (1979).

R.R. Douglas, "Conceptual Design of the 7 Megawatt MOD-5B Wind Turbine Generator," NASA Accession ID: 83N19272 (1982).

Dryden Flight Research Center Fact Sheets, "Centurion," NASA FS-056.

Dryden Flight Research Center Fact Sheets, "Helios Prototype," NASA FS-068.

Dryden Flight Research Center Fact Sheets, "Pathfinder Solar-Powered Aircraft," NASA FS-034 and FS-068.

Dryden Flight Research Center Fact Sheets, "ERAST," NASA FS-020.

Federal Energy Administration, "Project Independence Blueprint. Task Force. Report. Solar Energy," U.S. Department of Commerce, National Technical Information Service (1974).

General Electric Company, Space Division, "Mod-1 Wind Turbine Generator Analysis and Design Report," NASA CR-159497 (1979).

General Electric Company, "MOD-5A Wind Turbine Generator Program Design Report," vol. 1 (executive summary), NASA CR-174734 (1984).

John D. Hunley and Yvonne Kellogg, compilers, "ERAST: Scientific Applications and Technology Commercialization," *Proceedings of the NASA ERAST Exclusive Preview sponsored by NASA Dryden Flight Research Center, Edwards, CA, Oct. 13, 1999*, NASA CP-2000-209031 (2000).

Jay Levine, ed., "Special Helios Prototype Edition," *The X-Press*, vol. 44, issue 2, Dryden Flight Center, Edwards, CA (May 8, 2002).

Derek L. Lisoski and Mark B. Tischler, "Solar Powered Stratospheric Research Aircraft—Flight Test and System Identifications," paper presented at the RTO SCI Symposium held in Madrid, Spain, and published in RTO MP-11.

NASA, "Global Stratospheric Change: Requirements for a Very-High Altitude Aircraft for Atmospheric Research," NASA CP-10041 (1989).

NASA Facts, "ERAST: Environmental Research Aircraft and Sensor Technology," FS-2002-08-020 DFRC (2002).

Thomas E. Noll, et al., (Mishap Investigation Board) "Investigation of the Helios Prototype Aircraft Mishap," vol. 1, NASA (2004).

Richard L. Puthoff, "Fabrication and Assembly of the ERDA/NASA 100-Kilowatt Experimental Wind Turbine," NASA TM-X-3390 (1976).

Richard L. Puthoff, et al. "Installation and Checkout of the DOE/NASA MOD-1 2000-kW Wind Turbine Generator," NASA TM-81444 (1980).

Palmer C. Putnam, "Wind Power: Yesterday, Today, and Tomorrow," *Proceedings of a workshop sponsored by the U.S. Department of Energy and NASA Lewis Research Center held in Cleveland, OH, July 28–30, 1981*, NASA CP-2230 (1982).

S.M. Schoenung and S.S. Wegener, "Meteorological and Remote Sensing Applications of High Altitude Unmanned Aerial Vehicles," paper presented at the *Fourth International Airborne Remote Sensing Conference and Exhibition/21st Canadian Symposium on Remote Sensing, Ottawa, Ontario, Canada* (June 21–24, 1999).

D.A. Spera, et al., "Preliminary Analysis of Performance and Loads Data From the 2-Megawatt Mod-1 Wind Turbine Generator," paper prepared for the *Fourth Biennial Conference and Workshop on Wind Energy Conversion Systems, sponsored by the Department of Energy, Washington, DC, Oct. 29–31, 1979*, NASA TM-81408 (1979).

Edward H. Teets, Jr., et al., "Atmospheric Consideration for Uninhabited Aerial Vehicle (UAV) Flight Test Planning," NASA TM-1998-206541 (1998).

Ronald Thomas, et al., "Plans and Status of the NASA-Lewis Research Center Wind Energy Project," NASA TM-X-71701 (1975).

Ronald Thomas and Richard M. Donovon, "Large Wind Turbine Generators," NASA TM-73767 (1978).

Ronald Thomas and Joseph M. Savino, "Status of Wind-Energy Conversion," NASA TM-X-71523 (1973).

Larry A. Viterna, "The NASA-LeRC Wind Turbine Sound Prediction Code," presented at *Second DOE/NASA Wind Turbine Dynamics Workshop, Cleveland* (1981).

Larry A. Viterna and David C. Janetzke, "Theoretical and Experimental Power From Large Horizontal-Axis Wind Turbines," NASA TM-82944 (1982).

Books and Monographs:

Virginia P. Dawson, *Engines and Innovation: Lewis Laboratory and American Propulsion Technology*, ch. 10, NASA History Office, SP-4306 (Washington, DC: NASA, 1991).

Richard P. Hallion and Michael H. Gorn, *On the Frontier: Experimental Flight at NASA Dryden* (Washington, DC: Smithsonian Institution Press, 2004).

Alexander Laufer, et al., *Shared Voyage: Learning and Unlearning from Remarkable Projects*, NASA History Series, NASA SP-2005-4111 (Washington, DC: NASA, 2005).

Palmer C. Putnam, *Power from the Wind* (New York: Van Nostrand Reinhold Company, 1948).

Robert W. Righter, *Wind Energy in America: A History* (Norman, OK: University of Oklahoma Press, 1996).

David A. Spera, ed., *Wind Turbine Technology: Fundamental Concepts of Wind Turbine Engineering* (New York: ASME Press, 1994).

THIS PAGE INTENTIONALLY BLANK

THIS PAGE INTENTIONALLY BLANK

Index

201 Aircraft, 23n39
202 Aircraft, 23n39
367-80 Aircraft, 20
500E Helicopter, 165
707 Aircraft, 4, 20, 65, 123, 740, 804
727 Aircraft, 737, 745, 802, 804
737 Aircraft, 124, 616, 737, 745, 771
747 Aircraft: 747-400 winglet use, 89, 123–24; composite materials, 804; engine component improvements, 745; NASA's aeronautics role and, 787; Shuttle carrier aircraft, 661; simulators testing of PCA, 715; TAVs launched from, 337; titanium construction, 616; winglet use on, 89, 122, 123–24
757 Aircraft, 124, 715, 756, 795, 807–8
767 Aircraft, 124, 615
777 Aircraft: 777-300ER aircraft, 795; fly-by-wire systems, 722, 753; GE90 jet engine, 742, 743, 747; materials for, 620; noise from, 771; supercritical wing, 115; titanium construction, 616
787 Aircraft: electrical flight control surfaces and actuators, 722; engine for, 769, 795; materials for, 571, 620; noise from, 771; range of, 620; titanium construction, 616

A

A&P Technology, Inc., 768, 769
A-2 rocket, 635–36
A4D-1 Skyhawk, 45–46
A-4 Skyhawk, 65
A-12 Blackbird, 210, 570
A-17A aircraft, 734
Abbett, Mike, 444
Aberdeen Proving Ground, Maryland, 14
Ablation, heat of: measurement of, 305, 581; satellites, conditions for, 305, 581
Ablation cooling, 286–88, 417–18, 581–82, 591, 606–8
Accelerometers, 367, 376, 534
Ackeret, Jakob, 93, 182
Ackeret theory, 15
Acoustically Treated Nacelle program, 740
Acoustical Society of America (ASA): engine noise research, 736–37; research results, dissemination of through, 207; sonic boom conferences, 211, 217
Active cooling concepts, 420, 589–90, 605
Active-element computers, 468
Active flight control, 651–53, 805. *See also* Control Configured Vehicle (CCV)
Active laminar flow control, 806–8
Active Oscillation Suppression (AOS) system, 389
ACTOMP program, 556
ACTON program, 556
Actuators, 384–85
Adams, Michael J., 299, 373
Adaptive Engine Control System (ADECS) program, 711–12, 757
Adaptive flight control system (AFCS), 372–74, 375, 376, 379
Adaptive grid, 434
Ad Hoc Group for Structural Analysis, 474
Advanced Concepts Simulator, 715
Advanced Control Technology for Integrated Vehicles Experiment (ACTIVE) project, 258, 716–17
Advanced Fighter Technology Integration (AFTI) F-16 project: aircraft modifications

for, 672–73, 675; asynchronous digital flight control system, 672, 680–82, 683; Automated Maneuvering Attack System (AMAS), 672, 673, 674–75, 676; benefits of program, 632; control laws, 680–82; digital flight control system, 672–73, 674, 680–82; F-16 CCV as basis for, 648, 670, 672; flight-testing, 673–79; forward-looking infrared (FLIR) device, 673, 675; goals of, 672; ground collision avoidance system, 675; helmet-mounted target designation sight, 673, 675; Pilot Activated Recovery System, 676; power-by-wire testbed, 676–79; publication of technical papers on, 679–80; research activities, 630; Stores Management System (SMS), 672, 681; strike technology testbed, 675–76; terrain system with moving map, 675–76; Voice-Controlled Interactive Device (VCID)/Voice Command System, 672–73, 674, 675

Advanced Research and Technology, Office of, 212, 473, 654

Advanced Subsonic Combustion Rig (ASCR), 799

Advanced Subsonic Technology (AST) program, 69, 786, 790, 796–97, 798

Advanced Supersonic Technology (AST) program. *See* Supersonic Cruise Research (SCR)/Supersonic Cruise Aircraft Research (SCAR)/Advanced Supersonic Technology (AST) program

Advanced Tactical Fighter (ATF) program, 126–27

Advanced Tiltrotor Transport Technology (ATTT) program, 167–68

Advanced Turboprop (ATP) project, 744, 747–50, 786, 788, 789, 790–91, 799–803

Advancing Blade Concept (ABC) rotor, 152–53, 157–58

Adverse yaw, 19–20, 409

Advisory Group for Aerospace Research & Development (AGARD), 213

Aerion Corporation, 249

Aero Digest, 106, 106n48

Aerodynamic damping, 363–67

Aerodynamic heating, 414–15

Aerodynamics Committee, NACA, 59

Aerodynamic stability frequencies, 378

Aerodynamic structures, 805–6

Aeroelastic effects: aeroservoelastic effects, 496, 525, 528–29; aircraft structure, 387, 390–92, 397–98, 482–83, 521–22; B-47 aircraft, 20, 390, 409–10; B-52 Stratofortress, 390–91, 392; B-58 Hustler, 392; FB-111 aircraft, 392; rotary wing aircraft, 149–50; SR-71 Blackbird, 391–92; swept and delta wing aircraft, 390; X-29 aircraft, 496, 528–29, 694; XB-70 Valkyrie, 392

Aeroelastic Rotor Experimental System (ARES), 164

Aeroelastic tailoring, 521, 524–25, 694, 695

Aeroflightdynamics Directorate, 170

Aeronautical Research Associates of Princeton (ARAP), 213, 223

Aeronautical Research Institute of Sweden, 212, 222

Aeronautical Research Laboratory, 186

Aeronautical Research Radar Complex, 151

Aeronautics and Space Technology, Office of, 219–20

Aeronautics Research Mission Directorate, 770

Aeronautics System Division, 220

Aeroservoelastic effects, 496, 525, 528–29

Aerospace America, 340

Aerospace Industries Association, 248

Aerospace Maintenance and Regeneration Group (AMARC), 672

Index

Aerospaceplane (ASP) program, 334

Aerospace Vehicles Panel, Air Force Scientific Advisory Board, 585

Aerospatiale Dauphine helicopter, 165

Aerothermodynamic Elastic Vehicles (AEV), 308–10, 594, 595

Aerothermodynamics. *See also* hot structures: ablation cooling, 286–87, 417–18, 581–82, 591, 606–8; active cooling concepts, 420, 589–90, 605; aerodynamic heating, 414–15; blunt-body design, 291, 311, 415; ceramic tiles, 419; Dryden research activities, 495–96; heat-sink structures, 415–17, 579, 581, 607; LI-900 thermal protection, 311, 313–17, 600–601, 604, 605; LI-1500 thermal protection, 310–11, 313, 599–600, 601, 604; LI-2200 thermal protection, 317; localized heating, 416; materials selection and speed, 569–73; passive cooling, 606; reusable surface installation (RSI), 312–23, 596–605; satellites, conditions for, 305, 579, 581–82; speed limits, 572–73; Structural-Thermal-Optical (STOP) program, 501–2; thermal barrier coatings (TBC), 610–13; tiles, thermal-protection, 278, 288, 310–23, 419, 582, 598; tiles, thermal-protection, bonding of, 319, 320–23, 609; tiles, thermal-protection, coatings for, 315–16; tiles, thermal-protection, densification of, 320–21; tiles, thermal-protection, gaps between, 318, 319–20; tiles, thermal-protection, installation of, 319–23, 419; tiles, thermal-protection, shaping of, 314–15, 321; tiles, thermal-protection, testing of, 316–19; tiles, thermal-protection, thermal expansion and, 315; transpiration cooling, 579–80; water-wall cooling, 307, 589–90

Aerothermodynamic Structural Vehicle (ASV), 308–10, 594–95

AeroVironment, Inc., 853, 855, 856, 859, 863, 864, 868, 871

AFT-100 aircraft, 221

Agility, 651

AGM-129 FSW stealthy cruise missile, 22

Agusta Helicopter Company, 168

Aileron reversal, 409–10

Ailerons: aeroelastic effects, 390; flight control coupling, 409; flight control with, 633; swept wing aircraft, 19–20, 21

AIM-9 Sidewinder air-to-air missile, 378

Air, compressed, 324, 573

Airborne Real-Time Imaging System (ARTIS), 861

Air-breathing vehicles, 279

Airbus A310 aircraft, 615

Airbus A319 aircraft, 124

Airbus A320 aircraft, 124, 616, 722, 753

Airbus A330 aircraft, 124, 616

Airbus A340 aircraft, 124, 616, 752

Airbus A350 aircraft, 616

Airbus A380 aircraft, 571, 616, 643, 722, 795

Airbus Industrie, 692, 722

Air combat simulation, 703

Aircraft: challenges of developing, 361–62; evolution of, 3, 361–62; flight envelope limitations, 368, 632; handling qualities criteria, 145–46; nicknames for, 3–4; safety of, 506, 632; speed limits, 572–73; turbojet engine development and, 3, 90; unstable aircraft, 379–80, 383, 526, 631, 632, 633, 753; weight of, 461–62, 463

Aircraft and Armament Experimental Establishment (A&AEE), 638n16

Aircraft carrier landings, 33, 33n60

Aircraft design. *See also* area rule; computational fluid dynamics (CFD): aerodynamic predictions, 397–98; bullet shapes, 97; commercial aircraft, 208; DC-4 Generation, 4, 4n1; documentation and communication of design and system requirements, 683–84; evolution of, 3–4, 64; induced drag, 117, 118; NASA's role in, 65; noise reduction and, 770–72; optimization of, 533; profile drag, 117; refinement and efficiency improvements, 64–65, 89–90; safety factor, 387; stability and control and, 633; statically unstable aircraft, 633; supersonic aircraft design, 93, 96; transonic aircraft, 93–94, 362–63, 444–46; transonic drag, 94–100

Aircraft Energy Efficiency (ACEE) program: Advanced Turboprop (ATP) project, 744, 747–50, 786, 788, 789, 790–91, 799–803; advancements through, 758; Composite Primary Aircraft Structures project, 744, 804–5; Energy Efficient Engine (E Cubed) project, 742, 744, 745–47, 786, 788, 790–91, 792–95, 817; Energy Efficient Transport (EET) project, 744, 805–6; Engine Component Improvement project, 744–45, 746; Laminar Flow Control (LFC) project, 744, 784, 786, 806–8; purpose of, 743–44, 788; winglets development, 116–17, 120, 121, 123

Aircraft Engine Research Laboratory, 137, 786–87. *See also* Glenn Research Center/Lewis Research Center/Lewis Flight Propulsion Laboratory/Aircraft Engine Research Laboratory

Aircraft handling qualities criteria, 145–46

Aircraft noise. *See* noise

Aircraft structure: aerodynamic structures, 805–6; aeroelastic effects, 387, 390–92, 397–98, 409–10, 482–83, 521–22; aeroelastic tailoring, 521, 524–25; aeroservoelastic effects, 496, 525, 528–29; Computational Structural Mechanics (CSM) workshop, 497; electro-optical flight deflection measurement system, 521–22, 527–28; flutter, 306, 309, 387–89, 482–83, 525–26, 585–86, 595; redundant structures, 463–64; research activities, meeting about, 473; structural loads, 385–86, 390; structural resonances, 374–79; testing facilities, 553–55

Aircraft Synthesis (ACSYNT), 493

Air defense missions, 187–88

AiResearch, 335, 503

Airfoil design and theory. *See also* supercritical wing: evolution of, 109; NACA airfoil sections, 29n56, 109, 115

Air Force, U.S. (USAF): Aeronautical Systems Division, 337, 526; Aerospace Maintenance and Regeneration Group (AMARC), 672; Aerospaceplane (ASP) program, 334; AFTI/F-16 aircraft, 679; air defense missions, 187–88; Arnold Engineering Development Center (AEDC), 17n30, 158, 280, 291, 760, 787; ASSET vehicles, 308–10, 418–19; Atlas ICBM, 581, 582; ballistic missile accuracy, 284–85; biofuel research, 817; Bomarc missile, 289; BWB research, 71; C-17 Globemaster III, 127, 722; Century series fighters. *See* Century series fightersconcurrent development, 322; Digital Electronic Engine Control (DEEC) program, 709–10, 754–56; energy maneuverability concept, 651; Exceptional Service Medal, 107; F-22 Raptor, 126–27, 712; F-102/F102A Delta Dagger, 102–3; F-104 Starfighter, 104; F-111 program, 61–62; fighter aircraft program, 61; fly-by-wire systems, development of, 379–80; fly-by-wire systems, research activities, 645–48; fly-by-wire systems, transition to, 632; Goodyear Electronic Digital Analyzer (GEDA), 393, 394; hydrogen fuel research, 811–13; HyTech research program, 345; Integrated Propulsion Control System (IPCS) program, 754; International Test Organization (ITO), 700–701; KC-135 aerial tanker, 20, 120, 122–24, 204, 806; Lightweight Fighter (LWF) program, 370, 647, 664; missile tests, 288; NASTRAN, development of, 474; NASTRAN, Users' Colloquia, 480; National Aero-Space Plane Joint Program Office, 338–39; National Full-Scale Aerodynamics Complex, 158, 166; Navaho cruise missile, 289; 1954 Interceptor initiative, 50; Noise and Sonic Boom Impact Technology (NSBIT) program, 229–30, 231; Precision Aircraft Control Technology (PACT), 647–48; Project 680J Survivable Flight Control System (SFCS), 379–80, 384, 646–48, 658n59, 660; scramjet research, 334–35; Self-Repairing Flight Control System (SRFCS), 713–14; Shuttle study, 450; side stick flight controller, 658n59; sonic boom research, 194–203, 207–8; sonic booms, concerns and complaints about, 185–88; Space Division (SD), 337; stability augmentation system research, 365; STOL transport competition, 114; strain gages, 386; Strategic Defense Initiative (SDI), 279, 336–40; structural resonance research, 378; Suntan project, 811–13; supersonic aircraft research, 182; synthetic fuel research, 816–17, 816n114; Test Pilot School. *See* Test Pilot School, USAF Transonic Aircraft Technology (TACT) program, 113, 386; V-22 Osprey, 166; Vietnam war aircraft losses, 644n32; Wright Air Development Center, Flight Control Laboratory, 639; Wright Laboratory, 541; X-15 hypersonic flight, 290–91; X-20 Dyna-Soar program, 303–4, 400; X-29 aircraft, 526; X-31 aircraft, 707; X-51A vehicle, 345; YF-12 Blackbird project, 517; YF-102/YF102A Delta Dagger, 100–102

Air Force Academy, 216

Air Force Aero-Propulsion Laboratory, 335

Air Force Air Material Command, 59

Air Force Flight Dynamics Laboratory (AFFDL): Advanced Fighter Technology Integration (AFTI) F-16 project, 670; Control Configured Vehicle (CCV) F-16 project, 669–70; Digital Flight Control for Tactical Aircraft (DIGITAC), 382–83, 684–86; electronic flight control system, 372; FLEXSTAB program, 544–45; flutter suppression research, 388; fly-by-wire system research, 647; Highly Maneuverable Aircraft Technology (HiMAT), 686–89; JB-47E program, 645–46; reentry vehicle research, 404, 406, 415; supercritical wing refinement, 113; Transonic Aircraft Technology (TACT) program, 113, 386; X-24C computed surface streamlines, 299, 448–49; X-29 aircraft, 692

Air Force Flight Test Center (AFFTC): F-100A research, 44; inertia table, 396; limit-cycle oscillation research, 368; simulators and flight simulation, 403–4, 406–7; sonic boom research, 195–96, 201–3; sonic booms and Century series fighters, 185

Air Force Flight Test Center Museum, 686

Air Force Institute of Technology, 542

Air Force Museum, 648, 679, 709

Air Force Plant 42, 196

Air Force Research Laboratory (AFRL): Advanced Control Technology for

Integrated Vehicles Experiment (ACTIVE) project, 716–17; Air Vehicle Directorate, 388; Electrically Powered Actuator Design (EPAD) program, 384

Air Force Scientific Advisory Board, 585

Air Force Systems Command (AFSC), 29n56

Air France Concorde. *See* Concorde Supersonic Transport (SST)

Airlines, profitability of, 743–44, 788

AIRLOADS program, 556

Air Mobile Airborne Division, 139

Air pollution, 736. *See also* emissions

Airport noise. *See* noise

Air Proving Ground, 44

Air resonance, 149

AJ-37 Viggen, 391, 651n49

Alford, William J., Jr., 36, 60–61, 61n115, 638n16

Allee, Cody, 706, 707

Allen, Harry Julian "Harvey": blunt-body theory, 54, 285–86, 291, 311, 415, 440–41, 579, 599; flattop delta aircraft, 56; nickname for, 441

AlliedSignal Engines, 796

Allison Engines, 796, 800–801

Allison/Rolls-Royce, 762, 797

Allison turbo shaft engine, 800–801

All-moving tail research, 34–35, 35n65, 362–63, 363n4

Alloy 454, 614–15

Almroth, B.O. "Bo," 551

Alonso, Juan, 494

Alper, Marshall E., 474

Alumina, 590

Aluminum: ablation cooling and, 417; aircraft built with, 310, 571, 602, 606–9, 620; Alloy 454, 614–15; bonding to, 609; chemical milling, 569; composite materials compared to, 804; high-speed aircraft and, 569; sonic booms and, 210; Space Shuttle orbiter built with, 310, 602, 606–9; structural analysis for building with, 464; superalloy addition, 575, 583; temperature and strength of, 569

American Helicopter Society (AHS): Annual Forum *Proceedings*, 138; charter of, 135; Grover E. Bell Award, 169–70; *Journal*, 138; rotary wing noise research, 163, 164–65

American Institute of Aeronautics and Astronautics (AIAA): *Astronautics & Aeronautics*, 599; research results, dissemination of through, 138, 207; Sylvanus Albert Reed Award, 116

American Society of Mechanical Engineers, 759

American Technology Alliances (AmTech), 852n33, 853, 855

America's Cup yacht race, 124–25

Ames-Dryden AD-1 oblique wing testbed, 63–64, 515, 522–26, 530

Ames Research Center/Ames Aeronautical Laboratory: Advanced Concepts Simulator, 715; Advanced Tiltrotor Transport Technology (ATTT) program, 167–68; Advanced Turboprop (ATP) project, 800; Aeronautics Research Mission Center, transformation from, 166; aircraft handling qualities criteria, 145–46; Air Force F-111 program, 61–62; Army research presence at, 139, 147; blunt-body problem solution, 442; computational fluid dynamics (CFD), 234–36, 437–38, 451; computers, analog, 468; computers, ERA 1103 digital, 437; computers, IBM digital, 437–38; computers, Illiac IV, 438; CONMIN program, 492–93, 498, 512; delta wing development, 53; ENSAERO code,

493–94; finite element analysis activities, 491–94; flattop delta research, 55–56; FLEXSTAB program, 518, 544–45; Flight Simulator for Advanced Aircraft (FSAA), 151; Full-Scale Tunnel, 147, 153, 156–58; high-altitude aircraft workshop, 850, 851; Highly Maneuverable Aircraft Technology (HiMAT), 687, 689; Hypersonic Weapons Research and Development Supporting System (HYWARDS), 300–301; hypersonic wind tunnel, 283; Illiac IV computer, 325; Intelligent Flight Controller (IFC), 720; Intelligent Flight Control Systems (IFCS) project, 719; JF-100C Super Sabre, 402n62; Joint Sponsored Research Agreement (JSRA), 852n33; M-2 lifting body, 56; NASTRAN, development of, 474; NASTRAN, Users' Colloquia, 536; National Full-Scale Aerodynamics Complex, 158, 165–66; National Rotorcraft Technology Center, 168–69; OVERFLOW code, 452, 453; pitch-up research, 25–26; Propulsion Controlled Aircraft (PCA) program, 715; rotary wing aircraft research, 137, 147, 148, 149–50, 152–53, 155–61, 163–64, 492; rotary wing noise research, 163; Rotor Systems Research Aircraft (RSRA) program, 155, 163–64, 165; Rotor Test Apparatus (RTA), 153; RSI coatings, 315; Science Mission Center, transformation to, 166; sensor technology development, 861; sonic boom research, 214, 218, 223–24, 233, 235; Space Shuttle program, 56–57, 327; SST program, 66, 191; stability research activities, 637; Steep Approach program, 737; Structural Dynamics Laboratory, 553; tail configuration research, 36–37; thermal-protection tiles, 316–17, 319, 419; Tilt Rotor Research Aircraft (TRRA) program, 155, 156, 160–62, 165, 166; Transonic Aircraft Technology (TACT) program, 386; Ultra Efficient Engine Technology (UEET) program, 762, 797; Unitary Wind Tunnel, 214, 235, 586; variable stability aircraft, 401; Vertical Motion Simulator (VMS), 151, 167; wind energy program, 829; wind tunnel studies, 25, 51, 327, 524; X-29 aircraft, 526; YF-12 Blackbird project, 517

AmTech, 852n33, 853, 855

Analex Corporation, 539

Analog computers: active-element computers, 468; direct analog computers, 466–68; flight simulation and, 392–94, 405–7, 468; fly-by-wire systems, 648–49, 656, 656n57, 660, 664; hybrid direct-analog/digital computers, 468; indirect analog computers, 468

Anderson, Griffin, 336

Anderson, John D. Jr., 341–42

Angle of attack (AoA): aerodynamic damping and, 364, 366; arrow wing and, 69; deep stall and, 67n127; dihedral effects, 410; fly-by-wire systems and angle-of-attack limiters, 665–66; oblique wing concept, 524–25; pitch-up and swept and delta wing aircraft, 21, 33, 34, 34n61, 36, 50, 56; Shuttle reentry, 57; X-29 aircraft, 694, 695, 696–98; X-31 aircraft, 701–2, 706–7

Anglo-French Concorde SST. *See* Concorde Supersonic Transport (SST)

Anhedral effects, 29, 34, 34n62

ANSYMP program, 545

Antelope Valley, 247

Antenna Design and Analysis, 549–50

Antipodal aircraft. *See* Silver Bird (Silbervögel)

Antispin parachute, 666n70, 695, 696

Apollo 8 mission, 513
Apollo 9 mission, 513
Apollo 11 mission, 654
Apollo 13 mission, 513
Apollo program: ablation cooling, 417; centrifuge program, 400; Command and Service Module, 654; Display and Keyboard (DSKY), 655, 655n55; flight control systems, 644, 753; fly-by-wire systems, 653–54, 663; fuel cells, 814; hydrogen fuel, 814; Inertial Management System (IMS), 655; Lunar Landing Training Vehicles (LLTVs), 653–54; Lunar Module, vii, 654; moon landing mission, vii; pogo, 328; pogo oscillations, 513; reentry vehicles, 54; side stick flight controller, 658n59; silica fiber radome, 598; sonic boom research, 206, 226; success of, 64
Applications Engineering process, 488
Applied Physics Laboratory, Johns Hopkins University, 332, 333
Apt, Milburn "Mel," 44n87, 394–95, 407
Aramid fibers, 767
Arc-Heated Scramjet Test Facility (AHSTF), 343
Arc-jet, 316–17, 415
Area rule: aircraft built with, 104; announcement of, 106–7, 106n48; application of to aircraft design, steps in, 100; benefits of, 89–90, 126; development of, 95n14, 97–100; models, 88, 97, 99; naming of, 99; recognition of, 107–8; research memorandum on, 99–100; secret status of, 106; transonic transport fuselage, 127
Area rule validation in flight, 100–106; F8U Crusader series, 103–4, 105; F9F-9/F11F-1 Tiger, 104, 105, 106; F-102 Delta Dagger, 51, 52; F-104 Starfighter, 104; YF-102 Delta Dagger, 51, 89, 100–102, 103

Argonne National Laboratory, 540
Argyris, John, 471
Arlington, Virginia, 537
Arms race, 442
Armstrong, Neil, vii, 64, 276, 307, 591, 654
Armstrong-Whitworth flying wing, 70
Army, U.S. (USA): Aeroflightdynamics Directorate, 170; Air Mobile Airborne Division, 139; Armament Research & Development Command, 481–82, 540, 541; Atlas ICBM, 581, 582; Aviation Applied Technology Directorate, 146; aviation laboratory (AVLABS), 146, 150; Benet Laboratory, 541; Frankford Arsenal, 542; Harry Diamond Laboratory, 541; missile tests, 288; NASA, relationship with, 139, 146, 147, 152; NASTRAN Users' Colloquia, 480, 481–82, 540; National Rotorcraft Technology Center, 168–69; rotary wing aircraft research, 147, 150, 154–55, 166, 168–69, 170–71; Rotorcraft Division, Army-NASA, 166, 170; Rotor Systems Research Aircraft (RSRA) program, 154–55; Tactical Mobility Requirements Board (Howze Board), 139; Tank Automotive Command, 549; Tilt Rotor Research Aircraft (TRRA) program, 155, 165, 166; Vietnam war aircraft losses, 644n32
Army Aeronautical Activity at Ames (AAA–A), 147
Army Air Corps (AAC), 141–42
Army Air Forces (AAF): Dart, sponsorship of, 9–10; glomb (guided glide bomb), 11; swept wing research, 13–14; technical intelligence interrogations, 14–15, 17
Army Air Mobility Research & Development Laboratory, U.S., 492–93, 540
Army Air Service Engineering Division, 118
Army Research Laboratory, 612

Army Research Office, 169
Army Structures Laboratory, 162–63
Army Tactical Missile System (ATACMS) rocket booster, 345
Army Transportation School, Fort Eustis, Virginia, 146
Arnold, Henry H. "Hap," 14, 291
Arnold, Steven, 500
Arnold Engineering Development Center (AEDC), 17n30, 158, 280, 291, 760, 787
Arrowplane, 5
Arrow-Shaped Lifting Surface (Pfielförmiges Tragwerk), 6
Arrow wing, 65, 67–68, 69, 209, 225
Artificial feel devices, 635, 643
Aspect ratio of a wing, 119–20, 407
ASSIE-I program, 552
Astronautics & Aeronautics, 599
Asynchronous digital flight control system, 672, 680–82, 683
Atar 9K-50 engine, 691
Atlanta, 206
Atlas intercontinental ballistic missile (ICBM), 284–85, 286, 288, 293–94, 579–82
Atmospheric Effects of Aviation project, 797
Atmospheric Effects of Stratospheric Aircraft project, 795–96
Atomic bombs, 284
Atomic Energy Commission (AEC), 206–7, 213, 481
Augenstein, Bruno, 284
Augmented Longitudinal Control System (ALCS), 372
Ault, Stan, 861
Aurora Flight Sciences, 851, 853
Autogiros, 135, 137, 140, 141–42, 145
Automated Maneuvering Attack System (AMAS), 672, 673, 674–75, 676
Autonomous Frequency Domain Identification (AU-FREDI), 504, 505
Autopilots, 631, 687–88, 706
Autothrottle system, 706
Auxiliary power units (APUs), 295, 503, 815
Avery, William, 332
Aviation Applied Technology Directorate, Army Transportation School, Fort Eustis, Virginia, 146
Aviation industry: airlines, profitability of, 743–44, 788; certification of new aircraft, 683–84; design of commercial aircraft, 208; finite element structural analyses use by, 533–34; fly-by-wire systems, 632, 722–23; NASTRAN use by, 478, 480–81; NASTRAN Users' Colloquia, 480; supercritical wing use by, 114–15; winglet use by, 120–22, 806
Aviation laboratory (AVLABS), 146, 150
Aviation Partners, Inc., 124
Aviation Partners Boeing, 124
Aviation Week, 106, 106n48, 319, 639, 856
Aviation Week and Space Technology, 253, 260–61
Avro 707C aircraft, 643
Avro Canada, 641–42, 642n25
Avro CF-105 Arrow, 641–42, 642n25, 643
Avro Vulcan bomber, 45, 639

B

B-1 bomber, 62, 62n118, 378; B-1A bomber, 392; B-1B bomber, 392
B-2 Spirit, 722, 753
B-17 Flying Fortress, 736
B-26 Invader, 402n62
B-29 Superfortress, 29, 38, 92, 290, 736
B-36 bomber, 55
B-45 aircraft, 634–35
B-47 aircraft: aeroelastic effects, 20, 390, 409–10; B-47B aircraft, 645, 736; B-47E

aircraft, 658n59; JB-47E aircraft, 645–46; side stick flight controller, 658n59

B-50 aircraft, 290

B-52 Stratofortress: aeroelastic effects, 390–91, 392; DAST launch from, 389; flight control systems, 752; NB-52/NB-52B Stratofortress, 343, 369, 378–79, 568; vortex generators, 65; X-15 launch from, 197

B-57B bomber, 810, 811

B-58 Hustler: aeroelastic effects, 392; B-70 as replacement, 191; cross-country flight by, 188; design of, 45, 53, 103; sonic boom research, 194, 195–98, 199, 200, 202, 203, 204, 205, 209; sonic booms from, 188–89; supersonic flight by, 188, 189

B-70 aircraft, 191

Baals, Donald, 99

Babcock and Wilcox, 601

Backup Control System (BCS), 656, 657, 658, 660

Baer-Riedhart, Jenny, 854–55

Balaclava, 828

Baldwin, Darrell H., 830

Baldwin-Lomax algebraic model, 341

Ball bearings, 307, 590

Ballistic Missile Defense Organization, 850

Ballistic missiles. *See also* intercontinental ballistic missiles (ICBMs): development of, 284; intermediate-range ballistic missile (IRBM), 286, 442, 571–72; V-2 ballistic missiles, 183–84, 280, 282, 571, 635–36

Ballistic shock waves, 183–85

Bamford, Robert M., 474

Banas, Ronald, 310, 598

Bare Reactor Experiment Nevada (BREN) Tower, 206–7

Barger, Raymond L., 223–24, 227

Barlow, Robert, 647

Barrowsville, Virginia, 634

Barter, Steve, 678

Base Amplitude and Direction Sensor (BADS), 247

Batch operation, 437–38

Battelle Memorial Institute, 230

Bauer, Jeffrey, 855

Baughman, John, 791, 793

Beam, Richard M., 474

Beamont, Roland, 23–24, 24n40

Bearingless Main Rotor (BMR), 157

Bearings, 307, 590

Beasley, R.M., 310, 598

Becker, John V.: aerothermodynamic research, 93, 418, 573; flat-bottom delta aircraft, 55; hypersonic aircraft and flight research, 282–83, 291; HYWARDS, 300, 301; scramjet engines, 333; on Whitcomb's research techniques, 113; X-20 Dyna-Soar program, 307

Beech C-45 aircraft, 199

Beechcraft, 481

Beechcraft Starship, 121

Beechcraft Super King Air 200 twin turboprop aircraft, 481

Beechcraft T-34C Mentor, 481

Beech L-23 sailplane, 248, 257

Beggs, James, 192

Belgium, 184

Bell, Lawrence D., 93, 94

Bell Aerospace Company, 479

Bell AH-1 Cobra FLITE configuration, 170

Bell Aircraft Corporation, 17n30, 290

Bell Award (AHS), 169–70

Bell BA 609 aircraft, 168

Bell BoMi bomber-missile, 290, 300

Bell Brass Bell boost-glider, 300, 301, 302–3

Bell H-13 helicopter, 150

Bell Helicopter Company/Bell Helicopter Textron: Design Analysis Methods for

Vibration (DAMVIBS), 508, 509, 509n128; helicopter design and production, 136; RITA membership, 169; tilt rotor design, 153

Bell L-39 aircraft, 22–23, 23n37, 23n39

Bell Lunar Landing Research Vehicles (LLRVs), 653–54

Bell Model 222 rotary wing aircraft, 156

Bell P-39 Aircobra, 23n37

Bell P-63 Kingcobra fighters (L-39 aircraft), 22–23, 23n37, 23n39

Bell RoBo rocket bomber, 300, 301, 303

Bell UH-1 Huey, 147

Bell V-22 Osprey Tilt Rotor, 136, 155, 161, 166

Bell X-1 aircraft: Collier Trophy, 93; development and manufacture of, 108, 290, 494; engines for, 294; flight control systems, 362–63, 752; gliding landings, 402–3; lift-to-drag ratio (L/D), 403; Round One research aircraft, 55n106; supersonic capabilities, 94; tail configuration, 411; X-1A aircraft, 40–41, 40n74, 43, 292, 363, 407; X-1B aircraft, 40n74, 298; X-1D aircraft, 40n74; Yeager's flight, 40–41, 292, 363, 407

Bell X-2 aircraft: accident with, 44n87, 394–95, 407, 410; control coupling, 410; design of, 34; development and manufacture of, 17n30, 290; dihedral effects, 410; flight control systems, 298; inertial coupling, 44n87; pressure suit for, 297; simulator, 394–95; tail configuration, 411

Bell X-5 aircraft, 17n30, 58, 59–60, 413

Bell XS-1 aircraft: design of, 3, 28, 34, 34n62, 36, 184; falling body test, 38, 43; flight control systems, 635; inspiration for, 14n22; sonic boom from, 182, 182n3; supersonic flight by, 182, 183; testing of, 35; variable-geometry wing tests, 59; wingspan-to-fuselage length ratio, 37; XS-1 FSW planform, 21; Yeager's flight, 182, 182n3, 283, 635

Bell XS-2 aircraft, 34, 34n62

Bell XV-1 Convertiplane, 134, 147

Bell XV-15 Tilt-Rotor Research Aircraft (TRRA), 155, 156, 160–62, 165, 166

Bendix nose skid, 592

Bendix ramjets, 289

Benet Laboratory, 541

Benscoter, Stanley, 467

Berg, Moe, 332–33

Beryllia, 587

Beryllium, 607, 620–21

Beta-21S, 617–18, 621

Betz, Albert, 15

Beyond the Limits gallery, National Air and Space Museum, 427, 432

B.F. Goodrich, 481, 538

Big Boom, Project, 195

Bikle, Paul, 296

Bi-Level Integrated System Synthesis (BLISS), 511

Billig, Frederick, 332

Biofuel, 817

Biplanes, 4–5

Birchenough, Arthur, 830

Birds, 385, 386

Bisplinghoff, Raymond, 473

Blackburn Buccaneer, 638n16

BLAde SIMulation (BLASIM), 547

Blade-vortex interaction (BVI) noise, 163, 164–65

Blended wing-body (BWB): Boeing X-48B Blended Wing Body UAV, 2; BWB-450, 71; development of, 70–72; heavy-lift military missions, 71, 71n131; hybrid laminar flow control, 808; performance of,

415; thermodynamic characteristics, 415
Blériot, Louis, 385
Block Island, Rhode Island, 827
Blow-down design, 280, 283
Blue Angels, 104
Blue Ridge Electrical Membership Corporation, 835
Blunt-body theory: computational fluid dynamics and solution to, 439–44; development of, 54, 415, 441; limitations of, 415; numerical solution for, 442–44; reentry vehicles, 54, 285, 291, 441, 599; shock waves, 286, 415, 440–42; Star Clipper spacecraft, 311; thermodynamic characteristics, 291, 311, 415, 579
BMW 801 command equipment, 633n1
Bobbitt, Percy "Bud," 235, 246
Boccadoro, Charles H., 246
Bodies of revolution, 184
Boeing 367-80 aircraft, 20
Boeing 707 aircraft, 4, 20, 65, 123, 740, 804
Boeing 727 aircraft, 737, 745, 802, 804
Boeing 737 aircraft, 124, 616, 737, 745, 771
Boeing 747 aircraft: 747-400 winglet use, 89, 123–24; composite materials, 804; engine component improvements, 745; NASA's aeronautics role and, 787; Shuttle carrier aircraft, 661; simulators testing of PCA, 715; TAVs launched from, 337; titanium construction, 616; winglet use on, 89, 122, 123–24
Boeing 757 aircraft, 124, 715, 756, 795, 807–8
Boeing 767 aircraft, 124, 615
Boeing 777 aircraft: 777-300ER aircraft, 795; fly-by-wire systems, 722, 753; GE90 jet engine, 742, 743, 747; materials for, 620; noise from, 771; supercritical wing, 115; titanium construction, 616

Boeing 787 aircraft: electrical flight control surfaces and actuators, 722; engine for, 769, 795; materials for, 571, 620; noise from, 771; range of, 620; titanium construction, 616
Boeing B-17 Flying Fortress, 736
Boeing B-29 Superfortress, 29, 38, 92, 290, 736
Boeing B-47 aircraft: aeroelastic effects, 20, 390, 409–10; B-47B aircraft, 645, 736; B-47E aircraft, 658n59; JB-47E aircraft, 645–46; side stick flight controller, 658n59
Boeing B-50 aircraft, 290
Boeing B-52 Stratofortress: aeroelastic effects, 390–91, 392; DAST launch from, 389; flight control systems, 752; NB-52/NB-52B Stratofortress, 343, 369, 378–79, 568; vortex generators, 65; X-15 launch from, 197
Boeing Business Jet, 124
Boeing Company: Advanced Control Technology for Integrated Vehicles Experiment (ACTIVE) project, 717; biofuel research, 817; BWB research, 71–72; columbium coating, 585; composite materials for aircraft parts, 804; Cooperative Test Organization/VECTOR, 705–6; F-16XL modifications, 261; financial problems, 217; FLEXSTAB program, 518; high-altitude aircraft workshop, 851; High-Speed Civil Transport (HSCT), 231, 232, 243; hydrogen fuel cell aircraft, 815; Intelligent Flight Control Systems (IFCS) project, 719; McDonnell-Douglas merger, 127; NASA contract work, 120; NASTRAN Users' Colloquia, 480; nose cap design, 587, 588–89; Numerical Propulsion System Simulation (NPSS) program, 760; P&W turbo fan engines on jetliners, 710;

Phantom Works, 71; QSP proposal, 245; RITA membership, 169; Shaped Sonic Boom Demonstration (SSBD), 246; sonic boom research, 213, 224; SST program, contract for, 211, 217–18; SST program, data to support, 519; SST program, proposal for, 210; STOL transport competition, 114; supersonic business jet (SSBJ) concept, 249; Supersonics Project, 261; Trans Atmospheric Vehicles (TAV), 337; Ultra Efficient Engine Technology (UEET) program, 762, 797; Vertol Division, 136; wind energy program, 829, 835, 838–39, 840, 841; winglet design, 124; winglet use by, 121–22, 124

Boeing F-15 aircraft, 710. *See also* McDonnell-Douglas F-15 Eagle

Boeing FLEXSTAB, 392

Boeing Helicopters, 508, 509–10. *See also Boeing-Vertol entries*

Boeing JB-47E aircraft, 645–46

Boeing KC-135 aerial tanker: aerial refueling, 253; design of, 20; sonic boom research, 204; winglet use on, 120, 122–23, 806

Boeing NB-52/NB-52B Stratofortress, 343, 369, 378–79, 568

Boeing PB2-1S Superfortress (B-29), 29

Boeing Scientific Research Laboratories, 445

Boeing Spanloader, 70

Boeing SST concept, 63, 68

Boeing-Vertol, 540

Boeing-Vertol Bearingless Main Rotor (BMR), 157

Boeing-Vertol CH-47 Chinook, 151, 162, 163; CH-47C Chinook, 152; CH-47D Chinook, 509–10

Boeing-Vertol tilt rotor concept, 153

Boeing-Vertol XCH-62 Heavy Lift Helicopter (HLH) program, 158

Boeing WC-135B Stratotanker, 204

Boeing X-20 Dyna-Soar: cancellation of, 304, 308, 400, 592; centrifuge program, 400; contributions from program, 307–8; design confidence, 307–8; development of, 300–305; flight control system, 372; flutter-testing panels, 585–86; hot structures, 305–8, 418, 582–92; justification of, 302–5; landing skids, 592; materials for, 305–8; missiles to launch, 301–2; nose cap, 306, 308, 587–89, 598; Round Three research aircraft, 55–56; Round Two research aircraft, 55n106; schematic drawing, 583; simulator, 405–6; windows, 307

Boeing X-48B Blended Wing-Body UAV, 2, 71–72

Boeing X-51A vehicle, 279, 345–46

Boeing XB-52 aircraft, 20

Boeing YC-14 transport, 114–15, 670–72

Bomarc missile, 289

Bombardier CSeries aircraft, 752

BoMi bomber-missile, 290, 300

Bond, Alan, 124–25

Bongo, Operation, 197–98

Bongo II, Operation, 199–200

Bonneville Power Administration, 837–38

Boom Event Analyzer Recorder (BEAR), 229–30, 238

Boomfile, 229, 238

Boone, North Carolina, 835

Boosted (hydraulic) control-surface actuators, 384, 590, 631, 634–35, 643, 644

Boost-glider, 55

Boria, 315

Boron, 574, 583, 614

Boron, California, 182, 202

Borosilicate glass coatings, 315–16, 599–600

Boscombe Down, 638n16
Bossart, Karel J. "Charlie," 284
Boston, 536
Bosworth, John, 719
Boundary condition, 410, 431
Boundary layer: boundary layer control (BLC) experiments, 25, 147; laminar flow control research, 235, 495, 744, 784, 786, 806–8; sonic boom research and, 196
Bowles, Mark, 802
Bradshaw, Peter, 341
Brass Bell boost-glider, 300, 301, 302–3
BREN Tower, 206–7
Breuhaus, Waldemar "Walt," 401
Brewer, Jack D., 36
Bright, George W., 648
British Aeronautical Research Committee, 118
British Aerospace (BAe), 690
British Aircraft Corporation BAC 1-11, 67n127
British Airways Concorde. See Concorde Supersonic Transport (SST)
Brooklyn Polytechnic Institute, 332–33
Brooks, Peter, 4
Brown, Clinton E., 67
Brugh, Robert L. "Bob," 485, 536
Brush, Charles F., 825
BUCKY program, 507, 550
Bullet, Project, 103–4
Bulzan, Dan, 799, 816–17
Bureau of Reclamation, 839
Burgess Company, 5
Burgess-Dunne seaplanes, 5
Burke, Melvin E., 654, 655
Burning, Pieter, 452
Burroughs Corporation Illiac IV computer, 325, 438
Burstein, Adolph, 46, 46n90

Busemann, Adolf, 5–6, 15, 15n27, 93, 98, 99
Butler, Thomas G., 473, 474, 476, 532
Butler Analyses, 539
BWB-450 aircraft, 71

C
C-5 aircraft, 337, 407, 867
C-17 Globemaster III transport, 127, 384, 715, 722, 795
C-45 aircraft, 199
C-47 transport, 195, 411
C-131B aircraft, 204
C-141 aircraft, 324, 658n59
Cadmium, 570
Cadmium-plated tools, 570
California Institute of Technology (Caltech), 12, 17n30, 502
CALSPAN (Cornell Aeronautical Laboratory), 401–2, 402n62
CALSPAN NT-33 aircraft, 642
CALSPAN VISTA Lockheed-Martin NF-16, 402n62
Calverton, New York, 528
Cambridge, 736
Campbell, John, 63
Canada, 641
Canadair Challenger 600 aircraft, 115
Canadair Sabre, 24
Canards, 224, 391, 392, 519, 526, 647–48
Carbides, formation of, 574–75. See also silicon carbide; titanium carbide; tungsten carbide
Carbon, 614
Carbon braid engine fan casing, 768, 769
Carbon-carbon composite, 318, 576–78, 618–20
Carbon composites, 619–20
Carbon dioxide emissions, 790, 790n22, 797–98, 799, 813
Carbon fibers, 576, 578, 620, 768,

769, 866, 868
Carbon-phenolic composite, 576
Carderock Division, Naval Surface Weapons Center, 157
CARES. *See* Ceramics Analysis and Reliability Evaluation of Structures (SCARE/CARES)
Carl, Marion, 30, 297
Carlson, Harry W.: SCAT-15 arrow wing, 67, 209; sonic boom research, 193, 207, 218; sonic boom research, conference and symposia presentations, 210, 211, 212; sonic boom research, minimization concepts, 223–24, 225–26; sonic boom research, wind tunnel studies, 208–9, 208n66
Carlswell Air Force Base, 673
Carter, Alan, 518
Case, William R., 476
Castigliano, Carlo Alberto, 466
Castle Bravo nuclear test, 284–85
Caterpillar Tractor Company, 612
Caustics, 226–27, 228
Center for High-Speed Commercial Flight, 230
Center for Rotorcraft Innovation (CRI), 169–70
Center for Turbulence Research, Stanford University, 341
Center of gravity (cg), 363, 632
Central Intelligence Agency (CIA), 312, 582
Centrifuge, 398–400
Centurion, 850, 856, 859, 862, 864–66
Century series fighters. *See also* Convair F-102 Delta Dagger; Lockheed F-104 Starfighter; McDonnell F-101 Voodoo; North American F-100 Super Sabre; Republic F-105 Thunderchief: area rule aircraft, 103; damper systems, 365, 366; design of, 401, 408–9; management of program for, 53n103, 61; roll coupling (inertial coupling), 408–9; sonic booms from, 185, 187; speed of, 290; tail configuration, 34
Ceramics: ceramic fiber mats, 310–11, 598; ceramic matrix composite (CMC) components, 499, 798; ceramic tiles, 419; ceramic zirconia, 598; environmental barrier coatings, 612; limitations of, 598; thermal barrier coatings, 610; for X-20 Dyna-Soar, 587–88
Ceramics Analysis and Reliability Evaluation of Structures (SCARE/CARES), 499–500; CARES/LIFE, 499–500; CARES/PC, 499; CMC CARES (CCARES), 499
Ceria, 587
Cerium, 574
Cessna 150 aircraft, 204
Cessna 152 aircraft, 55
Cessna 182 Skylane, 720
Cessna supersonic business jet (SSBJ) concept, 249
CF-105 Arrow, 641–42, 642n25, 643
CFL criterion, 436
CFM International, 803
Challenger, 329–30
Chambers, Joseph R., 125
Chamis, Christos, 536
Champine, Robert, 30
Chance Vought nose cap, 306, 587–89
Charles, Prince of Wales, 371
Chemical milling, 569
Chemical vapor deposition (CVD) coatings, 613
Cheranovskiy, Boris Ivanovich, 5
Cheung, Samsun, 234–35
Chevron exhaust nozzle, 808–9
Chicago, 200, 206
Chlorine, 570

Christian Science Monitor, 107
Chromium, 306, 575, 583, 614
Chrysler Corporation, 498
Circulation Control Rotor (CCR), 156–57
Civil Aeronautics Administration (CAA), 135
Civil aviation. *See* aviation industry
Civil Tilt Rotor Development Advisory Committee (CTRDAC), 167
CL-400 aircraft, 812–13
CL-1200 Lancer, 34n61
Clark, Larry L., 111
Clark, Thomas L., 475
Clark Atlanta University, 71
Clarke, Mike, 647
Clark MC-2 full-pressure suit, 276, 297, 298
Clayton, New Mexico, 833
Clean Air Act, 792
Clemson University, 542
Cleveland, 105, 137, 191
Climactic Impact Assessment Program (CIAP), 789
Climate change research, 789–90, 790n22, 818
Clinton, William J., 243
Clipper Wind Power, 848
Closed-form analytical solutions, 433, 434
Clough, R.W., 471
Cobalt, 306, 575, 583
Cobalt-based alloys, 573, 574, 575, 610
Coca-Cola bottle design, 93, 106–7
Coen, Peter G., 235, 246, 248, 249, 251, 254
Cold jet hydrogen peroxide reaction control, 298
Cole, Julian, 445
Collier Research Corporation, 512
Collier Trophy: Gulfstream business jets, 121; McDonnell-Douglas innovations, 127; NACA for engine and aerodynamic technology, 736; supersonic flight by X-1, 96;

turboprop propulsion system, 748, 800; Whitcomb's work, 52, 93–94, 107
Colorado, 200
Colorado Springs, 537
Columbia. *See* Space Shuttle Columbia
Columbium, 312, 584, 585, 587, 595, 602–3
Columbium alloy, 306, 308
Combined-cycle engines, 337
Combustion-Heated Scramjet Test Facility (CHSTF), 343
Combustor liners, 798
Combustors, 794–95, 798–99
Command and Service Module, 654
Command Augmentation Systems (CAS), 366–67, 382
Commercial aircraft. *See* aviation industry; Supersonic Commercial Air Transport (SCAT) program
Commercial Remote Sensing program, 861
Committee on Technology, National Science and Technology Council, 171–72
Common Object Request Broker Architecture (CORBA), 760, 761
Community and Structural Response Program, 197–99
COmposite Blade STRuctural ANalyzer (COBSTRAN), 546–47
Composite materials: aircraft parts, 804–5; engine fan casings, 767–69; engine parts, 798; Long Range Planning Study for Composites (RECAST), 804; obstacles to use of, 804; UAV construction, 866, 868
Composite Primary Aircraft Structures project, 744, 804–5
Composite structures, 534
Compressed air, 324, 573
Compressible flows, 12, 14, 15
Compressors, 746–47, 793–94, 795
Computational fluid dynamics (CFD):

application of to aircraft design, 428; benefits of, 340, 341; blunt-body problem solution, 439–44; CFL criterion, 436; computer codes, 451–53; development of, 340, 427–28, 453; digital computer development and, 436–39; direct numerical simulation (DNS), 342, 439, 450–51; fluid dynamics theory and, 428–33, 434; limitations of, 340–42; MacCormack's method, 446–47; NASA's role in development of, 428, 453; NASP and, 279; Navier-Stokes equations solutions, 438–39, 446–47; propulsion system research, 803; Shuttle study, 450; sonic boom research, 234–37, 246, 248, 253, 260; three-dimensional flow-field calculations, 447–49; time-marching process, 443–44, 445–46; transonic flow and airfoils, 444–46; upwinding, 445

Computational fluid dynamics (CFD) grid points: computer technology development and, 438–39; direct numerical simulation (DNS), 342, 439, 450–51; Euler equations, 438; flow-field calculations, 433, 434, 435; flow-field calculations, three-dimensional, 446–47; grid generation, 434; numerical accuracy, 446–47

Computational simulations. *See also* simulators and flight simulation: analog computers and, 392–94, 405–7, 468; digital computers, 405–7; hybrid computers, 406–7; Space Shuttle orbiter, 325–26

Computational structural analysis. *See* finite element structural analyses

Computational Structural Mechanics (CSM) workshop, 497

Computer-aided design (CAD), 493, 532, 550

Computer codes for computational fluid dynamics, 451–53

Computerized Structural Analysis and Research (CSAR), 486

Computer Programs Abstracts Journal, 489

Computers: active-element computers, 468; direct analog computers, 466–68; equations of motion and, 393; flight control system use of, 631, 633n1; Goodyear Electronic Digital Analyzer (GEDA), 393, 394; indirect analog computers, 468; supercomputers, 438–39, 446; women mathematicians as, 465, 494

Computers, analog: active-element computers, 468; direct analog computers, 466–68; flight simulation and, 392–94, 405–7, 468; fly-by-wire systems, 648–49, 656, 656n57, 660, 664; hybrid direct-analog/digital computers, 468; indirect analog computers, 468

Computers, digital: batch operation, 437–38; development of, 436–39, 468–70; flight simulation and, 405–7; fly-by-wire systems, 659

Computers, hybrid: direct-analog/digital computers, 468; flight simulation and, 406–7

Computer Sciences Corporation (CSC), 475, 478, 538

Computer Software Management and Information Center (COSMIC), 477, 488–90, 490n62, 499, 508

Concorde Supersonic Transport (SST): aluminum and speed, 569; American research and, 208; demise of, 249; design of, 3, 45, 56, 66; economic condition, 229; flight control systems, 643; ozone layer damage from, 789; service to U.S., 227–28; sonic booms from, 227–28

Concurrent development, 322

Congress, U.S.: Civil Tilt Rotor Development Advisory Committee, 167; fuel efficiency initiatives, 788, 792; High-Speed Research (HSR) program, 232; Joint Planning and Development Office (JPDO) authorization, 171n53; ozone layer research, 789; SCR/SCAR/AST program funding, 221; sonic boom complaints, 187; SST program funding, 218, 219; supersonic research, 219

Conical camber, 51–53

CONMIN program, 492–93, 498, 511, 512

Conservation of energy principle, 432, 433, 443

Consortium for the Design and Analysis of Composite Materials, 500

Continenal Airlines, 123

Continuity equation, 430, 431, 433, 438, 443

Continuous Flow Hypersonic Tunnel (CFHT), 283

Contracted research, 153–54

Control-augmentation (CAS) mode, 659, 660

Control Configured Vehicle (CCV): British Jaguar Active Control Technology (ACT), 670, 690–91, 692; F-16 CCV, 648, 669–70; French Dassault Mirage 3NG, 691; German CCV F-104G, 670, 689–90, 692; Japanese CCV T-2, 670, 691–92; purpose and benefits of, 651–53, 692

Control coupling, 409–10

Control Data Corporation (CDC): 7600 computer, 324, 325; computers, 469; NASTRAN Users' Colloquia, 538

Controllable Twist Rotor (CTR) concept, 153, 156

Control laws, 381–82, 389, 402, 661, 662, 665, 680–82, 696

Control stick feedback, 370–71, 635, 638–39, 643

Control-surface actuators: electrical flight control surfaces and actuators, 384–85,
631, 635; electrohydrostatic actuator, 385, 676–79, 722; electro-mechanical actuator, 385; hydraulically actuated flight control surfaces, 384, 590, 631, 634–35; speed of, 383

Control surface commands, 365, 367

Convair Aerospaceplane proposal, 334

Convair B-36 bomber, 55

Convair B-58 Hustler: aeroelastic effects, 392; B-70 as replacement, 191; cross-country flight by, 188; design of, 45, 53, 103; sonic boom research, 194, 195–98, 199, 200, 202, 203, 204, 205, 209; sonic booms from, 188–89; supersonic flight by, 188, 189

Convair C-131B aircraft, 204

Convair delta wing aircraft, 46, 46n89, 50

Convair F-102 Delta Dagger: area rule and, 51, 52, 103; design of, 52–53, 103, 409; development and manufacture of, 290; F-102A Delta Dagger, 102–3; inertial coupling (roll coupling), 409; management of program for, 53n103, 61

Convair F-102 Delta Dagger (YF-102): area rule and, 51, 89, 100–102, 103; handling qualities, 53; sonic booms, 185; YF-102A prototype, 89, 101–2, 103

Convair F-106 Delta Dart: design of, 45, 53, 409; handling qualities, 50; inertial coupling (roll coupling), 409; purpose of, 103; sonic booms, 185, 196, 199, 201, 204, 261; variable stability aircraft, 402n62

Convair NC-131H Total In-Flight Simulator (TIFS), 402n62

Convair XF2Y-1 Sea Dart, 45, 103

Convair XF-92A aircraft, 45, 46n90, 49–51, 100

Convair XP-92 aircraft, 46

Convair YF-102 Delta Dagger: area rule and,

51, 89, 100–102, 103; handling qualities, 53; sonic booms, 185; YF-102A prototype, 89, 101–2, 103

Conventional Take-Off and Landing (CTOL) aircraft, 120

Convertible Engine Systems Technology (CEST) program, 158, 162

Cook, William L., 476

Cooper, Robert, 337, 339

Cooperative Test Organization, 705–6

Cooper-Harper Handling Qualities Rating Scale, 146

Coors Porcelain Company, 590

Copper Canyon research program, 337–38

Copper chill plate, 614

Copper heat shield, 579

Corliss, William, 99

Cornell Aeronautical Laboratory, 401–2, 402n62

Cornell University, 211, 217, 226

Corona spy satellite, 305

Corrigan, Bob, 846

COSMIC. *See* Computer Software Management and Information Center (COSMIC)

Couch, James P., 830

Counter Intelligence Agency (CIA), 811

Coupled Eulerian Lagrangian Finite Element (CELFE), 514

Courant, Richard, 435–36, 471

Courant Institute, New York University, 111, 471

Cowart, Robert A., 253

Cowling design, NACA, 734, 735–36, 809

Craig, Ray, 695

Cranfield Aerospace, Ltd., 71

Cranked arrow wing, 69

Crash-test activities, 152

Cray computers: Cray-1 computer, 325, 438; Cray X-MP computer, 438, 446; X-24C

computed surface streamlines, 449

Cristobalite, 600

Crocco, Arturo, 6

Crossfield, A. Scott, 26, 27–28, 30, 42, 50

Crow, David Ed., 793, 794

Cudnik, Jason, 260

Culebra, Puerto Rico, 827, 833

Cunningham, Herbert J., 474

Curtin, Robert, 856

Curtis, Al, 518

Curtiss, Glenn, 107

Curtiss XP-46 aircraft, 35

Cut-and-try research technique, 113–14

Cyber 205 computer, 446

Cylindrical Optimization of Rings, Skin and Stringers (CORSS), 513

Czysz, Paul, 572

D

D-36 columbium alloy, 306, 585, 591

D-558-1 Skystreak, 3, 28–29, 65

D-558-2 Skystreak, 21; engines for, 294; inertial coupling experiment, 39; nose-rise problems, 414; pitch-up problems, 28–32, 413; pressure suit for, 297; Round Two research aircraft, 55n106; tail configuration, 36, 411, 413; wing-stall research, 414

Daimler-Benz, 701, 706

Dallas, 206

Dallas Convention Center Heliport/Vertiport, 166

Damage-tolerant fan casing, 766–69

Damper systems, 364–67, 412

Dana, William H., 658, 674, 679, 714

Darden, Christine M., 224–25, 229, 235, 241–42

Dassault Falcon 50 aircraft, 115

Dassault Mirage series: development of, 53;

Mirage 3NG, 691; Mirage 2000 aircraft, 669; Mirage G, 62; Mirage I, 45

Dassault Rafale, 692

Dassault supersonic business jet (SSBJ) concept, 249

Data buses, 684–85

David G. Clark Company pressure suits, 276, 297, 298

David Taylor Naval Ship Research & Development Center, 481, 484, 540, 541

David Taylor Research Center, 541, 542

Da Vinci, Leonardo, 428, 429

Davis-Monthan Air Force Base, 672

Dawson, Virginia, 802

Day, Richard E., 394, 407, 409, 410

Dayton, 638n15

DC-2 aircraft, 6

DC-3 aircraft, 411

DC-4 Generation, 4, 4n1

DC-8 aircraft, 4, 737, 740, 817

DC-9 aircraft, 745

DC-10 aircraft: engine component improvements, 745; Sioux City airliner crash, 757, 766; winglet research, 119, 120, 122, 123

Decoupled (unconventional) control modes, 652, 669–70

Deep-stall conditions, 67, 67n127, 413, 666n70

Deets, Dwain A., 654

Defense, Department of: aircraft safety, 506; aircraft technology and systems to support, 171, 172; flight control systems, research activities on, 362–63; Jason advisory group, 227; software development, 489; SST program, 191, 210–11

Defense Advanced Research Projects Agency (DARPA): computational fluid dynamics, 340; Convertible Engine Systems Technology (CEST) program, 158, 162; Copper Canyon research program, 337–38; hypersonic aircraft, 230; NASA, relationship with, 139; powder metallurgy, 617; Quiet Supersonic Platform (QSP), 244–48, 249; Trans Atmospheric Vehicles (TAV), 337; X-29 aircraft, 526, 692; X-31 aircraft, 701; X-51A vehicle, 345; X-Wing Stoppable Rotor program, 157, 158, 163–64, 165

Degree of freedom: unit displacement method, 469, 470

De Haven, Hugh, 9n13

De Havilland, Geoffrey, 27

De Havilland D.H. 108 Swallow, 27

Del Frate, John, 855, 858, 858n48, 862

Delta wing aircraft. *See* swept and delta wing aircraft

Denmark, 828, 846–47, 848

Denver, 206

Desert Storm, 62

Design Analysis Methods for Vibration (DAMVIBS), 508–11

Deutsche Aerospace, 699–700, 706

Deutsches Museum, 707

Deutsch Metal Components, 539

DFS 346, 36

DFS DM-1, 13

Dielmann, Howard, 476

Differential Maneuvering Simulator (DMS), 666–67

Digital Array Scanned Interferometer (DASI), 861

Digital computers: batch operation, 437–38; development of, 436–39, 468–70; flight simulation and, 405–7; fly-by-wire systems, 659

Digital Electronic Engine Control (DEEC) program, 707–10, 754–57, 758–59, 766

Digital Electronic Flight Control System (DEFCS), 711

Digital Flight Control for Tactical Aircraft (DIGITAC), 382–83, 684–86
Digital flight control system (DFCS), 655, 672–73, 674, 680–82
Digital fly-by-wire (DFBW) flight control systems. *See also* Digital Fly-By-Wire F-8 program and aircraft: actuators, 384–85; development of, 380, 753; military use of, 383–84; redundant systems, 384; software programs, 381–82, 659–60, 671–72, 683–84; space flight program use of, 632, 653–54; testing of, 380–83
Digital Fly-By-Wire F-8 program and aircraft: aircraft modifications for, 380, 654–56, 658, 661; AP-101 computers, 659, 660; Backup Control System (BCS), 656, 657, 658, 660; benefits of program, 380–81, 632, 662–64; control laws, 381–82, 661, 662; exhibit showing, 663; flight control modes, 659–60; flight-testing, 647, 657, 661–62, 753; follow-on research testing, 381–82; gain, 656; implementation of Phase I, 655–58; implementation of Phase II, 658–62; Iron Bird simulator, 655, 656–57, 659, 660; Linear Differential Variable Transformers (LDVTs), 656; performance of, 663–64; preflight safety procedures, 383; redundant systems, 380, 655–56, 658, 659; Resident Backup System (REBUS), 662; side stick flight controller, 658, 658n59; simulated Shuttle landing flights, 661; software programs, 381–82, 659–60; transport delay, 661; YF-16 design validation, 665
Digital Matrix Abstraction Programming (DMAP), 476
Digital Systems Validation Handbook (FAA), 684
Digital terrain system with moving map, 675–76

Dihedral effects, 19, 34n62, 390, 410–11
Direct analog computers, 466–68
Direct Connect Supersonic Combustion Test Facility (DCSCTF), 343
Direct control system, 640, 653, 659, 660
Direct current power generation system, 677–78
Directionally solidified (d.s.) castings, 614–15
Directional stability: flattop delta aircraft, 55; fuselage length and, 37–38; inertial coupling (roll coupling) and, 38–39, 40–41; static, 408, 409, 410–11, 648, 651; swept and delta wing aircraft, 19; wedge-shaped tail and, 283
Direct numerical simulation (DNS), 342, 439, 450–51
Discoverer II vehicle, 305, 582
Discretization of structures, 466, 469, 470–72
Display and Keyboard (DSKY), 655, 655n55
Distinguished Service Medal, 107
Distortion Estimation System (DES), 717–18
Divone, Louis, 825
Donlan, Charles, 59, 603
Doolittle, James H., 737
Dornberger, Walter, 290, 300
Dornier Do-31 aircraft, 648–49
Dorsch, Robert, 331
Doublet-lattice model, 482–83, 526
Douglas, Frank J., 475
Douglas A4D-1 Skyhawk, 45–46
Douglas A-4 Skyhawk, 65
Douglas Aircraft Company. *See also* McDonnell-Douglas Corporation: delta wing development, 46–47; High-Speed Civil Transport (HSCT), 231, 232, 233; winglet use by, 121–22

Douglas B-26 Invader, 402n62
Douglas C-47 transport, 195, 411
Douglas D-558-1 Skystreak, 3, 28–29, 65
Douglas D-558-2 Skyrocket: engines for, 294; inertial coupling experiment, 39; nose-rise problems, 414; pitch-up problems, 28–32, 413; pressure suit for, 297; Round Two research aircraft, 55n106; tail configuration, 36, 411, 413; wing-stall research, 414
Douglas D-558 FSW planform, 21
Douglas DC-2 aircraft, 6
Douglas DC-3 aircraft, 411
Douglas DC-8 aircraft, 4, 740
Douglas F4D-1 Skyray, 48n94
Douglas F5D-1 Skylancer, 591
Douglass, R.R., 843–44
Douglas X-3 "Stiletto," 3, 4, 37, 40, 41–42, 43, 440
Douglas XF4D-1 Skyray, 32–33
Dow, Norris, 291, 292
Dow Corning, 314
Drake, Hubert, 63
Draper, Charles Stark, 38, 284
Draper Laboratory, Massachusetts Institute of Technology (MIT), 654
Drinkwater, Fred, 641
Driver, Cornelius "Neil," 220
Drones for Aerodynamic and Structural Testing (DAST), 388–89
Dryden, Hugh L., 101
Dryden Flight Research Center (DFRC)/Flight Research Center. *See also* High-Speed Flight Research Station/High-Speed Flight Station: ablation cooling research, 417; Advanced Fighter Technology Integration (AFTI) F-16 project, 673–79; Advanced Turboprop (ATP) project, 800; aeroservoelastic effects, 496; Boeing X-48B testbeds, 72; Centurion, 866; computers at, 468; Digital Electronic Engine Control (DEEC) program, 708–10, 754–56; Digital Fly-By-Wire F-8 program and aircraft, 655, 657; Electrically Powered Actuator Design (EPAD) program, 384; Environmental Research Aircraft and Sensor Technology (ERAST) program, 849, 853, 855, 856, 862, 866, 867, 868–69; finite element analysis activities, 494–96; FLEXSTAB program, 544–45; flutter suppression research, 388; fly-by-wire research activities, 383, 653, 654, 753; General Dynamics F-16XL research, 69, 72, 237, 242–43, 245; General Purpose Airborne Simulator, 402n62; Grumman F-14A Tomcat testing, 63; Helios Prototype, 867, 868–69; high-altitude aircraft workshop, 850; Highly Maneuverable Aircraft Technology (HiMAT), 686–89; High Temperature Loads Calibration Laboratory, 553; hypersonic flight research, 299; inertial coupling research, 408–9; Inlet Spillage Shock Measurement (ISSM), 246; Integrated Propulsion Control System (IPCS) program, 754; Intelligent Flight Control Systems (IFCS) project, 719; laminar flow control research, 495, 784, 808; limit-cycle oscillation research, 368–69; Lunar Landing Research Vehicles (LLRVs), 653; NASTRAN development, 474; Performance Seeking Control (PSC) program, 712; pilot-induced oscillations research, 381; Propulsion Controlled Aircraft (PCA) program, 714–15, 757, 758; rotary wing aircraft research, 164, 165, 169–70; Self-Repairing Flight Control System (SRFCS), 713–14; Shuttle concept development, 57; simulators at, 296;

solar cells, 862; sonic boom research, 195–97, 201–4, 227, 227n133, 236–39, 242–43, 245, 251, 252–53, 255–57, 258–60; Structural Analysis Routines (STARS), 556; structural resonance research, 376–77, 378–79; supercritical wing program, 115, 805–6; supercritical wing testing, 112–13; supersonic flight research, 193; synthetic fuel research, 816–17; Thermal Loads Laboratory, 553–55; thermal research, 495–96; tilt rotor aircraft, 160; Transonic Aircraft Technology (TACT) program, 386; variable-geometry wing research, 63; X-15 aircraft, 294, 395–97; X-29 aircraft, 526, 694–98, 699; X-31 aircraft, 701, 704–5, 707; X-43 Hyper-X vehicle, 344–45; X-Wing Stoppable Rotor testing, 164; YF-12 Blackbird project, 517; YF-12 Experiment Symposium, 519

Duits-Nederlandse Wind Tunnel (Dutch-German Wind Tunnel), 164

Dulles Airport, 166, 638n15

Dunagan, Steven, 861

Dunne, John, 4–5

DuPont, 316, 320–21, 767

DuPont, Anthony "Tony," 335, 337, 338, 339

Dutch-German Wind Tunnel (Duits-Nederlandse Wind Tunnel), 164

Dutch roll motions, 19, 50, 410–12, 637, 637n11

DYLOFLEX program, 545

Dynacs Engineering, 539

Dynamic Analysis System, 153

Dynamic Engineering, Inc. (DEI), 481

Dynamic Simulation of Controls & Structure (DISCOS), 548–49

Dyson, Ken, 700

E

Eagle Engineering/Aeronautics, 235, 238, 246, 249

EAL program, 498

Earth Resources Survey ER-2 aircraft, 850, 851, 861

Earth Science Enterprise, 849

EBASCO Services, 538

Eckert, J. Presper, 436

E Core engine, 803

Eddy viscosity, 341

Edison, Thomas A., 91, 98

Edisonian research technique, 113

Edwards, George, 56

Edwards, Thomas A., 234–35

Edwards Air Force Base. *See also* Dryden Flight Research Center (DFRC)/Flight Research Center: Air Force Flight Test Center (AFFTC), 44, 185, 195–96; deep-strike fighter competition, 69; fly-by-wire testing, 753; Helios Prototype, 866n61, 867; High-Speed Flight Research Station (HSFRS), 184, 185; landings at, 403; Pathfinder, 858; Shaped Sonic Boom Demonstration (SSBD), 247; sonic boom research, 193, 195–97, 200, 201–4, 206–7, 227, 229–30, 251; sonic booms over, 226; Survivable Flight Control System program (F-4 SFCS program), 646–47; X-2 aircraft, 394–95

Eggers, Alfred J., Jr.: B-70 design, 191; blunt-body theory, 55, 285–86, 291, 311, 415; flattop delta aircraft, 54–55, 56, 56n107; hypersonic tunnel, 283; HYWARDS, 300–301

Eglin Air Force Base, 187

Ehernberger, L.J., 236

Eindecker (Monoplane), 34

Eisenhower, Dwight D., 191, 288, 582

Electric airplane, 815
Electrical flight control surfaces and actuators, 384–85, 631, 635, 722
Electrically operated trim tabs, 633, 634–35
Electrically Powered Actuator Design (EPAD) program, 384–85
Electrohydrostatic actuators (EHAs), 385, 676–79, 722
Electro-mechanical actuator (EMA), 385
Electron beam physical barrier vapor deposition (EB-PVD) coatings, 610, 612, 613
Electronic damper systems, 364–67, 412
Electronic engine control systems, 631, 632, 633n1
Electronic flight control system. *See* flight control systems
Electro-optical flight deflection measurement system, 521–22, 527–28
Elevators, 362–63, 633, 635
Elevons, 9, 50, 278, 391
Elliptic behavior, 443
Elliptic partial differential equations, 435, 435n11
Emissions: Advanced Subsonic Technology (AST) program, 796–97, 798; Atmospheric Effects of Aviation project, 797; carbon dioxide emissions, 790, 790n22, 797–98, 799, 813; HSCT aircraft, 795–96; nitrogen oxides (NOx) emissions, 762, 790, 794, 795–99, 813; ozone layer damage from, 789–90, 795, 798; reduction in, 785–86, 788–89; regulations and standards, 745, 791–92, 794, 796, 797, 798, 799; Ultra Efficient Engine Technology (UEET) program, 762–66; Zero CO2 project, 813
Emissions Reduction project, 763
Enabling Propulsion Materials program, 767
Endplates, 118

Energy, conservation of, 432, 433, 443
Energy, Department of (DOE), 439, 824, 825, 836–39, 846, 847
Energy Efficient Engine (E Cubed) project, 742, 744, 745–47, 786, 788, 790–91, 792–95, 817
Energy Efficient Transport (EET) project, 123, 744, 805–6
Energy maneuverability concept, 651
Energy method, 466
Energy Research and Development Administration (ERDA), 824, 825
Energy Technology Division, 830
Enevoldson, Einar K., 658, 661
Engine Component Improvement project, 744–45, 746, 747
Engineering Dynamics, Inc., 551
Engine inlet unstarts, 336, 754
Engine mufflers, 737, 740
Engines. *See also* turbine blades; *specific types of engines*: combustor liners, 798; combustors, 794–95, 798–99; component parts, improvements to, 794, 798; compressors, 746–47, 793–94, 795; damage-tolerant fan casing, 766–69; design of, computer programs for, 746–47, 748; design of, NASA initiatives for, 791; digital engine control systems, 754–57, 758–59; electronic engine control systems, 631, 632, 633n1; emissions standards, 745; Energy Efficient Engine (E Cubed) project, 742, 744, 745–47, 786, 788, 790–91, 792–95; materials for, 584; noise research, 736–37; polymer composite engine fan casings, 767–69; Structural Tailoring of Engine Blades (STAEBL), 545–46; Ultra Efficient Engine Technology (UEET) program, 613, 762–66, 786, 790, 797–98; unstarts, 336, 708–9, 754

Engines, aircraft noise and improvements to, 737–40; Acoustically Treated Nacelle program, 740; Quiet Clean Short Haul Experimental Engine, 741–43; Quiet Engine program, 740, 741; Steep Approach program, 740

Engine Structural Dynamics Program, 555

England. *See* Great Britain

English Electric Lightning, 32

Enhanced Fighter Maneuverability (EFM) X-31 aircraft, 699–705

ENIAC, 436, 468–69

Eniwetok Atoll, 439

ENSAERO code, 493–94

Enterprise. *See* Space Shuttle Enterprise

Envia, Edmane, 770, 771, 772

Environmental barrier coatings (EBCs), 612–13

Environmental impact statement (EIS), 226

Environmental Protection Agency (EPA): emissions standards, 745, 791–92, 794; Environmental Science Services Administration, 206

Environmental Research Aircraft and Sensor Technology (ERAST) program. *See also* Unmanned aerial vehicles (UAVs): alliance organizational structure, 853–55, 872; budget and funding, 853, 854–55; goals of, 849–50, 852, 861, 872; high-altitude aircraft, requirements and capabilities of, 851–52; high-altitude aircraft, workshop on, 850–52; Joint Sponsored Research Agreement (JSRA), 849, 852–55, 852n33, 872; ownership of hardware developed, 854; participants and partners, 853–55; report requirements, 853; science missions, 859, 861; sensor technology development, 861; success of, 855, 871–72

Environmental Science Services Administration, 206

Epstein, Alan H., 750–51

Equations. *See also* Euler equations; Navier-Stokes equations: continuity, 430, 431, 433, 438, 443; elliptic partial differential, 435; hyperbolic partial differential, 435n11, 436, 443; integral, 433; linear partial differential, 431, 435–36; of motion, 393; nonlinear partial differential, 430–31, 432–33; parabolic partial differential, 435n11, 436; partial differential, 430–31, 432–33, 434–36, 435n11, 446–47

ER-2 aircraft, 850, 851, 861

Error Correction Information System (ECIS), 479

ESA 3560 HF ablative blanket, 607

Ethell, Jeffrey, 805

Ethylene, 346

Ettinger, Robert C., 647

Euler, Leonhard, 430–31

Euler equations: blunt-body problem solution, 443–44; continuity equation, 430, 431, 433, 438, 443; ENSAERO code, 493; friction and, 431; LAURA code, 451; momentum equation, 430, 431, 433, 438, 443; transonic flow and airfoils, 445

Eurofighter, 692

European countries interest in NASTRAN, 481, 484

Everest, Frank K. "Pete," 51, 102, 297

Exceptional Service Medal, 107

Exhaust nozzle chevron, 808–9

Exhaust nozzles, thrust-vectoring, 715–16, 717

Experimental Aircraft Program (EAP), 692

Experimental Air Vehicle (EAV), 720

Experimental Clean Combustor program, 794–95

External tanks, 326, 327, 329, 578

Extremely short takeoff and landing (ESTOL), 705–7

Exxon, 538

F

F-2 fighter, 692

F4D-1 Skyray, 48n94

F4H-1 Phantom II, 32, 33n60, 34, 195

F-4 Phantom, 644n32

F-4 Survivable Flight Control System program, 379–80, 384, 646–48, 658n59, 660

F4U-5 Corsair, 401

F5D-1 Skylancer, 591

F-5 fighter: F-5E Shaped Sonic Boom Demonstration, 245–48, 250, 252; sonic boom research, 245; tail configuration, 32, 413

F6F-5 Hellcat, 401, 637

F-8A fighter, 112–13

F-8 Digital Fly-By-Wire aircraft. *See* Digital Fly-By-Wire F-8 program and aircraft

F-8 Super Cruise Wing project, 655

F8U Crusader series. *See also* Digital Fly-By-Wire F-8 program and aircraft: area rule and redesign of, 103–4, 105; digital flight control system, 380–81, 753; F-8C, 380–81, 654–55, 753; F-8 Supercritical wing, 89; F-8 Super Cruise Wing project, 655; F8U-1, 32, 37, 43; F8U-3, 194; XF8U-1 Crusader, 37, 363

F9F-2 Panther, 638–39, 642, 658n59

F9F-5 Panther, 33

F9F-6 Cougar, 33

F9F-7 aircraft, 33n60

F9F-9 Tiger, 104, 105, 106

F11F-1 Tiger, 32, 104, 105

F-14 Tomcat, 44, 62, 413; F-14A Tomcat, 63

F-15 aircraft, 710. *See also* McDonnell-Douglas F-15 Eagle

F-15 Eagle: Advanced Control Technology for Integrated Vehicles Experiment (ACTIVE) project, 716, 717; Digital Electronic Engine Control (DEEC) program, 709–10, 754–56; Dryden research activities, 496; engines, 713; F-15A, 44, 711, 713; F-15E Strike Eagle, 69; fly-by-wire system, 383–84; Highly Integrated Digital Electronic Control (HIDEC), 711; High Stability Engine Control (HISTEC) system, 718; Intelligent Flight Control Systems (IFCS) project, 718–21; Propulsion Controlled Aircraft (PCA) program, 713, 757; Self-Repairing Flight Control System (SRFCS), 713–14; simulator, 407; sonic boom research, 238–39, 246–48, 252–53, 256, 258–60, 261; tail configuration, 413; TF-15, 715

F-15 Eagle (F-15B/NF-15B): Advanced Control Technology for Integrated Vehicles Experiment (ACTIVE) project, 716, 717; exhibit showing, 721; High Stability Engine Control (HISTEC) system, 718; Intelligent Flight Control Systems (IFCS) project, 718–21; Lift and Nozzle Change Effects on Tail Shocks (LaNCETS), 258–60, 261; modifications for experiments, 258–59; Quiet Spike device, 252–53, 256; Short Takeoff and Landing Maneuver Technology Demonstrator (S/MTD), 715–16; standard features, 259

F-16 Fighting Falcon: deep-stall conditions, 666n70; development of. *See* General Dynamics YF-16 Lightweight Fighter (LWF)Digital Electronic Engine Control (DEEC) program, 710, 756; F-16A, 378, 380, 669; F-16B, 380, 669; F-16C, 669; F-16 CCV, 648, 651–52, 669–70; F-16XL-1, 808; F-16XL-2, 784, 808; F-16XL SCAMP, 69, 69n129, 72, 237, 242–43, 245, 261; F-16XL

SCAMP model, 807; fly-by-wire systems, 380, 383–84, 669; pitch rocker technique, 666n70; production of, 669; success of, 668–69; tail configuration, 413, 666n70

F-18 Hornet, 237, 251, 255–57, 496, 676

F-22 Raptor, 126–27, 384, 402, 413, 712, 713, 722

F-35 Lightning II Joint Strike Fighter, 384, 413, 678, 713, 722

F-84 aircraft, 333

F-86 Sabre: accidents with, 24, 24n43; development of, 19n32, 23n39; F-86A, 26; F-86D Sabre, 102; F-86F, 26; Korean war use of, 4, 19, 26; pitch-up problems, 24–26, 36, 37, 413; reputation of, 23–24; sonic booms, 186–87; stability research activities, 637; tail configuration, 36, 413; variable stability aircraft, 401; XP-86 Sabre prototype, 18, 635; YF-86D low-tail testbed, 36–37

F-89 Scorpion, 102

F-94 Starfire, 102

F100 engine model derivative (EMD) engine, 711

F-100 Super Sabre: development and manufacture of, 290; F-100C Super Sabre, 409; F-100D Super Sabre, 409; flight control systems, 635; JF-100C Super Sabre, 402n62; landing simulator, 403; sonic boom research, 194, 196; tail configuration, 413; YF-100A Super Sabre, 34, 363; YF-100 Super Sabre, 37, 40, 41, 42, 185

F-100 Super Sabre (F-100A): design of, 41, 42–43, 393–94; roll-coupling accident, 41, 42, 44, 393; roll coupling research, 408–9; wingspan-to-fuselage length ratio, 37

F100 turbojet engine, 710, 712–13, 717, 754–56

F-101 Voodoo: development and manufacture of, 290; flight control system, 372; pitch-up problems, 33–34, 413; sonic boom research, 194, 199; sonic booms, 185; tail configuration, 33–34, 36, 413; turbofan engines, 741

F-102 Delta Dagger: area rule and, 51, 52, 103; design of, 52–53, 103, 409; development and manufacture of, 290; F-102A Delta Dagger, 102–3; inertial coupling (roll coupling), 409; management of program for, 53n103, 61

F-102 Delta Dagger (YF-102): area rule and, 51, 89, 100–102, 103; handling qualities, 53; sonic booms, 185; YF-102A prototype, 89, 101–2, 103

F-103 interceptor fighter, 569; XF-103 interceptor fighter, 289, 290, 333

F-104 Starfighter: design of, 104, 401, 409, 440; development and manufacture of, 290; F-104A Starfighter, 37, 440; F-104B, 409; F-104G, 688; F-104G (German CCV), 670, 689–90, 692; heating, structural, 413; inertial coupling, 43, 409; landing of, 405; landing simulator, 403–4; nickname for, 4, 104; pitch-up problems, 34n61, 413; reaction controls, 298; sonic boom research, 195–96, 199, 200, 204, 205, 206–7; speed of, 413; supersonic flight by, 3, 104; use of, 104; wingspan-to-fuselage length ratio, 407; XF-104 Starfighter, 185

F-105 Thunderchief: design of, 333, 409; development and manufacture of, 290; F-105B Thunderchief, 32, 44; inertial coupling (roll coupling), 409; sonic booms, 185, 216; tail configuration, 37, 413

F-106 Delta Dart: design of, 45, 53, 409; handling qualities, 50; inertial coupling (roll coupling), 409; purpose of, 103; sonic booms, 185, 196, 199, 201, 204, 261;

variable stability aircraft, 402n62

F-107 aircraft, 372, 642, 658n59

F-111 aircraft: adaptive flight control systems (AFCS), 373–74, 375; design of, 61–62, 62n118; digital engine control system, 754; F-111A variable-sweep aircraft, 113, 386, 404; F-111F, 61; fly-by-wire system, 639; management of program for, 61; mission-adaptive wing (MAW), 386; sonic boom research, 238

F117 engine, 795

F-117 stealth fighter, 70, 384, 721–22

F119 engine, 712, 713

F135 engine, 713

F404 engine, 802

F/A-18 Hornet: Active Oscillation Suppression (AOS) system, 389; autothrottle system from, 706; design of, 669; engines for, 706n152; F/A-18A Hornet, 44, 383–85; F/A-18B Hornet, 386; tail configuration, 413

FAA. *See* Federal Aviation Agency/Administration (FAA)

Faget, Maxime "Max," 56, 291, 602, 603

Fail-operational, fail-operational, fail-safe (FOFOFS) systems, 367, 384

Fail-operational, fail-operational (FOFO) system, 379–80

Fail-operational, fail-safe (FOFS) systems, 366

Fairey F.D.2 aircraft, 45, 53

Falling body test, 38, 43, 92

Fallon, 247

Fanjet engines, 747

Farmer, Crofton, 851

Farnborough, England, 161, 643

FASTOP program, 527

FB-111 aircraft, 392

Fedden, Roy, 35n65

Federal Aviation Agency/Administration (FAA): aircraft noise standards, 739–40, 771; aircraft safety, 506; aircraft technology and systems to support, 171, 172; certification of MD-11, 123; certification programs, 123, 683–84, 723; Civil Aeronautics Administration (CAA), 135; Digital Systems Validation Handbook, 684; emissions regulations, 745; International Helicopter Study Team (IHST), 172; NASA, relationship with, 139; National Rotorcraft Technology Center, 168; polymer composite engine fan casings, 768; recertification of DC-10, 123; rotary wing noise research, 163, 164; Skyshield air defense exercises, 188; SST program, 65–66, 191–92, 210–11; Supersonic Transport Development Office, 199; tilt rotor aircraft, civil transportation use of, 166–67; unmanned aerial vehicles (UAVs) issues, 854

Federal Aviation Agency/Administration (FAA) and sonic booms: noise level standards, 250; noise standards workshop, 248, 249; prohibition of by Concorde, 227–28; reactions to, 261; research activities, 199–201, 206, 207–8, 251; restrictions on, 248–49; supersonic flight ban and, 220–21, 220n112, 224

Federal Trade Commission (FTC), 486

Federal Wind Energy Program, 823–25, 827, 828–29, 831, 833, 840, 843, 846–49

Federation of American Scientists, 227

Ferri, Antonio, 213, 215–16, 217, 222, 223, 332–33, 334

Fiberglass, 804, 845

Fiber-optic controllers, 385, 685, 686

Finite-difference solutions, 434–36

Finite Element Composite Analysis Program (FECAP), 556

Finite Element Modeling of Piping Systems

program, 552

Finite element structural analyses. *See also* NASA Structural Analysis System (NASTRAN): Ad Hoc Group for Structural Analysis, 474; advancements in, 152, 530–34; aviation industry use of, 533–34; benefits of, 534–35; computational analysis or simulation, 515, 516; discretization of structures, 466, 469, 470–72; finite element method (FEM), 471–72, 485, 510; flight-testing, 514, 515, 534–35; HiMAT, 515, 520–22, 530; laboratory testing, 515; materials behavior and, 532–33; modeling process, 531–32, 534–35; NASA's contributions to, 535, 556; oblique wing concept, 515, 522–26, 530; optimization of aircraft design, 533; pre- and post-processing challenges, 531–32; rotary wing aircraft, 162–63, 515, 519–20, 529–30; software programs, 158, 490n62, 502, 544–52; Space Shuttle studies, 327–28, 506; technology development for, 486; testing facilities, 553–55; theory, 515, 516; validation of, 530, 534–35, 553–55; X-29 aircraft, 515, 526–29, 530; YF-12 Blackbird project, 515, 516–19, 529

Finite element structural analyses, projects and contributions of Centers, 490–91, 491n64; Ames Research Center, 491–94; Dryden Flight Research Center, 494–96; Glenn (Lewis) Research Center, 496–500, 532; Goddard Space Flight Center, 500–502; Jet Propulsion Laboratory, 502–5; Johnson Space Center, 505–7; Langley Research Center, 507–12, 532; Marshall Space Flight Center, 512–14

Finnegan, Patrick M., 830

Fischel, Eduard, 636

Fischer-Tropsch process, 816n114, 817

Fischetti, Thomas L., 105

Fisher Body Company, 91

Flaperons, 694, 697

Flat-bottom delta aircraft, 55–57, 301, 302

Flat-top delta aircraft, 55–56, 56n107, 300–301

Flechner, Stuart G., 119, 120

Fletcher, Edward, 331

Fletcher, James, 116

Flexibility matrix, 471

FLEXSTAB program, 392, 518, 544–45

Flight control coupling, 409–10

Flight control surfaces. *See also* control-surface actuators; *specific surfaces*: electrical flight control surfaces and actuators, 384–85, 631, 635, 722; hydraulically actuated flight control surfaces, 384, 590, 631, 634–35, 643, 644

Flight control systems. *See also* digital fly-by-wire (DFBW) flight control systems; fly-by-wire (FBW) systems: adaptive flight control system (AFCS), 372–74, 375, 376, 379; advancements in, 278, 752–53; aerodynamic damping and, 363–67; artificial feel devices, 635, 643; Command Augmentation Systems (CAS), 366–67, 382; computer use for, 631, 633n1; control stick feedback, 370–71, 635, 638–39, 643; design of, frequency spectrum of challenges, 377, 378; direct control system, 640, 653, 659, 660; electronic feedback flight control research, 636–37; evolution of, 632–35, 633n1; fly-by-light fiber-optic digital flight control system, 685, 686; gain, 656; hydraulically actuated flight control surfaces, 384, 590, 631, 634–35, 643, 644; hypersonic flight, 298; limit-cycle oscillation, 367–69, 377, 378; mechanical linkages, 525, 631, 633, 637–38, 639,

655; Minneapolis Honeywell MH-96 flight control system, 298–99, 372–74; model-following systems, 366–67; pilot-induced oscillations (PIO), 369–72, 378, 381, 409, 412, 661; redundant systems, 365–67; research activities, 362–63; resonant motions, 374–79, 513; Self-Repairing Flight Control System (SRFCS), 713–14; side stick flight controller, 298, 638, 638n15, 642, 658, 658n59, 664, 665, 668; Space Shuttle orbiter, 367; stability augmentation systems, 364–66, 399, 636–37; total loop gain, 367–69, 372; transport delay, 661; Vietnam war and, 644–45; X-1 aircraft, 362–63; X-2 aircraft, 298

Flight envelope limitations, 368, 632

Flight gain ground rule, 376–77

Flight Operations, 494

Flight Research Center. *See* Dryden Flight Research Center (DFRC)/Flight Research Center

Flight Simulator for Advanced Aircraft (FSAA), 151

Flight simulators. *See* simulators and flight simulation

Flittie, Kirk, 856

Floating point operations per second (FLOPS), 341, 437, 438–39

FLO codes, 445–46

Florida, 226

Flourine, 570

Flow direction fences, 414

Fluid dynamics, 428–33. *See also* computational fluid dynamics (CFD)

Flush Air Data System, 707

Flutter, 306, 309, 482–83, 525–26, 585–86, 595

Fly-by-light fiber-optic digital flight control system, 685, 686

Fly-by-wire (FBW) systems: acceptance of, 379; actuators, 384–85; advancements in, 631; advantages of, 631–32; aeroservoelastic effects, 496; analog computers, 648–49, 656, 656n57; angle of attack limiters, 665–66; asynchronous digital flight control system, 672, 680–82, 683; Augmented Longitudinal Control System (ALCS), 372; aviation industry use of, 632, 722–23; backup systems, 656, 657, 658, 660, 662, 688; benefits of, 650–53, 654, 721–23; certification of new aircraft, 683–84; control stick feedback, 370–71; development of, 379–80, 631, 632, 635–37, 753; digital flight control system, 655, 672–73, 674, 680–82; early applications, 637–43; events that supported need for, 643–45; nonaerospace applications, 631, 723; redundant systems, 379–80, 655–56, 658, 659, 671–72; research activities at Dryden, 383, 753; research activities at Edwards, 646–47; research activities at Langley, 638–39; research activities by Air Force, 645–48; research activities in Europe, 648–50; software programs, 381–82, 659–60, 671–72, 683–84; space flight program, transfer of technology from, 632, 653–54; Survivable Flight Control System program (F-4 SFCS program), 379–80, 384, 646–48, 658n59, 660; transition to, 632; transport delay, 661

Fly-by-wire (FBW) systems, digital: actuators, 384–85; development of, 380, 753; military use of, 383–84; redundant systems, 384; software programs, 381–82, 659–60, 671–72, 683–84; space flight program use of, 632, 653–54; testing of, 380–83

Fly-by-wire ship control system certification program, 723

Flying Laboratory for Integrated Test and Evaluation (FLITE), 170
Flying-quality criteria, 145–46
Flying wings (Nurflügeln), 70
Focke-Wulf FW 190 fighter, 633n1, 636, 703n144
Focke-Wulf Ta 183 aircraft, 36
Fokker, Anthony, 385
Fokker Eindecker (Monoplane), 34
Folland Gnat, 32
Forest Service, U.S., 200
Forman, Royce, 506
Fort Eustis, Virginia, 146
FORTRAN programming language, 437, 469, 476
Fort Worth, 667, 674, 678
Forward-looking infrared (FLIR) device, 673, 675
Forward-swept wing (FSW), 21–22, 524, 527, 693–94
France: Concorde Supersonic Transport (SST). *See* Concorde Supersonic Transport (SST)Control Configured Vehicle (CCV), 691; delta wing development, 45, 53; Paris Air Show, 161; Supersonic Commercial Air Transport (SCAT) program, 190; V-2 missile shock waves, 184; variable-geometry wing research, 62
Frankford Arsenal, 542
Franklin Institute, 142
Free body diagrams, 464–65
Freon gas, 149, 150
Frequencies and Modes of Shells of Revolution (FAMSOR), 506
Friction, 431–32
Friedrichs, K.O., 435–36
Fuel: alternative, 785, 815, 817; biofuel, 817; Fischer-Tropsch process, 816n114, 817; hydrogen, liquid, 809–14, 815–16; oil prices, 743, 788, 790–91, 816, 840–41; resonant motions from, 513; synthetic, 815–17, 816n114; transfer between tanks, 632
Fuel cells: for aircraft, 785, 814–15; hydrogen-air fuel cell system, 864, 869, 871; hydrogen fuel cell aircraft, 815; hydrogen-oxygen regenerative fuel cell system (RFCS), 863–64, 871; regenerative fuel cell system (RFCS), 863–64, 871, 872; for UAVs, 856, 863–64, 869, 871
Fuel efficiency programs, 785–86, 790–91, 817. *See also* Aircraft Energy Efficiency (ACEE) program
Full Authority Digital Engine Control program, 755
Full-cost accounting methods, 166
Fullerton, Gordon, 715
Fundamental Aeronautics Program, 254
FW 190 fighter, 633n1, 636, 703n144

G

Gain, 656
Galileo, 428, 429
Galileo probe, 277, 278, 573
Garrett AiResearch, 335
Garrett Turbine Company, 612
Garrison, Charles P. "Pete," 646
Gas-metal-arc (GMA) welding, 575
GASP 2.0 / 3.0, 341
Gas-tungsten-arc (GTA) welding, 575
Geared Turbofan engine, 750, 751–52, 800–801, 802
Gedser wind turbine, 827, 828, 846–47
Geisler, Frank, 645
Gemini program: ablation cooling, 417; centrifuge program, 400; flight control systems, 653; fuel cells, 814; reentry vehicles, 54; side stick flight controller, 658n59; spacesuits, 297; Titan II missile/

booster, 328
General American Research Division, 215
General Applied Science Laboratories (GASL), 215–16, 333, 336, 338, 343
General Atomics Aeronautical Systems, 853
General Aviation Manufactures Association, 248
General Dynamics, 480
General Dynamics F-16 Fighting Falcon: deep-stall conditions, 666n70; development of. *See* General Dynamics YF-16 Lightweight Fighter (LWF)Digital Electronic Engine Control (DEEC) program, 710, 756; F-16A, 378, 380, 669; F-16B, 380, 669; F-16C, 669; F-16 CCV, 648, 651–52, 669–70; F-16XL-1, 808; F-16XL-2, 784, 808; F-16XL SCAMP, 69, 69n129, 72, 237, 242–43, 245, 261; F-16XL SCAMP model, 807; fly-by-wire systems, 380, 383–84, 669; pitch rocker technique, 666n70; production of, 669; success of, 668–69; tail configuration, 413, 666n70
General Dynamics F-111 aircraft: adaptive flight control systems (AFCS), 373–74, 375; design of, 61–62, 62n118; digital engine control system, 754; F-111A variable-sweep aircraft, 113, 386, 404; F-111F, 61; fly-by-wire system, 639; management of program for, 61; mission-adaptive wing (MAW), 386; sonic boom research, 238
General Dynamics FB-111 aircraft, 392
General Dynamics Tactical Fighter Experiment (TFX), 61
General Dynamics YF-16 Lightweight Fighter (LWF), 371; angle of attack limiters, 665–66; flight-testing, 667–68; flutter suppression research, 388, 389; fly-by-wire system, 379–80, 647, 664–69; pilot-induced oscillations, 370; roll stick force gradient, 667; side stick flight controller, 658, 664, 665, 668; simulator studies, 666–67; strain gages, 385–86; structural resonance research, 378; success of, 668; unintentional first flight, 667
General Electric (GE): ablation cooling, 286, 581; Advanced Subsonic Technology (AST) program, 796; combustor concepts, 794–95, 798–99; Cooperative Test Organization/VECTOR, 705–6; E Cubed technology use, 742, 747, 792–95; high-bypass fanjet engines, 741–43; Numerical Propulsion System Simulation (NPSS) program, 760; open-rotor engines, 803; Quiet Engine program, 737; thermal barrier coatings, 612–13; Ultra Efficient Engine Technology (UEET) program, 762, 797; wind energy program, 829, 835, 836, 840, 841, 848
General Electric CF6-50C engine, 745
General Electric CF6 engine, 744
General Electric compressor, 793, 795
General Electric E Core engine, 803
General Electric F404 engine, 802
General Electric GE90-115B jet engine, 795
General Electric GE90 jet engine, 742, 743, 793
General Electric GE F404 engine, 706n152
General Electric GEnx engine, 769, 795
General Electric GP-7200 engine, 795
General Electric J47 jet engine, 736
General Electric J79 turbojet, 104
General Electric J85 engine, 686
General Electric Mark 3 nose cone, 288
General Electric mullite RSI, 313, 601, 604
General Electric RVX-1 nose cone, 288, 582
General Electric RVX-2 nose cone, 288, 294
General Electric TF30 turbofan engines, 754
General Electric Twin Annular Pre-mixing Swirler (TAPS) combustor, 798–99, 817

General Electric Ultra-High Bypass Turbofan engine, 750–51
General Electric Unducted Fan (UDF) engine, 749, 750, 801–2
General Motors, 481, 538
General Purpose Airborne Simulator, 402n62
General Purpose Structural Analysis program, 474–75. *See also* NASA Structural Analysis System (NASTRAN)
General Rotorcraft Aeromechanical Stability Program (GRASP), 492
Geophysical Finite Element Simulation (GeoFEST), 556
George, Albert R., 211, 217, 222, 223
George Washington University, 542
Georgia Institute of Technology, 169, 719
Germany: A-2 rocket, 635–36; A-4b (winged V-2 derivative), 280, 281; CCV F-104G, 670, 689–90, 692; Cooperative Test Organization/VECTOR, 706; Dutch-German Wind Tunnel (Duits-Nederlandse Wind Tunnel), 164; fly-by-wire system research, 648–49; forward-swept wing aircraft, 21, 22; high-speed flight program, 280–81; International Test Organization (ITO), 700; oblique wing concept, 522; pitch-up problems, 27, 27n51; rotary wing noise research, 164; side stick flight controller, electrical, 638n15; swept and delta wing aircraft, 7, 13, 13n20, 14–15, 15n27, 17, 72–73, 93; synthetic fuel process, 817; technical intelligence interrogations, 14–15, 17; turbojet engines, 92; V-2 ballistic missiles, 183–84, 280, 282, 571, 635–36; variable-geometry wing research, 63; wind energy, 828; X-31 aircraft, 699–700, 706, 707
Gessow, Alfred, 144

Giesing, J.P., 483
Gifu Air Base, 692
Gigaflops, 341, 438–39
G II aircraft, 404–5
Gimbaled nose sensor, 298
G-induced loss-of-consciousness (g-loc), 33
Giner, 864
Glacier National Park, 201
Glazer, Jack, 852n33
Glenn, John H., Jr., 103–4
Glennan, T. Keith, 192
Glenn Research Center/Lewis Research Center/Lewis Flight Propulsion Laboratory/Aircraft Engine Research Laboratory: Advanced Subsonic Combustion Rig (ASCR), 799; Advanced Tiltrotor Transport Technology (ATTT) program, 167–68; Advanced Turboprop (ATP) project, 800; aeronautics expertise, 787; Aircraft Engine Research Laboratory, 137, 786–87; Air Force F-111 program, 61–62; AIRLOADS program, 556; ANSYMP program, 545; Army research presence at, 139, 147; BWB research, 71; COmposite Blade STRuctural ANalyzer (COBSTRAN), 546–47; Computational Structural Mechanics (CSM) workshop, 497; computers at, 438; Coupled Eulerian Lagrangian Finite Element (CELFE), 514; Digital Electronic Engine Control (DEEC) program, 709–10, 754; engine design, computer programs for, 746–47, 748; engine design and aircraft speed, 105–6; engine fan case improvements, 767; finite element analysis activities, 496–500, 532; Glenn Research Center, 137; Greenlab Research Facility, 817; Helicopter Transmission Technology program, 158; higher-order theory for functionally graded materials

(HOTGFM), 512; High Stability Engine Control (HISTEC) system, 717–18; hydrogen fuel research, 809–14; impact analysis research, 497; Integrated Composite Analyzer (ICAN), 556; jet engine innovations, 105–6; Lewis Research Center, 137; Low Emissions Alternative Power (LEAP) program, 872; Micromechanics Analysis Code with Generalized Method Cells (MAC/GMC), 512; NASTRAN, development of, 474, 497; NASTRAN, Users' Colloquia, 536, 539; noise research, 738; Numerical Evaluation of Stochastic Structures under Stress (NESSUS), 498–99, 514; Numerical Propulsion System Simulation (NPSS) program, 759–62; propulsion system research, 786–87; Quiet Engine program, 737; rotary wing aircraft research, 156, 158, 162; rotary wing noise research, 163; Spin Rig, 555; SST program, 66, 191; Structural Performance and Redesign (SPAR), 495, 498, 512; Structural Tailoring of Engine Blades (STAEBL), 545–46; supersonic combustion testing, 331; thermal barrier coatings research, 613; TRansfer ANalysis Code to Interface Thermal Structural (TRANCITS), 546; TRASYS Radiative Heat Transfer program, 550; Ultra Efficient Engine Technology (UEET) program, 762, 797; Ultra-High Bypass Turbofan engine, 750–51; wind energy program, 824–25, 827, 829, 831, 836–39, 841, 844–49; Wind Power Office, 829–30; women mathematicians, 465; YF-12 Blackbird project, 517

Gliding landings, 402–3

Global Disaster Information Network, 861

Global Positioning System (GPS), 242, 706, 856

Global warming research, 789–90, 790n22, 818

Glomb (guided glide bomb), 11–12

Gloster Javelin, 45

Gluhareff, Michael E., 8–11, 13n20

Gluhareff-Griswold "Dart" concept, 8–11, 9n12

Gnoffo, Peter, 451, 451n26

Goddard Space Flight Center (GSFC): ACTOMP program, 556; ACTON program, 556; Dynamic Simulation of Controls & Structure (DISCOS), 548–49; finite element analysis activities, 500–502; Integrated Modeling of Optical Systems (IMOS), 548; NASTPLT program, 556; NASTRAN, development of, 152, 473–74, 476, 501–2; NASTRAN, maintenance of, 478; NASTRAN, use of, 477; NASTRAN, Users' Colloquia, 480, 536; NPLOT program, 547; Structural-Thermal-Optical (STOP) program, 501–2; Ultra Efficient Engine Technology (UEET) program, 762, 797; VIEW program, 502

Goldburg, Arnold "Bud," 445

Goldendale, Washington, 837–38

Goldstein, Howard, 315

Goodrich, 481, 538

Goodyear: Goodyear Electronic Differential Analyzer (GEDA), 393, 394; Micromechanics Analysis Code and tire design, 500; X-20 Dyna-Soar landing skids, 592

Goodyear blimp Mayflower, 204

Gossamer Albatross, 856

Gossamer Penguin, 856

Göthert, Bernard, 29n56

Göttingen aerodynamics, 17

Goward, C.W., 611

GP-7200 engine, 795

Graham, David H., 245–46

Graphite, 576–77, 587–88, 594
Graphite Car, 578
Graphite composite, 804
Graphite-epoxy composite, 576
Gray, George, 18
Great Britain: aeronautical community, ties to Langley Research Center, 60–61; Aircraft and Armament Experimental Establishment (A&AEE), 638n16; all-moving tail research, 35n65; Concorde Supersonic Transport (SST). *See* Concorde Supersonic Transport (SST)delta wing development, 45, 53; fly-by-wire system research, 649; Jaguar Active Control Technology (ACT), 670, 690–91, 692; Royal Air Force (RAF), 186, 188, 643; Supersonic Commercial Air Transport (SCAT) program, 190; swept wing aircraft, 27, 28; turbojet engines, 92; V-2 missile shock waves, 183–84; variable-geometry wing research, 62; wind energy, 828, 848
Great Lakes Industrial Technology Center (GLITeC), 500
Green, J.R., 641
Greenhouse gases, 790
Greenlab Research Facility, 817
Grid points: computer technology development and, 438–39; direct numerical simulation (DNS), 342, 450–51; Euler equations, 438; flow-field calculations, 433, 434, 435; numerical accuracy, 446–47; three-dimensional flow-field calculations, 447–49
Griffith, John, 30
Grindle, Thomas, 253
Griswold, Roger W., 9, 9n13
Ground collision avoidance system, 675
Ground Demonstrator Engine (GDE), 345
Ground resonance, 144–45

Ground vibration tests (GVT), 376, 526
Grover E. Bell Award (AHS), 169–70
Groves, Al, 701
Grumman: FASTOP program, 527; NASTRAN Users' Colloquia, 480, 538; Shaped Sonic Boom Demonstration (SSBD), 245–47
Grumman F6F-5 Hellcat, 401, 637
Grumman F9F-2 Panther, 638–39, 642, 658n59
Grumman F9F-5 Panther, 33
Grumman F9F-6 Cougar, 33
Grumman F9F-7 aircraft, 33n60
Grumman F9F-9 Tiger, 104, 105, 106
Grumman F11F-1 Tiger, 32, 104, 105
Grumman F-14 Tomcat, 44, 62, 413; F-14A Tomcat, 63
Grumman G II aircraft, 404–5
Grumman Gulfstream, 325
Grumman TBM Avenger, 695n125
Grumman X-29 aircraft: aeroelastic effects, 496, 528–29, 694; aeroelastic tailoring, 694, 695; angle of attack capabilities, 694, 695, 696–98; benefits of, 697–98; contract for, 692–93; control laws, 696; control surfaces, 694–95; difference between two X-29s, 693–94; exhibit showing, 698–99; finite element analysis, 460, 515, 526–29, 530; flight-testing, 695–98; fly-by-wire system, 383, 692–98, 753; forward-swept wing design, 22, 527, 693–94; NASTRAN model, 460; redundant systems, 695; sensor selection thresholds, 698; strain gages, 528; supercritical wing, 693, 694, 695, 695n126; vortex flow control, 698; X-29A aircraft, 460
Grumman XF10F-1 Jaguar, 59, 60
Guggenheim Aeronautical Laboratory, California Institute of Technology (GALCIT),

12, 17n30, 502
Guided glide bomb (glomb), 11–12
Guidonia Laboratory, 6, 332–33
Gulfstream Aerospace Corporation: Quiet Spike device, 251–53, 256, 261; Shaped Sonic Boom Demonstration (SSBD), 246, 249; sonic boom mitigation techniques, 251–53; sonic boom simulator, 257; supersonic business jet (SSBJ) concept, 249
Gulfstream II business jet, 124
Gulfstream III, 121
Gulfstream II turboprop modification, 801
Gulfstream IV, 121
Gulfstream V, 121
Gulfstream X-54A aircraft, 255
Gurevich, M.I., 19n32
Guruswamy, Guru, 494
Gyroplane, 141
Gyros, 298, 364–65, 366, 367

H

H-13 helicopter, 150
Haering, Edward A., 236, 242, 248, 255
Hafnia, 574, 587, 588, 614
Haise, Fred, 370–71, 381
Halaby, Najeeb E., 192
Hall, Charles F., 51–53
Hall, David B., 476
Hall, Eldon, 810, 811, 814
Hallion, Richard P., 115, 595
Hamburger Flugzeugbau HFB-320 Hansa Jet, 22
Hamilton Standard, 742, 800–801, 839
Hamilton Standard SR-7A prop fan, 800–801
Hamilton Sundstrand direct current power generation system, 677–78
Hamlin, Fred, 106n48
Hampton, Virginia, 90, 91, 126, 137, 146, 191
Hansen, J.R., 706
Harrier strike fighter, 60
Harris, Henry, 536
Harris, Roy, 67
Harry Diamond Laboratory, 541
Harting, David, 475
Harvard University, 850
Harvey the Rabbit, 441
Have Blue aircraft, 721
Have Coupling project, 685
Hawaii: Corona spy satellite, 305; Helios Prototype flights, 867, 869; Pathfinder flights, 858, 859; sonic booms over, 226n130; wind energy program, 827, 842
Hawaiian Electric Incorporated, 842
Hawker Hunter T.12 aircraft, 649
Hayes, Wallace D., 94, 211, 212, 217, 223
Heat shields, 579, 591, 596, 602–3
Heat-sink structures, 415–17, 579, 581, 607
Heavy Lift Helicopter program, 158
Heavy-lift military missions, 71, 71n131
Heinemann, Edward "Ed," 29, 46–47
Heitmeyer, John C., 51–52
Helicopter Association International (HAI), 163
Helicopters. *See* rotary wing aircraft and helicopters
Helicopter Subcommittee (Rotating Wing Subcommittee), 142
Helicopter Theory (Johnson), 159
Helicopter Transmission Technology program, 158
Helios Prototype, 850, 853, 856, 862, 863–64, 866–72, 866n61
Helium gas gun, 318
Helium gas launcher simulator, 214
Helmet-mounted target designation sight, 673, 675

Henderson, Herbert, 217
Hennrich, Carl, 476
Henry, John, 335
Henschel Hs 129 aircraft, 636–37, 637n11
Henschel Hs 293 aircraft, 638n15
Herbst, Wolfgang, 703n144
Herbst Maneuver, 703, 703n144
Hersman, Walt, 647
Hicks, John, 855
High-altitude aircraft, 850–52. *See also* Environmental Research Aircraft and Sensor Technology (ERAST) program
High-Altitude Long-Endurance (HALE) vehicles, 496, 849, 859, 868, 871
High-Altitude Solar (HALSOL) aircraft, 850
High aspect ratio aircraft, 6–7, 13, 13n20, 390
High aspect ratio wings, 120
High-bypass fanjet engines, 741–43
Higher-order theory for functionally graded materials (HOTGFM), 512
Highly Integrated Digital Electronic Control (HIDEC), 710–11, 712
Highly Loaded Turbomachinery project, 763
Highly Maneuverable Aircraft Technology (HiMAT), 22, 325, 515, 520–22, 530, 686–89
High Performance Computing and Communications program, 759
High-Speed Aircraft Industrial Project (HISAC), 250
High-Speed Civil Transport (HSCT) aircraft, 231–33, 236, 241, 242, 519, 795–96
High-Speed Flight Research Station/High-Speed Flight Station. *See also* Dryden Flight Research Center (DFRC)/Flight Research Center: aerodynamic refinement and efficiency improvements, 65; conical camber research, 53; Convair XF-92A aircraft, 49; establishment of, 184; flight control systems, research activities on, 363; organization of, 494–95; pitch-up research, 25; roll coupling research, 393; sonic boom research, 185
High-Speed Research (HSR) program, 232–43, 519, 795–96, 797
High-Speed Tunnel (HST), 94–97, 127–28
High Stability Engine Control (HISTEC) system, 258, 717–18
High Temperature Loads Calibration Laboratory (Dryden Flight Research Center), 553
High-temperature structures and materials, 569–73
High Temperature Tunnel (HTT), 342, 343, 573
Hill, G.T.R., 5
Hillaker, Harry J., 69, 69n129, 668
Hiller Aircraft, 136
HiMAT (Highly Maneuverable Aircraft Technology), 22, 325, 686–89
HL-10 aircraft, 56, 360, 368–69, 378–79
Holloman Air Force Base, 200
Honeywell: DISCOS, 549; Numerical Propulsion System Simulation (NPSS) program, 760; Propulsion Controlled Aircraft (PCA) program, 715; Ultra Efficient Engine Technology (UEET) program, 762, 797
Hoover, Herbert "Herb," 634
Horizontal-axis wind turbine (HAWT) generators, 823, 825
Horten, Reimar, 70
Horten flying wing, 70
Horton, Walter, 70
Hot Section Technology (HOST) program, 612
Hot structures: ASSET vehicles, 304, 308–9,

418–19, 592–96, 607; complexity of, 312–13, 603; concept and characteristics of, 305, 418–19; Dyna-Soar vehicles, 305–8, 418, 582–92; nose cones, 286, 579, 580–81; Space Shuttle orbiter, 310–13, 601–3; thermal protection tiles compared to, 278; titanium, 312
House Variable Intensity Boom Effect on Structures (House VIBES), 256
Houston, 217
Howze, Hamilton H., 139
Hrennikoff, Alexander, 471
Hubbard, Harvey H., 193, 210, 211, 229
Huckleberry Finn technique, 107
Huff, Dennis, 803
Hughes Falcon missiles, 102
Hughes Helicopter, 136
Hughes Offshore, 539
Hughes OH-6 helicopter, 163
Humes, Donald, 318
Hunter, Maxwell, 311, 599, 603
Hunton, Lynn, 218
Hütter, Ulrich, 826–27, 846
Hütter-Allgaier wind turbine, 826–27, 831, 846
Huygens, Christiaan, 429
Hybrid computers, 406–7, 468
Hybrid laminar flow control (HLFC), 235, 807–8
Hydraulically actuated flight control surfaces, 384, 590, 631, 634–35, 643, 644
Hydrocarbons, liquid, 817
Hydrogen, 617, 618
Hydrogen, liquid, 809–14, 815–16
Hydrogen-air fuel cell system, 864, 869, 871
Hydrogen-air supersonic combustion testing, 332
Hydrogen bomb, 439
Hydrogen fuel cell aircraft, 815

Hydrogen-oxygen regenerative fuel cell system (RFCS), 863–64, 871
Hydrogen peroxide reaction control, 298
Hyperbolic behavior, 443
Hyperbolic partial differential equations, 435n11, 436, 443
Hyperelastic structures, 503–4
HyperSizer program, 512
Hypersonic aircraft and flight. *See also* McDonnell Aerothermodynamic/elastic Structural Systems Environmental Tests (ASSET); North American X-15 aircraft; Space Shuttle orbiter: aircraft for, 279; challenges of, 361–62; delta wing development, 53–57; development of technology for, 278–79; flight control systems, 298; flutter, 306, 309, 388–89; heating, structural, 414–15; military use of, 279; record flights, 278, 293, 299, 642; reentry challenges, 285–89, 291–92; research activities, 231; scramjet engines, 279, 330–36, 340, 342–46, 343n126; stability issues, 292; support for, 230–31; wing configuration, 300–301
Hypersonic Research Engine (HRE), 335, 337
Hypersonic speed, 276
Hypersonic Weapons Research and Development Supporting System (HYWARDS), 300–301. *See also* Boeing X-20 Dyna-Soar
Hypersonic wind tunnels, 280–83, 573
Hyper-X vehicle, 343. *See also* X-43 Hyper-X vehicle
HYPULSE Shock Tunnel, 343
HyTech research program, 345

I

IBM. *See* International Business Machines (IBM)

Icing environment research, 162
Illiac IV computer, 325, 438
Illinois Center Plaza Hotel, 481
Ilyushin Badger, 102
Imaging sensors, 861
Impact analysis research, 497
Impact Dynamics Research Facility, 152
Incompressible flows, 12, 15
Inconel 625 nickel alloy, 345
Inconel X, 292, 293, 295, 416, 571, 592
Independence, Project, 823–24, 831
Indian Springs, 195
Indirect analog computers, 468
Induced drag, 117, 118
Industrial Application Centers, 488
Industry/University Cooperative Research Center (I/UCRC), 549
Inertial coupling (roll coupling), 37–45, 40n74, 44nn86–87, 393, 394–95, 407–9
Inertial Management System (IMS), 655
Inertia table, 396
In-Home Noise Generation/Response System (IHONORS), 240
Iniki, Hurricane, 859
Inlet Spillage Shock Measurement (ISSM), 246
Innis, Robert, 649
Institute for Computer Applications in Science and Engineering (ICASE), 446
Institute for Scientific Research, 719
Instrument flight rules (IFR), 151
Instrument meteorological conditions (IMC), 159
Integral equations, 433
Integrated Component Technology Demonstrations project, 764
Integrated Composite Analyzer (ICAN), 500, 547, 556; ICAN/PART, 500
Integrated Design and Evaluation of Advanced Spacecraft (IDEAS), 511–12, 550
Integrated Flight Control System (IFCS), 649
Integrated High Performance Turbine Engine Technology (IHPTET) program, 764
Integrated Modeling of Optical Systems (IMOS), 548
Integrated Propulsion Control System (IPCS) program, 754
Integrity Beacon Landing System (IBLS), 706
Intelligent Flight Controller (IFC), 720
Intelligent Flight Control Systems (IFCS) project, 259, 718–21
Intelligent Propulsion Controls Project, 764
Intercontinental ballistic missiles (ICBMs): accuracy of, 284–85; development of, 284, 439–40, 572; guidance systems, 284; materials for, 579–82; nose cones, 278, 285–88, 305, 417, 440–42, 572, 579–82; reentry challenges, 284–88
Interference studies, 53, 65
Intergraph, 539
Interior, Department of, 839
Intermediate-range ballistic missile (IRBM), 286, 442, 571–72
International Business Machines (IBM): AP-101 computers, 659, 660; blunt-body problem solution and, 442; computer advancements, 341; FORTRAN programming language, 469; IBM 360 time-sharing computer, 438; IBM 370-195 computer, 324; IBM 650 digital computer, 437; IBM 704 digital computer, 437; IBM 7090 mainframe computer, 437; IBM 7094 mainframe computer, 437–38; NASTRAN Users' Colloquia, 542; NASTRAN

version development, 484; 20-petaflop computer, 439
International Civil Aviation Organization (ICAO): emissions standards, 762, 791, 796, 797, 798, 799; noise level standards, 250; sonic booms, reaction to, 261
International Harvester Aerothermodynamic Structural Vehicles (ASV), 308–9, 595
International Helicopter Study Team (IHST), 172
International Nickel, 292
International Panel on Climate Change, United Nations, 790, 790n22
International Space Station (ISS), 243, 506
International Test Organization (ITO), 700–701
Inventions and Contributions Board, 116
Inviscid flows, 438, 444–45, 448
Iron Bird simulator, 655, 656–57, 659, 660
Iron Cross simulator, 298
Iron-nickel–based alloys, 573, 574, 575
Ishmael, Stephen D., 662, 679, 695
Israel, 481
Ivan C. Kincheloe Award, 649
Izu, Douglas, 310, 598

J

J29 fighter, 23n39
J32 Lancen (Lance), 23n39
J35 Draken (Dragon), 45
J47 jet engine, 736
J57 turbojet, 100, 101, 102
J58 engines, 708, 709–10
J75 turbojet, 103
J79 turbojet, 104
J85 engine, 686
Jackass Flats testing, 206–7, 213
Jacobs, Eastman, 17n30, 109

Jacobs, Joseph, 826
Jacobs, Marcellus, 826, 827
Jacobs, Peter F., 119, 120
Jaguar Active Control Technology (ACT), 670, 690–91, 692
Jameson, Antony, 341, 445–46, 494
Japan: CCV T-2, 670, 691–92; NASTRAN use by, 481, 484; Technical Research Development Institute (TRDI), 692
Japan Aerospace Exploration Agency (JAXA), 255
Jarvis, Calvin R., 654, 655
JAS-39 Gripen fighter, 706
JB-47E aircraft, 645–46
Jesperson, Dennis, 452
Jet engine aircraft: handling qualities, 634; hydraulically actuated flight control surfaces, 634–35; innovations in and aircraft speed, 105–6
Jet Propulsion Laboratory (JPL): Antenna Design and Analysis, 549–50; finite element analysis activities, 502–5; Geophysical Finite Element Simulation (GeoFEST), 556; high-altitude aircraft workshop, 851; history of, 502–3; Integrated Modeling of Optical Systems (IMOS), 548; NASTRAN development, 474, 503; SAMIS program, 506; System Identification/Parameter Identification (PID), 504–5, 504n113
Jettisonable assistance takeoff (JATO), 29
JF-100C Super Sabre, 402n62
John Deere, 539
Johns Hopkins University Applied Physics Laboratory, 332, 333
Johns Manville, 305, 314, 586, 600, 601
Johnson, Clarence "Kelly," 34n61
Johnson, James, 474
Johnson, Kelly, 812–13
Johnson, Lyndon B., 66

Johnson, Richard L. "Dick," 89
Johnson, Wayne, 159
Johnson Space Center/Manned Spacecraft Center: carbon-carbon composite study, 577; finite element analysis activities, 505–7; NASTRAN development, 474, 506; OVERFLOW code, 452, 453; Shuttle concept development, 56–57; structural test facility, 553; thermal-protection tiles, 318, 319, 419; TRASYS Radiative Heat Transfer program, 550
Johnston, A.M. "Tex," 22–23
Johnsville, Pennsylvania, 297, 398
Joint Acoustic Propagation Experiment (JAPE), 238–39
Joint Planning and Development Office (JPDO), 171, 171n53
Joint Sponsored Research Agreement (JSRA), 849, 852–55, 852n33, 872
Joint Strike Fighter (JSF) program, 675
Joint Strike Fighter/Integrated Subsystems Technology (J/IST) program, 676–77
Joint Task Force 2, 195
Jones, L.B., 223
Jones, Robert T.: delta wing development, 12–13, 13n20, 46, 48, 51, 53, 67; oblique wing concept, 63, 523, 523n169; Perl's association with, 17n30; swept and delta wing aircraft, 8, 11–17, 93, 108
Jordan, Gareth, 195–96
Joyce, David, 751
JP-7 fuel, 345–46
JT8D engine, 744
JT9D-7A engine, 745
JT9D engine, 611, 615, 744, 745
Junkers, Hugo, 70
Junkers Ju 88 bomber, 636
Junkers Ju 287 bomber, 21
Jupiter, 277, 288, 573. *See also* Galileo probe
Jupiter-C missile, 288
Jupiter intermediate-range ballistic missile (IRBM), 286, 287

K

Kalman, T.P., 483
Kaman Aerospace Corporation/Kaman Aircraft Corporation: helicopter design and production, 136; RITA membership, 169; wind energy program, 829, 835, 845
Kaman Circulation Control Rotor (CCR), 156–57
Kaman Controllable Twist Rotor (CTR), 153
Kane, Edward J., 224
Kansas City, Missouri, 537
Kantrowitz, Arthur, 14, 14n23, 17n30, 286
Karlskronavarvet (KKRV), 839
Kartveli, Alexander "Sasha," 333, 334, 569
Kauai, 858, 859
Kaufman, Stanley, 475
KC-135 aerial tanker: aerial refueling, 253; design of, 20; sonic boom research, 204; winglet use on, 120, 122–23, 806
Kellett KD-1 autogiro, 141
Kellett YG-1A autogiros, 141
Kellett YG-1B autogiros, 141
Kempel, Robert, 369
Kennedy, John F., 66, 109, 191, 193, 304, 644
Kennedy Space Center (KSC): finite element analysis activities, 491n64; mission of, 491n64; NASTRAN Users' Colloquia, 536; Space Shuttle Columbia, 321
Kenyon, George, 56
Kerosene, synthetic aviation, 815–16
Keyworth, George, 337
Kick angle, 527
Kiel probe, 701, 704, 705, 705n149
Kincheloe, Iven, 297, 298, 394
King Air 200 twin turboprop aircraft, 481

Kleber-Colombes, 538
Kleinknecht, Kenneth, 323
Knight, William J. "Pete," 64, 294, 418
Knöpfel, Rüdiger "Rudy," 706
Knox, Fred, 701
Kock, Berwin, 517
Korean war: F-86 Sabres in, 4, 19, 26; MiG 15s in, 26–27; rotary wing aircraft use, 138–39
Korycinski, Peter, 55, 301, 307
Kotcher, Ezra, 13–14, 14n22, 94
Krier, Gary E., 647, 657, 658, 660, 661
Küchemann, Dietrich, 13n20, 93, 107
Küssner, H.G., 15
Kvaternik, Raymond G., 508
Kwajalein Atoll, 607

L

L-23 sailplane, 248, 257
L-39 aircraft, 22–23, 23n37, 23n39
L-1011 aircraft, 120, 122
Lambert Field, 646
Laminar Flow Control (LFC) project, 235, 495, 744, 784, 786, 806–8
Lancaster, 196
Lanchester, Frederick W., 118
Lang, Karl, 701, 704, 705
Langley Aerothermodynamic Upwind Relaxation Algorithm (LAURA), 451–52, 453
Langley Research Center/Langley Aeronautical Laboratory/Langley Memorial Aeronautical Laboratory: Acoustically Treated Nacelle program, 740; Acoustics Division, 195, 234; Advanced Tiltrotor Transport Technology (ATTT) program, 167–68; Advanced Turboprop (ATP) project, 800; aerodynamic refinement and efficiency improvements, 90, 128–29; Aeroelastic Rotor Experimental System (ARES), 164; Aeronautics Research Mission Center, transformation to, 166; Aeronautics System Division, 220; Air Force F-111 program, 61–62; all-moving tail research, 35; Arc-Heated Scramjet Test Facility (AHSTF), 343; Army research presence at, 139, 146, 147; AST Project Office, 220; Avro CF-105 Arrow development, 641–42, 642n25; Bi-Level Integrated System Synthesis (BLISS), 511; British aeronautical community, ties to, 60–61; BWB research, 71; columbium heat shields, 602–3; Combustion-Heated Scramjet Test Facility (CHSTF), 343; computational fluid dynamics and transonic flow and airfoils, 445–46; computational fluid dynamics CFL3D code, 452–53; computational fluid dynamics LAURA code, 451–52, 453; computational fluid dynamics Shuttle study, 426; Computational Structural Mechanics (CSM) workshop, 497; computers, analog, 468; computers, IBM, 438; CONMIN program, 511; Continuous Flow Hypersonic Tunnel (CFHT), 283; Convair YF-102 Delta Dagger testing, 100–101; delta wing development, 53; delta wing research, 47–48, 48n94; Design Analysis Methods for Vibration (DAMVIBS), 508–11; Differential Maneuvering Simulator (DMS), 666–67; Direct Connect Supersonic Combustion Test Facility (DCSCTF), 343; DM-1 glider research full-scale tunnel studies, 47; finite element analysis activities, 507–12, 532; Finite Element Composite Analysis Program (FECAP), 556; flat-bottom delta research, 55–56; FLEXSTAB program, 544–45; Flight Research Branch,

154; flutter suppression research, 388, 389; fly-by-wire research activities, 638–39; forward-swept wing research, 21–22; Free Flight Tunnel studies, 63; fuel research, 815–16; fuel-saving aircraft structures, 787; Gas Dynamics Branch, 332–33; glomb (guided glide bomb), 11; Ground Demonstrator Engine (GDE), 345; high-altitude aircraft workshop, 850; High-Speed Tunnel (HST), 94–97, 127–28; high-speed tunnel studies, 36; high-temperature tunnels, 342, 343, 573; hingeless rotor concept full-scale tunnel studies, 150; HL-10 lifting body research, 56; hypersonic flight research, 299; Hypersonic Research Engine (HRE), 335; Hypersonic Tunnel, 282–83; Hypersonic Weapons Research and Development Supporting System (HYWARDS), 301; HYPULSE Shock Tunnel, 343; Impact Dynamics Research Facility, 152; inertial coupling research, 42–43; Institute for Computer Applications in Science and Engineering (ICASE), 446; instrument flight rules (IFR) investigation, 151; Integrated Design and Evaluation of Advanced Spacecraft (IDEAS), 511–12; laminar flow control research, 807; Low-Frequency Noise Facility, 215; Lunar Landing Research Facility, 152; Mach 6 tunnels, 343; Mach 10 tunnel, 343; multidisciplinary optimization (MDO) studies, 511; NASTRAN, development of, 152, 473–74; NASTRAN, management of, 477, 508; NASTRAN, Users' Colloquia, 480, 508, 536, 540, 542; noise research, 163, 164–65; Programming Structural Synthesis (PROSS), 511–12; rotary wing aircraft full-scale wind tunnel studies, 140, 143, 144; rotary wing aircraft research, 137, 139–45, 147–52, 154–55, 156, 161, 162–63, 164–65; Rotor Systems Research Aircraft (RSRA) program, 154–55; SCAT-15F, 67; scramjet engines, 342–44; scramjet research, 335–36, 343n126; SCR/SCAR/AST conference, 225–26, 228–29; Shaped Sonic Boom Demonstration (SSBD), 246; slotted transonic tunnel, 93, 95–96; sonic boom conferences, symposiums, and workshops, 209–10, 231, 233–34; sonic boom research, 192–200, 203, 208–9, 214–15, 217, 227, 239–40, 256–58; sonic boom research, publication of, 218–19, 221; sonic boom research simulators, 249, 257–58; sonic boom wind tunnel studies, 235; Space Shuttle concept development, 56–57; Space Shuttle orbiter heat shield, 312; SST program, 66–67, 191; Steep Approach program, 737; Structural Analysis of General Shells (STAGS), 514, 551–52; Structural Performance and Redesign (SPAR), 498; structural test facility, 553; ST-SIZE program, 512; supercritical wing program, 110–15, 805–6; Supersonic Pressure Wind Tunnel, 208, 208n66, 214; supersonic tunnel, 14; swept wing research, 6, 8, 13–15; tail configuration research, 36; thermal-protection tile testing, 318; Transonic Aerodynamic Branch, 94, 108; Transonic Aircraft Technology (TACT) program, 386; Transonic Dynamics Tunnel (TDT), 149–50, 161, 164, 388, 389; Transonic Pressure Tunnel (TPT), 109–10, 118, 119, 123, 127–28; Transonic Tunnel (High-Speed Tunnel), 95n16; Ultra Efficient Engine Technology (UEET) program, 762, 797; Unitary Wind Tunnel, 235, 585–86; variable-geometry wing research, 59, 60–61, 61n115, 63; Vehicle Technology

Directorate (VTD), 162–63; vortex lift research, 651n49; wind energy program, 829; wind tunnel studies, 665, 702; winglet program, 117–19, 120, 122–23; X-15 development, 291–93; X-43 Hyper-X vehicle, 343–45, 344n127; X-48B testing, 71; YF-12 Blackbird project, 517

Large Eddy Simulation (LES), 341–42

Larson, George, 117

Larson, Howard, 315

Larson, Nils, 259, 260

Lateral-directional stability, 19, 38–39

LAURA (Langley Aerothermodynamic Upwind Relaxation Algorithm), 451–52, 453

Laureate Award in Aeronautics and Propulsion, 253

Laurel Award, 856

Lavochkin fighters, 36

Leading edge extensions, 414

Lean Pre-mixed, Pre-vaporized (LPP) combustor concepts, 798–99

Lear, Bill, 115

Lear 28/29 aircraft, 481

Lear 35/36 aircraft, 480

Lear 55 business jet, 481

Learjet, NASTRAN use by, 480–81

Learjet Model 28 aircraft, 121

Learjet Model 55 aircraft, 121

Learjet variable stability aircraft, 402n62

Leduc, René, 50

Lemay, Curtis E., 812

Lewis, David, 11

Lewis, George W., 14, 142

Lewis Flight Propulsion Laboratory. *See* Glenn Research Center/Lewis Research Center/Lewis Flight Propulsion Laboratory/Aircraft Engine Research Laboratory

Lewis Research Center. *See* Glenn Research Center/Lewis Research Center/Lewis Flight Propulsion Laboratory/Aircraft Engine Research Laboratory

Lewy, Hans, 435–36

LI-900 thermal protection, 311, 313–17, 600–601, 604, 605

LI-1500 thermal protection, 310–11, 313, 599–600, 601, 604

LI-2200 thermal protection, 317

Lichtenstein, Jacob H., 36

Liebeck, Robert H., 70

Lift and Nozzle Change Effects on Tail Shocks (LaNCETS), 258–60, 261

Lifting bodies research, 56, 417, 418

Lifting line theory, 12

Lifting surface predictions, 483

Lift-to-drag ratio (L/D), 118, 403, 415, 595

Light emitting diodes (LEDs), 521–22, 527–28

Lightweight Fighter (LWF) program, 370, 647, 664

Limit-cycle oscillation, 367–69, 377, 378

Lindbergh, Charles A., 591

Lindbergh Field, 102

Linear Differential Variable Transformers (LDVTs), 656

Linear partial differential equations, 431, 435–36

Ling-Temco-Vought A-7D Corsair II DIGITAC, 382–83

Ling-Temco-Vought shock expansion tube, 214–15

Link Trainer, 295, 392

Lippisch, Alexander, 5, 7, 8, 13, 13n20, 14, 45, 46–47, 58

Lippisch DFS DM-1 aircraft, 13

Lippisch DM-1 aircraft, 47–48, 48n94

Lippisch P 13 aircraft, 13, 47

Liquid hydrogen, 809–14, 815–16
Liquid methane, 815–16
Liquid oxygen (LOX) lines, 513
Lithgow, Mike, 67n127
Little Boom, Project, 194–95
Littleman, Project, 199
Littleton Research & Engineering, 538
Load factors, 377
Loads Alleviation and Mode Suppression (LAMS) system, 392
Loads prediction, 461–62, 462n1, 465–66
Lockalloy, 620
Lockheed A-12 Blackbird, 210, 570
Lockheed C-5 aircraft, 337, 407, 867
Lockheed C-141 aircraft, 324, 658n59
Lockheed CL-400 aircraft, 812–13
Lockheed CL-1200 Lancer, 34n61
Lockheed coatings: LI-0042 coating, 315; LI-0050 coating, 315
Lockheed Corporation: Aerospaceplane proposal, 334; high-altitude aircraft workshop, 851; NASTRAN Users' Colloquia, 480; Palo Alto Research Laboratory, 551, 600; sonic boom research, 237; SST proposal, 210; supersonic business jet (SSBJ) concept, 249; titanium hot structures, 312; wind energy program, 829, 832, 834; winglet use by, 121–22
Lockheed F-16, 710. *See also* General Dynamics F-16 Fighting Falcon
Lockheed F-94 Starfire, 102
Lockheed F-104 Starfighter: design of, 104, 401, 409, 440; development and manufacture of, 290; F-104A Starfighter, 37, 440; F-104B, 409; F-104G, 688; F-104G (German CCV), 670, 689–90, 692; heating, structural, 413; inertial coupling, 43, 409; landing of, 405; landing simulator, 403–4; nickname for, 4, 104; pitch-up problems, 34n61, 413; reaction controls, 298; sonic boom research, 195–96, 199, 200, 204, 205, 206–7; speed of, 413; supersonic flight by, 3, 104; use of, 104; wingspan-to-fuselage length ratio, 407; XF-104 Starfighter, 185
Lockheed F-117 stealth fighter, 70, 384, 721–22
Lockheed Have Blue aircraft, 721
Lockheed L-1011 aircraft, 120, 122
Lockheed Loads Alleviation and Mode Suppression (LAMS) system, 392
Lockheed Martin: QSP proposal, 245; Shaped Sonic Boom Demonstration (SSBD), 246; supersonic business jet (SSBJ) concept, 250; Ultra Efficient Engine Technology (UEET) program, 762, 797
Lockheed Martin F-22 Raptor, 126–27, 384, 402, 413, 712, 713, 722
Lockheed Martin F-35 Lightning II Joint Strike Fighter, 384, 413, 678, 713, 722
Lockheed Martin X-33 Research Vehicle, 384
Lockheed Missiles and Space, 310
Lockheed NT-33A Shooting Star, 402
Lockheed P-80 Shooting Star, 634
Lockheed SR-71 Blackbird: accident with, 392; aerodynamics of, 70; aeroelastic effects, 391–92; digital engine control system, 754; engine inlet unstarts, 708–9, 754; heating, structural, 413; JP-7 fuel, 346; simulator, 407; sonic boom research, 202–3, 204, 205–6, 236–38, 239, 242–43, 245, 261; spacesuit for, 297; speed of, 413; thermal barrier coatings, 611; thermal loads project, 516; titanium construction, 312, 570–71, 602
Lockheed Star Clipper, 311, 599, 600

Lockheed TF-104G aircraft, 688
Lockheed thermal protection, 310–11, 419, 596–99, 602; borosilicate glass coatings, 315–16, 599–600; LI-900 thermal protection, 311, 313–17, 600–601, 604, 605; LI-1500 thermal protection, 310–11, 313, 599–600, 601, 604; LI-2200 thermal protection, 317; Space Shuttle tiles, 310–23, 419
Lockheed U-2 spy plane, 297, 811, 812–13
Lockheed X-7 testbed, 278, 289, 290, 572
Lockheed X-17 missile, 287–88
Lockheed X-27 aircraft, 34n61
Lockheed XF-104 Starfighter, 185
Lockheed XH-51N helicopter, 150
Lockheed X-Wing Stoppable Rotor, 157, 158
Lockheed YF-12 Blackbird: digital engine control system, 754; Loads Alleviation and Mode Suppression (LAMS) system, 392; sonic boom research, 202–3, 204, 210, 227, 227n133; thermal loads project, 515, 516–19, 529, 554; YF-12A, 202–23; YF-12C, 516, 708–9; YF-12 Experiment Symposium, 519
Lockheed YF-22 aircraft, 402
Lockheed YO-3A aircraft, 242–43
Loftin, Laurence K., Jr., 108, 111, 125
Long-Davis, Mary Jo, 254
Longhorn wing and winglet system, 121
Longitudinal stability: all-moving tail and, 34–35, 35n65, 362–63, 363n4; inertial coupling (roll coupling) and, 38–39; static, 648, 664, 669, 690, 692; swept and delta wing aircraft, 19, 21; T-tail aircraft, 36–37
Long Range Planning Study for Composites (RECAST), 804
Lord, W.T., 94
Los Alamos National Laboratory, 498, 541
Los Angeles, 206

Louisiana Tech University, 542
Love, Eugene S., 56
Low aspect ratio aircraft, 11–12, 13n20, 18, 43, 54
Low-boom demonstrator, 249, 250–51, 255, 261
Low boom/no boom technique, 255–57
Low Emissions Alternative Power (LEAP) program, 872
Low-Frequency Noise Facility, 215
LTV Aerospace, 576, 618
LTV YA-7D Digital Flight Control for Tactical Aircraft (DIGITAC), 684–86
Ludington-Griswold Company, 9
Ludox, 320–21
Lunar Landing Research Facility, 152
Lunar Landing Research Vehicles (LLRVs), 653–54
Lunar Landing Training Vehicles (LLTVs), 653–54
Lunar Module, vii, 654
Lux, David, 236
Lynntech, Inc., 864
Lyons, Valerie, 814

M

M2-F2 lifting body, 407
MacCormack, Robert, 446–47, 448
MacCormack's method, 446–47
Mach, Ernst, 97, 182
Mach 2+ supersonic transport, 11
Mach angle of wing, 15–16, 18
Mach number: definition, 182; of wing, 5, 12, 14, 18, 52, 53
Mack, Robert, 223–24, 225, 229, 235, 249
Mackall, Dale, 682, 683
MacKay, John, 331–32
Maclise, Dougal, 861
MacNeal, Richard H., 467, 475

Index

MacNeal Schwendler Corporation (MSC): cofounder of, 467; finite-element heat transfer program, 502; MSC/NASTRAM, 478, 484, 499, 546, 547; NASTRAN competitors, 486; NASTRAN development, 475; NASTRAN maintenance, 479

Maglieri, Dominic J.: AST project, 220; Shaped Sonic Boom Demonstration (SSBD), 246; sonic boom conferences, 210, 211, 212; sonic boom research, 193, 229, 236; supersonic aircraft proposals, 245–46; XB-70 records, 238

Magnesia, 587

Magnesia-stabilized zirconia, 611

Magnesium, 293, 571

Mahesh, Krishnan, 341

Makani Uwila Power Corporation, 842

Mallick, Donald L., 638

Management and Budget, Office of (OMB), 601–2

Manching, 690

Maneuver load control, 652

Manke, John A., 381, 661

Manned Orbiting Laboratory (MOL), 303

Manned Spacecraft Center (MSC). *See* Johnson Space Center/Manned Spacecraft Center

Marc Analysis & Research, 484, 546

Marching solutions, 435–36

Marconi Company, 671

MARC structural analysis code, 546

Marine Corps, U.S. (USMC): AFTI/F-16 aircraft, 679; V-22 Osprey, 166; Vietnam war aircraft losses, 644n32; X-31 aircraft, 707

Mariotte, Edme, 428–29

Marquardt ramjets, 289, 290

Marquardt scramjet flight-test program, 335, 336

Marshall Islands, 439

Marshall Space Flight Center: ASSIE-I program, 552; Coupled Eulerian Lagrangian Finite Element (CELFE), 514; Cylindrical Optimization of Rings, Skin and Stringers (CORSS), 513; finite element analysis activities, 498, 512–14; impact analysis research, 497; NASTRAN, development of, 474, 514; NASTRAN, Users' Colloquia, 536; Program for Analysis of Nonlinear Equilibrium and Stability (PANES), 552; Shuttle concept development, 57; Structural Analysis of General Shells (STAGS), 514, 551–52; structural test facility, 553; VLOADS launch loads and dynamics program, 513–14; wind energy program, 829

Martin, H.C., 471

Martin, Roy, 247

Martin B-57B bomber, 810, 811

Martin Baltimore, 475

Martin MA-25S ablative coating, 417–18

Martin Marietta TRASYS Radiative Heat Transfer program, 550

Martins, Joaquim, 494

Martin SV-5D (X-23) PRIME (Precision Recovery Including Maneuvering Entry), 27n51, 278, 288, 304, 417, 418, 606–8

Martin X-24A (SV-5P) lifting body, 27n51, 407, 607

Martin X-24B lifting body, 56, 404, 407

Martin X-33 Research Vehicle, 384

Martin X-38 aircraft, 27n51

Mascitti, Vincent R., 220

Mason, James B., 476

Massachusetts Institute of Technology (MIT), 471, 654

Master Builders, Inc., 500

Materials and Structures for High Performance project, 763

Matranga, Gene, 517

Matrix method, 466, 470–71
Matrix Methods in Structural Mechanics conference, 474–75
Mattingly, Kenneth, 658
Mauchly, John, 436
Maxwell, James Clerk, 433–34
Mayo Graduate School of Medicine, 542
McAleese, James D., 474
McComas, Robert L., 474, 536
McConnell, Richard D., 476
McCormick, Caleb W., 475
McDonnell Aerothermodynamic/elastic Structural Systems Environmental Tests (ASSET): Aerothermodynamic Elastic Vehicles (AEV), 308–10, 594, 595; Aerothermodynamic Structural Vehicle (ASV), 308–10, 594–95; flights in program, 309, 594–95; hot structures, 304, 308–10, 418–19, 592–96, 607; International Harvester Aerothermodynamic Structural Vehicles (ASV), 308–9, 595; nose cap, 594–95; performance of, 309, 596–97
McDonnell Aircraft Corporation, 592
McDonnell-Douglas 500E helicopter, 165
McDonnell-Douglas Aerospace, 494
McDonnell-Douglas BWB-450 aircraft, 71
McDonnell-Douglas C-17 Globemaster III transport, 127, 384, 715, 722, 795
McDonnell-Douglas Corporation: Advanced Control Technology for Integrated Vehicles Experiment (ACTIVE) project, 717; Boeing merger, 127; BWB transport, 70–71; Collier Trophy, 127; lifting surface predictions, 483; NASA contract work, 120; NASTRAN Users' Colloquia, 480; sonic boom research, 237; STOL transport competition, 114; winglet research, 123, 125
McDonnell-Douglas DC-8 aircraft, 737, 817
McDonnell-Douglas DC-9 aircraft, 745
McDonnell-Douglas DC-10 aircraft: engine component improvements, 745; Sioux City airliner crash, 757, 766; winglet research, 119, 120, 122, 123
McDonnell-Douglas F-4 Phantom, 644n32
McDonnell-Douglas F-15 Eagle: Advanced Control Technology for Integrated Vehicles Experiment (ACTIVE) project, 716, 717; Digital Electronic Engine Control (DEEC) program, 709–10, 754–56; Dryden research activities, 496; engines, 713; F-15A, 44, 711, 713; F-15E Strike Eagle, 69; fly-by-wire system, 383–84; Highly Integrated Digital Electronic Control (HIDEC), 711; High Stability Engine Control (HISTEC) system, 718; Intelligent Flight Control Systems (IFCS) project, 718–21; Propulsion Controlled Aircraft (PCA) program, 713, 757; Self-Repairing Flight Control System (SRFCS), 713–14; simulator, 407; sonic boom research, 238–39, 246–48, 252–53, 256, 258–60, 261; tail configuration, 413; TF-15, 715
McDonnell-Douglas F-15 Eagle (F-15B/NF-15B): Advanced Control Technology for Integrated Vehicles Experiment (ACTIVE) project, 716, 717; exhibit showing, 721; High Stability Engine Control (HISTEC) system, 718; Intelligent Flight Control Systems (IFCS) project, 718–21; Lift and Nozzle Change Effects on Tail Shocks (LaNCETS), 258–60, 261; modifications for experiments, 258–59; Quiet Spike device, 252–53, 256; Short Takeoff and Landing Maneuver Technology Demonstrator (S/MTD), 715–16; standard features, 259
McDonnell-Douglas F-18 Hornet, 237, 251,

255–57, 496, 676

McDonnell-Douglas F/A-18 Hornet: Active Oscillation Suppression (AOS) system, 389; autothrottle system from, 706; design of, 669; engines for, 706n152; F/A-18A Hornet, 44, 383–85; F/A-18B Hornet, 386; tail configuration, 413

McDonnell-Douglas Harrier strike fighter, 60

McDonnell-Douglas Helicopter Company, 508, 510

McDonnell-Douglas MD-11 aircraft, 123, 715, 757, 758, 806

McDonnell-Douglas Project 680J Survivable Flight Control System (SFCS), 379–80, 384, 646–48, 658n59

McDonnell-Douglas YC-15 aircraft, 114, 127

McDonnell-Douglas YF-4E Phantom Survivable Flight Control System program, 379–80, 384, 646–48, 658n59, 660

McDonnell F-4E Phantom II, 195

McDonnell F4H-1 Phantom II, 32, 33n60, 34

McDonnell F-101 Voodoo: development and manufacture of, 290; flight control system, 372; pitch-up problems, 33–34, 413; sonic boom research, 194, 199; sonic booms, 185; tail configuration, 33–34, 36, 413; turbofan engines, 741

McDonnell XF-88 aircraft, 33

McKee, William F. "Bozo," 192

McKinney, Linwood, 651n49

McLean, F. Edward, 67, 210, 211, 217, 220, 222

McLellan, Charles, 292

McMurtry, Thomas C. "Tom," 89, 112, 113, 657, 658, 661, 714

McNamara, Robert S., 192, 211, 303–4, 400, 592

MD-11 aircraft, 123, 715, 757, 758, 806

MDBOOM, 237

Mechanical linkages in flight control systems, 525, 631, 633, 637–38, 639, 655

Medal for Exceptional Scientific Achievement, 116

Medicine Bow, Wyoming, 839, 848

Mercury ballast, 595

Mercury program: ablation cooling, 417; centrifuge program, 400; flight control systems, 653; heat shields, 587; McDonnell Aircraft spacecraft, 592; reaction controls, 298; reaction thrusters, 278; Redstone missile/booster, 636; reentry vehicles, 54; side stick flight controller, 658n59; spacesuits, 297

Messerschmitt-Bölkow-Blohm/Messerschmitt: advanced project group, 14, 48n94; Military Aircraft Division, 703n144; variable-geometry wing aircraft, 58–59

Messerschmitt Me 163 Komet (Comet), 7, 13, 27, 27n51

Messerschmitt Me 262 aircraft, 27n51

Messerschmitt P 1101 aircraft, 58–59

Metallurgy: directionally solidified (d.s.) castings, 614–15; powder metallurgy, 616–17; rapid-solidification technique, 279, 617, 620–21

Metal matrix composites (MMCs), 617–18

Methane, 342, 815–16

Methyl methacrylate (Plexiglas), 310–11, 598

Meyer, Corwin "Corky," 23, 59, 104

Michel, Doug, 473–74

Microcraft X-43 vehicle, 279, 341, 343–45, 344n127

Micromechanics Analysis Code (MAC), 500

Micromechanics Analysis Code with Generalized Method Cells (MAC/GMC), 512

Middleton, Wilbur, 67
MiG aircraft. *See Mikoyan and Gurevich entries*
Mighty Mouse rockets, 102
Mikoyan, Anushavan "Artem," 26
Mikoyan, Stepan, 26
Mikoyan and Gurevich MiG-15 aircraft: development of, 19n32, 23n39; Korean war use of, 4, 19, 26; pitch-up problems, 26–27; tail configuration, 36
Mikoyan and Gurevich MiG-17 aircraft, 27
Mikoyan and Gurevich MiG-21 aircraft, 32, 45
Mikoyan and Gurevich MiG-23/27 Flogger, 62
Miles M.52 aircraft, 35n65
Miller, Robert, 611, 612–13
Millikan, Clark, 14
Minneapolis, 206
Minneapolis Honeywell MH-96 Adaptive Flight Control System (AFCS), 298–99, 372–74
Minneapolis Honeywell MH-96 flight control system, 376, 379
Minuteman intercontinental ballistic missile (ICBM), 286, 303
Mirage IIIE aircraft, 691
Mishap Investigation Board (MIB), 870–71
Missile Defense Agency, 850
Mission-adaptive wing (MAW), 385–86
Mistle (Mistletoe) aircraft, 636
Mitsubishi F-2 fighter, 692
Mitsubishi Regional Jet, 752
Mitsubishi T-2 aircraft, 691–92
Mod-0A wind turbine, 831, 833–35
Mod-0 wind turbine testbed, 827, 830, 831–33, 846
Mod-1A wind turbine, 836
Mod-1 wind turbine, 831, 835–36, 845
Mod-2 wind turbine, 830, 831, 836–39, 840, 843, 847, 848
Mod-3 wind turbine, 840
Mod-4 wind turbine, 840
Mod-5A wind turbine, 840, 841
Mod-5B wind turbine, 840, 841–44, 847
Mod-5 wind turbine, 831, 840
Mod-6 wind turbine, 840
MODE-ID program, 504
Model-following systems, 366–67
Modern Rotor Aerodynamic Limits survey, 515, 519–20, 529–30
Modlin, C. Thomas, Jr., 474, 506
Moes, Timothy R., 236, 259
Moffett Field, 137
Molybdenum, 306, 575, 583, 587
Molybdenum alloy, 306, 308, 584
Molzahn, Leslie M., 253
Momentum equation, 430, 431, 433, 438, 443
Monoplane, 387
Moon: mission to, 787; reentry from, 278, 288; return to, 765; training to land on, 653–54
Moore School of Electrical Engineering, University of Pennsylvania, 436
Morane brothers, 385
Morane Bullet, 34, 385
Moretti, Gino, 444
Morgan, David, 24
Morgan, Ray, 855, 856, 868–69
Morgenstern, John, 246
Moskalev, Alexandr Sergeevich, 9n12
MOSTAB computer code, 834
Motion, equations of, 393
MSC-040 concept, 602
MSC/NASTRAM, 478, 484, 499, 546, 547
Mufflers, 737, 740
Mullite reusable surface installation,

313, 604
Multhopp, Hans, 27n51
Multidisciplinary optimization (MDO) studies, 511
Munk, Max, 11, 12
Murman, Earll, 445
Muroc Army Airfield, 182, 184
Muroc Dry Lake, 23–24
Muroc Flight Test Unit, 494. *See also* Dryden Flight Research Center (DFRC)/Flight Research Center
Mussolini, Benito, 332–33
Mutterperl, William, 17n30
Myasishchev Bison, 102
Myers, Dale, 313, 603–4
Myers, Garry, 144

N

N-156 aircraft, 32
N aircraft concepts: N+1, 771, 772, 808; N+2, 771, 772; N+3, 771, 772; N2A, 808; N2B, 808
Nap-of-the-Earth (NOE) flight, 159, 170
NASA Structural Analysis System (NASTRAN): aeroelastic effects research, 392, 521–22; capabilities of, 328, 472, 476–77, 531, 535; commercial versions, 484–87, 531; competitor organizations, 486; development of, 152, 497, 501–2, 503, 506, 535, 556; distribution of, 477, 489, 508; documentation, 477; early use of, 477–83; elements development, 478, 479, 485, 501–2, 514; equations of, 328; error corrections, 478, 479; government agency use of, 481–82, 540–41; maintenance of, 478–79; management of, 477, 508; nonaerospace applications, 481, 535, 538–39; organization-specific versions, 484–86; origins and development of, 152, 468, 472–77; price of, 477; reference sources, 543; requirements of, 475; Space Shuttle studies, 327–28; success of, 486–87; Systems Management Office (NSMO), 477, 478, 480, 484, 508; team members, 475–76; upgrades to, 478, 479, 482–83, 531, 535; user community profile, 478, 480–81, 535; Users' Colloquia, 479–82, 484, 491n64, 497, 508, 536–37; X-29 aircraft model, 460
NASGRO program, 506
NASTPLT program, 556
NASTRAN. *See* NASA Structural Analysis System (NASTRAN)
National Academy of Sciences, 211, 215
National Advisory Committee for Aeronautics (NACA): aerodynamic research and development, 17, 17n30; aeronautics contributions, viii–x, 108, 171–72, 735–36; aeronautics staff, 146; airfoil sections, 29n56, 109, 115; computation tasks, 137–38; cowling design, 734, 735–36, 809; Distinguished Service Medal, 107; formation of, viii, 361; Jones as aide, 11; rotary wing aircraft contributions, 136–37; Rotating Wing Subcommittee (Helicopter Subcommittee), 142; Special Subcommittee on Aircraft Noise, 737; technical intelligence collection target, 17n30; transformation into NASA, 146, 361, 737
National Aeronautics and Space Act, 487, 852, 852n33
National Aeronautics and Space Administration (NASA): aeronautics contributions, viii–x, 64–65, 817–18; aeronautics research, return to, 253–54; Army, relationship with, 139; budget for aeronautics, 785; Executive Order to support aeronautics research, 171–72; full-cost accounting methods, 166; Headquarters meeting,

472–74; Medal for Exceptional Scientific Achievement, 116; Outstanding Mechanical Engineer Award, 828n6; rotary wing aircraft contributions, 136–37; Software of the Year Award, 761; Space Act Award, 846; space initiatives, vii–viii, 787; transformation of NACA to, 146, 361, 737

National Aeronautics Association, 93, 107, 116

National Aeronautics Research and Development Policy, 171–72

National Aero-Space Plane (NASP), 230–31, 241, 279, 336–40, 342, 419, 615–21

National Aero-Space Plane Joint Program Office, 338–39

National Air and Space Museum: Beyond the Limits gallery, 427, 432; delta test model donation, 15n26; HiMAT vehicle, 689; Udvar-Hazy Center, 166, 638n15

National Business Aircraft Association, 121

National Carbon, 587

National Full-Scale Aerodynamics Complex (NFAC), 158, 165–66

National Hypersonic Flight Research Facility (NHFRF), 299, 605

National Medal of Science, 116

National Museum of the United States Air Force, 638n15, 698–99

National Oceanic and Atmospheric Administration (NOAA), 206–7

National Opinion Research Center, University of Chicago, 199

National Park Service, 201

National Register of Historic Places, 127–28

National Research Council, 248

National Rotorcraft Technology Center (NRTC), 168–69

National Science and Technology Council, Committee on Technology, 171–72

National Science Foundation (NSF), 549, 823–24, 825, 827

National Sonic Boom Evaluation Office (NSBEO), 203

National Sonic Boom Evaluation Program, 203–5

National Technology Transfer Center (NTTC), 490

NATO (North Atlantic Treaty Organization), 213, 365

Naumann, Eugene C., 508

Navaho cruise missile, 289, 290, 569–70

Naval Air Development Center (NADC), 297, 398–400, 540

Naval Air Warfare Center, 384

Naval Ocean Systems Center, 541

Naval Research Laboratory, 227

Naval Ship Research & Development Center, 481, 484, 540, 541

Naval Surface Weapons Center, Carderock Division, 157, 541

Naval Technical Mission, 29n56

Naval Underwater Systems Center, 481, 540, 541

Naval Weapons Center, 540

Navier, Claude-Louis-Marie-Henri, 431–32

Navier-Stokes equations: Beyond the Limits gallery display, 427, 432; concepts governed by, 324, 432–33, 446; ENSAERO code, 493; LAURA code, 451; origins of, 431–32; solution to, 323, 432–33, 434, 436, 438–39, 446–47; turbulence, 450–51; X-24C computed surface streamlines, 448–49

Navy, U.S. (USN): Cooperative Test Organization/VECTOR, 705–6; Douglas D-558-2 Skyrocket, 28; electronic ship control systems, 631; F-8C Crusaders, 654–55; fighter aircraft program, 61; fly-by-wire ship control systems, 723; Full Authority Digital Engine Control program, 755; Grumman F9F-9/F11F-1 Tiger, 104, 105; helicopter specifications, 145–46; International Test Organization (ITO), 700–701; NASA, relationship with, 139; NASTRAN Users' Colloquia, 480, 481, 541; National Aero-Space Plane Joint Program Office, 338–39; National Rotorcraft Technology Center, 168; Pacific Missile Range Facility, 867, 869; sonic boom research, 206; spacesuits, 297; stability augmentation system research, 365; swept wing research, 22–23; Talos missile, 289; Tilt Rotor Research Aircraft (TRRA) program, 155; Vanguard satellite project, 500; Vietnam war aircraft losses, 644n32; X-15 hypersonic flight, 290; X-31 aircraft, 707

NB-52/NB-52B Stratofortress, 343, 369, 378–79, 568

NC-131H Total In-Flight Simulator (TIFS), 402n62

Nellis Air Force Base, 194–95

Netherlands: Dutch-German Wind Tunnel (Duits-Nederlandse Wind Tunnel), 164; V-2 missile shock waves, 184; wind energy, 827

Neural network project, 718–21

Nevada, 202

New Mexico, 827

Newton, Isaac, 429–30, 432

Newton-Raphson Parameter Identification, 397

Newton's fluid dynamics laws, 429–30; second law, 429–30; sine-squared law, 429

New York Times, 216

New York University: Courant Institute, 111, 471; sonic boom research, 213, 217, 222

Next Generation Clean Aircraft Power (NEXCAP) program, 815

NFK Engineering Associates, 484

Nibe A wind turbine, 848

Nicholas-Beazley, 11

Nicholls, J. Arthur, 332

Nickel, 575, 583

Nickel-based alloys. *See also* René 41 alloy: Alloy 454, 614–15; fasteners from, 619; gamma-prime, 574; oxygen and, 575; stress to cause failure, 613–14; superalloys, 573–76; thermal barrier coatings (TBC), 610; welds on, 575

1954 Interceptor initiative, 50

Niobium, 575

Nippon Univac Kaisha Ltd. (NUK), 484

Nitrogen oxides (NOx) emissions, 762, 790, 794, 795–99, 813

Nixon, Richard M., 66–67, 107, 116, 217, 219, 313, 603

NKF Engineering, 538, 539

Noise. *See also* sonic booms: airport noise studies, 206; design of aircraft and, 770–72; engine noise research, 736–37; exhaust nozzle chevron, 808–9; FAA regulations and standards, 739–40, 741, 771; open-rotor engines, 791; research activities at Lewis, 738; rotary wing aircraft, 163, 164–65, 167–68; turbojet engines, 737; turbopropeller engines, 800, 802

Noise and engine improvements, 737–40; Acoustically Treated Nacelle program, 740; Quiet Clean Short Haul Experimental Engine, 741–43; Quiet Engine program, 740, 741; Steep Approach program, 740

Noise and Sonic Boom Impact Technology (NSBIT) program, 229–30, 231

Nomex, 767

Nomex nylon, 316

Nonlinear partial differential equations,

430–31, 432–33
Nonplanar lifting system, 118
Nordic Windpower, 848
Northam, Burton, 336
North American A3J-1 Vigilante, 32
North American A-5 Vigilante, 639
North American Air Defense Command (NORAD), 188
North American Aviation/North American Rockwell: B-70 contract, 191; business jet concept, 228–29; Space Shuttle orbiter contract, 313, 604; SST proposal, 210; supercritical wing contract, 112
North American B-45 aircraft, 634–35
North American B-70 aircraft, 191
North American F-86 Sabre: accidents with, 24, 24n43; development of, 19n32, 23n39; F-86A, 26; F-86D Sabre, 102; F-86F, 26; Korean war use of, 4, 19, 26; pitch-up problems, 24–26, 36, 37, 413; reputation of, 23–24; sonic booms, 186–87; stability research activities, 637; tail configuration, 36, 413; variable stability aircraft, 401; XP-86 Sabre prototype, 18, 635; YF-86D low-tail testbed, 36–37
North American F-100 Super Sabre: development and manufacture of, 290; F-100C Super Sabre, 409; F-100D Super Sabre, 409; flight control systems, 635; JF-100C Super Sabre, 402n62; landing simulator, 403; sonic boom research, 194, 196; tail configuration, 413; YF-100A Super Sabre, 34, 363; YF-100 Super Sabre, 37, 40, 41, 42, 185
North American F-100 Super Sabre (F-100A): design of, 41, 42–43, 393–94; roll-coupling accident, 41, 42, 44, 393; roll coupling research, 408–9; wingspan-to-fuselage length ratio, 37

North American F-107 aircraft, 372, 642, 658n59
North American JF-100C Super Sabre, 402n62
North American P-51D Mustang, 37
North American P-51 Mustang, 92
North American Sabre 45 aircraft, 36
North American simulators, 296
North American SM-64 Navaho cruise missile, 289, 290, 569–70
North American T-2C Buckeye, 114–15
North American X-15 aircraft: ablation cooling research, 417–18; accident with, 299, 373–74; centrifuge program, 399–400; contracts for, 293; contributions from program, 299; design confidence, 307, 308; development of, 278, 283, 290–93; at Dryden, 294; Dutch roll motions, 411–12; emergencies, practice for, 296; flight control system, 298–99, 372–74, 376, 379, 642; flight-testing, 294–95; flight time, 299; funding for, 291; heating, structural, 414–15; heat-sink structures, 415–17; inertial coupling (roll coupling), 409; instrumentation development, 298; landing of, 405; landings of, 403–4; lift-to-drag ratio (L/D), 403; limit-cycle oscillation research, 368; materials for, 292, 293, 295, 571, 592; pressure suits, 276, 278, 297, 298; reaction controls, 298; record hypersonic flight, 278, 293, 299, 642; redundant systems, 365–66; roll damper, 412; Round Two research aircraft, 54, 55, 290; side stick flight controller, 642, 658n59; simulator, 295–97, 395–97, 403–4, 405, 409, 411–12; sonic boom research, 197, 226; speed of, 415, 642; Stability Augmentation Systems, 365–66, 399; stability issues, 292, 411–12; structural resonances, 375–76;

tail configuration, 278, 411; thermal barrier coatings, 611; thermal damage, 294–95, 416, 418; thermal data from, 517–18; X-15A-2 aircraft, 64, 335, 406, 417–18

North American XB-70 Valkyrie: aeroelastic effects research, 392; SCAT program and, 108; sonic boom research, 201–2, 201n49, 203, 204, 205, 210; XB-70-1 testbed, 201, 202, 204; XB-70-2 testbed, 201–2, 204; XB-70A, 56n107, 191

North American XP-86 Sabre, 18, 635

North American YF-86D low-tail testbed, 36–37

North American YF-100A Super Sabre, 34, 363

North American YF-100 Super Sabre, 37, 40, 41, 42, 185

North Atlantic Treaty Organization (NATO), 213, 365

Northrop, John "Jack," 70

Northrop A-17A aircraft, 734

Northrop B-2 Spirit, 722, 753

Northrop F-5 fighter: F-5E Shaped Sonic Boom Demonstration, 245–48, 250, 252; sonic boom research, 245; tail configuration, 32, 413

Northrop F-89 Scorpion, 102

Northrop Grumman/Northrop Corporation: flying wings concept, 70; NASTRAN Users' Colloquia, 480; Shaped Sonic Boom Demonstration (SSBD), 245–47

Northrop HL-10 aircraft, 56, 360, 368–69, 378–79

Northrop M2-F2 lifting body, 407

Northrop N-156 aircraft, 32

Northrop Q-ball, 296, 298

Northrop Snark cruise missile, 52n100

Northrop T-38 aircraft, 32, 238, 239, 241

Northrop X-4 aircraft, 27–28

Northrop YB-49 flying wing jet bomber, 19, 637

Northrop YF-17 Lightweight Fighter (LWF), 388, 389, 664, 669

Nose cones/caps: ablation cooling, 286–88, 581–82; ASSET vehicles, 594–95; General Electric Mark 3 nose cone, 288; General Electric RVX-1 nose cone, 288, 582; General Electric RVX-2 nose cone, 288, 294; heat sinks, 579, 581; hot structures, 286, 579, 580–81; ICBMs, 278, 285–88, 305, 417, 440–42, 572, 579–82; Space Shuttle orbiter, 318, 577; transpiration cooling, 579–80; V-2 ballistic missiles, 571; X-20 Dyna-Soar, 306, 308, 587–89, 598

Nose sensor, 298

No-slip condition, 431

Notch filter, 376, 377, 378

NPLOT program, 547

NT-33 aircraft, 642

NT-33A Shooting Star, 402

Nuclear Metals, 617

Numerical accuracy, 446–47

Numerical Evaluation of Stochastic Structures under Stress (NESSUS), 498–99, 514

Numerical Propulsion System Simulation (NPSS) program, 759–62

Numerical solutions, 442–44, 446–47

Nurflügeln (flying wings), 70

O

Oahu, Hawaii, 827, 842

Obama, Barack, 818

Oberammergau, 14, 48n94, 59

Oberschleißheim, 707

Oblique wing concept, 63–64, 515, 522–26, 523n169, 530

Oblique Wing Research Aircraft (OWRA) program, 526

Oestricher, Philip, 370, 658, 663, 665, 667
Ohio Aerospace Institute, 512
Ohio State University, 437, 768, 845
Oil embargo, 116, 743, 787–88, 817
Oil prices, 743, 788, 790–91, 816, 840–41
Oklahoma City sonic boom study, 199–200
Old Dominion University, 71, 542
Ontario Hydro, 538
Open Channel Foundation, 486, 490, 490n62
Open Channel Software (OCS), 489–90, 490n62
Open-rotor engines, 750–51, 786, 788, 791, 800–803, 817
Optical data links, 659
Optimum Trajectory Research Experiment (OPTRE), 662
Orbital Sciences Pegasus rocket/booster, 343–45
Organization of the Petroleum Exporting Countries (OPEC), 116, 743, 787–88
Orient Express, 230, 339. See also National Aero-Space Plane (NASP)
Orion spacecraft, 506
Orlando, 536
Orton, George, 344
Oscillations: aerodynamic damping, 363–67; limit-cycle oscillation, 367–69, 377, 378; pilot-induced oscillations (PIO), 369–72, 378, 381, 409, 412, 661; pogo oscillations, 328, 513; resonant motions, 375
Outstanding Mechanical Engineer Award, 828n6
OVERFLOW code, 452, 453
Ozone layer research, 789–90, 795, 798

P

P-39 Aircobra, 23n37
P-47 aircraft, 333
P-51D Mustang, 37
P-51 Mustang, 92
P-63 Kingcobra fighters (L-39 aircraft), 22–23, 23n37, 23n39
P-80 Shooting Star, 634
P 1101 aircraft, 58–59
Pacemaker reentry vehicle, 600
Pacific Gas and Electric, 839
Pacific Missile Range Facility, 417, 867, 869
Painter, Weneth, 369
Palmdale, 196, 707
Pamidi, P.R., 485
Parabolic partial differential equations, 435n11, 436
Parameter Identification (PID), 504–5, 504n113
PAREDYM program, 510
Paris Air Show, 161, 707, 750
Paris engine noise and sonic boom conference, 213
Parker Aerospace, 677
Parker Pen Company, 616
Parks, William, 856
Parrish, Philip, 617
Partial differential equations, 430–31, 432–33, 434–36, 435n11, 446–47
Passive cooling, 606
Passive laminar flow control, 806–8
Pasteur, Thomas, Jr., 36
Pathfinder, 850, 855–59, 861, 862, 863
Pathfinder Plus, 850, 859–60, 862
PATRAN finite element preprocessor, 514
Patuxent River Naval Air Station, 706, 707
PB2-1S Superfortress (B-29), 29
PCA-2 autogiro, 137, 140
PCBoom, 230, 237, 257
PCBoom4 modeling, 255
Peabody, Paul R., 475
Pegasus rocket/booster, 343–45

Pennsylvania State University, 169, 251, 850–51
Performance Seeking Control (PSC) program, 712–13, 757
Performance Test Engine (PTE), 345
Perl, William, 17n30
Perseus, 850, 853
Petaflop computers, 341, 439
Peterson, Bruce, 369
Pfeilförmiges Tragwerk (Arrow-Shaped Lifting Surface), 6
Phenolic fiberglass, 607
Phenolic glass, 581
Phillips, William Hewitt, 38–39, 41, 42, 44n86, 408
Phoenix Integration, Inc., 493
Piasecki Helicopter, 136
Pilot Activated Recovery System, 676
Pilot-induced oscillations (PIO), 369–72, 378, 381, 409, 412, 661
Pilot-induced oscillations (PIO) suppression filter, 661–62
Pilotless Aircraft Research Division, Wallops Island, Virginia, 39–40
Pima Air and Space Museum, 672
Pinckney, Shimer, 335
Pinkel, Irving, 331
Piper Colt, 199
Piston engines, 634, 736
Pitcairn PCA-2 autogiro, 137, 140
Pitcairn YG-2 autogiro, 141
Pitch rocker technique, 666n70
Pitch-up: Douglas D-558 Skyrocket program, 28–32; Douglas XF4D-1 Skyray, 32–33; Grumman F9F-6 Cougar, 33, 33n60; Lockheed F-104 Starfighter, 34n61, 413; McDonnell F-101 Voodoo, 33–34; stick-kicker, 34, 34n61; swept and delta wing aircraft, 19, 20–28, 27n51, 32–33, 33n59, 50, 53; tail configuration and, 32–37, 33n60, 34nn61–62, 413
Pitot nose inlet, 29, 50
Pitot tube, 701, 704, 705
Plasma Jet Subsonic Splash Facility, 588
Plasma Rotating Electrode Process, 617
Plexiglas (methyl methacrylate), 310–11, 598
Plotkin, Kenneth J., 226–27, 246
Plum Brook Station, 831, 846
Pogo oscillations, 328, 513
Polaris missiles, 303
Polentz, Perry P., 474, 518
Polhamus, Edward C., 60–61, 61n115, 651n49
Polymer composite engine fan casings, 767–69
Popson, Raymond, 59–60
Porter, Lisa J., 253–54
Portland, Oregon, 537
Poststall maneuvering, 702–3, 703n144
Powder metallurgy, 616–17
Powell, C.W., 647
Power, Thomas, 300
Power-by-wire AFTI F-16 testbed, 676–79
Powered Flight Control Units (PFCUs), 642–43, 677
Prandtl, Ludwig, 15
Pratt & Whitney compressor, 794
Pratt & Whitney F100 engine model derivative (EMD) engine, 711
Pratt & Whitney F100 turbojet engine, 710, 712–13, 717, 754–56
Pratt & Whitney F117 engine, 795
Pratt & Whitney F119 engine, 712, 713
Pratt & Whitney F135 engine, 713
Pratt & Whitney Geared Turbofan engine, 750, 751–52
Pratt & Whitney J57 turbojet, 100, 101, 102

Pratt & Whitney J58 engines, 708, 709–10
Pratt & Whitney J75 turbojet, 103
Pratt & Whitney JT8D engine, 744
Pratt & Whitney JT9D-7A engine, 745
Pratt & Whitney JT9D engine, 611, 615, 744, 745
Pratt & Whitney PW1000G engine, 752
Pratt & Whitney PW2000 engine, 795
Pratt & Whitney PW2037 turbofan engines, 756, 795
Pratt & Whitney Rocketdyne/Pratt & Whitney: Advanced Control Technology for Integrated Vehicles Experiment (ACTIVE) project, 716–17; Advanced Subsonic Technology (AST) program, 796; combustor concepts, 795, 798, 799; Digital Electronic Engine Control (DEEC) program, 709–10; digital engine control system, 754–56; E Cubed technology use, 747, 792–95; geared turbofan engine, 800–801, 802; HyTech research program, 345; merger of Pratt & Whitney and Rocketdyne, 345; Numerical Propulsion System Simulation (NPSS) program, 760; Propulsion Controlled Aircraft (PCA) program, 715; scramjet engines, 345; thermal barrier coatings, 611–13; turboprop engine development, 800–802; Ultra Efficient Engine Technology (UEET) program, 762, 797
Pratt & Whitney Technology for Advanced Low Nitrogen Oxide (TALON) combustor, 799, 817
Pratt & Whitney TF34 turbofan engines, 154, 158, 162, 163
Precision Aircraft Control Technology (PACT), 647–48
Precision Recovery Including Maneuvering Entry (PRIME). *See* Martin SV-5D (X-23)
PRIME (Precision Recovery Including Maneuvering Entry)
Prescribed Velocity Distribution (PVD) blades, 612
Presidential Advisory Committee (PAC), 192, 210–11
Presidential Executive Order, 171–72
Pressure suits, 276, 278, 297
Prien, Austria, 47
Princeton University, 211, 217
Principia (Newton), 429
Probabilistic Structural Analysis Methods (PSAM), 498
Profile drag, 117
Program for Analysis of Nonlinear Equilibrium and Stability (PANES), 552
Programming Structural Synthesis (PROSS), 511–12
Project 680J Survivable Flight Control System (SFCS), 379–80, 384, 646–48, 658n59
Project Independence, 823–24, 831
Project Independence Interagency Solar Task Force, 823–24
Projekt 100LDU testbed, 649–50, 692
Propeller, folding, 91
Propulsion-Airframe Integration project, 763–64
Propulsion Controlled Aircraft (PCA) program, 714–15, 757, 758, 766
Propulsion-Only Flight Control System, 711
Propulsion systems. *See also* engines: advancements in, 3, 735, 769–70; computational fluid dynamics (CFD) and, 803; computers and technology improvements, 757–59; propulsion efficiency, 794
Propulsion Systems Integration and Assessment project, 763
Proteus, 850, 853
Puccinelli, Edward F., 476
Puerto Rico, 827, 833

Pulse-type maneuvers, 397
Purifoy, Dana, 679
Puthoff, Richard L., 828, 830
Putnam, Palmer Cosslett, 826, 827, 846
PW1000G engine, 752
PW2000 engine, 795
PW2037 turbofan engines, 756, 795
Pyle, Richard S., 475
Pyrolysis, 577

Q

Q-ball, 296, 298
Q-felt insulation, 305, 586–87, 589, 590
Quarles, Donald A., 106
Quartz, 600
Quesada, Elwood R. "Pete," 66, 192
Quiet Clean Short Haul Experimental Engine, 741–43, 758
Quiet Engine program, 740, 741
Quiet Spike device, 251–53, 256, 258, 261
Quiet Supersonic Platform (QSP), 244–48, 249

R

R&D Magazine, 499
R-4 helicopter, 135, 142
Radar systems, 159–60
Radiative cooling, 588
Ragusa, Vivian, 706
Rahn, Robert O. "Bob," 33
Ramjet engines: capabilities of, 278, 279, 289, 571; scramjet engines, 279, 330–36, 340, 342–46, 343n126
Rand Corporation, 284
Raney, J. Philip, 536
Rapid-solidification technique, 279, 617, 620–21
RAPTOR/Pathfinder UAV, 850
Rate gyros, 298, 364–65, 366, 367

Rausch, Vincent, 343
Rauscher, Manfred, 38
RAVFAC program, 502
Rayleigh, Lord, 440, 466
Rayon, 577
Raytheon, 246, 249
Reaction controls, 298
Reaction-cured glass, 315
Reaction Motors XLR11 engines, 294
Reaction thrusters, 278
Reactor Centrum Nederland & Hazameyer, 538
Reagan, Ronald, 229, 230, 279, 336, 339
Reclamation, Bureau of, 839
Recoverable Orbital Launch System (ROLS), 334
Redner, Keith, 475
Redstone missile/booster, 288, 636
Redundant structures, 463–64
Redundant systems, 671–72; Boeing YC-14 transport, 671–72; Digital Fly-By-Wire F-8 program and aircraft, 380, 655–56, 658, 659; flight control systems, 365–67; fly-by-wire systems, 379–80, 384, 655–56, 658, 659, 671–72; Grumman X-29 aircraft, 695; North American X-15 aircraft, 365–66; unmanned aerial vehicles (UAVs), 855–56
Reeder, John P. "Jack," 641
Reentry vehicles: Air Force Flight Dynamics Laboratory research, 404, 406, 415; blunt-body theory, 54, 285–86, 291, 599; Dryden research activities, 495–96
Reeves Electronic Analogue Computer (REAC), 42
Regenerative fuel cell system (RFCS), 863–64, 871, 872
Reid, H.J.E., 142
Reinhardt, Walter, 325
Relaxation method, 435

Relaxed stability aircraft, 647–48, 651
Remington Rand ERA 1103 digital computer, 437
Remotely piloted vehicle (RPV), 236
Remotely Piloted Vehicle Facility, 662
René 41 alloy: ASSET vehicles, 594; composition of, 583–84; Dyna-Soar vehicles, 306, 307, 308, 582–84, 586, 590, 591, 592, 597; jet engine use, 584; Space Shuttle orbiter, 602; strength of, 583; strength versus temperature, 597
Renegar, William W., 474
Republic Aviation, 333, 334, 569
Republic F-84 aircraft, 333
Republic F-103 interceptor fighter, 569; XF-103 interceptor fighter, 289, 290, 333
Republic F-105 Thunderchief: design of, 333, 409; development and manufacture of, 290; F-105B Thunderchief, 32, 44; inertial coupling (roll coupling), 409; sonic booms, 185, 216; tail configuration, 37, 413
Republic P-47 aircraft, 333
Republic XF-103 interceptor fighter, 289, 290, 333
Research Airplane Projects Panel (RAPP), 14n23
Research Applied to National Needs (RANN) project, 824
Research Applied to National Needs (RANN) Symposium, 827–28
Resident Backup System (REBUS), 662
Resonances, structural, 374–79, 513
Reusable surface installation (RSI), 312–23, 596–605
Reuther, James, 494
Reynolds number flows, 438
Rhoades, Carrie, 260
Rhode Island, 827

Rich, Quick Mix, Lean (RQL) combustor concept, 798, 799
Richardson, Holden C., 5
Richardson, L.F., 434–35
Richwine, David, 248
Ritchie, Virgil S., 95
Ritz, Walter, 466
RM-1 engine, 706
Robbins, Karen Risa, 853
Robbins, William, 830
Robert C. Byrd National Technology Transfer Center (NTTC), 490
Robert J. Collier Trophy. *See* Collier Trophy
Robins, A. Warner, 67
RoBo rocket bomber, 300, 301, 303
Roché, Jean, 13
Rocket Assisted Take Off solid rocket motors, 736
Rocketdyne, 345. *See also* Pratt & Whitney Rocketdyne/Pratt & Whitney
Rockwell Aerospace/Rockwell International: International Test Organization (ITO), 700–701; NASTRAN, development of, 484; NASTRAN, Users' Colloquia, 480; X-31 aircraft, 700–701, 706
Rockwell B-1 bomber, 62, 62n118, 378; B-1A bomber, 392; B-1B bomber, 392
Rockwell Highly Maneuverable Aircraft Technology (HiMAT), 22, 325, 686–89
Rockwell Sabreliner 65 aircraft, 115
Rockwell Space Shuttle contract: awarding of, 313; thermal-protection tiles, 318, 319, 321, 419; wind tunnel studies, 326
Rockwell X-31 Enhanced Fighter Maneuverability (EFM) aircraft, 699–707
Rocky Mountains, 200
Rodden, W.P., 483
Rodgers Dry Lake, 182, 394, 403, 688, 858
Rokide coatings, 611

Roll coupling (inertial coupling), 37–45, 40n74, 44nn86–87, 393, 394–95, 407–9

Roll damper, 364–67, 412

Rolling tail, 278

Rolls-Royce: Allison/Rolls-Royce, 762, 797; combustor concepts, 799; Numerical Propulsion System Simulation (NPSS) program, 760; turbine blades, 609

Rolls-Royce RB 108 engines, 64

Roll stability, 37

Roll stick force gradient, 667

Ronald, Terence, 621

Root, L. Eugene, 27n51, 48n94

Rose, Peter, 572

Rosecrans, Richard J., 474

Rosenberg espionage investigation, 17n30

Rotary wing aircraft and helicopters: aircraft handling qualities criteria, 145–46; airfoil design for, 144; Ames as lead center for research, 156; Army research on, 147; birth and growth of industry, 135–36; changes in program priorities and, 165–66; Design Analysis Methods for Vibration (DAMVIBS), 508–11; development of, 135; flying-quality criteria, 145–46, 148–49; ground resonance, 144–45; instrument flight rules (IFR), 151; Korean war use of, 138–39; military interest in, 141, 142, 168–69; NACA/NASA contributions, 136–37; National Aeronautics Research and Development Policy, 171–72; National Rotorcraft Technology Center, 168–69; piloting techniques, 145; tilt rotor aircraft, civil transportation use of, 166–68; Vietnam war aircraft losses, 644n32

Rotary wing aircraft and helicopter research activities: Advanced Tiltrotor Transport Technology (ATTT) program, 167–68; Advancing Blade Concept (ABC) rotor, 152–53, 157–58; aeroelastic stability, 149–50; Airloads Program, 170–71; at Ames, 137, 147, 148, 149–50, 152–53, 155–61, 163–64, 492; by Army, 147, 150, 154–55, 166, 168–69, 170–71; Bearingless Main Rotor (BMR), 157; blade motion studies, 141–42; blade pressure research, 144, 147–48; Circulation Control Rotor (CCR), 156–57; contracted research, 153–54; Controllable Twist Rotor (CTR) concept, 153, 156; Convertible Engine Systems Technology (CEST) program, 158, 162; crash-test activities, 152; at Dryden, 164, 165, 170; Executive Order to support, 171–72; finite element analysis, 152, 162–63, 515, 519–20, 529–30; Flying Laboratory for Integrated Test and Evaluation (FLITE), 170; Heavy Lift Helicopter program, 158; hingeless rotor concept, 150, 152–53; icing environment, 162; at Langley, 137, 139–45, 147–52, 154–55, 156, 161, 162–63; at Lewis/Glenn, 156, 158, 162; by manufacturers, 146; nap-of-the-Earth (NOE) flight, 159, 170; noise research, 163, 164–65, 167–68; radar systems, 159–60; resources for research, 155–56, 165–66, 169, 171; results and findings, publication or distribution of, 138, 143, 144, 154; rotary dynamics, 144–45; rotor efficiency, 159; rotor-flow studies, 146; Rotor Systems Research Aircraft (RSRA) program, 154–55, 163–64, 165; safety research, 167–68, 172; Tilt Rotor Research Aircraft (TRRA) program, 155, 156, 160–62, 165; tilt rotor simulations, 151; Vehicle Technology Directorate (VTD), 162–63; X-Wing Stoppable Rotor, 157, 158, 163–64; X-Wing Stoppable Rotor program, 165

Rotating-Wing Aircraft Meeting, Franklin

Institute, 142
Rotating Wing Subcommittee (Helicopter Subcommittee), 142
ROTONET, 164–65
Rotor Aerodynamic Limits survey, 515, 519–20, 529–30
Rotorcraft Aircrew Systems Concepts Airborne Laboratory (RASCAL), 170
Rotorcraft Centers of Excellence (RCOE) program, 168–69
Rotorcraft Division, Army-NASA, 166, 170
Rotorcraft Industry Technology Association (RITA), 169–70
Rotor Systems Research Aircraft (RSRA) program, 154–55, 163–64, 165
Rotor Test Apparatus (RTA), 153
Round One research aircraft, 53–54, 55n106, 301. *See also* Bell X-1 aircraft; Bell X-2 aircraft; Douglas D-558-2 Skyrocket
Round Three research aircraft, 55–56, 55n106, 301. *See also* Boeing X-20 Dyna-Soar
Round Two research aircraft, 54, 55n106, 290, 301. *See also* North American X-15 aircraft
Royal Aeronautical Establishment, 94
Royal Aeronautical Society (RAeS), 333
Royal Aircraft Establishment (RAE): Hawker Hunter T.12 aircraft, 649; Jaguar Active Control Technology (ACT), 670, 690–91, 692; Short S.C.1 aircraft report, 641; swept and delta wing aircraft, 13n20
Royal Air Force (RAF): Skyshield air defense exercises, 188; sonic booms, 186; VC10K tanker, 643
Rudder, 411, 412, 632, 633, 637, 703–4
Ruhrstahl Fritz X missiles, 638n15
Runge-Kutta time integration, 445
Runyan, Harry L., 217, 473

Rural Electrification Act, 826
Russell, Philip, 851
Russia, 22, 827–28
Rutan, Burt, 120–21, 619
Rutan, Dick, 121n106
Rutkowski, Mike, 524
RVX-1 nose cone, 288, 582
RVX-2 nose cone, 288, 294
Ryder, Jim, 647

S

S-61 Sea King, 154, 163–64
S-72 Rotor Systems Research Aircraft (RSRA), 155
S-76 helicopter, 156
Saab, 706
Saab 201 aircraft, 23n39
Saab 202 aircraft, 23n39
Saab AJ-37 Viggen, 391, 651n49
Saab J29 fighter, 23n39
Saab J32 Lancen (Lance), 23n39
Saab J35 Draken (Dragon), 45
Saab JAS-39 Gripen fighter, 706
Saab Safir aircraft, 23n39
Saab seat restraint, 9n13
Sabre 45 aircraft, 36
Sabreliner 65 aircraft, 115
Safe Slide, Project, 201
Safety factor, 387
Safety of aircraft, 506, 632
Safir aircraft, 23n39
Saint Augustine, 246–47
Saint Germain, France, 14
Saint Louis, 197–99
SAMIS program, 506
SAMSAN program, 549
San Antonio, 537
Sandusky, 831
San Francisco, 536

Sänger, Eugen, 54, 300
Sänger-Bredt, Irene, 54, 300
Sänger-Bredt Silbervögel (Silver Bird), 54, 300
Sargent & Lundy, 538
Satellites, 305, 579, 581–82
Saturn launch vehicle, 814
Saturn V launch vehicle, 328, 513
Savannah, 252, 253, 801
Savino, Joseph M., 827–28, 828n6
Scaled Composites, 619, 853
Scherr, S.J., 448
Schkolnik, Gerard, 253, 714, 719
Schlichting, Hermann, 15
Schlieren photography, 97, 239, 241
Schneider, Edward, 662, 663
Schramm, Wilson, 310, 598
Schwartz, Ira R., 213, 216, 217
Science and Technology, Office of, 230–31
Science Museum, 641
Scott, H.G., 611
Scott, William C., 474
Scott Air Force Base, 197–98
Scramjet engines, 279, 330–36, 340, 342–46, 343n126, 620
Seat restraint, three-point, 9n13
Seawolf submarines, 723
Seebass, A. Richard, 211, 212, 217, 218, 222, 223, 226–27, 244, 244n181
Seeck, Dietrich, 700, 701
Self-adaptive flight control system (AFCS), 372–74, 375, 376, 379
Self-Repairing Flight Control System (SRFCS), 713–14
Semispan model, 118, 119
Serafini, John, 331
Sewell, Charles A. "Chuck," 695, 695n125
SH-3 Sea King, 151
Shaker vanes, 519

Shang, Joseph S., 448
Shaped Sonic Boom Demonstration (SSBD), 244–48, 249, 250, 252, 258
Shaped Sonic Boom Experiment (SSBE), 248
Sharkey, John, 855
Shell Oil Company, 503
Shepard, Alan, 636, 653
Shepherd, Kevin P., 240, 257
Ship control systems, 723
Shock waves. *See also* sonic booms: blunt-body theory, 286, 415, 440–42; three-dimensional flow-field calculations, 448; transonic flow, 94, 97, 98–99, 111, 445; unstarts, 336, 708–9, 754
Short Brothers, 641
Short S.C.1 aircraft, 639–41
Short Take-Off and Landing (STOL) aircraft, 670–72; Air Force transport competition, 114; engine innovations for, 743; F-15B/NF-15B, 258, 715–16; research on, 146
Short Takeoff and Landing Maneuver Technology Demonstrator (S/MTD), 258, 715–16
Shrout, Barrett L., 211
Shuttle Training Aircraft (STA), 404–5
Sideslip, 39–40, 42–43, 408, 409–10, 681–82, 700, 701
Side stick flight controller, 298, 638, 638n15, 642, 658, 658n59, 664, 665, 668
Sight, helmet-mounted target designation, 673, 675
Sikorsky, Igor I., 9, 135, 136
Sikorsky Advancing Blade Concept (ABC) rotor, 152–53, 157–58
Sikorsky Aircraft Corporation: Design Analysis Methods for Vibration (DAMVIBS), 508, 509n128, 510; helicopter design and

production, 136; PAREDYM program, 510; RITA membership, 169
Sikorsky R-4 helicopter, 135, 142
Sikorsky S-61 Sea King, 154, 163–64
Sikorsky S-72 Rotor Systems Research Aircraft (RSRA), 155
Sikorsky S-76 Spirit, 156
Sikorsky SH-3 Sea King, 151
Sikorsky UH-60 Blackhawk: Airloads Program, 170–71; RASCAL configuration, 170; UH-60A Blackhawk, 510, 519–20, 529–30
Sikorsky VS-300 helicopter, 9, 135, 136, 142
Sikorsky X2 helicopter, 158
Sikorsky XH-59A Advancing Blade Concept (ABC) helicopter, 152–53, 157–58
Sikorsky YR-4B helicopter, 142, 143–44
Silane, 344–45
Silbervögel (Silver Bird), 54, 300
Silica ablative coating, 417–18
Silica fiber insulation, 305, 598–99, 600–601
Silica thermal-protection tiles, 310–18, 599–605
Silicides, 603
Silicon carbide, 315, 317, 587, 599
Silicon dioxide, 599–600
Siliconized graphite, 306, 308, 587, 588
Silicon tetraboride (SiB4), 315–16
Silver Bird (Silbervögel), 54, 300
Silverstein, Abe, 810, 811, 814
Simonov, Mikhail Petrovich, 650n46
Simplifications, 463
Simulators and flight simulation: accuracy of, 396–98, 406; analog computers and, 392–94, 405–7, 468; centrifuge, 398–400; digital computer simulation, 405–7; evolution of, 392–96; finite element analysis, 515, 516; hybrid computer simulator, 406–7; importance of, 392; Iron Bird simulator, 655, 656–57, 659, 660; Iron Cross simulator, 298; landing-pattern trainers, 402–5; Link Trainer, 295, 392; moving-base cockpits, 398; reaction controls, 298; Space Shuttle orbiter, 371; variable stability aircraft, 401–2; visual displays, 407; X-15 aircraft, 278–79, 295–97, 395–97, 403–4, 405, 409, 411–12; X-20 Dyna-Soar, 405–6
SINDA codes, 546
Sine-squared law, 429
Single-crystal blades, 614–15
Single-stage-to-orbit (SSTO) vehicles, 279, 337
Single-string damper systems, 365, 412
Sioux City airliner crash, 757, 766
Skantze, Lawrence "Larry," 337, 338
Skoog, Mark, 679
Skyshield air defense exercises, 188
SkyTower, 853
Slender-beam theory, 466–67
Slender wing theory, 12, 15
Slots, fixed and retracting, 414
SM-64 Navaho cruise missile, 289, 290, 569–70
Small Business Innovation Research (SBIR) grant, 768
Smidinger, Peter A., 474
Smith, Apollo M.O., 29n56, 48n94, 70
Smith, Beauchamp, 826, 827
Smith, Burwell, 826
Smith, Rogers, 662, 679, 695, 701
Smith, Ron, 524
Smith-Putnam wind turbine, 826, 827, 828, 835, 846
Smithsonian Institution. *See* National Air and Space Museum
Smolka, James W. "Jim," 253, 255, 259, 260, 679, 714, 719
S. Morgan Smith Company, 826

Snark cruise missile, 52n100

Snecma, 803

Sobieszczanski-Sobieski, Jaroslaw, 511

Society of Automotive Engineers (SAE)-Aerospace, 138

Society of Experimental Test Pilots, 649

Software programs. *See also* NASA Structural Analysis System (NASTRAN): Backup Flight System software, 661; *Computer Programs Abstracts Journal*, 489; Computer Software Management and Information Center (COSMIC), 477, 488–90, 490n62, 499, 508; digital fly-by-wire, 381–82, 659–60, 671–72, 683–84; documentation, 489; finite element structural analyses, 158, 490n62, 502, 544–52; NASA-developed software, 490; Open Channel Software (OCS), 489–90, 490n62; prices, 489, 490n62; Robert C. Byrd National Technology Transfer Center (NTTC), 490

Solano County, California, 839

Solar cells, 862–63, 868, 872

Solar Energy Research Institute, 836

Sol-gel deposition coatings, 613

Solid-propellant rocket boosters, 326, 327, 329, 606

Solid rocket fuels, 503–4

Solovyev, Yevgeny, 650, 650n46

Sonic bang, 184

Sonic booms: complaints and damage claims, 186–89, 190, 197–98, 199–200, 204–6, 210–11, 216–17; conferences, symposiums, and workshops, 209–10, 211–14, 217, 231, 233–34, 241–42; control and reduction of, 181, 211–12, 218, 221–26, 231–32; database on, 229, 238; double-boom sounds, 184–85, 186; duration of, 188–89; impact of, 185–88; low-boom/no-boom technique, 255–57; magnification of, 187; National Sonic Boom Evaluation Program, 203–5; natural, 182–83; prediction of, 212–13; prohibition of by Concorde, 227–28; Quiet Spike device, 251–53, 256, 258, 261; secondary booms, 227–28; signature of, 196–97, 198, 222–23; signature of, evanescent waves, 256; signature of, N-shaped, 185, 186, 197, 206, 223, 232, 255–56; SST program research, 192–207; super booms, 187; survey of civilians exposed to, 206, 207, 240–41, 242, 249

Sonic Boom Mitigation Project, 251, 261

Sonic boom research activities: by Air Force, 194–203, 207; at Ames, 214, 223–24; arrow-wing configurations, 209, 225; by Boeing, 224; caustics, 226–27, 228; Community and Structural Response Program, 197–99; computational fluid dynamics (CFD), 234–37, 246, 248, 253, 260; computer program development and use, 224–25, 226, 230, 237, 255, 257; at Dryden, 227, 227n133, 236–39, 242–43, 245, 251, 252–53, 255–57, 258–60; at Edwards, 193, 195–97, 200, 201–4, 206–7, 229–30, 251; by FAA, 199–201, 206, 207; flight-testing, 227; High-Speed Research (HSR) program, 232–43; at Jackass Flats test site, 206–7; at Langley, 192–200, 203, 208–9, 214–15, 217, 227, 239–40, 249, 256–58; Lift and Nozzle Change Effects on Tail Shocks (LaNCETS), 258–60, 261; low-boom demonstrator, 249, 250–51, 255, 261; low-boom/no-boom technique, 255–57; by Navy, 206; results and findings, publication or distribution of, 207–8, 213–14, 218–19, 221, 229, 233, 243, 243n179; Shaped Sonic Boom Demonstration (SSBD), 244–48, 249, 250, 252, 258; Shaped Sonic Boom Experiment

(SSBE), 248; simulators, 214–16, 240, 249, 257–58; SST program research, 192–207, 210; Supersonics Project, 254–60; techniques for, 218; at Wallops Island Station, 194, 198, 206; wind tunnel studies, 208–9, 208n66, 214, 222, 235
Sonic Boom Symposium, 211
Sorensen, Hans, 222
Soulé, Hartley A., 14, 14n23
Sound, speed of, 182
South Kensington, London, 641
Southwest Research Institute (SwRI), 498, 514
Soviet Union (USSR): aerodynamic research and development, 17, 17n30; arms race with, 442; delta wing development, 45–46; fly-by-wire system research, 649–50; Projekt 100LDU testbed, 649–50, 692; space race against, 632, 644, 786; Sputnik, 288, 301, 361, 786, 818; Supersonic Transport study, 66; swept and delta wing aircraft, 9n12; technical intelligence collection, 17, 17n30; variable-geometry wing research, 62; Vietnam war, 644
Space Act, 487, 852, 852n33
Space Act Agreement, 761
Space Act Award, 846
Spacecraft Design Division, NASA, 56–57
Space race, 632, 644, 786
Space Shuttle Challenger, 329–30
Space Shuttle Columbia: accident with, 578; delay in launch of, 322–23; descent of, 322; launch of, 323; record hypersonic flight, 278; significance of return of, 289; structural testing, 329–30; thermal-protection tiles, 313, 319–23
Space Shuttle Enterprise: approach and landing evaluations, 661; pilot-induced oscillations (PIO), 370–71, 381, 661; structural testing, 329; wind tunnel studies, 327
Space Shuttle Main Engine (SSME), 337
Space Shuttle orbiter: aluminum construction, 310, 602, 606–9; AP-101 computers, 659; approach and landing evaluations, 661; Backup Flight System software, 661; computational fluid dynamics study, 426, 448, 450; computational simulations, 325–26; concept development, 56–57; contract for, 313, 604; control stick feedback, 370–71; design development tests, 317–18; external tanks, 326, 327, 329, 578; finite element analysis, 327–28, 506; flight control system, 278; flight control systems, 367; fly-by-wire system, 753; fuel cells, 814; heating, structural, 415, 572; hot structures thermal protection, 310–13, 601–3; hypersonic flow pattern, 426; insulating foam, 578; landing simulator, 404–5; nose cap, 318, 577; pilot-induced oscillations (PIO), 370–71, 381; pilot-induced oscillations (PIO) suppression filter, 661–62; pogo (oscillation), 328; record hypersonic flight, 278, 299; Shuttle Orbiter 036 series, 57; Shuttle Orbiter 040C design, 57; side stick flight controller, 658n59; solid-propellant rocket boosters, 326, 327, 329, 606; speed of, 415; strain isolator pad (SIP), 316, 319, 320, 321; structural verification study, 554–55; thermal modeling, 495; thermal protection, 576, 577–78; thermal-protection tiles, 278, 288, 310–23, 419, 582, 598; thermal-protection tiles, bonding of, 319, 320–23, 609; thermal-protection tiles, coatings on, 315–16; thermal-protection tiles, densification of, 320–21; thermal-protection tiles, gaps between, 318, 319–20; thermal-protection tiles,

installation of, 319–23, 419; thermal-protection tiles, shaping of, 314–15, 321; thermal-protection tiles, testing of, 316–19; thermal-protection tiles, thermal expansion and, 315; three-dimensional shock wave shape, 448; titanium construction, 608–9; wind tunnel studies, 323–30; wing configuration, 301; wing leading edge protection, 577–78

Space Shuttle program: approval of, 313, 603; cancellation of, 765; concurrent development, 322–23; cost overruns, 229, 602; Digital Fly-By-Wire F-8 program to support, 661; environmental impact statement (EIS), 226; Johnson Space Center activities, 506; sonic booms from, 226, 226n130; spacesuits for, 297

Space station, 243, 311–12, 506

Spacesuits, 276, 278, 297

SPAD XIII fighter, 37

Spark discharge system, 214

Special Subcommittee on Aircraft Noise, 737

Specific fuel consumption (SFC), 745–46, 792–93

Spectralab, 862

Speed of sound, 182

Spera, David A., 830

Sperry Marine, 538

Sperry Support Services, 484, 656, 656n57, 660

Spinoff, 512

Spin Rig (Glenn [Lewis] Research Center], 555

Spirit of St. Louis, 591

Splat-cooling, 617

Split-S maneuvers, 702, 702n139

Sputnik, 288, 301, 361, 786, 818

SR-71 Blackbird: accident with, 392; aerodynamics of, 70; aeroelastic effects, 391–92; digital engine control system, 754; engine inlet unstarts, 708–9, 754; heating, structural, 413; JP-7 fuel, 346; simulator, 407; sonic boom research, 202–3, 204, 205–6, 236–38, 239, 242–43, 245, 261; spacesuit for, 297; speed of, 413; thermal barrier coatings, 611; thermal loads project, 516; titanium construction, 312, 570–71, 602

STA-099, 329–30. *See also* Space Shuttle Challenger

Stabilator tail, 34–35, 670

Stability augmentation system research, 636–37

Stability augmentation systems (SASs), 364–66, 399, 659, 660

Stability Management Control (SMC), 717–18

Stabilizer, 26, 35, 362–63

Stack, John P.: aeronautics contributions, 107; Anglo-American partnership and Swallow aircraft, 60, 61n115; area rule and, 99, 101; Bell X-1 aircraft, 108; Collier Trophy, 93, 96; SCAT program, 193; slotted transonic tunnel, 93, 95–96; supersonic aircraft research, 94, 108; Supersonic Transport Research Committee (STRC), 65–66

Stage 1 aircraft, 739

Stage 2 aircraft, 739, 741

Stage 3 aircraft, 739

Stage 4 aircraft, 739

Stainless steel, 210, 590

Stall conditions, 67, 67n127, 412–14, 666n70, 711

Stall margin, 711n163

Stanford Research Institute (SRI), 203, 205, 215

Stanford University, 71, 244, 315, 341
Star Clipper, 311, 599, 600
Star Clipper spacecraft, 311, 599, 600
Starship, 121
Star Wars (Strategic Defense Initiative), 279, 336–40
Static Nonlinear Analysis of Shells of Revolution (SNASOR), 506–7
Static stability: directional, 408, 409, 410–11, 648, 651; longitudinal, 648, 664, 669, 690, 692; relaxed stability aircraft, 647–48, 651
Statically unstable aircraft, 633
Stealth aircraft, 70, 384, 721–22
Stecura, Stefan, 611
Steel, 569, 620
Steel, stainless, 210, 590
Steep Approach program, 740
Stefan-Boltzmann law, 585
Stennis Space Center: finite element analysis activities, 491n64; Finite Element Modeling of Piping Systems program, 552; mission of, 491n64
Stephenson, Geoffrey, 41
Stewart, James, 855
Stick kicker, 34, 34n61, 413
Stick shaker, 413
Stiffness matrix, 470–71
Stinger, 261
Stockpile Stewardship program, 498
Stokes, George Gabriel, 431–32
Stores Management System (SMS), 672, 681
Strain equations, 464–65
Strain gages, 370, 385–86, 528, 534, 554, 555
Strain isolator pad (SIP), 316, 319, 320, 321
Strake flaps, 526, 694
Strategic Air Command (SAC) missions, 188, 197
Strategic Defense Initiative (SDI), 279, 336–40
Stress distribution, 471
Stress equations, 464–65
Structural analysis. *See also* finite element structural analyses; NASA Structural Analysis System (NASTRAN): aircraft built without, 464; digital computer development and, 468–72; direct analog computers, 466–68; discretization of structures, 466, 469, 470–72; energy method, 466; flexibility matrix, 471; free body diagrams, 464–65; goal of, 461; hybrid direct-analog/digital computers, 468; loads prediction, 461–62, 462n1, 465–66; matrix method, 466, 470–71; NASA's contributions to, 461; principles of, 462–63; prior to computers, 462–66; redundant structures, 463–64; rotary wing aircraft, 152; simplifications, 463; stiffness matrix, 470–71; stress distribution, 471; topics covered by, 462n1; unit displacement method, 469, 470
Structural analysis groups, 461
Structural Analysis of General Shells (STAGS), 514, 551–52; **STAGS-C1**, 551
Structural Analysis Routines (STARS), 496, 525–26, 528, 556
Structural Dynamics Laboratory (Ames Research Laboratory), 553
Structural Dynamics Research Corporation, 550
Structural Performance and Redesign (SPAR), 495, 498, 512
Structural resonances, 374–79, 513
Structural Tailoring of Engine Blades (STAEBL), 545–46
Structural-Thermal-Optical (STOP) program, 501–2
ST-SIZE program, 512
Subsonic aircraft, 109, 117

Index

Sukhoi: Experimental Design Bureau (OKB), 650n46; supersonic business jet (SSBJ) concept, 249
Sukhoi Su-7U aircraft, 649–50
Sukhoi Su-17/22 Fitter, 62
Sukhoi Su-24 Fencer, 62, 62n118
Sukhoi Su-27 Flanker, 692
Sukhoi Su-37 FSW prototype, 22
Sukhoi T-4 aircraft, 649
Sukhoi T-10 aircraft, 650, 650n46
Sullivan, Brenda M., 257
SunPower Corporation, 860, 862, 872
Suntan project, 811–13
Superalloys, 573–76, 602. *See also* cobalt-based alloys; nickel-based alloys
Superchargers, 736
Supercomputers, 438–39, 446
Supercritical wing: aviation industry use of, 114–15; benefits and importance of, 89–90, 115–16, 126, 805–6, 817; development of, 65, 68, 109–12, 786, 805; F-8 Supercritical wing, 89; models, 110–11; naming of, 111; patent for, 114; publication of findings, 111, 113; recognition of, 116; testing of, 112–14; X-29 aircraft, 693, 694, 695, 695n126
Supercritical Wing Technology, 113
Supermarine Swift, 28
Supersonic Aerospace International (SAI), 249–50
Supersonic aircraft and flight. *See also* area rule: Air Force research activities, 182; ban of supersonic flight by FAA, 220–21, 220n112, 224; challenges of, 262, 361–62; design of, 93, 96, 100–105, 440; engine fan casings, 769; future of, 260–62; materials and structures for, 569–73; ozone layer damage from, 789–90; passenger plane development, 181, 189–90; proposals for, 245–46
Supersonic business jets (SSBJ): companies working on concepts, 249; concept for, 228–29; consortiums for exploring, 249–50; economic conditions and, 260–61; FAA sonic boom restrictions and, 249; future of, 260–62; sonic boom research, 256–57
Supersonic combustion testing, 331–32
Supersonic Commercial Air Transport (SCAT) program: cranked arrow wing development, 69; deep stall, 67, 67n127; design of aircraft, 67, 108–9, 108n61; goals of, 108–9, 189–91; SCAT-4, 67, 108–9, 191; SCAT-15, 67, 191; SCAT-15F, 67, 69, 221; SCAT-15F model, 180; SCAT-16, 191; SCAT-17, 191; sonic boom research, 193
Supersonic Cruise Aircraft Research program. *See* Supersonic Cruise Research (SCR)/Supersonic Cruise Aircraft Research (SCAR)/Advanced Supersonic Technology (AST) program
Supersonic Cruise and Maneuver Prototype (SCAMP), 69, 807. *See also* General Dynamics F-16 Fighting Falcon
Supersonic Cruise Industry Alliance (SCIA), 250
Supersonic Cruise Research (SCR)/Supersonic Cruise Aircraft Research (SCAR)/Advanced Supersonic Technology (AST) program: AST Project Office, 220; cancellation of, 228; conferences on, 225–26, 228–29; design of aircraft, 221; establishment of, 219; funding for, 221; goals of, 69, 219, 220; name changes, 219; sonic boom research, 220, 224, 227
Supersonic Pressure Wind Tunnel, 208, 208n66, 214 Supersonics Project, 254–60, 261

Supersonic Transport (SST): cancellation of, 68–69, 217, 218, 519, 789; Concorde Supersonic Transport (SST), 3, 45, 56, 66; contract for, 211, 217–18; data to support, 519; design of, 217–18; flight locations, 218; flutter suppression research, 388; funding for, 218, 219; opposition to, 216–17; Presidential Advisory Committee (PAC), 192; request for proposals (RFP), 210; research activities, 192; research activities, sonic booms and, 192–207, 210; research activities, Whitcomb's work and, 108–9, 108n61; research activities at Langley, 65–69; responsibility for, 191–92; XB-70A testbeds, 191

Supersonic Transport Development Office, FAA, 199

Supersonic Transport Research Committee (STRC), 65–66, 67, 108, 108n61

Supersonic Vehicles Technology program, 249

Supersonic wind tunnels, 280–81

Survivable Flight Control System program (F-4 SFCS program), 379–80, 384, 646–48, 658n59, 660

Sutton, George, 286, 581

SV-5D (X-23) PRIME (Precision Recovery Including Maneuvering Entry), 27n51, 278, 288, 304, 417, 418, 606–8

SV-5P (X-24A) lifting body, 27n51, 407, 607

Swallow aircraft, 60–61

Sweden: AFTI/F-16 aircraft, 679; Cooperative Test Organization/VECTOR, 706; Swedish Air Force, 679; wind energy, 839, 848; X-31 aircraft, 706

Sweep theory, 12, 15, 15n27

Swept and delta wing aircraft: aeroelastic effects, 390; American research and designs, 7–19, 72–73, 93, 108; Blended Wing-Body (BWB), 2, 70–72; British aircraft, 27, 28; challenges of, 18–28, 27n51, 50; conical camber, 51–53; delta wing development, 12–13, 13n20, 45–49, 51, 53, 67; Dutch roll motions, 19, 50, 410–12; early history of, 4–7, 414; flat-bottom delta aircraft, 55–57, 301, 302; flat-top delta aircraft, 55–56, 300–301; forward-swept wing (FSW), 21–22, 524, 527; German development of, 7, 8, 13, 13n20, 14–15, 15n27, 72–73, 93; hypersonic aircraft, 53–57; iconic status of, 4; inertial coupling (roll coupling), 37–45, 40n74, 44nn86–87, 393, 394–95, 407–9; Langley Memorial Aeronautical Laboratory testing, 13–15; Mach angle of wing, 15–16, 18; Mach number of wing, 5, 12, 14, 18, 52, 53; nose-rise problems, 414; pitch-up problems, 19, 20–28, 27n51, 32–33, 33n59, 50, 53; sections, thickness of, 13; stall, 414; Supersonic Transport (SST), 65–69; test model, 14, 15n26; variable-geometry wing aircraft, 57–64, 61n115, 62n118; vortex lift research, 651n49

Sylvanus Albert Reed Award, 116

Synthetic aviation kerosene, 815–16

Synthetic fuel research, 815–17, 816n114

System Identification/Parameter Identification (PID), 504–5, 504n113; Autonomous Frequency Domain Identification (AUFREDI), 504, 505; MODE-ID program, 504

Systems Research Aircraft (SRA), 384–85, 386, 676

System verification units (SVU), 839

Syvertson, Clarence A., 54–55, 56, 56n107, 300–301

Szalai, Kenneth J., 654, 659, 663

Index

T

T-2 aircraft, 691–92
T-2C Buckeye, 114–15
T-4 aircraft, 649
T-10 aircraft, 650, 650n46
T-34C Mentor, 481
T-38 aircraft, 32, 238, 239, 241
Ta 183 aircraft, 27n51, 36
Tactical Fighter Experiment (TFX), 61
Tactical Fighter Experimental (TFX), 61
Tactical Mobility Requirements Board (Howze Board), 139
Tail configuration. *See also* T-tail aircraft: adjustable horizontal tail, 362–63, 635, 666n70; all-moving tail, 34–35, 35n65, 73, 362–63, 363n4, 665–66; Dutch roll motions and, 411; pitch-up and, 32–37, 33n60, 34nn61–62, 413; rolling tail, 278; stabilator tail, 34–35, 670
Tailless aircraft, 4–5, 13, 27, 27n51, 32–33, 703–4. *See also* swept and delta wing aircraft
TALON (Technology for Advanced Low Nitrogen Oxide) combustor, 799, 817
Talos missile, 289
Tampa, 537
Tantalum, 575, 584
TAPS (Twin Annular Pre-mixing Swirler) combustor, 798–99, 817
TBM Avenger, 695n125
Technical Report 1007, 465–66
Technical Report server, 168
Technical Research Development Institute (TRDI), 692
Technology for Advanced Low Nitrogen Oxide (TALON) combustor, 799, 817
Technology transfer activities: Applications Engineering process, 488; Computer Software Management and Information Center (COSMIC), 477, 488–90, 490n62; at Dryden, 496; Industrial Application Centers, 488; NASA-developed software, 490; Open Channel Foundation, 486, 490, 490n62; Open Channel Software (OCS), 489–90, 490n62; priority of, 487; publications, 487–88; Robert C. Byrd National Technology Transfer Center (NTTC), 490; Technology Utilization (TU) program for, 487–88; Technology Utilization Officers (TUOs), 488
Technology Utilization (TU) program, 487–88
Technology Utilization Officers (TUOs), 488
Teflon, 305, 582
Tel Aviv University, 512
Teledyne-Ryan Aeronautical, 688, 760
Teledyne-Ryan autopilots, 687–88
Teledyne-Ryan BQM-34E Firebee II RPV, 245
Teledyne-Ryan BQM-34 Firebee II RPV, 236, 242, 388–89
Teller, Edward, 439
Temperature effects. *See* aerothermodynamics; *high-temperature entries; thermal entries*
Tennessee Eastman Company, 481, 538, 539
Teraflop computers, 439
Terrain system with moving map, 675–76
Test Pilot School, USAF: Digital Flight Control for Tactical Aircraft (DIGITAC), 382–83, 684–86; Have Coupling project, 685; sonic boom research, 248, 257
Tether, Anthony "Tony," 337
Texas Instruments, 481, 539
TEXLESP-S, 503–4
TF30 turbofan engines, 754
TF34 turbofan engines, 154, 158, 162, 163
TF-104G aircraft, 688

Theodorsen, Theodore, 15
Thermal Analyzer, NASTRAN, 501–2
Thermal barrier coatings (TBC), 610–13
Thermal Loads Laboratory (Dryden Flight Research Center), 553–55
Thermal loads project, 515, 516–19, 529
Thermal protection. *See* aerothermodynamics
Thermal Radiation Analysis System (TRASYS), 506
Thermodynamics, first law of, 432, 433
Theseus, 850
Thigpin, David J., 670
Thinking sessions, 98
Thiokol XLR99 engines, 294
Thomas, Arthur, 337–38
Thomas, Charles L., 223, 226
Thomas, Jim, 452
Thomas, Ronald, 827–28, 830
Thomas, Russell, 772
Thor-Able missiles/boosters, 288, 594
Thor-Delta missiles/boosters, 308, 309, 310, 594
Thoria, 587–88
Thorium, 308
Thorium oxide, 595
Thor missiles/boosters, 308, 309, 310, 592, 594
Three-axis flight control system, 633
Three-dimensional flow-field calculations, 447–49
Thrust vectoring, 700, 701, 703–4, 705–7, 716–17
Thrust-vectoring exhaust nozzles, 715–16, 717
Thrust-vectoring paddles, 699, 700, 706
Tiernay Turbines, 539
Tilt rotor aircraft, civil transportation use of, 166–68

Tilt Rotor Research Aircraft (TRRA) program, 155, 156, 160–62, 165, 166. *See also* Bell V-22 Osprey Tilt Rotor
Time-marching process, 443–44, 445–46
Tinker Air Force Base, 199
Titan III-C missile, 302
Titan III missile/booster, 400
Titan II missile/booster, 302, 328
Titan I missile, 301
Titan intercontinental ballistic missile (ICBM), 286
Titanium: aircraft built with, 570–71, 616, 620; B-120 version, 570; Beta-21S, 617–18, 621; bonding to, 609; characteristics of, 615–16; chemical incompatibilities, 570–71; chemical milling, 569; experience working with, 312, 602; high-speed aircraft and, 569–71; hot structures, 312; hypersonic aircraft and flight, 293; rapid-solidification technique and, 279; sonic booms and, 210; Space Shuttle orbiter, 608–9; SR-15 Blackbird construction, 312, 570–71, 602; superalloy addition, 575, 583; superalloys compared to, 574; X-15 aircraft, 295
Titanium carbide, 307, 590
Titanium Metals, 617
Toll, Thomas, 291
Topp, L.J., 471
Tornado interceptor and strike fighter program, 62, 649
Total In-Flight Simulator (TIFS), 402n62
TRACOR, Inc., 205–6
Transatmospheric flight: aircraft for, 279; development of technology for, 277–79; hypersonic flight and, 278–79; reentry challenges, 285–89, 291–92; stability issues, 292
Trans Atmospheric Vehicles (TAV), 337

Index

Trans Atmospheric Vehicles (TAV) Program Office, 337, 339

TRansfer ANalysis Code to Interface Thermal Structural (TRANCITS), 546

Transonic Aerodynamic Branch, 94, 108, 125

Transonic aerodynamic effects, 5

Transonic Aircraft Technology (TACT) program, 113, 386

Transonic Dynamics Tunnel (TDT), 149–50, 161, 164

Transonic flow and airfoils, 444–46

Transonic Pressure Tunnel (TPT), 109–10, 118, 119, 123, 127–28

Transonic regime: aircraft design, 93–94, 362–63, 444–46; challenges of, 91–92; importance of operation in, 94; Mach limits, 92; research tools, 92–93; shock waves in, 94, 97, 98–99, 111, 445; transonic drag research, 94–100; transonic flow, visualization of, 98; wind tunnel research, 92–97

Transonic transport fuselage, 127

Transonic transport wing configuration, 52, 52n100, 68

Transonic Tunnel (High-Speed Tunnel), 95n16

Transpiration cooling, 579–80

Transportation, Department of: Climactic Impact Assessment Program (CIAP), 789; NASTRAN use by, 481; ozone layer research, 789–90; SST program, 192; supersonic research, 219

Transport Canada, 251

Transport delay, 661

TRASYS Radiative Heat Transfer program, 550

Tremaine, Stanley A., 337

Tridymite, 600

Trim tabs, electrically operated, 633, 634–35

Trippensee, Gary, 701, 705

Truckee, California, 850

Tsien, Hsue-shen, 12, 14, 571

T-tail aircraft: deep stall, 67n127, 413; F-101 Voodoo, 33–34; longitudinal stability, 36–37; performance and, 33, 413

Tullahoma, Tennessee, 280, 291, 787

Tungsten, 308, 575, 584, 587, 595

Tungsten carbide, 592

Tupolev Bear, 102

Tupolev Tu-22M Backfire, 62

Tupolev Tu-144 aircraft, 789

Tupolev Tu-144 Charger Supersonic Transport (SST), 66

Tupolev Tu-160 Blackjack, 62, 62n118

Turbine blades: environmental barrier coatings, 612–13; fabrication of, 613–15; materials for, 609–10; thermal barrier coatings, 610–13

Turbofan engines: Digital Electronic Engine Control (DEEC) program, 709–10; materials for, 571; Quiet Engine program, 737, 741; TF30 turbofan engines, 754; TF34 turbofan, 154, 158; TF34 turbofan engines, 162, 163

Turbojet engines: General Electric J79 turbojet, 104; invention of, 3, 90; noise from, 737; Pratt & Whitney J57 turbojet, 100, 101, 102; Pratt & Whitney J75 turbojet, 103; Quiet Engine program, 737, 741; speed capabilities, 289, 571; transonic flight with, 92

Turbopropeller engines: Advanced Turboprop (ATP) project, 744, 747–50, 786, 788, 789, 790–91, 799–803; Collier Trophy for, 748, 800; noise from, 800, 802; operation of, 747–48; public acceptance of, 747

Turbulence, 340–42, 439, 450–51

Turner, M.J., 471

951

Twin Annular Pre-mixing Swirler (TAPS) combustor, 798–99, 817
Twining, Nathan F., 101
Typhoon, 692
TZM molybdenum: ASSET vehicles, 594; X-20 Dyna-Soar, 584–87, 588, 594

U

U-2 spy plane, 297, 811, 812–13
Udvar-Hazy Center, National Air and Space Museum, 166, 638n15
UH-1 Huey, 147
UH-60 Blackhawk: Airloads Program, 170–71; RASCAL configuration, 170; UH-60A Blackhawk, 510, 519–20, 529–30
Ulster Folk and Transport Museum, Flight Experience exhibit, 641
Ultra Efficient Engine Technology (UEET) program, 613, 762–66, 786, 790, 797–98
Ultra-High Bypass Turbofan engine, 750–51
Unconventional (decoupled) control modes, 652, 669–70
Unducted Fan (UDF) engine, 749, 750, 801–2
Union Carbide, 601
Unitary Wind Tunnel: at Ames, 214, 235, 586; at Langley, 235, 585–86
Unit displacement method, 469, 470
United Aircraft Corporations, Vought-Sikorsky Aircraft Division, 8–9
United Airlines Flight 232 crash, 757, 766
United Nations International Panel on Climate Change (IPCC) report, 790, 790n22
United States Air Force. *See* Air Force, U.S. (USAF)
United States Air Force Museum, 202
United States Army. *See* Army, U.S. (USA)
United States Army Air Forces (USAAF). *See* Army Air Forces (AAF)
United States Ballistic Missile Defense Organization, 850
United States Capital, 11
United States Forest Service, 200
United States House of Representatives. *See* Congress, U.S.
United States Marine Corps. *See* Marine Corps, U.S. (USMC)
United States National Park Service, 201
United States Navy (USN). *See* Navy, U.S. (USN)
United States Senate. *See* Congress, U.S.
United Technologies Corporation, 799, 839
Universal Analytics, 484, 486, 538
Universal Automatic Computer (UNIVAC) computers, 436, 469, 484
University of Akron, 768
University of California at Berkeley, School of Engineering, 162, 471
University of Chicago, 199, 490
University of Colorado, 244
University of Florida, 71
University of Georgia, 488, 542
University of Iowa, 549
University of Manchester, 94
University of Maryland at College Park, 169, 452, 542
University of Michigan, 332, 406
University of Missouri, 542
University of Naples, 542
University of New Brunswick, 542
University of Pennsylvania, 436, 468–69
University of Southern California, 71
University of Texas at San Antonio, 498–99
University of Toronto, 225, 240
University of Virginia, 512, 538, 542
University of Wichita, 845
Unmanned aerial vehicles (UAVs): advantages of UAVs, 861–62; Boeing X-48B,

2, 71–72; Centurion, 850, 856, 859, 862, 864–66; control systems, 856; development of, 71, 823, 849–50, 852, 853, 870–71; flight termination, 858, 858n48; fuel cells, 856, 863–64, 869, 871; hardware reliability, 859, 869; Helios Prototype, 850, 853, 856, 862, 863–64, 866–72, 866n61; Mishap Investigation Board (MIB) report, 870–71; Pathfinder, 850, 855–59, 861, 862, 863; Pathfinder Plus, 850, 859–60, 862; record altitude flights, 859, 860; redundant systems, 855–56; rotary wing, 136; sensors and communication equipment, 854; solar cells, 862–63, 868, 872; solar-powered, 850, 857, 858, 860, 872; weather and, 858–59

Unstable aircraft, 379–80, 383, 526, 631, 632, 633, 651–52, 753

Unstarts, 336, 708–9, 754

Upwinding, 445

USSR. *See* Soviet Union (USSR)

V

V-2 ballistic missiles, 183–84, 280, 282, 571, 635–36

V-22 Osprey Tilt Rotor, 136, 155, 161, 166

VAK-191B aircraft, 649

Vandenberg Air Force Base, 607

Vanderplaats, Garret, 493

Van Dyke, Milton O., 442

Vanguard satellite project, 500

Variable cycle engine (VCE), 220

Variable-geometry wing aircraft, 57–64, 61n115, 62n118

Variable stability aircraft, 401–2

Vari-Eze, 120–21

Vectoring, Extremely Short Takeoff and Landing Control and Tailless Operation Research (VECTOR), 705–7

Vehicle Integration Branch, 220

Vehicle Systems program, 765

Vehicle Technology Directorate (VTD), 162–63

Venona signals intelligence decryption program, 17n30

Vermont wind power program, 826, 828, 831

Vertical Motion Simulator (VMS), 151, 167

Vertical/Short Take-Off and Landing (V/STOL) aircraft, 25, 60, 146

Vertical Take-Off and Landing (VTOL) aircraft: Dornier Do-31 aircraft, 648–49; research activities, 148–49; Short S.C.1 aircraft, 639–41; VAK-191B aircraft, 649; VJ-101fighter, 648

Vertical Take-Off and Landing (VTOL) Approach and Landing Technology (VALT) program, 151

Vertol Aircraft Corporation, 136

Vertol CH-47 Chinook, 151, 162, 163; CH-47C Chinook, 152; CH-47D Chinook, 509–10

Vertol Division, Boeing Company, 136

Vertol YHC-1A, 148

Vestas wind turbine, 822, 848

Vibration testing, 328–29

Vickers Swallow aircraft, 60–61

Vickers VC10K tanker, 643

Vickers VC10 tranport, 642–43, 677

Vickers/Vickers-Supermarine: F-86 intelligence collection, 24; variable-geometry wing research, 60–61

Vietnam war, 289, 632, 644–45, 644n32

View factors: RAVFAC program, 502; VIEW program, 502

Virgin Atlantic, 619

Virginia-class submarines, 723

Virginia Polytechnic Institute, 493

Virginia Tech, 494
Visco-Elastic and Hyperelastic Structures (VISCEL), 503
Viterna, Larry A., 846
VJ-101fighter, 648
VLOADS launch loads and dynamics program, 513–14
Voice-Controlled Interactive Device (VCID)/Voice Command System, 672–73, 674, 675
Voigt, Waldemar, 14, 48n94, 58–59
Völkenrode, 14
Volta Congress on High Speeds in Aviation, 5–7, 15n27, 93
Volvo, 706
Volvo RM-1 engine, 706
Von Braun, Wernher, 286, 512, 636
Von Kármán, Josephine "Pipa," 17n30
Von Kármán, Theodore, 6, 14, 17n30, 216, 291, 502
Von Kármán Institute, 244
Von Neumann, John, 284
Vortex flow control (VFC), 698
Vortex generators, 65
Vortex lift, 651, 651n49
Vortex ring state (VRS), 168
Vought, 480
Vought F4U-5 Corsair, 401
Vought F-8A fighter, 112–13
Vought F8U Crusader series. *See also* Digital Fly-By-Wire F-8 program and aircraft: area rule and redesign of, 103–4, 105; digital flight control system, 380–81, 753; F-8C, 380–81, 654–55, 753; F-8 Supercritical wing, 89; F-8 Super Cruise Wing project, 655; F8U-1, 32, 37, 43; F8U-3, 194; XF8U-1, 37, 363
Vought Missiles and Space Company, 576–77
Vought nose cap, 306

Vought-Sikorsky Aircraft Division, United Aircraft Corporation, 8–9
Vought XF8U-1 Crusader, 37, 363
Voyager aircraft, 121, 121n106, 576, 578
VS-300 helicopter, 9, 135, 136, 142
VTOL Approach and Landing Technology (VALT) program, 151
Vulcan bomber, 45, 639

W

Walkden, Frank, 187, 196
Walker, Harry, 695
Walker, Joseph A. "Joe," 42, 294, 298, 412
Wall, Steven D., 475
Wallis, Barnes, 60–61, 61n115
Wallops Flight Center/Wallops Island Station: Aeronautical Research Radar Complex, 151; instrument flight rules (IFR) investigation, 151; noise research, 165; Rotor Systems Research Aircraft (RSRA) program, 154–55; sonic boom research, 194, 198, 206, 239; transonic research, 92
Walton, William C., Jr., 508
Ward, G.N., 94
Washington, DC, 217, 427
Washington Dulles International Airport, 166, 635n15
Waspalloy, 592
Wasp-waist fuselage design, 89, 99, 107
Wasserfall (Waterfall) missiles, 638n15
Waterman, Waldo, 5
Water-wall cooling, 307, 589–90
WC-135B Stratotanker, 204
Weapon systems, 22
Weber, Richard, 331–32
Wedge-shaped vertical tail, 283
Wegener, Steven, 851, 861
Weidman, Deene J., 536
Weinberger, Caspar, 336, 338

Weinstein, Leonard M., 239
Welch, George "Wheaties," 41, 42, 393
Westenhoff and Novick, 538
Western USA Sonic Boom Survey, 240–41, 242
Westinghouse, 538
Westinghouse Electric Corporation, 829, 834, 842
Westland Pterodactyls, 5
West Virginia University, 719
Whitcomb, Kenneth F., 90
Whitcomb, Richard Travis "Dick": aerodynamic refinement and efficiency improvements, 89–90, 128–29; area rule, announcement of, 106–7; area rule, benefits of, 89–90, 126; area rule, development of, 51, 52, 95n14, 97–100; area rule, models, 88, 99; area rule, recognition for, 107–8; area rule, transonic transport fuselage, 127; area rule, validation in flight, 100–106; character of, 115–16, 125–26; Collier Trophy, 52, 93–94, 107; contributions and legacy, 125–29; cut-and-try research technique, 113–14; Distinguished Service Medal, 107; early life of, 90–91; education of, 91; Exceptional Service Medal, 107; influences on, 91; interference studies, 53, 65; intuitive reasoning and imagination, 95n14, 98, 114, 126; Medal for Exceptional Scientific Achievement, 116; National Medal of Science, 116; reputation of, 94; resignation of, 125–26; SCAT-4, 67; SST program, 65, 67; supercritical wing, aviation industry use of, 114–15; supercritical wing, benefits of, 89–90, 126, 805–6, 817; supercritical wing, development of, 805; supercritical wing, invention of, 68, 109–12; supercritical wing, models of, 110–11; supercritical wing, naming of, 111; supercritical wing, patent for, 114; supercritical wing, publication of findings, 111, 113; supercritical wing, recognition for, 116; supercritical wing, testing of, 112–14; Supersonic Transport study, 108–9, 108n61; Sylvanus Albert Reed Award, 116; thinking sessions, 98; Transonic Aerodynamic Branch, 108, 125; transonic drag research, 94–100; transonic transport wing configuration, 52, 52n100, 68; visualization capabilities, 98; winglets, announcement of, 120; winglets, aviation industry use of, 120–22, 806; winglets, benefits of, 89–90, 119–20, 123–25, 126, 806, 817; winglets, development of, 68, 117–19; winglets, flight-testing of, 122–23; winglets, models of, 118, 119; winglets, placement and design of, 119; winglets, publication of findings, 118; winglets, wind tunnel studies of, 118, 119; Wright Brothers Memorial Trophy, 116

White, Robert, 294
White, Thomas D., 285
Whitehead, Robert, 340
White House: Executive Order to support aeronautics research, 171–72; National Sonic Boom Evaluation Program, 203–4; Presidential Advisory Committee (PAC) on SST, 192, 210–11; Science and Technology, Office of, 203, 230–31; supersonic research, 219
White Sands Missile Range, 200, 238–39
Whitman, Gerald B., 185
Whitten, James, 291
Wierzbanowski, Theodore Ted," 695
Wieselsberger, Carl, 280
Williams, Robert "Bob," 337, 338, 340
Williamsburg, Virginia, 233, 537
Williams International, 760, 762, 797

Windcharger Corporation, 826

Wind energy: cost of turbines, 843–44; Federal Wind Energy Program, 823–25, 827, 828–29, 831, 833, 840, 843, 846–49; history of wind turbines, 825–28; horizontal-axis wind turbine (HAWT) generators, 823, 825; manufacturers of turbines, 848; Mod-0A wind turbine, 831, 833–35; Mod-0 wind turbine testbed, 827, 830, 831–33, 846; Mod-1A wind turbine, 836; Mod-1 wind turbine, 831, 835–36, 845; Mod-2 wind turbine, 830, 831, 836–39, 840, 843, 847, 848; Mod-3 wind turbine, 840; Mod-4 wind turbine, 840; Mod-5A wind turbine, 840, 841; Mod-5B wind turbine, 840, 841–44, 847; Mod-5 wind turbine, 831, 840; Mod-6 wind turbine, 840; Nibe A wind turbine, 848; objectives of program, 828–29; RANN symposium, 827–28; research activities at Lewis/Glenn, 844–49; rotor blade research, 844–46, 847; system verification units (SVU), 839; Vestas wind turbine, 822, 848; wind-electric dynamo, 825; Wind Energy Project Office, 830; Wind Power Office, 829–30

Windows, 307, 590–91

Wind tunnel and shock tunnel research: aerodynamic predictions, 397–98; conical camber, 51; DM-1 glider research, 47; hingeless rotor concept, 150; pitch-up, 25; rotary wing aircraft and helicopters, 140, 143, 144; SCAT-15F, 67; sonic booms, 208–9, 208n66, 214, 222, 235; Space Shuttle studies, 323–30, 327; speed limits, 573; winglets, 118, 119; X-48B, 71

Wing aspect ratio, 119–20, 407

Wing box research, 618–19

Wing dropping, 21

Wing fences, 414

Wing flaperons, 694, 697

Wing flaps, 414, 526, 595

Wing-flow model, 92

Winglets: announcement of, 120; aviation industry use of, 120–22, 806; benefits of, 89–90, 119–20, 123–25, 126, 806, 817; blended winglets, 124; concept of, 117; development of, 65, 68, 117–19, 786; flight-testing, 122–23; models, 118, 119; placement and design of, 119; publication of findings, 118; wind tunnel studies, 118, 119

Wingspan-to-fuselage length ratio, 3–4, 37, 40n74, 407

Wingtip stall, 414

Wingtip vortexes, 117, 118, 119

Wing twist, 414

Wing warping, 386, 633

Wisneski, Jim, 701

Wlezien, Richard W., 244, 250–51, 254

Wolfe, Alex V., 673

Wolverine Ventures, Inc., 761

Women mathematicians, 465, 494

Wood, Drury W., Jr., 649

Woods, Robert J., 59, 290

Worcester Polytechnic Institute, 91, 107–8

World War I, 17, 70, 387, 463, 633

World War II: aluminum use, 571; flight control systems, 631, 633–34, 633n1, 635–37; Link Trainer, 295, 392; piston engines, problems with, 736; rotary wing aircraft use, 142; V-2 missile shock waves, 183–84

Wright, Orville, 107, 364, 385, 386, 632–33

Wright, Ray, 95

Wright, Wilbur, 364, 385, 386, 632–33

Wright Aeronautical ramjets, 289

Wright Air Development Center, Flight Control Laboratory, 639

Wright Brothers Memorial Trophy, 116

Wright Field Aircraft Laboratory, 9–10
Wright Flyer, 37, 364, 385, 391, 633, 752
Wright Laboratory, 541
Wright-Patterson Air Force Base: Aeronautical Research Laboratory, 186; Air Force Museum, 648, 679, 709; Flight Dynamics Laboratory, 592; Matrix Methods in Structural Mechanics conference, 474–75; NASTRAN development, 474; NASTRAN Users' Colloquia, 540, 541; Numerical Propulsion System Simulation (NPSS) program, 760
Wright R-1820 Cyclone engine, 736
Wright R-3350 Duplex Cyclone engine, 736
Wyle Laboratories, 240, 242, 246
Wyle Research, 226

X

X-1 aircraft: Collier Trophy, 93; development and manufacture of, 108, 290, 494; engines for, 294; flight control systems, 362–63, 752; gliding landings, 402–3; lift-to-drag ratio (L/D), 403; Round One research aircraft, 55n106; supersonic capabilities, 94; tail configuration, 411; X-1A aircraft, 40–41, 40n74, 43, 292, 363, 407; X-1B aircraft, 40n74, 298; X-1D aircraft, 40n74; Yeager's flight, 40–41, 292, 363, 407

X-2 aircraft: accident with, 44n87, 394–95, 407, 410; control coupling, 410; design of, 34; development and manufacture of, 17n30, 290; dihedral effects, 410; flight control systems, 298; inertial coupling, 44n87; pressure suit for, 297; simulator, 394–95; tail configuration, 411

X2 helicopter, 158

X-3 "Stiletto," 3, 4, 37, 40, 41–42, 43, 440

X-4 aircraft, 27–28

X-5 aircraft, 17n30, 58, 59–60, 413

X-7 testbed, 278, 289, 290, 572

X-15 aircraft: ablation cooling research, 417–18; accident with, 299, 373–74; centrifuge program, 399–400; contracts for, 293; contributions from program, 299; design confidence, 307, 308; development of, 278, 283, 290–93; at Dryden, 294; Dutch roll motions, 411–12; emergencies, practice for, 296; flight control system, 298–99, 372–74, 376, 379, 642; flight-testing, 294–95; flight time, 299; funding for, 291; heating, structural, 414–15; heat-sink structures, 415–17; inertial coupling (roll coupling), 409; instrumentation development, 298; landing of, 405; landings of, 403–4; lift-to-drag ratio (L/D), 403; limit-cycle oscillation research, 368; materials for, 292, 293, 295, 571, 592; pressure suits, 276, 278, 297, 298; reaction controls, 298; record hypersonic flight, 278, 293, 299, 642; redundant systems, 365–66; roll damper, 412; Round Two research aircraft, 54, 55, 290; side stick flight controller, 642, 658n59; simulator, 295–97, 395–97, 403–4, 405, 409, 411–12; sonic boom research, 197, 226; speed of, 415, 642; Stability Augmentation Systems, 365–66, 399; stability issues, 292, 411–12; structural resonances, 375–76; tail configuration, 278, 411; thermal barrier coatings, 611; thermal damage, 294–95, 416, 418; thermal data from, 517–18; X-15A-2 aircraft, 64, 335, 406, 417–18

X-17 missile, 287–88

X-20 Dyna-Soar: cancellation of, 304, 308, 400, 592; centrifuge program, 400; contributions from program, 307–8; design confidence, 307–8; development of, 300–305;

flight control system, 372; flutter-testing panels, 585–86; hot structures, 305–8, 418, 582–92; justification of, 302–5; landing skids, 592; materials for, 305–8; missiles to launch, 301–2; nose cap, 306, 308, 587–89, 598; Round Three research aircraft, 55–56; Round Two research aircraft, 55n106; schematic drawing, 583; simulator, 405–6; windows, 307

X-23 (SV-5D) PRIME (Precision Recovery Including Maneuvering Entry), 27n51, 278, 288, 304, 417, 418, 606–8

X-24A (SV-5P) lifting body, 27n51, 407, 607

X-24B lifting body, 56, 404, 407

X-24C hypersonic lifting body proposal, 299, 448–49

X-27 aircraft, 34n61

X-29 aircraft: aeroelastic effects, 496, 528–29, 694; aeroelastic tailoring, 694, 695; angle of attack capabilities, 694, 695, 696–98; benefits of, 697–98; contract for, 692–93; control laws, 696; control surfaces, 694–95; difference between two X-29s, 693–94; exhibit showing, 698–99; finite element analysis, 460, 515, 526–29, 530; flight-testing, 695–98; fly-by-wire system, 383, 692–98, 753; forward-swept wing design, 22, 527, 693–94; NASTRAN model, 460; redundant systems, 695; sensor selection thresholds, 698; strain gages, 528; supercritical wing, 693, 694, 695, 695n126; vortex flow control, 698; X-29A aircraft, 460

X-30 National Aero-Space Plane, 230–31, 241, 279, 336–40, 342, 419, 615–21

X-31 aircraft: accident with, 704–5; air combat simulation, 703; angle of attack, 701–2, 706–7; autothrottle system, 706; Enhanced Fighter Maneuverability, 699–705; exhibit showing, 707; flight-testing, 701–7; Flush Air Data System, 707; fly-by-wire system, 700–705; Kiel probe, 701, 704, 705, 705n149; poststall maneuvering, 702–3, 703n144; tail configuration, 703–4; thrust vectoring, 700, 701, 703–4, 705–7; thrust vectoring paddles, 699, 700, 706; VECTOR program, 705–7

X-31 Enhanced Fighter Maneuverability (EFM) aircraft, 699–707

X-33 Research Vehicle, 384

X-38 aircraft, 27n51

X-43 Hyper-X vehicle, 279, 341, 343–45, 344n127, 568

X-48B Blended Wing-Body UAV, 2, 71–72

X-51A vehicle, 279, 345–46

XB-52 aircraft, 20

XB-70 Valkyrie: aeroelastic effects research, 392; SCAT program and, 108; sonic boom research, 201–2, 201n49, 203, 204, 205, 210; XB-70-1 testbed, 201, 202, 204; XB-70-2 testbed, 201–2, 204; XB-70A, 56n107, 191

XCH-62 Heavy Lift Helicopter (HLH) program, 158

XF2Y-1 Sea Dart, 45, 103

XF4D-1 Skyray, 32–33

XF8U-1 Crusader, 37, 363

XF10F-1 Jaguar, 59, 60

XF-88 aircraft, 33

XF-92A aircraft, 45, 46n90, 49–51, 100

XF-103 interceptor fighter, 289, 290, 333

XF-104 Starfighter, 185

XH-51N helicopter, 150

XH-59A Advancing Blade Concept (ABC) helicopter, 152–53, 157–58

XP-86 Sabre, 18, 635

XP-92 aircraft, 46

XS-1 aircraft: design of, 3, 28, 34, 34n62, 36,

184; falling body test, 38, 43; flight control systems, 635; inspiration for, 14n22; sonic boom from, 182, 182n3; supersonic flight by, 182, 183; testing of, 35; variable-geometry wing tests, 59; wingspan-to-fuselage length ratio, 37; XS-1 FSW planform, 21; Yeager's flight, 182, 182n3, 283, 635

XS-2 aircraft, 34, 34n62

XV-1 Convertiplane, 134, 147

XV-15 Tilt-Rotor Research Aircraft (TRRA), 155, 156, 160–62, 165, 166

X-Wing Stoppable Rotor, 157, 158

X-Wing Stoppable Rotor program, 157, 158, 163–64, 165

Y

Yaw, 19, 407–8, 410

Yaw, adverse, 19–20, 409

Yaw damper, 637

YB-49 flying wing jet bomber, 19, 637

YC-14 transport, 114–15, 670–72

YC-15 aircraft, 114, 379–80

Yeager, Charles E. "Chuck": Collier Trophy, 93; flight control systems, 752; supersonic aircraft research, 94; X-1A flight, 40–41, 292, 363, 407; XS-1 flight, 182, 182n3, 283, 635

YF-4E Phantom Survivable Flight Control System program, 379–80, 384, 646–48, 658n59, 660

YF-12 Blackbird: digital engine control system, 754; Loads Alleviation and Mode Suppression (LAMS) system, 392; sonic boom research, 202–3, 204, 210, 227, 227n133; thermal loads project, 515, 516–19, 529, 554; YF-12A, 202–23; YF-12C, 516, 708–9; YF-12 Experiment Symposium, 519

YF-16 Lightweight Fighter (LWF), 371; angle of attack limiters, 665–66; flight-testing, 667–68; flutter suppression research, 388, 389; fly-by-wire system, 379–80, 647, 664–69; pilot-induced oscillations, 370; roll stick force gradient, 667; side stick flight controller, 658, 664, 665, 668; simulator studies, 666–67; strain gages, 385–86; structural resonance research, 378; success of, 668; unintentional first flight, 667

YF-17 Lightweight Fighter (LWF), 388, 389, 664, 669

YF-22 aircraft, 402

YF-86D low-tail testbed, 36–37

YF-100A Super Sabre, 34, 363

YF-100 Super Sabre, 37, 40, 41, 42, 185

YF-102 Delta Dagger: area rule and, 51, 89, 100–102, 103; handling qualities, 53; sonic booms, 185; YF-102A prototype, 89, 101–2, 103

YG-2 autogiro, 141

YO-3A aircraft, 242–43

York, Herbert, 301

YR-4B helicopter, 142, 143–44

Yttria, 587

Yttria-stabilized zirconia, 610, 611–12, 613

Z

Zahm, Albert, 11

Zero CO2 project, 813

Zhu, Dong-ming, 613

Zirconia: magnesia-stabilized zirconia, 611; nose cones/caps, 587, 588, 589, 594, 595, 598; reusable surface installation (RSI), 601; yttria-stabilized zirconia, 610, 611–12, 613

Zirconium, 574, 584, 588, 614

Zirconium Corporation of America, 588

Zirconium oxide, 306, 308

THIS PAGE INTENTIONALLY BLANK

THIS PAGE INTENTIONALLY BLANK

www.ingramcontent.com/pod-product-compliance
Lightning Source LLC
Chambersburg PA
CBHW082102230426
43671CB00015B/2585